*Edited by Matthias Beller, Albert Renken,
and Rutger van Santen*

Catalysis

Related Titles

Hashmi, A. S. K., Toste, D. F. (eds.)
Modern Gold Catalyzed Synthesis

2012
ISBN: 978-3-527-31952-7

Behr, A., Neubert, P.
Applied Homogeneous Catalysis

2012
ISBN: 978-3-527-32641-9

Steinborn, D.
Fundamentals of Organometallic Catalysis

2011
ISBN: 978-3-527-32716-4

Zecchina, A., Bordiga, S., Groppo, E. (eds.)
Selective Nanocatalysts and Nanoscience
Concepts for Heterogeneous and Homogeneous Catalysis

2011
ISBN: 978-3-527-32271-8

Bullock, R. M. (ed.)
Catalysis without Precious Metals)

2010
ISBN: 978-3-527-32354-8

Caprio, V., Williams, J.
Catalysis in Asymmetric Synthesis
Series: Postgraduate Chemistry Series

Second edition
2009
ISBN: 978-1-4051-9091-6

Rothenberg, G
Catalysis
Concepts and Green Applications

2008
ISBN: 978-3-527-31824-7

*Edited by Matthias Beller, Albert Renken,
and Rutger van Santen*

Catalysis

From Principles to Applications

WILEY-VCH Verlag GmbH & Co. KGaA

The Editors

Prof. Dr. Matthias Beller
Universität Rostock
Leibniz-Institut für Katalyse e.V.
Albert-Einstein-Str. 29a
18059 Rostock
Germany

Prof. Dr. Albert Renken
École Polytechnique Fédérale
EPFL-ISISC-LGRC
Station 6
1015 Lausanne
Switzerland

Prof. Dr. Rutger van Santen
Eindhoven University of Technology
Schuit Institute of Catalysis
Den Dolech 2
5612 AZ Eindhoven
The Netherlands

All books published by **Wiley-VCH** are carefully produced. Nevertheless, authors, editors, and publisher do not warrant the information contained in these books, including this book, to be free of errors. Readers are advised to keep in mind that statements, data, illustrations, procedural details or other items may inadvertently be inaccurate.

Library of Congress Card No.: applied for

British Library Cataloguing-in-Publication Data
A catalogue record for this book is available from the British Library.

Bibliographic information published by the Deutsche Nationalbibliothek
The Deutsche Nationalbibliothek lists this publication in the Deutsche Nationalbibliografie; detailed bibliographic data are available on the Internet at <http://dnb.d-nb.de>.

© 2012 Wiley-VCH Verlag & Co. KGaA, Boschstr. 12, 69469 Weinheim, Germany

All rights reserved (including those of translation into other languages). No part of this book may be reproduced in any form – by photoprinting, microfilm, or any other means – nor transmitted or translated into a machine language without written permission from the publishers. Registered names, trademarks, etc. used in this book, even when not specifically marked as such, are not to be considered unprotected by law.

Print ISBN: 978-3-527-32349-4

Cover Design Formgeber, Eppelheim
Typesetting Laserwords Private Limited, Chennai, India
Printing and Binding Markono Print Media Pte Ltd, Singapore

Printed in Singapore
Printed on acid-free paper

Contents

List of Contributors   *XVII*
Preface   *XXI*

Part I   Basic Concepts   *1*

1   Catalysis in Perspective: Historic Review   *3*
Rutger van Santen
1.1   History of Catalysis Science   *3*
1.1.1   General Introduction   *3*
1.1.2   Heterogeneous Catalysis: the Relationship between a Catalyst's Performance and its Composition and Structure   *4*
1.1.3   Homogeneous and Enzyme Catalysis   *8*
1.1.4   Important Scientific Discoveries   *9*
1.2   The Development of Catalytic Processes: History and Future   *11*
1.3   Fundamental Catalysis in Practice   *13*
1.4   Catalyst Selection   *13*
1.5   Reactor Choice   *16*
1.6   Process Choice   *17*
           References   *19*
           Further Reading   *19*

2   Kinetics of Heterogeneous Catalytic Reactions   *20*
Rutger van Santen
2.1   Physical chemical principles   *20*
2.1.1   The Catalytic Cycle   *20*
2.2   The Lock and Key Model, the Role of Adsorption Entropy   *27*
2.3   Equivalence of Electrocatalysis and Chemocatalysis   *30*
2.4   Microkinetics; the Rate-Determining Step   *32*
2.5   Elementary Rate Constant Expressions for Surface Reactions   *34*
2.6   The Pressure Gap   *36*
2.6.1   Surface Reconstruction   *37*
2.6.2   Altered Surface Reactivity   *38*
2.7   The Materials Gap   *39*

2.7.1	Structure Sensitivity 39
2.7.2	Catalyst Activation or Deactivation 40
2.7.3	Inhomogeneous Site Distribution 40
2.8	Coupling of Catalytic Reaction and Inorganic Solid Chemistry 42
2.9	*In situ* Generation of Organo-Catalyst 42
2.10	The Compensation Effect 44
	References 46

3 Kinetics in Homogeneous Catalysis 48
Detlef Heller

3.1	Principles of a Catalyst and Kinetic Description 48
3.2	Catalyst Activity 54
3.3	Catalyst Activation and Deactivation 58
3.3.1	Induction Periods as Catalyst Activation 59
3.3.2	Catalyst Deactivation due to Formation of Non-Reactive Complexes 61
3.3.3	Catalyst Deactivation due to Formation of Multinuclear Complexes 61
3.3.4	Catalyst Deactivation due to Irreversible Reactions 63
	References 64

4 Catalytic Reaction Engineering Principles 67
Albert Renken and Lioubov Kiwi-Minsker

4.1	Preface 67
4.2	Formal Kinetics of Catalytic Reactions 68
4.2.1	General Definitions 69
4.2.2	Heterogeneous Catalytic Reactions 70
4.2.3	The Langmuir Adsorption Isotherms 72
4.2.4	Reaction Mechanisms 73
4.2.4.1	Langmuir–Hinshelwood Model 74
4.2.4.2	The Quasi-Surface Equilibrium Approximation 75
4.2.4.3	The Masi Approximation 75
4.2.4.4	Bimolecular Catalytic Reactions 76
4.3	Mass and Heat Transfer Effects 77
4.3.1	Introduction 77
4.3.2	External Mass and Heat Transfer 78
4.3.2.1	Isothermal Pellet 78
4.3.2.2	Non-Isothermal Pellet 84
4.3.3	Internal Mass and Heat Transfer 85
4.3.3.1	Isothermal Pellet 87
4.3.3.2	Non-Isothermal Pellet 94
4.3.4	Combination of External and Internal Transfer Resistances 96
4.3.4.1	Internal and External Mass Transport in Isothermal Pellets 96
4.3.4.2	Implication of Mass Transfer on the Temperature Dependence 98
4.3.4.3	External and Internal Temperature Gradient 100

4.3.5	Criteria for the Estimation of Transport Effects 101
4.4	Homogenous Catalysis in Biphasic Fluid/Fluid Systems 103
	References 108

Part II   The Chemistry of Catalytic Reactivity 111

5   Heterogeneous Catalysis 113
Rutger van Santen

5.1	General Introduction 113
5.2	Transition Metal Catalysis 114
5.2.1	Ammonia Synthesis 114
5.2.1.1	The Mechanism of the Reaction 114
5.2.1.2	Structure Sensitivity, Composition Dependence 114
5.2.2	Methane Reforming 120
5.2.2.1	The Mechanism of the Reaction 120
5.2.2.2	Structure Sensitivity and Composition Dependence 120
5.2.3	Hydrogenation, Dehydrogenation, and C–C Bond Cleavage 126
5.2.3.1	Mechanism of Hydrogenation and Dehydrogenation 126
5.2.3.2	Kinetics of Olefin Hydrogenation 126
5.2.3.3	The Mechanism of Ethane Hydrogenolysis 127
5.3	Solid Acids and Bases 132
5.3.1	Introduction 132
5.3.2	Proton Activation by Zeolites 135
5.3.3	General Mechanistic Considerations 139
5.3.3.1	Direct Alkane Activation 139
5.3.3.2	Hydride Transfer 141
5.3.3.3	Isomerization Catalysis 141
5.4	Reducible Oxides 143
5.4.1	Comparison of the Relative Stabilities of Some Oxides 143
5.4.2	Structure Sensitivity 145
5.4.3	Mechanism of Important Oxidation Reactions 148
5.4.3.1	The Selective Oxidation of Propylene 148
5.4.3.2	Propane Oxidation 150
	References 150

6   Homogeneous Catalysis 152
Matthias Beller, Serafino Gladiali, and Detlef Heller

6.1	General Features 152
6.1.1	Acid and Base Catalysis 155
6.1.2	Nucleophilic and Electrophilic Catalysis 157
6.1.3	Transition Metal-Centered Homogeneous Catalysis 159
	References 169

7	**Biocatalysis** *171*
	Uwe Bornscheuer
7.1	Introduction *171*
7.1.1	Choice of Reaction Strategy: Kinetic Resolution or Asymmetric Synthesis *174*
7.1.2	Choice of Reaction Systems *175*
7.2	Examples *176*
7.2.1	Oxidoreductases (EC 1) *176*
7.2.1.1	Dehydrogenases (EC 1.1.1.-, EC 1.2.1.-, EC 1.4.1.-) *176*
7.2.1.2	Oxygenases *178*
7.2.2	Hydrolases (EC 3.1) *182*
7.2.2.1	Lipases (EC 3.1.1.3) and Esterases (EC 3.1.1.1) *182*
7.2.2.2	Peptidases, Acylases, and Amidases *185*
7.2.2.3	Nitrilases (EC 3.5.5.1) and Nitrile Hydratases (EC 4.2.1.84) *186*
7.2.2.4	Hydantoinases (EC 3.5.2.-) *187*
7.2.3	Lyases (EC 4) *188*
7.2.3.1	Hydroxynitrile Lyases (EC 4.1.2.-) *188*
7.2.3.2	Aldolases (EC 4.1.2.-; 4.1.3.-) *190*
7.2.4	Transaminases *193*
7.3	Summary/Conclusions *194*
	References *194*
8	**Electrocatalysis** *201*
	Timo Jacob
8.1	Introduction *201*
8.2	Theory *203*
8.2.1	Electrochemical Potentials *203*
8.2.2	Electric Double Layer *204*
8.3	Application to the Oxygen Reduction Reaction (ORR) on Pt(111) *207*
8.4	Summary *212*
	References *213*
9	**Heterogeneous Photocatalysis** *216*
	Guido Mul
9.1	Introduction *216*
9.1.1	What Is Photocatalysis? *216*
9.1.2	What Is the Principle of Photocatalysis? *217*
9.2	Applications of Photocatalysis *219*
9.3	Case Studies *220*
9.3.1	Water Purification: the Quest for the Structure–Activity Relationship of TiO_2 *220*
9.3.2	Energy Conversion: Advanced Materials to Go Thermodynamically Uphill! *222*
9.3.2.1	Design of Crystalline Catalysts *222*
9.3.2.2	The Quest for Visible Light-Sensitive Systems *223*

9.3.2.3	Supported Chromophores	223
9.3.3	Photocatalysis in Practice: Some Reactor Considerations	225
9.3.3.1	Microreactors	227
9.4	Concluding Remarks	228
	References	228

Part III   Industrial Catalytic Conversions   231

10   Carbonylation Reactions   233
Matthias Beller
10.1   General Aspects   233
10.2   Hydroformylation   234
10.3   Other Carbonylations of Olefins and Alkynes   238
10.4   Carbonylations of Alcohols and Aryl Halides   244
         References   246

11   Biocatalytic Processes   250
Uwe Bornscheuer
11.1   Introduction   250
11.1.1   How to Choose the Best Route?   250
11.2   Examples   253
11.2.1   General Applications   253
11.3   Case Study: Synthesis of Lipitor Building Blocks   257
11.4   Conclusions   259
         References   259

12   Polymerization   261
Vincenzo Busico
12.1   Introduction   261
12.2   Polyolefins in Brief   262
12.3   Olefin Polymerization Catalysts   264
12.3.1   The Catalytic Species: Structure and Reactivity   264
12.3.2   Polymerization Kinetics: Active, 'Dormant' and 'Triggered' (?) Sites   269
12.4   Olefin Polymerization Process Technology   273
12.4.1   Heterogeneous Catalysis   273
12.4.2   Homogeneous Catalysis   278
12.5   The Latest Breakthroughs   280
         References   285

13   Ammonia Synthesis   289
Jens Rostrup-Nielsen
13.1   Ammonia Plant   289
13.2   Synthesis   291
13.2.1   Technology Development   291

13.2.2	The Catalysis	*292*
13.2.3	Process Optimization	*295*
13.3	Steam Reforming	*295*
13.3.1	Technology	*295*
13.3.2	The Catalysis	*296*
13.3.3	Secondary Phenomena	*297*
13.4	Conclusions	*299*
	Abbreviations	*299*
	References	*299*
14	**Fischer–Tropsch Synthesis in a Modern Perspective**	*301*
	Hans Schulz	
14.1	Introduction	*301*
14.2	Stoichiometry and Thermodynamic Aspects	*304*
14.2.1	Stoichiometry	*304*
14.2.1.1	Thermodynamic Aspects	*305*
14.2.1.2	Rate Equations and Operation Ranges	*306*
14.2.1.3	Operating Ranges (Pichler)	*306*
14.3	Processes and Product Composition	*308*
14.3.1	Commercial FT-Synthesis	*308*
14.3.1.1	Low-Temperature Synthesis	*309*
14.3.1.2	Slurry Reactors	*310*
14.3.1.3	High-Temperature Fischer–Tropsch Synthesis	*310*
14.3.1.4	Synthesis Gas	*311*
14.4	Catalysts, General	*311*
14.4.1	Cobalt	*312*
14.4.2	Iron	*312*
14.5	Reaction Fundamentals	*313*
14.5.1	Ideal Polymerization Model	*313*
14.5.1.1	Chain Growth	*314*
14.5.1.2	Alternative Reactions on Growth Site	*315*
14.5.1.3	Branching	*315*
14.5.1.4	Alcohols in FT-Synthesis	*316*
14.5.1.5	Desorption (Olefins/Paraffins)	*316*
14.5.1.6	Catalyst Formation *in situ*	*319*
14.6	Concluding Remarks	*323*
	References	*323*
15	**Zeolite Catalysis**	*325*
	Rutger van Santen	
15.1	Introduction	*325*
15.2	The Hydrocracking Reaction; Acid Catalysis	*325*
15.2.1	The Dependence of Cracking Selectivity and Activity on Hydrocarbon Chain Length	*326*

15.2.2	Symmetric versus Asymmetric Cracking Patterns. Stereoselectivity, Pore Size, and Topology Dependence *328*	
15.3	Lewis Acid–Lewis Base Catalysis; Hydrocarbon Activation *332*	
15.4	Selective Oxidation; Redox Catalysis *333*	
15.4.1	The Reactivity of Extra-Framework Single-Site versus Two-Center Fe Oxycations *334*	
15.5	Framework-Substituted Redox Ions *335*	
15.5.1	Ti-Catalyzed Epoxidation *335*	
15.5.2	Thomas Chemistry; Redox Cations in the $AlPO_4$ Framework *339*	
	References *339*	
16	**Catalytic Selective Oxidation – Fundamentals, Consolidated Technologies, and Directions for Innovation** *341*	
	Fabrizio Cavani	
16.1	Catalytic Selective Oxidation: Main Features *341*	
16.2	Catalytic Selective Oxidation: What Makes the Development of an Industrial Process More Challenging (and Troublesome) than Other Reactions *353*	
16.3	Catalytic Selective Oxidation: the Forefront in the Continuous Development of More-Sustainable Industrial Technologies *355*	
16.4	The Main Issue in Catalytic Oxidation: the Control of Selectivity *356*	
16.5	Dream Reactions in Catalytic Selective Oxidation: a Few Examples (Some Sustainable, Some Not Sustainable) *359*	
16.6	A New Golden Age for Catalytic Selective Oxidation? *361*	
16.7	Conclusions: Several Opportunities for More Sustainable Oxidation Processes *363*	
	References *363*	
17	**High-Temperature Catalysis: Role of Heterogeneous, Homogeneous, and Radical Chemistry** *365*	
	Olaf Deutschmann	
17.1	Introduction *365*	
17.2	Fundamentals *366*	
17.2.1	Heterogeneous Reaction Mechanisms *367*	
17.2.2	Homogeneous Reactions *369*	
17.2.3	Coupling of Chemistry with Mass and Heat Transport *369*	
17.2.4	Monolithic Catalysts *370*	
17.2.5	Experimental Evaluation of Models Describing Radical Interactions *371*	
17.2.6	Mathematical Optimization of Reactor Conditions and Catalyst Loading *372*	
17.3	Applications *372*	
17.3.1	Turbulent Flow through Channels with Radical Interactions *372*	
17.3.2	Synthesis Gas from Natural Gas by High-Temperature Catalysis *373*	

17.3.3	Olefin Production by High-Temperature Oxidative Dehydrogenation of Alkanes *373*	
17.3.3.1	Formulation of an Optimal Control Problem *375*	
17.4	Hydrogen Production from Logistic Fuels by High-Temperature Catalysis *378*	
17.5	High-Temperature Catalysis in Solid Oxide Fuel Cells *380*	
	References *385*	
18	**Hydrodesulfurization** *390*	
	Roel Prins	
18.1	Introduction *390*	
18.2	Hydrodesulfurization *391*	
18.3	The C–X Bond-Breaking Mechanism *393*	
18.4	Structure of the Sulfidic Catalyst *393*	
18.4.1	Structure of Mo *393*	
18.4.2	Structure of the Promoter *394*	
18.4.3	DFT Calculations *395*	
18.5	Hydrodenitrogenation *397*	
18.6	Determination of Surface Sites *398*	
	References *398*	
	Part IV Catalyst Synthesis and Materials *399*	
19	**Molecularly Defined Systems in Heterogeneous Catalysis** *401*	
	Fernando Rascón and Christophe Copéret	
19.1	Introduction *401*	
19.2	Single Sites: On the Border between Homogeneous and Heterogeneous Catalysis *402*	
19.2.1	Taking Homogeneous Catalysis to the Heterogeneous Phase via a Molecular Approach: the Case of Single-Site Alkene Metathesis Catalysts *404*	
19.2.2	Bridging the Gap with Classical Heterogeneous Systems by a Molecular Approach: the Case of Re_2O_7/Al_2O_3 vs $MeReO_3/Al_2O_3$ *408*	
19.2.3	Toward New Reactivity: the Case of Supported Transition-Metal Hydrides *410*	
19.2.4	Beyond a Molecular Viewpoint: a Closer Look at the Role of the Surfaces *413*	
19.3	Conclusion and Perspectives *415*	
	References *415*	
20	**Preparation of Supported Catalysts** *420*	
	Krijn P. de Jong	
20.1	Introduction *420*	
20.2	Support Surface Chemistry *422*	

20.3	Ion Adsorption	423
20.4	Impregnation and Drying	425
20.5	Deposition Precipitation	427
20.6	Thermal Treatment	428
	References	429

21 Porous Materials as Catalysts and Catalyst Supports 431
Petra de Jongh

21.1	General Characteristics	431
21.2	Sol-gel and Fumed Silica	433
21.3	Alumina and Other Oxides	436
21.4	Carbon Materials	438
21.5	Zeolites	440
21.6	Ordered Mesoporous Materials	442
21.7	Metal-Organic Frameworks	442
21.8	Shaping	443
	References	444

22 Development of Catalytic Materials 445
Manfred Baerns

22.1	Introduction	445
22.2	Fundamental Aspects	446
22.3	Micro-Kinetics and Solid-State Properties as a Knowledge Source in Catalyst Development	448
22.3.1	Reaction Mechanism and Kinetics of the Catalytic OCM Reaction	448
22.3.2	Surface Oxygen Species in Methane Conversion	449
22.3.3	Kinetic Analysis	450
22.3.4	Physico-Chemical Properties of Catalytic Solid Materials for the OCM Reaction	451
22.3.5	Structural Defects	451
22.3.6	Surface Acidity and Basicity	452
22.3.7	Redox Properties, Electronic Conductivity, and Ion Conductivity	452
22.3.8	Supported Catalysts	453
22.3.9	Conclusions	453
22.4	Combinatorial Approaches and High-Throughput Technologies in the Development of Solid Catalysts	453
22.4.1	Combinatorial Design of Catalytic Materials for Optimal Catalytic Performance	453
22.4.2	High-Throughput Technologies for Preparation and Testing of Large Numbers of Catalytic Materials	456
22.4.2.1	Preparation of Catalytic Materials	457
22.4.2.2	Testing and Screening of Catalytic Materials	457
22.4.3	Data Analysis	458
	References	459

Part V Characterization Methods 463

23 *In-situ* Techniques for Homogeneous Catalysis 465
Detlef Selent and Detlef Heller
- 23.1 Introduction 465
- 23.2 *In-situ* Techniques for Homogeneous Catalysis 466
- 23.3 Gas Consumption and Gas Formation 467
- 23.4 NMR Spectroscopy 470
- 23.5 IR-Spectroscopy 481
- 23.6 UV/Vis Spectroscopy 486
- 23.7 Summary 490
- References 490

24 *In-situ* Characterization of Heterogeneous Catalysts 493
Bert Weckhuysen
- 24.1 Introduction 493
- 24.2 Some History, Recent Developments, and Applications 495
- 24.3 *In situ* Characterization of a Reactor Loaded with a Catalytic Solid 497
- 24.3.1 A Reactor Loaded with a Catalytic Solid Probed by One Characterization Method 497
- 24.3.2 A Reactor Loaded with a Catalytic Solid Probed by Multiple Characterization Methods 499
- 24.4 *In situ* Characterization at a Single Catalyst Particle Level 501
- 24.4.1 *In situ* Micro-Spectroscopy of a Catalytic Solid 501
- 24.4.2 Single-Molecule *in-situ* Spectroscopy of a Catalytic Solid 504
- 24.4.3 *In-situ* Nano-Spectroscopy of a Catalytic Solid 509
- 24.5 Concluding Remarks 511
- Acknowledgments 511
- References 511

25 Adsorption Methods for Characterization of Porous Materials 514
Evgeny Pidko and Emiel Hensen
- 25.1 Introduction 514
- 25.2 Physical Adsorption 514
- 25.3 Classification of Porous Materials 517
- 25.4 Adsorption Isotherms 517
- 25.5 The Application of Adsorption Methods 518
- 25.6 Theoretical Description of Adsorption 519
- 25.6.1 Langmuir Isotherm 519
- 25.6.2 BET Theory 521
- 25.6.3 Standard Isotherms and the t-Method 522
- 25.7 Characterization of Microporous Materials 524
- 25.7.1 Dubinin–Radushkevich and Dubinin–Astakhov Methods 524
- 25.7.2 Horvath–Kawazoe (HK) Equation 525

25.8	Characterization of Mesoporous Materials	527
25.8.1	The Kelvin equation	528
25.8.2	BJH Method	529
25.8.3	Nonlocal Density Functional Theory (NL–DFT)	530
25.9	Mercury Porosimetry	533
25.10	Xenon Porosimetry	533
	References	534

26 A Critical Review of Some "Classical" Guidelines for Catalyst Testing 536
Frits Dautzenberg

26.1	Introduction	536
26.2	Encouraging Effectiveness	536
26.3	Ensuring Efficiency	537
26.3.1	Apply Effective Experimental Strategies	538
26.3.2	Collect Meaningful Data	540
26.3.3	Select the Most Appropriate Laboratory Reactor	543
26.3.4	Establish Ideal Flow Pattern	545
26.3.5	Ensure Isothermal Conditions	546
26.3.6	Diagnose and Minimize Effects of Transport	549
26.3.7	Assess Catalyst Stability Early	551
26.4	Concluding Remarks	552
	Appendix A: Three-Phase Trickle-Bed Reactors	552
	List of Symbols and Abbreviations	558
	References	559

Part VI Catalytic Reactor Engineering 561

27 Catalytic Reactor Engineering 563
Albert Renken and Madhvanand N. Kashid

27.1	Introduction	563
27.2	Types of Catalytic Reactors	564
27.2.1	Single-Phase Reactors	564
27.2.1.1	Stirred-Tank Reactor	564
27.2.1.2	Tubular Reactors	567
27.2.2	Fluid–Solid Reactors	568
27.2.2.1	Fixed-bed Reactors	568
27.2.2.2	Fluidized-bed Reactors	569
27.2.3	Fluid–Fluid Reactors	571
27.2.3.1	Liquid–Liquid–Gas System	573
27.2.4	Three-Phase Gas–Liquid–Solid Systems	573
27.2.4.1	Fixed-Bed Reactors	574
27.2.4.2	Slurry–Suspension Reactors	574
27.2.4.3	Structured Catalysts for Multiphase Reactions	575
27.3	Ideal Reactor Modeling/Heat Management	575

27.3.1	Mass and Energy Balances	*576*
27.3.2	Batchwise-Operated Stirred-Tank Reactors	*578*
27.3.3	Continuously Operated Ideal Stirred Tank Reactors	*580*
27.3.4	Ideal Plug Flow Reactor	*581*
27.4	Residence Time Distribution	*587*
27.4.1	Experimental Determination of the Residence Time Distribution	*589*
27.4.1.1	Step Function	*589*
27.4.1.2	The Pulse Function	*590*
27.4.2	RTD for Ideal Reactors	*591*
27.4.2.1	Ideal Plug Flow Reactor	*591*
27.4.2.2	Ideal Continuously-Operated Stirred Tank Reactor	*591*
27.4.2.3	Cascade of Ideally Stirred Tanks	*592*
27.4.2.4	Laminar Flow Reactor	*593*
27.4.3	RTD Models for Real Reactors	*595*
27.4.3.1	Dispersion Model	*595*
27.4.3.2	Cell Model	*596*
27.4.4	Estimation of the Residence Time Distribution in Tubular Reactors	*597*
27.4.5	Influence of RTD on Performance of Real Reactors	*599*
27.5	Microreaction Engineering	*602*
27.5.1	General Criteria for Reactor Selection	*602*
27.5.2	Types of Microstructured Reactors	*604*
27.5.2.1	Single-Phase MSR	*604*
27.5.2.2	Fluid–Solid MSR	*607*
27.5.3	Fluid–Fluid MSR	*610*
27.5.3.1	Gas–Liquid Systems	*611*
27.5.3.2	Liquid–Liquid Systems	*613*
27.5.3.3	Three-Phase Reactors	*616*
27.5.4	Heat Management in Microstructured Reactors	*622*
	References	*625*

Index *629*

List of Contributors

Manfred Baerns
Fritz-Haber-Institute of
Max-Planck Society
Department of Inorganic
Chemistry
Faradayweg 4–6
14195 Berlin
Germany

Matthias Beller
Leibniz Institut für Catalysis
Rostock University
A.-Einstein-Str. 29a, room 240
18059 Rostock
Germany

Uwe Bornscheuer
University of Greifswald
Institute of Biochemistry
Department of Biotechnology &
Enzyme Catalysis
Felix-Hausdorff-Str. 4
17487 Greifswald
Germany

Vincenzo Busico
Laboratory of Stereoselective
Polymerizations (LSP)
"Paolo Corradini" Department of
Chemistry
Federico II University of Naples
Via Cintia - Complesso di Monte
S. Angelo
80126 Napoli
Italy

Fabrizio Cavani
Dipartimento di Chimica
Industriale e dei Materiali
ALMA MATER STUDIORUM
Università di Bologna, INSTM
Viale Risorgimento 4
40136 Bologna
Italy

Christophe Copéret
Université de Lyon, Institut de
Chimie de Lyon, C2P2 UMR5265
(CNRS – CPE – Université
Lyon 1) ESCPE Lyon
43, Bd. du 11 Novembre
69616 Villeurbanne
France

Frits Dautzenberg
Serenix Corporation
5008 Chelterham Terrace
San Diego, CA 921309
United States of America

Olaf Deutschmann
Karlsruhe Institute of Technology
Institute for Chemical Technology and Polymer Chemistry
Engesserstr. 20
76128 Karlsruhe
Germany

Serafino Gladiali
Università di Sassari
Dipartimento di Chimica e Farmacia
Via Vienna 2, 07100 Sassari
Italy

and

Leibniz Institut für Catalysis, Rostock University
A.-Einstein-Str. 29a
18059 Rostock
Germany

Emiel Hensen
Eindhoven University of Technology
Inorganic Materials Chemistry
Den Dolech 2, STW 3.33
5612 EZ Eindhoven
The Netherlands

Timo Jacob
Universität Ulm
Institut für Elektrochemie
Albert-Einstein-Allee 47
89081 Ulm
Germany

Krijn P. de Jong
Utrecht University
Department of Inorganic Chemistry and Catalysis
Postbus 80083, David de Wied Building, room 4.80
3508 TB Utrecht
The Netherlands

Petra de Jongh
Utrecht University
Department of Inorganic Chemistry and Catalysis
Postbus 80083, David de Wied Building room 4.84
3508 TB Utrecht
The Netherlands

Detlef Heller
Leibniz-Institut für Katalyse an der Universität Rostock e.V.
Albert-Einstein-Str. 29a
18059 Rostock
Germany

Madhvanand N. Kashid
Ecole Polytechnique Fédérale de Lausanne
Institut des sciences et ingénierie chimiques
EPFL SB-DO, Office: CH J2 500
CH J2 500 (Bât. CH), Station 6
CH-1015 Lausanne

Lioubov Kiwi-Minsker
Ecole polytechnique fédérale de Lausanne
Institut des sciences et ingénierie chimiques
EPFL SB ISIC GGRC, Office: CH H3 594
CH H3 594 (Bât. CH), Station 6
CH-1015 Lausanne
Switzerland

Guido Mul
University of Twente
Photo Catalytic Synthesis Group
MESA + Institute for
Nanotechnology
Faculty of Science and
Technology
Meander 225
P.O. Box 217
7500 AE Enschede
The Netherlands

Evgeny Pidko
Eindhoven University of
Technology
Inorganic Materials Chemistry
Den Dolech 2, room STW 3.27
5612 AZ Eindhoven
The Netherlands

Roel Prins
ETH-Hönggerberg, Inst.für
Chemie-/Bio-ingenieur
Wissenschaft
Wolfgang-Pauli-Str. 10,
HCI E 125
CH-8093 Zürich
Switzerland

Fernando Rascón
Department of Chemistry
ETH Zurich/HCI H 229
Wolfgang-Pauli-Str.
108093 Zurich
Switzerland

Albert Renken
École Polytechnique Fédérale de
Lausanne
Institut des sciences et ingénierie
chimiques
EPFL SB-DO, Office: CH J2 500
CH J2 500 (Bât. CH), Station 6
CH-1015 Lausanne
Switzerland

Jens Rostrup-Nielsen
Haldor Topsoe A/S
Director R&D Division and
Member of Executive Board
Nymoellevej 55
DK 2800 Lyngby
Denmark

Rutger van Santen
NRSC-C Director
Honorary Professor at Eindhoven
University of Technology
Molecular Heterogeneous
Catalysis
Den Dolech 2, STW 3.28
5612 AZ Eindhoven
The Netherlands

Hans Schulz
University of Karlsruhe
Engler-Bunte Institute
Kaiserstrasse 12
76131 Karlsruhe
Germany

Detlef Selent
Leibniz-Institut für Katalyse an
der Universität Rostock e.V.
Albert-Einstein-Str. 29a
18059 Rostock
Germany

Bert Weckhuysen
Utrecht University, Debye
Institute for Nano Materials
Science
Inorganic Chemistry and
Catalysis Group
Sorbonnelaan 16, room 4.82
3584 CA Utrecht
The Netherlands

Preface

This book aims to provide material of use to advanced catalysis courses. The basic idea is to present a collection of chapters that explains the fundamentals of catalysis as developed over the past decades and also introduces new catalytic systems that are of increasing current importance.

There is no doubt that the science of catalysis is currently being transformed by the growth in our understanding of the molecular chemistry underlying catalytic phenomena. Not only are our techniques of manipulating catalysis on a molecular level becoming increasingly sophisticated, but so also is our ability to fully grasp the complexity of the changes in the catalytic systems as they occur in time and space.

The various disciplines: heterogeneous catalysis, homogenous catalysis and biocatalysis as well as reactor engineering are integrated in the different chapters on the six different themes around which the book is organized.

Part I deals with basic concepts. The book is organized in six Parts that cover physical chemistry, mechanism, synthesis, catalyst characterization and catalyst testing. There are also chapters on industrial applications and reactor engineering aspects. The first chapter gives an historic introduction to the field. It views catalytic science and technology as parallel developments driven by the needs of society, leading to the invention of new catalytic materials, and corresponding new developments in science and engineering. Important recently developed insights and concepts are emphasized, these mainly deriving from advances in computational catalysis and the design of new molecular chemistry.

This chapter is followed by three chapters that deal with the physical chemistry and kinetics of heterogeneous and homogeneous catalytic systems, complemented by kinetic considerations in reactor engineering, again with emphasis on recent molecular understanding of the fundamental complexity of catalytic systems, and is illustrated with the aid of some relevant examples.

Part II deals with the mechanism of catalytic reactions. This part contains five chapters on homogeneous, heterogeneous, and biocatalysis, with further sections on electrocatalysis and photocatalysis.

Part III deals with the full cycle of industrial catalytic systems. Novel homogeneous, biocatalytic, and polymerization catalytic reactions have been selected for presentation, and five heterogeneous catalytic systems are also discussed.

Essential to catalysis is the discovery, synthesis, and preparation of catalytically active systems, and in Part IV the molecular surface chemistry and details of the preparation of heterogeneous catalytic systems are presented. This part concludes with a chapter on combinatorial approaches and high-throughput methods of value in the development and discovery of new heterogeneous catalytic systems.

Part V deals with modern techniques for the (*in-situ*) characterization of catalytically working systems. The modern approach to the characterization of heterogeneous and homogeneous catalysts is described, including a section on the correct way to test the reactivity of heterogeneous catalysts.

The final part deals with catalytic reactor engineering aspects and specifically discusses microreactor systems.

Many authors of high reputation in their respective fields have written outstanding contributions to the different chapters. We are extremely grateful to them for being willing to find the time to contribute to the project, which was initiated by the Integrated Design of Catalysis (IDECAT), a European Community Network of Excellence, during the period 2006–2011. The IDECAT network has created the opportunity for many meetings and workshops jointly organized by European catalytic scientists. Within this network, it was felt that an integrated catalysis textbook of the type realized here would be extremely useful and was therefore highly desirable, and IDECAT provided funding to initiate the book. We believe that it will prove invaluable in workshops organized by the recently founded sister network, the European Institute of Catalysis (ERIC), in which we expect that many of the important IDECAT activities will continue.

May 2012

Matthias Beller
Albert Renken
Rutger van Santen

Part I
Basic Concepts

Physical chemical aspects of the kinetics of catalytic reactions are introduced. Chapter 1 can be viewed as an introduction to catalysis science and technology. The history of catalytic innovation is given and an historic context is provided of the development of fundamental understanding of catalytic reactivity. The three chapters that follow provide detailed discussions of the kinetics and physical chemistry of heterogeneous-, homogeneous- and bio-catalytic systems.

1
Catalysis in Perspective: Historic Review

Rutger van Santen

1.1
History of Catalysis Science

1.1.1
General Introduction

Catalysis as a scientific discipline originated in the early part of the last century. Earlier, the unique feature of a catalytic substance, namely that when added in small quantities to a reaction it affects its rate and selectivity but is not consumed, had become widely recognized, and many applications had been developed. Only after chemical thermodynamics had been defined did a rational approach to discover new catalytic processes become possible. Thermodynamics would define the proper conditions at which a material should be tested as a catalyst and catalytic turnover would be expected. Ostwald, one of the founding fathers of chemical thermodynamics, introduced thermodynamics into the physical chemical definition of a catalyst, specifying that it is a material that will leave the equilibrium of a reaction unchanged.

The past century can be viewed as the age of the molecularization of the sciences. It took nearly a century before the molecular basis of catalytic processes, now widely applied at very large scale, became understood. The Haber–Bosch process of ammonia synthesis was discovered early in the twentieth century once the thermodynamics of this process had become properly understood. The Nobel Prize to Ertl in 2007 recognized his discovery of the molecular principles of this reaction.

The three scientific disciplines that are essential to catalysis: chemical engineering, inorganic chemistry, and organic chemistry, which developed in the past fairly independently, now have a common basis (see Figure 1.1).

The chemical tradition of the nineteenth century had culminated 100 years earlier in the Nobel Prize for Sabatier for catalytic hydrogenation, useful because coal had made the production of hydrogen cheap. Sabatier formulated the principle that the reaction intermediates formed at the surface of a catalytic material should have an intermediate stability. When too stable they would not decompose, when too unstable they would not be formed. This molecular view of the catalytic

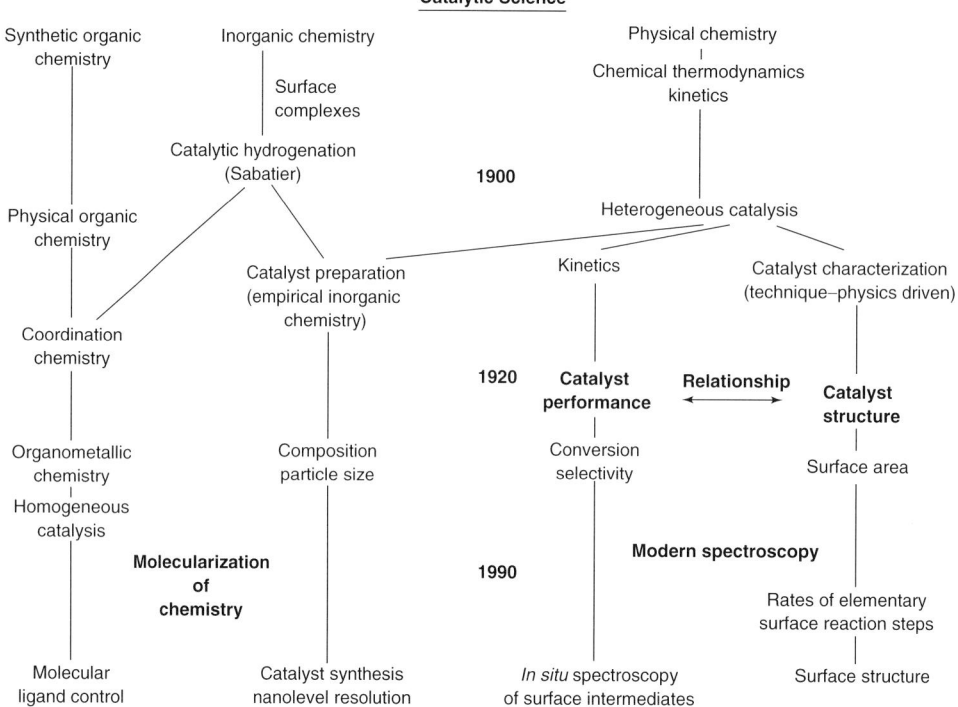

Figure 1.1 The three branches of catalysis history.

reaction, not as a single reaction but as a cycle of reaction steps, in which intermediate complexes between a catalyst and a reagent are formed and then decay, was particularly modern. Sabatier's principle is the formulation of the molecular basis of catalytic action and complements Ostwald's physical chemical view.

Current state-of-the-art physical chemical instrumentation and computational and molecular chemistry provide a basis for the formulation of a molecular theory of catalysis. This basis will be presented in the first three chapters.

1.1.2
Heterogeneous Catalysis: the Relationship between a Catalyst's Performance and its Composition and Structure

Kinetics provides the basis to the physical chemical description of catalytic reactivity. Reaction engineering is the discipline that connects reactor performance with the chemical reactivity of the catalyst. The level of accuracy to be obtained is highly dependent on the accuracy, sensitivity, and compatibility of measuring technique with catalytic reactor operation. It is essential to understand the difference between performance parameters, which are caused by such phenomena as heat and mass transfer, and features of the catalytic reaction, which in turn are due to the chemical reactivity of the catalytic material. This is the difference between extrinsic and intrinsic kinetics, to be discussed in Part I.

The link between chemical kinetics and molecular activation at the catalyst surface is through the mechanism of the catalytic reaction. This implies a physical organic chemistry type of understanding of the reaction paths of molecular transformations that happen when the reaction proceeds in its catalytic cycle. Transient and isotope-labeled kinetic studies as well as *in situ* spectroscopic measurements are the experimental tools. To obtain detailed molecular information, such studies often have to limit themselves to the investigation of a single elementary step of the overall catalytic reaction. The integration of such information in order to understand the kinetics of the complete catalytic cycle usually involves the lumping together of properties of many elementary reaction events, which causes the physical chemistry of kinetics and molecular understanding to be often indirectly related. Agreement between prediction and experiment is important as a tool, but cannot be considered the ultimate proof of the validity of a proposed mechanism.

This state of affairs is dramatically improved by advances in computational modeling of catalytic chemistry, which currently enables the prediction of rate constants of elementary reaction steps and discrimination between mechanistic options of reaction routes. The mechanistic discussions in later chapters will demonstrate this.

A central theme in the science of heterogeneous catalysis has been and still is the characterization of the catalytic material at the level of detail relevant to its performance.

Similarly to the direct access to the mechanism of reactions, molecular characterization of the catalyst surface also has long been beyond direct reach of experimentalists. Due to the heterogeneous nature of a catalyst's particles this is still not always completely possible, but for model type catalysts a molecular description is becoming possible.

It is becoming evident that ultimately transformations of the catalyst structure are intimately related to the chemical transformations they induce and the influence of the reaction medium. Thus surface changes and surface chemistry are intimately coupled. Again, advanced spectroscopy in combination with computational approaches is getting to a stage where definitive study becomes possible.

Probing the molecular basis of catalysis generated the need to study model systems to validate theory and experiment at their respective time and length scales. It gave rise to the surface science approach in catalysis, with a corresponding generation of catalytic surfaces and surfaces of increasing molecular definition. This, coupled with the considerable parallel advances in coordination chemistry, metal organic chemistry, and homogeneous and molecular biocatalysis, gives rise to new generations of catalysts of increasing complexity, but also increasingly better molecular. This not only leads to catalysts of increased selectivity, stability, and energy efficiency, but also helps fundamental progress in catalysis. Definition of reactivity descriptors, which are of molecular refinement, becomes possible. The history of catalysis shows this in the progression of techniques for characterization and kinetics modeling approaches.

The first important characterization tool next to chemical composition was structural characterization. The Brunauer-Emmett-Teller (BET) and later T-plot techniques were developed to determine catalyst surface area and porosity structure. These methods were based on thermodynamics, and hence could not provide molecular information. X-ray scattering techniques were developed to characterize particle size and kind. A major advance was the use of electron microscopy. The ongoing development of spectroscopic techniques, such as photo-emission and synchrotron-based methods, has only recently made it possible to analyze complex working catalysts at a nanometer scale.

Truly molecular information needs access to model systems and model conditions in order to apply spectroscopic techniques with atomic resolution. This very important branch of catalysis found its early start in Langmuir's surface science studies in a vacuum. Later this was brought to high levels of sophistication especially by the schools of Somorjai and Ertl. It is the data from such studies that provide validation of the now also well-established discipline of computational catalysis. In the next chapters the fundamental insights on catalyst reactivity provided by these approaches will be given.

The formulation of kinetics underwent similar changes. A proper description of the kinetics of a catalytic system is crucial to the design of catalytic reactor systems. An understanding of the relationship between the volume of a catalyst and its performance is fundamental. Knowledge of porosity and surface area is of direct engineering relevance. Whereas the equations can be mathematically complex, the basis of kinetics up to the present is empirical. Over the past century kinetic models, useful for engineering purposes, were designed by fitting parameters to experimental measurements. To improve such models, mechanistic assumptions about the reaction network had to be made. These became more and more refined with the increasing understanding of the physical chemistry of catalytic reactivity.

The simplest kinetic equation is the power law rate expression that relates the rate of the reaction with the concentration of reactants and products. As we will see in the next chapter the parameters of such power law expressions depend in principle also on the concentration of reacting molecules and can be expected to be only valid over a limited concentration, pressure, and temperature regime.

The Langmuir–Hinshelwood–Watson–Hougson (LHWH) expressions, which do not use the assumption of the power law concentration dependence, can be used to derive the simpler and more easy-to-use kinetic expressions. The LHWH expressions need the explicit assumption of a reaction mechanistic network. They also involve the assumption that most reaction steps are equilibrated and only one reaction is not equilibrated (often called the *rate-limiting step*). In addition, the surfaces are usually assumed to be uniform, and, most importantly, in the models the total number of reaction sites is maintained constant. For this reason the LHWH expressions are the equivalent of the Michaelis–Menten expression used in enzyme and homogeneous catalysis. The parameters in the LHWH expressions that are fitted to experiment are to be considered lumped parameters, since many details of the elementary surface reactions are ignored.

Microkinetics is the kinetics approach that takes full account of the elementary reaction rate expressions deduced from catalysis studies at the molecular level. Eyring's transition state expressions are used. No assumption is made about equilibrated and non-equilibrated reaction steps.

Ultimately, Dynamic Monte Carlo methods have been developed that can include non-uniformity of surfaces, overlayer phase formation, and even surface reconstruction effects. No mean field approximation of uniform surface concentrations, as made in the other methods, is necessary in this case. Such methods are useful to deduce kinetic parameters valid in particular concentration regimes to be used in other simpler-to-handle approaches.

The ongoing molecular understanding of the chemistry and characterization is also impacting the molecular level design of heterogeneous catalytic systems. Most important was the insight, which gained general support, that the reactivity of a surface and that of related molecular complexes can be quite similar. This brought heterogeneous catalysis into the heart of molecular catalysis. Here we now see an important cross-fertilization of the intrinsically molecular approach of coordination complex chemistry and the increasing importance and refinement of homogeneous catalysis in the second half of the last century.

Whereas the early heterogeneous catalysts were simply metal powders, not much later they became materials with highly dispersed catalytically active components on high-surface supports. Control of composition, size, and shape of catalytically active particles and of catalyst support morphology gave rise to important improvements in the catalytic performance. This can be viewed as one of the main benefits of continued research aimed at understanding the relationship between catalyst performance and its structure.

There are many examples of the sustained gradual improvement in catalyst performance that occurred over many years. The introduction of microporous solid acidic zeolites instead of non-structured clays as catalysts for catalytic cracking of oil saved society a significant fraction of crude oil, that otherwise would have been converted into residual coke. In chemicals production a familiar example is the production of ethylene epoxide by a silver-based catalyst. In the course of 50 years the selectivity improved from initially 40% to the current 90% with large savings in ethylene that otherwise would have been combusted.

Catalyst preparation methods were based on increasing understanding of the chemistry of the often complex reaction mixtures and of the inorganic chemistry of their reaction with catalyst supports. Molecular catalysis gave rise to molecularly defined immobilized systems and the use of well-defined clusters or complexes in catalyst preparation. Zeolites, with their microscopic channel structure that is atomistically defined, can be considered to be an example of the ultimate molecular heterogeneous catalyst.

1.1.3
Homogeneous and Enzyme Catalysis

The first molecularly defined catalyst was the Co carbonyl hydroformylation catalyst discovered by Roelen in 1938. Its mechanism, defined in physical organic terms, was unraveled in the 1960s by Heck and Breslow, and it was later developed commercially by Shell.

Earlier, mercury sulfate had been industrially applied for the conversion of acetylene to acetaldehyde. Later, in the 1950s, the Wacker process of selective oxidation of ethylene catalyzed by the Pd/Cu system was introduced.

The Ziegler–Natta invention of an ethylene and propylene polymerization catalyst in the 1950s, based on $TiCl_3$, signaled the beginning of well-defined (immobilized) coordination complexes serving as catalytically active species, in parallel with the development of metal–organic chemistry.

This development was crowned by Wilkinson's discovery of homogeneous hydrogenation in 1965. The catalyst, $RhCl(PPh_3)_3$, consists of a single metallic center stabilized by triarylphosphines. The unique feature of such organometallic complexes is that they can be manipulated molecularly by variation of the ligands. With their invention the field of molecular catalysis has been expanded from the organic chemist's realm into metal-based catalysis. Catalyst design through development of physical chemical approaches, ligand synthesis, and computational modeling techniques has become one of the outstanding features of this branch of catalysis.

These developments have provided the basis of several large-scale homogeneous bulk industrial processes. Examples are the Rh-based carbonylation of methanol and hydroformylation processes. More recently we see the development of metathesis applied, for instance, in the ring-opening polymerization process of Huels, and enantiomeric catalysis due to the invention of highly enantiomeric ligand systems, as for the production L-Dopa by Monsanto.

A special issue in homogeneous catalysis is separation of catalyst from product after reaction, and for this there are unique developments and applications of biphasic systems and membrane reactors.

Whereas biocatalysis has been used widely in fermentation processes from the early beginning of mankind, the science of biocatalysis only started when Sumner and Northrop were able to crystallize an enzyme, the molecule active as a biocatalyst in the living system, and identified it as a protein. The protein acts as the complex ligand of the catalytically active center, that can be an organic acid or base, a metal, or an inorganic metallic complex. Variation of the protein composition far from the actual catalytic site can have a major effect on catalyst performance. Unique to enzyme catalysis is multipoint contact and activation of a substrate molecule when this is adsorbed into the interior of the enzyme.

Very early in the nineteenth century, Willstätter discovered catalases and peroxidases that activate hydrogen peroxide, and Summer concentrated on urease that decomposes urea.

One of the early bulk processes that employ a hydrolase enzyme is the Mitsui Toatsu process that converts acetonitrile into the corresponding amide.

Modern biomolecular chemistry has a major impact on the design and improvement of biocatalytic systems through the use of combinatorial and recombinative techniques that allow for DNA reshuffling. Such evolutionary molecular biological techniques have been developed especially for application to fine chemical synthesis. Mutations are introduced through the biochemical polymer chain reaction or other random chain reactions. This approach has led to the development of bacterial lipases with significantly enhanced enantioselectivity.

Differently from the design approaches in homo- and heterogeneous catalysis, in this approach to catalysis no mechanistic information on the catalytic reaction is used to optimize the system. The desired catalyst is found by feedback of the information obtained by screening into the selection of the bacteria possessing the desired gene sequences [1].

1.1.4
Important Scientific Discoveries

To give a historic illustration of the scientific advances that gave rise to the science of catalysis as we now know it, we have listed in this section the relevant Nobel prizes. The Nobel prizes for heterogeneous catalysis are followed by the recognition of discoveries in coordination chemistry and biochemistry.

- 1909 – W. Ostwald, for his work on catalysis, and for his investigations into the fundamental principles governing chemical equilibria and rates of reactions.
- 1912 – P. Sabatier, for his method of hydrogenating organic compounds in the presence of finely divided metals whereby the progress of organic chemistry has been greatly advanced in recent years.
- 1918 – F. Haber, for the synthesis of ammonia from its elements.
- 1931 – C. Bosch and F. Bergius, in recognition of their contributions to the invention and development of chemical high-pressure methods.
- 1932 – I. Langmuir, for his discoveries and inventions in surface chemistry.
- 1956 – C. N. Hinshelwood and N. N. Semenov, for their researches into the mechanism of chemical reactions.
- 1963 – K. Ziegler and G. Natta, for their discoveries in the field of chemistry and technology of high polymers.
- 1973 G. Wilkinson and E. O. Fischer for pioneering work on the chemistry of the organometallic so-called sandwich compounds.
- 1983 – H. Taube, for his work on electron transfer reactions, especially in metal complexes.
- 1989 – S. Altman and T. Cech, for their discovery of the catalytic properties of RNA.
- 1993 – K. B. Mullis, for his invention of the polymerase chain reaction.
- 1994 – G. A. Olah, for his contributions to carbocation chemistry.

- 2001 – W. S. Knowles, R. Noyori, and K. Barry Sharpless, for their work on chirally catalyzed hydrogenation reactions and for the work of KBS on chirally catalyzed oxidation reactions.
- 2005 – Y. Chauvin, R. H. Grubbs, and R. R. Schrock, for the development of the metathesis method in organic synthesis.
- 2007 – G. Ertl, for his studies of chemical processes on solid surfaces.
- 2010 – R. F. Heck, E. Negishi, and A. Suzuki, for palladium-catalyzed cross-couplings in organic synthesis.

The word catalysis is mentioned four times during a period of nearly 100 years – first, in the context of heterogeneous catalysis, then biocatalysis and homogeneous catalysis, and recently for application in organic synthesis. Before 1963 no research award was given for the use of coordination complexes in catalysis. In biocatalysis the first Nobel Prize was in 1946 for crystallization of an enzyme (Sumner, Northrop, and Stanley), followed by Kendrew and Perutz's (1962) Nobel Prize for crystallographic studies of hemoglobin.

It is interesting to realize that the increased understanding of the molecular mechanism in homogeneous and biocatalysis came earlier than it did in heterogeneous catalysis.

There are many factors that contribute to the recognition of a Nobel Prize award. It is interesting to ask the question whether for catalysis there are other breakthrough developments that possibly are equivalent at least in impact and originality. Below, a list of such discoveries essential to the advancement in catalysis can be found. They are important because they often are basic to the development of new industries or new catalytic processes:

- F. Fischer and H. Tropsch (1926). Oligomerization of hydrocarbons from CO – heterogeneous catalysis.
- V. Ipatieff and E. Houdry (1930). Amorphous solid acid catalysts – catalytic cracking.
- (1930) Reducible vanadium oxide catalysts for benzene oxidation, earlier discovered for SO_2 oxidation.
- T. E. Lefort (1931). Silver-catalyzed epoxidation of ethylene – a heterogeneous process.
- H. Pines (1940). Superacid catalysis (liquid phase) – alkylation.
- O. Roelen (1938) and W. Reppe (1941). Hydroformylation, carbonyl chemistry, and homogeneous catalysis.
- V. Haensel (1940). Bifunctional heterogeneous catalysts (catalytic reforming).
- J. Smidt, WACKER (1959). Homogeneous oxidation of ethylene by oxygen by Pd-Cu oxidation redox couple.
- R. K. Graselli, SOHIO (1955). Mixed oxides for selective oxidation and ammoxidation.
- R. L. Banks and G. C. Bailey (1964). Heterogeneous metathesis by supported oxide clusters.
- C. J. Plank and E. J. Rosinski, MOBIL (1968). Zeolite catalytic cracking.

- H. P. Wulff and F. Wattimena, SHELL (1969). Heterogeneous Ti-catalyzed epoxidation.
- W. Keim, SHELL (1972). Shell Higher Olefins Process.
- C. D. Chang, A. J. Silvestri, and W. H. Lang, MOBIL (1972). Methanol to gasoline, ZSM-5 zeolite catalyst.

Before 1900 we have two important inorganic chemicals produced by catalytic processes:

- Sulfuric acid by the lead chamber process (NO_x) or contact process (Pt).
- Chlorine from HCl by the Deacon process ($Cu/ZnCl_x$).

Early in the nineteenth century we have the invention of electrocatalysis, for which Faraday gave the first scientific basis:

- The invention of metal-catalyzed hydrolysis, oxygen reduction, and the fuel cell.

Also in the early part of the nineteenth century there are important inventions of applications of heterogeneous catalysis. Famous are Davy's miner's lamp (based on catalytic oxidation by Pt) and the lamp by Döbereiner (this was a Kipp's apparatus for hydrogen generation combined with a Pt catalyst for oxidation). Berzelius was inspired by acid- and base-catalyzed hydrolysis reactions.

The first half of the twentieth century is dominated by the development of heterogeneous hydrogenation processes (NH_3 and hydrocarbon-related chemistry). Hydrogen was now available from coal. High-pressure continuous processes were developed through the use of this technology. This can be considered to be the origin of reactor engineering as we now know it.

Note the early discovery of heterogeneous catalytic hydrogenation and the discovery of homogeneous catalytic hydrogenation 60 years later by Wilkinson – similarly the early discovery by Banks *et al.* of the heterogeneous catalysis of the metathesis reaction and the 30-years-later discovery of homogeneous metathesis.

1.2
The Development of Catalytic Processes: History and Future

The technological development of catalysis shows an intimate relationship between important political and societal developments and the exploitation of new catalytic technologies. The relationship between scientific discovery and its use in society is complex. Earlier we discussed highlights of catalytic advances; here we will provide a historic list of important industries based on catalysis (Table 1.1).

We have three columns. In the first column we indicate the major societal issues of that moment, for instance, the changing uses of raw materials. We recognize the transitions from coal to oil and natural gas. There also the two World Wars that had a major impact on the development of particular technologies. At the end of the last century there were oil crises and needs for environmentally friendly techniques. In the early part of this century the issues of climate change generated

Table 1.1 The history of catalytic processes.

	Catalyst	Process
1900	Noble metal	Hydrogenation
1910 World War I	Promoted iron	Nitrogen to ammonia
		Synthesis gas to methanol gasoline
1920	Sulfides	Desulfurization–denitrogenation
1930 Automobiles	Solid acids	Catalytic cracking
1940 World War II	New acid catalysis	Synthetic kerosene
	Superacids	Alkylates
	Anionic catalysis	Synthetic rubber
1950	Coordination catalysis	Polymers
1960 Petrochemical industry	Bifunctional catalysis	Hydrocracking
	Zeolites	Catalytic cracking
	Reducible oxidic systems	Selective oxidation
1970 Energy crisis	Methanol to gasoline	Novel synthetic acidic zeolites
	Synthesis gas to chemicals	Organometallic complexes
1980 Environment	Noble metal alloys	Exhaust catalysis
	Mixed oxides	Stack gas treatment
	Molecular heterogeneous catalysis	Fine chemical catalysis
	Catalytic organic chemistry	
1990 Environment	Immobilized enzymes	Enantiomeric catalysis
	Organometallic complexes in nano/mesoporous materials	
	Supported reducible oxides	NO_x, SO_2 reduction
	Zeolitic redox systems	N_2O utilization (Panov reaction)
Raw materials: natural gas and coal	Reducible mixed oxides	Selective alkane oxidation and ammoxidation
	Pt/Rh	Alkane dehydrogenation; synthesis gas
	Ga, Zn in zeolites	Alkane to olefins and aromatics
	$Si_xAl_{1-x}(PO_4)_2$ zeolitic systems	Methanol to olefins
	Co nanoparticles	Synthesis gas to hydrocarbons
2000 Climate	Electrocatalysts	Fuel cell
	Cr/molten salts	Glucose to diesel
	Early transition metals	Hydrogen storage
	Inorganic oxides	CO_2 storage and activation
	Photocatalysis	
	Hybrid systems	

a demand for processes based on renewable energy. The wide recognition of these issues inspired goal-oriented research. Inventions were made that could not have been foreseen but had considerable consequences.

In the second column the technologies are listed. We recognize several of these from the previous sections. It appears that very often the new catalytic technology relates to utilizing the discovery of a new catalytically active material or complex. These are listed in the final column.

Whereas heterogeneous catalysis gave rise to the construction of highly efficient large-scale processes, we see in our age a shift to smaller scales and also to the use of catalysis in various devices.

The automotive exhaust catalyst illustrates this. The reactor is embedded in the exhaust of the automotive engine and operates by integration with gas sensors and computer control. This trend persists in fuel cell development and devices for hydrogen storage and solar energy conversion.

1.3
Fundamental Catalysis in Practice

There are economic constraints on catalyst cost, catalytic performance, and process selection that are very relevant. Catalyst improvement does not necessarily lead to a practically relevant system. Interestingly, as the matter of catalyst choice will illustrate, there is a direct relationship between molecular predictive understanding and new options for catalyst choice. Catalytic performance and reactor choice are determined, among other things, by extrinsic kinetics, in which mass and heat transfer play a dominant role. In energy conversion technology, process selection of preferred conversion processes is largely determined by how the energy content of product and raw material relate.

1.4
Catalyst Selection

A major contribution of computational catalysis is the development of a computational approach to evaluating and predicting the catalytic activity of heterogeneous catalytic systems. This method uses the dependence of a measured rate of some reaction as a function of a reactivity descriptor that can be calculated. The method exploits the physical chemical interpretation of the volcano type behavior that is found in this way. This is a consequence of the Sabatier principle. If one uses the interaction strength of a reactant with a catalyst as a reactivity parameter the maximum in the volcano curve occurs where the rate of reactant activation and that of product desorption are the same (see Figure 1.2).

To the left of the volcano the rate of reaction increases with increasing interaction energy, to the right it decreases with increasing reaction energy. The overall catalytic rate shows a maximum rate at an optimum value of the reactivity descriptor.

The second ingredient in the extrapolative prediction of the rate is the use of a Brønsted type linear activation energy–reactivity descriptor relationship for elementary surface reactions. As reactivity descriptor a thermodynamic quantity as the reaction energy of an essential elementary reaction step is used. For surface reactions this is called the Brønsted–Evans–Polanyi relationship (BEP).

An example of such a relationship is shown in Figure 1.3a. The computed activation energy for CO dissociation, that is, for C–O bond cleavage, is plotted against the sum of the adsorption energies of adsorbed C and O. The surface structure is corrugated.

The slope of the curve, the BEP proportionality constant, is nearly one. It indicates that the structure of transition state and dissociated state are very similar and that in the transition state the C–O bond is very weak. Once the validity of the BEP relationship for an essential reaction step of a catalytic reaction step has been established the relevant descriptor can be used to construct a volcano curve when a large set of experimental data is available. One can also use microkinetic modeling techniques of the intrinsic catalytic rate to create such a data set computationally.

Figure 1.3b shows a plot of the experimentally measured rate constants of the catalytic reaction:

$$CO + 3H_2 \rightarrow CH_4 + H_2O$$

against a reactivity descriptor that is the sum of the adsorption energies of the dissociation products E_{diss}. A volcano-type reaction rate dependence is found in accordance with the Sabatier principle. The optimum value of the reactivity descriptor is close to 0.

Access to a volcano curve such as that in Figure 1.3b creates the possibility to computationally screen many metal combinations. One then has to calculate the sum of the adsorption energies of C and O (E_{diss}) for different metal compositions. In this way the reaction rate for many metal combinations can be computationally analyzed.

The method has been used to find alternatives to the expensive Co and Ru metals that show maximum performance for the methane reforming reaction. Figure 1.4 illustrates the way these data can be used to arrive at an economic decision as to which is the material to be preferred.

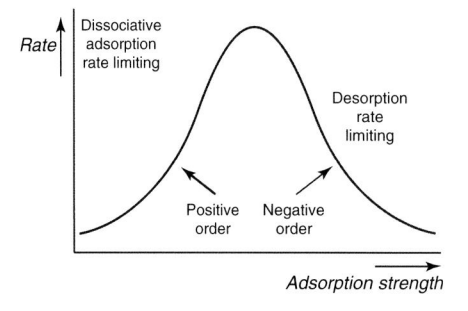

Figure 1.2 Sabatier's catalytic reactivity principle; rate reaches a maximum at optimum interaction strength of reagent and catalyst.

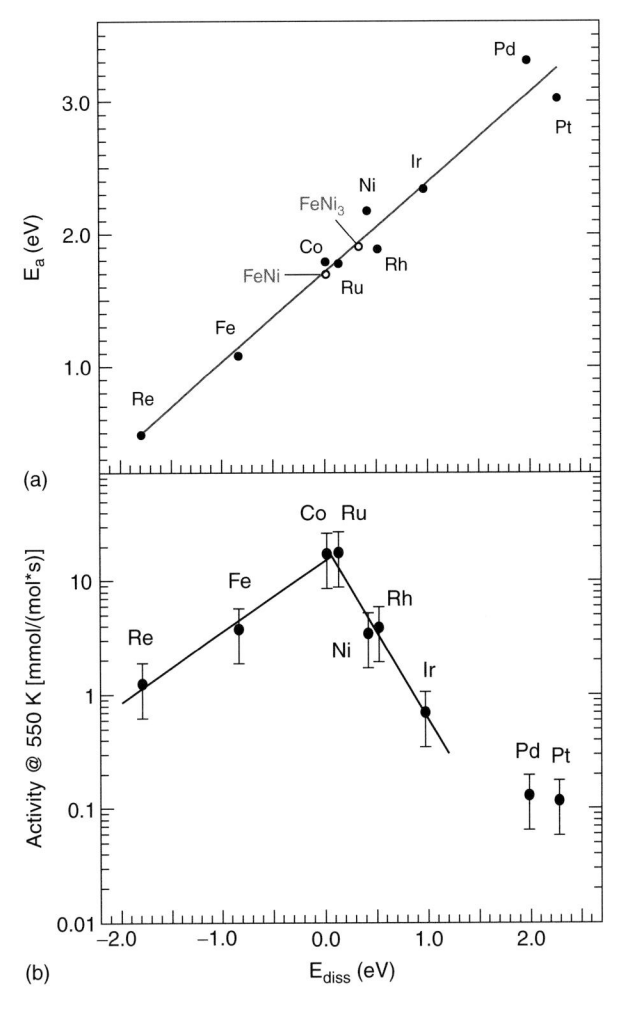

Figure 1.3 (a) BEP relationship for the activation energy for CO dissociation vs the CO dissociation energy, E_{diss}. (b) Measured catalytic activities for supported metal catalysts are shown vs E_{diss}. The calculations were performed using the RPBE exchange correlation on periodically repeated stepped fcc (211) metal slabs with 12 layers in the (211) direction (adapted from Ref. [2]).

Shown is a so-called Pareto plot of E_{diss} for a particular metal combination against the price of this combination. When ΔE_{diss} is zero the performance of the material is maximal. Only in exceptional cases will this coincide with the lowest price. The Pareto plot allows for a choice in which there is a trade-off between lowest process and best performance. According to the Pareto plot the lowest price with close to maximum performance is the alloy FeNi$_3$.

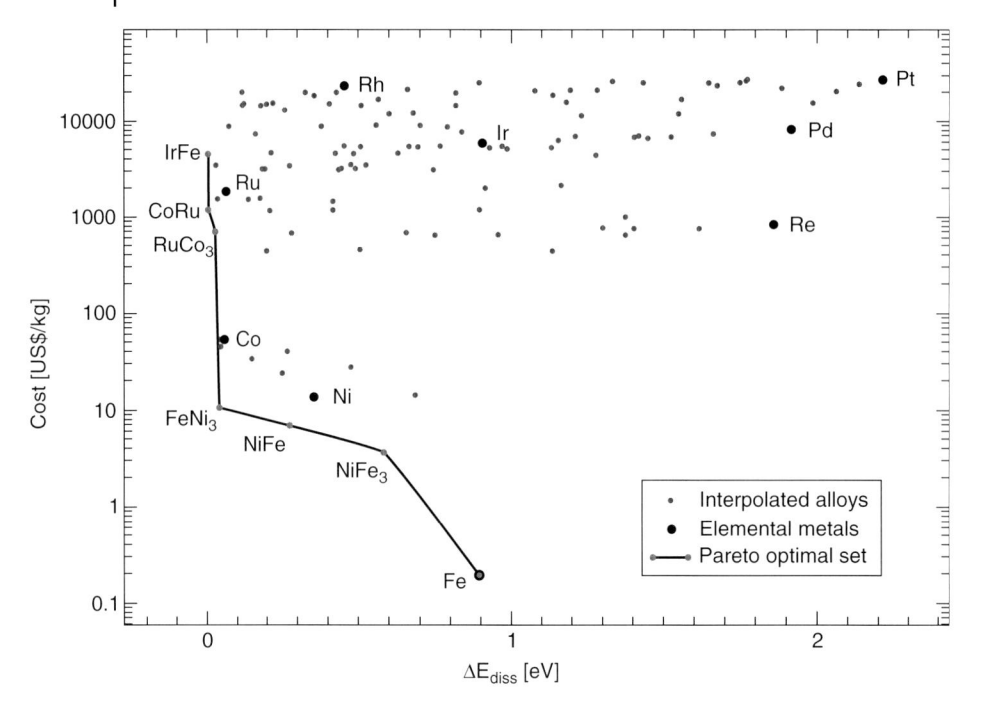

Figure 1.4 Pareto plot of the activity measure $\Delta E_{diss} = \left| E'_{diss} - E^{opt}_{diss} \right|$ and the cost for 117 elemental metals and bimetallic alloys (adapted from Ref. [2]).

1.5
Reactor Choice

The catalyst of maximum activity is not always the catalyst that is useful in practice. The overall rate of a catalytic reaction also depends on extrinsic kinetic parameters such as rates of mass and heat transfer. In Figure 1.5 reactivity regimes are compared for three processes. The petroleum geochemical processes generate the oil and gas reservoirs. The space–time yield of such processes is very low. There is also a comparison of biochemical processes for fermentation and the optimum process window for industrial catalysis. Note the three orders of

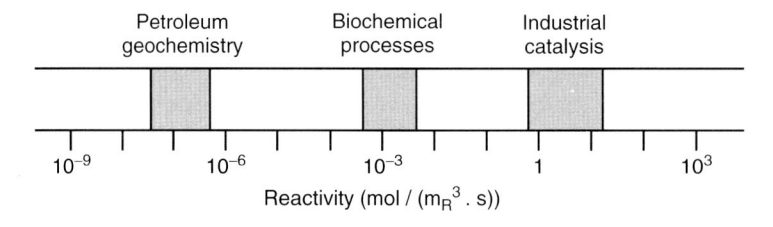

Figure 1.5 Reactivity regimes for different chemical processes.

magnitude difference between industrial catalysis and biochemical processes. The reason for the optimum window of the industrial process is the trade-off between two parameters. One is the intrinsic rate of a catalyst, which preferably is maximal. However, when the rate of a reaction becomes very high other factors can become limiting, such as the rate at which mass can be transported to or from the catalyst or the amount of heat to be supplied or removed.

Reactor design sets these limitations. The process is typically executed at the optimum condition where mass and heat transfer limit production rate. Intrinsic kinetics then sets the limiting values that can be used. Clearly the better the extrinsic kinetic parameters are controlled and can be increased, the higher the production.

1.6
Process Choice

The thermodynamic and material efficiency of a process are critical to its selection. A measure of material efficiency is the concept of Atom Utilization introduced by R. Sheldon [3]. The essential idea is to evaluate the production of waste material by simply counting the ratio of the number of atoms in the reactant material to the number of atoms in the product. For practical purposes it is useful to convert the Atom Utilization number into a weight ratio, so as to evaluate the efficiency on a weight basis.

The effective Atom Utilization can be influenced by the choice of catalyst and reactor. It depends on the selectivity of a reaction. We illustrate this for epoxidation processes to ethylene (Figure 1.6).

The classical route proceeds through the intermediate chlorohydrin using Cl_2 and $Ca(OH)_2$ as reactants. Atom Utilization is 25% with $CaCl_2$ as waste product. The catalytic process based on Ag ideally produces ethylene epoxide with 100% selectivity. The Atom Utilization is 100%. In practice the reaction is run with 90% selectivity giving an effective Atom Utilization of 77%. It illustrates the importance of catalytic processes. This number is still significantly higher than that of the chlorohydrin route.

In dealing with such transformation processes as the conversion of coal or gas to liquid fuels for transportation thermal efficiency is critically important. This is the heat of combustion of the products divided by that of (all) the feedstock used. For example, the thermal efficiency of oil refineries is typically 90%. This is to be contrasted to the thermal efficiency of the production of a non-fuel product such as methanol, which has a thermal efficiency of only 67%.

De Jong [4] mentions two important principles for efficient processes:

- minimizing the difference in hydrogen content between feed and product, and
- minimizing the number of process steps.

The importance of hydrogen content is illustrated by the conversion of natural gas or coal into liquid energy carriers. As illustrated in Figure 1.7, conversion

Classical chlorohydrin route

$$H_2C=CH_2 + Cl_2 + H_2O \longrightarrow ClCH_2CH_2OH + HCl$$

$$ClCH_2CH_2OH + Ca(OH)_2 \longrightarrow H_2C\overset{O}{-}CH_2 + CaCl_2 + 2H_2O$$

Overall:
$$H_2C=CH_2 + Cl_2 + Ca(OH)_2 \longrightarrow H_2C\overset{O}{-}CH_2 + CaCl_2 + H_2O$$
Mol. wt.   44   111   18

Atom utilization = 44/173 = 25%

Mobern pertochemical route

$$H_2C=CH_2 + \tfrac{1}{2} Cl_2 \xrightarrow{catalyst} H_2C\overset{O}{-}CH_2$$

Atom utilization = 100%

Figure 1.6 Atom utilization for the classical chlorohydrin route to ethylene oxide is 25%, whereas that of the modern petrochemical route is 100%.

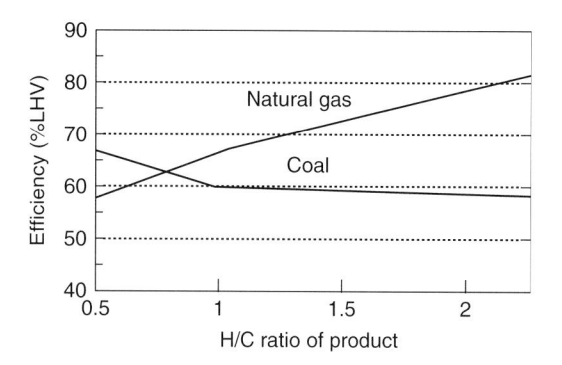

Figure 1.7 Maximum thermal efficiency for synthetic fuels: synthesis gas to hydrocarbons (reproduced from Ref. [4]).

of natural gas into hydrogen-rich paraffinic molecules has a substantially higher thermal efficiency than the conversion of coal.

Aromatic gasoline (H/C = 1) has a substantially lower thermal efficiency than paraffin (H/C = 2). There is a very low thermal efficiency of fuel production via the synthesis gas route because of the energy cost of coal gasification. Production of liquid fuels via direct hydrogen addition (Bergius process) is more efficient.

De Jong also gives an interesting example of the effect of the introduction of several process steps. The example is the production of methyl *tert*-butyl ether.

Direct etherification from isobutane requires one step, and reaction with *n*-butene requires an additional isomerization step, whereas reaction with butane needs an

Table 1.2 Energy use in MTBE manufacturing [4].

Process/feedstok[a]	Process/name	Energy consumed (GJ/t MTBE)[b]
iso-C_4^{2-} + MeOH → MTBE	Etherification	<0.1
n-C_4^{2-} + MeOH → MTBE	Isomerization	1.9
iso-C_4^0 + MeOH → MTBE + H_2	Dehydrogenation	5.5

[a]C_4^{2-} = butene, C_4^0 = butane, and MeOH = methanol.
[b]Energy consumptions excluding low-pressure steam.

additional dehydrogenation step. Table 1.2 shows the increasing use of energy with number of process steps.

References

1. Reetz, M. and Jaeger, K.-E. (1999) Superior biocatalysts by directed evolution. *Biocatal. Discov. Appl.*, **200**, 31–57.
2. Andersson, M.P. et al. (2006) Toward computational screening in heterogeneous catalysis: pareto-optimal methanation catalysts. *J. Catal.*, **239** (2), 501–506.
3. Sheldon, R.A. (1994) Consider the environmental quotient. *Chemtech*, **24** (3), 38–47.
4. De Jong, K.P. (1996) Efficient catalytic processes for the manufacturing of high-quality transportation fuels. *Catal. Today*, **29** (1–4), 171–178.

Further Reading

Jencks, W. (1969) *Catalysis in Chemistry and Enzymology*, McGraw-Hill, New York.

Moulijn, J.A., Makkee, M., and van Diepen, A. (2001) *Chemical Process Technology*, Wiley-VCH Verlag GmbH, Chichester.

Santen, R.Av. and Neurock, M. (2006) *Molecular Heterogeneous Catalysis*, Wiley-VCH Verlag GmbH, Weinheim.

van Leeuwen, P.W.M. (2004) *Homogeneous Catalysis*, Kluwer Academic Publishers, Dordrecht.

2
Kinetics of Heterogeneous Catalytic Reactions

Rutger van Santen

2.1
Physical chemical principles

2.1.1
The Catalytic Cycle

The Sabatier Principle states that a catalytic reaction has its maximum rate when there is an optimum interaction of reactant or product molecules with the catalyst. This is because the catalytic reaction is a cycle of consecutive elementary reaction steps. The rate of the reaction has a maximum when the rate of reactant activation to give a surface reaction intermediate equals the rate of its removal (see Figure 1.2; Chapter 1). The intermediate complex of reactant and surface should be stable enough that chemical bonds in the reactant molecule become activated, but have enough instability so that it can desorb. It is useful to analyze this kinetic argument using elementary kinetics equations of a simple reaction.

We will initially consider the activation of a diatomic molecule that reacts upon adsorption to a surface. In a consecutive reaction, it dissociates. The adsorbed reaction fragments transform to desorbed product. An example of such a reaction would be the hydrogenation of nitrogen to give ammonia or CO hydrogenation to give methane.

The elementary kinetic scheme of such a reaction is shown in Eq. (2.1).

$$A_{2,gas} \underset{k_{des}}{\overset{k_{ads}}{\rightleftarrows}} A_{2,ads} \quad (2.1a)$$

$$A_{2,ads} \overset{k_{diss}}{\longrightarrow} 2A_{ads} \quad (2.1b)$$

$$A_{ads} \overset{k_1}{\longrightarrow} B_{gas} \quad (2.1c)$$

To dissociate molecule A_2 fragments into two A_{ads}, two vacant surface site positions are needed, whereas to adsorb A_2 one surface site is used.

The kinetic equations corresponding to Eq. (2.1) can be easily formulated (see Chapter 4). Using the steady-state approximations the following equations for the

---

Catalysis: From Principles to Applications, First Edition. Edited by Matthias Beller, Albert Renken, and Rutger van Santen.
© 2012 Wiley-VCH Verlag GmbH & Co. KGaA. Published 2012 by Wiley-VCH Verlag GmbH & Co. KGaA.

rate R_B of product B formation are found. They are given by Eq. (2.2):

$$R_B = k_1 \cdot \theta_A \qquad (2.2)$$

$$\theta_A = 1 + \tfrac{1}{2}\lambda - \tfrac{1}{2}\sqrt{\lambda^2 + \lambda} \qquad (2.3a)$$

$$\approx \frac{1}{1+\lambda} \qquad (2.3b)$$

$$\lambda = \frac{k_1}{k_{diss}} \frac{(K_{eq}^{ads}[A_2]+1)^2}{K_{eq}^{ads}[A_2]} \qquad (2.3c)$$

In Eq. (2.2) θ_A is the surface coverage with component A, K_{eq}^{ads} is the equilibrium constant of A_2 adsorption.

Using the approximate Eq. (2.3), Eq. (2.2) can be converted to the useful equations of:

$$R_B = \frac{k_1}{1 + \frac{k_1}{k_{diss}} \frac{(K_{eq}^{ads}[A_2]+1)^2}{K_{eq}^{ads}[A_2]}} \qquad (2.4a)$$

$$R_B \approx k_{diss} \frac{K_{eq}^{ads}[A_2]}{(K_{eq}^{ads}[A_2]+1)^2}, \; k_{diss} \ll k_1 \qquad (2.4b)$$

$$R_B \approx k_1, \; k_{diss} \gg k_1 \qquad (2.4c)$$

Eq. (2.3) contains examples of rate equations that are used in microkinetic modeling. No *a priori* assumptions on the relative size of kinetic parameters have been made that would imply a choice of the slowest elementary rate constant.

Eq. (2.4b) is the kinetic expression for the limiting case when the rate constant of dissociation is small compared to the rate of product formation; Eq. (2.4c) is the result for the opposite case.

Eq. (2.4b) is a very common Langmuir–Hinshelwood–Watson–Hougen (LHWH) type equation that could have been directly deduced by only assuming the dissociation step to be non-equilibrated. It implies that this is the smallest rate constant. Such equations are extensively discussed in Chapter 4.

Note that we do not use the term *rate-limiting step*. This denomination is in fact erroneous because at steady state the rates that correspond to intermediate product formation steps of Eq. (2.2) are the same. As we discuss below it is better to denote such slow elementary rate steps as rate controlling.

Eq. (2.4b) will lead to the volcano type behavior, if the equilibrium constant K_{eq}^{ads} of molecule A_2 adsorption is considered to be essentially the reactivity descriptor.

If one ignores variation of k_{diss}, Eq. (2.4b) will show a maximum at $K_{eq}^{ads} \cdot [A_2] = 1$. In this case the volcano curve behavior is due to the need for two empty vacant sites to be present for dissociation. The initial increase in overall rate is due to increasing coverage of the surface. The maximum rate is found as a compromise between increasing coverage and decreasing probability of finding neighboring vacant sites.

Note that in this particular case the volcano curve is not due to a change in rate-controlling step, which is the basis of the Sabatier Principle. It is due to changes in surface coverage by change of conditions. In this case the pressure changes the surface coverage. Maxima in rate of reaction as a function of pressure or temperature are very common and are not necessarily due to the Sabatier Principle.

Generally the relative rates of r_{diss} versus r_1 change much more then the relative rate of molecular adsorption and desorption.

A very important basic rule of surface reactivity is that the molecular adsorption behaves very differently from dissociative adsorption when one studies this as a function of changing interaction with a surface. The changes in adsorption energies of molecules are much less than those of atoms or the molecular fragments that result from dissociation. In molecules, electronic changes in the donating and back-donating chemical bonding interactions tend to counteract each other. In adatoms the changes in donating chemical bonding interactions tend to dominate. Therefore the activation energies for molecular dissociation or recombination of adatoms relate primarily to changes in the adsorption energies of the adatoms. This is due to the second important surface reactivity principle we have already met in Chapter 1: the linear activation energy–reaction energy relation for elementary surface reaction rate activation energies. This is the Brønsted–Evans–Polanyi relationship (BEP):

$$\delta E_{act} = \alpha \cdot \delta E_{reaction}, \quad 0 \leq \alpha \leq 1 \tag{2.5}$$

δE_{act} is the change in activation energy and $\delta E_{reaction}$ the difference in energy between the adsorbed molecule and the dissociated fragments. α is a proportionality constant with values between zero and one. Because $\delta E_{reaction}$ is dominated by the change in the adsorption energies of the adsorbed adatoms, this also will dominate the change in the activation barriers.

This holds clearly for weakly adsorbing molecules such as N_2 or methane, but also, though more approximately, is also found for a molecule such as CO [1].

The consequence of this is that the relevant reactivity descriptor is not the molecular adsorption energy but is the atomic interaction energy. Actually compared to the variation in the activation energies the energy of molecular adsorption can often be assumed constant. With this assumption we find the dependence on the rate of product formation versus λ (see Eq. (2.2)) that now is shown in Figure 2.1.

The maximum in the volcano curve is now found for the value of λ:

$$\lambda_{max} = \frac{\alpha_1}{2\alpha_{diss}} \tag{2.6}$$

λ_{max} is proportional to the ratio of the two BEP constants of the rate-controlling elementary rate constants. This determines at which rate constants of reactant dissociation and surface intermediate to product formation are in balance. We note (compare Eqs. (2.4b) and (2.4c)) that to the left and right of the Sabatier maximum surface coverage changes.

For many reactions computational catalysis or detailed experimental model catalyst studies have generated reaction energy diagrams that follow the changes in

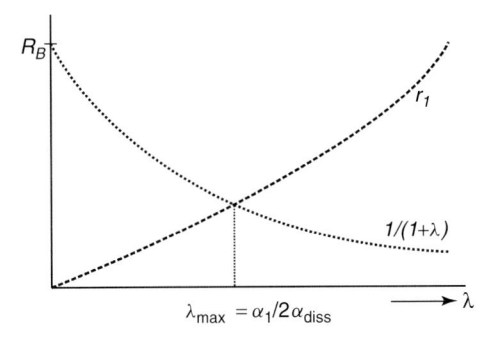

Figure 2.1 Schematic representation of the Sabatier effect. The parameter λ is defined in Eq. (2.2c). The decrease in surface coverage with decreasing relative rate of dissociation $1/(1+\lambda)$ is compensated by the increasing rate of product desorption r_1.

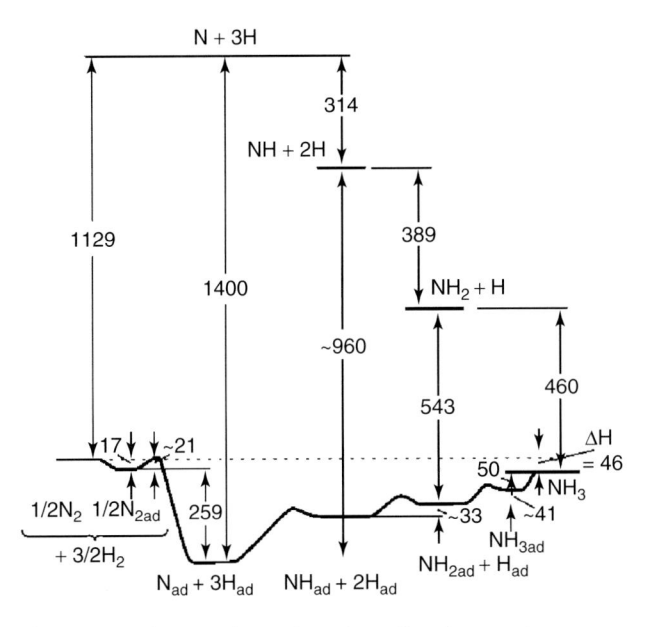

Figure 2.2 Schematic thermodynamic profile of ammonia synthesis on Fe(111) [2]. Energies are in kilojoules per mole.

energy of the reacting catalytic system through the catalytic cycle. They contain the energies of reaction intermediates as well the activation energies of the elementary reaction steps.

A well-known example is the famous reaction energy diagram produced by Ertl on ammonia synthesis (Figure 2.2) catalyzed by Fe:

In this figure the corresponding energy differences in the gas phase and on the surfaces are compared.

The interesting dominating feature of such reaction energy diagrams is that initially the total energy of the system decreases and that the cycle is closed by an upward movement of the thermodynamic energy of the reaction. Of course the difference in initial energy of reactants and the final product energy has to be equal to the overall reaction energy of the reaction.

In order to elucidate the use of such graphs to deduce kinetic information we show in Figure 2.3 schematically the reaction energy scheme for the catalytic conversion of reactant A into product B. In Figure 2.3a the energy change is sketched at zero temperature, in Figure 2.3b the changes in free energy are sketched at a finite temperature.

We will first analyze Figure 2.3a. One notes the lowering in energy due to adsorption of A, the activation energy for the transformation of A into B, and the adsorption energy of product B.

The difference in energy of A and B is the overall reaction energy. At finite temperature (Figure 2.3b) equilibria and rates are controlled by free energy differences.

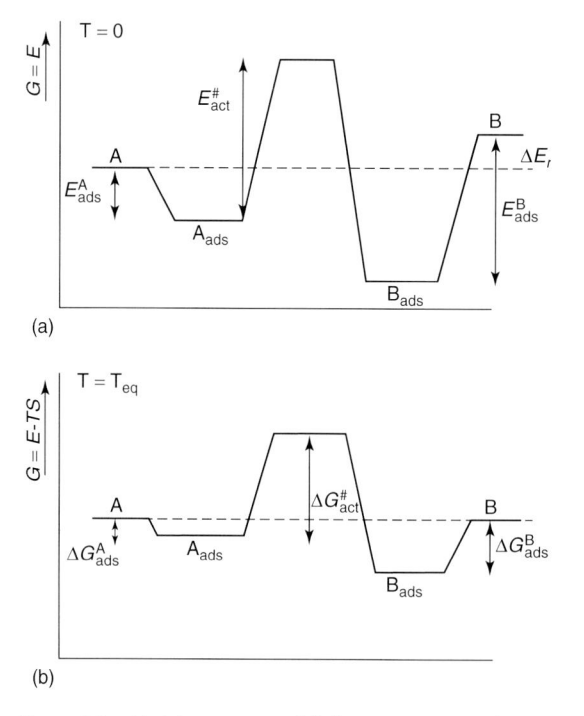

Figure 2.3 (a) Schematic simplified reaction energy diagram at 0 K. (b) The reaction free energies at the temperature where product and reactant are equilibrated.

Free energy differences affect the relative stability of adsorbed species significantly because of the high entropy of the free moving molecules in the gas phase and reduced entropies of molecules adsorbed on surfaces. At higher temperature this tends to destabilize the adsorbate with respect to the gas phase. It is reflected in the relative upward shift of the free energies of adsorption. The entropies of adsorbed molecules or corresponding adsorbed molecular fragments intermediates do not vary significantly (see Section 2.6). Therefore changes with temperature in the activation free energy of surface reactions tend to be small compared to those between adsorbed state and gas phase.

The optimum temperature for a reaction is the temperature at which the reaction free energy is zero. Reaction then proceeds with optimum free energy efficiency. This situation is sketched in Figure 2.3b.

The rate of the reaction is determined by the relative position of the free energy of the transition state. Important to kinetics is the question whether this barrier has to be compared with the free energy of the adsorbed phase A or with A in the gas phase.

This question is resolved considering the corresponding LHWH equations (see Chapter 4). Their general form is given in Eq. (2.7). The form they take depends on an assumption concerning the smallest rate constant. The other steps then are assumed to be equilibrated.

$$r = k \frac{K_{eq}^A P_A}{1 + K_{eq}^A P_A + \sum_{i \neq A} K_{eq}^i P_i} \tag{2.7}$$

In this case the rate constant of the transformation A_{ads} to B_{ads} is small compared to the respective rates of adsorption and desorption of reaction products and reagents. Expression (2.7) reduces to expression (2.8) for the case illustrated by Figure 2.3:

$$R_B = k_{A \to B} \frac{K_{eq}^A [A]}{K_{eq}^A [A] + 1} \tag{2.8a}$$

If the rate of desorption of product B is slow the rate of product formation becomes:

$$R_B = k_{des}^B \frac{\frac{k_{A \to B}}{k_{B \to A}} K_{eq}^A [A]}{\frac{k_{A \to B}}{k_{B \to A}} K_{eq}^A [A] + 1} \tag{2.8b}$$

$$\approx k_{des}^B \tag{2.8c}$$

We have assumed in Eq. (2.8) that gas phase composition of B remains low.

According to Eq. (2.8a) the equation for the apparent activation energy becomes:

$$E_{app}^{act} = E_{act} + E_{ads}^A (1 - \theta_A) \tag{2.9}$$

The activation energy of the reaction now depends on surface coverage θ_A. At low surface coverage the activation energy of the reaction equals the activation energy with respect to A in the gas phase. The rate of reaction now is first order in A. At high coverage of A ($\theta_A = 1$) the activation energy of the reaction is to be measured with respect to the adsorbed state of A. The reaction now is zero order in A.

The other limit with rapid equilibration of A_{ads} and B_{ads} and slow desorption is given in Eq. (2.8c). Now the activation energy is that of product B desorption.

The maximum rate of a reaction is given by the condition that:

$$G_{app}^{act} = 0 \tag{2.10a}$$

$$E_{app}^{act} = T_m \Delta S_{app}^{act} \tag{2.10b}$$

We can introduce a catalytic efficiency (CE) measure of how efficient the reaction is energetically:

$$CE = \frac{E^{react}}{E_{app}^{act}} \times 100\%, \text{ when } E^{react} > 0 \tag{2.10c}$$

$$CE = \frac{E^{react}}{E_{app}^{act}} \times 100\%, \text{ when } E^{react} > 0 \tag{2.10d}$$

These can also be expressed in terms of temperatures. Similarly to the equations for the case of the maximum rate, the reaction equilibrium is given by the condition that:

$$\Delta G^{react} = 0 \tag{2.10e}$$

$$E^{react} = T_{eq} \Delta S^{react} \tag{2.10f}$$

This gives:

$$CE = \frac{T_{eq}}{T_m} \frac{\Delta S^{eq}}{\Delta S^{act}} \times 100\%, \text{ when } E_{react} > 0 \tag{2.10g}$$

$$CE = \frac{T_{eq} \Delta S^{eq}}{T_{eq} \Delta S^{eq} + T_m \Delta S^{act}}, \text{ when } E_{react} < 0 \tag{2.10h}$$

If the entropy of the reactants and that of the transition state are close, these can be rewritten:

$$CE = \frac{T_{eq}}{T_m} \times 100\%, \text{ when } E_{react} > 0 \tag{2.10i}$$

$$CE = \frac{T_{eq}}{T_{eq} + T_m}, \text{ when } E_{react} < 0 \tag{2.10j}$$

$\Delta T = T_m - T_{eq}$. In this case Eq. (2.10i) becomes:

$$CE = \frac{1}{1 + \frac{\Delta T}{T_{eq}}} \times 100\%, \text{ when } E_{react} > 0 \tag{2.10k}$$

It follows from Eqs. (2.9) and (2.10b) that the temperature of maximum rate of the catalyst sensitively depends on the coverage of the surface. At low surface coverage (or low temperature) the activation energy of the conversion A_{ads} to B_{ads} is lowered by the adsorption energy of A. At high coverage (and high temperature) it equals the activation energy of the conversion A_{ads} to B_{ads}.

According to Eqs. (2.10c) and (2.10d), the higher the apparent activation energy the lower is the thermodynamic efficiency of the catalytic process.

A general rule for maximum catalytic rate and optimum thermodynamic efficiency is that differences in intermediate free energies of adsorption and activation should be minimal with respect to the free energy line that connects the free energy of the reactants with that of the products.

2.2
The Lock and Key Model, the Role of Adsorption Entropy

The LHWH model is a kinetic rate equation, mainly used in heterogeneous catalysis, that is quite analogous to the Michaelis–Menten equation (Eq. (2.11)) as used in homogeneous catalysis. This will be discussed more extensively in Chapter 3. Here we will emphasize the relationship with molecular heterogeneous catalysis:

$$r = k_2 E_0 \frac{[S]}{K_m + [S]} \quad (2.11)$$

where

$E_0 \equiv$ total concentration of enzyme catalyst
$[S] \equiv$ concentration of substrate
$K_m = \frac{1}{K_C} \equiv$ Michaelis constant
$k_2 \equiv$ the rate measured when all enzyme molecules are complexed with substrate.

We recognize the similarity of this equation with Eq. (2.7). There are two parameters, K_m, the Michaelis constant, which is the inverse of the complexation constant, and k_2, the conversion rate constant of adsorbed reagent. One is the equilibrium constant for complexation; the other is the rate constant of conversion of the substrate complexed with enzyme or organometallic complex. In the case of the enzyme the activity of the catalyst is often dominated by the complexation constant.

Important is the fit of reactant or product with the shape and size of the cavity of the peptide enzyme molecule. This is schematically illustrated in Figure 2.4.

It appears that the shape of the catalyst cavity adapts to the transformation of reagent(s). This is the induced fit model. According to Rose [3, 4], when reaction progresses subsequent enzyme cavity transformations assist optimum rate of adsorption of reagent as well as rate of product removal.

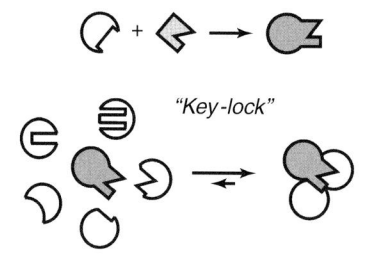

Figure 2.4 Schematic illustration of the lock and key match of cavity and substrate.

Entropy differences of transition state intermediates can have a large effect on catalyst selectivity.

A very interesting illustration of this is the selectivity model of enantiomeric hydrogenation by Halpern et al., discussed more extensively in Part II. The catalyst in this case is a chiral organometallic complex. When the difference in catalytic reactivity is controlled by the lock and key model, differences in rate should be controlled by the most favorable reaction intermediate interaction energy. In this particular case the reverse is found.

Figure 2.5a illustrates the effect of bulky ligands on the preferred orientation of the coordinated reactant molecules upon enantioselective hydrogenation to produce L-Dopa. Figure 2.5b presents the respective two reaction energy diagrams that illustrate the differences in barrier heights for the two enantiomers [5].

It appears that that the preferred conversion route is the one that proceeds with the most unfavorable misfit. This is the anti-lock and key model. This at the same time implies that selectivity of this reaction is determined by the gain in entropy for the system with the more unfavorable interaction energy. Such entropy effects appear quite general, also in heterogeneous catalysis.

We will give as an example in heterogeneous catalysis a comparison of conversion rates in different zeolites. Zeolites are microporous systems with well-defined pores of the order of a nanometer. These materials are used as solid acid catalysts for the cracking, isomerization, or dehydrogenation of aliphatic hydrocarbons.

We will discuss the example of the hydro-isomerization of n-pentane to iso-pentane by acidic zeolites activated with Pt.

This is an example of a bifunctional catalyst. The Pt is used to equilibrate alkane with alkene. The reaction is executed in excess H_2 at a temperature where equilibrium is on the alkane side. It implies that only a minute fraction of the hydrocarbons is present as alkene.

The reaction is performed under the condition that the rate of proton-induced isomerization has the lower rate constant k_1. Then the LHWH rate equation for the conversion of n-alkane to isoalkane becomes Eq. (2.12):

$$\frac{1}{N_p}\frac{d}{dt}[iCH] = \frac{1}{N_p}K_{eq}[H_2]\frac{d}{dt}[iCH^=] \qquad (2.12a)$$

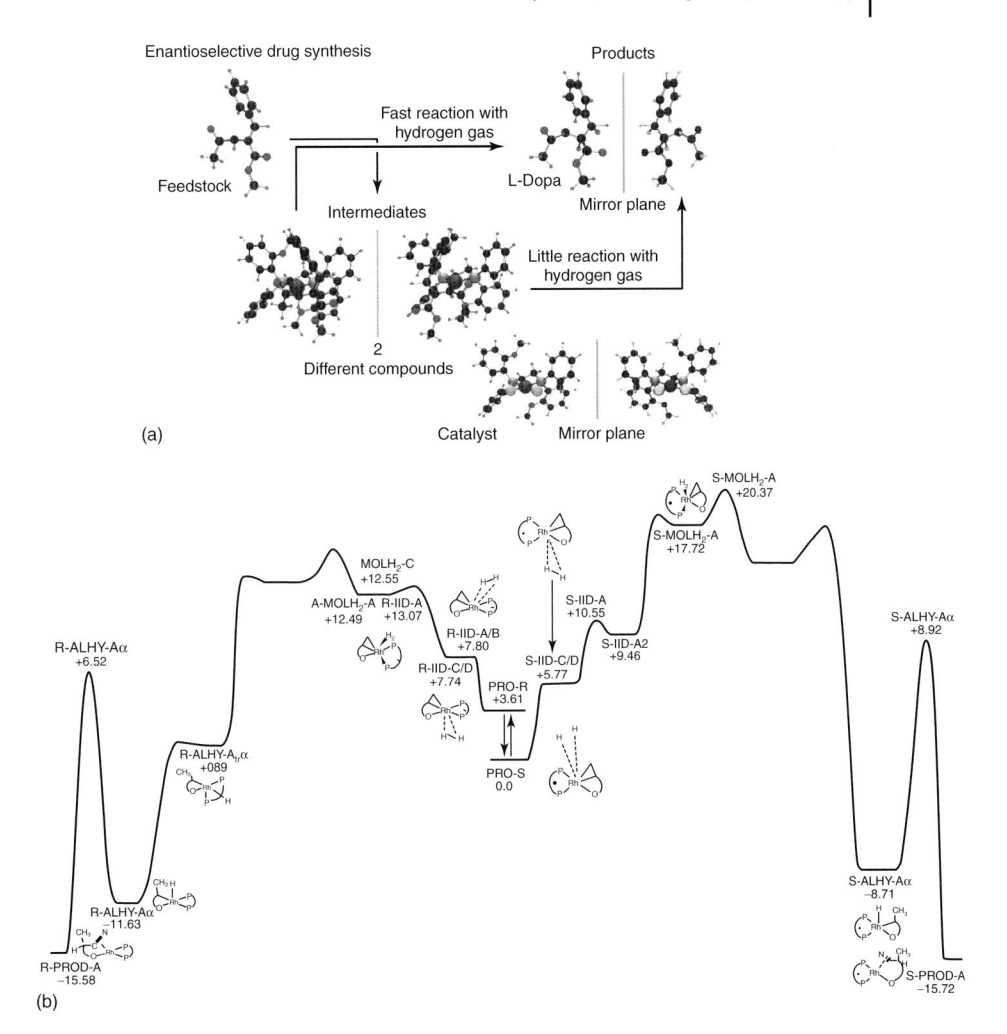

2.2 The Lock and Key Model, the Role of Adsorption Entropy

Figure 2.5 (a) The enantioselective hydrogenation by homogeneous Rh–BINAP catalyst – effect of bulky ligands. (b) The enantioselective hydrogenation by homogeneous Rh–BINAP catalyst – two reaction energy diagrams (adapted from S. Feldgus and C.R. Landis [5]).

where N_p is the number of reactive protons in zeolite

$$\frac{d}{dt}[i\text{CH}^=] = k_1 \frac{K_{\text{ads}}[n\text{CH}^=]}{1 + K_{\text{ads}}[n\text{CH}^=] + K_{\text{ads}}[i\text{CH}^=]}, \quad (2.12b)$$

$$\approx k_1 \frac{K_{\text{ads}}[n\text{CH}^=]}{1 + K_{\text{ads}}[n\text{CH}^=]}. \quad (2.12c)$$

For our purpose Eq. (2.12c) is relevant. We observe that the rate of production of i-alkene, which is proportional to the rate of i-alkane production (Eq. (2.12a)), has now the (for us) familiar form of a rate equation that depends on the rate of activation of adsorbed reagent and its concentration on the catalyst surface.

When zeolites are compared with one-dimensional channels and with cross sections containing 10- and 12-membered rings, experimental study demonstrates that k_1 changes only slightly. The differences in reactivity are dominated by differences of the adsorption constants K_{ads}. This is more extensively discussed in Part III. It appears that at this temperature entropic effects dominate differences of K_{ads}, the adsorption constant of alkene adsorption to the zeolite.

The molecules loose entropy upon adsorption in a micropore channel from the gas phase. The more limited mobility of the molecules in the smaller micropore causes the concentration of hydrocarbons adsorbed to be suppressed and generates preference of adsorption of molecules in the wider channels. The decrease in entropy counteracts the increase in the adsorption energy.

Comparison of mordenite with the 12-ring channels and ferrierite with the 10-ring channels gives a more than sixfold increase in turnover frequency (the number of molecules converted per unit time per site) for the zeolite with the wider pore channel.

2.3
Equivalence of Electrocatalysis and Chemocatalysis

The role temperature plays in chemocatalysis is taken over in electrocatalysis by the electrode potential. The rate of the reaction is determined by the overpotential. An interesting illustration is provided by an elementary computational study of oxygen reduction on the Pt(111) surface [6].

This exploits the thermodynamic equivalence of the electrocatalytic reaction steps:

$$O^* + H^+ + e^- \longrightarrow OH^* \tag{2.13a}$$

and

$$OH^* + H^+ + e^- \longrightarrow H_2O \tag{2.13b}$$

with the corresponding chemocatalytic elementary reaction steps:

$$O^* + \frac{1}{2}H_2 \longrightarrow OH^* \tag{2.14a}$$

$$HO^* + \frac{1}{2}H_2 \longrightarrow H_2O + * \tag{2.14b}$$

The free energies of reaction steps (Eq. (2.14)) are available from gas phase or liquid phase quantum-chemical calculations.

We will deduce the electrochemical overpotential for oxygen reduction and water hydrolysis by setting the reference potential equal to that of the standard hydrogen electrode. We then can relate the chemical potential for the reaction $H^+ + e^-$ to that of the chemical potential of $1/2$ H_2.

This means that the reaction energies of Eq. (2.14) are equal to the reactions Eq. (2.13) at electrode potential $U = 0$.

Dissociation of O_2 to O^* occurs without electron transfer. Figure 2.6 shows the relative energies of the different intermediates as a function of potential. Use is made of the Faraday relation that relates free energy with electrode potential.

$$\Delta\mu = Z.F.\Delta V \tag{2.15}$$

$\Delta\mu$ is the change in chemical potential, Z the number of electron transferred; F the Faraday constant, and ΔV the electrochemical potential.

In Figure 2.6 the curve $U = 0$ shows the computed relative energies of adsorbed intermediates at zero potential U. According to the Faraday relation the equilibrium potential corresponding to the reaction free energy of 2.46 eV corresponds to an electrode potential of 1.23 V (two electrons are transferred).

In Figure 2.6 the curve $U = 1.23$ V gives a comparison of the relative energies at this potential. The energy of adsorbed oxygen (O^*) is shifted downward by the same amount as that of O_2 (no difference in electron transfer), but that of OH^* by half that amount because in the latter case only one electron is transferred.

One notes that at equilibrium O^* is the dominant species. An upward shift of the potential by 0.45 V makes the energies of O^* and H_2O equivalent and hence predicts this to be the overpotential of this reaction. This appears to be close to the experimental value.

Note that so far we have made no estimates of activation energies, which are small compared to the energetic differences.

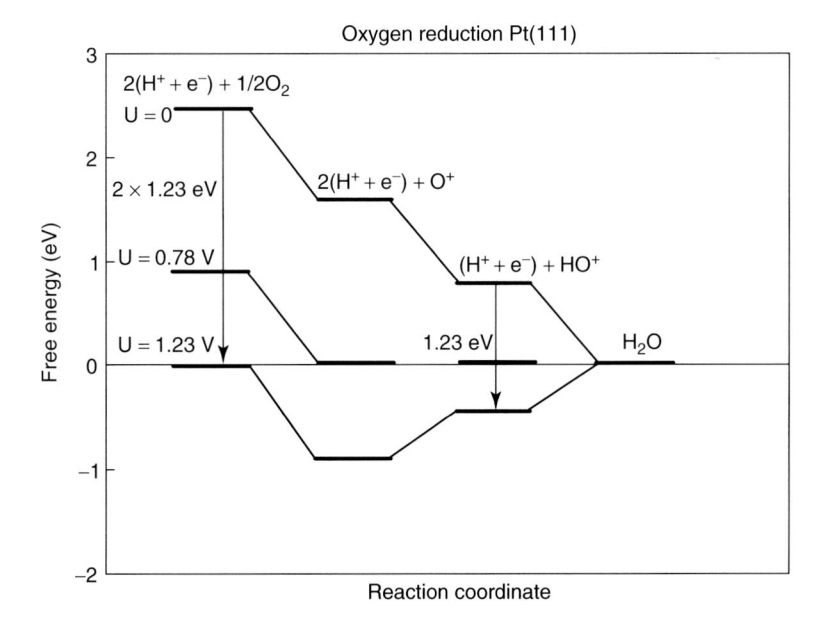

Figure 2.6 Free-energy diagram for electrochemical oxygen reduction over Pt(111) [6].

For the reverse reaction of hydrolysis we find that the first step of H_2O dissociation to produce H and surface OH is thermodynamically neutral by shifting the $U = 0$ curve downward by 0.78 eV. This then represents the electrode potential for the onset of H_2O dissociation.

The diagram of Figure 2.6 can also be used to illustrate the condition of optimum CE. The overpotential is predicted to be zero when at zero potential the relative energies of O^* and OH^* differ by a factor of 2 and dissociative adsorption of O_2 is energetically neutral. Then at equilibrium potential all the energies will be located on the equilibrium energy line. There is now no fluctuation of the energies around this equilibrium line and the overpotential is predicted to be zero.

2.4
Microkinetics; the Rate-Determining Step

Since at steady state the rates of formation and removal of reaction intermediates are the same, the notion of rate-determining step does not directly apply. In Section 2.1 we have seen that to transform microkinetics equations into the easier-to-handle LHWH equations one has to determine which reaction steps are close to equilibrium and which steps are far from equilibrium. This identifies the type of LHWH equation to use (compare for instance Eq. (2.3)).

Dumesic et al. [7] and Campbell [8] have discovered an important method that enables the identification of the out-of-equilibrium steps, which they call rate-controlling.

In order to do so they exploit the de Donder concept of the affinity of a reaction:

$$A = -RT \ln \frac{\prod_j (a_j^{v_{ij}})}{K_{eq}} \qquad (2.16)$$

The affinity of the reaction is zero when it is at equilibrium. In Eq. (2.16) v_{ij} are the stoichiometric coefficients for species j and K_{eq} is the overall equilibrium constant. a_j is the activity of species j. A is a measure of the difference between equilibrium and the actual state of a reaction. A provides the driving force of the reaction.

Similarly one can also define the affinity of a species i. This defines its closeness to equilibrium:

$$A_i = -RT \ln \left(\frac{\left[\prod_j (a_j^{v_{ij}}) \right]}{K_{eq,i}} \right) \qquad (2.17)$$

The reversibility (closeness to equilibrium) has been defined as:

$$z_i = \exp(-A_i/RT) = \frac{\left[\prod_j (a_j^{v_{ij}}) \right]}{K_{eq,i}} \qquad (2.18)$$

2.4 Microkinetics; the Rate-Determining Step

When rate of change of component i is at equilibrium z_i equals 1, and when there are no products z_i equals 0. The net rate of step i then can be written as:

$$r_i = \vec{r}_i(1 - z_i) \tag{2.19}$$

With $\vec{r}_i$ the ratio of forward and reverse reaction steps.

Whereas this provides a recipe to determine with the LHWH which steps in a kinetic scheme can be considered equilibrated, it is not yet completely satisfactory since calculation of z_i implies insight into the steady-state concentrations of the reaction intermediates, which one has to deduce from an estimate of the relative rate constants as we did earlier in Section 2.1.

Campbell [8] defined an operational definition that does not necessitate this and can be readily computerized. Campbell defined degree of rate control of step i thus:

$$X_{rc,i} = (k_i/R) \cdot (\partial R/\partial k_i) \tag{2.20}$$

The partial derivative $\partial R/\partial k_i$ is calculated maintaining the equilibrium constant for step i constant and keeping all other rate constants equal. Keeping $K_{eq,i}$ constant means that the that the rate of forward and reverse steps have to be changed equally:

$$\frac{\partial k_i}{k_i} = \frac{\partial k_i^{-1}}{k_i^{-1}} \tag{2.21}$$

For example, by increasing both the forward and reverse rate constants by 1% one can calculate the resulting fractional increase in the overall rate. The degree to which the step controls the overall rate is just the percentage increase in the overall rate divided by this 1%.

Equation (2.20) is the differential calculus definition normalized such that the rate-determining step is that step in a mechanism which has a degree of rate control $X_{rc,i}$ equal to unity. A *rate-controlling step* or *slow step* is defined as any step with a non-zero $X_{rc,i}$. This equation has the explicit form:

$$X_{rc,i} = \frac{k_i}{R}\left(\frac{\partial R}{\partial k_i}\right)_{K_{i,eq},k_j} = \frac{k_i}{R}\left[\left(\frac{\partial r}{\partial k_i}\right)_{k_j} + \left(\frac{\partial r}{\partial k_{-i}}\right)_{k_j}\frac{k_{-i}}{k_i}\right] \tag{2.22}$$

$X_{rc,i}$ is the sum of the two terms:

$$X_{rc,i} = s_i + s_{-i} \tag{2.23}$$

s_i is the sensitivity of the overall rate with respect to k_i. It can be shown that [7] the sum of the rate-controls of the different rate constants is a constant:

$$\sum_i X_{rc,i} = 1 \tag{2.24}$$

An elementary rate has maximum rate control if $X_{rc,i} = 1$.

Following Dumesic et al. [7] we can now analyze the concept of rate-determining step more deeply. We will do this for three coupled reactions:

$$C \underset{k_{-1}}{\overset{k_1}{\rightleftharpoons}} D \qquad (2.25a)$$

$$B \xrightarrow{k_2} C \qquad (2.25b)$$

$$C \xrightarrow{k_3} D \qquad (2.25c)$$

Reagent A and intermediate B are assumed to be quasi-equilibrated. The other reaction steps are irreversible.

The following expressions for $X_{rc,i}$ can be deduced:

$$X_{rc,1} = (1 - z_1)s_1 = 0 \qquad (2.26)$$

$$X_{rc,2} = z_1(1 - z_2)s_1 = s_1 = 1 \qquad (2.27)$$

$$X_{rc,3} = z_1 z_2 (1 - z_3)s_1 = 0 \qquad (2.28)$$

Remember that z_i approaches 0 as step i is irreversible; z_i equals 1 when i is quasi-equilibrated. Step 2 (Eq. (2.25b)) controls the overall rate of the reaction, and hence it can be called *rate-limiting*, whereas the net rates of the three reaction steps are equal at steady state.

2.5
Elementary Rate Constant Expressions for Surface Reactions

The transition state reaction rate equation depends not only on the activation energy but also on the transition state entropy. In Eq. (2.29) the Eyring equation is given for the reaction rate based on the transition state model.

$$k = \frac{k_B T}{h} e^{-\frac{\Delta G^\#}{RT}} = \frac{k_B T}{h} e^{-\frac{\Delta H^\#}{RT}} e^{\frac{\Delta S^\#}{T}}, \qquad (2.29)$$

where k is the rate constant, k_B the Boltzmann constant, $\Delta H^\#$ and $\Delta S^\#$ are transition state or activation enthalpy and entropy, respectively. The transition state free energy is:

$$\Delta G^\# = RT \ln K^\# = \Delta H^\# - T\Delta S^\#, \qquad (2.30)$$

When a molecule adsorbs and desorbs from a surface the entropy of the molecule increases due to its increased mobility in the gas phase. Typical values for a desorbing molecule of the pre-exponential frequencies are given in Table 2.1. These numbers are to be compared with the pre-exponential frequencies for a surface dissociation reaction. We will illustrate this for CO dissociation.

Figure 2.7 shows the changes in CO geometries as well as changes in electron density distribution upon dissociation on an Ru surface. We observe the change of the CO molecule from its perpendicular position to side-on and finally dissociated state. Electrons localized around C and O rehybridize with the stretching of the C–O bond and the late character of the CO transition state, where the C–O bond is

Table 2.1 Experimental activation energies and pre-exponential factors for CO and NO desorption from a range of clean and well-defined single surfaces [9].

System	Prefactor (s^{-1})	Activation energy (kJ mol^{-1})
CO/Co(0001)	10^{15}	118
CO/Ni(111)	10^{15}	126
CO/Ni(100)	10^{14}	130
CO/Cu(100)	10^{14}	67
CO/Ru(001)	10^{16}	160
CO/Rh(111)	10^{14}	134
CO/Pd(111)	10^{14}	143
CO/Pd(100)	10^{16}	160
CO/Pd(211)	10^{14}	147
CO/Ir(110)	10^{13}	155
CO/Pt(111)	10^{14}	134
NO/Pt(111)	10^{15}	126
CO/Rh(111)	10^{14}	134

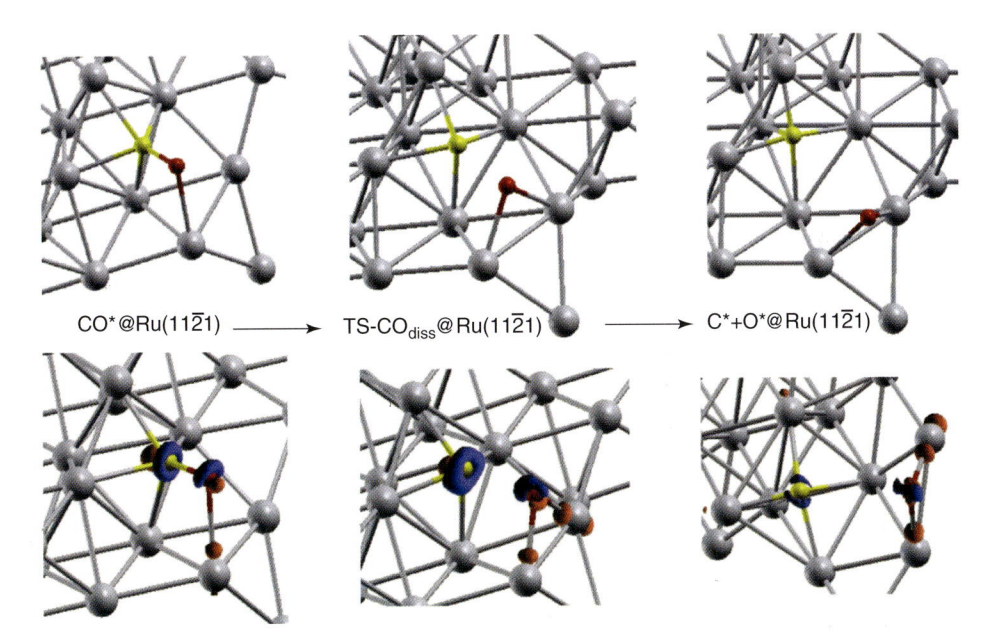

Figure 2.7 Product, transition state, and final state geometries (top panels) of CO dissociating on an open Ru (11–21) surface (Ru atoms are shown in gray, carbon and oxygen atoms are shown in yellow and red, respectively). The corresponding difference electron densities (DEDs) are shown at the bottom panels [10].

already weak and the positions of the C and O atoms are already close to the final dissociated state. Clearly there is close contact between dissociating molecule and the surface in the transition state.

Vibrational frequencies can be computed and hence the corresponding partition functions Q.

The partition function is a statistical mechanical function that is related to the entropy of a system:

$$S = k \ln Q \tag{2.31}$$

From harmonic vibrational frequencies the expression for Q_{vib} is:

$$Q_{vib} = \prod_i \left(\frac{1}{1 - e^{-\frac{h\nu_i}{kT}}} \right) \tag{2.32}$$

2.6
The Pressure Gap

An important question is whether one can use kinetic data obtained far from reaction conditions, for instance in a surface science experiment, to predict the rates of the reaction at catalytically practical conditions. The same question is relevant for calculated rate parameters in computational catalysis studies to be used in simulations at finite temperature and pressure.

A typical experimental example that illustrates this issue is the difference in product distribution of the reaction of CO and H_2 at low pressure compared to the rate of this reaction at high pressure.

At low temperature a typical surface science experiment would saturate the surface with pre-adsorbed CO. One would study the rate of methane formation as a function of temperature with a low pressure of H_2 in the gas phase. In such an experiment often a densely packed surface is used such as the (111) surface of an face centered cubic (fcc) transition metal. For methanation Ni would be useful. Whereas in this experiment no methane will be observed, a very different result would be found if this experiment were repeated in the presence of high pressure of CO.

We discussed this reaction in the previous section, where we presented the microkinetics Eq. (2.2) and deduced the corresponding LHWH equations in Eq. (2.4).

Useful for our discussion is Eq. (2.4b), which shows that at low CO pressure the rate of reaction is proportional to the rate constant of CO dissociation and the surface coverage of CO.

The activation energies of CO dissociation and energies of CO adsorption are well known from theory and experiment. Computed values for low coverage are given in Tables 2.2 and 2.3:

One notes for the group VIII metals the previously mentioned large variation in activation energies of CO activation and minor variation in the CO adsorption energies. An important observation is also that the barriers of CO dissociation decrease comparing metals moving from left to right in a row of the periodic

Table 2.2 DFT computed adsorption energies of CO on dense (111) fcc surfaces (in kJ mol^{-1}) of group VIII and IB metals [1].

Co −176	Ni −183	Cu −85
Rh −193	Pd −189	Ag −18
Ir −162	Pt −168	Au −29

Table 2.3 DFT computed activation energies of CO dissociation on (111) fcc of group VIII and IB metal surfaces.

Co 251	Ni 355	Cu 517
Rh 315	Pd 424	Ag 592
Ir 336	Pt* 416	Au 581

system. This increase in reactivity, which also reflects increased interactions of reagent and product atoms with the metals, is a general trend that relates to increasing reactivity of metals with a decrease of electrons in the d-valence electron band.

We can now understand the reason why when one performs the CO methanation experiment with preadsorbed CO and no CO in the gas phase there will be no reaction. The CO desorbs before it can dissociate.

The high concentrations of surface overlayers can have two additional important effects. It can lead to surface reconstruction and even to changes in the chemical reactivity of atoms in the overlayer.

2.6.1
Surface Reconstruction

Adatoms adsorbed onto the surface of group VIII metals form strong chemical bonds. When an adsorbate strongly interacts with a surface, bonds between the metal atoms involved in the surface complex and the bulk metal will weaken as the metal atom electrons that are part of the adsorbate−surface complex become less available for bonding to the bulk metal atoms. In this case an overlayer forms an excess of metal−metal bonds, and this can provide a driving force for reconstruction.

An example is the reconstruction caused by the deposition of atomic carbon on a Co surface. The relevant energetics is illustrated in Figure 2.8.

In this figure, computed surface energies are shown as a function of surface coverage. The densest surface is initially the most stable. Formation of the C overlayer weakens the energies of bonds between the metal atoms of the surface. However, the clean Co(100) surface has a higher surface energy then the Co(111)

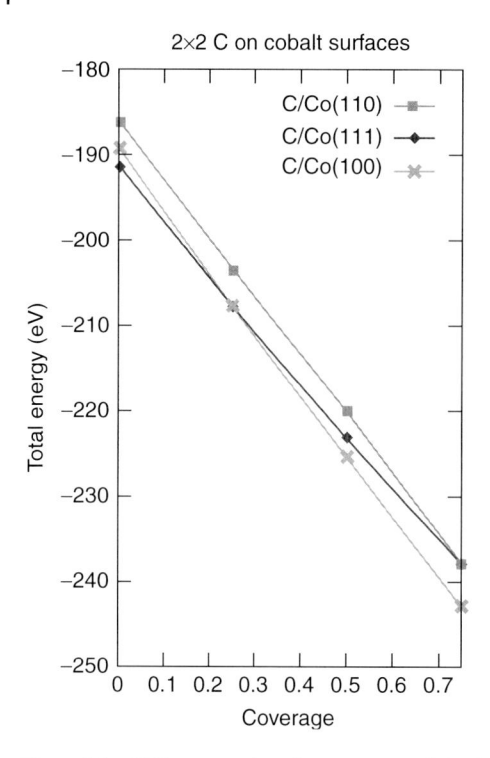

Figure 2.8 DFT-computed surface energies of Co (111), (100), and (110) surfaces as a function of atomic carbon coverage [11].

surface. On the Co(100) surface C_{ad} will coordinate with four Co metal atoms in a plane. Whereas adsorbed carbon interacts with only three surface atoms at the (111) surface of the fcc Co, it has a substantially stronger interaction with the more reactive Co(100) surface. This effect further stabilizes the (100) surface due the preferential binding of C_{ad} to four Co atoms in a plane. At a surface coverage of 30% C the Co(111) surface reconstructs into the Co (100) surface. Since the (100) surface has a lower atom density than the (111) surface excess Co atoms are created. These are pushed out of the surface and cause formation of surface step and kink sites.

2.6.2
Altered Surface Reactivity

The reactivity of surface adatoms can also change with overlayer composition. An example is the selectivity of the ethylene epoxidation reaction catalyzed by Ag. At low surface concentration the selectivity to the epoxide is low, at higher O concentration the chemical nature of the oxygen atom changes from nucleophilic to electrophilic.

On a silver surface with low O_{ads} coverage, the reaction of ethylene will predominantly lead to the formation of acetaldehyde, which will be readily oxidized to CO_2 and H_2O.

When the surface is oxidic, as in Ag_2O, the dominant reaction product is the epoxide.

2.7
The Materials Gap

2.7.1
Structure Sensitivity

The materials gap refers to the difference between the structure of catalytically reactive materials and the very often used dense transition metal surfaces in surface science.

Whereas the heat of adsorption of molecules does not vary significantly with surface, the activation energies behave very differently. This is illustrated in Table 2.4 in which a comparison is given of the activation energies of CO as a function of surface structure.

One may note the dramatic decrease in activation energy for stepped compared to non-stepped surfaces. A reaction where these differences are very relevant is the Fischer-Tropsch reaction. This reaction, which produces long-chain hydrocarbons from synthesis gas, requires CO dissociation with low barriers, since otherwise the surface coverage of "C1" species to be inserted in the growing hydrocarbon chains will be too low. It therefore only takes place on unique sites with very low activation barriers for CO dissociation.

In practice, heterogeneous transition metal catalysts consist of small (5–100 nm) particles dispersed on high-surface-area alumina or silica type materials. Such particles will have surfaces more similar to the stepped surface than to the corresponding terraces. Hence direct comparison with the reactivity of dense surfaces is often not useful.

Table 2.4 DFT computed activation energies for CO dissociation on different Ru surfaces (D.A.J.M. Ligthart and E.J.M. Hensen, manuscript in preparation) [10].

Surfaces	Activation energy (kJ mol^{-1})
Ru(0001)	227
Ru(1015) steps	89
Ru(1121)	65
Ru(1010)	46

2.7.2
Catalyst Activation or Deactivation

Furthermore, the working catalyst may have a very different chemical composition from that of the material that was originally present. For instance, in the start-up phase of the Fischer-Tropsch reaction, initially reduced Fe is converted to the less reactive Fe_2C_5 phase, which provides the sites for this reaction.

Small metal particles distributed on a support are metastable. The stable phase is always the bulk phase. The catalyst surface becomes destabilized by interaction with reagent molecules or reaction intermediates. For instance, reaction with CO and formation of intermediate carbonyl species can increase the mobility of the surface atoms.

Changes that can lead to deactivation of the catalyst are illustrated in Figure 2.9:

Figure 2.9 illustrates the important effect of time of reaction on the reactivity of a small Rh particle distributed over a high-surface-area support. Whereas the larger particles do not deactivate in time, the smaller particle becomes more rapidly deactivated. In this case this is most likely because of increased sensitivity to oxidation.

2.7.3
Inhomogeneous Site Distribution

Most heterogeneous catalysts have an intrinsically inhomogeneous distribution of reactive centers.

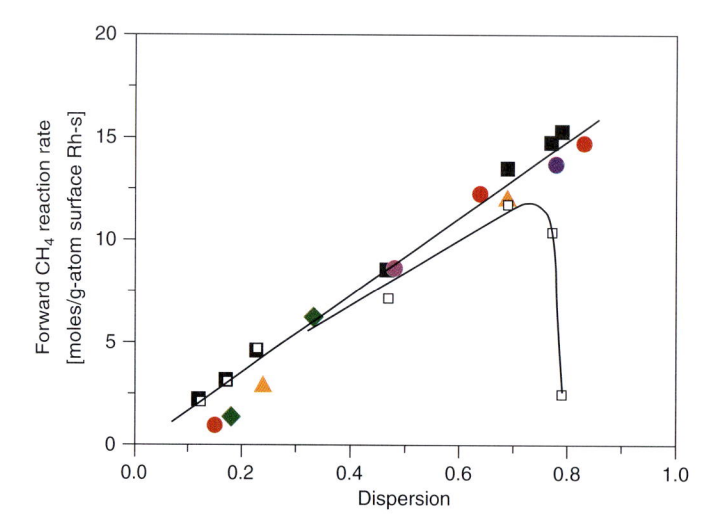

Figure 2.9 Reaction rate as a function of dispersion for $CH_4 - H_2O$ reforming (773 K). Rh supported on ZrO_2 (■), CeO_2-poly (●), CeO_2-rod (●), CeO_2-cube (●), $CeZrO_2$ (▲), SiO_2 (♦) symbols: initial rate (closed) vs. rate after 15 h (open). A comparison of initial reaction rate is with those after 15 h. The extremely small particles deactivate. (D.A.J.M. Ligthart and E.J.M. Hensen, manuscript in preparation.)

2.7 The Materials Gap

Even a uniform surface may be inhomogeneous. An example is provided by a combined computational and experimental study of the metathesis reaction catalyzed by monomeric Mo oxide sites on the surface of γ-Al$_2$O$_3$ by Handzlik and Sautet [12].

Even on a single well-defined surface, anchoring of MoO$_x$ clusters can happen in many different ways. A cluster can attach to a single bridging O between two Al ions or via an O attached end-on to Al. Also, two of the O atoms of Mo can connect to the surface via two Al cations or, alternatively, Mo may attach through surface O, while one of the Mo oxygen atoms attaches to the other Al atoms.

These different ways of attachment are illustrated in Figure 2.10 for a Mo-methylidene species. The detailed structure of this intermediate controls the rate of the metathesis reaction. Reaction energies strongly depend on the O–M–O angle with the oxygen atoms of the alumina support.

The great majority of sites are blocked by the stability of intermediate molybdocyclobutane, and their catalytic activity is low. The species active at low temperature has reduced stability because of the constraint imposed by the surface that deforms

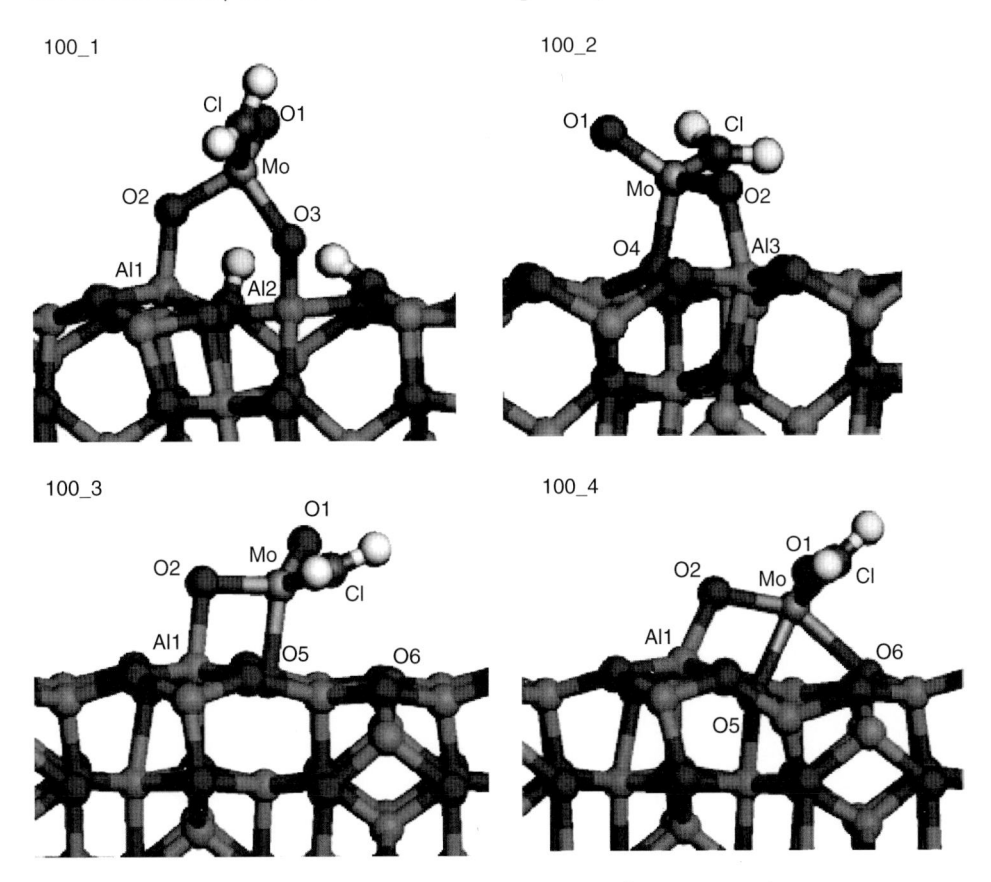

Figure 2.10 Mo-methylidene species grafted to γ-Al$_2$O$_3$ (100) surface.

the Mo complex. The various Mo species show very different activities due to the different constraints imposed by the wide variety of ways in which the Mo complex can be attached to the alumina support. As a result, only a small fraction of the Mo sites are catalytically active.

2.8
Coupling of Catalytic Reaction and Inorganic Solid Chemistry

Surface reconstruction and phase changes are often dynamically coupled to the conversion of reactants. The prototype examples are the spatially organized patterns on Pt or Pd surfaces generated by Nettesheim et al. [13] for the oxidation of CO. In this case, CO adsorption causes surface reconstruction. The appearance and disappearance of time-dependent surface structural changes has been observed with varying surface overlayer patterns. The chemistry of the surface reaction and surface structure transformations are closely related. To properly describe the chemistry of such process simulations, surface phase transformations as well as the chemical transformations have to be included.

Hence there is sometimes an additional constraint on the rate of catalytic reactions that relates to the rate of the surface transformation.

Butane oxidation by $VOPO_4$ is another reaction that illustrates this system, which oxidizes butane to maleic acid with high efficiency. Oxidation of butane reduces V^{5+} to a lower valency, and the reduced cationic vanadium is reoxidized by oxygen.

Based on extensive *in situ* and model studies Blühm et al. [14] have shown that this ongoing change of composition at the surface of the oxide implies complex inorganic chemistry. Maintainance of reversibility of the inorganic chemistry is a second important reaction condition requirement. The proposals of these authors for the changes in catalyst structure are summarized in Figure 2.11:

2.9
In situ Generation of Organo-Catalyst

Catalysis by zeolites appears to occur on an organocatalyst that is formed *in situ* during reaction. As an example, we discuss here the conversion of methanol to aromatics by a solid acid zeolite ZSM-5. This reaction proceeds via the formation of olefins followed by consecutive cyclization and aromatization.

The overall reaction is

$$CH_3OH \longrightarrow C_6H_6 + H_2O + H_2 \quad (2.33)$$

The dimensions of the zeolite micropores inhibit formation of higher aromatics and hence catalyst deactivation. The difficult step in the catalysis of this reaction is the first C–C bond-forming step.

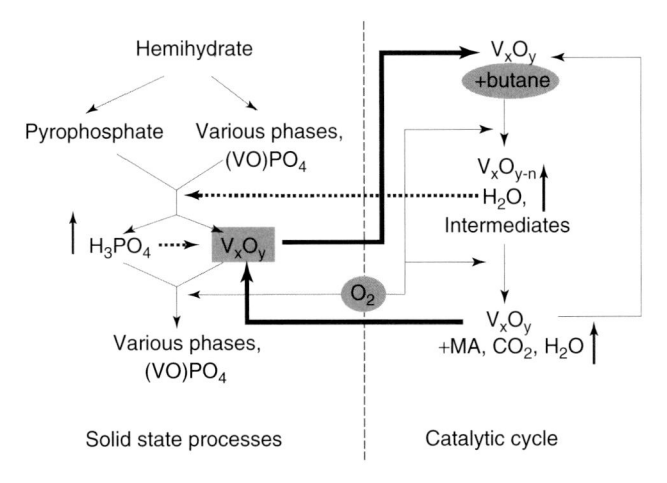

Figure 2.11 Structural changes in the VOPO$_4$ catalyst under reaction conditions. H$_2$O that is produced by oxidation of butane segregates a VO$_2$ phase with vanadium +4 cations. The particles in this phase remain small because phosphoric acid inhibits crystallite growth. Reduction of butane prevents over-oxidation of the V^{4+} overlayer [14].

Many computational studies investigating this reaction step are available. They show that initially the C–C bond forming reaction is a reaction between zeolite surface-alcoxyl intermediate and methanol (Figure 2.12).

A surface methoxy species is easily formed from reaction of methanol with zeolite protons. The activation energy of reaction with methanol is at least 200 kJ mol^{-1}.

Once ethanol or its corresponding ether is formed, chain growth can occur by further reaction of surface methoxy species with olefins formed by dehydration.

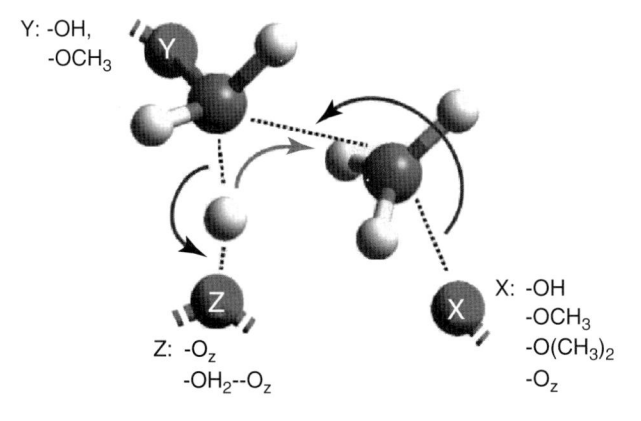

Figure 2.12 Concerted C–C bond formation and C–H cleavage during the reaction of a surface methoxy-group and methanol [15].

Consecutive aromatics formation then occurs by olefin oligomerization and hydrogen transfer reactions, to be discussed in more detail in Parts II and III. All these reactions have reaction barriers of 100 kJ mol^{-1} or less. At the temperature of reaction, around 350 °C, the direct C–C bond formation reaction will be extremely slow. Therefore it cannot be responsible for the steady state of the catalytic reaction.

The main evidence from experiment [7] is the formation of a hydrocarbon pool in the micropore of the zeolite. Product formation occurs through participation of reactants in this hydrocarbon pool. Solid-state NMR [8] demonstrated the presence of polymethylnaphthalenes and cyclopentadienyl species that are efficient catalysts for C–C bond formation with low activation energies of approximately 100 kJ mol^{-1}.

Such molecules can be the result of slow initiation of the reaction by direct C–C combination and subsequent transformation to cyclopentadienyl type species stabilized by contact with the zeolite. They may also be due to the presence of ethylene impurities in the methanol feed. The actual catalysts then are the organocatalytic intermediates formed *in situ*. The two reaction paths proposed to give intermediate olefins are shown in Figure 2.13.

2.10
The Compensation Effect

When reaction conditions or catalytic reactivity of a surface changes, the surface coverage of the catalyst is modified. This change in surface coverage changes the rate by changing the order of the reaction. It leads to a compensation effect.

According to the compensation effect there is a linear relationship between apparent activation energy and entropy of a catalytic reaction.

In Eq. (2.4) the surface coverage given by θ_A and θ_B is related to parameter λ of Eq. (2.3c). Equation (2.4a) can be rewritten to show explicitly its dependence on gas phase concentration. Equation (2.34) gives the result. This equation can be rewritten in the form of power law type expressions, as is done in Eq. (2.34b). Power law type rate equations present the rate of a reaction as a function of the reaction order. In Eq. (2.34b) the reaction order is m in H_2 and $-n$ in CO.

$$R_H = \frac{r_H k_{diss} K_{ads}^{CO}[CO]}{k_{diss} K_{ads}^{CO}[CO] + r_H \left(K_{ads}^{CO}[CO] + 1\right)^2} \quad (2.34a)$$

$$\approx k_H^l k_{diss}^{-n} \left(K_{ads}^{CO}\right)^{-n} [CO]^{-n} [H_2]^m$$

$$l \leq 1; \quad (2.34b)$$

with:

$$r_H = k_H \cdot [H_2]^t ; \ m = t \cdot l \quad (2.35)$$

Power law equations are useful as long as the approximate orders on reactant concentration are constant over a particular concentration course. A change in the order of the reaction corresponds to a change in surface concentration of a particular

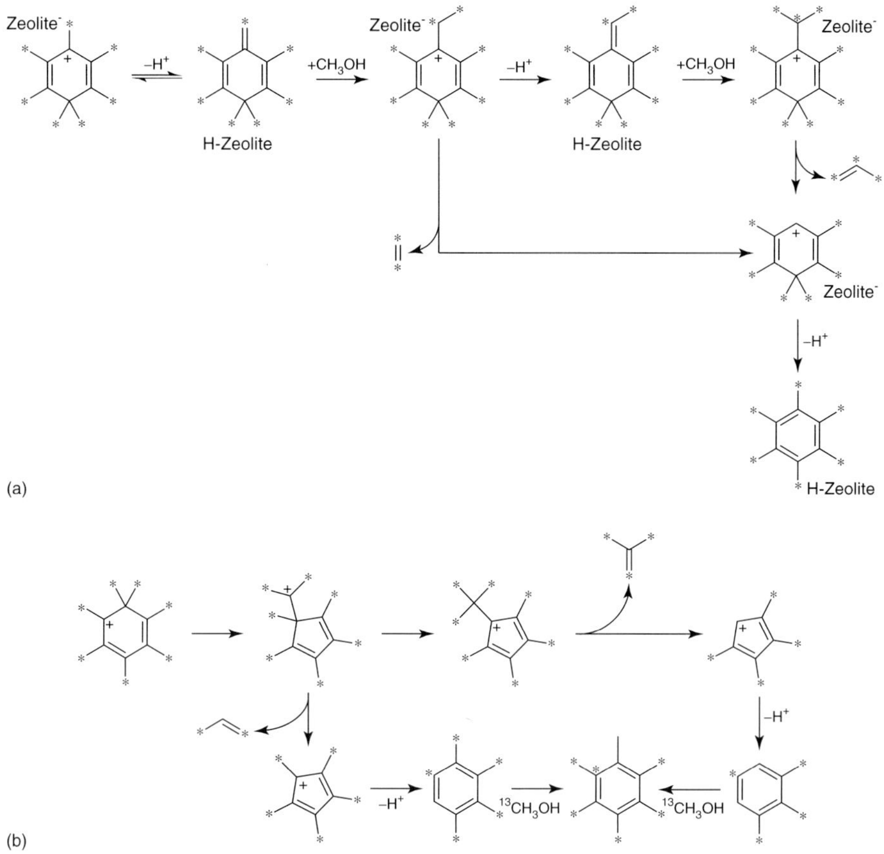

Figure 2.13 (a) The exocyclic methylation reaction in an acidic zeolite (adapted from Ref. [16]). (b) The pairing reaction (*denotes ^{13}C) (adapted from Ref. [16]).

reactant. A low reaction order usually implies a high surface concentration, and a high reaction order a slow surface reaction of the corresponding adsorbed intermediates.

From Eq. (2.34b) the apparent activation energy as well as the pre-exponent can be readily deduced. They are given by Eq. (2.36).

$$E_{app} = lE^H_{act} - n\{E_{act}(\text{diss}) + E_{ads}(\text{CO})\} \tag{2.36a}$$
$$= lE^H_{act} - n\Delta E_{app}$$

$$\ln A_{app} = \ln \Gamma \cdot \frac{ekT}{h} + \frac{l}{k}\Delta S^H_{act} - \frac{n}{k}\Delta S_{app} \tag{2.36b}$$

$$\Delta S_{app} = \Delta S_{act}(\text{diss}) + \Delta S_{ads}(\text{CO}) \tag{2.36c}$$

The orders of the reaction appear as coefficients of activation energies and adsorption energies and their corresponding entropies.

A consequence of the compensation effect is the presence of an isokinetic temperature. For a particular reaction the logarithm of the rate of a reaction measured at different conditions versus T^{-1} should cross at the same (isokinetic) temperature. For conditions with varying n this isokinetic temperature easily follows from Eq. (2.36) and is given by Eq. (2.37):

$$T_{iso} = \frac{\Delta E_{app}}{k \Delta S_{app}} \tag{2.37}$$

It is important to realize that the compensation effect in catalysis refers to overall catalytic reactions.

Sometimes the activation energies of elementary reaction steps also show a relationship between activation energy changes and activation entropies.

A reaction with a high activation energy tends to have a weaker interaction with the surface and hence will have enhanced mobility that is reflected in a larger activation entropy. For this reason the pre-exponents of surface desorption rate constants are 10^4–10^6 larger than the pre-exponents of surface reaction rates.

In classical reaction rate theory equations this directly follows from the frequency–pre-exponent relationship:

$$k_{class} = \nu \frac{r_t^2}{r_i^2} e^{-\frac{E_{act}}{kT}} \tag{2.38}$$

Higher vibration frequencies for r_i and r_t, the initial and transition state radii between surface and reactant imply a stronger bond, which will give a higher activation energy for desorption. Increase of pre exponent and corresponding activation energies are seen to counteract this. Equation (2.38) is the rate equation for a weakly bonded complex.

Compensation type behavior is quite general and has been extensively studied especially in transition metal catalysis [17], sulfide catalysis [18], and zeolite catalysis [19].

References

1. van Santen, R.A. and Neurock, M. (2006) *Molecular Heterogeneous Catalysis*, Wiley-VCH Verlag GmbH, Weinheim.
2. Ertl, G. (1983) in *Catalysis, Science and Technology* (eds J.R. Anderson and M. Boudart), Springer, Berlin, p. 273.
3. Rose, I.A. (1997) Restructuring the active site of fumarase for the fumarate to malate reaction. *Biochemistry*, **36** (40), 12346–12354.
4. Rose, I.A. (1998) How fumarase recycles after the malate →fumarate reaction. Insights into the reaction mechanism. *Biochemistry*, **37** (51), 17651–17658.
5. Feldgus, S. and Landis, C.R. (2000) Large-scale computational modeling of Rh(DuPHOS) (+)-catalyzed hydrogenation of prochiral enamides: reaction pathways and the origin of enantioselection. *J. Am. Chem. Soc.*, **122** (51), 12714–12727.
6. Norskov, J.K., Rossmeisl, J., Logadottir, A., Lindqvist, L., Kitchin, J.R., Bligaard, T., and Jonsson, H. (2004) Origin of the overpotential for oxygen reduction at a fuel-cell cathode. *J. Phys. Chem. B*, **108** (46), 17886–17892.
7. Dumesic, J.A., Huber, G.W., and Boudart, M. (2008) in *Handbook of*

Heterogeneous Catalysis (eds G. Ertl, H. Knözinger, F. Shüth, and J. Weitkamp), Wiley-VCH Verlag GmbH, Weinheim, pp. 1445–1461.
8. Campbell, C.T. (2001) Finding the rate-determining step in a mechanism – comparing DeDonder relations with the 'degree of rate control'. *J. Catal.*, **204** (2), 520–524.
9. Zhdanov, V.P., Pavlicek, J., and Knor, Z. (1988) Preexponential factors for elementary surface processes. *Catal. Rev. – Sci. Eng.*, **30** (4), 501–517.
10. Shetty, S., Jansen, A.P.J., and van Santen, R.A. (2008) Dissociation on the Ru(11(2)$\bar{1}$) surface. *J. Phys. Chem. C*, **112** (36), 14027–14033.
11. Ciobica, I.M., van Santen, R.A., van Berge, P.J., and de Loosdrecht, J.V. (2008) Adsorbate induced reconstruction of cobalt surfaces. *Surf. Sci.*, **602** (1), 17–27.
12. Handzlik, J. and Sautet, P. (2008) Active sites of olefin metathesis on molybdena-alumina system: a periodic DFT study. *J. Catal.*, **256** (1), 1–14.
13. Nettesheim, S., Vonoertzen, A., Rotermund, H.H., and Ertl, G. (1993) Reaction-diffusion patterns in the catalytic cooxidation on Pt(110) – front propagation and spiral waves. *J. Chem. Phys.*, **98** (12), 9977–9985.
14. Bluhm, H., Havecker, M., Kleimenov, E., Knop-Gericke, A., Liskowski, A., Schlogl, R., and Su, D.S. (2003) In situ surface analysis in selective oxidation catalysis: n-butane conversion over VPP. *Top. Catal.*, **23** (1–4), 99–107.
15. Lesthaeghe, D., Van Speybroeck, V., Marin, G.B., and Waroquier, M. (2006) Understanding the failure of direct C-C coupling in the zeolite-catalyzed methanol-to-olefin process. *Angew. Chem. Int. Ed.*, **45** (11), 1714–1719.
16. Olsbye, U., Bjorgen, M., Svelle, S., Lillerud, K.P., and Kolboe, S. (2005) Mechanistic insight into the methanol-to-hydrocarbons reaction. *Catal. Today*, **106** (1–4), 108–111.
17. Bligaard, T., Honkala, K., Logadottir, A., Norskov, J.K., Dahl, S., and Jacobsen, C.J.H. (2003) On the compensation effect in heterogeneous catalysis. *J. Phys. Chem. B*, **107** (35), 9325–9331.
18. Toulhoat, H. and Raybaud, P. (2003) Kinetic interpretation of catalytic activity patterns based on theoretical chemical descriptors. *J. Catal.*, **216** (1–2), 63–72.
19. Bond, G.C., Keane, M.A., Kral, H., and Lercher, J.A. (2000) Compensation phenomena in heterogeneous catalysis: general principles and a possible explanation. *Catal. Rev. – Sci. Eng.*, **42** (3), 323–383.

3
Kinetics in Homogeneous Catalysis
Detlef Heller

3.1
Principles of a Catalyst and Kinetic Description

Kinetic investigations produce quantitative relationships between concentration and time data of reactants and thus are some of the most important approaches for the elucidation of reaction mechanisms. Knowledge of substantiated mechanistic ideas opens up the possibility of aimed manipulations of activity and selectivity, two very important parameters in catalysis. Also, scale-up from laboratory to industrial scale application requires consideration of kinetic findings.

Even though kinetics is one of the most important methodical approaches in catalysis, it should be emphasized that it is in principle not possible to prove reaction mechanisms solely from kinetic investigations! Thus, by means of kinetics only certain reaction sequences can be excluded, for example, if the experimentally determined reaction order for one component is not in accordance with a model. Nevertheless, rejection of this model does not enable any of the remaining possibilities to be preferred. It is possible that kinetically equivalent reaction sequences differ chemically, that is, they cannot be distinguished from one another using only kinetic methods. Only additional findings such as the detection or even isolation of intermediates, the interpretation of isotope labeling studies, or computational chemistry allow us to describe the experimental findings in the form of a closed catalytic cycle that is the most probable reaction mechanism based on the available findings.

Although the outstanding relevance of kinetics is indisputable, several sets of problems have to be considered when applying this method in homogeneous catalysis. One arises from the nature of catalysis itself; only the continual repetition of the catalytic cycle leads to an advantage over the simple stoichiometric reaction. A good catalyst thus has to effectively lead to the desired product. Due to the high substrate-to-catalyst ratio, however, detailed kinetic investigations are difficult, as the interesting intermediates of the catalytic cycle have to be detected or quantified in the presence of huge amounts of substrates and/or products. With common techniques such as GC or HPLC this is hardly possible. In addition, the applied transition metal complex is distributed between several intermediates. And finally, in the case of stereoselective catalysis, the number of intermediates can easily multiply. Above all, due to an unfavorable equilibrium position intermediates may not be detectable at all under quasi-stationary catalytic conditions.

The application of transition metal complexes as homogeneous catalysts introduces further challenges. Owing to the anaerobic methods of operation mostly necessary here [1], classical methods for the determination of concentration–time data are available to only a limited extent. For example, it is practically impossible to monitor a hydrogenation in a cell for UV/Vis spectroscopy under isobaric conditions.

Furthermore, the differential equations that result from a catalytic cycle and are used to quantitatively describe the time dependence of individual species cannot, in most cases, be solved directly. Approximate solutions are necessary, such as the equilibrium approximation [2], the Bodenstein principle [3], or the more general quasi-steady-state approach [4]. A discussion of differences and similarities of several approximations can be found in Refs. [5–7].

Finally, a challenge exists in that kinetic investigations in the field of homogeneous catalysis require not only experience in chemistry and methodical research but also in physical chemistry. However, the essential additional expense in a time of increasing financial pressure on research is often not readily agreed to or considered a necessity [1].[1)]

The efficiency of a catalyst can be estimated from the ratio of rate constants of the catalyzed (k_{cat}) to the uncatalyzed (k) reaction, the activation energy being smaller for a catalyzed reaction than for an uncatalyzed one, which is why the rate of reaction increases. Ostwald commented in 1895 that catalysts do change the rate of a chemical reaction but do not change its energy, and that a catalyst increases the rate of a chemical reaction but does not appear in the reaction product [9, 10]. This corresponds to the actual IUPAC definition of catalysis.

Using the example of a bimolecular reaction, the general principles of catalysis and important fundamental terms are shortly explained, Scheme 3.1.

Two limiting cases can be formulated for the rate of product formation, depending on the rate-determining step [11]. If k_2 is rate determining and provided the catalyst concentration is constant, a second-order reaction results; the reaction order in substrates S and R is 1 in each case.

$$k_2 \cdot [R] \ll k_{-1} \; r_{cat} = \frac{k_2 \cdot k_1 \cdot [cat] \cdot [S] \cdot [R]}{k_{-1}} = k_{cat} \cdot [cat] \cdot [S] \cdot [R] \tag{3.1}$$

The intermediate [catS] is called the '*Arrhenius intermediate*' (Figure 3.1a). In the other case the formation of the intermediate is rate determining ('van't Hoff intermediate'); the intermediate reacts faster than it is formed (Figure 3.1b).

$$k_2 \cdot [R] \gg k_{-1} \; r_{cat} = k_1 \cdot [S] \cdot [cat] \tag{3.2}$$

---

1) "Quite often, kinetic studies are not performed because they appear time-consuming, expensive (with respect to the price or availability of most of chiral ligands), and not rewarding" [8].

3 Kinetics in Homogeneous Catalysis

$$S + R \longrightarrow P \qquad \text{Exergonic stoichiometric bimolecular reaction}$$

$$\text{cat} + S \underset{k_{-1}}{\overset{k_1}{\rightleftharpoons}} \text{catS} \qquad \text{Pre-equilibrium with the catalyst (cat)}$$

$$\text{catS} + R \xrightarrow{k_2} P + (\text{cat}) \qquad \text{Irreversible reaction of the catalyst-substrate complex with R}$$

$$r_{\text{cat}} = \frac{d[P]}{dt} = k_2 \cdot [\text{catS}] \cdot [R] \qquad \text{Reaction rate}$$

$$\frac{d[\text{catS}]}{dt} \cong 0 = k_1 \cdot [\text{cat}] \cdot [S] - k_{-1} \cdot [\text{catS}] - k_2 \cdot [\text{catS}] \cdot [R]$$

$$[\text{catS}] = \frac{k_1 \cdot [\text{cat}] \cdot [S]}{k_{-1} + k_2 \cdot [R]} \qquad \text{Steady state for the intermediate [catS]}$$

$$r_{\text{cat}} = \frac{d[P]}{dt} = \frac{k_2 \cdot k_1 \cdot [\text{cat}] \cdot [S] \cdot [R]}{k_{-1} + k_2 \cdot [R]} \qquad \text{Reaction rate}$$

Scheme 3.1 Principle of a catalyst and reaction rate using the example of a bimolecular reaction. (S, R = starting materials, P = product, and cat = catalyst).

The free enthalpy as a function of the reaction coordinate is each shown in Figure 3.1.

If k_2 is rate determining (Figure 3.1a), the catalyst–substrate complex is characterized by a distinct minimum in free enthalpy. The free activation enthalpy of the catalyzed reaction ($\Delta G_{\text{cat}}^{\neq}$) equals the sum of all $\Delta G_i^{\neq}$, including the back reaction, considering the algebraic sign convention for the stoichiometry. $\Delta G_{\text{cat}}^{\neq}$ is smaller (indicative of stronger catalytic activity) when ($\Delta G_1^{\neq} - \Delta G_{-1}^{\neq}$) and $\Delta G_2^{\neq}$ decrease, in other words, the more stable and reactive is the catalyst–substrate complex [catS]. If, however, k_1 is rate determining (Figure 3.1b) the intermediate is relatively unstable (high reactivity of [catS] since $\Delta G_2^{\neq} < \Delta G_{-1}^{\neq}$) and thus difficult to detect. The free activation energy of the catalyzed reaction ($\Delta G_{\text{cat}}^{\neq}$) corresponds to $\Delta G_1^{\neq}$.

The true problem in determining the individual rate constants is the acquirement of the free catalyst concentration [cat] under quasi-stationary catalytic conditions. Michaelis–Menten kinetics, originally published in the field of enzyme kinetics at the beginning of the last century [2], solves the problem considering the catalyst's balance and is thus of vital importance for homogeneous catalysis. The active catalyst reacts with the substrate in a pre-equilibrium, a characteristic of homogeneous catalytic cycles, to form a catalyst–substrate complex, which – usually irreversibly – reacts to give the product in the rate-determining step, releasing the catalyst (Scheme 3.2).[2)]

2) Many hydrogenations under isobaric conditions follow exactly this pattern. The classical example is the asymmetric hydrogenation of prochiral dehydroamino acid derivatives with rhodium or ruthenium catalysts.

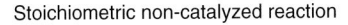

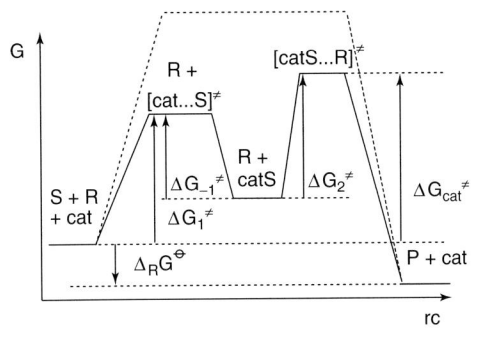

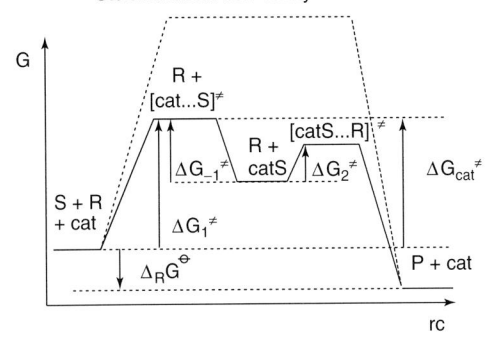

3.1 Principles of a Catalyst and Kinetic Description | 51

Figure 3.1 Reaction diagrams of two catalyzed bimolecular reactions having each a different rate-determining step; (a) rate-determining product formation and (b) rate-determining intermediate formation. The energy profile for the non-catalyzed reaction is shown in both cases for reasons of comparison. (According to Ref. [12].)

$$\text{cat} + S \underset{k_{-1}}{\overset{k_1}{\rightleftharpoons}} \text{cat}S \overset{k_2}{\longrightarrow} P + (\text{cat}) \qquad [\text{cat}]_0 = [\text{cat}] + [\text{cat}S]$$

Scheme 3.2 Reaction sequence and catalyst's balance of the simple Michaelis–Menten kinetics. cat = catalyst, S = substrate, catS = catalyst–substrate complex as intermediate, and P = product.

For the rate of product formation, the so-called Michaelis–Menten equation, Eq. (3.3), results.[3] Its derivation can, for example, be found in Ref. [13].

$$r = \frac{dP}{dt} = \frac{k_2 \cdot [\text{cat}]_0 \cdot [S]}{K_M + [S]} = \frac{r_{\text{sat}} \cdot [S]}{K_M + [S]} \qquad (3.3)$$

3) Today, the name is also used synonymously for the result of the more general quasi-steady state approach by Briggs and Haldane [4].

$$K_M = \frac{k_{-1} + k_2}{k_1} = \frac{[cat] \cdot [S]}{[cat\, S]} \quad K_M \cong \frac{k_{-1}}{k_1} \quad \text{if} \quad k_2 \ll k_{-1};$$

$$\text{if} \quad k_{-1} \ll k_2 \; K_M \cong \frac{k_2}{k_1} \quad \text{if} \quad k_{-1} \ll k_2;$$

k_i rate constants, K_M Michaelis constant, and r_{sat} saturation rate.

With this equation, which besides independent variables and the known initial catalyst concentration only contains the relatively easily accessible concentrations of substrate or product, the desired constants can be determined in a straightforward way.

A more detailed examination shows that in the case of the quasi-equilibrium approximation the value for K_M corresponds to the reverse stability constant; in the case of the steady-state approach the term additionally contains the rate constant for the (irreversible) product formation. Since one cannot macroscopically determine whether the equilibrium constant for a given system is reasonably valid, one should be careful with the interpretation of the K_M value as an inverse stability constant. At best the inverse K_M represents a lower limiting case of a 'stability constant'! In other words, the stability constant quantifying the pre-equilibrium can never be smaller than the inverse value of the Michaelis constant, though it may be larger.

The Michaelis–Menten equation is described by two constants: the rate constant for the reaction of the catalyst–substrate complex forming the product (k_2) and the Michaelis constant (K_M). The Michaelis constant has the dimension of a concentration and describes – independently of the approximation for the intermediate concentration – the substrate concentration at which the ratio of free catalyst to catalyst–substrate complex ([cat]/[catS]) is exactly 1. In other words, one half of the catalyst is coordinated to the substrate. Analogously it follows that, for example, at [S] = 20 K_M the ratio [cat]/[catS] is 0.05, that is, practically 95% of the applied catalyst ([cat]$_0$) is present as catalyst–substrate complex.

Figure 3.2 schematically shows the rate of product formation as a function of substrate concentration (Eq. (3.3)).

Due to the complexity of biological systems, Eq. (3.3) as the differential form of Michaelis–Menten kinetics is usually analyzed using the method of initial rates. By limiting the analysis to the initial range of conversion, unwanted influences such as reversible product formation, enzyme inhibition, or side reactions can be reduced to a minimum in its effect. The disadvantage of this method, however, is the fact that many experiments may be necessary to determine the desired constants.

The analysis of product conversion, however, is in principle not limited to the range of initial rates. Laidler investigated the problem of the valid range of the Michaelis–Menten equation as a function of time under the presumption of quasi-steady-state conditions for the catalyst–substrate complex [14]. As long as one of the conditions of Eq. (3.4) is fulfilled, Eq. (3.5) will be valid up to high conversions. Equation (3.5) corresponds to Eq. (3.3) – due to the free choice of experimental conditions in most cases [S]$_0 \gg$ [cat]$_0$ holds.

$$[S]_0 \gg [cat]_0 \quad [cat]_0 \gg [S]_0 \quad k_{-1} + k_2 \gg k_1 \cdot [cat]_0 \quad k_{-1} + k_2 \gg k_1 \cdot [S]_0 \quad (3.4)$$

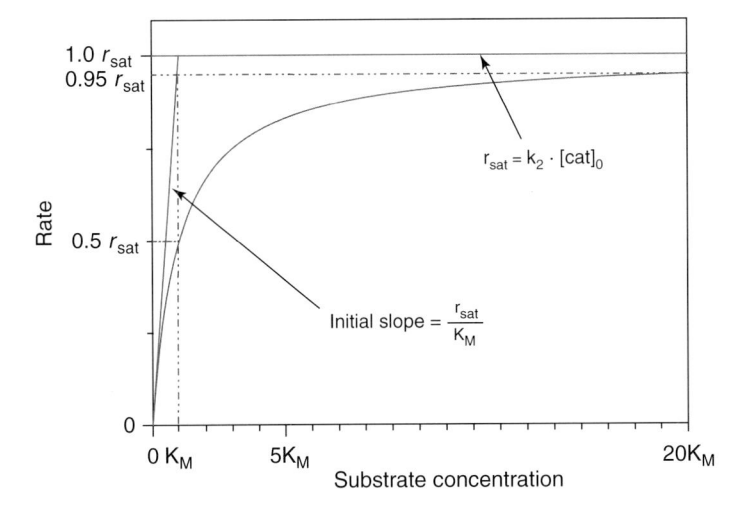

Figure 3.2 Plot of rate of product formation as a function of substrate concentration, cp. Eq. (3.3).

$$\frac{d[P]}{dt} = \frac{k_2 \cdot [cat]_0 \cdot [S]}{K_M + [cat]_0 + [S]} \quad (3.5)$$

After all, every point on a conversion–time plot can be interpreted as an 'initial-rate experiment.' Thus, by analyzing a large range of conversion a huge number of single initial-rate experiments can be saved.

There are two limiting cases for Michaelis–Menten kinetics. Based on Eq. (3.3), Eq. (3.6a) results at a very high excess of substrate (and/or a very small Michaelis constant). This corresponds to a zero order reaction; the rate of product formation is independent of the substrate concentration. In contrast, at very small substrate concentrations (and/or very high Michaelis constants) a first order reaction results (Eq. (3.6b)).

$$r = k_2 \cdot [cat]_0 = r_{sat} \text{ }^{4)} \quad (3.6a)$$

$$r = \frac{k_2 \cdot [cat]_0}{K_M} \cdot [S] = k_{obs} \cdot [S] \quad (3.6b)$$

For the analysis of the Michaelis–Menten equation both the differential (Eq. (3.3)) and the integrated form (Eq. (3.10)) can be used. Because differential values such as flowmeter data are available and time-dependent substrate or product concentrations (or proportional measures) can be easily differentiated numerically, the differential form of the Michaelis–Menten equation is often applied.

4) The term 'maximum rate' (r_{max}) commonly used in earlier reports should not now be used, as suggested by the International Union of Biochemistry, since in a mathematical sense it is not a maximum but rather a limit.

Initial values for a non-linear analysis of Eq. (3.3) can be determined from linearization. The most common linearizations result from transformations of the Michaelis–Menten equation and are plots according to:

Lineweaver and Burk: [15] $\quad \dfrac{1}{r} = \dfrac{K_M}{r_{sat}} \cdot \dfrac{1}{[S]} + \dfrac{1}{r_{sat}} \quad$ plot $1/r$ versus $1/[S]$ (3.7)

Eadie-Hofstee: [16–18] $\quad r = r_{sat} - \dfrac{r}{[S]} \cdot K_M \quad$ plot r versus $r/[S]$ (3.8)

Hanes: [19] $\quad \dfrac{[S]}{r} = \dfrac{K_M}{r_{sat}} + \dfrac{1}{r_{sat}} \cdot [S] \quad$ plot $[S]/r$ versus $[S]$ (3.9)

An analysis of the influence of errors clearly shows that the Lineweaver-Burk double reciprocal plot[5] is the least suitable. ('Although it is by far the most widely used plot in enzyme kinetics, it cannot be recommended, because it gives a grossly misleading impression of the experimental error: for small values of r small errors in r lead to enormous errors in $1/r$; but for large values of r the same small errors in r lead to barely noticeable errors in $1/r'$ [13].) For a considerably more constant error distribution, the Hanes plot, Eq. (3.9), is preferred.

The integrated form of the simple Michaelis–Menten kinetics is suitable for the analysis of time-dependent data of a progressing substrate conversion or the corresponding product formation (Eq. (3.10)).

$$\dfrac{1}{t} \cdot \ln \dfrac{[S]_0}{[S]_t} = -\dfrac{[P]_t}{K_M \cdot t} + \dfrac{r_{sat}}{K_M} \qquad (3.10)$$

A detailed discussion of further possibilities to analyze (Eq. (3.3)) can be found in Ref. [13].

Beside the simple Michaelis–Menten reaction sequence shown in Scheme 3.2, there are other, more complex, models. By analyzing the experimental conditions such as the applicability of the Bodenstein approximation or of pseudo-first-order rate coefficients (linear algebra) one can distinguish between 'simple' and 'non-simple' pathways and networks [7]. Other algorithms for complex models that are described in the literature are the procedure of King and Altman and also the analogous, older works of Christiansen [20–22].

3.2
Catalyst Activity

The best way to quantify catalyst activity is to specify the rate constants for the reaction sequence. Since these are reliably known in only a few cases, most specifications of catalyst activity refer to the dimensionless 'turnover number'

---

5) The original work of Lineweaver and Burk [15] is the most cited paper in the Journal of the American Chemical Society according to Chem. Eng. News **2003**, *81*, issue 48.

(TON; defined as mol substrate per mol of catalyst) or 'turnover frequency' (TOF; defined as TON per time). Note that in enzyme kinetics TON is referred to as TOF.[6] Another common way to quantify catalyst activity is by means of half-life or initial rate.[7]

From an industrial point of view it is the productivity of a catalyst that is of interest, that is, the maximum amount of product formed relative to catalyst. This is also called TON, but this assumes that the catalyst – in contrast to its definition – is consumed by the end of the reaction. Consequently, the macroscopically observed activity not only describes the product formation but also a catalyst deactivation which is usually difficult to quantify.

From a kinetic standpoint, comparison of several catalyst systems is only reasonable if, specifying a TOF, the reaction proceeds zero order, or, if specifying a half-life, the reaction proceeds first order in substrate. Only in these cases is the quantification of catalyst activity independent of the applied substrate concentration.

On the other hand, for the above-mentioned Michaelis–Menten kinetics it could be shown that, dependent on substrate concentration, both a zero-order and a first-order reaction are possible as a limiting case. Thus, for hydrogenations of various substrates with pre-equilibria literature precedents, we find reaction orders of 1 or 0, respectively, in substrate.

If the substrate concentration is sufficiently high and/or the Michaelis constant is sufficiently small (stable catalyst–substrate complexes) a change in the reaction order from zero (beginning of reaction) to one (high substrate conversion) can be observed for one single catalysis! This clearly shows the limits of the common quantification of activity by means of TOF or half-life. Much more informative, therefore, is the specification of catalyst activity on the basis of the Michaelis–Menten equation.

In spite of the advantages of describing the catalyst activity on the basis of Michaelis–Menten kinetics compared to TOF or half-life, however, one has to consider the following. If a reaction follows Michaelis–Menten kinetics and if the reaction order is 1 in substrate, the observed rate constant (k_{obs}) is – according to Eq. (3.6b) – a composite. Without further information a conclusion as to the value of the single constants k_2 and K_M, respectively, cannot be drawn. Conversely, the limiting case of a zero order reaction in substrate does not allow the determination of the Michaelis constant.

However, specific objectives such as the solid comparison of activities of several catalytic systems or a catalytic system in different solvents require both parameters.

---

6) 'It is also sometimes called the *turnover number*, because it is a reciprocal time and defines the number of catalytic cycles (or 'turnovers') the enzyme can undergo in unit time or the number of molecules of substrate that one molecule of enzyme can convert into products in one unit of time [13].'

7) Besides the difficulty to determine the initial rate, one disadvantage is that right at the beginning of the catalysis stationary conditions are not yet reached. Furthermore, the beginning of a reaction is usually the least understood state in terms of thermal equilibration, vapor pressure equilibration, gas solubilities, and so on.

If they are unknown, the comparison of catalyst activities for given experimental conditions can lead to completely false conclusions. Using an example from asymmetric hydrogenation the comparison of a catalyst in different solvents will be discussed below.

Diphosphinites – ligands derived from chiral pool sugar derivatives – were intensively investigated by Selke et al. [23, 24]. The asymmetric hydrogenation of dimethyl itaconate (ItMe$_2$) in MeOH under standard conditions with the catalyst [Rh(Ph-β-glup-OH)(MeOH)$_2$] BF$_4$ (Ph-β-glup-OH : phenyl -2, 3-bis-(o-diphenyl-phosphino)-β-d-glycopyranoside) leads to a first-order reaction up to very high conversions as shown in Figure 3.3 where ln(δc) is plotted as a function of time (Guggenheim plot) [25].

The reaction order of 1 in the sense of Michaelis–Menten kinetics characterizes the hydrogenation in the initial range of Eq. (3.6b), that is, $K_M \gg [S]_0$ is valid. The concentration of diastereomeric catalyst–substrate complexes as intermediates should be small due to the shift of the pre-equilibrium toward the starting material (solvate complex as free catalyst). This can be proven UV/Vis spectroscopically, cp. Figure 3.4. To the solvate complex [Rh(Ph-β-glup-OH)(MeOH)$_2$]BF$_4$ (blue) is added – the usual 100-fold excess of substrate; the hydrogenation is started by exchange of argon by hydrogen (green). Although the stereoisomeric substrate complexes absorb in the measurement range (small picture in Figure 3.4), the UV/Vis spectrum virtually does not change. A GC analysis after termination of the reaction shows that the hydrogenation proceeds, and the usual values for conversion and selectivity result. Apparently, practically all catalyst in solution is

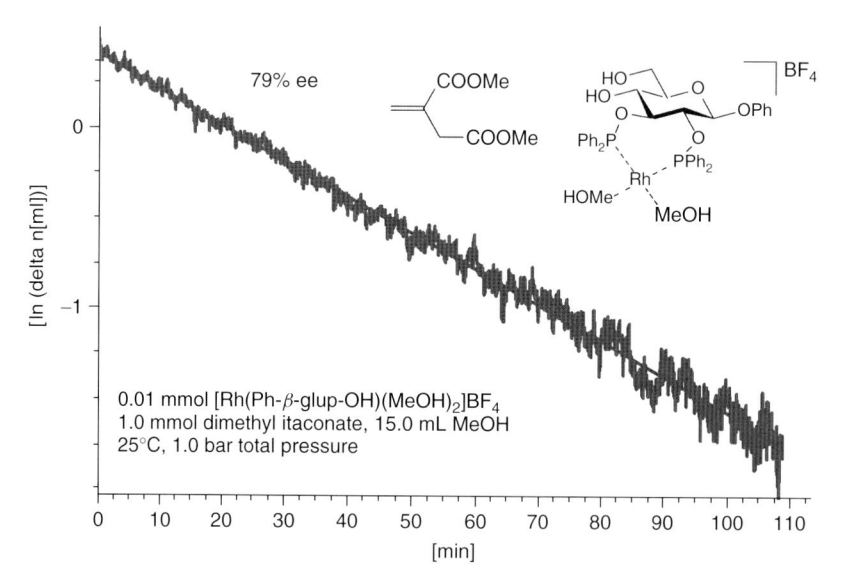

Figure 3.3 Guggenheim plot as linearization of a first order reaction for the asymmetric hydrogenation of ItMe$_2$ with [Rh(Ph-β-glup-OH)(MeOH)$_2$]BF$_4$.

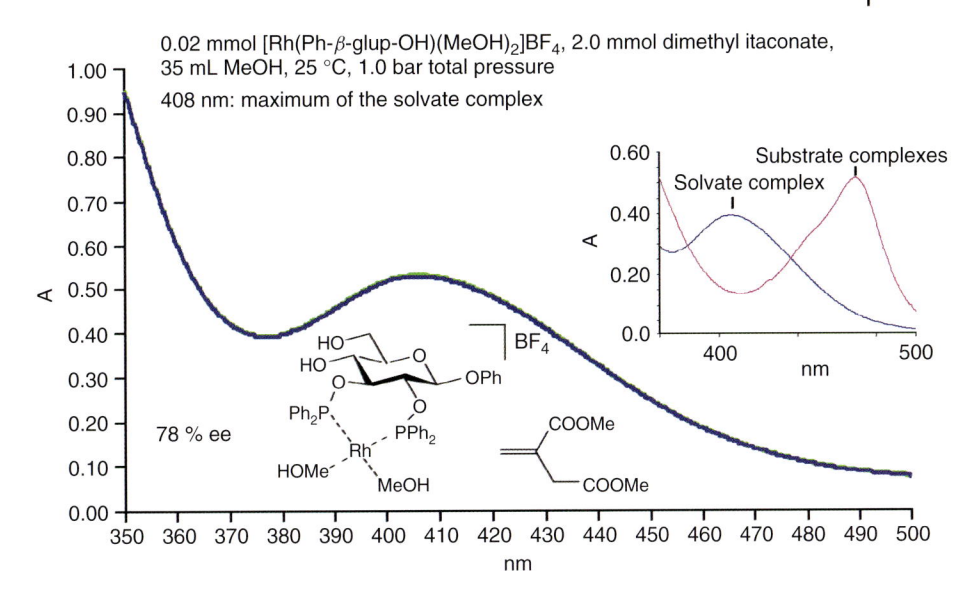

Figure 3.4 UV/Vis spectrum of [Rh(Ph-β-glup-OH)(MeOH)2]BF4 in MeOH under argon (blue), and five spectra (cyclic; 6.0 min) after addition of an about 100-fold excess of ItMe2 and exchange of argon by hydrogen (green). The small picture shows the spectrum of the ItMe2-complexes for comparison (magenta).

present in form of solvate complex during the hydrogenation of ItMe$_2$, which is consistent with the model.

If instead of MeOH tetrahydrofuran (THF) is used as the reaction medium, the situation changes entirely [26]. The activity increases considerably while the enantioselectivity increases only slightly (84% ee). The main difference from MeOH as solvent, however, is that the hydrogenation in THF cannot be described as a first-order reaction. The direct, non-linear regression to the Michaelis–Menten equation (3.3) is shown in Figure 3.5.

The simple exchange of solvent from MeOH to THF thus causes a different stability of the catalyst–substrate complexes under otherwise identical conditions. While there is an extremely low concentration of the catalyst–substrate complex in MeOH under hydrogenation conditions, in THF a value that is in the range of half the applied catalyst concentration results under the given experimental conditions – at least in the beginning of the reaction ($r_{sat}/2 = 0.24$). The higher activity that is macroscopically observed as reaction rate under defined reaction conditions can hence partly be ascribed to the higher concentration of catalyst–substrate complex.

One has to be aware of this effect if several solvents are compared with respect to activity. At the same time, different 'standard conditions' – such as substrate concentrations – significantly influence the levels of activity.

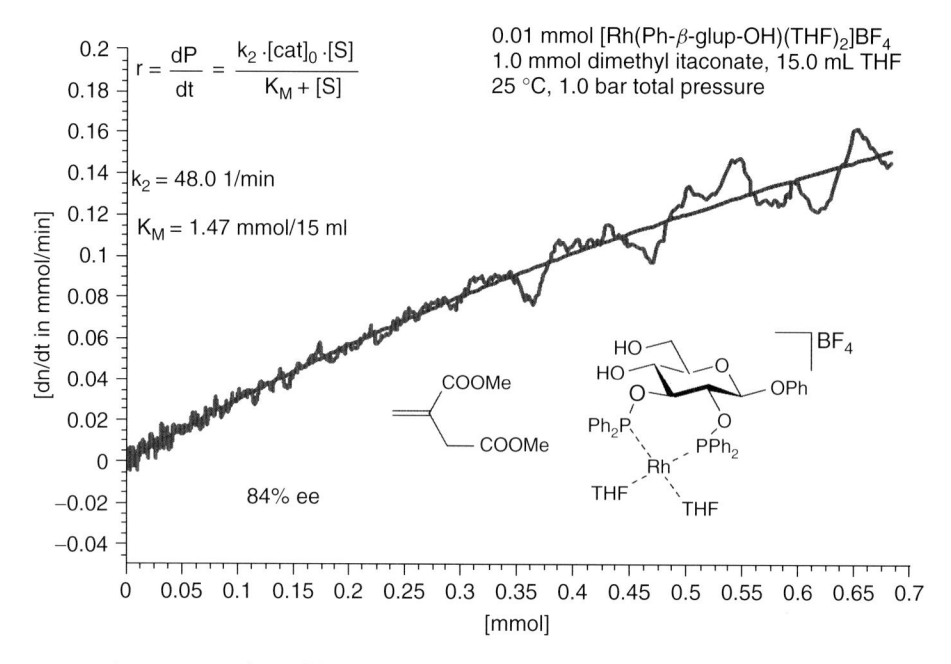

Figure 3.5 Analysis of the asymmetric hydrogenation of ItMe$_2$ with [Rh(Ph-β-glup-OH)(THF)$_2$]BF$_4$ under standard conditions in THF according to Eq. (3.3).

$$r = \frac{dP}{dt} = \frac{k_2 \cdot [cat]_0 \cdot [S]}{K_M + [S]}$$

0.01 mmol [Rh(Ph-β-glup-OH)(THF)$_2$]BF$_4$
1.0 mmol dimethyl itaconate, 15.0 mL THF
25 °C, 1.0 bar total pressure

$k_2 = 48.0$ 1/min
$K_M = 1.47$ mmol/15 ml

84% ee

Using the example of asymmetric hydrogenation (a further example is given in Ref. [25]) it could be shown that pre-equilibria under stationary catalytic conditions, which considerably co-determine the macroscopic activity, are strongly dependent on solvent, catalyst (ligand), and on the substrate and its concentration. The results emphasize that 'gross activities' on the basis of TOF's or half-lives are hardly suitable to reliably compare activities of several catalytic systems, even with very similar catalyst structures. 'Optimal' catalytic conditions determined under specific experimental conditions can by no means be relied upon to still be optimal if experimental conditions, such as substrate concentration or solvent, are slightly changed.

3.3
Catalyst Activation and Deactivation

As important as an understanding of the reaction mechanism is the knowledge of phenomena which effect the catalyst activity. The thermal denaturation of enzymes provides an example of irreversible catalyst deactivation. In other cases phenomena take place which reduce the concentration of the active catalyst. For instance, strongly coordinating ligands in organometallic complex catalysis can lead to such a decrease in catalyst concentration.

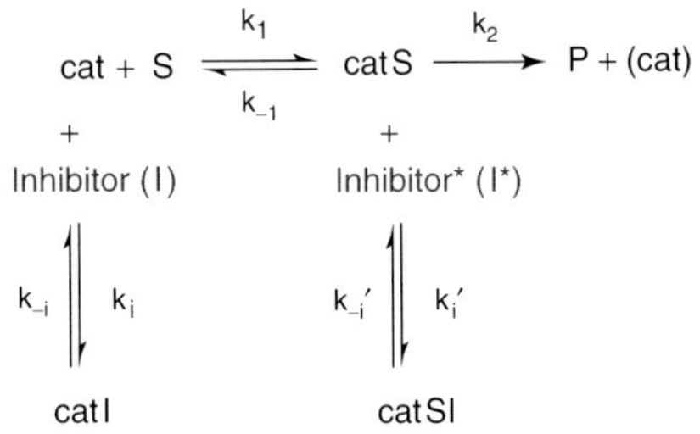

$$\text{cat} + \text{S} \underset{k_{-1}}{\overset{k_1}{\rightleftharpoons}} \text{catS} \xrightarrow{k_2} \text{P} + (\text{cat})$$

$$+ \qquad\qquad\qquad +$$

$$\text{Inhibitor (I)} \qquad\qquad \text{Inhibitor* (I*)}$$

$$k_{-i} \updownarrow k_i \qquad\qquad k_{-i}' \updownarrow k_i'$$

$$\text{catI} \qquad\qquad\qquad \text{catSI}$$

Scheme 3.3 Competitive, non-competitive, and mixed inhibition in enzyme catalysis (for simplification a further reaction of inhibitor complexes catI and catSI is not assumed).

Deactivation phenomena are especially important in industrial applications. Due to very high TONs, even traces of impurities can have huge effects. For example, at a substrate-to-catalyst ratio of 1000, 1/1000 of an undesirable strongly complexing substance can be enough to completely poison the catalyst. Inhibitions by the product are possible as well.

From enzyme catalysis, competitive inhibition (the inhibitor blocks the free enzyme), non-competitive inhibition (the inhibitor blocks the enzyme-substrate complex), and mixed inhibition (both free enzyme and enzyme-substrate complex are blocked) are known. For the simple Michaelis–Menten kinetics, this is shown in Scheme 3.3. By varying the inhibitor concentration and kinetic analysis, differentiation between the three cases is possible [13].

A special case of non-competitive inhibition is inhibition by the substrate itself; a characteristic of substrate inhibition is that the catalytic activity at first increases.

In the following sections activation and deactivation phenomena in homogeneous catalysis are discussed using examples of organometallic complex catalysis, or more precisely asymmetric hydrogenation with cationic rhodium complexes.

3.3.1
Induction Periods as Catalyst Activation

'Active species' mediating a catalytic process are characterized by their high reactivity, as is implied in their name. Due to labile ligands or free valences, free coordination sites are available which ultimately cause the reactivity. Because of the high reactivity of these 'active species' they are generally difficult to handle. Thus, in many cases it is not the active species that is employed – if they are known at all. Instead, the actual catalysts are formed by pre-activation with so-called *co-catalysts*. A typical example is Ziegler-Natta polymerization with MAO (methylaluminoxane) as activator.

The application of precatalysts (catalysts stabilized by suitable ligands) or their formation by *in situ* techniques is another very common approach. Stabilization of active catalysts can be realized by so-called 'spectator ligands'. These include, for example, cyclic diolefins such as COD (1,5-cyclooctadiene) or NBD (2,5-norbornadiene), ethylene, and CO.

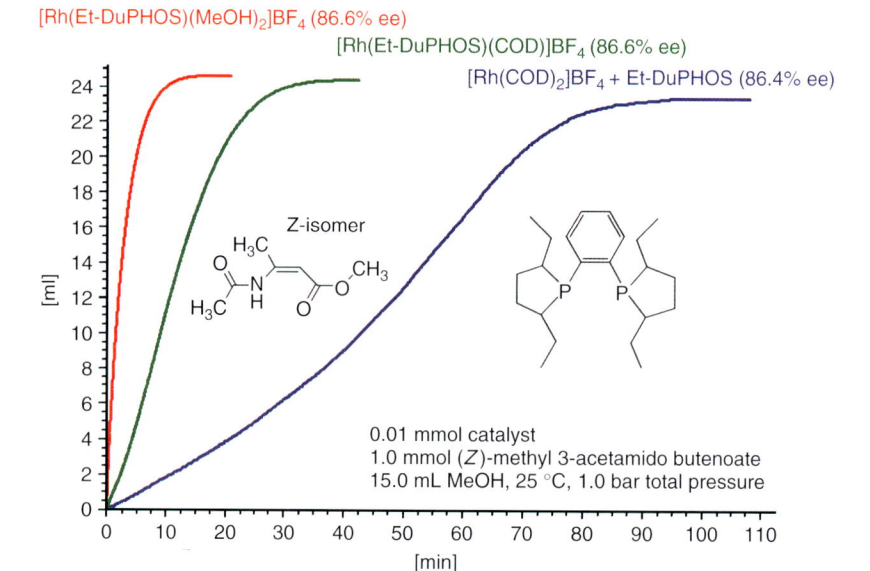

Figure 3.6 Hydrogenation of (Z)-methyl 3-acetamido butenoate with the Rh-Et-DuPHOS catalyst by several methods under standard conditions: 0.01 mmol [Rh(COD)$_2$]BF$_4$ + 0.01 mmol Et-DuPHOS (*in situ*, blue), 0.01 mmol [Rh(Et-DuPHOS)(COD)]BF$_4$ (green), or 0.01 mmol [Rh(Et-DuPHOS)(MeOH)$_2$]BF$_4$ (red).

In most cases both the application of the *in situ* technique and of precatalysts has consequences in terms of catalyst activity, at least at the beginning of the reaction. The different methods are contrasted in Figure 3.6 using the asymmetric hydrogenation of a β-dehydroamino acid derivative with the Rh-Et-DuPHOS catalyst (Et-DuPHOS = 1,2-bis(2,5-diethylphospholano)benzene) in MeOH as example.

It should be noted that the enantioselectivity was not affected by the different methods used for the preparation of the catalyst. For both the use of [Rh(Et-DuPHOS)(COD)]BF$_4$ and the *in situ* technique, distinct *induction periods* are characteristic; that is, the hydrogenation accelerates in the initial stages; an activation seems to occur. These induction periods are especially noticeable as a maximum in the rate profile. They are caused by the fact that part of the catalyst concentration is blocked (by diolefin) and thus not accessible for the hydrogenation of the prochiral olefin. Due to the rather slowly proceeding COD hydrogenation the amount of active catalyst increases at first; induction periods result. To eliminate the undesirable induction periods but also to maximize the use of the 'intrinsic activity' of the catalyst it is best to apply the respective solvate complexes, which are obtained by pre-hydrogenation of diolefin complexes [27, 28]. Under these conditions, the catalytic reaction profits from the whole amount of the 'catalyst' added at the beginning of the reaction.

3.3.2
Catalyst Deactivation due to Formation of Non-Reactive Complexes

Common solvents for homogeneous hydrogenations are simple alcohols, THF, dichloromethane, and aromatic solvents such as toluene or benzene. The sometimes considerable stability of Rh(I)-η^6-arene complexes with chelating diphosphines is often disregarded, although it has been known for quite some time [29, 30]. Only a few literature precedents exist that describe an inhibition of the asymmetric hydrogenation if aromatic solvents are used [31]. Even substrates that bear phenyl rings in their structure are able to form stable arene complexes with rhodium species. For example, the complex in which the hydrogenation product of methyl (Z)-N-acetylamino cinnamic acid (N-acetyl phenylalanine methyl ester) coordinates to the rhodium atom with its phenyl ring was shown to exist by NMR spectroscopy [32]. By means of the PHIP method, Rh complexes with styrene derivatives have been described [33]. X-ray structures of stable arene complexes are also known, for example, the benzene complex of (1R,2R)-trans-1,2-bis((diphenylphosphine)-methyl)cyclobutane-Rh(I) and [Rh((S,S)-Me-DuPHOS)(toluene)]BF$_4$ (Me-DuPHOS = 1,2-bis(2,5-dimethylphospholano)benzene) [34, 35].

Owing to the apparent stability of such η^6-arene complexes one has to take into consideration the fact that asymmetric hydrogenations only utilize one part of the applied rhodium concentration if they are carried out in the presence of aromatic solvents. The other part is blocked for the asymmetric hydrogenation because of the formation of stable Rh(I)-arene complexes [34], as also the following example proves.

A further representative example of the inhibiting solvent effect is the hydrogenation of (Z)-methyl-3-acetamido butenoate with the Rh/Me-DuPHOS catalyst. While Zhang et al. reported complete conversion after 24 h (63.7% ee) with the in situ prepared catalyst in toluene at 20 bar hydrogen pressure and ambient temperature [36], the hydrogenation of the same substrate in MeOH using [Rh(Me-DuPHOS)(MeOH)$_2$]BF$_4$ under comparable conditions takes only 4 min under normal pressure (87.8% ee) [37]. This differing behavior is simply caused by the inhibiting effect of the aromatic solvent. In toluene, a large amount of catalyst is present as stable and inactive arene complex. In MeOH, on the other hand, the total amount of applied rhodium is active for the asymmetric hydrogenation.

Note that due to the fact that the formation of arene complexes can be reversed by other complexing agents by ligand exchange, it is the concentrations and ratios of stability constants of all complexes in solution that decide the degree of a possible activity loss.

3.3.3
Catalyst Deactivation due to Formation of Multinuclear Complexes

As early as 1977 it was reported that the addition of the base triethylamine to the solvate complex [Rh(DPPE)(MeOH)$_2$]BF$_4$ (DPPE = 1,2-bis-(diphenylphosphino) ethane) leads to the formation of trinuclear rhodium complex [Rh$_3$(DPPE)$_3$ (μ_3-OMe)$_2$]$^+$ [29]. An analogous structure of [Rh$_3$(BINAP)$_3$(μ_3-OH)$_2$]ClO$_4$ (BINAP = (2,2′-bis-(diphenylphosphino)-1,1′-binaphthyl)) was published by Saito et al. [38]. As the structural principle for both complexes it was found that three rhodium cores form a regular triangle, with each rhodium atom coordinated a bidentate ligand. The planes P-Rh-P are perpendicular to the Rh$_3$ plane. Above and below the Rh$_3$ plane two μ_3-bridging anions (methoxy or hydroxy) are located. The trinuclear rhodium complexes show a rather high formation tendency and stability; thus, a number of different complexes have been synthesized since then [39]. Figure 3.7a (left) shows the molecular structure of trinuclear rhodium complex [Rh$_3$((S,S)-Me-DuPHOS)$_3$(μ_3-OH)$_2$]BF$_4$.

Additionally, it could be shown that certain olefinic substrates can be basic enough by themselves and without any additive to induce the formation of trinuclear complexes and thus decrease the activity of their hydrogenation [39].

In solvents such as dichloromethane, dinuclear complexes are often observed. Due to the fact that dichloromethane is in practice a non-coordinating solvent, diolefin complexes [Rh(ligand)(diolefin)]anion as common precatalysts applied in asymmetric hydrogenation are stabilized by dimerization in the presence of hydrogen (hydrogenation of the precatalyst's diolefin), and arene-bridged dimers are formed. Figure 3.7b (right) shows the molecular structure with DIPAMP (1,2-bis[(methoxyphenyl)(phenylphosphino)]ethane) as ligand [40]. Since these sometimes quite stable dimers are inactive catalysts, their formation can also lead to a decrease in catalytic activity.

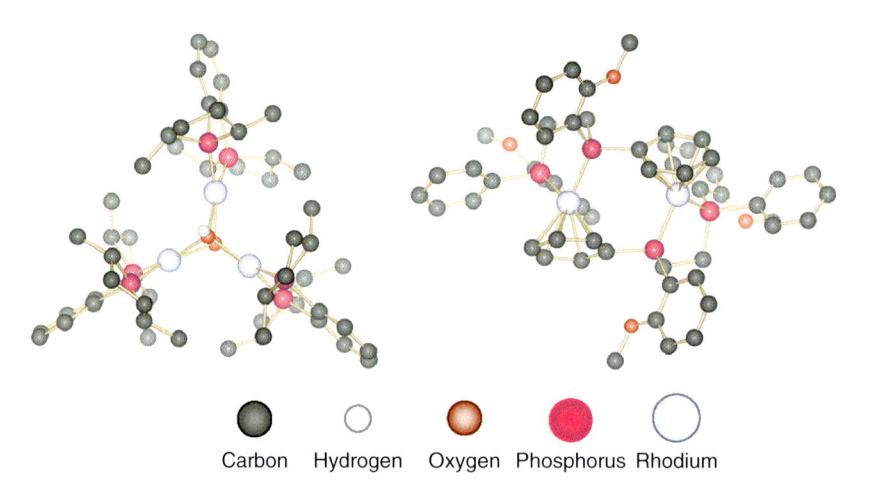

Figure 3.7 X-ray structures of [Rh$_3$((S,S)-Me-DuPHOS)$_3$(μ_3-OH)$_2$]BF$_4$ and of [Rh((S,S)-DIPAMP)]$_2$(BF$_4$)$_2$. Hydrogen atoms except those of the OH-bridge and the anion BF$_4^-$ are omitted for clarity.

3.3.4
Catalyst Deactivation due to Irreversible Reactions

The asymmetric hydrogenation of itaconic acid (ItH$_2$) with the Rh/DIPAMP catalyst in MeOH surprisingly showed that conversions continuously decreased with increasing substrate concentration under otherwise identical conditions [41, 42]. Thus, a substrate inhibition is practically observed. Single crystals could be isolated from the solution of [Rh(DIPAMP)(MeOH)$_2$]BF$_4$ and itaconic acid. The X-ray structure surprisingly revealed that the substrate was not coordinated via the C=C double bond and one carboxylate oxygen as usual. In fact, a tridentate complexation was observed. The substrate coordinates via the two carboxylate oxygens and the quarternary carbon atom. Formally, the acidic functions of the substrate were completely deprotonated; the methylene group was already converted into a methyl group via H-transfer. The tetrafluoroborate anion was no longer present in the crystal as the counterion. Thus, the complex is a neutral Rh(III)-alkyl species and corresponds to the complex present in large excess in solution.

It could be proven that the isolated Rh(III)-alkyl complex is not involved in the hydrogenation owing to the fact that the oxidative addition of hydrogen is formally impossible on an Rh(III) complex. By ^{31}P-NMR spectroscopy it could be shown that the alkyl complex formation is indeed responsible for the above-mentioned substrate concentration-dependent deactivation phenomenon.

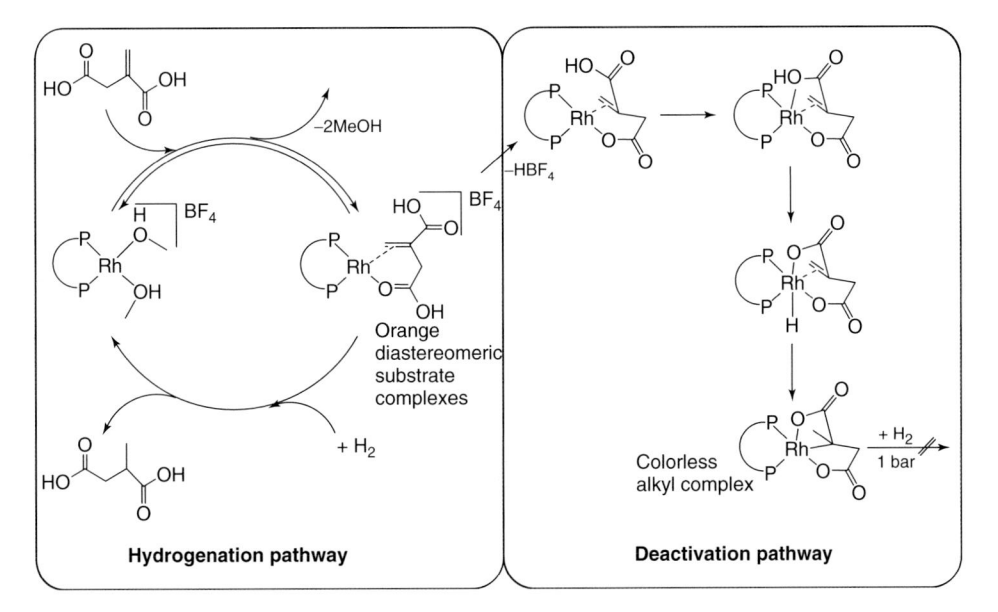

Figure 3.8 Reaction sequence for the usual hydrogenation and the parallel-running deactivation pathway.

In Figure 3.8 the reaction sequence is shown in which, besides the usual hydrogenation that can be accelerated by increasing the hydrogen pressure, the parallel formation of the inactive Rh(III)-alkyl complex takes place. This irreversible catalyst deactivation is the reason for the apparent substrate inhibition. At lower substrate concentrations, however, the overall reaction time is shorter, and the deactivation is less obvious.

As the examples have shown, there are several factors that influence the catalytic activity. Already the transformation of precatalysts into catalytically active species is a factor that is often underestimated. Furthermore, there are a number of complexing agents, which can – in most cases reversibly – decrease the catalyst concentration available for the catalysis. Coordinating functional groups or structural elements can be part of the solvent, the substrate, the product, or additives. Here, concentration and complex stability play an important role. However, a stabilization of unstable intermediates, for example, by formation of multinuclear complexes, decreases the catalytic activity as well. Finally, irreversible transformations of the active catalyst are also unwanted.

After all, the knowledge of activation and deactivation phenomena is as important as the elucidation of the reaction mechanism. Kinetics is especially useful for such investigations if used in combination with other methods such as the isolation of intermediates and their characterization by X-ray analysis.

References

1. Herrmann, W.A. and Salzer, A. (1996) Literature, laboratory techniques and common starting materials, in *Herrmann/Brauer: Synthetic Methods of Organometallic and Inorganic Chemistry* (ed. W.A. Herrmann), Georg Thieme Verlag, Berlin.
2. Michaelis, L. and Menten, M.L. (1913) The kenetics of the inversion effect. *Biochem. Z.*, **49**, 333–369.
3. Bodenstein, M. (1913) Eine theorie der photochemischen reaktionsgeschwindigkeiten. *Z. Phys. Chem.*, **85**, 329–397.
4. Briggs, G.E. and Haldane, J.B.S. (1925) A note on the kinetics of enzyme action. *Biochem. J.*, **19** (2), 338–339.
5. Connors, K.A. (1990) *Chemical Kinetics: The Study of Reaction Rates in Solution*, VCH Publishers Inc., New York.
6. Espenson, J.H. (2003) in *Encyclopedia of Catalysis* (ed. I.T. Horváth), Wiley-VCH Verlag GmbH, pp. 490–508.
7. Helfferich, F.G. (2001) *Kinetics of Homogeneous Multistep Reactions, Comprehensive Chemical Kinetics*, Elsevier, Amsterdam.
8. de Bellefon, C., Pestre, N., Lamouille, T., Grenouillet, P., and Hessel, V. (2003) High throughput kinetic investigations of asymmetric hydrogenations with microdevices. *Adv. Synth. Catal.*, **345** (1–2), 190–193.
9. Ostwald, W. (1894) 98. Über den Wärmewert der Bestandteile der Nahrungsmittel. *Z. Phys. Chem.*, **15**, 705–706.
10. Ostwald, W. (1910) Über Katalyse. *Ann. Naturphilos.*, **9**, 1–25.
11. Laidler, K.J. (1988) Rate-controlling step – a necessary or useful concept. *J. Chem. Educ.*, **65** (3), 250–254.
12. Taube, R. (1988) *Homogene Katalyse*, Akademie-Verlag, Berlin.
13. Cornish-Bowden, A. (2004) *Fundamentals of Enzyme Kinetics*, 3rd edn, Portland Press Ltd, London.
14. Laidler, K.J. (1955) Theory of the transient phase in kinetics, with special

reference to enzyme systems. *Can. J. Chem.*, **33** (10), 1614–1624.
15. Lineweaver, H. and Burk, D. (1934) The determination of enzyme dissociation constants. *J. Am. Chem. Soc.*, **56**, 658–666.
16. Eadie, G.S. (1942) The inhibition of cholinesterase by physostigmine and prostigmine. *J. Biol. Chem.*, **146** (1), 85–93.
17. Hofstee, B.H.J. (1952) Specificity of esterases.1. Identification of 2 pancreatic aliesterases. *J. Biol. Chem.*, **199** (1), 357–364.
18. Hofstee, B.H.J. (1959) Non-inverted versus inverted plots in enzyme kinetics. *Nature*, **184** (4695), 1296–1298.
19. Hanes, C.S. (1932) Studies on plant amylases. I. The effect of starch concentration upon the velocity of hydrolysis by the amylase of germinated barley. *Biochem. J.*, **26**, 1406–1421.
20. Christiansen, J.A. (1931) Ein Versuch zur Anwendung der Methode der stationären Geschwindigkeiten auf die Reaktion $CH_3OH + H_2O \rightarrow 3H_2 + CO_2$. *Z. Phys. Chem. Bodenstein Festband*, 69–77.
21. Christiansen, J.A. (1935) Some comments on the application of the Bodenstein method of stationary concentrations of intermediary materials in reaction kinetics. *Z. Phys. Chem.*, **28** (4), 303–310.
22. King, E.L. and Altman, C. (1956) A schematic method of deriving the rate laws for enzyme-catalyzed reactions. *J. Phys. Chem.*, **60** (10), 1375–1378.
23. Harthun, A., Kadyrov, R., Selke, R., and Bargon, J. (1997) Proof of chiral dihydride complexes including catalyst and substrate during the bis(phosphinite)rhodium(I)-catalyzed hydrogenation of dimethyl itaconate. *Angew. Chem. Int. Ed. Engl.*, **36** (10), 1103–1105.
24. Harthun, A., Selke, R., and Bargon, J. (1996) Proof of a reversible, pairwise hydrogen transfer during the homogeneously rhodium(I)-catalyzed hydrogenation of alpha,beta-unsaturated carbonic acid derivatives with in situ NMR spectroscopy and parahydrogen. *Angew. Chem. Int. Ed. Engl.*, **35** (21), 2505–2507.
25. Drexler, H.-J., Preetz, A., Schmidt, T., and Heller, D. (2007) Kinetics of homogeneous hydrogenation: measurement and interpretation, in *Handbook of Homogeneous Hydrogenation*, Chapter 10 (eds H.G. de Vries and C. Elsevier), Wiley-VCH Verlag GmbH, 257–293.
26. Heller, D. (1999) Habilitation, Greifswald.
27. Heller, D., de Vries, A.H.M., and de Vries, J.G. (2007) Catalyst inhibition and deactivation in homogeneous hydrogenation, in *Handbook of Homogeneous Hydrogenation*, Chapter 44 (eds H.G. de Vries and C. Elsevier), Wiley-VCH Verlag GmbH, 1483–1516.
28. Preetz, A., Drexler, H.J., Fischer, C., Dai, Z., Borner, A., Baumann, W., Spannenberg, A., Thede, R., and Heller, D. (2008) Rhodium-complex-catalyzed asymmetric hydrogenation: transformation of precatalysts into active species. *Chem. – Eur. J.*, **14** (5), 1445–1451.
29. Halpern, J., Riley, D.P., Chan, A.S.C., and Pluth, J.J. (1977) Novel coordination chemistry and catalytic properties of cationic 1,2-bis(diphenylphosphino)ethanerhodium(I) complexes. *J. Am. Chem. Soc.*, **99** (24), 8055–8057.
30. Landis, C.R. and Halpern, J. (1983) Homogeneous catalysis of arene hydrogenation by cationic rhodium arene complexes. *Organometallics*, **2** (7), 840–842.
31. Burk, M.J., Kahlberg, C.S., and Pizzano, A. (1998) Rh-DuPHOS-catalyzed enantioselective hydrogenation of enol esters. application to the synthesis of highly enantioenriched r-hydroxy esters and 1,2-diols. *J. Am. Chem. Soc.*, **120** (18), 4345–4353.
32. Gridnev, I.D., Yasutake, M., Higashi, N., and Imamoto, T. (2001) Asymmetric hydrogenation of enamides with Rh-BisP* and Rh-MiniPHOS catalysts. Scope, limitations, and mechanism. *J. Am. Chem. Soc.*, **123** (22), 5268–5276.
33. Hübler, P. and Bargon, J. (2000) In situ transfer of parahydrogen-induced nuclear spin polarization - Structural

34. Heller, D., Drexler, H.J., Spannenberg, A., Heller, B., You, J.S., and Baumann, W. (2002) The inhibiting influence of aromatic solvents on the activity of asymmetric hydrogenations. *Angew. Chem. Int. Ed.*, **41** (5), 777–780.
35. Townsend, J.M. and Blount, J.F. (1981) Crystal structure of (eta-6-hexadeuteriobenzene) (1R,2R)-trans-1,2-bis((diphenylphosphino) methyl)cyclobutane rhodium(i) perchlorate - a model for the resting state of an asymmetric hydrogenation catalyst. *Inorg. Chem.*, **20** (1), 269–271.
36. Zhu, G.X., Chen, Z.G., and Zhang, X.M. (1999) Highly efficient asymmetric synthesis of beta-amino acid derivatives via rhodium-catalyzed hydrogenation of beta-(acylamino)acrylates. *J. Org. Chem.*, **64** (18), 6907–6910.
37. Heller, D., Holz, J., Drexler, H.J., Lang, J., Drauz, K., Krimmer, H.P., and Borner, A. (2001) Pressure dependent highly enantioselective hydrogenation of unsaturated beta-amino acid precursors. *J. Org. Chem.*, **66** (20), 6816–6817.
38. Yamagata, T., Tani, K., Tatsuno, Y., and Saito, T. (1988) A new rhodium trinuclear complex containing highly protected hydroxo groups, (rh(binap))3(mu-3-oh)2 cio4, responsible for deactivation of the 1,3-hydrogen migration catalyst of allylamine binap = 2,2′-bis(diphenylphosphino)-1, 1′-binaphthyl. *J. Chem. Soc., Chem. Commun.*, (7), 466–468.
39. Preetz, A., Baumann, W., Drexler, H.J., Fischer, C., Sun, J.T., Spannenberg, A., Zimmer, O., Hell, W., and Heller, D. (2008) Trinuclear rhodium complexes and their relevance for asymmetric hydrogenation. *Chem. Asian J.*, **3** (11), 1979–1982.
40. Preetz, A., Baumann, W., Fischer, C., Drexler, H.J., Schmidt, T., Thede, R., and Heller, D. (2009) Asymmetric hydrogenation. Dimerization of solvate complexes: synthesis and characterization of dimeric Rh(DIPAMP) (2)(2+), a valuable catalyst precursor. *Organometallics*, **28** (13), 3673–3677.
41. Schmidt, T., Baumann, W., Drexler, H.J., and Heller, D. (2011) Unusual deactivation in the asymmetric hydrogenation of itaconic acid. *J. Organomet. Chem.*, **696** (9), 1760–1767.
42. Schmidt, T., Drexler, H.J., Sun, J., Dai, Z., Baumann, W., Preetz, A., and Heller, D. (2009) Unusual deactivation in the asymmetric hydrogenation of itaconic acid. *Adv. Synth. Catal.*, **351** (5), 750–754.

4
Catalytic Reaction Engineering Principles

Albert Renken and Lioubov Kiwi-Minsker

4.1
Preface

Every industrial chemical process is designed to produce economically a desired product from suitable starting materials through a succession of treatment steps. The design of equipment for the physical treatment steps belongs to the field of unit operations. Understanding how chemical reactors work lies at the heart of almost every chemical process. The appropriate choice and dimensioning of the chemical reactor to carry out a heterogeneous catalytic process is the main subject of 'Catalytic Reaction Engineering.' This particular topic is always closely associated with the catalyst design and represents a true integration of the skills of the chemist and the chemical engineer. To be successful, one uses knowledge from a variety of areas: material and surface science, thermodynamics, chemical kinetics, fluid mechanics, and heat and mass transfer. Design of the catalytic reactor is not a routine matter since many alternatives can be proposed for a process. In searching for the optimum it is not just the cost of the reactor and of the substances (raw materials, catalyst, additives, etc.) that must be minimized. One design may have low cost, but the products obtained may require supplementary treatments involving much higher cost than that for alternative designs. Hence, the economics of the overall process must also be considered. Catalytic reaction engineering applies this knowledge to produce the best design of a chemical reactor together with the most suitable (selective/active/stable) catalyst.

This text is intended to introduce the principles of chemical engineering of heterogeneous catalytic reactions. It is supposed that the students already have a basic grounding in chemical kinetics and thermodynamics including reaction equilibria, heats of reaction, and mass/energy balances. The main objective is to integrate a fundamental understanding of catalytic reaction kinetics with its application to the design and analysis of chemical reactors used to carry out catalytic processes. This will be discussed in more detail in Part 6.

4.2
Formal Kinetics of Catalytic Reactions

Kinetics describes the rate of a chemical reaction in relation to concentrations, pressure, and temperature and is key knowledge for the reactor design [1, 2]. Starting at the beginning of the twentieth century as a purely empirical discipline, it nowadays uses different spectroscopic and computational tools linking molecular dynamics (reaction mechanism at the level of individual molecules) to macroscopic description of the chemical process used in industrial reaction engineering.

Any heterogeneous catalytic reaction involves at least two phases (fluid(s) and solid), with the reaction occurring on the surface of a solid catalyst. Therefore, the reacting molecules must be supplied from the bulk fluid to the fluid/solid interface, introducing complications in overall kinetics which are not present in homogeneous systems. Heterogeneous catalytic reactions involve by their nature a combination of reaction and transport processes. The combined reaction and transport processes are schematically presented in Figure 4.1.

We suppose a porous catalyst particle with a large specific surface area surrounded by the liquid or gaseous reaction mixture. For the transformation of the reactant A_1 to the product A_2 the following 7 steps, marked in Figure 4.1, are necessary:

1) transfer of A_1 from the bulk to the outer surface of the catalyst particle
2) diffusion of A_1 through the pores of the catalyst to a catalytically active site
3) adsorption of the reactant
4) transformation of A_1 to A_2 by chemical reaction
5) desorption of the product

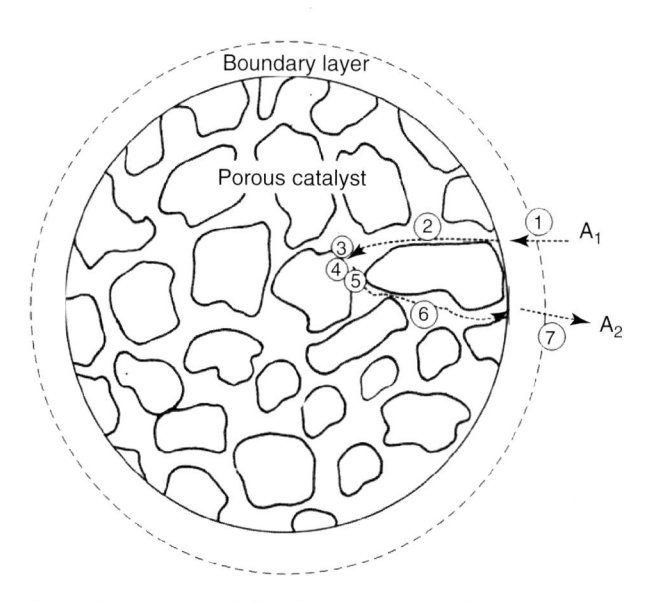

Figure 4.1 Steps involved in heterogeneous catalytic reactions.

6) diffusion of A_2 through the pores to the outer surface
7) transfer of the product from the outer surface into the bulk of the fluid.

If the rates of the chemical steps 3–5 are slow as compared to the transport processes 1, 2 and 6, 7, the overall rate of the A_1 transformation to A_2 is said to be *'kinetically controlled.'* This means that the transformation rate can be described by the rate equation derived from the elementary steps involved in the catalytic cycle (reaction mechanism). Often the reaction mechanism is *a priori* not known; therefore, for the practical purpose of the reactor design, it is convenient to present the reaction rate as a function of reactant concentrations or partial pressures in the form of simple power laws.

4.2.1
General Definitions

The rate of A_i transformation (R_i), (mol · m^{-3} · s^{-1}), is the sum of the rates (r_j) of the reactions in which A_i participates:

$$R_i = \sum_j v_{i,j} r_j \tag{4.1}$$

The reaction rate r_j can be defined in terms of change in the number of moles of A_i per unit of reactor volume (V) and per unit of time:

$$r_j = \frac{1}{V} \cdot \frac{d\xi}{dt}, \text{ with } d\xi = \frac{1}{v_i} dn_i = \text{extent of the reaction} \tag{4.2}$$

A *reversible reaction* occurring in the reactor can be presented in the following way:

$$j = 1: v_{1,1}A_1 + v_{2,1}A_2 = v_{3,1}A_3 + v_{4,1}A_4$$
$$j = 2: v_{3,2}A_3 + v_{4,2}A_4 = v_{1,2}A_1 + v_{2,2}A_2 \tag{4.3}$$

The transformation rates for the species A_1, A_3 are given by:

$$R_1 = v_{1,1}r_1 + v_{1,2}r_2; \quad R_3 = v_{3,1}r_1 + v_{3,2}r_2 \tag{4.4}$$

The reaction rate depends on the concentrations of the reacting species and is described by a *power rate law*:

$$r_j = k_j \cdot \prod_i c_i^{n_i} \tag{4.5}$$

where r_j is the rate of the reaction j, c_i is the concentration of reactant A_i, n_i is called the *reaction order* with respect to A_i and can have all values including integer and fractional, positive and negative, or zero. The reaction rate constant, k_j in Eq. (4.5) is independent of the composition of the reaction mixture, but is strongly influenced by the temperature, as described by the *Arrhenius law*:

$$k_j = k_{0,j} \cdot \exp\left(\frac{-E_{a,j}}{RT}\right) \tag{4.6}$$

where $k_{0,j}$ is the pre-exponential or frequency factor and $E_{a,j}$ is the *apparent activation energy* which virtually may assume any value including negative. But, for a chemical reaction, the activation energy is positive and lies usually in the range of 20–200 kJ · mol^{-1}.

The expression of power rate law may be determined empirically or may, in part, be based upon the understanding of the reaction mechanism. *Reaction mechanism* is a sequence of elementary steps. A reaction is considered as *elementary* if it occurs in a single pathway without any sub steps. If, for example, the reactions (Eq. (4.3)) are elementary with $|v_i| = 1$, and the reactions are of first order for all the reactants ($n_i = 1$), than Eq. (4.4) can be rewritten as:

$$R_1 = -1 \cdot k_1 c_1 c_2 + 1 \cdot k_2 c_3 c_4; \quad R_3 = +1 \cdot k_1 c_1 c_2 - 1 \cdot k_2 c_3 c_4 \tag{4.7}$$

A chemical reaction proceeds in the direction in which the Gibbs free energy, G, of the reaction mixture diminishes. When equilibrium is reached, $R_1 = R_3 = 0$:

$$k_1 c_{1,eq} c_{2,eq} = k_2 c_{3,eq} c_{4,eq}; \quad K_c = \frac{k_1}{k_2} = \frac{c_{3,eq} c_{4,eq}}{c_{1,eq} c_{2,eq}} \tag{4.8}$$

If the activities of the reactants correspond to their concentrations, the equilibrium constant K_c can be estimated from:

$$K_c(T) \cong K(T) = \exp\left(\frac{-\Delta G^0}{RT}\right) \tag{4.9}$$

Taking the derivative of Eq. (4.9), where $\Delta G^0 = \Delta H^0 - T\Delta S^0$, the van't Hoff equation is obtained:

$$\frac{d \ln K}{dT} = \frac{d}{dT}\left(\frac{-\Delta G^0}{RT}\right) = \frac{\Delta H^0}{RT^2} \tag{4.10}$$

By integrating from the standard temperature (273 K) to the desired temperature T, we obtain the dependence of the equilibrium constant on the reaction enthalpy:

$$\ln K = \ln K(273) + \int_{273}^{T} \frac{\Delta H^0}{RT^2} dT' \tag{4.11}$$

The superscript '0' indicates standard conditions.

4.2.2
Heterogeneous Catalytic Reactions

Any catalytic reaction, due to its nature, cannot be elementary as at least three steps are involved: adsorption of reactant, surface reaction, and desorption of a product. For a simple monomolecular reaction, for example, an isomerization, the steps involved in the catalytic process are depicted in Figure 4.2.

The catalyst must be included in the kinetic equation as a participating species. The simplest way to do so is to consider a solid catalyst as an ensemble of single active sites (*).

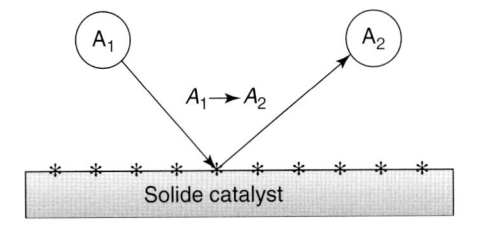

Figure 4.2 Schematic presentation of a catalytic reaction.

The transformation from A_1 to A_2 can be represented as a sequence of elementary steps (Eq. (4.12)):

$$A_1 + * \underset{k_{-1}}{\overset{k_1}{\rightleftarrows}} A_1^* \quad \text{reactant adsorption/desorption}$$

$$A_1^* \underset{k_{-2}}{\overset{k_2}{\rightleftarrows}} A_2^* \quad \text{surface reaction}$$

$$A_2^* \underset{k_{-3}}{\overset{k_3}{\rightleftarrows}} A_2 + * \quad \text{product desorption/adsorption} \tag{4.12}$$

The adsorption of the reactant is herein considered as a reaction with an empty site (*) to give an intermediate A_1^*. All sites are equivalent and each can be occupied by a single species only.

Considering all steps as elementary processes expressions for the rates can be directly obtained:

$$R_{ad} = r_1 - r_{-1} = k_1 p_1 Z_{tot} \theta_v - k_{-1} Z_{tot} \theta_1 \tag{4.13}$$

$$R_{rx} = r_2 - r_{-2} = k_2 Z_{tot} \theta_1 - k_{-2} Z_{tot} \theta_2 \tag{4.14}$$

$$R_{des} = r_3 - r_{-3} = k_3 Z_{tot} \theta_2 - k_{-1} p_2 Z_{tot} \theta_v \tag{4.15}$$

The total concentration of active sites is represented by Z_{tot} and $\theta_1, \theta_2, \theta_v$ are the fractions occupied by A_1, A_2, and the fraction of vacant sites, respectively.

In order to derive the rate equation of this sequence of elementary steps, one should know the fraction of the sites occupied by each species, θ_1 and θ_2, which is called *fractional surface coverage*:

$$\theta_i = \frac{Z_i}{Z_{tot}} \tag{4.16}$$

where Z_i is the amount of A_i adsorbed at a given pressure and temperature and Z_{tot} is the maximum amount of surface sites at this temperature and pressure.

The coverage of a catalyst surface by gaseous molecules at constant temperature depends on the partial pressure of this gas above the surface. The quantitative relationships are called *isotherms*.

4.2.3
The Langmuir Adsorption Isotherms

For describing the kinetics of heterogeneous catalytic reactions, mainly the *Langmuir adsorption isotherms* are used. We will now derive them for associative, dissociative, and competitive adsorption. The main assumptions are the following:

- The solid surface is uniform and contains a number of equivalent sites, each of which can be occupied by only one species of adsorbate;
- A dynamic equilibrium exists between the gas and the adsorbed layer at constant temperature and pressure; adsorbate molecules from the gas phase are continually colliding with the surface. If they impact a vacant adsorption site, they may form a bond with the surface and stick. If they strike a filled site, they are reflected back into the gas phase.

Once adsorbed, the molecules are localized.

The enthalpy of adsorption per site remains constant irrespective of coverage (no lateral interaction between the adsorbed species).

When molecules are hitting the surface, they can interact by bonding with an active site, being attached for some time. This process can be considered as chemical reaction and are characterized by the rates of adsorption, r_a and desorptions, r_d.

$$A + * \longrightarrow A^* \quad r_a = k_a p_A (1 - \theta)$$
$$A^* \longrightarrow A + * \quad r_d = k_d \theta \tag{4.17}$$

When equilibrium is attained, $r_d = r_a$ and after introducing the adsorption equilibrium constant $K = k_a/k_d$, we can write the *Langmuir adsorption isotherms* for associative adsorption of gas A (without any dissociation upon interaction with the surface):

$$\theta_1 = \frac{K \cdot p_1}{1 + K \cdot p_1} \tag{4.18}$$

An illustration of the Langmuir isotherm is given in Figure 4.3.

The common form of presenting this equation as a linear dependence of $1/\theta_1$ against $1/P_1$ allows us to find experimentally the constant of adsorption equilibrium and to verify the consistency of the Langmuir assumptions:

$$\frac{1}{\theta_1} = \frac{1}{K \cdot p_1} + 1 \tag{4.19}$$

For dissociative adsorption (when molecules break their bonds upon interaction with the surface) the same considerations can be applied, leading to the corresponding isotherm:

$$A_2 + 2* \rightleftharpoons 2 A_1^*$$

$$\theta_1 = \frac{\sqrt{K \cdot p_2}}{1 + \sqrt{K \cdot p_2}} \tag{4.20}$$

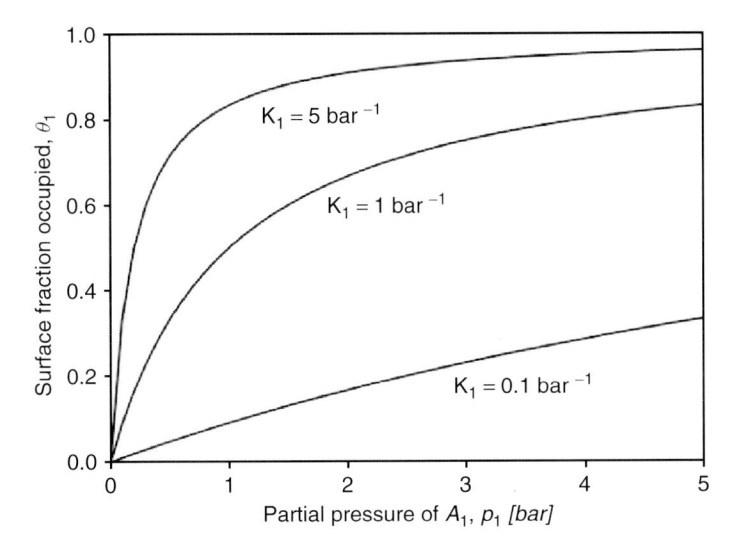

Figure 4.3 Examples of adsorption isotherms.

This is a very important case, since molecules like H_2 or O_2 (participating in catalytic hydrogenation and oxidation reactions) often dissociate on the catalytic surface, adsorbing with fragmentation. This brings some consequences for the observed kinetics order and optimization of the reaction conditions.

Competitive adsorption takes place when two (or more) different molecules compete for the same sites. If each species adsorbs on one site only without dissociation, the corresponding Langmuir isotherm is as follows:

$$\theta_i = \frac{K_i p_i}{1 + \sum_i K_i p_i} \quad (4.21)$$

where θ_i is the fractional coverage and K_i the constant of adsorption of molecule A_i.

Data for the isotherm can be obtained experimentally from the equilibrium coverage of the surface at a particular temperature over a range of pressures and then presented in a linear form allowing to find K_i. From the K_i at different temperatures, the heat of adsorption ($\Delta H_{a,i}$) can be estimated:

$$K_i = \frac{k_{a,i}}{k_{d,i}} = \frac{k_{a,i,0}}{k_{d,i,0}} \cdot \frac{\exp(-E_{a,i}/RT)}{\exp(-E_{d,i}/RT)} = \frac{k_{a,i,0}}{k_{d,i,0}} \exp(-\Delta H_{a,i}/RT) \quad (4.22)$$

Since the adsorption is always exothermic, K_i decreases with temperature.

4.2.4
Reaction Mechanisms

In general, a mechanism for any *complex reaction* (catalytic or not) is defined as a sequence of elementary steps involved in the overall transformation. To determine these steps and especially to find their kinetic parameters is very rarely possible. It requires sophisticated spectroscopic methods and/or computational tools. Therefore, a common way to construct a micro-kinetic model describing the overall transformation is to assume a reaction mechanism that is based on some experimental data. Once the model is chosen, a rate expression can be obtained and fitted to the kinetics observed.

Below, some basic models often used for heterogeneous catalytic reactions are described and the overall rate expressions are developed.

4.2.4.1 Langmuir–Hinshelwood Model

The main assumption of this model is that the catalytic reaction proceeds only via chemical adsorption of all reactants on the catalytic surface and the transformation takes place as a series of surface reactions ending up with a desorption of the products.

Let us first consider a monomolecular transformation, like the catalytic isomerization of hydrocarbons, as a simple example.

$$A_1 \rightleftharpoons A_2 \tag{4.23}$$

The transformation can be described by considering three surface processes as shown in Eq. (4.12): adsorption of the reactant, surface reaction, and desorption of the product. If the reaction is carried out in an open reactor under constant conditions, for example, in a catalytic packed-bed reactor, the fractions occupied by A_1 and A_2 are time invariant ($\frac{d\theta_1}{dt} = \frac{d\theta_2}{dt} = 0$). With Eqs. (4.13–4.15) we obtain:

$$\frac{d\theta_1}{dt} = k_1 Z_{tot} p_1 \theta_v - k_{-1} Z_{tot} \theta_1 - k_2 Z_{tot} \theta_1 + k_{-2} Z_{tot} \theta_2 = 0$$

$$\frac{d\theta_2}{dt} = k_2 Z_{tot} \theta_1 - k_{-2} Z_{tot} \theta_2 - k_3 Z_{tot} \theta_2 + k_{-3} Z_{tot} p_2 \theta_v = 0 \tag{4.24}$$

As the open reactor operates under stationary conditions, accumulation of products and reactant are excluded. In consequence, the transformation rate of A_1 corresponds to the production rate of A_2.

$$-R_1 = R_2$$

or

$$-k_1 p_1 \theta_v + k_{-1} \theta_1 = k_3 \theta_2 - k_{-3} p_2 \theta_v \tag{4.25}$$

With

$$\theta_1 + \theta_2 + \theta_v = 1 \tag{4.26}$$

We can eliminate the different occupied fractions of active sites and we finally get the the production rate as a function of the partial pressures of A_2 and A_1.

$$R_2 = \frac{k Z_{tot} (p_1 - p_2/K)}{1 + k_I p_1 + k_{II} p_2} \tag{4.27}$$

With:

$$k = \frac{k_1 k_2 k_3}{k_{-1}(k_{-2}+k_3) + k_2 k_3}$$

$$k_I = \frac{k_1(k_2 + k_{-2} + k_3)}{k_{-1}(k_{-2}+k_3) + k_2 k_3}$$

$$k_{II} = \frac{k_{-3}(k_{-1} + k_2 + k_{-2})}{k_{-1}(k_{-2}+k_3) + k_2 k_3}$$

$$K = \frac{k_1 k_2 k_3}{k_{-1} k_{-2} k_{-3}} \tag{4.28}$$

It is evident that the six individual rate constants cannot be obtained under steady-state reaction conditions. To estimate their values, independent measures of the sorption and reaction behavior under transient (non-steady-state) conditions are necessary.

4.2.4.2 The Quasi-Surface Equilibrium Approximation

If we suppose that the adsorption and desorption processes are fast compared to the surface reaction, we can estimate the surface concentrations from the equilibrium constants. With the Langmuir adsorption isotherm the following relationships result for the simple monomolecular reaction presented in Eq. (4.23).

$$\theta_1 = \frac{K_1 p_1}{1 + K_1 p_1 + K_2 p_2}$$

$$\theta_2 = \frac{K_2 p_2}{1 + K_1 p_1 + K_2 p_2}$$

$$\text{with: } K_1 \simeq \frac{k_1}{k_{-1}}; \; K_2 \simeq \frac{k_3}{k_{-3}} \tag{4.29}$$

The transformation rate is than simply given by:

$$-R_1 = k_2 Z_{\text{tot}} \theta_1 - k_{-2} Z_{\text{tot}} \theta_2 = \frac{k_2 Z_{\text{tot}} (p_1 - p_2/K_{\text{eq}})}{1 + K_1 p_1 + K_2 p_2} \tag{4.30}$$

4.2.4.3 The Masi Approximation

Catalytic transformations may include the formation of many intermediates which are difficult to identify. In these cases, it is impossible to formulate a kinetic model based on elementary steps. Often, one of the intermediates adsorbs much more strongly in comparison to the other surface species, thus occupying nearly the whole of the active sites. This intermediate is called the *most abundant surface intermediate* or 'masi.' For a simple monomolecular reaction $A_1 \rightarrow A_2$ the situation can be illustrated by the following scheme:

$$A_1 + * \xrightarrow{k_1} A_1^*$$

$$A_1^* \xrightarrow{k_2} I_2^*$$

·

·

$$I_{n-1}^* \xrightarrow{k_{n-1}} I_n^* \text{ (masi)}$$

$$I_n^* \xrightarrow{k_n} A_2 + * \tag{4.31}$$

Neglecting all intermediates having a very short lifetime on the catalyst results in:

$$\theta_v + \theta_{\text{masi}} \simeq 1 \tag{4.32}$$

The transformation rate of reactant A_1 corresponds to the first step in Eq. (4.31):

$$-R_1 = k_1 Z_{\text{tot}} p_1 \theta_v = k_1 Z_{\text{tot}} p_1 (1 - \theta_{\text{masi}}) \tag{4.33}$$

The final product is formed in the nth step and corresponds to the transformation to A_2 and its desorption:

$$R_2 = k_n Z_{\text{tot}} \theta_{\text{masi}} \tag{4.34}$$

At steady state $R_2 = -R_1$ holds, θ_v can easily be calculated, and the final expression for describing the production is given by:

$$R_2 = \frac{k_1 Z_{\text{tot}} p_1}{1 + K p_1}; \text{ with } K = \frac{k_1}{k_n} \tag{4.35}$$

It is important to emphasize that the mathematical form of the obtained kinetic models (Eqs. (4.27), (4.30), and (4.35)) are quite similar, whereas the interpretations of the corresponding model parameters are very different.

4.2.4.4 Bimolecular Catalytic Reactions

Supposing that the surface reaction is rate determining, we obtain for an irreversible bimolecular reaction the following relations:

$$A_1 + A_2 \rightarrow A_3$$

$$-R_1 = k Z_{\text{tot}} \theta_1 \theta_2 \tag{4.36}$$

The surface fractions occupied by the reactants A_1 and A_2 are given by the Langmuir isotherm, supposing competitive adsorption and neglecting the product adsorption.

$$\theta_1 = \frac{K_1 p_1}{1 + K_1 p_1 + K_2 p_2}$$

$$\theta_2 = \frac{K_2 p_2}{1 + K_1 p_1 + K_2 p_2}$$

$$\theta_3 \simeq 0 \tag{4.37}$$

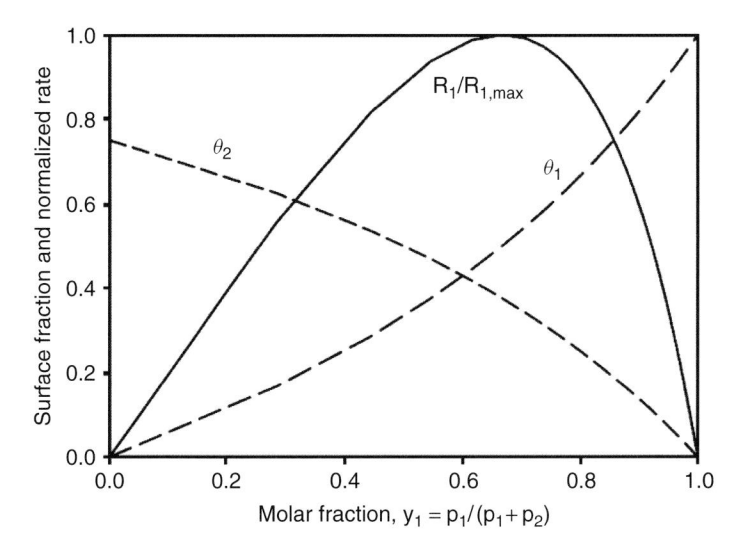

Figure 4.4 Surface coverage of A_1 and A_2 and the transformation rate as function of the mole fraction of A_1 ($K_1 = 2$ bar^{-1}, $K_2 = 3$ bar^{-1}, $p_2 = 1$ bar).

$$-R_1 = R_3 = k_3 \cdot Z_{tot}\theta_A \cdot \theta_B = \frac{k_3 \cdot Z_{tot} K_1 p_1 \cdot K_2 p_2}{(1 + K_1 p_1 + K_2 p_2)^2} \tag{4.38}$$

At constant pressure, the transformation rate as a function of p_1 passes through a maximum. The maximum depends on p_2 and the values of the adsorption constants K_1 and K_2. The optimum pressure for A_1 can easily be calculated with Eq. (4.39).

$$p_{1,op} = \frac{1 + K_2 p_2}{K_1} \tag{4.39}$$

The transformation rate as function of the mole fraction of A_1 is illustrated in Figure 4.4 for given adsorption constants.

4.3
Mass and Heat Transfer Effects

4.3.1
Introduction

As pointed out in Section 4.2, heterogeneous catalytic reactions involve by their nature a combination of reaction and transport processes, since the reactants must be first transferred from the bulk of the fluid phase to the catalyst surface, where the reaction occurs [3]. The combined reaction and transport processes are shown schematically in Figure 4.1. We picture a porous catalyst particle with a large specific surface area surrounded by the liquid or gaseous reaction mixture.

If the rates of the intrinsic steps 3–5 are comparable with or higher than the transport processes 1, 2 and 6, 7 significant concentration profiles of A_1 and A_2 inside the catalyst particle or in the surrounding layer will occur. The concentration of the reactant A_1 on the outer surface is smaller than that in the reaction mixture and a pronounced concentration profile within the pellet will develop. Finally, if the intrinsic rates are very fast compared to the diffusion process in the pores, the reaction will take place only near the external surface and the observed transformation rate is only influenced by the external mass transfer. The same situation is observed for non-porous pellets or so called 'eggshell' catalysts, where the active phase is placed in a layer near the outer pellet surface (see Chapter 21 (Preparation of Supported Catalysts)). Simultaneously with the chemical transformation, heat is released or consumed in the case of exothermic or endothermic reactions respectively. Consequently, temperature gradients inside and outside the catalyst pellet will develop. The different situations are illustrated in Figure 4.5.

As a consequence of the concentration profiles caused by the transfer phenomena the observed (effective) reaction rates are modified compared to the rate which would occur at constant bulk phase concentration. This effect is commonly characterized by an effectiveness factor as defined in Eq. (4.40).

$$\eta_{ov} = \frac{\text{observed rate of reaction}}{\text{rate of reaction at bulk concentration and temperature}} \quad (4.40)$$

Besides the modification of the overall reaction rate, the product selectivity may be changed. This will be discussed in detail in the following subchapters.

4.3.2
External Mass and Heat Transfer

The first step in heterogeneous catalytic processes is the transfer of the reactant from the bulk phase to the external surface of the catalyst pellet. If a non-porous catalyst is used, only external mass and heat transfer can influence the effective rate of reaction. The same situation will occur for very fast reactions, where the reactants are completely exhausted at the external catalyst surface. As no internal mass and heat transfer resistances are considered, the overall catalyst effectiveness factor corresponds to the external effectiveness factor, η_{ex}. For a simple irreversible reaction of nth order the following relation results:

$$\eta_{ov} = \eta_{ex} = \frac{k(T_s) \cdot c_{i,s}^n}{k(T_b) \cdot c_{i,b}^n} = \frac{k(T_s)}{k(T_b)} \left(\frac{c_{i,s}}{c_{i,b}}\right)^n \quad (4.41)$$

4.3.2.1 Isothermal Pellet

The external transfer process can be described by the so-called film model as shown in Figure 4.6. According to the film model a stagnant fluid layer of thickness δ surrounds the external surface, where the total resistance to mass transfer is

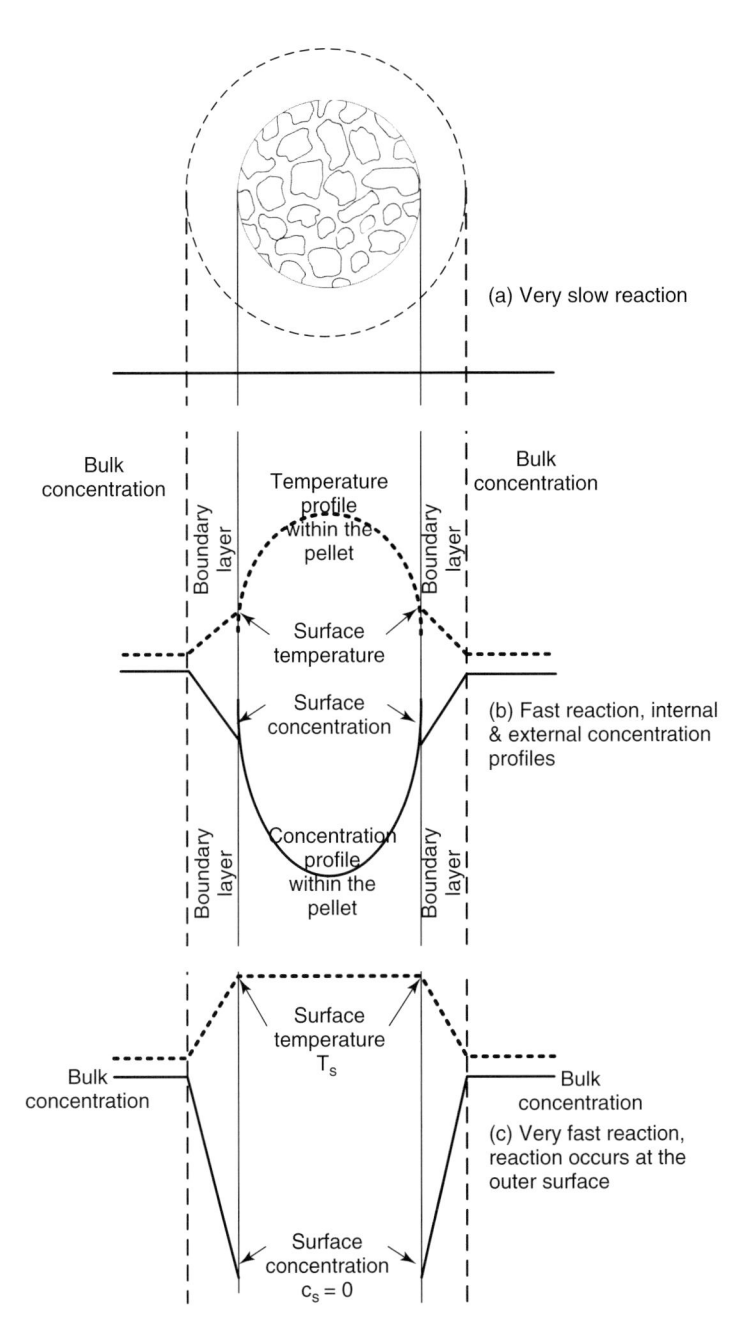

Figure 4.5 Concentration profiles in porous catalysts for different reaction regimes.

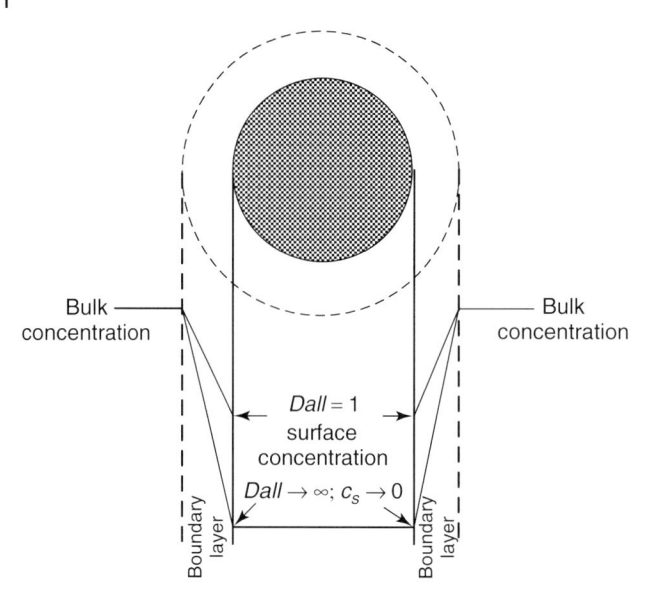

Figure 4.6 External concentration profile according to the film model.

located. Accordingly, the concentration profile is confined to this layer. The molar flux of reactant A_i is proportional to the driving force as given in Eq. (4.42).

$$J_i = k_m \left(c_{i,b} - c_{i,s} \right) \tag{4.42}$$

with k_m the mass transfer coefficient and $c_{i,s}$, $c_{i,b}$ the concentration at the external surface and the bulk of the fluid, respectively. At steady-state condition, the molar flux of A_i is equal to the rate of transformation at the outer surface:

$$J_i = k_m \left(c_{i,b} - c_{i,s} \right) = -R_i \cdot \frac{V_p}{A_p} = \frac{-R_i}{a_p};$$

$$\text{with} \quad R_i = \frac{v_i r_p}{a_p} \tag{4.43}$$

with R_i the transformation rate of A_i, r_p, the reaction rate per volume of catalyst pellet, V_p, A_p the pellet volume and outer surface respectively, and a_p the specific external surface area of the pellet.

$$a_p = \frac{A_p}{V_p}; \quad a_p = \frac{6}{d_p} \text{(sphere)} \tag{4.44}$$

For an irreversible first-order reaction we obtain for the reactant A_1:

$$J_1 = k_m \left(c_{1,b} - c_{1,s} \right) = \frac{-R_1}{a_p} = \frac{k_r c_{1,s}}{a_p} \tag{4.45}$$

The reactant concentration on the surface is given by:

$$c_{1,s} = \frac{k_m a_p}{k_m a_p + k_r} \cdot c_{1,b} \tag{4.46}$$

and for the observable, effective reaction rate:

$$-R_{1,\text{eff}} = k_r \frac{k_m a_p}{k_m a_p + k_r} \cdot c_{1,b} \tag{4.47}$$

A similar development can be done for irreversible nth order reactions. At steady state:

$$k_m a_p (c_{1,b} - c_{1,s}) = -R_{1,p} = k_r c_{1,s}^n \tag{4.48}$$

Dividing by $k_m a_p c_{1,b}$ leads to

$$1 - \frac{c_{1,s}}{c_{1,b}} = \frac{k_r c_{1,b}^{n-1}}{k_m a_p} \left(\frac{c_{1,s}}{c_{1,b}}\right)^n = DaII \left(\frac{c_{1,s}}{c_{1,b}}\right)^n \tag{4.49}$$

The *second Damköhler number*, $DaII$ is defined as the ratio of the characteristic mass transfer time $t_m = 1/(k_m a_p)$ to the characteristic reaction time, $t_r = 1/\left(k_r c_{1,b}^{n-1}\right)$.

$$DaII = \frac{t_m}{t_r} = \frac{k_r c_{1,b}^{n-1}}{k_m a_p} \tag{4.50}$$

With Eq. (4.41) we find for the external effectiveness under isothermal conditions:

$$\eta_{\text{ex}} = \left(\frac{c_{1,s}}{c_{1,b}}\right)^n \tag{4.51}$$

The external effectiveness factors as a function of the second Damköhler number are obtained by solving Eq. (4.49). This is done for reaction orders $n = 1, 2, 1/2;$, and -1 and displayed in Figure 4.7 [4].

$$n = 1: \eta_{\text{ex}} = \frac{1}{1 + DaII}$$

$$n = 2: \eta_{\text{ex}} = \left(\frac{\sqrt{1 + 4DaII} - 1}{2DaII}\right)^2$$

$$n = \frac{1}{2}: \eta_{\text{ex}} = \left[\frac{2 + DaII^2}{2}\left(1 - \sqrt{1 - \frac{4}{(2 + DaII^2)^2}}\right)\right]^{\frac{1}{2}}$$

$$n = -1: \eta_{\text{ex}} = \frac{2}{1 + \sqrt{1 - 4DaII}}; \text{ for } DaII < 0.25 \tag{4.52}$$

From Figure 4.7 we see that

- The effectiveness factor diminishes for the same $DaII$ with increasing reaction order
- An effectiveness factor higher than one is obtained for reactions with educt inhibition (negative reaction order)
- For large values of $DaII$ the effectiveness is inversely proportional to $DaII$ ($\eta_{\text{ex}} \simeq 1/DaII$) for all reactions with positive reaction order.

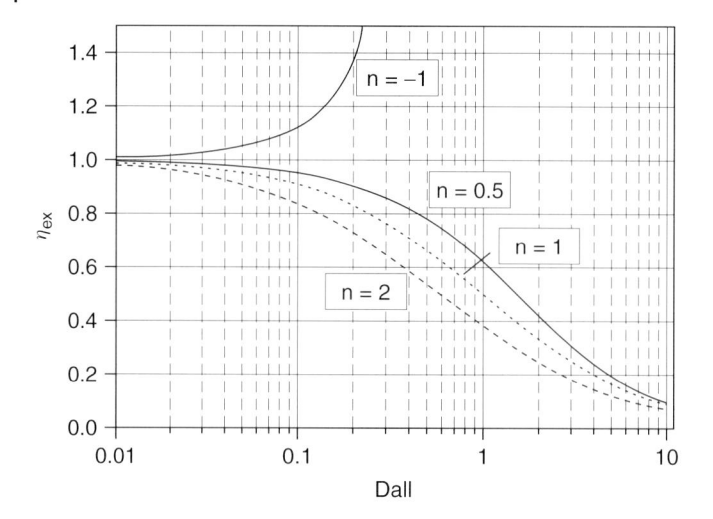

Figure 4.7 Isothermal external effectiveness factor as function of the Damköhler number.

The observable reaction rate is given by:

$$r_{p,eff} = k_{eff} c_{1,b}^{n'} = \eta_{ex} k_r c_{1,b}^{n} \tag{4.53}$$

With increasing rate of the intrinsic reaction (increasing $DaII$) the observed rate constant approaches the volumetric mass transfer coefficient ($k_{eff} \to k_m a_p$) and the reaction order changes from n to unity.

Whereas Figure 4.7 is quite instructive, it is not of practical use for estimating the importance of the mass transfer influence from experimental data, since the intrinsic rate constant is normally unknown. Replotting the effectiveness factor as a function of the ratio of the observed reaction rate to the maximum mass transfer rate allows the external effectiveness factor to be estimated.

$$\frac{r_{p,eff}}{k_m a_p c_{1,b}} = \eta_{ex} \frac{k_r c_{1,b}^{n-1}}{k_m a_p} = \eta_{ex} DaII = Ca \text{ (Carberry number)} \tag{4.54}$$

This relationship is plotted in Figure 4.8.

Isothermal Yield and Selectivity For a network of parallel and/or consecutive reactions, mass transfer may drastically influence the obtainable product yield. For *consecutive first-order reactions* and in the absence of mass transfer influence we obtain for the transformation rate of the reactant and the production rate of the intermediate:

$$A_1 \xrightarrow{k1} A_2 \xrightarrow{k2} A_3 \tag{4.55}$$

$$-R_1 = k_1 c_{1,b}$$

$$R_2 = k_1 c_{1,b} - k_2 c_{2,b} \tag{4.56}$$

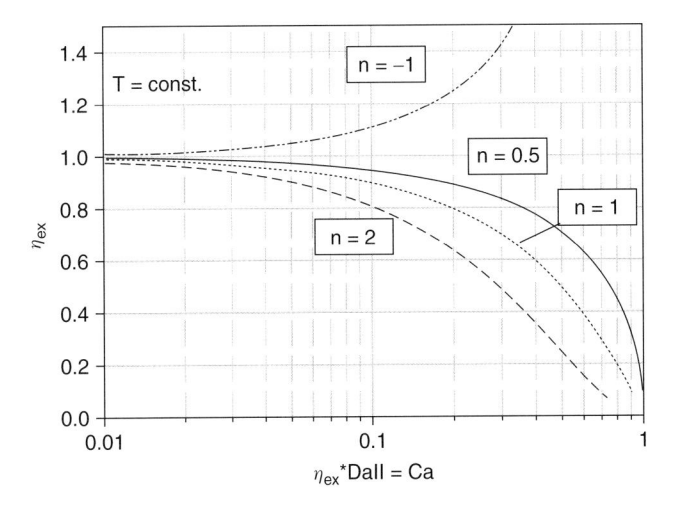

Figure 4.8 Effectiveness factor as function of the observable variable: the Carberry number.

The instantaneous or point selectivity for the intermediate product is obtained by dividing R_2 by $(-R_1)$:

$$s_{2,1} = -\frac{R_2}{R_1} = 1 - \frac{k_2 c_{2,b}}{k_1 c_{1,b}} \quad (4.57)$$

Anticipation of mass transfer phenomena leads to:

$$k_m a_p (c_{1,b} - c_{1,s}) = k_1 c_{1,s} \quad (4.58)$$

$$k_m a_p (c_{2,s} - c_{2,b}) = k_1 c_{1,s} - k_2 c_{2,s} \quad (4.59)$$

Solving for the surface concentrations $c_{1,s}$ and $c_{2,s}$, we obtain for the instantaneous selectivity the following relations:

$$\begin{aligned}(s_{2,1})_{\text{eff}} &= -\frac{R_{2,s}}{R_{1,s}} = 1 - \frac{k_2 c_{2,s}}{k_1 c_{1,s}} \\ (s_{2,1})_{\text{eff}} &= \frac{1}{1 + DaII_2} - \frac{k_2 (1 + DaII_1)}{k_1 (1 + DaII_2)} \frac{c_{2,b}}{c_{1,b}}\end{aligned} \quad (4.60)$$

with $DaII_1 = k_1/(k_m a_p)$ and $DaII_2 = k_2/(k_m a_p)$.

Under initial conditions, at the reactor entrance the product concentrations are zero and the instantaneous selectivity becomes:

$$(s_{2,1})_{\text{eff},0} = \frac{1}{1 + DaII_2} = \frac{1}{1 + (k_2/k_m a_p)} \quad (4.61)$$

Obviously, the effective selectivity of the intermediate product depends on the ratio of the escape rate from the surface to its rate of destruction on the surface. Low mass transfer rate compared to the rate of the consecutive reaction is detrimental for the selectivity and yield of the intermediate product.

For *parallel reactions* the influence of mass transfer depends on the individual reaction orders:

$$A_1 \xrightarrow{k1} A_2; \quad R_2 = k_1 c_1^{n1} \tag{4.62}$$

$$A_1 \xrightarrow{k2} A_3; \quad R_3 = k_2 c_1^{n2} \tag{4.63}$$

The ratio between the products A_2 and A_3 depends on the rate constants and the reactant concentration.

$$\frac{R_2}{R_3} = \frac{k_1}{k_2} c_1^{(n1-n2)} \tag{4.64}$$

As the concentration gradient around the catalyst leads to a lower surface concentration compared to that in the bulk, the observed alteration of rate ratio depends on the individual reaction order:

$$\frac{(R_{2,s}/R_{3,s})}{(R_{2,b}/R_{3,b})} = \left(\frac{c_{1,s}}{c_{1,b}}\right)^{(n1-n2)} \tag{4.65}$$

As $c_{1,s} < c_{1,b}$ we see that diffusion intrusion leads to

- a reduced selectivity for A_2, if $n_1 > n_2$
- an increased selectivity for A_2, if $n_1 < n_2$
- no change of the selectivity for $n_1 = n_2$.

4.3.2.2 Non-Isothermal Pellet

For highly endothermic or exothermic reactions, the temperature of the catalyst surface can be considerably different from the temperature of the surrounding fluid.

We evaluate the surface temperature by the heat balance:

$$(-\Delta H_r) \cdot r_{p,\text{eff}} = h \cdot a_p (T_s - T_b) \tag{4.66}$$

with h being the heat transfer coefficient.

We divide Eq. (4.66) by $k_m a_p c_{1,b}$

$$\frac{h \cdot a_p}{k_m a_p c_{1,b}} (T_s - T_b) = (-\Delta H_r) \cdot \frac{r_{p,\text{eff}}}{k_m a_p c_{1,b}}$$

or

$$\frac{h \cdot a_p}{k_m a_p c_{1,b}} (T_s - T_b) = (-\Delta H_r) \cdot \eta_{\text{ex}} DaII = (-\Delta H_r) \cdot Ca \tag{4.67}$$

Invoking the Chilton–Colburn analogy between heat and mass transfer:

$$\frac{h}{\rho \cdot c_p} \Pr^{2/3} = k_m Sc^{2/3} \tag{4.68}$$

we obtain the ratio between heat and mass transfer coefficient:

$$\frac{h}{k_m} = \rho c_p \left(\frac{Sc}{\Pr}\right)^{2/3} \tag{4.69}$$

and finally with Eq. (4.67):

$$T_s = T_b + \Delta T_{ad} \left(\frac{Pr}{Sc}\right)^{2/3} \cdot Ca$$

$$\frac{T_s}{T_b} = 1 + \frac{\Delta T_{ad}}{T_b}\left(\frac{Pr}{Sc}\right)^{2/3} \cdot Ca = 1 + \beta_{ex} \cdot Ca \qquad (4.70)$$

With $\Delta T_{ad} = \frac{(-\Delta H_r)c_{1,b}}{\rho c_p}$ the adiabatic temperature rise, $Pr = \frac{\nu}{\alpha} = \frac{\nu}{\lambda/(\rho c_p)}$ the Prandtl number, and $Sc = \frac{\nu}{D_m}$ the Schmidt number.

For a given system the temperature difference between bulk and surface depends on the reactant concentration via ΔT_{ad}, the ratio between Prandtl and Schmidt numbers, and the Carberry number. The temperature difference is maximal for reactions limited by mass transfer ($Ca \to 1$). As for gases, where the Schmidt and Prandtl numbers are approximately unity ($Pr \simeq Sc \simeq 1$), the temperature difference can reach the adiabatic temperature ($T_s - T_b \simeq \Delta T_{ad}$).

The non-isothermal external effectiveness factor is

$$\eta_{ex} = \frac{r_{p,eff}}{r_{p,b}} = \frac{k(T_s)}{k(T_b)}\left(\frac{c_{1,s}}{c_{1,b}}\right)^n \qquad (4.71)$$

Based on Eqs. (4.51) and (4.52) we can estimate the surface concentration and obtain for a first order reaction: $c_{1,s}/c_{1,b} = (1 + DaII)^{-1}$.

The rate constant at the surface temperature, $k(T_s)$, is given by the Arrhenius law:

$$k(T_s) = k(T_b)\exp\left(-\frac{E}{RT_b}\left(\frac{T_b}{T_s} - 1\right)\right) = k(T_b)\exp\left(-\gamma\left(\frac{T_b}{T_s} - 1\right)\right) \qquad (4.72)$$

with $\gamma = \frac{E}{RT_b}$, the Arrhenius number. The surface temperature is determined by the adiabatic temperature rise and the ratio of the Schmidt to the Prandtl number as shown in Eq. (4.70). In summary, the external effectiveness factor for a given Ca number depends on the Arrhenius number, γ, and the parameter $\beta_{ex} = \frac{\Delta T_{ad}}{T_b}\left(\frac{Pr}{Sc}\right)^{2/3}$.

The non-isothermal effectiveness as function of the observable Ca number for different Arrhenius numbers and β_{ex} are shown in Figures 4.9 and 4.10.

We see that

- the effectiveness factor can be greater than unity for exothermic reactions
- the Arrhenius number, γ, is more important than the parameter β_{ex} in determining η_{ex}.
- at high values of the Carberry number the effectiveness falls well below unity.

4.3.3
Internal Mass and Heat Transfer

For most catalytic processes porous catalysts with a high inner specific surface area are used. Therefore, the reactant has to be transported through the pores to the

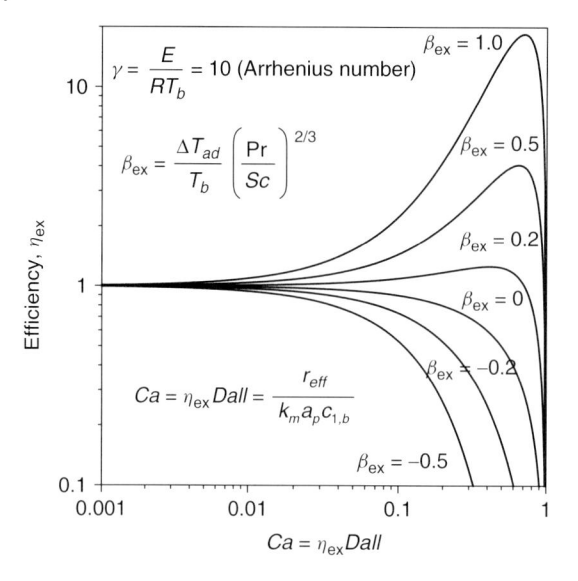

Figure 4.9 Non-isothermal external effectiveness factor as function of the parameter β_{ex} and the Arrhenius number ($\gamma = 10$).

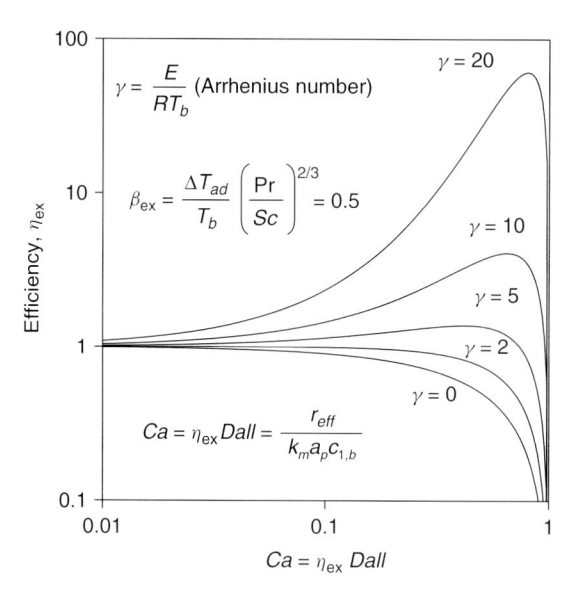

Figure 4.10 Non-isothermal external effectiveness factor as function of the Arrhenius number, γ, and the Carberry number ($\beta_{ex} = 0.5$).

catalytically active sites as described in Section 4.3.1. Due to the chemical reaction, a concentration gradient develops from the outside to the center of the pellet. For the following discussion we assume isotropic particles, and that the transport process can be represented by molecular diffusion. The molar flux of reactant A_1 can be described by:

$$J_1 = -D_{1,e} \frac{dc_1}{dz} \tag{4.73}$$

where $D_{1,e}$ is the effective diffusion coefficient for reactant A_1, and z is the particle coordinate, defined as the distance from the center. Formally, Eq. (4.73) corresponds to Fick's law. As the diffusion does not occur in a homogeneous phase, an effective diffusion coefficient is introduced. The effective diffusion coefficient takes into account the fact that pores only occupy a fraction, ε_p, of the particles, and that the pores are not all oriented in the z direction. As a consequence the diffusion path through the pores is longer than dz. This is accounted for by introducing a tortuosity factor τ_p. With both corrections the effective diffusion coefficient can be estimated with the following expression:

$$D_{1,e} = D_1 \frac{\varepsilon_p}{\tau_p} \tag{4.74}$$

with D_1 the molecular diffusion coefficient of reactant A_1.

The particle porosity is in the order of $0.3 < \varepsilon_p < 0.6$, and the tortuosity is found to be in the range of $2 < \tau_p < 5$.

4.3.3.1 Isothermal Pellet

To illustrate the simultaneous diffusion/reaction processes occurring in a porous catalyst, we consider a catalyst in the form of a flat slab of semi-infinite dimensions on the outer surface and of a half thickness L as shown in Figure 4.11.

An irreversible first-order reaction takes place in the porous matrix. The mass transport is represented by a normal diffusion. A steady-state mass balance over a differential volume element yields:

$$D_e \frac{dc_1}{dz} - (-R_1) = 0 \tag{4.75}$$

The boundary conditions define the concentration on the outer surface and have a symmetry at the slab center.

$$z = L : c_1 = c_{1,s}; \quad z = 0 : \frac{dc_1}{dz} = 0 \tag{4.76}$$

For first order kinetics with $-R_1 = k_r c_1$ Eq. (4.75) can be rewritten in a non-dimensional form as follows:

$$\frac{d^2 f}{dZ^2} - L^2 \left(\frac{k_r}{D_e} \right) f = 0 \tag{4.77}$$

with $f = \frac{c_1}{c_{1,s}}; Z = \frac{z}{L}$.

The group $L^2 \frac{k_r}{D_e}$ corresponds to the ratio between the characteristic diffusion time, t_D in the catalyst and the characteristic reaction time. This ratio is commonly

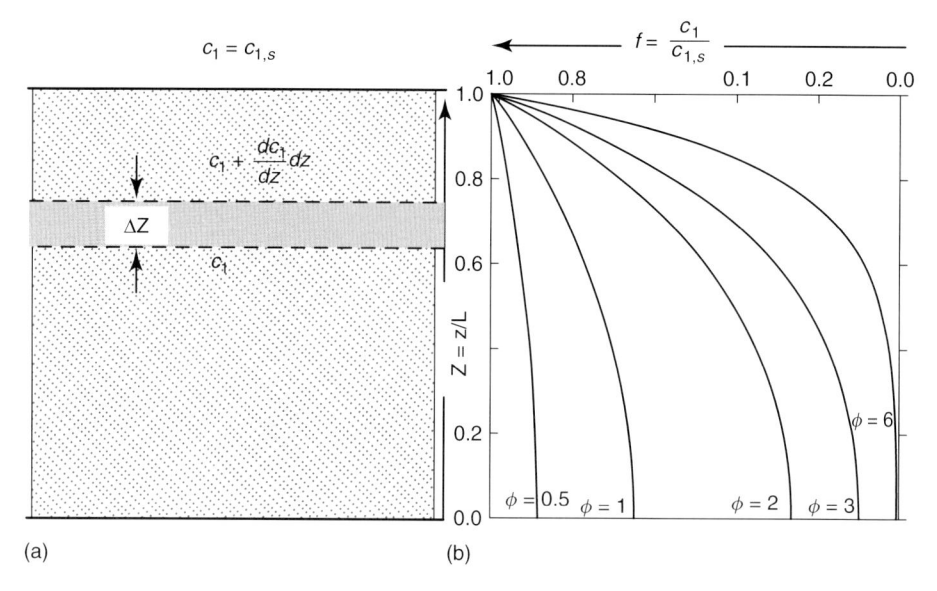

Figure 4.11 Diffusion and reaction in a semi-infinite flat slab.

called *Thiele modulus*, ϕ.

$$\phi^2 = \frac{t_D}{t_r} = \frac{L^2}{D_e}k_r; \phi = L\sqrt{\frac{k_r}{D}}; \text{ first order reaction} \tag{4.78}$$

The solution of Eq. (4.77) for the concentration profile in the slab is:

$$f = \frac{c_1}{c_{1,s}} = \frac{\cosh(\phi Z)}{\cosh(\phi)} \tag{4.79}$$

The effective rate of reaction corresponds the molar flux at the external surface $J_{1,L}$. Using the concentration profile evaluated for $Z = 1$ from Eq. (4.79) we obtain.

$$J_{1,L} = -D_e \left(\frac{dc_1}{dz}\right)_{z=L} \tag{4.80}$$

which results in an effective reaction rate per external area of:

$$J_{1,\text{eff}} = \frac{D_e c_{1,s}}{L}\phi \tanh(\phi) \tag{4.81}$$

The overall reaction rate per external area in the absence of an internal concentration profile is $J_{1,s} = k_r c_{1,s} L$. From Eq. (4.40) it follows that the effectiveness factor in the porous catalyst, η_p, is given by:

$$\eta_p = \frac{J_{1,\text{eff}}}{J_{1,s}} = \frac{D_e c_{1,s}/L \cdot \phi \tanh(\phi)}{k_r c_{1,s} L} = \frac{\tanh \phi}{\phi} \tag{4.82}$$

The effective reaction rate per volume of the catalyst is then given by

$$-R_{1,\text{eff}} = \eta_p k_r c_{1,s} = \frac{\tanh \phi}{\phi} k_r c_{1,s} \tag{4.83}$$

If the diffusional influence is pronounced ($L^2/D_e \gg \frac{1}{k_r}$), the Thiele modulus becomes large and $\tanh \phi \to$ unity. Therefore, the effectiveness factor for strong diffusional resistances is

$$\eta_p \simeq \frac{1}{\phi} \qquad (4.84)$$

Concentration profiles in slabs for different values of ϕ are shown in Figure 4.11b. The results presented above are specific for a first-order reaction and a catalyst in the form of a slab. For spherical particles the corresponding equation is

$$\eta_p = \frac{3}{\phi_s} \left[\frac{1}{\tanh \phi_s} - \frac{1}{\phi_s} \right] \qquad (4.85)$$

The corresponding solution for a cylinder is

$$\eta_p = \frac{2}{\phi_c} \frac{I_1(\phi_c)}{I_0(\phi_c)} \qquad (4.86)$$

where $I_1(\phi)$ and $I_0(\phi)$ denote the modified Bessel functions of first and zero order, respectively.

In Figure 4.12 the effectiveness factor as a function of the Thiele modulus for different pellet shapes is shown. For small values the effectiveness factor reaches unity in all cases. The reaction is controlled by the intrinsic kinetics, and the reactant concentration is identical to the concentration at the outer surface. This situation may be observed for low catalyst activity or very small particles as used in fluidized beds or suspension reactors. For large values of the Thiele modulus the dependence of η_p approaches an asymptotic solution: $\eta_p = m/\phi$ with $m = 1, 2, 3$ for a slab, a cylinder and a sphere, respectively. This situation may occur for very fast

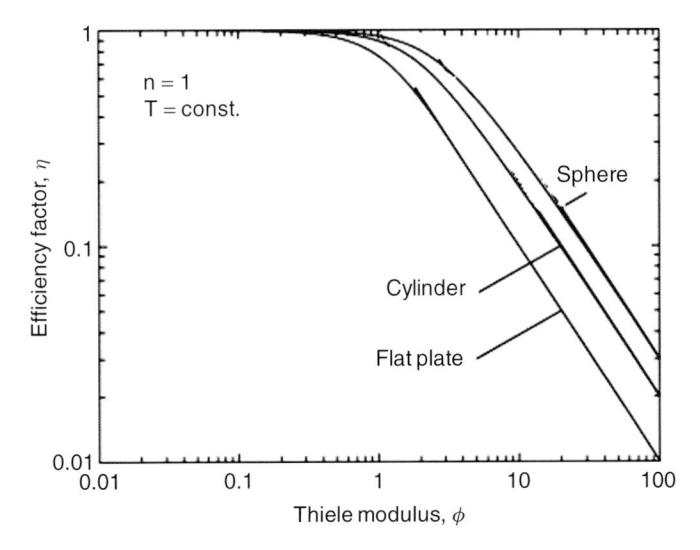

Figure 4.12 Effectiveness factor as function of the Thiele modulus for different pellet shapes.

reactions or large catalyst particles. The concentration in the center of the catalyst particles approaches zero as indicated for $\eta_p < 0.2$.

The observation that the slope of the asymptotic solution for $\eta(\phi)$ becomes independent of the particle geometry suggests that the dependence of the effectiveness factor on the Thiele modulus can be described by a generalized relationship, valid for arbitrary pellet shapes. This was in fact demonstrated by Aris by defining a general Thiele modulus ϕ_{gen}, based on the ratio of pellet volume to external surface as characteristic diffusion length. A further correction was proposed by Petersen to get a general effectiveness factor for an nth-order reaction with a characteristic reaction time $t_r = \left(k_r c_{1,s}^{(n-1)}\right)^{-1}$. The final definition is given in Eq. (4.87).

$$\phi_{gen} = \frac{V_p}{A_p} \sqrt{\frac{k_r c_{1,s}^{(n-1)}}{D_e}} \cdot \sqrt{\frac{n+1}{2}} \qquad (4.87)$$

The effectiveness factor as a function of the generalized Thiele modulus is shown for a slab and a sphere. Both curves exactly coincide for $\phi_{gen} \to \infty$. The maximum deviations are in the order of 10–15% (Figure 4.13).

In general, the intrinsic kinetic parameters are unknown. Therefore, the relationships based on the Thiele modulus cannot be used to estimate the influence of inner mass transfer on the reaction. Observed is the experimentally accessible efficient reaction rate, $r_{p,eff}$, is observed. In addition, the characteristic diffusion time in the porous pellet can be estimated. This allows us to define a new modulus based on the characteristic effective reaction time $t_{r,eff}$ and the characteristic diffusion time in the particle, t_D. The ratio of these two values is known as *Weisz modulus*. We obtain for spherical pellets:

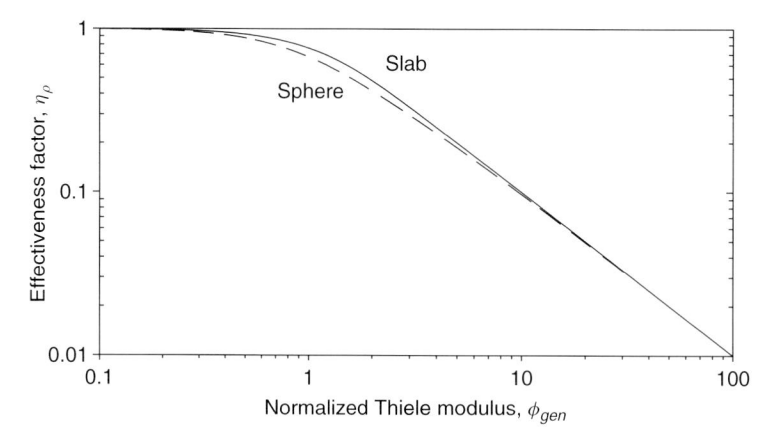

$$\psi_s^2 = \frac{t_D}{t_{r,eff}} = \frac{R_{sphere}^2}{D_e} \frac{c_s}{r_{p,eff}} = \eta_p \phi_s^2 \qquad (4.88)$$

Figure 4.13 Effectiveness factor as function of the generalized Thiele modulus for different pellet geometries.

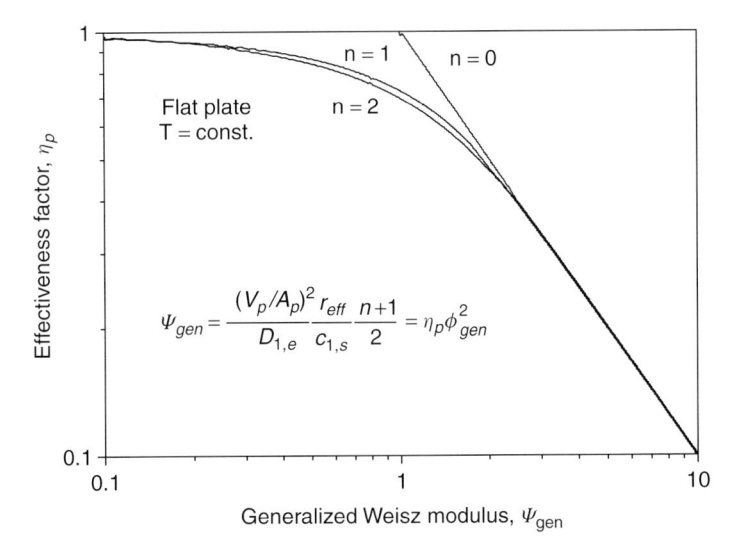

Figure 4.14 Effectiveness factor as function of the generalized Weisz modulus for different reaction orders.

In analogy with the generalized Thiele modulus we can define a Weisz modulus which applies to arbitrary pellet shapes and different reaction order:

$$\psi_{gen}^2 = \frac{t_D}{t_{r,eff}} = \left(\frac{V_p}{A_p}\right)^2 \frac{n+1}{2} \frac{r_{p,eff}}{D_e c_s} = \eta_p \phi_{gen}^2 \qquad (4.89)$$

In Figure 4.14 a plot of the effectiveness factor against the generalized Weisz module for different reaction orders is shown. Using this relation, the effectiveness factor can be estimated based on the experimental results and the estimated diffusion coefficient.

Isothermal Yield and Selectivity The influence of transport phenomena on selectivity and yield is often more important than its influence on the effective catalyst activity. The following analysis is restricted to two important schemes for multiple reactions [5]:

$$\text{Parallel reactions: } A_3 \xleftarrow{k_2} A_1 \xrightarrow{k_1} A_2, \text{ and}$$
$$\text{consecutive reactions: } A_1 \xrightarrow{k_1} A_2 \xrightarrow{k_2} A_3.$$

We also neglect the effect of external mass transfer resistances and assume that the concentration at the pellet surface is identical to the bulk concentration ($c_{i,s} = c_{i,b}$).

In the case of *parallel reactions* the rate equations for the disappearance of the reactant A_1 and for the formation of the desired product A_2 are given by:

$$-R_1 = k_1 c_1^{n_1} + k_2 c_1^{n_2}$$
$$R_2 = k_1 c_1^{n_1} \qquad (4.90)$$

with k_1 and k_2 the intrinsic rate constants. The *instantaneous* or *point selectivity* is defined as the ratio of the A_2 production rate to the rate of reactant disappearance:

$$s_{2,1} = \frac{R_2}{-R_1} = \frac{1}{1 + k_2/k_1 \cdot c_1^{(n2-n1)}} = \frac{1}{1 + \kappa c_1^{(n2-n1)}} \tag{4.91}$$

There is no influence of the concentration profile on selectivity in the case of equal order kinetics for the two reaction paths. If $n_1 \neq n_2$, the effective selectivity will be influenced by intraparticle diffusion. As the influence of the internal concentration profile becomes more pronounced with increasing reaction order, the product selectivity will diminish, whenever the desired reaction has a higher order than that of the undesired reaction. Otherwise the selectivity will be improved with increasing internal mass transfer resistance.

To discuss the influence of internal transport processes on *consecutive reactions*, we assume simple irreversible first-order reactions. With k_1 and k_2 being the intrinsic rate constants, the production rate of A_2 and the disappearance of A_1 are given by:

$$R_2 = k_1 c_1 - k_2 c_2$$
$$-R_1 = k_1 c_1 \tag{4.92}$$

We obtain for the instantaneous selectivity in the kinetic regime:

$$s_{2,1} = \frac{R_2}{-R_1} = 1 - \frac{k_2}{k_1}\frac{c_{2,b}}{c_{1,b}} = 1 - \kappa \frac{c_{2,b}}{c_{1,b}}; \text{ with } \kappa = \frac{k_2}{k_1} \tag{4.93}$$

If transport resistances can be neglected, the concentration inside the catalyst pellet corresponds to the bulk concentration $c_i = c_{i,b}$. The instantaneous selectivity decreases with increasing conversion, X. In a catalytic fixed-bed reactor with plug flow behavior (see Chapter 28) the product yield can be determined by integration:

$$Y_{2,1} = \int_0^X s_{2,1} dX$$

$$Y_{2,1} = \frac{1}{1-\kappa}[(1-X)^\kappa - (1-X)] \tag{4.94}$$

The yield increases up to a maximum and finally reaches zero for $X = 1$ as shown in Figure 4.15. The maximum depends on the ratio of the two rate constants and is given by

$$Y_{2,1,\max} = \kappa^{\kappa/(1-\kappa)} \text{ at } X_{\text{op}} = 1 - \kappa^{1/(1-\kappa)}; \text{ for } \kappa \neq 1 \tag{4.95}$$

To evaluate the influence of internal mass transfer on the product selectivity and yield, we have to solve the material balance for A_1 and A_2 in the porous catalyst. Assuming a flat plate and equal diffusion coefficient ($D_{e,1} = D_{e,2} = D_e$), we obtain

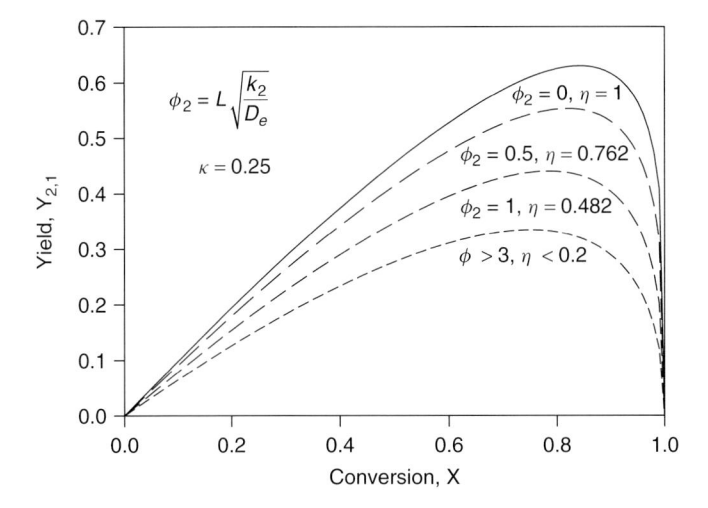

Figure 4.15 Effective product yield as function of conversion for consecutive reaction. Influence of internal mass transfer resistance. $\kappa = k_2/k_1 = 0.25$, initial product concentration $c_{2,0} = 0$.

with $c_{i,s} = c_{i,b}$:

$$\frac{d^2 f_1}{dZ^2} = \phi_1^2 f_1; \quad f_1 = \frac{c_1}{c_{1,s}} = \frac{c_1}{c_{1,b}}; \quad Z = \frac{z}{L}; \quad \phi_1 = L\sqrt{\frac{k_1}{D_e}}$$

$$\frac{d^2 f_2}{dZ^2} = \phi_2^2 \left(f_2 - \frac{1}{\kappa}\frac{c_{1,s}}{c_{2,s}} f_1 \right); \quad \kappa = \frac{k_2}{k_1} \quad \phi_2 = L\sqrt{\frac{k_2}{D_e}} \qquad (4.96)$$

With the concentration profile for the reactant A_1 given by Eq. (4.79), the solution of the differential equation leads to:

$$f_1 = \frac{\cosh(\phi_1 Z)}{\cosh(\phi_1)}; \quad \phi_1 = L\sqrt{\frac{k_1}{D_e}}$$

$$f_2 = \left(1 + \frac{c_{1,s}}{c_{2,s}}\frac{1}{1-\kappa}\right)\frac{\cosh(\phi_2 Z\kappa)}{\cosh(\phi_2\kappa)} - \frac{c_{1,s}}{c_{2,s}}\frac{1}{1-\kappa}\frac{\cosh(\phi_1 Z)}{\cosh(\phi_1)}; \quad \phi_2 = \sqrt{\kappa} \cdot \phi_1 \qquad (4.97)$$

The efficient instantaneous catalyst selectivity is given by the ratio of the efficient production rate of A_2 and the rate of reactant disappearance (see also Eq. (4.80)).

$$S_{2,1,\text{eff}} = \frac{R_{2,\text{eff}}}{-R_{1,\text{eff}}} = -\frac{(dc_2/dZ)_{Z=1}}{(dc_1/dZ)_{Z=1}} = \frac{1}{1-\kappa} - \left(\frac{c_{2,s}}{c_{1,s}} + \frac{1}{1-\kappa}\right) \cdot \sqrt{\kappa}\,\frac{\tanh(\phi_2\sqrt{\kappa})}{\tanh(\phi_2)} \qquad (4.98)$$

For strong diffusional resistance in the catalyst pellet, Eq. (4.98) can be simplified to:

$$s_{2,1,\text{eff}} = \frac{1}{1+\sqrt{\kappa}} - \sqrt{\kappa}\frac{c_{2,s}}{c_{1,s}} \quad \text{for } \phi_2\sqrt{\kappa} > 3 \tag{4.99}$$

The overall product yield is obtained by integration of Eq. (4.98) over a range of conversion. For a product concentration at the reactor inlet $c_{2,b,0} = 0$ the result is:

$$Y_{2,1,\text{eff}} = \frac{1}{1-\kappa}\left[(1-X)^{\Delta\phi} - (1-X)\right]; \quad c_{2,b,0} = 0$$

$$\text{with } \Delta\phi = \sqrt{\kappa}\frac{\tanh(\phi_2)}{\tanh(\phi_1)} = \sqrt{\kappa}\frac{\tanh(\phi_2)}{\tanh(\phi_2/\sqrt{\kappa})} \tag{4.100}$$

For very strong diffusional resistance, Eq. (4.100) can be simplified to yield:

$$Y_{2,1,\text{eff}} = \frac{1}{1-\kappa}\left[(1-X)^{\sqrt{\kappa}} - (1-X)\right]; \quad c_{2,0} = 0 \tag{4.101}$$

The integral product yield as a function of conversion for different values of the Thiele modulus is shown in Figure 4.15 for $\kappa = k_2/k_1 = 1/4$. It is obvious that internal diffusional resistance leads to a drastic decrease in the product selectivity and yield. In the domain of practical interest, with $\kappa < 1$, the maximum obtainable yield for strong diffusion resistance ($\phi_2 \geq 3$, Eq. (4.102)) drops to roughly 50% of the value reached in the kinetic regime (Eq. (4.95)). This demonstrates the disastrous influence of pore diffusion limitation on the overall productivity of the catalytic process.

$$\left(Y_{2,1,\text{max}}\right)_{\phi_2 \geq 3} = \frac{\kappa^{\left[0.5\sqrt{\kappa}/(1-\sqrt{\kappa})\right]}}{1+\sqrt{\kappa}}; \quad \text{at } X_{\text{op}} = 1 - \kappa^{\left[0.5/(1-\sqrt{\kappa})\right]} \tag{4.102}$$

4.3.3.2 Non-Isothermal Pellet

A large number of catalytic reactions are accompanied by thermal effects due to the heat of reaction. For relatively fast reactions compared to the mass and heat transfer phenomena, the development of internal temperature gradients can be expected. Heat and mass transfer balances have to be solved simultaneously to estimate concentration and temperature profiles under steady-state conditions. As the reaction rate depends exponentially on temperature the resulting temperature and concentration profiles have to be calculated by numerical methods.

$$D_e\frac{d^2c_1}{dz^2} - (-R_1) = 0$$

$$\lambda_e\frac{d^2T}{dz^2} - (-R_1)(-\Delta H_r) = 0 \tag{4.103}$$

where $(-\Delta H_r)$ is the reaction enthalpy and λ_e is the effective thermal conductivity in the porous pellet. As the reaction rate is the same in both balances, we obtain:

$$\frac{D_e(-\Delta H_r)}{\lambda_e}\frac{d^2c_1}{dz^2} = \frac{d^2T}{dz^2} \tag{4.104}$$

4.3 Mass and Heat Transfer Effects

With the surface concentration and temperature $c_{1,s}$ and T_s we obtain after integration a linear relationship between internal temperature and reactant concentration:

$$T - T_s = (-\Delta H_r) \frac{D_e}{\lambda_e} (c_{1,s} - c_1) \qquad (4.105)$$

The largest possible temperature difference in the particle is attained when the concentration in the particle center becomes $c_{1,\text{center}} \to 0$.

$$(T_{\text{center}} - T_s)_{\max} = (-\Delta H_r) c_{1,s} \frac{D_e}{\lambda_e} \qquad (4.106)$$

Obviously, the maximum temperature difference will depend on the reaction enthalpy and the ratio between effective diffusion and effective thermal conductivity.

If we refer the maximum temperature difference to the surface temperature, we get the dimensionless so-called Prater number.

$$\beta = \frac{\Delta T_{\max}}{T_s} = \frac{(-\Delta H_r) c_{1,s}}{T_s} \frac{D_e}{\lambda_e} \qquad (4.107)$$

For exothermic reactions the temperature inside the pellet will be higher than the surface temperature. Due to the exponential increase of the reaction rate, the

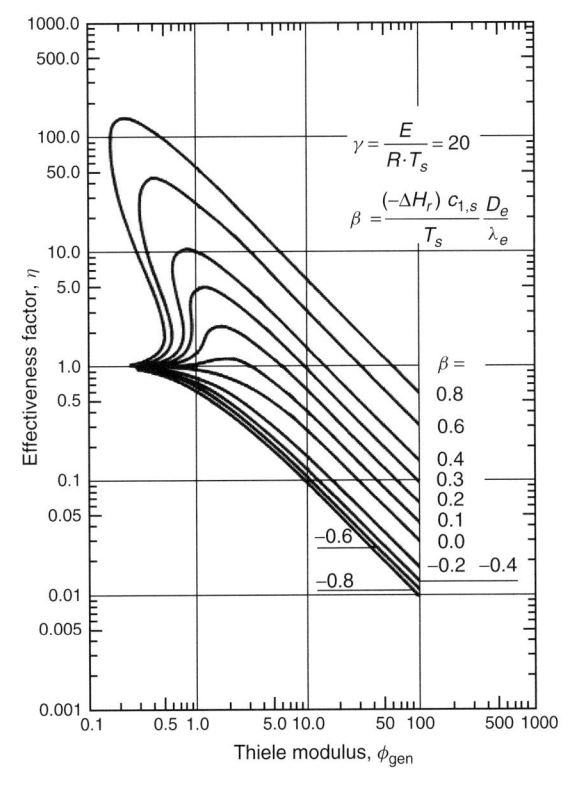

Figure 4.16 Effectiveness factor as function of the Thiele modulus. Non-isothermal sphere, first-order reaction. (Adapted from Ref. [5])

temperature effect can overcompensate the lower concentration in the pellet. An example is shown in Figure 4.16 where the effectiveness factor is plotted versus the Thiele factor for an Arrhenius number of $\gamma = 20$ and different Prater numbers.

The curves shown were obtained by numerical integration by Weisz and Hicks. Effectiveness factors higher than one can be expected at relatively low Thiele modulus and high Prater and Arrhenius numbers. At large values of ϕ, the effectiveness factor becomes inversely proportional to the Thiele modulus, as observed at isothermal conditions. Besides the increase of the observed reaction rate due to the high internal temperature, multiple steady states are predicted for reactions with high Arrhenius numbers and high Prater numbers. In the region of multiple steady states different temperature and concentration profiles for the same Thiele modulus may exist leading to different effectiveness factors. This behavior is shown in Figure 4.16 for $0.2 < \phi < 1$ and $\beta > 0.2$.

For industrially important catalysts the effective heat conductivity is in the order of $0.2 < \lambda_e < 0.5$ and effective diffusion coefficients for gas phase reactions are in the order of 10^{-5}–10^{-6} m^2 s^{-1}. Therefore Prater numbers seldom exceed values of $\beta = 0.1$ and the maximum temperature in the pellet center is seldom higher than $\Delta T_{\max} = 10$ K. In summary, temperature differences between gaseous bulk and catalyst surface are much more important as discussed in Section 4.3.4.3.

4.3.4
Combination of External and Internal Transfer Resistances

In the previous chapters we discussed the influence of internal mass and heat transfer by neglecting external transport phenomena. Hence, we assumed that concentrations and temperature at the outer surface of the catalyst particle and the bulk of the fluid are the same. But this assumption is not justified under certain conditions and concentration and temperature profiles inside and outside the porous catalyst must be considered.

4.3.4.1 Internal and External Mass Transport in Isothermal Pellets

If the efficient reaction rate is high enough, the reactant concentration drops significantly across the external boundary layer as indicated in Figure 4.17. In this case the surface concentration is lower than that in the bulk of the fluid phase ($c_{1,s} < c_{1,b}$). At first we will neglect potential heat effects and assume equal temperatures in the fluid and the catalyst particle ($T = T_s = T_b$). To determine the concentration profile in the particle, we first have to calculate the concentration at the external surface. This will be done based on the mass balance for the reactant A_1. At steady state, the molar flux of A_1 from the bulk to the external surface must be equal to the effective rate of transformation (see Eq. (4.45)).

$$J_1 = k_m \left(c_{1,b} - c_{1,s} \right) = \frac{-R_{1,p,\text{eff}}}{a_p} \qquad (4.108)$$

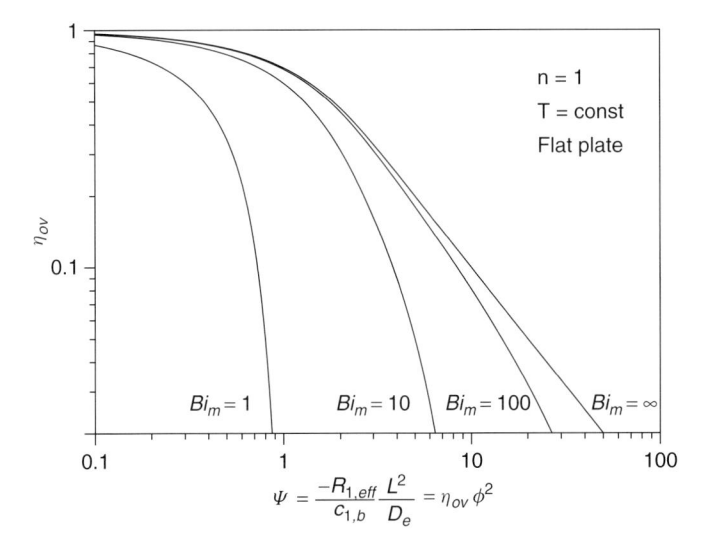

Figure 4.17 Overall effectiveness factor as a function of the Weisz modulus for different mass Biot numbers (isothermal, irreversible first-order reaction in a porous slab).

For a simple irreversible first order reaction we obtain:

$$k_m (c_{1,b} - c_{1,s}) = \frac{\eta_p k_r c_{1,s}}{a_p} \quad \text{with } \eta_p \text{ the internal effectiveness factor.} \quad (4.109)$$

Solving Eq. (4.109) for the unknown surface concentration:

$$c_{1,s} = \frac{c_{1,b}}{1 + \eta_p k_r / (k_m a_p)} \quad (4.110)$$

If we introduce the ratio between the characteristic diffusion time in the pellet t_D and the external mass transfer time t_m we will get a clearer physical interpretation of the relationship. The mentioned ratio is known as the *mass Biot number* Bi_m.

$$Bi_m = \frac{t_D}{t_m} = \frac{L_c^2}{D_e} k_m a_p \quad \text{with } L_c \text{ the characteristic length of the pellet} \quad (4.111)$$

Introducing the Biot number into Eq. (4.110) yields:

$$c_{1,s} = \frac{c_{1,b}}{1 + \eta_p \frac{k_r L_c^2}{D_e} \cdot \frac{1}{Bi_m}} = \frac{c_{1,b}}{1 + \eta_p \frac{\phi^2}{Bi_m}} \quad (4.112)$$

The *overall effectiveness factor* is defined as the ratio between the effective transformation rate and the rate at constant bulk concentration.

$$\eta_{ov} = \frac{-R_{1,\text{eff}}}{-R_{1,b}} = \frac{-R_{1,\text{eff}}}{k_r c_{1,b}} = \frac{\eta_p c_{1,s}}{c_{1,b}} \quad (4.113)$$

In combination with Eq. (4.112) we obtain:

$$\eta_{ov} = \frac{\eta_p}{1 + \eta_p \frac{\phi^2}{Bi_m}} = \frac{1}{\frac{1}{\eta_p} + \frac{\phi^2}{Bi_m}} \quad (4.114)$$

For a catalyst in the form of a flat plate the effectiveness factor is given by $\eta_p = \tanh\phi_L/\phi_L$ (Eq. (4.82)) and the overall effectiveness factor can be expressed as a function of the Thiele modulus and the Biot number.

$$\eta_{ov} = \frac{\tanh\phi}{\phi\left(1 + \frac{\phi\cdot\tanh\phi}{Bi_m}\right)} \quad (4.115)$$

The relationship shown in Eq. (4.115) suffers from the fact that the Thiele modulus must be specified to estimate the catalyst efficiency. This is, in general, not possible as the intrinsic kinetics is not known. It is therefore, more convenient to relate the overall effectiveness factor to the Weisz modulus, which is based only on observable parameters.

The catalyst efficiency decreases strongly at small mass Biot numbers as seen in Figure 4.17. This is due to the reduced reactant concentration on the external pellet surface. In contrast, external mass transfer influences can be neglected at $Bi_m > 100$. In practice, catalytic particles are in the range of several millimeters and the mass Biot numbers are in the order of 100–200. Hence, the overall effectiveness factor is almost entirely determined by the intraparticle diffusion.

4.3.4.2 Implication of Mass Transfer on the Temperature Dependence

As pointed out, the influence of mass transfer on the observed reaction rate depends on the ratio between the characteristic reaction time and the characteristic time for mass transfer. By increasing the temperature, the intrinsic reaction rate increases more strongly than the rates of inter- and intraparticle mass transfer. Consequently, the Thiele modulus and the second Damköhler number increase with increasing temperature and transport phenomena become more and more important and will finally control the transformation process. In addition, the temperature dependence of the observed reaction rate will change as indicated.

At low temperatures the process is controlled by the intrinsic chemical kinetics and the rate constant increases exponentially following Arrhenius's law:

$$k = k_0 \exp\left(\frac{-E}{RT}\right) \quad (4.116)$$

with k_0 the frequency factor and E the true activation energy.

The temperature dependence of the diffusion process can also be represented by an Arrhenius equation:

$$D_e = D_{e,0} \exp\left(\frac{-E_D}{RT}\right); \text{ with } 5 < E_D < 10\,\text{kJ/mol} \quad (4.117)$$

This is not a theoretical dependence of D_e, but the approximation is useful for the following discussions. At strong influence of internal diffusion on the reaction rate, the effectiveness factor was found to be inversely proportional to the Thiele modulus (e.g., Eq. (4.84)). Accordingly, the effective rate constant is given by:

$$k_{\text{eff}} = \frac{k}{L\sqrt{k/D_e}} = \frac{1}{L}\sqrt{k \cdot D_e} \tag{4.118}$$

The temperature dependence is given by:

$$k_{\text{eff}} = \frac{\sqrt{k_0 D_{e,0}}}{L} \exp\left(-\frac{E + E_D}{2 \cdot RT}\right) \tag{4.119}$$

Normally $E \gg E_D$, since diffusion is not very temperature sensitive, so the observed apparent activation energy is about one-half the true value when pronounced internal concentration profiles are present.

Further temperature increase will diminish the reactant concentration on the outer pellet surface as the influence of external mass transfer becomes important. Finally, interphase mass transfer will be the rate-controlling step and the surface concentration drops to zero. Under those conditions, the apparent activation energy corresponds to E_D. The influence of mass transfer on the observed temperature dependence is shown schematically in Figure 4.18.

Besides the apparent activation energy, the effective reaction order changes during the transition from kinetic to diffusion control. First-order reaction will be observed under external mass transfer control. The effective reaction order observed approaches $n_{\text{app}} = (n + 1)/2$ for severe influence of intraparticle diffusion.

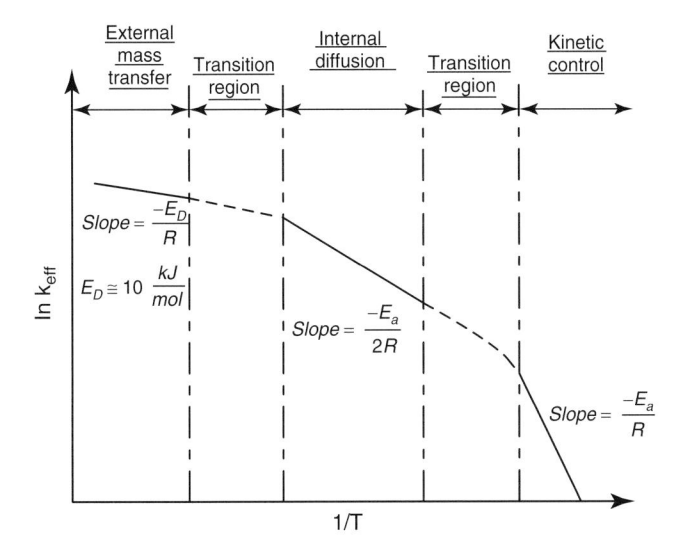

Figure 4.18 Arrhenius plot for heterogeneous catalytic reactions. Transition from the kinetic regime to mass transfer controlled regime.

4.3.4.3 External and Internal Temperature Gradient

In the case of fast highly exothermic or endothermic reactions, temperature gradients inside the porous catalyst and temperature differences between the fluid phase and catalyst surface cannot be neglected. Depending on the physical properties of the fluid and the solid catalyst important temperature gradients may occur. The relative importance of internal to external temperature profiles can be estimated based on the relationships presented in Sections 4.3.2.2 and 4.3.3.2. According to Eq. (4.66) the temperature difference between bulk and outer pellet surface is:

$$T_s - T_b = \frac{(-\Delta H_r)}{h \cdot a_p} \cdot r_{eff} \quad \text{with } r_{eff} = k_m a_p (c_{1,b} - c_{1,s})$$

$$T_s - T_b = (-\Delta H_r) \frac{k_m}{h} (c_{1,b} - c_{1,s}) \tag{4.120}$$

With the Chilton–Colburn analogy we can replace the ratio k_m/h and obtain (see Eq. (4.69)).

$$T_s - T_b = (-\Delta H_r) \frac{1}{\rho c_p} \left(\frac{Pr}{Sc}\right)^{2/3} (c_{1,b} - c_{1,s}) \tag{4.121}$$

For large internal diffusional resistance, the concentration of the reactant in the pellet center drops to zero. In this situation the temperature difference between the outer surface and the center of a porous catalyst pellet is maximal and given by Eq. (4.106).

In Eq. (4.122) the temperature difference between bulk and pellet surface is compared with the maximum internal temperature gradient. The ratio between these two temperature differences depends on the ratio of the mass to the thermal Biot numbers.

$$\frac{T_s - T_b}{(T_{center} - T_s)_{max}} = \frac{k_m}{h} \frac{\lambda_e}{D_e} \frac{c_{1,b} - c_{1,s}}{c_{1,s}} = \frac{Bi_m}{Bi_{th}} \frac{c_{1,b} - c_{1,s}}{c_{1,s}}$$

$$\text{with } Bi_{th} = \frac{h \cdot L}{\lambda_e}; \quad Bi_m = \frac{k_m \cdot L}{D_e} \tag{4.122}$$

In Table 4.1 the order of magnitude of some physical properties of fluid/solid systems are summarized. Based on these values we can conclude that for gas/solid systems the ratio of Bi_m/Bi_{th} is in the range of $10-10^4$. Hence, the temperature gradient in the external boundary layer is much more important than that within the pellet under usual conditions:

$$(T_s - T_b) >> (T_{center} - T_s); \text{ gas/solid system} \tag{4.123}$$

In contrast, we expect a higher temperature difference within the pellet in liquid/solid systems.

Table 4.1 Physical properties of fluid/solid systems [1].

	Gas	Liquid	Porous solid	Gas/solid	Liquid/solid
D or D_e (m$^2 \cdot$s^{-1})	10^{-5} to 10^{-4}	10^{-10} to 10^{-9}	10^{-7} to 10^{-5}	–	–
λ or λ_e (W$\cdot$m^{-1} K^{-1})	10^{-3} to 10^{-1}	10^{-2}–10	10^{-2}–1	–	–
$\rho\, c_p$ (J$\cdot$m^{-3}K^{-1})	10^2–10^5	10^5–10^7	10^6–10^7	–	–
Bi_m/Bi_{th}	–	–	–	10–10^4	10^{-4} – 0.1
$\beta = \dfrac{(-\Delta H_r)\, c_{1,s}}{T_s}\, \dfrac{D_e}{\lambda_e}$	–	–	–	10^{-3}–0.3	10^{-3}–0.1
$\gamma = \dfrac{E}{RT_s}$	–	–	–	5–30	5–30

4.3.5
Criteria for the Estimation of Transport Effects

For the catalyst development and optimization as well as for the correct reactor design, it is important to ascertain the influences of transport phenomena on the reaction kinetics. It is essential that criteria for estimating transport effects are based on what is measurable or observable [6, 7].

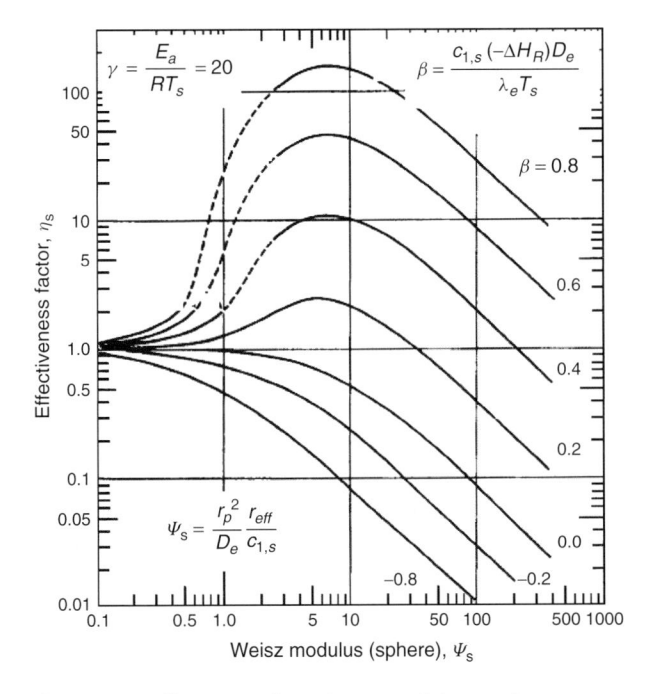

Figure 4.19 Effectiveness factor in terms of the experimentally observable Weisz modulus. First-order reaction, sphere. (Adapted from Ref. [5].)

One way to estimate the influence of transport processes is to use directly experimental results observed under given experimental conditions. In general, the experimentalist has information concerning observed reaction rates, reactant concentration and temperature, and catalyst form and dimensions. With this information to hand, the Weisz module can be estimated.

$$\psi_s^2 = \frac{t_D}{t_{r,\text{eff}}} = \frac{R_{\text{sphere}}^2}{D_e} \frac{c_s}{r_{p,\text{eff}}} = \eta_p \phi_s^2 \tag{4.87}$$

Each of the quantities in ψ_s is experimentally accessible and the effectiveness factor can be computed from results such as those shown in Figure 4.19. Hence a set of graphs can be prepared relating ψ_s to η with the Arrhenius and Prater numbers as parameters allowing the effectiveness to be directly estimated from experimental results. An example is given in Figure 4.18.

A number of important criteria for estimating the influence of transport phenomena on catalytic reaction rates are published in the open literature. In general, these criteria are derived assuming that transport effects do not alter the true rate by more than ±5%. Due to the uncertainty involved in estimating the different parameters, the application of the criteria should be done in a conservative manner.

Table 4.2 Experimental criteria for the absence of inter and intra transport phenomena ($0.95 < \eta < 1.05$) for simple irreversible reactions.

1	Absence of interphase concentration gradients in isothermal systems	$\dfrac{r_{p,\text{eff}} d_p}{2 k_m c_{i,b}} < \dfrac{0.15}{|n|}$
2	Absence of interphase temperature gradients. The criterion is independent whether intraparticle gradients exist or not.	$\dfrac{(-\Delta H_r) r_{p,\text{eff}} d_p}{2h \cdot T_b} \dfrac{E}{RT_b} < 0.15$
3	Absence of intraparticle/interphase gradients	$\dfrac{r_{p,\text{eff}} d_p^2}{4 c_{i,b} D_e} < \dfrac{1 + \gamma \cdot \chi}{|n - \gamma \cdot \beta|(1 + 0.33 nw)}$ $\chi = \dfrac{(-\Delta H_r) r_{p,\text{eff}} d_p}{2h \cdot T_b}, w = \dfrac{r_{p,\text{eff}} d_p}{2 c_{i,b} k_m}$
4	Absence of concentration profiles in an isothermal porous catalyst pellet	$\dfrac{r_{p,\text{eff}} d_p^2}{4 D_e c_{i,s}} \begin{cases} < 6 & \text{for } n = 0 \\ < 0.6 & \text{for } n = 1 \\ < 0.2 & \text{for } n = 2 \end{cases}$ $< 0.2 \ n = 2$
5	Absence of intraparticle temperature profile	$\dfrac{(-\Delta H_r) r_{p,\text{eff}} d_p^2}{4 \lambda_e T_s} < \dfrac{RT_s}{E}$
6	Absence of combined effect of temperature and concentration gradients	$\dfrac{r_{p,\text{eff}} d_p^2}{4 c_{1,b} D_e} < \dfrac{1}{|n - \gamma \cdot \beta|}; \gamma = \dfrac{E}{RT_b}, \beta = \dfrac{(-\Delta H_r D_e c_{1,b})}{\lambda T_b}$

Adapted from Baerns et al. [1].

The observed values should be at least several times or even an order of magnitude better than those proposed.

The most general of the criteria (Table 4.2, last equation) ensures the absence of any internal and external concentration and temperature gradient. But a problem may arise due to compensation between mass transport and heat transport. This situation will occur, if $\gamma \cdot \beta \cong n$. Therefore, it may be better to respect separately the criteria for isothermicity.

It is disturbing that criteria for the absence of heat effects are based on the true activation energy, which is not observable if mass transfer affects the rate of reaction. It is therefore indispensable to assess the experimental results critically and to apply the criteria prudently.

4.4
Homogenous Catalysis in Biphasic Fluid/Fluid Systems

In homogeneously catalyzed systems the selectivity of the reaction can be controlled by the appropriate choice of ligands on the catalytic metallic center. Combining a catalytic active metal with ligands often allows the synthesis of organic compounds that are otherwise only accessible through complex multistep synthesis. A drawback of homogeneously catalyzed processes is often the complex and costly separation and recycling of the catalyst. Therefore, considerable efforts are made to combine the easy separation of heterogeneous catalysts with the high potential reactivity and selectivity of homogeneous molecular catalysts.

Different methods are proposed to facilitate the recovery of the catalyst. A very successful way is to use biphasic systems of two immiscible liquids. The catalyst should be soluble in only one phase, in which the transformation takes place, while the products and sometimes the reactants should be preferentially soluble in the second phase. The catalyst is thus 'immobilized' in a 'liquid support.' Liquid supports may be water, supercritical fluids, ionic liquids, organic liquids, or fluorous liquids [8]. The Shell Higher Olefin Process (SHOP) and the Oxo synthesis (hydroformylation) are important industrial processes based on the biphasic catalytic system.

The immiscible liquids can be easily separated after the reaction and the catalyst is recycled to the reactor. This can be done without any thermal or chemical treatment. As the reaction is carried out in the presence of the homogeneously dissolved catalyst the advantages of homogeneous catalysis are fully preserved.

Special attention has to be paid to the mixing and mass transfer efficiency in the reactor. As the reaction takes place in the catalyst-containing phase, the reactants must first be transferred from the second, possibly gaseous, phase to the reaction phase. The mass transfer rate between the different phases depends on the area of the interface and the mass transfer coefficient. Whether the reaction will take place in the bulk of the reaction phase or near the interface depends on the ratio of the characteristic reaction time (t_r) to the characteristic time for mass transfer (t_m). This ratio is known as the *Hatta number (Ha)*.

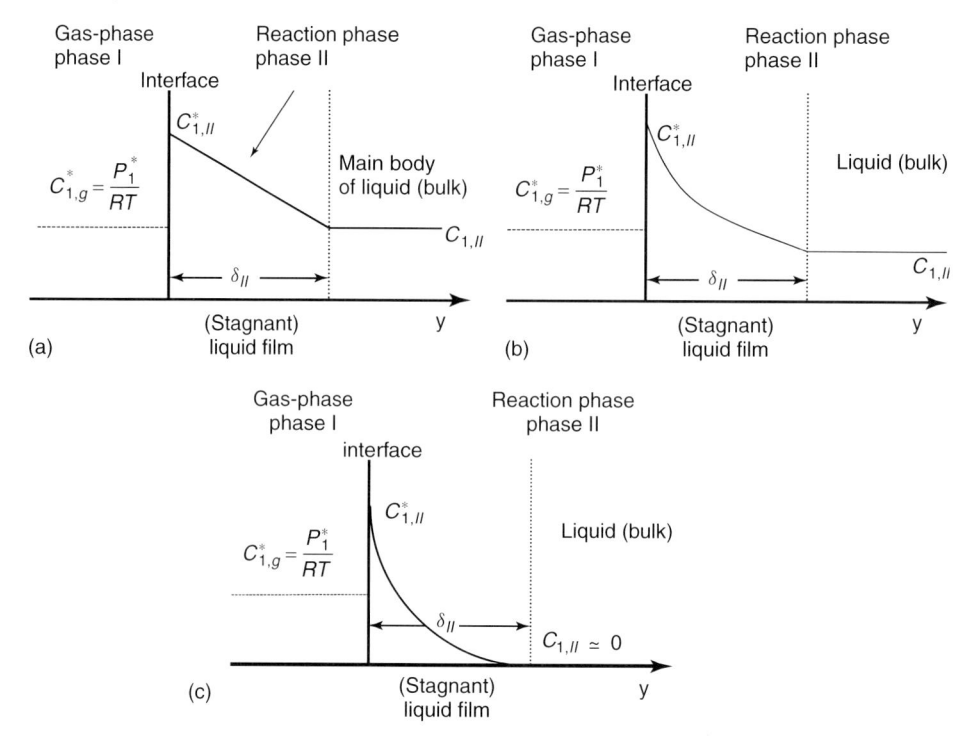

Figure 4.20 Concentration profiles for mass transfer with pseudo-first-order chemical reaction (film model) (a) slow chemical reaction: $Ha \leq 0.3$; (b) moderate chemical reaction: $0.3 \leq Ha \leq 3.0$; and (c) fast chemical reaction: $Ha \geq 3.0$.

The discussion can be facilitated on the basis of the film model and by supposing a first-order irreversible reaction in the reaction phase [9–11].

$$Ha = \sqrt{\frac{t_m}{t_r}} = \delta_{II}\sqrt{\frac{k'_r}{D_{1,II}}} = \frac{\sqrt{k'_r D_{1,II}}}{k_{L,II}} \tag{4.124}$$

With δ_{II}, the thickness of the boundary layer; k'_r the reaction rate constant, which is a function of the catalyst concentration ($k'_r = k_r \cdot c_{cat}$), $D_{1,II}$ the diffusion coefficient for the compound 1 in the second liquid phase, and $k_{L,II}$ the mass transfer coefficient in this liquid phase.

Depending on the value of Ha, different regimes can be distinguished (Figure 4.20):

For $Ha \leq 0.3$ the reaction rate is slow compared to the mass transfer and the reaction takes place in the bulk phase, and the mass transfer can be considered as an additional resistance in series to the reaction. The effective (observed) reaction rate is given by (Figure 4.20a):

$$r_{ov} = \left(\frac{1}{k_{L,II}a} + \frac{1}{k'_r}\right)^{-1} c_{1,II} \tag{4.125}$$

For values of the Hatta number of $0.3 \leq Ha \leq 3$ the reaction takes place partially in the boundary layer. This leads to deformation of the concentration profile as indicated in Figure 4.20b. The overall rate of reaction is given by the reaction in the bulk at reactant concentration $c_{1,L}$ and in the boundary layer. This leads to the following expression for the observable effective rate:

$$r_{\text{eff}} = \frac{Ha}{\tanh Ha} \left[1 - \frac{c_{1,\text{II}}}{c_1^*} \cdot \frac{1}{\cosh Ha} \right] \cdot k_L a \cdot c_1^* \tag{4.126}$$

The bulk concentration $c_{1,\text{II}}$ is a rather complex function of the intrinsic reaction rate, the mass transfer coefficient, and the area of the interface [11].

A further increase in the intrinsic reaction rate at constant volumetric mass transfer coefficient ($k_{L,\text{II}} \cdot a$) results in Hatta numbers greater than 3 ($Ha > 3$). The reaction rate can be considered as very fast compared to the mass transfer rate. As a consequence, the reactants do not reach the bulk phase ($c_{1,\text{II}} \approx 0$), and the reaction takes place only in the boundary layer (Figure 4.20c). Under these conditions, the reaction rate increases proportionally with the specific interfacial area between the phases (a), and the square root of the reaction rate constant and of the catalyst concentration as indicated in Eq. (4.127).

$$r_{\text{eff}} = k_{L,\text{II}} \cdot a \cdot Ha \cdot c_1^* = \sqrt{k_r' \cdot D_{1,\text{II}}} \cdot a \cdot c_1^* = \sqrt{k_r \cdot c_{\text{cat}} \cdot D_{1,\text{II}}} \cdot a \cdot c_1^* \tag{4.127}$$

In summary, increasing values of Ha lead to a decreasing reactant concentration in the reacting phase, and, as a consequence, the available volume of the reacting phase is less and less utilized. This situation can be characterized by introducing an efficiency factor η. The *efficiency factor* is defined as the ratio between the observed, effective rate and the maximal production rate referred to the reactor volume (V_R) corresponding to the maximum concentration in the reacting phase ($c_{1,\text{II}} = c_{1,\text{II}}^*$).

$$\eta = \frac{r_{\text{eff}}}{r_{\text{max}}}; \quad r_{\text{max}} = k_r' \frac{V_{\text{II}}}{V_R} c_{1,\text{II}}^* \tag{4.128}$$

The reactor efficiency depends on the Ha number and the specific interfacial area. For a first-order irreversible reaction the following relationship is obtained

$$\eta = \frac{B}{Ha} \left[\frac{\tanh(Ha) + (B^{-1} - 1) Ha}{1 + (B^{-1} - 1) Ha \tanh(Ha)} \right] \quad \text{with } B = \frac{a \cdot V_R \cdot D_{1,\text{II}}}{V_{\text{II}} \cdot k_{L,\text{II}}} \tag{4.129}$$

The parameter B can be interpreted as the ratio between the film volume (V_f) and the volume of the reacting phase (V_{II}). In Figure 4.21 the efficiency factor as a function of Ha and B is shown. The Figure clearly demonstrates that the reactor efficiency decreases with increasing Ha and with decreasing specific interfacial area a.

The presented relations above are strictly valid only for simple irreversible first-order reactions, but appropriate models for more complex kinetics can accordingly be developed based on the film model.

As shown above, fast chemical transformations characterized by $Ha \geq 3$ mainly occur near the interphase and are thus limited by the relatively low interfacial

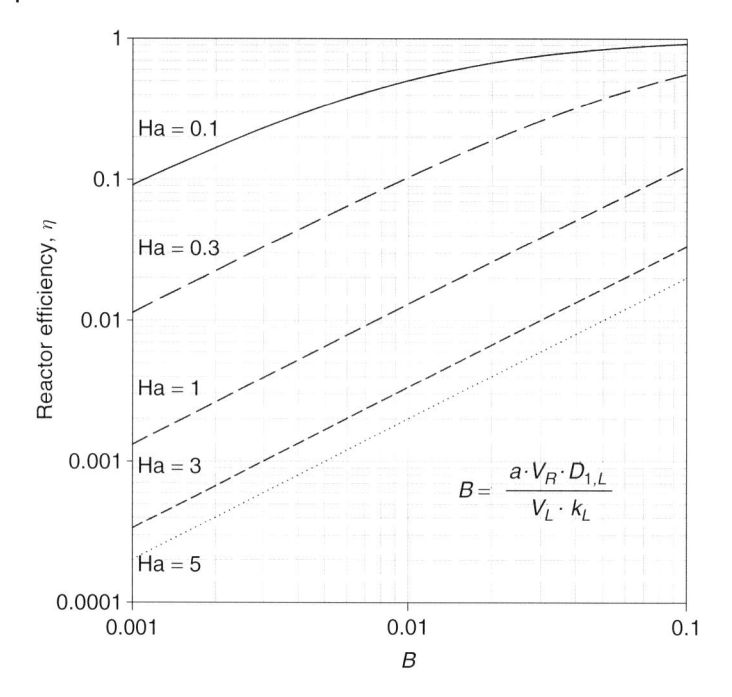

Figure 4.21 Effectiveness factor as for fluid/fluid reactions as function of Ha and B.

$$B = \frac{a \cdot V_R \cdot D_{1,L}}{V_L \cdot k_L}$$

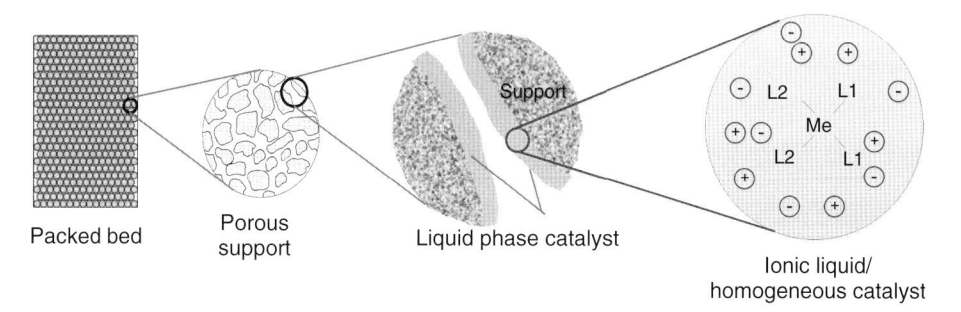

Figure 4.22 Schematic presentation of the concept of supported liquid-phase catalysis.

area which must be continuously generated by vigorous stirring of the multiphase mixture.

A possibility to overcome this drawback consists in immobilizing the liquid on a highly porous support. In this way a thin liquid layer is formed on the solid support, leading to the desired high fluid/fluid interface. This so called Supported Liquid-Phase Catalyst (SLPC) combines homogeneous catalysis with the advantages of a heterogeneous fluid/solid system. SLPCs can be used like traditional heterogeneous contacts in packed-bed reactors or even in fluidized beds. A schematic representation is shown in Figure 4.22.

The main problem related to SLP catalysts is the loss of solvent due to evaporation in a continuously operated catalytic reactor. This problem can be overcome by using ionic liquids as solvents [12–15]. Ionic liquids are molten salts, and their partial pressure is low under conditions commonly used for hydroformylation and hydrogenation reactions. As generally observed for SLPCs, the catalytic activity and product selectivity depends on the liquid loading and the nature of the porous support [16]. The *liquid loading* (α) is defined as the fraction of the pore volume occupied by the liquid:

$$\alpha = \frac{V_L}{\varepsilon \cdot V_p} \tag{4.130}$$

with V_L and V_p the liquid and particle volume, respectively, and ε the particle porosity. In consequence, the liquid loading varies between $\alpha = 0$ and $\alpha = 1$.

Typical experimental results for the determination of the effective reaction rate as a function of the liquid loading (α) of SLPCs are shown in Figure 4.23 for two different catalysts [17]. For the first catalyst (Figure 4.23a) a support with a low BET surface area of $7.5\,m^2\,g^{-1}$ was impregnated with RhCl(CO)(PPh$_3$)$_3$ as the active compound dissolved in PPh$_3$ and dioctylphthalate. The hydrogenation of ethene was used as a model reaction. The course of the curve with a pronounced maximum suggests that the effective reaction rate is influenced by two counteracting phenomena: (i) at low loadings, the reaction rate increases with α due to the growing amount of the homogeneous catalyst, but (ii) the accessibility of the gaseous educt diminishes at high loadings with liquid due to increasing diffusion resistances in the porous network and eventual pore blocking. The observed behavior was qualitatively confirmed for the hydroformylation of

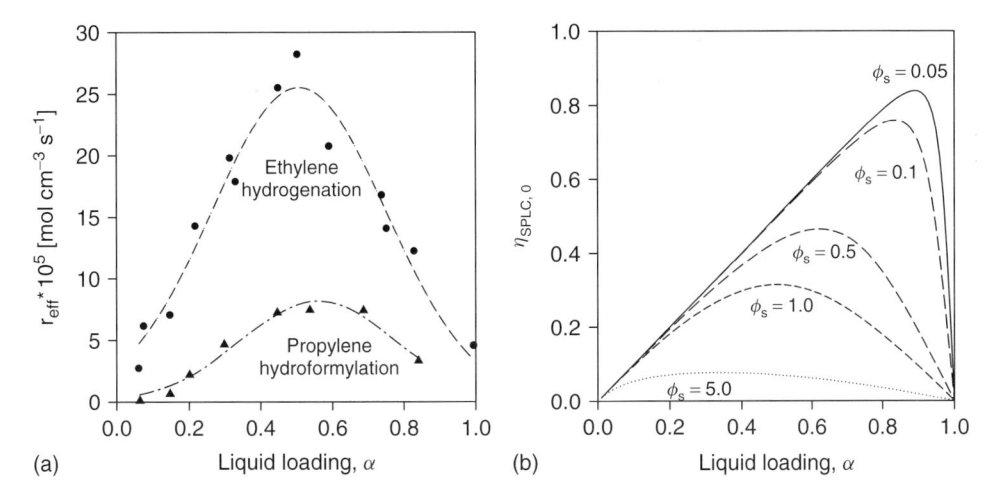

Figure 4.23 (a) Reaction rate of the ethane hydrogenation and the hydroformylation of propene as function of loading over different SLPCs. (b) The dependence of the global effectiveness factor on liquid loading and molecular Thiele modulus. (See text for details, data taken from Ref. [17].)

propene over RhH(CO) (PPh$_3$)$_3$ dissolved in dimethylglycolphthalate. The BET surface of the support was found to be in the order of 60 m^2 g^{-1}. This corresponds to values an order of magnitude higher compared to the first SLPC. In both cases the observed reaction rate passes through a maximum at loadings of $0.4 < \alpha < 0.6$.

Different models are proposed to explain the experimental observations, at least qualitatively [18]. The model proposed by Abed and Rinker [19] allows the estimation of the effectiveness factor as a function of the liquid loading and the Thiele modulus based on the intrinsic reaction rate and the characteristic diffusion time in the porous support.

The maximum reaction rate (r_{max}) per particle volume is reached for negligible diffusion resistance and complete pore filling ($\alpha = 1$).

The global effectiveness factor is based on the maximum rate:

$$\eta_{SLPC,0} = \frac{r_{eff}}{r_{max}} \qquad (4.131)$$

In Figure 4.23b the variation of the global effectiveness factor is shown as a function of the liquid loading and the Thiele modulus (ϕ_s). For all values of ϕ_s the effectiveness factor as function of the loading passes through a maximum. This corresponds essentially to published experimental results (see Figure 4.23). For comparable characteristic reaction and diffusion time ($\phi_s \approx 1$) the maximum effective rate is attained at loadings of $\alpha \approx 0.5$.

With decreasing intrinsic reaction rate, corresponding to decreasing Thiele modulus, the mass transfer resistance diminishes and the maximum effective rate is shifted to higher loadings as indicated in Figure 4.23b. Finally, for very low intrinsic reaction rates, the reaction will occur uniformly within the supported liquid and the optimal loading approaches unity ($\phi_m \to 0; \alpha \to 1$).

Although the presented model is able to afford some basic understanding of the influence of liquid loading on the observed chemical reactions in SLPC, more sophisticated models are necessary for a more quantitative description of the experimental findings. On the other hand, considerable experimental efforts exist to diminish diffusion resistances by using micro-structured supports with high porosity foams or fibrous materials in SLPC [20].

References

1. Baerns, M., Hofmann, H., and Renken, A. (1999) *Chemische Reaktionstechnik*, 3rd edn, Wiley-VCH Verlag GmbH, Weinheim.
2. Baerns, M. and Renken, A. (2004) Chemische reaktionstechnik, in *Winnacker Küchler: Chemische Technik* (eds W.K.R. Dittmeyer, G. Kreysa, and A. Oberholz), Wiley-VCH Verlag GmbH, Weinheim, **1**, pp. 453–643.
3. Satterfield, C.N. (1970) *Mass Transfer in Heterogeneous Catalysis*, M.I.T. Press, Cambridge.
4. Cassiere, G. and Carberry, J.J. (1973) The interphase catalytic effectiveness factor. *Chem. Eng. Ed.*, **7** (1), 22–26.
5. Dittmeyer, R. and Emig, G. (2008) in *Handbook of Heterogeneous Catalysis* (eds G. Ertl, H. Knözinger, F. Schüth, and J. Weitkamp), Wiley-VCH Verlag GmbH, Weinheim, pp. 1727–1784.

6. Mears, D.E. (1971) Tests for transport limitations in experimental catalytic reactors. *Ind. Eng. Chem. Process Des. Dev.*, **10** (4), 541–547.
7. Mears, D.E. (1971) Diagnostic criteria for heat transport limitations in fixed bed reactors. *J. Catal.*, **20** (2), 127–131.
8. Cornils, B., Herrmann, W.A., Horváth, I.T., Leitner, W., Mecklenburg, S., Olivier-Bourbigou, H., and Vogt, D. (2005) *Multiphase Homogeneous Catalysis*, vol. 1, Wiley-VCH Verlag GmbH, Weinheim.
9. Baerns, M., Hofmann, H., and Renken, A. (2002) *Chemische Reaktionstechnik*, 3rd edn, Wiley-VCH Verlag GmbH, Weinheim.
10. Levenspiel, O. (1999) *Chemical Reaction Engineering*, 3rd edn, John Wiley & Sons, Inc., New York.
11. Trambouze, P. and Euzen, J.-P. (2002) *Les Réacteurs Chimiques*, TECHNIP, Paris.
12. Mehnert, C.P., Cook, R.A., Dispenziere, N.C., and Afeworki, M. (2002) Supported ionic liquid catalysis – A new concept for homogeneous hydroformylation catalysis. *J. Am. Chem. Soc.*, **124** (44), 12932–12933.
13. Mehnert, C.P., Mozeleski, E.J., and Cook, R.A. (2002) Supported ionic liquid catalysis investigated for hydrogenation reactions. *Chem. Commun.*, **8** (24), 3010–3011.
14. Riisager, A., Eriksen, K.M., Wasserscheid, P., and Fehrmann, R. (2003) Propene and 1-octene hydroformylation with silica-supported, ionic liquid-phase (SILP) Rh-phosphine catalysts in continuous fixed-bed mode. *Catal. Lett.*, **90** (3–4), 149–153.
15. Riisager, A., Wasserscheid, P., Van Hal, R., and Fehrmann, R. (2003) Continuous fixed-bed gas-phase hydroformylation using supported ionic liquid-phase (SILP) Rh catalysts. *J. Catal.*, **219** (2), 452–455.
16. Riisager, A., Fehrmann, R., Haumann, M., and Wasserscheid, P. (2006) Supported Ionic Liquid Phase (SILP) catalysis: an innovative concept for homogeneous catalysis in continuous fixed-bed reactors. *Eur. J. Inorg. Chem.*, (4), 695–706.
17. Hoffmeister, M. and Hesse, D. (1990) The influence of the pore structure of the support on the properties of supported liquid-phase catalysts. *Chem. Eng. Sci.*, **45** (8), 2575–2580.
18. Renken, A. (2010) Chemical reaction engineering aspects for heterogenized molecular catalysis, in *Heterogenized Homogeneous Catalysts for Fine Chemical Production. Materials and Processes* (eds P. Barbaro and F. Liguori), Springer.
19. Abed, R. and Rinker, R.G. (1973) Diffusion-limited reaction in supported liquid-phase catalysis. *J. Catal.*, **31**, 119–126.
20. Kashid, M.N., Renken, A., and Kiwi-Minsker, L. (2011) Microstructured reactors and supports for ionic liquids. *Chem. Eng. Sci.*, **66** (7), 1480–1489.

Part II
The Chemistry of Catalytic Reactivity

This part focusses on the mechanism of catalytic reactions. It contains five chapters that deal with the different catalytic subdisciplines. Mechanistic aspects are discussed of heterogeneous catalytic, homogeneous catalytic and biocatalytic reactions. Chapters have also been included covering electrocatalysis and heterogeneous photocatalysis.

5
Heterogeneous Catalysis

Rutger van Santen

5.1
General Introduction

Heterogeneous catalysis science aims to relate catalytic performance to catalyst structure and composition. This can be reformulated as a search for the catalyst performance indicators that determine activity and selectivity of the overall reaction.

As we will see, a direct relation between a reactivity index and a rate constant can be uniquely defined only for elementary rate constants. However, for the overall reaction it is the combination of surface coverage and a balance of rate-controlling elementary steps that determines overall performance, which in turn strongly relates to reaction conditions.

The first example of CH activation that we will discuss illustrates this. We take the steam reforming reaction of methane as an example. The dependence on choice of the metal catalyst cannot be understood without explicit consideration of the surface composition of the catalyst.

Generally there are three reactivity-determining catalyst parameters to consider:

- particle size and shape
- catalyst composition
- structure and composition of surface during reaction.

Kinetics is very important when the selectivity of a reaction depends on the concentration of surface reaction intermediates formed successively in the course of the reaction.

We start this mechanistic chapter with the ammonia synthesis reaction. Surface scientists and computational catalytic scientists have especially used this reaction to further our understanding of the molecular basis of heterogeneous catalytic reactivity.

Many of the molecular catalytic chemistry concepts discussed in this book have their origin in such studies, recognized in 2008 by the award of the Nobel prize to Ertl.

The ammonia synthesis reaction is catalyzed by transition metal catalysts. In the section on transition metals, reactions involving C–H and C–C bond activation will also be discussed.

Throughout we emphasize topological surface structure requirements for low-activation energy barriers of surface chemical reactions.

The other two catalytic systems that we discuss are the reactivity of solid acids such as zeolitic materials. We will see that physical parameters as adsorption energies are essential to understand differences in acidity.

The third catalytic material that will be discussed is the group of reducible oxides. Structure and composition significantly affect the reactivity of the surface atoms.

Here we are especially interested in the tendency for substrate CH activation versus O insertion into the hydrocarbon π-bond.

5.2
Transition Metal Catalysis

5.2.1
Ammonia Synthesis

5.2.1.1 The Mechanism of the Reaction

The ammonia synthesis reaction corresponds to the following stoichiometry:

$$N_2 + 3H_2 \longrightarrow 2NH_3 \tag{5.1}$$

The currently preferred catalyst consists of Fe promoted with several metals, of which potassium is the most important. Recently Ru has been explored as an alternative catalyst.

In Chapter 2, Figure 2.2, the reaction energy diagram of this reaction is shown as constructed by Ertl for the Fe catalyst. Dissociation of N_2 takes place to give adsorbed N atoms and of H_2 forming hydrogen atoms. In consecutive surface reaction steps the adsorbed N atoms react with the adsorbed H atoms to give subsequently adsorbed NH, NH_2, and NH_3. The latter molecule desorbs as the product molecule.

5.2.1.2 Structure Sensitivity, Composition Dependence

In Figure 5.1 the results of an elegant experiment by Somorjai are presented. In this early surface science experiment he studied ammonia synthesis on single-crystal surfaces of different orientation. One observes that the overall catalytic conversion reaction changes by orders of magnitude when the activity of different surfaces is compared. The surface with the highest density of most coordinatively unsaturated atoms is found to have the highest reactivity.

The more bonds between metal surface atoms broken per unit surface area, the more reactive is the surface. The activity of the catalyst increases with increase of surface energy.

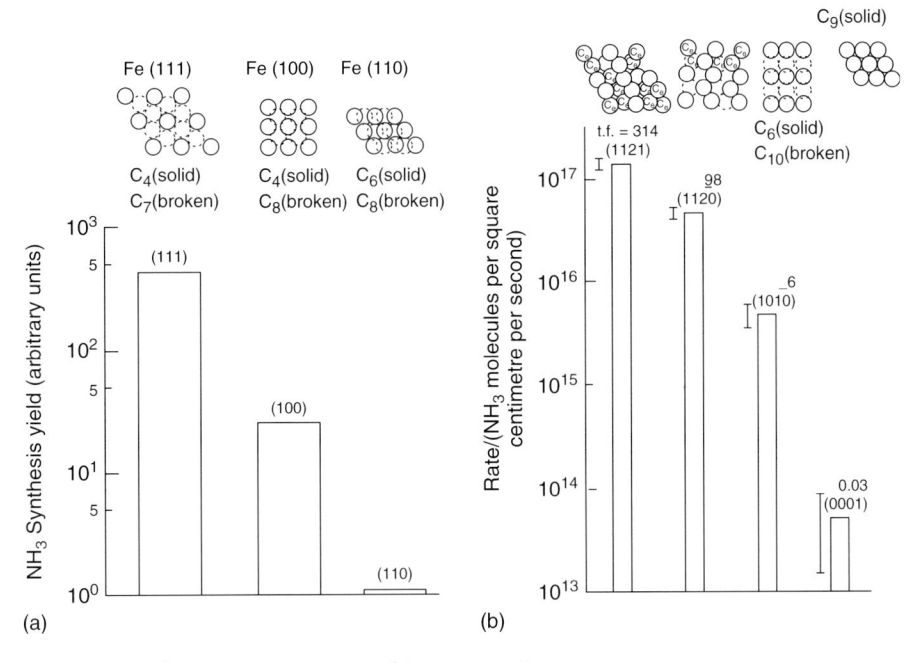

Figure 5.1 Surface science experiment of Somorjai on Fe and Re [1]; a comparison of the rate of ammonia formation catalyzed by different surfaces.

We will now show how this information can help to determine where on the Sabatier curve of rate versus substrate interaction energy these Fe surfaces are located.

As we can deduce from the corresponding reaction energy reaction diagram in Figure 2.2, there is an activation energy involved with N_2 dissociation (H_2 activation is non-activated), and also hydrogenation of adsorbed N atoms costs energy.

Dissociation of N_2 is exothermic. Hydrogenation of surface nitrogen is energetically uphill and endothermic. So there are two competing reactions:

- activation of N_2
- hydrogenation of adsorbed N.

The next task is to determine how the activation energies of the surface elementary steps depend on the reactivity of the surface atoms.

In Chapter 1 we briefly discussed the Brønsted–Evans–Polanyi relationship between the activation energy and the adsorption energy of adatoms generated upon dissociation.

The corresponding relationship between activation energy of N_2 and adsorption energy changes of adsorbed nitrogen atoms on two surfaces with different topologies is shown in Figure 5.2. The adsorption energies of N_{ads} vary according to choice of metal and is taken for the dense surface in which the surface atoms have a high coordination number.

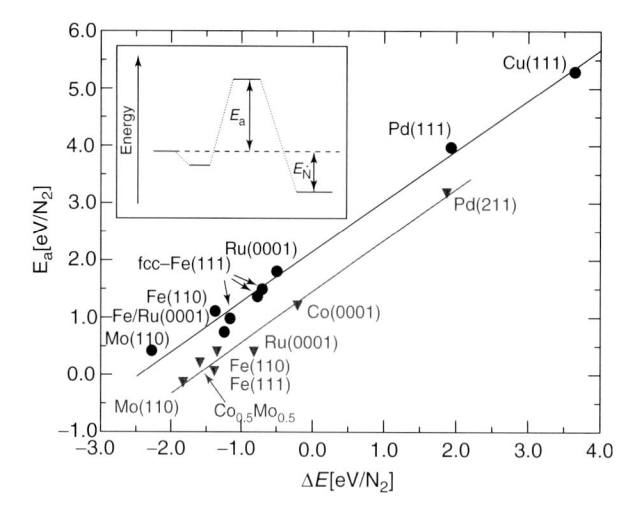

Figure 5.2 The calculated activation energies (transition state potential energies) for N_2 dissociation (E_a) on a range of metal surfaces plotted as a function of the adsorption energy for two nitrogen atoms (E_N). All energies are relative to $N_2(g)$. Results for both close packed surfaces (circles) and more open surfaces are shown (triangles). The inset shows a sketch of the energetics for the N_2 dissociation reaction [2].

As for CO (Figure 1.3) a linear dependence of the activation energy depending on the change in adsorption energy of the nitrogen atoms is found. The more open surface has the lower activation energy when plotted against the energy of the nitrogen atoms on the dense surface.

The change in the adsorption energy of the adatom with metal relates to chemical bonding properties of the metals. The adsorption energies increase along a row of the periodic system moving from the IB metals to the group VIII metals.

The changes in adsorption energies of the corresponding molecules are significantly less than those in the corresponding adatom adsorption energies. The reactivity of metal surfaces therefore relates with adsorption energies of the generated surface atoms instead of the molecular adsorption energies.

The lower activation energy for dissociation on the more open surface has two causes:

Surface metal atoms tend to interact more strongly with adsorbed atoms when their surface coordination number is less. When one compares the chemical bonding properties of surface atoms as a function of number of metal atom neighbors, the interaction energy with adsorbed adsorbate atoms generally tends to increase with decreasing neighbor surface metal atom coordination number. This is due to increased localization of the metal surface atom electrons. As we will see in the next section when we discuss methane activation, this dependence of surface reactivity on coordination with neighboring surface atoms is responsible for decreases in activation energy when bonds are activated by contact with a single metal atom. However, dissociation of diatomic molecules such as CO or N_2 requires a surface ensemble of atoms to dissociate.

(a)   (b)

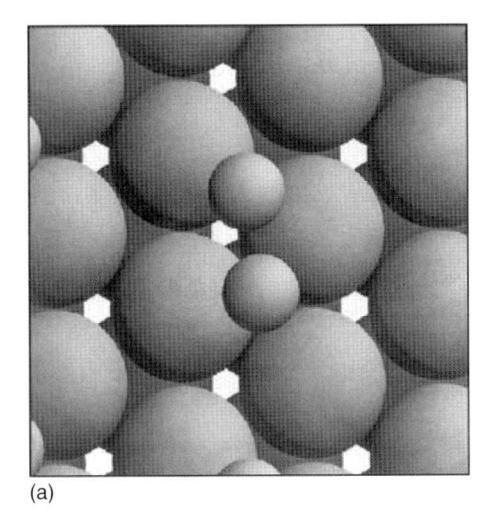

Figure 5.3 Configurations for the transition state for N_2 dissociation on Ru(0 0 0 1) surface (a) on a terrace and (b) at a step. The nitrogen atoms are smallest. Dark gray Ru atoms are surface atoms in deeper layers. On the terrace, the N_2 transition state binds to four Ru atoms but at the step the binding site consists of five surface atoms [2].

The requirement of surface sites that consist of an ensemble of several metal atoms to dissociate a diatomic molecule such as N_2 leads to extreme sensitivity of the activation energy of this molecule to surface topology. The general rule for diatomic molecules with π-bonds is that those surface sites are favored that allow for transition states in which the atoms in the molecule do not share bonding with the same metal atom. This explains, for instance, the high reactivity of the fcc (100) surface, with the arrangement of the surface atoms in squares, compared to that of the fcc (111) surface with the triangular arrangement of surface atoms. This is illustrated in Figure 5.3 for a (111) surface with steps in the (100) direction.

The reason for this is that adsorbate interaction with a surface metal atom weakens its reactivity. Electrons become consumed in the additional adsorbate–metal bond. Hence, when a second adsorbate interacts, bonding is weaker then when no additional adsorbate atom is attached to the same metal atom. The resulting increases in activation energies can be substantial, between 50 and 100 kJ mol^{-1}. Recombination of N adatoms is particularly favorable on (100) surfaces since they favor threefold coordination and hence are more strongly attached to a (111) type surface instead of to the (100) surface.

An even more favorable situation arises when the metal surface atoms are not in a plane but have the step-edge structure. Then in the transition state not only sharing of bonds with surface metal atoms is absent, but also the molecular bond has to stretch very little. Step-edges tend to have lower activation energies for dissociation.

We are now in a position to predict for the Fe or Re surfaces that Somorjai studied how the activation energies of the competing surface reactions depend on surface topology and coordinative unsaturation of the surface atoms.

The interaction energy of an N adatom with the surface metal atoms will increase when the surface metal becomes increasingly less surrounded by surface atom neighbors. Hence the energy cost to remove the N adatom from the surface by hydrogen atom addition will increase.

On the other hand, the activation energy of N_2 will decrease on the more open metal surfaces.

Let us now return to the Sabatier curve of Figure 1.2. Catalyst reactivity is at the left of the Sabatier maximum when increasing interaction energy of reactant with catalyst surface increases its rate. To the left when the trend is opposite.

One concludes that the Fe and Re surfaces both are at the left of the Sabatier maximum and hence the dissociation of N_2 in the Somorjai experiment [1] is rate limiting. Improvements in the catalyst relate to increases in the rate of N_2 dissociation.

The Temkin kinetic expression developed for the commercial process, employing a promoted Fe catalyst, illustrates how this conclusion depends on reaction conditions. At very high pressures NH_3 adsorption tends to compete with the N_2 activation step:

$$v_t = \vec{k}[N_2]\left\{\frac{[H_2]^3}{[NH_3]}\right\}^m - \overleftarrow{k}\left\{\frac{[H_2]^3}{[NH_3]}\right\}^{1-m} \tag{5.2}$$

Ruthenium has been proposed as an alternative catalyst. In Figure 5.2 we compared the activation energies for N_2 dissociation on a stepped and non-stepped surface. This preference of dissociation leads to an interesting particle size dependence of this reaction. This is illustrated in Figure 5.4.

In order to stabilize such sites on a metal particle a terrace has to be overgrown by a second overlayer so that step-edge sites are created. Such step-edge sites are also called *B5 sites*. It indicates the number of metal surface atoms involved in the

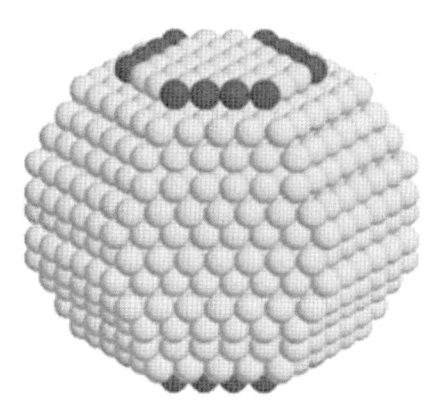

Figure 5.4 Particle with B5 sites [3].

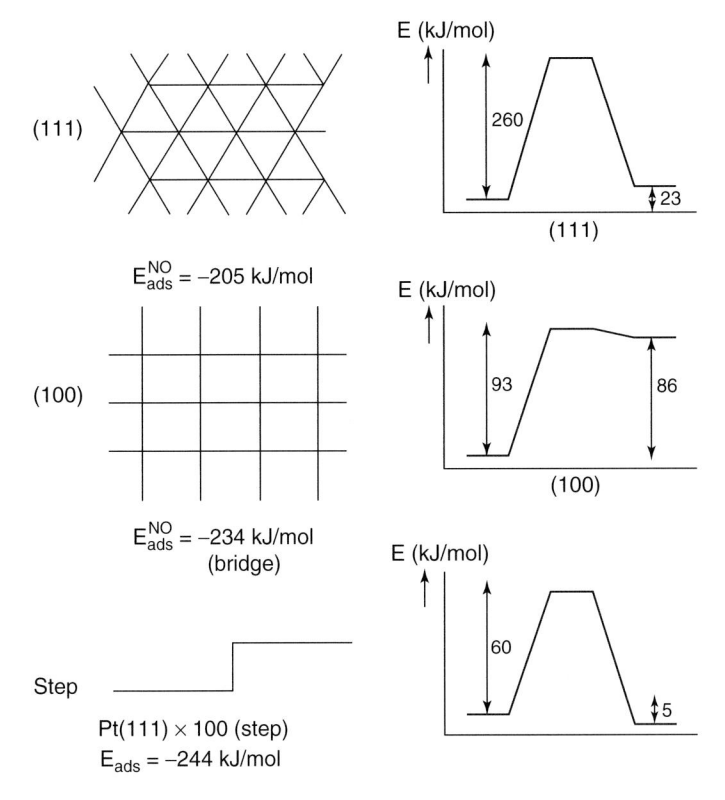

Figure 5.5 Reaction, transition state, and adsorption energies of NO on different Pt surfaces [4].

reaction site. When a particle becomes too small (typically less than 2 nm) such step edge sites cannot be supported and the activation energy of the molecules becomes the high barrier energy of the terraces.

Since the steric requirement for activation of π-valent electronic diatomic molecules is very similar, rates of reactions in which activation of molecules such as CO, N_2, or NO are rate controlling steeply decrease when the particle size is less than the size that can support the B5 type sites.

The dependence on surface topology for low-barrier transition states for dissociation as well as recombination is shown in Figure 5.5.

We see that recombination of the adatoms occurs with extremely low energy on a (100) surface, therefore cubic particle exposing only (100) surfaces have the preferred shapes for such reactions. Such reactions will not show a significant particle size dependence in contrast to the particle size dependence of the dissociation reaction. It relates to the very different thermodynamics of the two surface reactions for the two systems.

5.2.2
Methane Reforming

5.2.2.1 The Mechanism of the Reaction
The stoechiometry of the wet reforming reaction of methane is:

$$CH_4 + H_2O \longrightarrow CO + 3H_2 \tag{5.3}$$

It is related to the dry reforming reaction:

$$CH_4 + CO_2 \longrightarrow 2CO + 2H_2 \tag{5.4}$$

Whereas these two reactions are highly endothermic (ΔH^0_{298} = 205 and 247 kJ mol^{-1} respectively) the related partial oxidation reaction:

$$CH_4 + 1/2 O_2 \longrightarrow CO + 2H_2 \quad (\Delta H^0_{298} = -38 \text{ kJ mol}^{-1}) \tag{5.5}$$

is exothermic.

The water gas shift reaction:

$$CO + H_2O \longrightarrow CO_2 + H_2 (\Delta H^0_{298} = -41 \text{ kJ mol}^{-1}) \tag{5.6}$$

is strongly coupled to the steam reforming reaction, since the corresponding conversion steps are usually fast. Equilibrium will determine the product distribution of the reaction.

In this section we discuss molecular reactivity aspects of the activation of the initial step of the wet methane reforming reaction.

The mechanism of the reaction proceeds by first decomposition of CH_4 into CH_x fragments and adsorbed H atoms, and H_2O into first OH and subsequently into atomic hydrogen.

Initial decomposition of CH_4 and H_2O to dissociated surface fragments CH or C and OH and O respectively is exothermic. Subsequent CO formation and hydrogen desorption are the endothermic reactions. This reflects the fact that methane decomposition competes with CO formation to be rate controlling.

5.2.2.2 Structure Sensitivity and Composition Dependence
Figure 5.6 shows the very important result that the methane steam reforming conversion rate increases linearly with metal particle dispersion.

This linear increase is very different from what we would have expected based on the assumption that C_{ads} and O_{ads} recombination is rate controlling.

From the earlier discussion on the formation of π valence electron diatomic molecules we learned that for particles less than 2 nm the activation energies steeply increase compared to the activation energies typical for surface step sites. Hence if this reaction step were rate controlling beyond a particular dispersion the rate of reaction should decrease. This indicates that methane or H_2O activation is rate controlling.

The activation energy of bond cleavage of methane to produce adsorbed CH_3 and H atoms has been studied intensively as a function of surface atom coordination number. Some representative results are shown in Table 5.1.

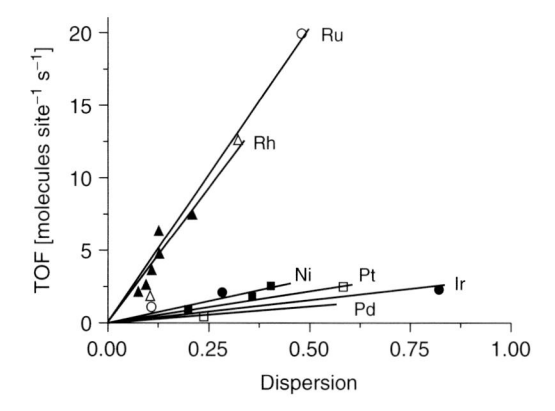

Figure 5.6 Rate of methane conversion as a function of dispersion for several metals [5].

Table 5.1 Activation energies of methane activation as a function of surface coordination number [4].

	E_{diss} (kJ mol^{-1})
Ru (0001)	76
Ru (1120)	56
Rh (111)	67
Rh step	32
Rh kink	20
Pd (111)	66
Pd step	38
Pd kink	41
Pd atom	5

The reactivity of such a surface atom increases with decreasing coordination number.

This leads to the uniform decrease of the methane activation energy with decreasing particle size, since methane activation occurs over a surface site that consists of a single metal atom. Such a transition state is illustrated for one of the Ru surfaces in Figure 5.7:

When the size of a metal particle decreases the ratio of corner to edge and terrace atoms changes. There is a uniform increase in the relative concentration of corner atoms and a uniform decrease in the relative concentration of terrace atoms. The relative concentration of the edge atoms goes through a maximum.

Since methane dissociation predominantly occurs over single atom sites, the relative concentration of the most active atoms increases with decreasing particle size. As a consequence the activation energy for methane activation decreases and the rate of methane activation increases with decreasing particle size. This

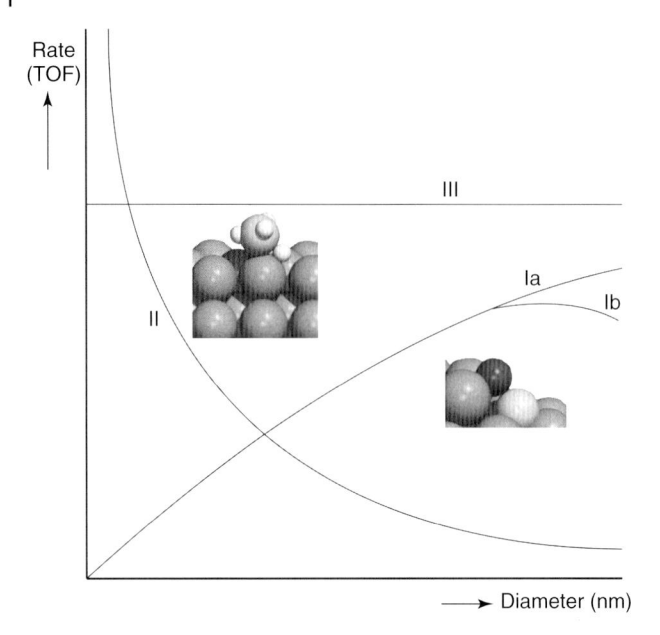

Figure 5.7 Schematic drawing of the dependence of rate of reaction as a function of particle diameter. The three types of particle size dependence relate to the differences in transition state intermediates. Inserts show the transition state for methane activation and CO dissociation respectively [4]. TOF is Turn Over Frequency; Number of molecules converted per site per second.

is consistent with the particle size dependence of the methane steam reforming reaction assuming methane activation to be rate controlling.

The uniform decrease in activation energy as found for methane is general for reactions of molecules in which σ-type bonds are broken.

Three types of particle size dependence have experimentally been found. They are illustrated in Figure 5.7.

An increase in normalized rate with decreasing particle size as in the methane reforming reaction is illustrated by curve II. Curve I, the decrease in rate of reaction when particle size becomes too small, corresponds to ammonia synthesis or the methanation reaction. Curve III, the rate independent of particle size belongs to the formation CH type σ-bonds.

This can be understood in the following way: as illustrated in Figure 5.8b in the case of methane the activation energy change is proportional to the reaction energy change:

$$\delta E_{act} = \alpha \cdot \delta E_{reaction} \tag{5.7}$$

For surface dissociation reactions in this BEP relation the value of α is typically 0.9. As a consequence, the BEP value for the proportionality constant of the reverse reaction follows from:

$$\delta E_{act}^{rev} = -(1 - \alpha) \cdot \delta E_{reaction} \tag{5.8}$$

5.2 Transition Metal Catalysis

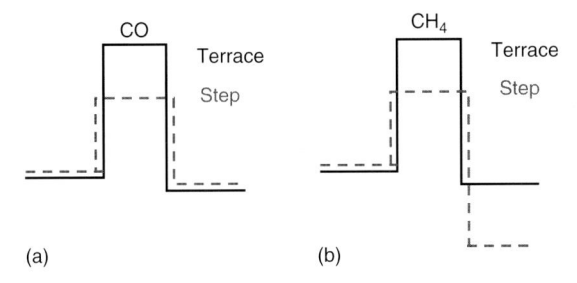

Figure 5.8 Relationship between activation energies (the difference between position of bottom horizontal lines and top horizontal lines) and reaction energies (the difference between the positions of horizontal lines to the left and right in figures a and b respectively) when reactions at step edge (---) and surface terrace (–) are compared (schematic); (a) CO dissociation and formation and (b) CH cleavage or formation.

and is equal to -0.1. So in Figure 5.7 the dependences I and II are complementary. The strong increase in rate of methane activation with particle size is consistent with the reverse hydrogenation reaction being independent of particle size.

If the rate of the methane steam reforming reaction were determined only by the activation energy of methane activation, the trend in rates of reaction as a function could be determined from known trends in the activation energies of methane.

Table 5.2 summarizes computed activation energies of methane activation of the dense surfaces of the group VIII metals.

One observes that Ir and Pt have the lowest barriers; next are Ru, Rh, and Pd, and Co and Ni are the least reactive. This sequence in reactivity is very different from that observed experimentally (see Figure 5.6).

The reason for this discrepancy is that in the steam reforming reaction ultimately CO has to be formed. This implies that there is an essential surface reaction between an adsorbed C1 fragment and adsorbed O.

The surface has to be reactive enough for H_2O to decompose on the metal surface. This depends on the adsorption energy of O. It behaves very differently as a function of the metal than the methane activation energies. This is shown in Table 5.3.

Table 5.2 DFT-computed activation energies in kJoules per mole for CH_4 decomposition to CH_3 computed on densily packed of some group VIII transition metals [4].

	100 Co(0001)	118 Ni(111)
77 Ru(0001)	67 Rh(111)	66 Pd(111)
	32 ≈ 40 Ir(111)	75 Pt(111)

Table 5.3 Adsorption energies of O atoms on the densily packed surfaces of group VIII transition metals. Critical values for E_{ads} of O for H_2O or CO_2 to decompose are 480 and 520 kJ mol^{-1} [6].

Fe (hcp)	Co	Ni	Cu
−714	−550	−496	−429
Ru	Rh	Pd	Ag
−557	−469	−382	−321
Os	Ir	Pt	Au
−533	−428	−354	−270

Interestingly, the adsorption energies of the oxygen atom to metals such as Pt, Ir, or Pd are too low. It has to be at least 480 kJ/atom to overcome the water dissociation energy.

Ru metal appears to be the best compromise between H_2O adsorption energy and methane activation barrier. Rh is second. Note that the computed O adsorption energy to the Rh (111) surface is borderline. It implies that more reactive surfaces than the (111) surface have to be involved.

Dry methane reforming requires dissociation of CO_2 into CO and adsorbed O. It requires a minimum interaction energy of 520 kJ/atom. Now Ru, Ni, and Co become interesting candidates.

Table 5.2 gives an indication of the activation barriers of methane activation. Comparison of the activation energies of C_{ads} and O_{ads} should give a consistent result. The activation energy for C_{ads} and O_{ads} recombination is independent of the metal because of the strong dependence of the activation energy of CO dissociation energy on the metal (principle of microscopic reversibility, see Eq. (5.4)). Recombination on step edges has an activation barrier of 90 kJ mol^{-1}. The activation energies for CH_4 activation and CO formation are comparable.

The endothermicity of the reaction requires a high temperature of reaction. For a reaction between gas phase molecules and surface this implies a significant loss in entropy resulting in a substantial activation entropy. At a temperature of 800 K, $T \cdot S_{act}$ can be as much as 100 kJ mol^{-1}. Figure 5.9 summarizes the trends in relative free activation energies as a function of adsorbate atom adsorption energies.

For a dense surface a typical value is 180 kJ mol^{-1}; for a more open surface with B5 sites a value of 90 kJ mol^{-1} for the overall activation energies of the reaction of CH_{ads} and O_{ads} to give CO and H_{ads} is found. The activation entropies of surface reactions are close to zero. The surface atoms are immobile. The transition states require strong electronic contact between formation of the chemical bond and surface electrons. It is also found to be immobile. The transition state can be considered tight. The free energy of activation to form CO_{ads} from the corresponding surface adatoms can be considered independent of temperature. As for methane activation the free energy of H_2O decomposition is strongly temperature dependent.

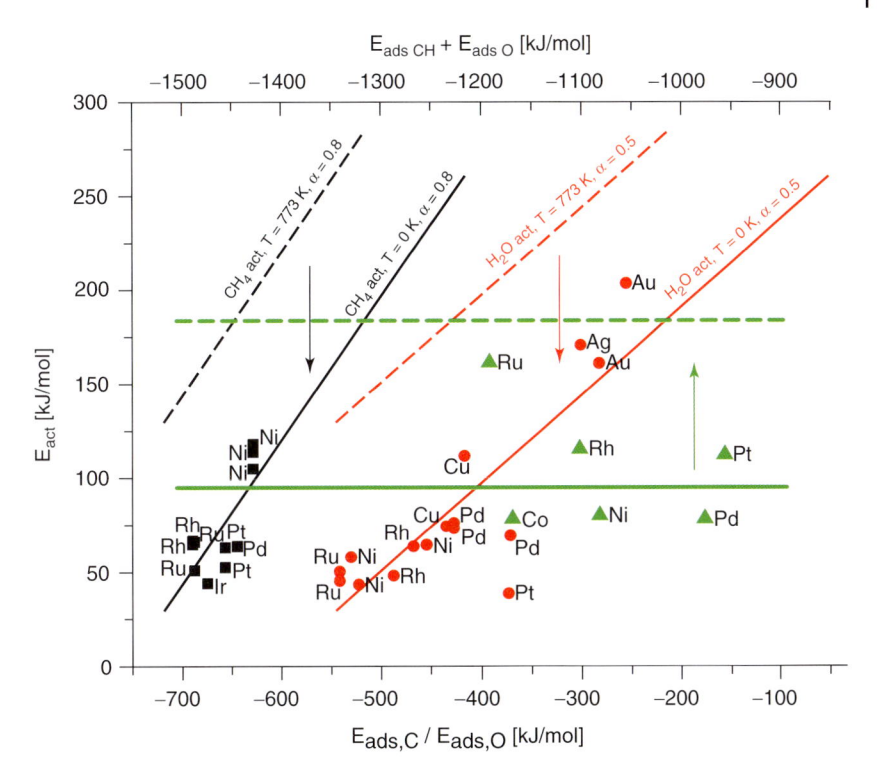

Figure 5.9 Comparison of Gibbs free energy activation energies for CH_4 activation (black), water activation (red), and C_{ads} and O_{ads} recombination (green) (schematic) [7].

We deduce from Figure 5.9 three different particle size dependence regimes. For high activation free energies of methane activation, corresponding to temperatures of reaction as used technologically, there is regime I. The free energy of methane activation is higher than the activation energies for CO formation from CH_{ads} and O_{ads} even on a dense surface. The rate of methane decomposition is rate controlling. There is a uniform increase in rate of the reaction with decrease in particle size.

In regime II (the next lower temperature regime) the activation free energy of methane activation is higher than that of the formation of CO on a step surface but lower than the corresponding activation energy on the dense surface. As long as particles are large enough to support B5 sites, methane activation is again rate controlling. For very small particles when B5 sites are no longer present the activation energy of the CO-forming surface reaction becomes higher that the activation free energy of methane decomposition, and hence the rate of reaction should decrease. Now a maximum in rate is expected as a function of decreasing particle size.

Finally there is regime III where the activation free energy of methane is lower than that of recombination of CH_{ads} and O_{ads} to give CO. Also in this case a maximum in the rate of reaction as a function of decreasing particle size is expected.

5.2.3
Hydrogenation, Dehydrogenation, and C–C Bond Cleavage

5.2.3.1 Mechanism of Hydrogenation and Dehydrogenation

Hydrogenation of olefin occurs in two steps:

$$C_2H_{4ads} + H_{ads} \longrightarrow C_2H_{5ads} \tag{5.9}$$

$$C_2H_{5ads} + H_{ads} \longrightarrow C_2H_{6ads} \tag{5.10}$$

This is the classical Horiuti-Polanyi mechanism.

Representative reaction diagrams for hydrogenation of ethylene on a Pd(111) surface are given in Figure 5.10:

Ethylene adsorbs strongly to the metal surface through its interaction with the π-bonds.

The activation energies for ethylene hydrogenation are a strong function of concentration. On the highly covered surface a very low activation barrier for the first hydrogen atom addition step is found, as illustrated in Figure 5.10.

At high adsorbate concentration before reaction, steric interactions between the ethylene molecules adsorbed to the surface have pushed the ethylene molecule from the strongly adsorbing di-σ state to the more weakly interacting π state.

Ethylene can in principle adsorb in these two different modes to the metal surface (Figure 5.11).

The lower activation energy of the π adsorbed complex is due to the weaker interaction of metal and ethylene carbon atom in the π adsorbed state. There is less energy needed to create a transition state with weakened M–C interaction than for reaction of di-σ adsorbed ethylene.

The two transition-state structures for hydrogen addition are shown in Figure 5.12.

In the respective transition states, H atom addition proceeds through recombination with a carbon atom over the top of a metal atom. Since σ CH bonds are formed, alkene hydrogenation is a structure-insensitive reaction.

5.2.3.2 Kinetics of Olefin Hydrogenation

The rate of hydrogenation as a function of substrate molecules is schematically shown in Figure 5.13.

Figure 5.13 shows a volcano type relationship between adsorption energy of unsaturated hydrocarbon and overall rate of hydrogenation. When the adsorption energy is too weak, there is a low surface coverage and rate is low. When the adsorption energy is too high the overall reaction rate is suppressed by reactant poisoning of the surface.

Differences in hydrogenation rate are seen not to be due to differences in intrinsic hydrogenation rates, but rather by differences in physical effects such as adsorption.

This has an interesting implication for the hydrogenation of large molecules as a function of metal particle size. When particle size becomes too small, the metal particle terraces cannot sustain adsorption of molecules larger than the terraces.

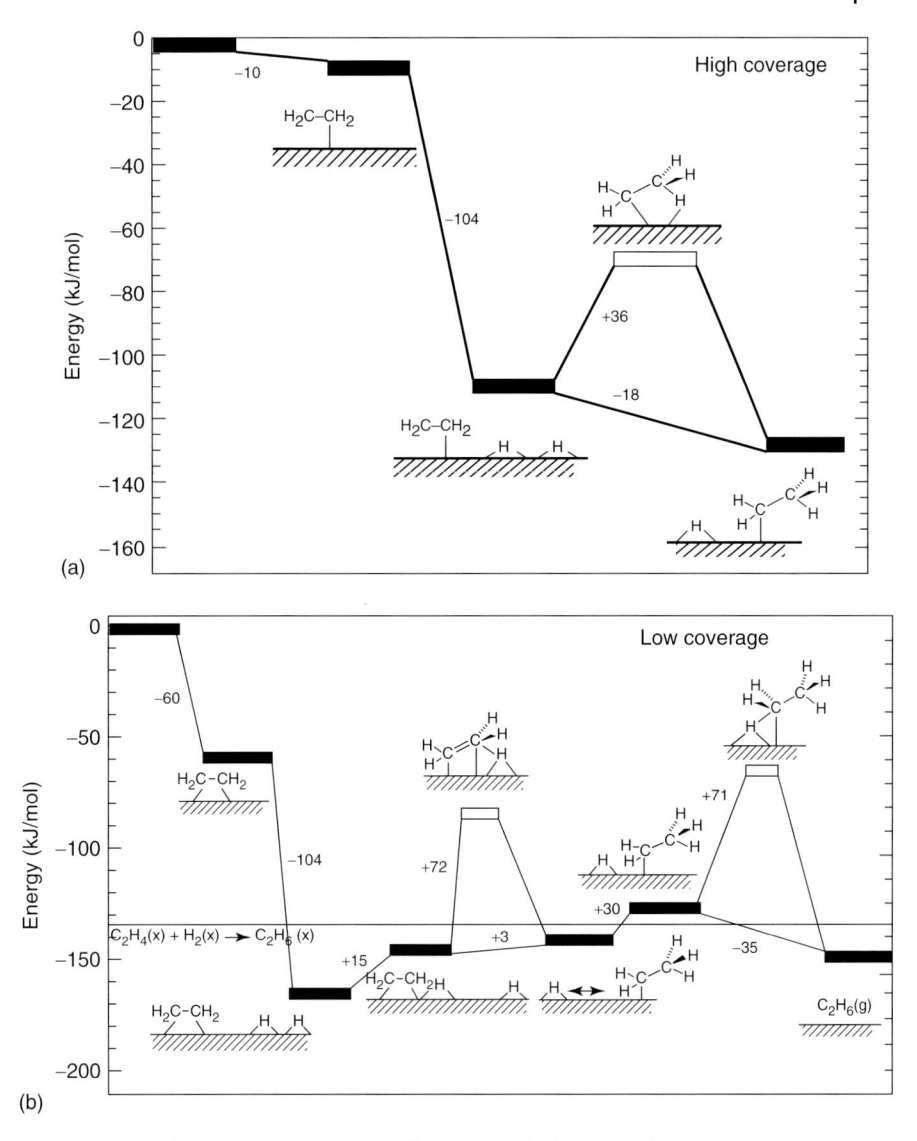

Figure 5.10 Ethylene hydrogenation over Pd(111) at (a) high and (b) low coverage [8].

The overall rate of reaction will decrease because the local surface concentration of adsorbed molecules is not sustained.

5.2.3.3 The Mechanism of Ethane Hydrogenolysis

The kinetic rate expression of ethane hydrogenolysis, in which a C–C bond is broken, is given by expression:

$$V = [L] k_2 [H_2] \frac{k_1 [C_2H_6][H_2]^{(x-6)/2}}{1 + k_1 [C_2H_6][H_2]^{(x-6)/2}} \quad (5.11)$$

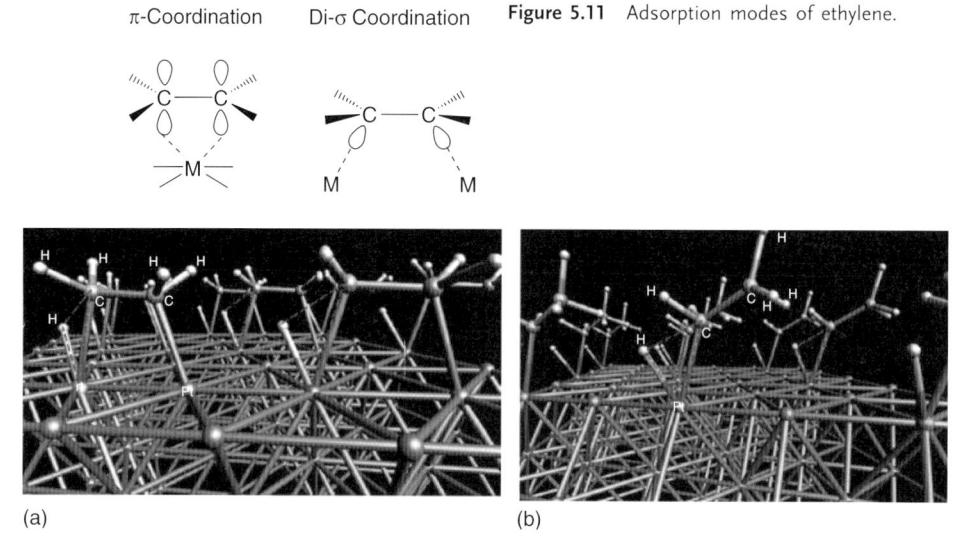

π-Coordination   Di-σ Coordination   **Figure 5.11** Adsorption modes of ethylene.

(a)   (b)

Figure 5.12 Three-center (M-C-H) transition states: (a) hydrogen atom addition to ethylene and (b) hydrogen addition to ethyl [9].

Figure 5.13 The rate of hydrogenation plotted as a function of the respective absorption energies (schematic) [10].

with [L] as surface area, k_2 rate constant for hydrogenation, k_1 equilibrium constant for absorption. Dependent on the value of x there can be a maximum in the overall rate.

Reaction is assumed to follow a two-step sequence: [11]

$$C_2H_6 \longrightarrow C_2H_x + \frac{(6-x)}{2}H_2 \tag{5.12}$$

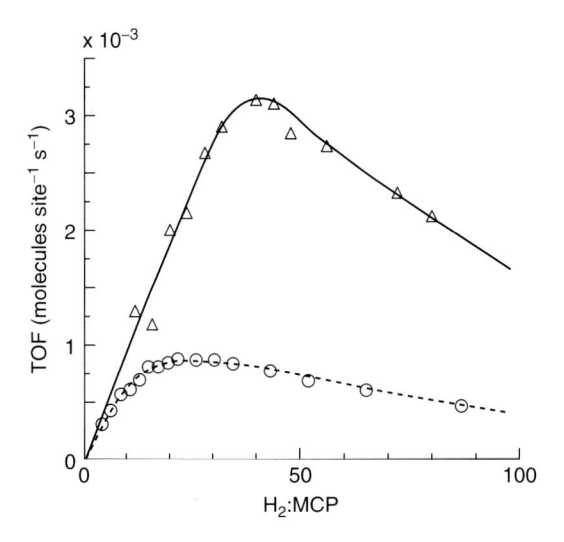

Figure 5.14 Ring-opening catalysis of methylcyclopentane to hexane catalyzed by Pt; [12] △: Pt supported on zeolite LTL; and ○: Pt supported on γ-Aluminia.

$$C_2H_x + \frac{1}{2}(8-x)H_2 \longrightarrow 2CH_4 \qquad (5.13)$$

The order of the reaction in H_2 pressure varies from positive for low hydrogen pressures to negative for high hydrogen pressures. Also for longer hydrocarbons this is generally found for the bond cleavage of σ type C–C bonds, as illustrated in Figure 5.14.

The change in order of the reaction with H_2 pressure has led to a variety of explanations.

The most important is the proposal that for alkane adsorption, next to any site for the adsorption of alkane itself vacant sites have to be available for hydrogen adsorption. It has also been postulated that surface hydrogen directly reacts with hydrocarbon hydrogen atoms to give H_2 in an associative reaction. These postulates imply the need for a surface ensemble of metal atoms of considerable size.

The interaction of an alkane with the metal surface is through van der Waals interactions. Therefore molecules such as methane and ethane have weak adsorption energies. On a metal surface, hydrogenolysis (the cleavage of C–C bonds in hydrocarbons) is initiated by activation of the CH bond, notwithstanding the stronger CH bond versus C–C bond (look up values). The CH_3 or CH_2 groups prevent sterically direct contact of metal surface with the σ C–C bond.

We have seen for methane activation that σ CH bond cleavage requires a substantial activation energy (of the order of $70\,kJ\,mol^{-1}$) and strongly depends on coordinative unsaturation of surface metal atoms.

Not surprisingly, hydrogenolysis of hydrocarbons is generally considered to be structure sensitive. Its rate is found to increase uniformly with decrease in particle

$$H_3C-CH_3 \longrightarrow H_2C-CH_{3\,(ad)} + H_{(ad)}$$

$$H_2C-CH_{3\,(ad)} \longrightarrow H_2C-CH_{2\,(ad)} + H_{(ad)}$$

$$H_2C-CH_{2\,(ad)} \longrightarrow 2\,CH_{2\,(ad)}$$

$$CH_{2\,(ad)} + 2H_{(ad)} \longrightarrow CH_4$$

Figure 5.15 α–β bond cleavage involving activation of a bond between two neighboring atoms [13].

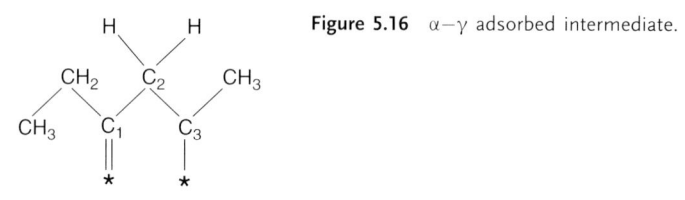

Figure 5.16 α–γ adsorbed intermediate.

size. The edge atoms provide the enhanced reactivity for CH bond activation. The terraces will provide the necessary ensemble sizes.

The C–C bond cleavage reaction step in ethane is very different from that in a longer hydrocarbon. In ethane, C–C bond cleavage takes place via an α–β adsorbed intermediate (see Figure 5.15). Dehydrogenation of the ethyl intermediate competes with C–C bond cleavage.

Along the step edge of Pt the activation energy to cleave the C–C bond in $CH_2CH_{3\,ads} \approx 100$ kJ mol^{-1} [14].

These activation energies are slightly higher than the corresponding values typical for CH activation.

In longer alkanes C–C bond cleavage occurs via α–γ adsorbed intermediates. Such an intermediate is sketched in Figure 5.16.

Activation of this reaction requires an ensemble of surface metal atoms or can occur at a single atom, as on the edge of a particle, through formation of intermediate metallocyclobutane intermediates analogous to the metathesis intermediates found in homogeneous complexes. This is illustrated in Figure 5.17.

Figure 5.17 Single-center route for hydrogenolysis of isopentane [15].

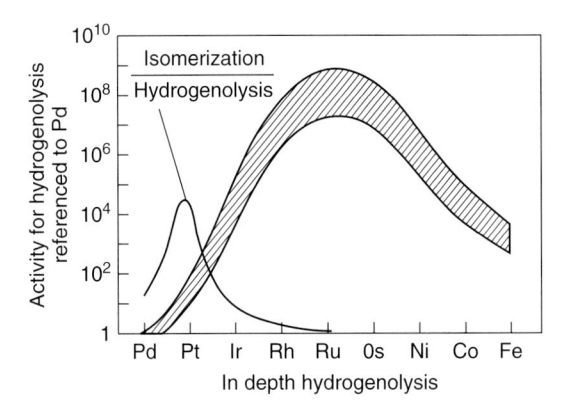

Figure 5.18 Schematic evolution of the activity in hydrogenolysis and the ratio isomerization/hydrogenolysis [15].

Figure 5.18 illustrates the differences in reactivity of the transition metals with respect to the activation of larger hydrocarbons. Pt is the metal with maximum isomerization and aromatization selectivity. The other metals show mainly deep hydrogenolysis with dominant production of methane. Remember that Pt is the metal with one of the lowest barriers of CH activation.

The rate of hydrogenolysis is strongly enhanced for small particles. Studies with alloys also indicate strong structure sensitivity of this reaction. A classical experiment in metal catalysis is the comparison of the reactivity of metals such as Pt or Ni with alloys of group VIII and IB metals.

Results for a Ni-Cu alloy are shown in Figure 5.19.

A dramatically different behavior is observed for cyclohexane dehydrogenation and hexane hydrogenolysis. The IB metal atoms, such as Cu or Au, do not activate C–H or C–C bonds. Alloying decreases the ensemble size of reactive group VIII metal atoms.

Cyclohexane dehydrogenation is essentially independent of surface alloy composition, whereas hexane hydrogenolysis rapidly decreases with increasing Cu content, the rate decreasing by several orders of magnitude. These results have been interpreted as proof of the action of a surface ensemble effect in the case of the hydrogenolysis reaction. Dehydrogenation is seen to be independent of ensemble size and hence should slightly decrease upon alloying. The logarithmic decrease in rate of the hydrogenolysis reactions with increasing Cu concentration is due to the dilution of the surface Ni ensembles by non-reactive Cu atoms. Decomposition of hexane may require ensembles of 10–12 reactive metal atoms. Dilution with Cu will steeply decrease their concentration.

Alternatively, hydrogenolysis occurs on single centers as illustrated in Figure 5.17. Because of the lower melting point of the IB metals they tend to be located at low surface atom coordination sites, as for instance step edge atoms. These are the preferred

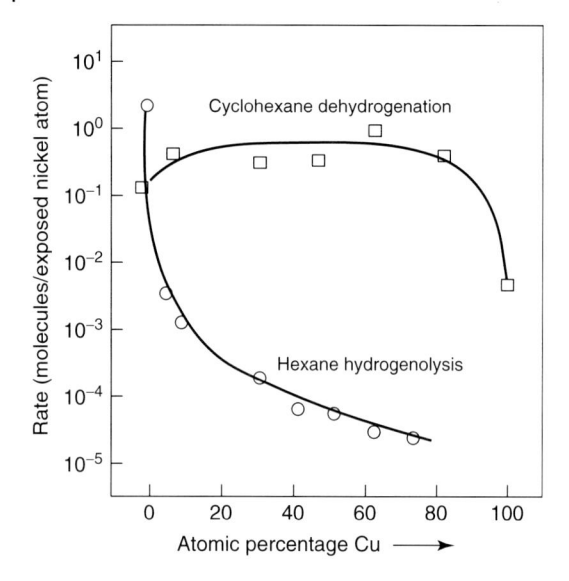

Figure 5.19 Cyclohexane dehydrogenation and hexane cracking conversion as a function of the Cu/Ni ratio of the catalyst [16].

sites for single-center hydrogenolysis. According to the single-center mechanism their occupation by non-reactive metal atoms would also suppress hydrogenolysis.

Carbon deposition also tends to suppress the hydrogenolysis reaction in favor of dehydrogenation reactions. This has been especially found for systems with very small particles. Such particles are used in bifunctional catalytic reactions, to be discussed in sections dealing with solid acid catalysis. Since both alkane isomerization and dehydrogenation are little effected by a decrease in the reaction center ensemble size, the ensemble effect can be considered mainly responsible for the decrease in hydrogenolysis activity.

5.3
Solid Acids and Bases

5.3.1
Introduction

The surface of a solid that consists of cations and anions contains Lewis acid or basic centers. Upon hydroxylation such centers become converted into Brønsted basic or Brønsted acidic centers.

Acid–base type reactivity can be considered to be due to local imbalances of cationic and anionic interactions, whereas an approximate view helps to distinguish Brønsted basic from Brønsted acidic protons on hydroxylated surfaces and also to understand differences in acid-base properties as a function of composition.

We will introduce the valancy concept of Pauling, who used this to deduce the relative stability of minerals as a function of structure and composition. We will use this to analyze acidity and basicity of oxidic surfaces.

The Pauling valency or the strength of an electrostatic bond with a cation or anion is defined as:

$$S^{\pm} = \left| \frac{\text{formal ion charge}}{\text{number of nearest neighbor ions}} \right| \quad (5.14)$$

Pauling valencies or charge excesses to be defined later are useful indicators of the relative interaction energies of coordinatively unsaturated surface ions.

These indicators can be used as a measure of acidity or basicity. We will use this to deduce rules on the effect of surface topology on surface reactivity.

We will use as an example the reactivity of the MgO(100) surface. In the bulk, each cation or anion has six neighbors. For the Mg^{2+} cation the Pauling bond strength equals:

$$S^+ = \frac{2}{6} = \frac{1}{3} \quad (5.15)$$

And for the O^{2-} ion, the Pauling bond strength becomes:

$$S^- = \left| \frac{-2}{6} \right| = \frac{1}{3} \quad (5.16)$$

Let us now consider the MgO(100) surface, in which we have an equal number of cations and anions and each has lost one neighbor, see Figure 5.20.

On the MgO surface the anion and cation have a charge excess. The respective charge excesses can be calculated from expressions:

$$e^+ = Q^+ - \sum_i S_i^- \quad (5.17)$$

$$e^- = Q^- + \sum_i S_i^+ \quad (5.18)$$

On the MgO(100) surface, the excess ionic charges present on the cation and anion are respectively:

$$e^{\pm} = \pm \frac{1}{3} \quad (5.19)$$

A positive excess ionic charge implies Lewis acidity whereas a negative excess ionic charge implies Lewis acidity.

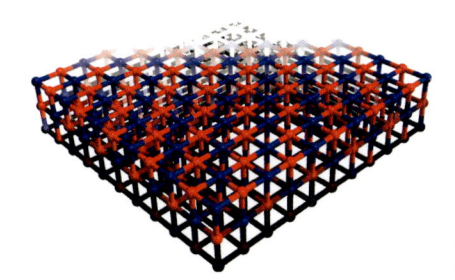

Figure 5.20 The structure of the MgO(100) surface.

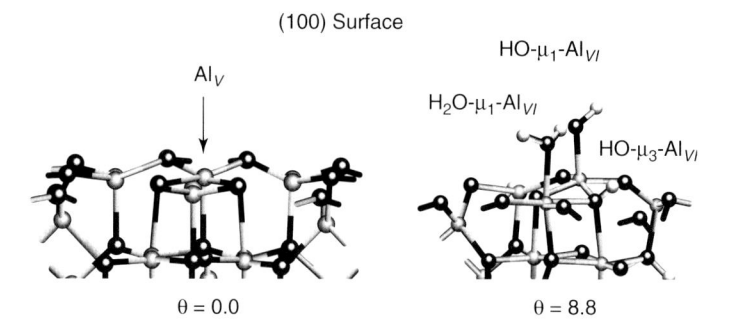

Figure 5.21 Relaxed configurations of γ-Al$_2$O$_3$ (100) surface for different hydroxyl coverages (in OH nm^{-2}). The most relevant surface sites are quoted. Al$_n$ stands for aluminum atoms surrounded by n oxygen atoms, and HO–μ$_m$ for OH groups linked to m aluminum atoms. Oxygen atoms are black, aluminum atoms are shown in gray whereas hydrogen atoms are white [17].

Brønsted acidity or basicity develops when a molecule or water dissociates. It turns out that the MgO surface is resistant with respect to dissociative adsorption of H$_2$O.

Alumina and titania surfaces are more reactive.

Figure 5.21 illustrates the three different situations that may arise when water adsorbs. On the γ-Al$_2$O$_3$ (100) surface an Al cation is five-coordinated; it is six-coordinated in the bulk. Its charge excess can be calculated and is found to be:

$$e_{Al}^+ = +\frac{1}{2} \tag{5.20}$$

The surface oxygen atoms each have three Al neighbors, one of which is below the surface, leading to an excess ion charge of:

$$e_O^- = -\frac{1}{2} \tag{5.21}$$

As is observed from Figure 5.21 three situations arise. H$_2$O can non-dissociatively adsorb and will attach to the surface through the Al cation. The oxygen atom of adsorbed H$_2$O now obtains a excess charge +1/2. This implies a charge imbalance on the water oxygen atom. The acidity of the water OH groups increases upon contact with the Lewis acidic Al cation.

Under the influence of the Al cation and basic oxygen anion the H$_2$O molecule can dissociate.

As is shown in Figure 5.21 a μ$_1$ and μ$_3$ proton appear on the surface. The charge excess of the O atom involved in the μ$_1$ coordinated hydroxyl is −1/2. The initially Lewis acidic site is converted into a Brønsted basic center.

The charge excess of the oxygen atom on the μ$_3$ site is +1/2. The originally basic Lewis basic acid site is converted into a Brønsted acidic site.

A hydroxylated oxide surface will contain Brønsted acidic and Lewis acidic sites. Their strength depends on the coordination of the oxygen anion with neighboring cations.

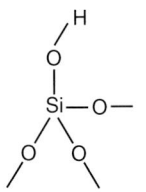

Figure 5.22 Surface silanol (schematic).

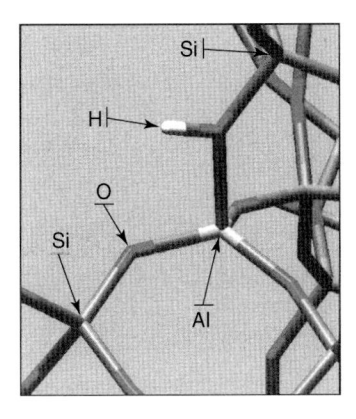

Figure 5.23 The structure SiOHAl of the zeolitic acidic site.

The higher the cation coordination the more acidic is the corresponding proton.

The ionic description of oxides is an approximation; one therefore has to be careful to apply this analysis to compare the reactivity of different oxidic surfaces.

An interesting example is the acidity of the silanol adsorbed to a silica surface (see Figure 5.22).

The oxygen atom that is part of the surface hydroxyl has charge excess zero, whereas the silanol proton behaves as chemically slightly acidic. This acidity would be different if Si is replaced by Ti. The weaker acidity that results requires a more refined model than Pauling's charge excesses. The Ti hydroxyl is more weakly acidic because of the larger Ti radius.

When one calculates the charge excess of the zeolitic protonic site (see Figure 5.23), one deduces for the oxygen anion a charge excess of $+3/4$. The corresponding large charge imbalance implies an acidity larger than we estimated for the alumina surface site.

5.3.2
Proton Activation by Zeolites

Zeolites are crystalline aluminosilicates that contain microchannels with dimensions that allow for accessibility of organic molecules. The framework consists of tetrahedrally coordinated cations. The tetrahedra are connected through the vertices (see Figure 5.24):

Replacement of four-valent Si by three-valent Al causes a charge imbalance on the framework, which gives it formally a negative charge. This negative framework charge is compensated by cations located in the zeolite micropores.

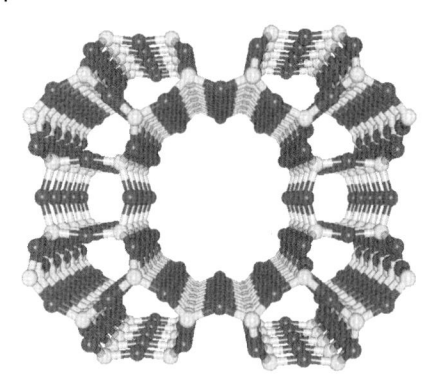

Figure 5.24 Zeolite framework; the mordenite one-dimensional channel.

Acidity is generated when such a cation is ammonium. Upon evolution of ammonia, protons are left attached to the framework that give it solid acid properties.

The local structure of such a site we discussed earlier and is given in Figure 5.23.

NMR measurements as well as quantum chemical calculations indicate that the undisturbed zeolitic proton has a very low charge. A positive charge develops only upon contact of the proton with a (Lewis base) reagent. Acidity develops because of the ease of polarizing the zeolitic hydroxyl. In contrast to the ionic picture in which the solid is composed of cations and anions, the zeolite framework instead appears to be covalent. In proton transfer reactions full charge separation develops as part of the transition states of protonated intermediates. This is stabilized by the developing negative charge on the zeolite framework.

In Figure 5.25 the computed reaction energy diagrams are compared for protonation of propylene reacting with a proton in ferrierite.

Whereas the reaction energies, the difference in energies between π adsorbed propylene and the two respective alcoxy intermediates, do not show a difference, there is a large difference in activation energies. Protonation on the secondary carbon atom is substantially easier than that on the primary C atom.

The shapes of the transition states are planar, as in the free carbenium ions.

The difference in activation energies relates to the difference in stabilities of the respective primary and secondary carbenium ions.

Differences in the energies of proton transfer relate to differences in the energies of the corresponding carbenium ions. In the case of solid acids they are part of transition states. This also holds for reaction intermediates formed by proton donations. In the transition state they are present as carbenion ions that convert according to well-known acid-type reactivity.

There is an essential difference between proton activation at the interphase of solid and gas and a liquid phase superacid. Proton transfer occurs within the reactive complex of zeolitic proton and reagent. The overall charge of the complex is neutral.

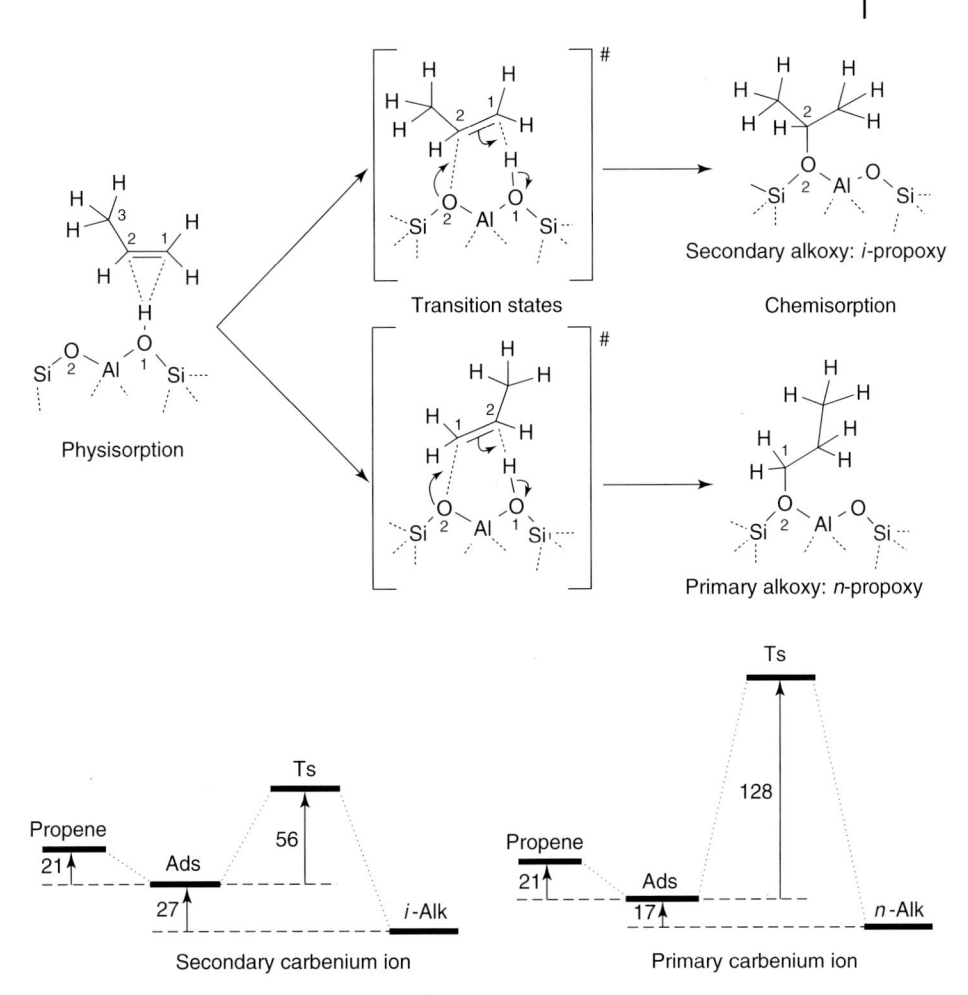

Figure 5.25 Comparison of primary and secondary proton attachment. To propylene by a proton of the ferrierite zeolite microchannel [18].

In the case of reactions in the liquid phase, there is equilibrium between the neutral monomers and dissolved ionic intermediates. For instance, in neat sulfuric acid there is the equilibrium:

$$2H_2SO_4 \rightleftharpoons H_2SO_4^{+1} + HSO_4^{-1} \tag{5.22}$$

Charge separation is stabilized by the high dielectric constant of H_2SO_4 medium.

The unstable cation intermediate easily releases H^{+1} to a reagent molecule. Protonated organic molecules become stable ground state intermediates. See Figure 5.26,b, which shows computed protonation reaction energy diagrams for a protonated and non-protonated sulfuric acid molecule:

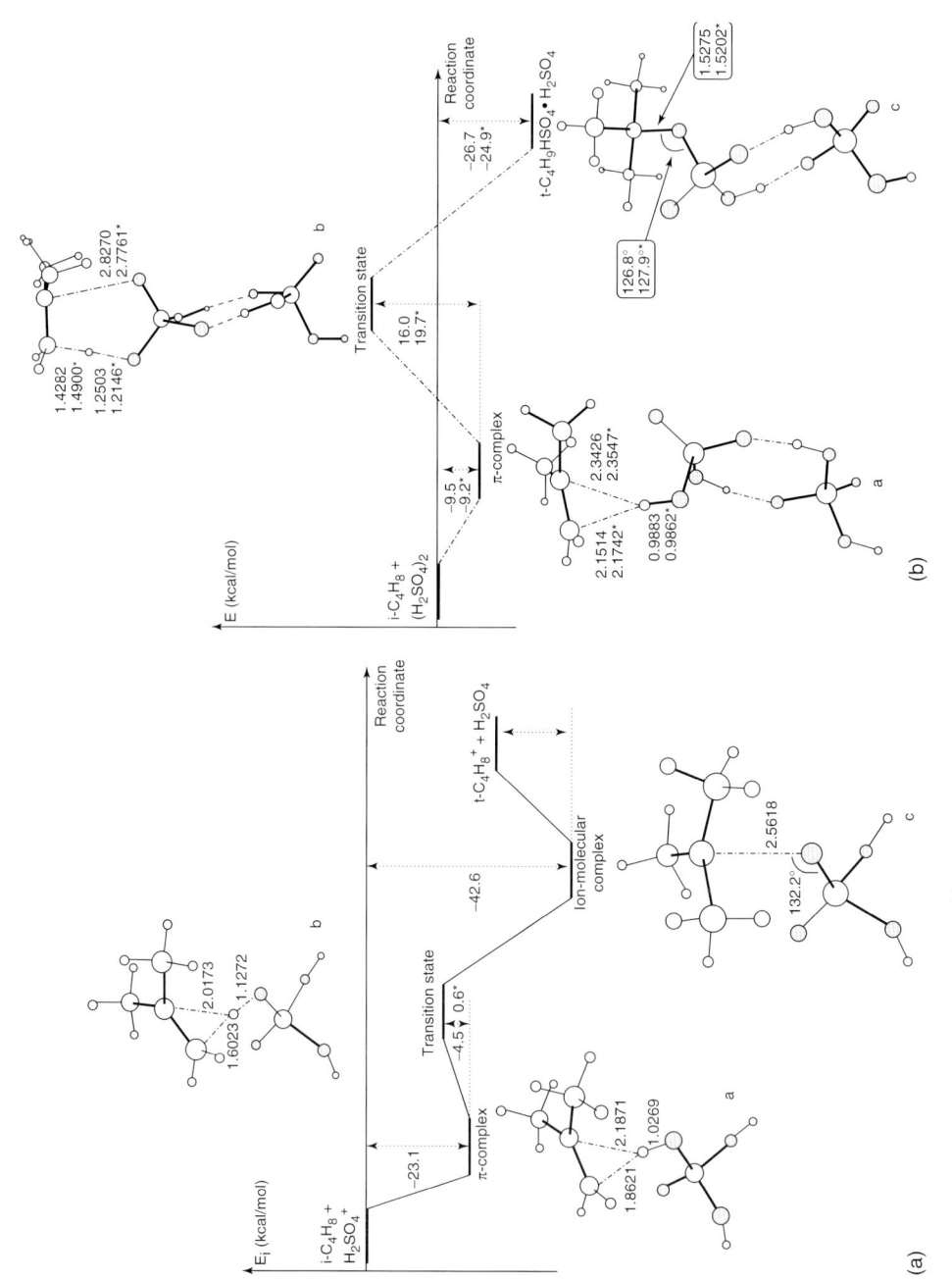

Figure 5.26 Proton transfer of (a) $H_3SO_4^{+1}$ and (b) H_2SO_4 with isobutene is compared [19].

Protonation by H_2SO_4 follows energetics similar to that of the zeolite. The barrier is reduced for the protonated case. Now the carbenium is a stable intermediate.

As long as molecules or intermediates fit well in the microcavities of the zeolite, stereochemical effects due to spatial constraints will be absent.

However, the weak van der Waals interactions that control the energies of adsorption and diffusion due to the interaction with the zeolite wall can considerably contribute to the overall adsorption energies of reaction intermediates due to their additivity. Since this will depend on structure as well as size and shape of molecules, overall kinetics may be strongly zeolite structure dependent.

Stereochemical effects may play a role when significant differences in rate of diffusion of reagent or product occur due to differences in the match of shape and size of molecules with that of the zeolite cavities and channels. This may also affect the adsorption energies of intermediates in ground and transition states. This will lead to transition state selectivity.

5.3.3
General Mechanistic Considerations

5.3.3.1 Direct Alkane Activation

When an olefin is activated by a proton in the transition state a carbenium ion is formed. The positive charge is located at a CH_2^{+1} type planar part of the molecule with sp^2 hybridization of the organic molecule chemical bonds.

When alkanes are activated the protonated intermediate transition states are the analogs of non-classical organic carbonium ions with five-coordinated C-atoms.

This is illustrated in Figure 5.27. The protonated alkane can decompose either to form a carbenium ion with the same C–C skeleton as the original alkane and H_2, or to a smaller alkane and carbenium ion. Olefins are formed upon deprotonation of the carbenium ions. Equal amounts of alkane and alkene molecules are then formed. The different reaction channels proceed through different transition states, as indicated in Figure 5.28.

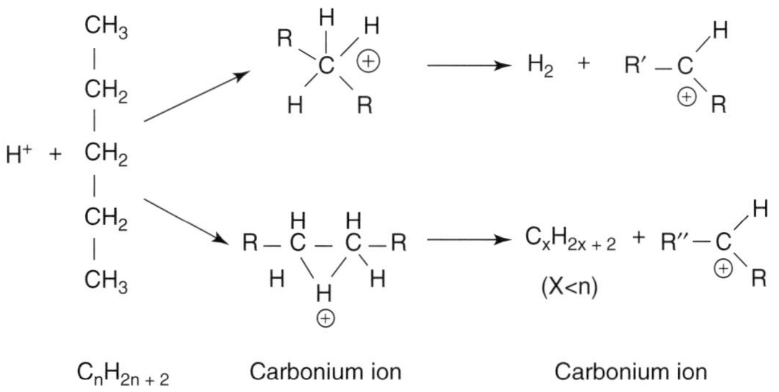

Figure 5.27 Proton activation of pentane. Initial carbonium ion formation and subsequent carbenium ion formation.

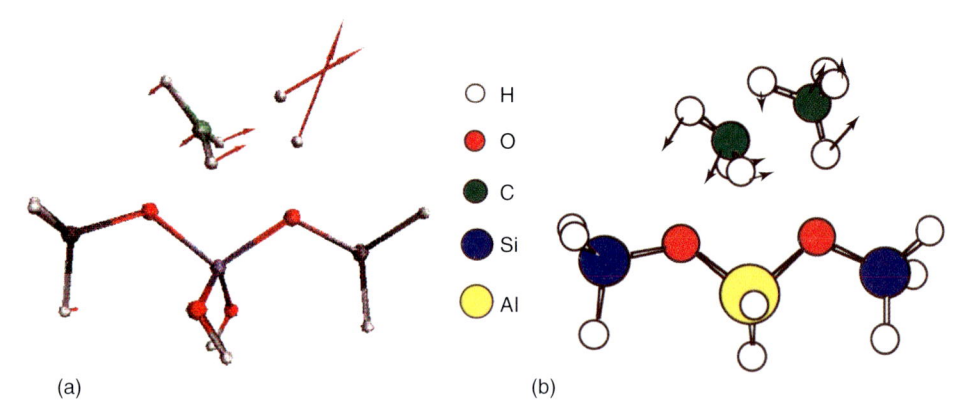

Figure 5.28 (a) Transition state of dehydrogenation of methane with formation of H_2 and CH_3^{+1} and (b) the quantum chemically calculated transition state for ethane cracking that produces methane [20].

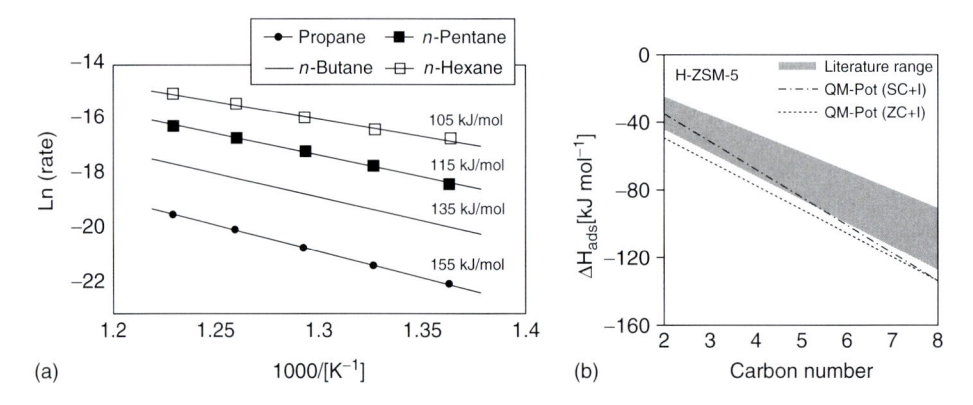

Figure 5.29 (a) Arrhenius plot for protolytic cracking of the n-alkanes catalyzed by H-ZSM-5 [21]. (b) Adsorption enthalpy range as obtained for experimental literature data and computed quantum chemical results obtained by different methods [22].

Dehydrogenation follows attack on the C–H bond. C–C bond cleavage follows attack on the corresponding C–C bond. Alkane cracking and hydrogenolysis has been extensively investigated. It is found that the activation energy of this reaction decreases with hydrocarbon chain length. As discussed in detail in Chapter 4, this relates to the increasing adsorption energy of hydrocarbons with chain length.

Here we will discuss its consequences for the reactivity of proton-activated alkene cracking as a function of chain length. The measured apparent activation energy decreases with increasing chain length. (see Figure 5.29)

This is not because of intrinsic higher reactivity of the alkane molecules with their chain length, but relates to the linear increase in the alkane absorption energy with increasing number of carbon atoms. This is illustrated by Eq. (5.23). At the high temperature of the alkane cracking reaction the rate is linear in pressure P:

$$r = k_{crack} \cdot K_{eq}^{ads}(C_n) \cdot P(C_n) \tag{5.23}$$

In this expression k_{crack} is the elementary rate constant of proton-activated cracking, and K_{eq}^{ads} is the equilibrium constant for adsorption. The apparent activation energy E_{app}^{act} becomes:

$$E_{app}^{act} = E_{crack}^{act} + E_{ads}(C_n) \tag{5.24}$$

E_{crack}^{act} is the activation energy of the elementary proton-activated cracking reaction and E_{ads} is the corresponding adsorption energy of the corresponding hydrocarbons as observed in Figure 5.29b.

5.3.3.2 Hydride Transfer

The reaction path, with relatively high activation energy for carbenium ion formation by direct proton addition, becomes bypassed by a secondary reaction with a substantially lower activation energy, the hydride transfer path illustrated in Figure 5.30:

Due to hydride transfer propagation, the cracking reaction occurs by reaction of intermediate carbenium ion with incoming alkane. Transfer of H- from a linear alkane will generate from this alkane molecule a secondary carbenium ion and will also generate a reaction of H- through a primary carbenium ion. This process prevents proton back-donation to the lattice and hence renewed formation of energy-costing carbenium ion transition state.

Whereas initiation by alkane activation of this overall reaction requires an initial high activation energy, propagation of the reaction by hydrotransfer proceeds with relatively low activation energies.

5.3.3.3 Isomerization Catalysis

Carbenium transformation reactions have been extensively studied in superacids. In solid acids similar transformations occur, but now as parts of transition states.

As illustrated in Figure 5.31a the pentene isomerization reaction occurs in three steps. When pentene adsorbs on a proton of a zeolite an alkoxy species is formed. Starting with the alkoxy species as the initial state the energies of the successive transformations are shown in this figure. Initially an activation energy of 49 kJ mol^{-1} has to be overcome to generate an activated free alkene carbenium ion. This alkene carbenium ion is transformed into a protonated dimethyl cyclopentane intermediate with a barrier of 98 kJ mol^{-1}. Its structure is shown in Figureb 5.31. Finally one of the C–C bonds in the cyclopropyl ring is opened to give the adsorbed isopentylalkoxy species. In the case of intermediate cyclopentene, this ring opening generates a secondary carbenium ion. When the same mechanism is studied for the isomerization of n-butene to iso-butene, in the corresponding ring-opening step a primary carbenium ion is generated. The resulting increased activation energy tends to inhibit this reaction. In case there is no steric constant for isomerization of

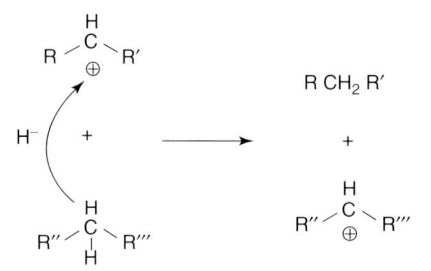

Figure 5.30 Chain propagation processes mediated via hydride transfer.

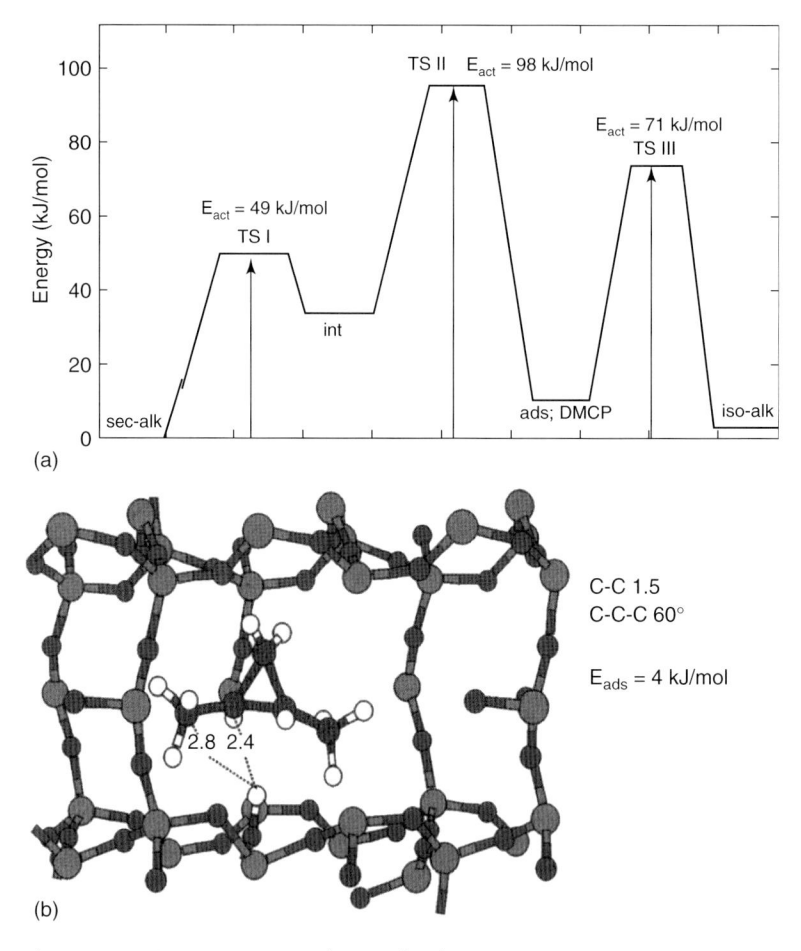

Figure 5.31 (a) Reaction energy diagram for the isomerization of pentyl alkoxy intermediate to isopentyl alkoxy intermediate in ferrierite and (b) adsorbed dimethyl cyclopentene in H-ZSM22(ferrierite) [23].

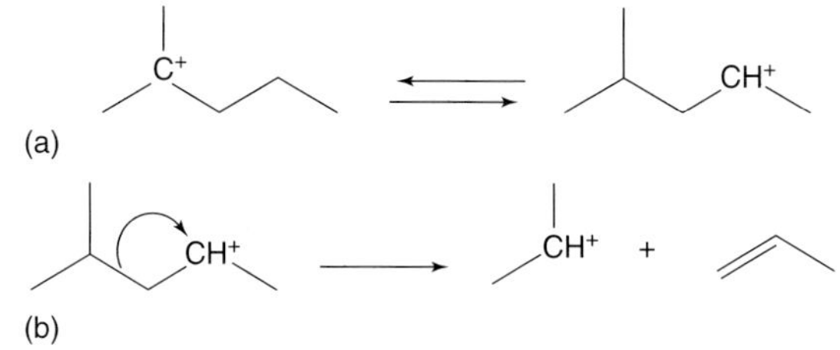

Figure 5.32 (a) isomerization of *tert*-isohexylcarbenium ion to secondary isohexylcarbenium ion and (b) β-C–C cleavage reaction.

butene, a more complex path is preferred: two molecules will dimerize; the octane molecule isomerizes and subsequently cracks to the now isomerized monomers.

A longer molecule such as pentene can undergo consecutive cracking reactions through beta C–C cleavage.

A C–C bond is broken through a secondary carbenium ion transition state. The reaction proceeds in two steps:

- initial isomerization of tertiary carbenium alkoxy species to secondary alkoxy species. When the analog reaction would be studied for a pentyl alkoxy species a primary alkoxyspecies would be generated which has significantly higher barrier, see Figure 5.32.
- An olefin is split off; a transition between secondary carbenium and tertiary carbon atom takes place. This is the dominating mechanism for cracking of olefins in zeolites and explains the formation of olefins such as ethylene or propylene and the absence of methane.

5.4
Reducible Oxides

5.4.1
Comparison of the Relative Stabilities of Some Oxides

Whether oxidation by a reducible oxide will be selective or will lead to total combustion depends on the strength of the M–O bond in the oxide. This is illustrated in Figure 5.33, where the conversion, as well as selectivity, of butane to butadiene is shown as a function of the heat of formation of the corresponding oxides.

Maximum selectivity is found of an optimum M–O bond strength for Fe_2O_3. Figure 5.33b and c show a linear increase in reaction temperature with increasing M–O bond strength.

Table 5.4 compares the relative stability of reducible oxides. The relative stability of the oxides increases as their position in the row of the periodic system becomes

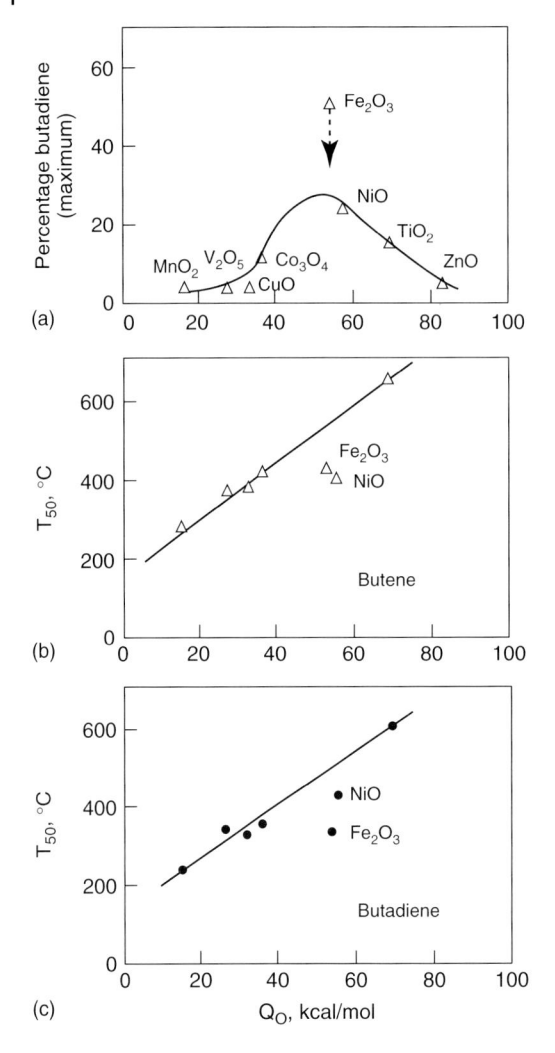

Figure 5.33 Relative activities of oxide catalysis for oxidation of butane and reaction intermediates in a pulse reactor. (a) selectivity of butadiene formation; (b) butene conversion temperature; and (c) butadiene conversion temperature [24].

lower. It decreases on moving toward a very stable metal such as Mn with its five d-electrons.

It is also interesting to have a look at the transition energies for release of oxygen. The reduction energy of V_2O_5 to give V_2O_4 and MoO_3 to give MoO_2 are -29 and -50 kcal mol^{-1} respectively, indicating substantially higher reactivity of V. On the other hand MoO_3 is 15 kcal mol^{-1} more reactive than WO_3. The reasons for non-reducibility of Nb and Ta are obvious. Cr and Mn are highly reactive, possibly too reactive, leading only to total oxidation.

Table 5.4 Heats of formation of important catalytically active reducible oxides.

V	Energy (kcal mol^{-1})	Cr	Energy (kcal mol^{-1})	Mn	Energy (kcal mol^{-1})	Nb	Energy (kcal mol^{-1})
V_2O_5	−373	–	–	–	–	Nb_2O_5	−463.2
V_2O_4	−344	–	–	–	–	Nb_2O_4	−367.8
V_2O_3	−290	Cr_2O_3	−209	–	–	–	–
V_2O_2	−200	–	–	MnO_2	−124	–	–

Mo	Energy (kcal mol^{-1})	Ta	Energy (kcal mol^{-1})	W	Energy (kcal mol^{-1})
–	–	Ta_2O_5	−550	–	–
MoO_3	−180	–	–	WO_3	−201
MoO_2	−130	–	–	WO_2	−136

Table 5.5 Calculated V = O bond dissociation energies for some complexes [25].

	V = O bond dissociation energy (kJ mol^{-1})
$V_{20}O_{62}H_{24}$ (a1)	113 (286)[a]
$OV(OCH_3)_3$ (b1)	312
$[OV(OCH_3)_3]_2$ (b2)	210

[a] Before structure relaxation.
a1, b1, and b2 refer respectively to the structures presented in Figure 5.34.

V_2O_5 is the preferred catalyst for selective benzene or butene oxidation to maleic acid. BiO_2/MoO_3 is the preferred catalyst for reaction with the more reactive olefins. This system oxidizes propylene with high selectivity to acrolein. The V_2O_5/MO_3 mixed oxide system is the preferred catalyst for acrylic acid production.

5.4.2
Structure Sensitivity

There is a large difference in reducibility of bulk oxides and monomers. This is illustrated in Table 5.5.

There is a large decrease in the VO bond energy calculated for the bulk-like system (Figure 5.34a) and that of the clusters (Figure 5.34, b1, b2).

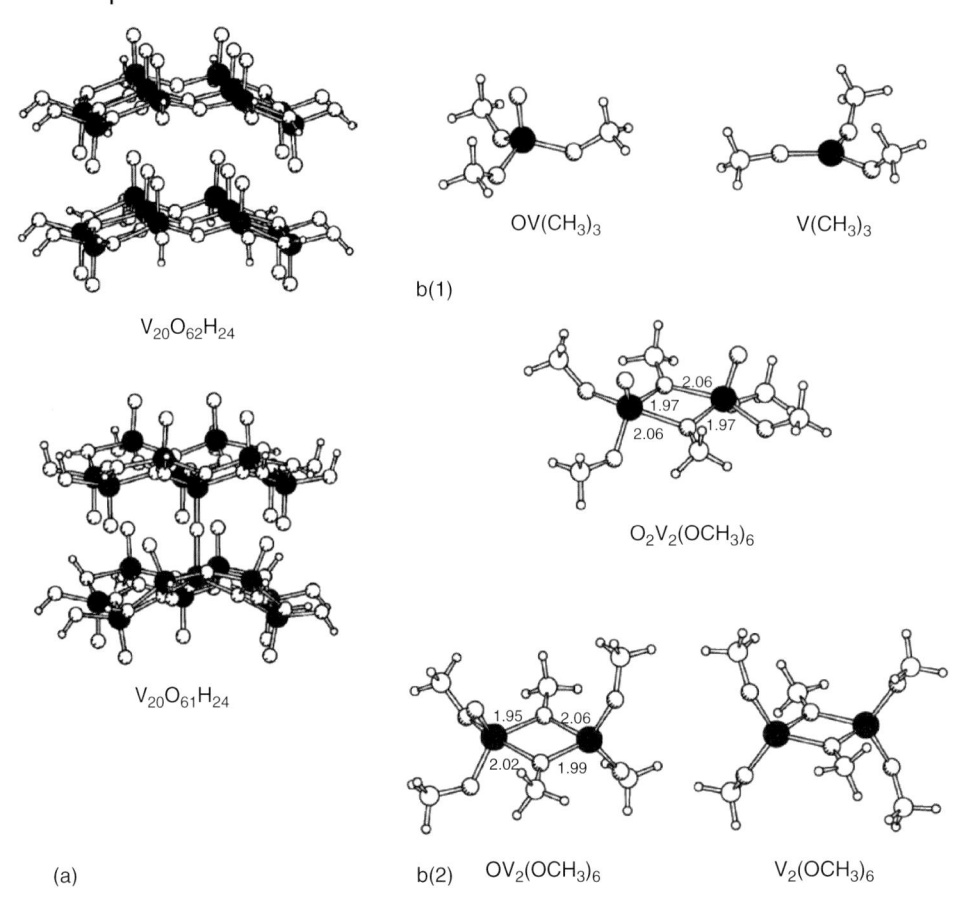

Figure 5.34 (a) Two-layer $V_{20}O_{62}H_{24}$ cluster model for the (001) crystal surface of V_2O_5 (left) and relaxed structure after formation of a vanadyl O defect. (b) The $O=V(OCH_3)_3$ molecule (b1) and its dimer (b2). Relaxed structures after removal of one and two vanadyl O atoms are also shown.

This is due to the stabilizing interaction between VO in one layer and reduce V in the other layer, as illustrated by scheme (eq. 5.25).

$$O=V(d^0) \quad O=V(d^0) \longrightarrow \left[V^{III}(d^2)\,O-V^{V}(d^0)\right] + O \longrightarrow V^{IV}(d^1) + O \quad (5.25)$$

The increased VO bond energy of the monomers implies a substantially lower reactivity. The monomeric state can be present when vanadium oxide is distributed over an inert oxidic support.

The vanadyl oxygen atoms are most reactive. Activation of butane is a radical reaction. Important intermediates are OH and alcoxy species, both formed by reaction with the vanadyl oxyatom. This explains the significantly lower reactivity of isolated vanadyl species, stabilized by silica or different support.

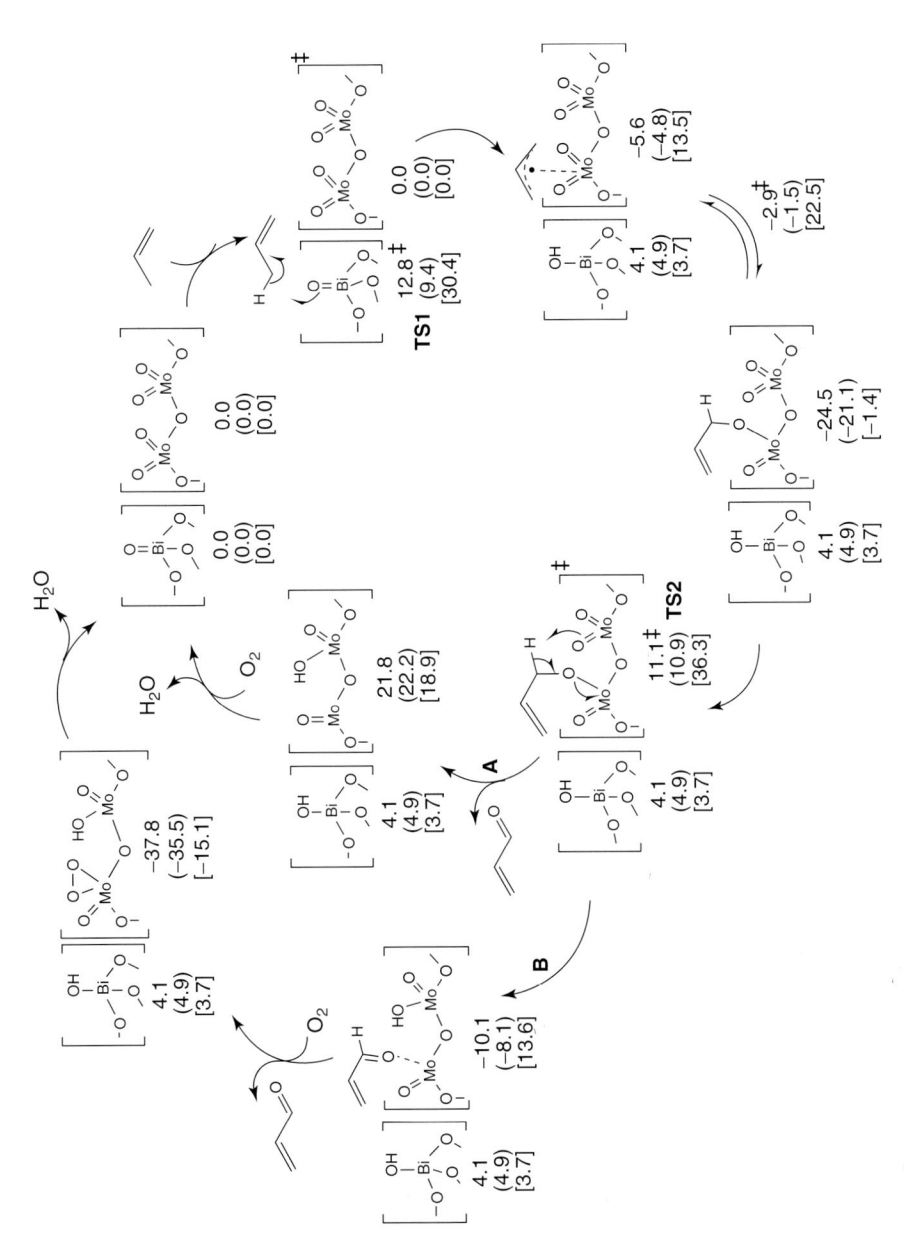

Figure 5.35 Proposed mechanism and energetics for propene oxidation over bismuth molbydate catalyst. The top energy is the ΔE from QM, the middle is $\Delta H_{0K} = \Delta E + \Delta ZPE$, and the bottom is ΔG_{593K}. All reported values are in kilocalories per mole. Results from molecular dynamics simulations [26].

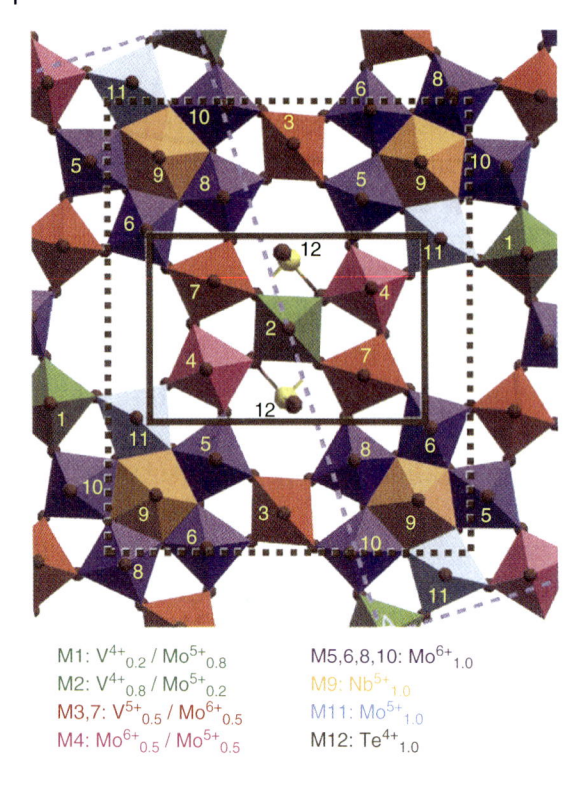

M1: $V^{4+}_{0.2}$ / $Mo^{5+}_{0.8}$
M2: $V^{4+}_{0.8}$ / $Mo^{5+}_{0.2}$
M3,7: $V^{5+}_{0.5}$ / $Mo^{6+}_{0.5}$
M4: $Mo^{6+}_{0.5}$ / $Mo^{5+}_{0.5}$
M5,6,8,10: $Mo^{6+}_{1.0}$
M9: $Nb^{5+}_{1.0}$
M11: $Mo^{5+}_{1.0}$
M12: $Te^{4+}_{1.0}$

Figure 5.36 Catalytically active center of Mo7.5V1.5NbTe29 in [001] projection and schematic depiction of the active site [27].

The increased reactivity of vanadyl oxygen that is part of a large particle relates to its increased reducibility.

5.4.3
Mechanism of Important Oxidation Reactions

5.4.3.1 The Selective Oxidation of Propylene

The mechanism of selective oxidation of propylene to acrolein by a bismuth-molydate catalyst is summarized in Figure 5.35.

Essential steps are the formation of allylic intermediate by reaction of a propylene C–H bond with the reactive Lewis basic oxygen atom from Bi.

The allyl radical adsorbs onto Mo and reacts in subsequent steps to Mo = O to give MoOH and a Mo oxygen vacancy.

The acidic BiOH proton than reacts with MoOH to give H_2O. Since one additional oxygen atom will be used to be inserted into propylene, two oxygen atom vacancies will be created on Mo. These are removed by reoxidation with O_2.

Isotope labeling studies indicate that the primary O_2 atoms consumed in this reaction are from the bulk oxide. The rate of O_2 atom diffusion is high. The sites

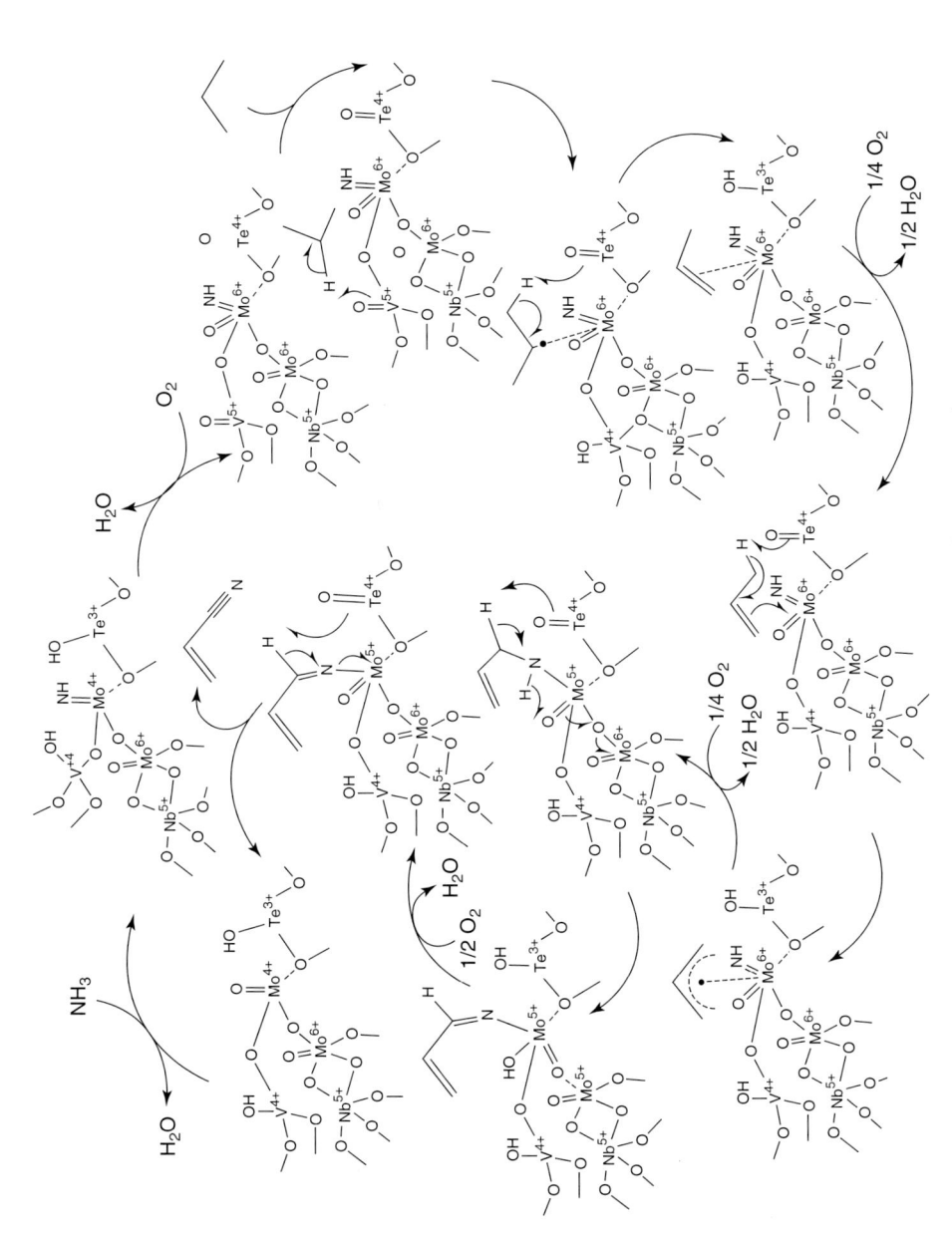

Figure 5.37 Proposed propane ammoxidation mechanism over Mo–V–Nb–Te–Ox catalysts [27].

where O_2 dissociates does not have to be the same sites where acrolein formation takes place. This is the so-called Mars–van Krevelen mechanism.

5.4.3.2 Propane Oxidation

Whereas ammoxidation of propylene to acrylonitrile requires different promotors than Bi, the catalyst remains Mo based. Now also NH_3 has to be activated. For activation of propane the more reactive V is also needed. A complex system, the MoVNbTeOx catalyst, has been found to have optimum activity and selectivity. The proposed structure of the reactive center of Mo and V is shown in Figure 5.36.

The Nb 5+ centers are thought to isolate the Mo and V centers. Vanadium is needed to activate propane, Mo for NH insertion.

Site isolation appears to be a prerequisite for selective oxidation. The presence of excess oxygen would otherwise tend to lead to total combustion.

This is very similar to the proposed action of phosphate in the vanadium phosphate catalyst used for the butane to maleic acid reaction (See Chapter 2). It is assumed also here that phosphate isolates vanadium oxide clusters on the catalyst surface so as to reduce the local concentration of reactive oxygen.

The proposed mechanism of this multicomponent reaction and the reaction cycle in which several reaction centers cooperate are shown in Figure 5.37.

As in the case of propylene oxidation we note the multifunctionality of the catalyst. A tandem of consecutive reactions occurs placed at different reaction centers. Remarkably, they all occur at the same temperature.

References

1. Strongin, D.R. et al. (1987) The importance of C7 sites and surface roughness in the ammonia synthesis reaction over iron. *J. Catal.*, **103** (1), 213–215.
2. Dahl, S. et al. (2001) Electronic factors in catalysis: the volcano curve and the effect of promotion in catalytic ammonia synthesis. *Appl. Catal. A: Gen.*, **222** (1–2), 19–29.
3. Honkala, K. et al. (2005) Ammonia synthesis from first-principles calculations. *Science*, **307** (5709), 555–558.
4. van Santen, R.A., Neurock, M., and Shetty, S.G. (2009) Reactivity theory of transition-metal surfaces: a Brønsted–Evans–Polanyi linear activation energy free-energy analysis. *Chem. Rev.*, **110** (4), 2005–2048.
5. Jones, G. et al. (2008) First principles calculations and experimental insight into methane steam reforming over transition metal catalysts. *J. Catal.*, **259** (1), 147–160.
6. van Santen, R.A. and Neurock, M. (2007) *Molecular Heterogeneous Catalysis*, Wiley-VCH Verlag GmbH & Co. KGaA, pp. I–XIII.
7. van Grootel, P.W., Hensen, E.J.M., and van Santen, R.A. (2010) The CO formation reaction pathway in steam methane reforming by rhodium. *Langmuir*, **26** (21), 16339–16348.
8. Neurock, M. and van Santen, R.A. (2000) A first principles analysis of C-H bond formation in ethylene hydrogenation. *J. Phys. Chem. B*, **104** (47), 11127–11145.
9. Neurock, M. (2003) Perspectives on the first principles elucidation and the design of active sites. *J. Catal.*, **216** (1–2), 73–88.
10. Hub, S., Hilaire, L., and Touroude, R. (1988) Hydrogenation of but-1-yne and

but-1-ene on palladium catalysts: particle size effect. *Appl. Catal.*, **36**, 307–322.
11. Boudart, M. and Djega-Mariadasson, G. (1984) Kinetics of heterogeneous catalytic reactions, in *Physical Chemistry: Science and Engineering* (eds J.M. Prausnitz and L. Brewer), Princeton University Press, p. 173–177.
12. Vaarkamp, M. (1993) *The Structure and Catalytic Properties of Supported Platinum Catalysts*, Eindhoven University of Technology.
13. Masel, R.I. (1996) *Principles of Adsorption and Reaction on Solid Surface*, John Wiley & Sons, Inc., New York.
14. Watwe, R.M. et al. (2000) Theoretical studies of stability and reactivity of CHx species on Ni(111). *J. Catal.*, **189** (1), 16–30.
15. Maire, G.L.C. and Garin, F.G. (1984) Catalysis, in *Science and Technology* (eds J.R. Anderson and M. Boudart), Springer, Heidelberg, p. 214.
16. Sinfelt, J.H., Carter, J.L., and Yates, D.J.C. (1972) Catalytic hydrogenolysis and dehydrogenation over copper-nickel alloys. *J. Catal.*, **24** (2), 283–296.
17. Digne, M. et al. (2002) Hydroxyl groups on [gamma]-alumina surfaces: a DFT study. *J. Catal.*, **211** (1), 1–5.
18. Rozanska, X. et al. (2002) A periodic structure density functional theory study of propylene chemisorption in acidic chabazite: effect of zeolite structure relaxation. *J. Phys. Chem. B*, **106** (12), 3248–3254.
19. Zhurko, D.A., Frash, M.V., and Kazansky, V.B. (1998) A quantum-chemical study of the formation mechanism and nature of tert-butyl carbenium ions in 100% sulfuric acid. *Catal. Lett.*, **55** (1), 7–14.
20. Blaszkowski, S.R., Nascimento, M.A.C., and van Santen, R.A. (1996) Activation of C-H and C-C bonds by an acidic zeolite: a density functional study. *J. Phys. Chem.*, **100** (9), 3463–3472.
21. Narbeshuber, T.F., Vinek, H., and Lercher, J.A. (1995) Monomolecular conversion of light alkanes over H-ZSM-5. *J. Catal.*, **157** (2), 388–395.
22. De Moor, B.A. et al. (2011) Adsorption of C2-C8 n-alkanes in zeolites. *J. Phys. Chem. C*, **115** (4), 1204–1219.
23. Demuth, T. et al. (2003) Catalytic isomerization of 2-pentene in H-ZSM-22-a DFT investigation. *J. Catal.*, **214** (1), 68–77.
24. Gates, B.C., Katzer, J.R., and Schuit, G.C.A. (1979) in *Chemistry of Catalytic Processes* (ed. J.H. Sinfelt), McGraw-Hill, p. 346.
25. Sauer, J. and Dobler, J. (2004) Structure and reactivity of V_2O_5: bulk solid, nano-sized clusters, species supported on silica and alumina, cluster cations and anions. *Dalton Trans.*, (19), 3116–3121.
26. Goddard, W. et al. (2008) Structures, mechanisms, and kinetics of selective ammoxidation and oxidation of propane over multi-metal oxide catalysts. *Top. Catal.*, **50** (1), 2–18.
27. Grasselli, R.K. et al. (2003) Multi-functionality of active centers in (Amm)oxidation catalysts: from Bi–Mo–O to Mo–V–Nb–(Te, Sb)–O. *Top. Catal.*, **23** (1), 5–22.

6
Homogeneous Catalysis

Matthias Beller, Serafino Gladiali, and Detlef Heller

6.1
General Features

Basically the definition of homogeneous catalysis applies to those transformations where catalyst, starting materials, and reaction products all reside in the same phase, almost always a liquid phase. In heterogeneous catalysis, reagents and products are in a different phase from that of the catalyst, which is almost always in a solid phase. On comparing homogenous with heterogeneous catalysis, some common characteristics and also some significant differences become apparent:

- Homogeneous catalysts (and often pre-catalysts) are in nature molecularly well-defined compounds in which the structure of the active center can be determined even *in situ* with reasonable reliability by normal spectroscopic techniques. This fact combined with the possibility to synthesize potential intermediates within a given catalytic cycle allows an easier understanding of the reaction mechanism. Given this mechanistic understanding, the development of catalysts with improved performances can in general be addressed on a rational basis.
- Homogeneous catalysis often takes place under much milder reaction conditions than those associated with heterogeneous catalysis. Typically, reactions take place at temperatures between ambient and 120 °C – conditions where even sensitive fine chemicals can be handled. Also, reaction pressures are comparatively low, and, although some high pressure processes are known, reactions generally proceed at atmospheric pressure unless gases or low-boiling materials are involved. Hence, the conversion of complicated organic building blocks with different functional groups is better performed with homogeneous catalysts because chemical reactions are more selective and catalysts are more tolerant at low rather than at high temperature. On the other hand, heterogeneous catalysts tend to be more stable than their homogeneous counterparts and retain their activity for a longer time even at high temperatures. Hence, reactions such as CH-activation processes, which require high temperatures (>250 °C) in order to take place at a significant rate, are mostly based on heterogeneous catalysts.

Catalysis: From Principles to Applications, First Edition. Edited by Matthias Beller, Albert Renken, and Rutger van Santen.
© 2012 Wiley-VCH Verlag GmbH & Co. KGaA. Published 2012 by Wiley-VCH Verlag GmbH & Co. KGaA.

- In the case of metal-catalyzed reactions, homogeneous metal-based catalysts are more efficient than heterogeneous ones in the utilization of the metal because in principle 100% of the metal sites are available for catalysis in homogeneous conditions, whereas only a limited number of active metal sites are available in a heterogeneous catalyst.
- In general, reactions driven by homogeneous catalysts are carried out in the liquid phase and in a batch-wise mode. Thus, in research laboratories homogeneous catalysis is associated with batch reactions in organic solvents. It is, however, also possible to operate in a continuous flow mode, as in fact takes place in industry in some large scale processes, for example, carbonylation reactions and oxidations, among others. Due to the presence of a solvent, control of the temperature is quite precise and the heat of the reaction can easily be dissipated.
- Isolation of the products and recovery and recycle of the catalyst are more complicated in homogeneous than in heterogeneous catalysis. Separation of the catalyst via filtration is straightforward in heterogeneous catalysis, whereas in homogeneous catalysis separation of starting materials and products by distillation, chromatography, crystallization, or modern techniques of multiphasic catalysis have to be applied.
- In the case of reactions mediated by transition metal complexes in homogeneous phase, the performance of the catalyst is basically dictated by the ligands coordinated to the metal center. Following advances in organometallic chemistry and organic ligand synthesis, a plethora of ligands (P-, N-, and recently C-ligands) is nowadays theoretically available (10 000–100 000). It is well known that activity, productivity, and selectivity of a homogeneous catalyst change dramatically on changing the ligands bonded to the metal, and in fact 'ligand tailoring' constitutes an extremely powerful tool to control all kinds of selectivity in a given catalytic reaction and to influence stability and activity of a homogeneous catalyst. A selection of the most popular ligands for metal-centered homogeneous catalysis is shown in Scheme 6.1. Apart from the well-known aryl and alkyl

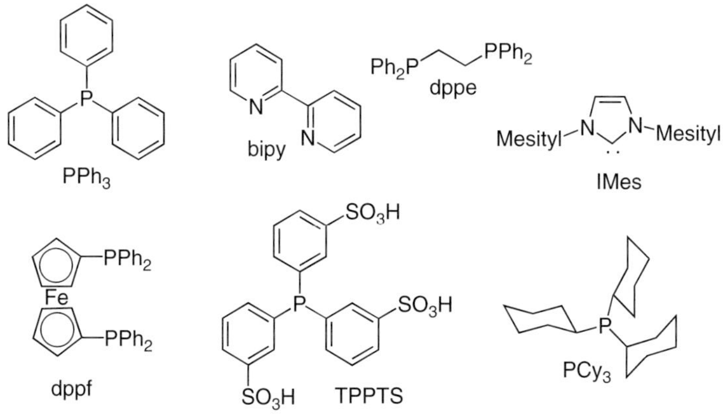

Scheme 6.1 Typical ligands for homogeneous catalysis.

phosphines and amines, carbenes [1] have also recently become increasingly important. In addition, mixed ligand systems with two different donors capable of chelate coordination to the metal as well as hemilabile ligands find increasing interest.
- The role of the ligand in homogeneous metal catalysts is essential in asymmetric catalysis [2] where the stereoselectivity of a given reaction is induced by a soluble metal complex modified with a chiral ligand, most frequently a phosphine or an amine. Although in general not recognized, there is an interesting analogy between the ligands for homogeneous catalysis and the supports and modifiers used in heterogeneous catalysis.

Due to the specific characteristics of homogeneous and heterogeneous catalysis, these two technologies are most often applied in different chemical communities. Although noteworthy efforts are being made to bridge this gap, a lot of work still has to be done in order to make full use of the synergism of the two techniques.

To date, homogeneous catalysts dominates in organic synthesis and in the fine chemical industry, while heterogeneous catalysis is by far the most used technology in large-scale petrochemical processes. There are also, however, some high-volume processes operated in the chemical industry which make use of homogeneous catalysis. Nearly all industrial carbonylation reactions, including the important production of aliphatic aldehydes by hydroformylation of olefins and of acetic acid by carbonylation of methanol, are based on homogeneous catalysts. However, despite enormous efforts, the leaching of volatile carbonyl clusters and complexes from the corresponding heterogeneous catalysts prevents the use of this technology in those processes where carbon monoxide is one of the reactants. Other large-scale processes operated with homogeneous catalysts, including oxidation reactions such as the Wacker oxidation, are performed with low-molecular-weight transition metal salts.

Clearly, homogeneous catalysis cannot be discussed comprehensively within the framework of this single chapter. Therefore this chapter deals only with a selection of important achievements of the past and some new developments in homogeneous catalysis which may be important in the future. Special focus is directed toward sub-areas of homogeneous catalysis such as carbonylations and transition metal-catalyzed coupling reactions for fine chemicals. Other homogeneous catalytic reactions such as oxidations, cyanations, olefin dimerization and oligomerization, hydrogenations, and asymmetric catalysis as well as 'simple' acid and base catalysis also constitute important areas in academic and industrial chemistry.

Four types of catalytic reactions occurring in the homogeneous phase can be distinguished:

1) (Brønsted)-acid/base catalysis (activation of substrate by means of protonation or deprotonation)
2) Nucleophilic/electrophilic catalysis (activation of substrate by means of Lewis base via electron pair donor complexes or Lewis acid via electron pair acceptor complexes)

3) Transition metal-centered homogeneous catalysis (activation of substrate via coordinative interaction, redox, and insertion reactions) [3]
4) Enzyme catalysis (activation of substrate by multifunctional means) [4].

6.1.1
Acid and Base Catalysis

Acid/base catalysis is the oldest form of homogeneous catalysis. According to Brønsted–Lowry theory, acids are proton donors and bases are proton acceptors [5].

Activation of the substrate occurs through protonation/deprotonation of suitable – mostly organic – molecules. Although proton affinities of organic compounds are in fact very high, proton transfer reactions proceed at remarkable rates. This is due to the nature of hydrogen bonds. Pioneering work in this field has been carried out by M. Eigen (Nobel Prize in Chemistry 1968).

Basically the course of a bimolecular, acid-catalyzed reaction involves a rapid protonation, followed by a slow activated reaction (Scheme 6.2). The neutralization proceeds very quickly due to the low activation barrier and is thus usually controlled by diffusion.

$$HA + S \underset{k_{-1}}{\overset{k_1}{\rightleftarrows}} HS^+ + A^- \quad \text{Pre-equilibrium with the Brønsted acid (HA) as catalyst}$$

$$HS^+ + R \xrightarrow{k_2} P + H^+ \quad \text{Irreversible reaction of the protonated substrate with R}$$

$$H^+ + A^- \rightleftarrows HA \quad \text{Very fast neutralization}$$

Scheme 6.2 Bimolecular, acid catalyzed reaction. (S, R = starting materials, P = product, and HA = Brønsted acid as catalyst.)

If product formation is rate determining (fast pre-equilibrium) we speak of *specific* acid catalysis since the rate only depends on the concentration of protons $[H_3O^+]$, (Eq. (6.1)).

$$r_{cat} = k_2[HS][R] \text{ with } [HS] = \frac{k_1}{k_{-1}}[S]\frac{[H_3O^+]}{K_a}{}^{1)} \text{ gives}$$

$$r_{cat} = \frac{k_2 k_1}{k_{-1} K_a}[S][H_3O^+][R] = k'[S][H_3O^+][R] \tag{6.1}$$

If on the other hand the protonation of substrate is rate determining we speak of *general* acid catalysis because each Brønsted acid in the system increases the catalytic activity, Eq. (6.2).

$$r_{cat} = k_1[HA][S] \tag{6.2}$$

1) Steady state for the pre-equilibrium: $k_1 \cdot [HA] \cdot [S] = k_{-1} \cdot [HS^+] \cdot [A^-]$, [HA] can be determined from acidity constant K_a ($K_a = [H_3O^+] \cdot [A^-]/[HA]$).

These two types of acid catalysis can be distinguished by isotope effects: in the case of special acid catalysis $k_{H_2O}/k_{D_2O} < 1$ (because D^+ is more acidic than H^+) and for general acid catalysis $k_{H_2O}/k_{D_2O} > 1$ (because H^+ is transferred faster than D^+).

Typical examples of special acid catalysis are ester hydrolysis,[2] alcoholysis of esters, inversion of sucrose, and hydrolysis of acetals. The mutarotation of glucose and the dehydration of acetaldehyde hydrate, on the other hand, are examples of general acid catalysis.

The counterpart of acid catalysis is base catalysis, Scheme 6.3.

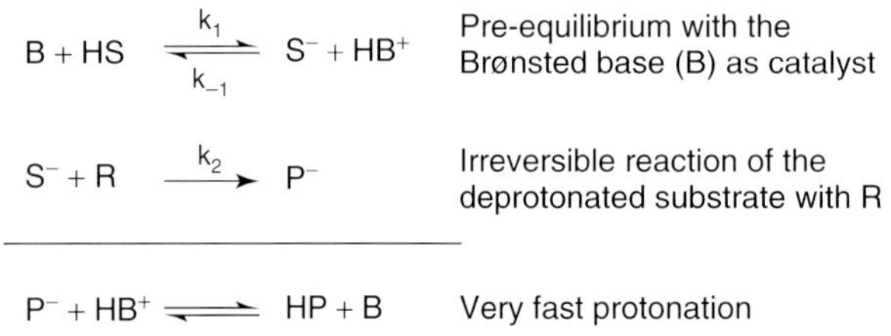

$$B + HS \underset{k_{-1}}{\overset{k_1}{\rightleftarrows}} S^- + HB^+ \quad \text{Pre-equilibrium with the Brønsted base (B) as catalyst}$$

$$S^- + R \overset{k_2}{\longrightarrow} P^- \quad \text{Irreversible reaction of the deprotonated substrate with R}$$

$$P^- + HB^+ \rightleftarrows HP + B \quad \text{Very fast protonation}$$

Scheme 6.3 Bimolecular, base catalyzed reaction. (HS, R = starting material, HP = product, and B = Brønsted base as catalyst.)

In analogy to acid catalysis we distinguish between *specific* (Eq. (6.3a)) and *general* base catalysis (Eq. (6.3b)), depending on the rate-determining step, or rather if only [OH$^-$] or also other bases such as anions of a buffer increase the rate of reaction by deprotonation of the substrate.

(a) $r_{cat} = k_2 [S^-][R] \; r_{cat} = k'[HS][OH^-][R]$[3] or

(b) $r_{cat} = k_1 [HS][B]$ (6.3)

A typical example of specific base catalysis – the second step in Scheme 6.3 is rate determining – is the formation of diacetone alcohol from acetone through aldol addition. In contrast, the analogous aldol reaction with acetaldehyde proceeds as general base catalysis: the deprotonation of acetaldehyde is rate determining.

The activation of substrate by protonation or deprotonation on one hand is not limited to aqueous systems; on the other hand it is strongly connected to acid or base strength, respectively. At this point the Hammett acidity function [6], superacids [7] such as Olah's 'magic acid' (FSO_3H/SbF_5 – Nobel prize in Chemistry 1994), and superbases such as Schlosser's base (BuLi/KO*t*Bu) are only mentioned here.

A number of reactions, for example, the hydration of acetaldehyde or the mutarotation of glucose, can be catalyzed by either acids or bases. Of special

---

2) In-depth mechanistic details (Sac2 mechanism: bimolecular acyl cleavage; or Sal1 mechanism: monomolecular alkyl cleavage) can be found in Ref. [3].

3) Steady state for the pre-equilibrium: $k_1 \cdot [HS] \cdot [B] = k_{-1} \cdot [S^{-1}] \cdot [HB^+], [B]$ can be determined from basicity constant K_b ($K_b = [HB^+ \cdot OH^-]/[B]$).

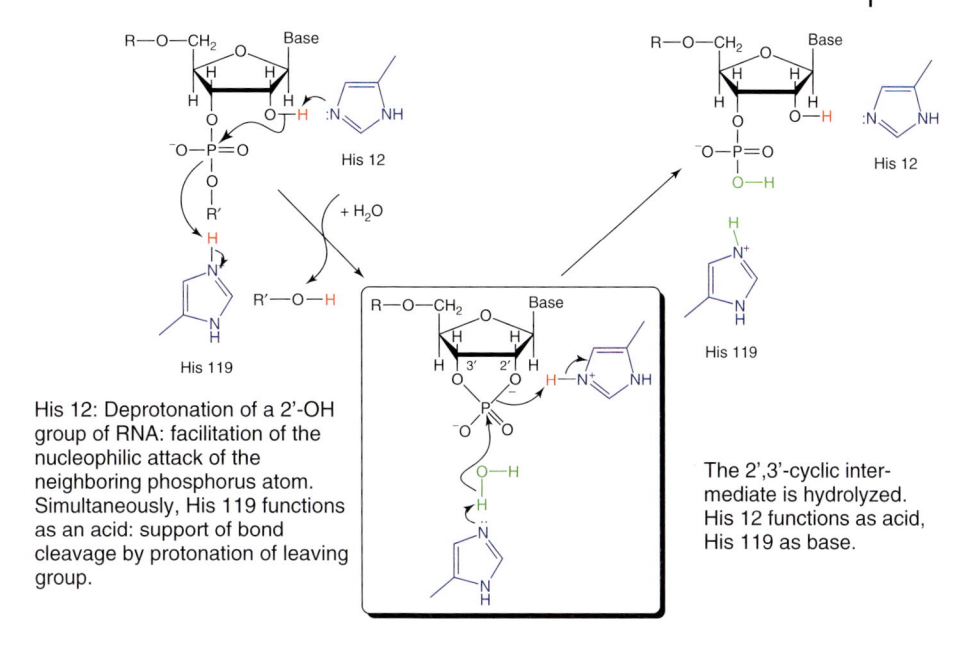

Figure 6.1 Acid/base catalyzed RNA hydrolysis catalyzed by enzyme RNase-A.

His 12: Deprotonation of a 2'-OH group of RNA: facilitation of the nucleophilic attack of the neighboring phosphorus atom. Simultaneously, His 119 functions as an acid: support of bond cleavage by protonation of leaving group.

The 2',3'-cyclic intermediate is hydrolyzed. His 12 functions as acid, His 119 as base.

interest and virtue is so-called 'bifunctional catalysis': the catalyst contains both an acidic and a basic function.

The special effectiveness can be shown impressively using the example of the mutarotation of tetramethyl-α-D-glucose in benzene [8]. Applying 2-hydroxypyridine as a bifunctional catalyst increases the rate of reaction by a factor of 7000 compared to simply phenol as acid or pyridine as base catalyst. Clearly, an important requirement for bifunctional catalyses – besides matching acidity or basicity of the catalyst – is a suitable steric arrangement between substrate and catalyst.

An example of an enzyme-catalyzed general acid/base catalysis is the hydrolysis of RNA into nucleotide building blocks by enzyme RNase-A (127 amino acid moieties, 13.7 kDa), which is present in the pancreas. The pH dependence of the enzyme catalysis as well as structural analyses point to two essential histidine side chains, His 12 and His 119, which work together as acid/base catalysts, Figure 6.1 [9].

6.1.2
Nucleophilic and Electrophilic Catalysis

Adequate substrates such as heteroatom-containing organic compounds – along with protonation/deprotonation – can also be activated through a covalent bond to another donor or acceptor forming a Lewis acid–base complex. From a mechanistic viewpoint the donor atom of the Lewis acid is an electrophilic center and the Lewis base a nucleophilic center [10]. While the thermodynamics of a Lewis acid or a

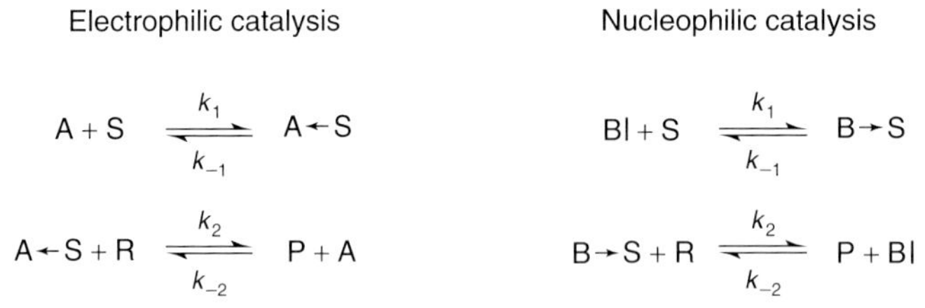

Electrophilic catalysis

$$A + S \underset{k_{-1}}{\overset{k_1}{\rightleftharpoons}} A{\leftarrow}S$$

$$A{\leftarrow}S + R \underset{k_{-2}}{\overset{k_2}{\rightleftharpoons}} P + A$$

Nucleophilic catalysis

$$B| + S \underset{k_{-1}}{\overset{k_1}{\rightleftharpoons}} B{\rightarrow}S$$

$$B{\rightarrow}S + R \underset{k_{-2}}{\overset{k_2}{\rightleftharpoons}} P + B|$$

Scheme 6.4 Electrophile and nucleophile catalyzed bimolecular reaction. (S, R = starting materials, P = product, A = Lewis acid, and B| = Lewis base as catalysts.) The formation of the Lewis acid-base adduct can occur via addition (as shown) or via substitution.

Lewis base are defined by its acidity or basicity, respectively [11], electrophilicity and nucleophilicity characterize the *rate* of reaction with a Lewis acidic or Lewis basic substrate. Thus, electrophilicity and nucleophilicity stand for the kinetic aspect of reactivity.

The relationship between basicity and nucleophilicity on one hand and acidity and electrophilicity on the other hand is rather complex. In addition to basicity and acidity, such effects as steric effects, ion charge, and solvation (small solvation increases reactivity, whereas stabilization through solvation leads to decreased reactivity [12]) are relevant [13]. The following examples demonstrate this: n-BuS⁻ and PhO⁻ possess similar basicity; however, face to face with saturated carbon, n-BuS⁻ is 1000 times more nucleophilic than PhO⁻. Conversely, PhO⁻ and Br⁻ are approximately equally nucleophilic, but PhO⁻ is 1000 times more basic than Br⁻.

The reaction sequence of a bimolecular, electrophilic, or nucleophilic reaction is shown in Scheme 6.4. Along with the easy formation of intermediate Lewis acid–base adducts A ← S and B → S a further prerequisite is that the Lewis acid A or the Lewis base B| as catalyst reacts faster with the substrate than with the reaction product. Nucleophilic catalysis and general base catalysis as well as electrophilic catalysis and general acid catalysis cannot be distinguished from one another kinetically.

Typical electrophilic catalysts are Lewis acidic metal halides (Friedel–Crafts alkylation of aromatic carbohydrates) and derivatives such as alkoxides, but also non-metallic Lewis acids such as halogens or carbonyl derivatives. Representative nucleophilic catalysts are π-electron-rich N-heterocycles like pyridine and imidazole, but also halides and alkoxides.[4]

In comparison to protonation/deprotonation in acid or base catalysis, electrophilic and nucleophilic catalysis provide much more mechanistic possibilities with regard to substrate activation. Furthermore, both nucleophilic and electrophilic catalysis profit from milder conditions compared to catalysis with superacids or superbases.

---

4) A number of examples of electrophilic and nucleophilic catalysis are discussed in detail in Ref. [3]

Owing to the fact that nucleophilicity and electrophilicity stand for reactivity, this field did not lack attempts to systematically investigate relative reactivity gradations by comparing rate constants for a reaction of a nucleophile with an electrophile relative to a reference system (including the solvent). For reactions of carbocations and diazonium ions with a range of nucleophiles, Ritchie was able to show that a nucleophilic system can be characterized by a constant parameter N_+ that does not depend on the type of electrophile [14].

$$\ln k - \ln k_0 = N_+ \tag{6.4}$$

This correlation is also called *'constant selectivity relationship,'* because the relative rate of reaction of two nucleophiles (= selectivity) does not depend on the absolute rate of reaction of the electrophile. Correspondingly, the relative reactivity of two electrophiles is independent of the strength of the nucleophile. On this basis, nucleophilicity and electrophilicity scales have been developed that allow for a quick overview of expected reactivities [15].

6.1.3
Transition Metal-Centered Homogeneous Catalysis

Metal-centered catalysis provides one of the most economic and powerful tools for steering selectivity in chemical reactions, since in these processes the kinetic control of the reaction is transferred from the substrate–reagent interaction to the metal center. As consequence, it is the nature of the metal and its environment that basically regulate the outcome of the reaction. This is immediately evident in, but not restricted to, asymmetric catalysis whereby one single chiral center embedded in the catalyst can be reproduced a large number of times. In this manner one molecule of chiral catalyst can generate millions of molecules of chiral product (chiral multiplication). This is the way in which the enzymes act in biological systems.

Coordination chemistry pioneered by Werner (Nobel Prize for Chemistry in 1923) forms the basis of organometallic catalysis mediated by transition metal complexes in homogeneous conditions. In these catalytic processes the metal center is the site where things occur: the substrate is activated toward the reagent(s) by interaction with the metal and the reagents also may or must require activation by the metal in order to interact efficiently with the substrate. The activation path proceeds through a precise sequence of events (elementary steps) that provide the conditions for the formation of the new bond(s) necessary for the reaction product to be obtained. The sequence of the elementary steps needed for one mole of product to be formed is the catalytic cycle of the process. Detailed knowledge of the catalytic cycle is mandatory for the selectivity of the reaction to be adjusted and optimized on rational grounds.

Activation of substrates and reagents can take place in different ways. Coordinative interactions, concerted transformations, redox, or radical reactions can be involved. Frequently these reactions result in the intermediate formation of organometallic species featuring a reactive metal–carbon bond. The structure and the stability of the resulting complexes depend on both the central atom and the ligands coordinated to the metal. It is well known that in metal-mediated reactions, both catalytic and stoichiometric, the reactivity of the metal promoter is critically dependent on the chemical environment in its surroundings as it is determined by the ligands coordinated.

For practical reasons transition metal catalysts are mostly introduced in the reaction vessel in the form of a pre-catalyst that in the early stages of the reaction gives rise to the 'real' catalyst through a more or less simple sequence of transformations. The catalyst precursor is most frequently a shelf-stable complex, the metal being surrounded by suitable ancillary ligands that have been designed in such a way as to generate a long-lived and easy to handle compound. For the metal to display its catalytic potential, these ancillary ligands must be removed and replaced by new ligands (substrate and reagents or solvent molecules), thus enabling the reactants to be brought into close proximity around the metal. This stage, which corresponds to the 'pre-activation step' of the catalyst, can be easily identified in the time–conversion plot of the reaction, being indicated by the presence of an initial induction period in which the reaction rate is lower than in the following steady-stage interval.

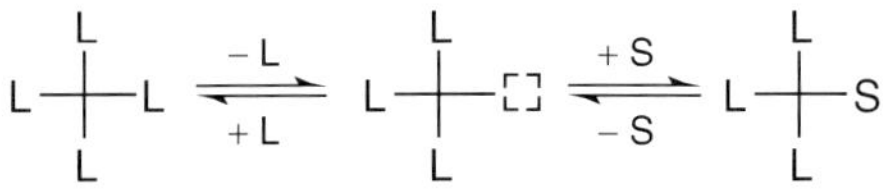

Scheme 6.5 Ligand substitution. Dissociative pathway.

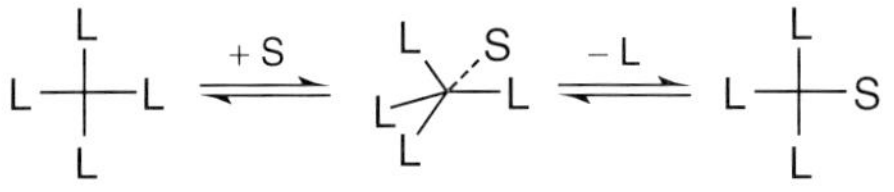

Scheme 6.6 Ligand substitution. Associative pathway.

Ligand substitution at the metal center can proceed according to two limiting mechanisms. In the first one (dissociative mechanism, Scheme 6.5) the cleavage of the coordinative bond between the metal and the leaving ligand is the rate-controlling step. The reaction is unimolecular and must precede the binding of the entering ligand. Opening a vacant coordination site that can be temporarily saturated by a solvent molecule is the most easily encountered activation path, given that the starting pre-catalysts are frequently complexes featuring a closed-shell electronic configuration that precludes the binding of one more ligand. Thus, metal catalysis is most frequently triggered by the production of a coordinative vacancy.

In the second case (associative mechanism, Scheme 6.6), breaking of the existing bond and formation of the new one at the metal occur simultaneously. The displacement is a bimolecular process resembling an $S_N 2$ reaction and conforms to a second-order rate equation. This mechanism is better suited for, but not restricted to, those pre-catalysts featuring fewer than 18 electrons in the valence orbitals. These complexes, however, can equally undergo ligand substitution via a dissociative mechanism since unsaturated species with two vacant coordination sites are kinetically accessible for several metals.

In some cases, when the entropy of activation of the reaction is fairly high, the same substitution process can occur through different pathways depending on the temperature, the dissociative path being favored at high temperatures [16].

The change of a ligand in the inner coordination sphere can affect remarkably the electronic and steric situation at the metal and an increasing reactivity is what is expected when a stabilizing ancillary ligand is substituted by a different one. Thus, when the entering ligand is a molecule of substrate, as normally occurs in the early stages of a catalytic reaction, the activation of the substrate usually goes along with the formation of a more reactive complex. This can easily undergo further transformations so that this ligand substitution opens up the way to the sequence of reactions that proceeds in the catalytic cycle.

The structure of the ligands employed in coordination chemistry is extremely varied but there is no doubt that phosphorus–metal ligands are by far the most important. The P-atom may act toward the metal as a strong σ-donor, as in the case of tertiary phosphines, or as a strong π-acceptor, as in the case of phosphites and phosphoramidites. The latter ligands are strong π-acids. This notwithstanding, they are capable of forming stable complexes not only with low-valent metals but also with metals in a high oxidation state. Strong σ-donors tend to increase the electron density at the metal, while strong π-acceptors display the opposite effect. Based on this, several methods have been devised in order to express the electronic effect of phosphorus ligands in such a way as to provide a tool useful for a predictive assessment to be formulated. One of the most popular is the electronic parameter χ as devised by Tolman [17] that reflects the differences induced in the IR frequencies of the symmetric stretching of $NiL(CO)_3$ complexes (L = phosphorus-donor ligand) upon changing the ligand L. According to the oxidation state of the phosphorus-donor, the ranges of χ-values can range between 0 and 20 for tertiary phosphines and between 20 and 40 for phosphites, increasing to 59 in the case of fluoro-substituted derivatives. Allowance should be made, however, for the fact that the electronic parameters of P-ligands may change somewhat from metal to metal.

In a methodology more recently devised, the electron density at phosphorus in isostructural phosphorus-donor ligands can be evaluated on the free ligand by measuring the $^1J_{P,Se}$ of the corresponding selenide. This can be prepared *in situ* by heating the ligand and selenium in $CDCl_3$. This method is among the most reliable ones for assessing the donating ability of the phosphorus donors, a smaller coupling constant corresponding to a more basic phosphorus-donor and *vice versa* [18].

Next to the electronic properties, the steric bulk of a phosphine ligand is also of prime importance. This can be estimated semi-quantitatively from Tolman's parameter θ, which corresponds to the cone angle [17] of the ligand as determined from a CPK model of a suitable complex. The θ-values obtained with this procedure are generally greater than the angles experimentally measured by X-ray diffraction, and this should be taken into account when using the cone angles. For instance, a recent survey in the database of crystal structures [19] has shown that the average P–Pd–P angle in a series of square planar *bis*-triphenylphosphine Pd(II)-complexes is 100.5°, while the θ-value is 145°.

A more reliable method for assessing the steric bulk of the phosphorus ligand is based on the dissociation constants of isostructural complexes containing homologous ligands. The values of these constants increase with the steric demand of the ligand [20], but the data available are restricted to a limited number of cases.

R-X + 2M ⇌ R-M-M-X or M-R + M-X   One electron process

R-X + M ⇌ R-M-X   Two electron process

Scheme 6.7 Modes of oxidative addition.

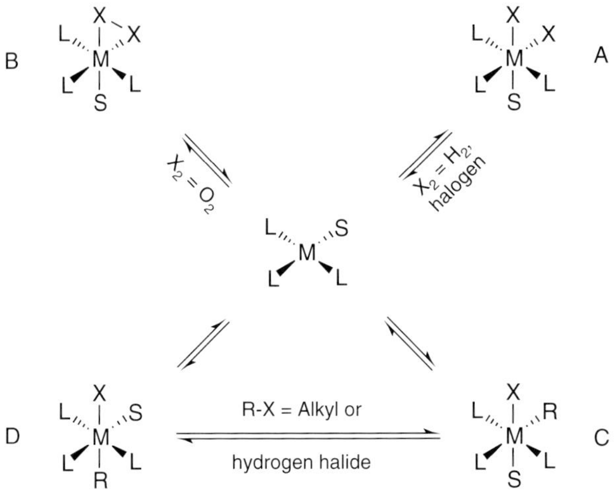

Scheme 6.8 Selected examples of oxidative addition adducts.

The activation of the substrate and of the reagents onto the metal can be achieved via oxidative addition, a path that provides an alternative for the substrates that are reluctant to participate in a direct binding to the metal. This is a ubiquitous reaction in which a reagent R–X (addendum) adds to a metal center M in such a way as to form in one shot two new bonds M–R and M–X. Both one- and two-electron processes are known: in the first one the formation of the new bonds occurs at two distinct metal centers, while in the second one, which is by far the most frequently encountered, the two bonds are formed at the same metal site (Scheme 6.7).

Depending on the type of addenda and of the operative conditions some differences can be noticed in the structure of the adducts (Scheme 6.8). In most cases the fragments A and B of the addendum are split into two separate formally anionic η^1-ligands (Scheme 6.8, **A**), but there are also addenda such as oxygen that retain at least one bond in the adducts and add in the form of dianionic

η^2-ligands (Scheme 6.8, **B**). Whichever the fate of the addendum, the net result is that the oxidation state and the coordination number of the metal both increase by two units, unless dissociation of a pre-existing ligand is required for the oxidative addition to occur.

The structure of the addenda R–X is very varied: they may be polar (protic acids, organic halides, etc.) as well apolar (Ar–H, R_3Si–H, etc.) compounds, simple symmetrical or non-symmetrical biatomic molecules (hydrogen, halogens, hydrogen halides, etc.), or single- or multiply-bonded species (oxygen, alkynes, etc.).

Oxidative addition is of paramount importance in transition metal catalysis since by adding in this way to the metal a wide variety of substrates, including alkyl and aryl halides, arenes, alkenes, alkynes, and so on, produce organometallic derivatives with highly reactive carbon–metal bonds that will react further to give the products of the reaction. Furthermore, a number of inorganic reagents such as hydrogen, oxygen, halogens, and so on, that in their fundamental state are poorly reactive toward organic substrates, upon binding to the metal in the form of anions become easily transferable to an organic functional group. Thus it is via oxidative addition that the basic conditions for a number of celebrated catalytic reactions mediated by homogeneous metal complexes such as hydrogenation, cross coupling, alkylation (Heck), arylation (Suzuky, Negishi), alkinylation (Sonogashira), and so on, are established.

The oxidative addition of non-polar addenda normally proceeds in apolar solvents with a complete *cis*-stereochemistry that implies a unimolecular concerted process where the two new bonds are formed simultaneously (Scheme 6.8, **A–C**).

The adduct with the opposite *trans*-stereochemistry (Scheme 6.8, **D**) is, however, the main or exclusive product when the addition of polar addenda such as alkyl and hydrogen halides is performed in polar solvents. Under these conditions the reaction is no longer concerted, but a step-wise process is operating where addition of a proton (or nucleophilic attack of the metal to the carbon bearing the halogen) produces in the first step an (incipient) cationic species. In a next separated step this then binds the halide forming the *trans*-adduct **D** (Scheme 6.8). Notably, the same polar solvents that favor the *trans*-addition may favor the interconversion between the two isomers **C** and **D**. This equilibrium can thus be shifted toward the *cis*-adduct **C** that is the thermodynamically more stable derivative, whereas the *trans*-derivative is the kinetically favored product.

The involvement of radical mechanisms in the oxidative addition of some alkyl halides to low-valent metal complexes has been noticed in several circumstances with halides that are poorly reactive in S_N2 chemistry. This radical chain mechanism (Scheme 6.9) proceeds in a way similar to the one-electron oxidative addition sketched in Scheme 6.5 and in some cases takes over the ionic mechanism. The lack of stereoselectivity of the reaction is an indication of the prevalence of radical chain mechanisms in oxidative addition [21].

Both coordinatively saturated and unsaturated complexes can react in oxidative addition, but the former are less reactive and usually only react with strongly oxidizing addenda. A low oxidation state and a high electron density at the metal

$$R\text{-}X \longrightarrow R\cdot + X\cdot$$

$$R\cdot + L_nM \longrightarrow L_{n-1}RM\cdot + L$$

$$L_{n-1}RM\cdot + R\text{-}X \longrightarrow L_{n-1}RMX + R\cdot$$

or/and

$$X\cdot + L_nM \longrightarrow L_{n-1}XM\cdot + L$$

$$L_{n-1}XM\cdot + R\text{-}X \longrightarrow L_{n-1}RMX + X\cdot$$

Scheme 6.9 Oxidative addition via radical chain mechanism.

center are additional positive factors for increasing the reactivity of the complex toward oxidative addenda.

Reductive elimination is regarded as the reverse of oxidative addition and is another most important elementary reaction in catalysis. For instance, the formation of the final C–H or C–C bond in the last step of the catalytic cycle of hydrogenation and cross coupling reactions takes place through a reductive elimination undergone by an alkylmetal hydride or a dialkylmetal derivative, respectively. The energetic conditions that are in favor of reductive elimination are just the opposite of those of the corresponding oxidative addition. Thus, the formation of an energetically stable bond, such as a C–H or a C–C bond as in the examples cited above, a relatively high oxidation state and a low electron density of the metal are all factors that contribute to speeding up the reductive elimination process. On the other hand, the principle of microscopic reversibility implies that the stereochemical constraints encountered in oxidative addition must also be met in reductive elimination. This means that for the elimination to be promptly accomplished, the leaving fragments must be *cis*-disposed around the metal for a concerted reaction path to be feasible. If by any reason this geometry is not easily accessible to the immediate precursor, a retardation or even a suppression of the process can be anticipated.

Migratory insertion provides one more routes for promoting the formation of the bond(s) between the substrate and the reagents onto the metal. The reaction proceeds with the formal insertion of an unsaturated molecule into a metal–anion bond. For such a process to be feasible the inserting molecule must be coordinated in advance to the metal in a site *cis* to the metal–anion bond where the insertion has to take place.

Alkenes and carbon monoxide are by far the most important and most frequently encountered inserting molecules, while metal–hydrogen and metal–carbon are the bonds that are more likely to undergo insertion. The insertion of olefins into M–H or M–C bonds (hydrometalation or carbometalation, respectively) are fundamental steps in many homogeneous catalytic reactions such as hydrogenation, carbonylation, and stereoselective Ziegler–Natta polymerization. Note that hydrometalation of an olefin gives an alkyl metal derivative whose reactive carbon–metal bond has to undergo some further transformation (reductive elimination or an additional insertion) for the reaction product to be formed.

With respect to the inserting molecule, the migratory insertion can be regarded as a 1,2-addition of an M–X bond into a C–C double bond or a 1,1-addition of M–X to a carbenic species, :C=O (Scheme 6.10).

There is compelling evidence that the reaction proceeds via migration of the anion to the *cis*-coordinated substrate rather than via a true insertion into the metal–anion bond. Notably, as observed in most the organic reactions involving migration of an anionic species, these reactions are strictly stereoconservative. This means that if the migrating anion and/or the metal are chiral, their configuration is preserved and no racemization should be noticed.

The insertion of olefins into M–H or M–C bonds is also a stereoselective process and gives the *cis*-adduct with complete stereoselectivity. The reaction has been regarded as a 2 + 2 cycloaddition involving a four-centered transition state (Scheme 6.11).

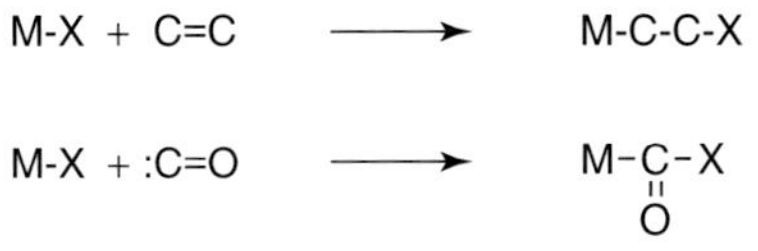

M-X + C=C $\longrightarrow$ M-C-C-X

M-X + :C=O $\longrightarrow$ M-C-X
$\|$
O

Scheme 6.10 Migratory insertion. 1,2- vs 1,1-addition.

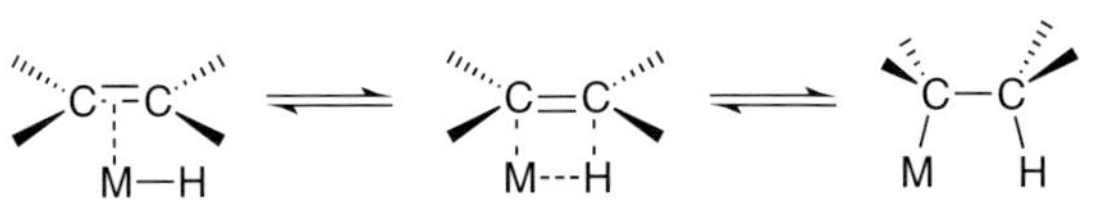

$$\text{C=C} \atop \text{M—H} \quad \rightleftharpoons \quad \text{C=C} \atop \text{M---H} \quad \rightleftharpoons \quad \text{C–C} \atop \text{M} \quad \text{H}$$

Scheme 6.11 Hydrometalation of alkenes. Concerted transition state.

The same reaction path should hold for the reverse reaction, namely the β-hydride elimination from an alkyl or an alkoxy metal complex, a process that can frequently be observed in homogeneous catalysis. The stereospecific character of migratory insertion reactions is of fundamental importance for developing efficient enantioselective processes based on homogeneous transition metal catalysts.

Catalyst precursors and catalytic intermediates are most conveniently characterized by multinuclear NMR spectroscopy. This spectroscopic technique has made possible the identification and sometimes even the *in-situ* observation of the organometallic species formed as intermediates in the course of the catalytic cycle of a number of diverse reactions. It must be stressed, however, that for practical purposes most NMR experiments are run at metal concentrations much higher and at substrate-to-metal ratios much lower than the ones used in catalysis. These differences in the operating conditions must be borne in mind in order to avoid wrong conclusions. Thus, even if the level of structural information is lower than with NMR, IR spectroscopy is a more reliable tool for the *in-situ* observation of catalytic intermediates since it is informative even in the range of concentrations used in catalytic runs. Furthermore, IR cells can be adapted so as to operate as on-line detectors, enabling where necessary continuous monitoring of the progress of a reaction even when it is carried out at high gas pressures.

With the introduction of a chiral monodentate phosphine ligand into Wilkinson's catalyst ([Rh(PPh)$_3$]$_3$Cl) (chlorotris(triphenylphosphine)rhodium(I)) [22] in 1968, Knowles (Nobel Prize in Chemistry in 2001) and Horner [23, 24] succeeded in catalytically and asymmetrically hydrogenating prochiral olefins for the first time. However, only the application of *chelating* phosphine ligands such as Kagan's DIOP (2,3-O-isopropylidene-2,3-dihydroxy-1,4-bis(diphenyl-phosphino)-butane) [25] made the breakthrough. Shortly after this the first industrial application of a transition metal complex-catalyzed hydrogenation was carried out: the synthesis of L-DOPA (treatment of Parkinson's desease) with a rhodium/DIPAMP complex (DIPAMP = 1,2-bis[(2-methoxyphenyl)(phenyl)phosphino]ethane).

Pioneering work to elucidate the reaction mechanism of asymmetric hydrogenation was accomplished by the research groups of Halpern (kinetics) and Brown (NMR). In contrast to Wilkinson's monodentate catalyst, two diastereomeric catalyst–substrate complexes are formed from the solvate complex and the prochiral dehydroamino acid derivative in a fast, reversible pre-step. Typically, one of the two complexes is present in considerable excess ('major substrate complex'), Figure 6.2. The substrate coordinates as a chelate, also, via the olefinic double bond and via the amide oxygen, which was revealed by ^{31}P NMR and ^{13}C NMR spectroscopy and partly isotope-labeled substrates [26]. In a sequence of elementary steps, namely the rate-determining oxidative addition of hydrogen, insertion, and reductive elimination, the substrate complexes react to form the enantiomeric products.

The key to the understanding of very high achievable enantioselectivities was provided by the X-ray analysis of a diastereomeric catalyst–substrate complex [27]. The oxidative addition of hydrogen to the isolated compound should have given the (S) enantiomer based on classical ideas; however, the main enantiomer observed experimentally was the (R) enantiomer! By elaborate kinetic investigations it could be shown that the minor catalyst–substrate complex dominantly determined the enantioselectivity with its high reactivity, Figure 6.2 [28].[5] Additional evidence that the reaction pathway via the minor complex leads to the excess product was deduced from NMR investigations. The hydrido-alkyl complex of the minor catalyst–substrate complex was thereby identified as the apparent transient intermediate of the main reaction pathway [29].

The ratio of reaction rates for product formation in Figure 6.2 (Michaelis–Menten kinetics) under isobaric conditions leads to Eq. (6.5). Hence, selectivity as er-value (enantiomeric ratio) is the result of two factors: the ratio of reactivities of intermediates as well as the stationary ratio of intermediates themselves.

---

5) Six rate constants that were determined temperature-dependently from six linearly independent equations describe the whole system. The conditional equations resulted from (i) gross stability constants (UV/vis), (ii) stability ratios of diastereomeric catalyst–substrate complexes (^{31}P NMR), (iii) measurements of the sum of rates of catalyst–substrate complex formation (UV/vis, stopped flow), (iv) determination of the dissociation constant for the major catalyst–substrate complex from substitution experiments with competing ligands (UV/vis, stopped flow), and finally (v) the product-proportional rate of hydrogen consumption of catalytic hydrogenations or – more precisely – the rates of formation of the single enantiomers, which are easily accessible from the conversion-independent, constant enantioselectivity.

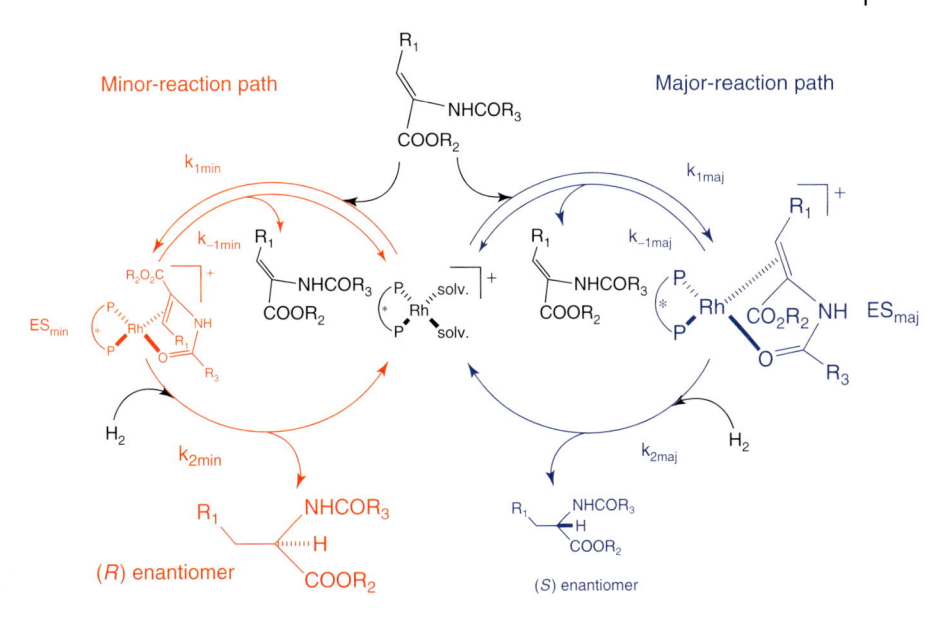

Figure 6.2 Reaction sequence of the asymmetric hydrogenation of α-dehydroamino acid derivatives with cationic rhodium complexes (PP* = chiral chelating ligand, ES_{min}, ES_{maj} = diastereomeric catalyst–substrate complexes).

$$\frac{[(S)\text{ enantiomer}]}{[(R)\text{ enantiomer}]} = \frac{K_{Mmaj}}{K_{Mmin}} \frac{k_{2min}}{k_{2maj}} = \frac{[ES_{min}]}{[ES_{maj}]} \frac{k_{2min}}{k_{2maj}} = \left(\frac{\frac{k_{1min}}{k_{-1min}+(k_{2min}[H_2])}}{\frac{k_{1maj}}{k_{-1maj}+(k_{2maj}[H_2])}} \right) \frac{k_{2min}}{k_{2maj}} \quad (6.5)$$

The discovery that the less stable and therefore lower concentration intermediate in solution is determining the stereochemical course of the reaction and therefore the overall selectivity, predominantly through its high reactivity, came as a big surprise at the time. It entered the literature as the so-called *major/minor concept*, also known as *anti-lock-and-key principle*. The major/minor concept can be expressed by the following three inequalities that must be valid simultaneously [30]:

$$k_{2min} > k_{2maj} \; [ES_{maj}] > [ES_{min}] \; (k_{2min}/k_{2maj}) > [ES_{maj}]/[ES_{min}] \quad (6.6)$$

The initial idea of an extreme reactivity of one intermediate is a basic principle in homogeneous catalysis and is reflected for instance in the concept of *ligand-accelerated catalysis* (LAC) [31].

> The goal of channeling the catalysis through one particular complex is usually achieved by an overwhelming *kinetic activity* favoring that one complex over the many other complexes which assemble in solution. For ligand-accelerated processes with early transition metal complexes this *in-situ* selection of a highly reactive (and selective) metallic complex from a variety of thermodynamically dictated assemblies is a crucial requirement.

Besides the *major/minor concept* it was recently recognized that the major substrate complex can *also* lead to the major enantiomer as the main hydrogenation product [32]. For example, the oxidative addition of H_2 to the catalyst–substrate complex shown in Figure 6.3 – formed from a protected β-dehydroamino acid derivative and the rhodium-(*S*,*S*)-DIPAMP catalyst – surprisingly leads to the (*S*) enantiomer, as was also observed experimentally.

Due to the fact that low-temperature NMR measurements proved this complex to be the major catalyst–substrate complex it could be unequivocally proven that, in this case, in contrast to classical ideas, the major catalyst–substrate complex predominantly determines the selectivity of the asymmetric hydrogenation [32b]. This behavior is already known from enzyme catalysis: the *lock-and-key principle* was introduced by Fischer (Nobel Prize in Chemistry 1902) [33].

The previous examples show that it is the interplay between kinetic investigations and other methods such as the structural elucidation of intermediates that leads to the mechanism. However, before one applies such methods it is important to know if the concentration of these intermediates under stationary catalytic conditions in the reaction solution is suitable for measurements at all. To answer this question, kinetic investigations are ideal. In homogeneous hydrogenation, for instance, the time-dependent hydrogen consumption reveals on which side pre-equilibria lie. The limiting case of a first order reaction of Michaelis–Menten kinetics means

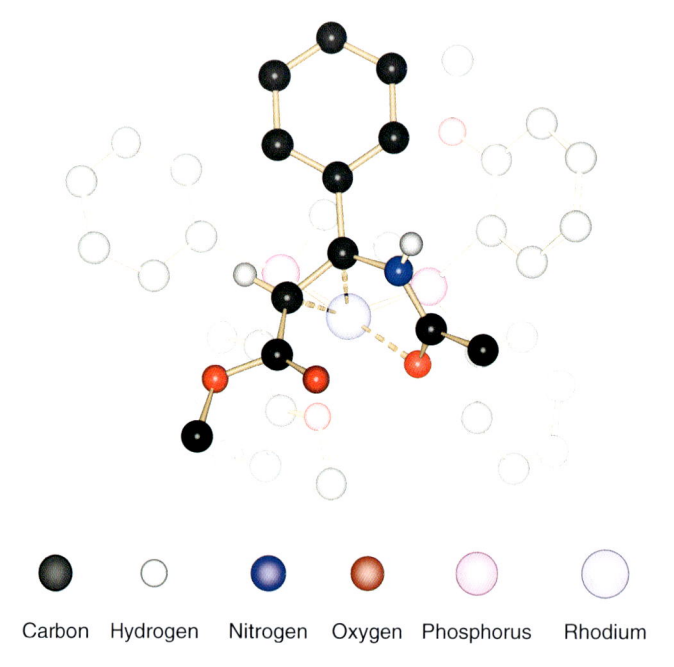

Figure 6.3 X-ray structure of [Rh((S,S)-DIPAMP)((Z)-3-N-acetylamino-3-(phenyl)-methyl acrylate)]BF_4 (foreground: coordinated substrate, background: rhodium–ligand skeleton).

that an intermediate cannot be detected because the equilibrium lies on the side of starting materials. On the other hand, if the hydrogen consumption shows a zero order reaction for the substrate the pre-equilibrium lies on the side of the intermediate.

References

1. (a) Herrmann, W.A. (2002) *Angew. Chem. Int. Ed.*, **41**, 1290; (b) Hillier, A.C. and Nolan, S.P. (2002) *Platinum Met. Rev.*, **46**, 50.
2. See for example: Jacobsen, E.N., Pfaltz, A., and Yamamoto, H. (eds) (1999) *Comprehensive Asymmetric Catalysis*, Springer, Berlin.
3. (a) Taube, R. (1988) *Homogene Katalyse*, Akademie-Verlag, Berlin; (b) Steinborn, D. (2012) *Fundamentals of Organometallic Catalysis*, Wiley-VCH Verlag & Co. KGaA, Weinheim.
4. The modern field of organocatalysis (that is homogeneous catalysis with entirely organic compounds) is not discussed here since it is acid–base catalysis in most cases.(a) Dalko, P.I. and Moisan, L. (2004) *Angew. Chem. Int. Ed.*, **43**, 5138–5175; (b) special edition (2004) *Acc. Chem. Res.*, **37** (8), 487–631.
5. (a) Brønsted, J.N. (1923) *Recl. Trav., Chim. Pays-Bas*, **42**, 718; (b) Lowry, T.M. (1923) *Chem. Ind. (London)*, **42**, 43.
6. Hammet, L.P. and Dreyrup, A.J. (1932) *J. Am. Chem. Soc.*, **54**, 2721–2739.
7. The term super acid traces back to Conant: Hall, N.F. and Conant, J.B. (1927) *J. Am. Chem. Soc.*, **49**, 3047–3061.
8. (a) Swain, C.G. and Brown, J.F. (1952) *J. Am. Chem. Soc.*, **74**, 2538–2543; (b) Swain, C.G. and Brown, J.F. (1952) *J. Am. Chem. Soc.*, **74**, 2534–2537.
9. Voet, D. and Voet, J.G. (1994) *Biochemie*, VCH Verlagsgesellschaft mbH.
10. The concepts of nucleophilicity and electrophilicity were introduced by Ingold: Ingold, C.K. (1934) *Chem. Rev.*, **15**, 225–274.
11. The so-called donor and acceptor numbers introduced by Gutmann are a relative quantification of Lewis basicity and acidity in several solvents:(a) Gutmann, V. (1977) *Chemtech*, 255–263; (b) Gutmann, V. (1976) *Electrochim. Acta*, **21**, 661–670.
12. Reichardt, C. (2003) *Solvents and Solvent Effects in Organic Chemistry*, 3rd updated and enlarged edn, Wiley-VCH Verlag GmbH, Weinheim.
13. According to Bunnet, J.F. (1963) *Annu. Rev. Phys. Chem.*, **14**, 271–290, no less than 17 factors influence nucleophilicity and electrophilicity!
14. (a) Ritchie, C.D. (1972) *Acc. Chem. Res.*, 348–354. Earlier, analog approaches can be found in (b) Swain, C.G. and Scott, C.B. (1953) *J. Am. Chem. Soc.*, **75**, 141–147; (c) Edwards, J.O. (1954) *J. Am. Chem. Soc.*, **76**, 1540–1547; (d) Edwards, J.O. (1956) *J. Am. Chem. Soc.*, **78**, 1819–1820.
15. (a) Mayr, H., Kempf, B., and Ofial, A.R. (2003) *Acc. Chem. Res.*, **36**, 66–77; http://www.cup.lmu.de/oc/mayr/ReactScalesPoster.pdf.; (b) Pike, R.D. and Sweigart, D.A. (1999) *Coord. Chem. Rev.*, **187**, 183–222; (c) Mayr, H. and Patz, M. (1994) *Angew. Chem. Int. Ed. Engl.*, **33**, 938–957; (d) Kane-Maguire, L.A.P., Honig, E.D., and Sweigart, D.A. (1984) *Chem. Rev.*, **84**, 525–543.
16. Good, R. and Merbach, A.E. (1975) *Inorg. Chem.*, **14**, 1030.
17. Tolman, C.A. (1977) *Chem. Rev.*, **77**, 313.
18. Allen, D.A. and Taylor, B.F. (1982) *J. Chem. Soc. Dalton Trans.*, 51–54.
19. van Leeuwen, P.W.N.M., Zuideveld, M.A., Swennehuis, B.H.G., Freixa, Z., Kamer, P.C.J., Goubitz, K., Fraanje, J., Lutz, M., and Spek, A.L. (2003) *J. Am. Chem. Soc.*, **125**, 5523.
20. Troegler, W.C. and Marzilli, L.G. (1975) *Inorg. Chem.*, **14**, 2942.
21. Kramer, A.V., Labinger, J.A., Bradley, J.S., and Osborn, J.A. (1974) *J. Am. Chem. Soc.*, **96**, 7145.

22. (a) Osborn, J.A., Jardine, F.H., Young, J.F., and Wilkinson, G. (1966) *J. Chem. Soc. A*, 1711–1732; (b) Young, J.F., Osborn, J.A., Jardine, F.H., and Wilkinson, G. (1965) *J. Chem. Soc., Chem. Commun.*, 131–132.
23. Knowles, W.S., Sabacky, M.J. (1968) *J. Chem. Soc., Chem. Commun.*, 1445–1446.
24. Horner, L., Siegel, H., Büthe, H. (1968) *Angew. Chem.*, **24** 1034–1035. (1968) *Angew. Chem. Int. Ed. Engl.* **7** 942.
25. (a) Kagan, H.B. and Dang, T.P. (1972) *J. Am. Chem. Soc.*, **94**, 6429–6433; (b) Dang, T.P. and Kagan, H.B. (1971) *J. Chem. Soc., Chem. Commun.*, 481.
26. Brown, J.M. and Chaloner, P.A. (1980) *J. Am. Chem. Soc.*, **102**, 3040–3048.
27. (a) Chan, A.S.C., Pluth, J.J., and Halpern, J. (1980) *J. Am. Chem. Soc.*, **102**, 5952–5954. Analogue results for different catalyst-substrate complexes can be found in (b) McCulloch, B., Halpern, J., Thomas, M.R., and Landis, C.R. (1990) *Organometallics*, **9**, 1392–1395; (c) Schmidt, T., Baumann, W., Drexler, H.-J., Arrieta, A., Heller, D., and Buschmann, H. (2005) *Organometallics*, **24**, 3842–3848; (d) Schmidt, T., Dai, Z., Drexler, H.-J., Baumann, W., Jäger, C., Pfeifer, D., and Heller, D. (2008) *Chem. Eur. J.*, **14**, 4469–4471; (e) Preetz, A., Baumann, W., Fischer, C., Drexler, H.-J., Schmidt, T., Thede, R., and Heller, D. (2009) *Organometallics*, **28**, 3673–3677.
28. Landis, C.R. and Halpern, J. (1987) *J. Am. Chem. Soc.*, **109**, 1746–1754.
29. Brown, J.M. and Chaloner, P.A. (1980) *J. Chem. Soc., Chem. Commun.*, 344–346.
30. The results if other inequations are valid are extensively described in: Schmidt, T., Dai, Z., Drexler, H.-J., Hapke, M., Preetz, A., and Heller, D. (2008) *Chem. Asian J.*, **3**, 1170–1180.
31. Berrisford, D.J., Bolm, C., and Sharpless, K.B. (1995) *Angew. Chem., Int. Ed. Engl.*, **34**, 1059–1070.
32. (a) Evans, D., Michael, F.E., Tedrow, J.S., and Campos, K.R. (2003) *J. Am. Chem. Soc.*, **125**, 3534–3543; (b) Drexler, H.-J., Baumann, W., Schmidt, T., Zhang, S., Sun, A., Spannenberg, A., Fischer, C., Buschmann, H., and Heller, D. (2005) *Angew. Chem. Int. Ed.*, **44**, 1184–1188; (c) Reetz, M.T., Meiswinkel, A., Mehler, G., Angermund, K., Graf, M., Thiel, W., Mynott, R., and Blackmond, D.G. (2005) *J. Am. Chem. Soc.*, **127**, 10305–10313; (d) Mori, S., Vreven, T., and Morokuma, K. (2006) *Chem. Asian J.*, **1**, 391–403.
33. (a) Fischer, E. (1894) *Ber. Dtsch. Chem. Ges.*, **27**, 2985–2993; (b) Koshland, D.E. Jr. (1994) *Angew. Chem., Int. Ed. Engl.*, **33**, 2375–2378.

7
Biocatalysis

Uwe Bornscheuer

7.1
Introduction

Biocatalysis refers to use the use of enzymes as biological catalysts. This discipline is part of industrial biotechnology, also known as *white biotechnology*. In a broader context, processes using whole micro-organisms are also included and these are often referred to as *biotransformations*, not only because entire cells are used, but also because often multistep reactions take place in the microorganism such as in the synthesis of the antibiotics Penicillin G or Cephalosporin C.

Indeed, many applications in organic synthesis use whole-cell systems, a well-known example being the use of baker's yeast (*Saccharomyces cerevisiae*) for stereoselective reduction of ketones to the corresponding chiral alcohols. The isolated enzymes used in biocatalysis are also produced by micro-organisms. Historically, the wild-type strains were used as a source of the enzyme of interest, but nowadays the majority of commercially produced biocatalysts are recombinantly expressed in host organisms such as *Escherichia coli*, *Pichia pastoris*, *Bacillus* species, *Aspergillus* species, and so on. The enzymes are then isolated, sometimes purified, and then used either directly in biocatalysis or immobilized to facilitate isolation and recycling.

Due to the fast progress made in molecular biology since the mid 1980s, enzymes isolated from plants, such as hydroxynitrile lyases (HNLs) or from mammalian tissues, for instance pig liver esterase (PLE), or aldolases, are used to a much lesser extent. One reason is the low concentration and often high price, but the major driving force to use recombinant microbial expression systems is the stable product quality and the substantially lowered risk (allergenic potential, viruses, pathogenic factors, etc.) especially associated with enzymes from mammalian sources.

The first applications of enzymes in organic chemistry date back a century. As early as 1908, Rosenthaler used an HNL-containing extract for the preparation of (*R*)-mandelonitrile from benzaldehyde and hydrogen cyanide (HCN). Since then more and more enzymes have been identified and in parallel their use in organic chemistry has steadily increased. Especially since the mid-1970s, the number of

Catalysis: From Principles to Applications, First Edition. Edited by Matthias Beller, Albert Renken, and Rutger van Santen.
© 2012 Wiley-VCH Verlag GmbH & Co. KGaA. Published 2012 by Wiley-VCH Verlag GmbH & Co. KGaA.

publications utilizing enzymes as well as the number of industrialized processes has increased substantially.

Several reasons can be identified for this development:

- More organic chemists accept the use of biocatalysts.
- Biocatalysis can save additional reaction steps compared to organic synthesis.
- Enzymes are often highly chemo-, regio-, and stereospecific.
- Biocatalysis is a safer and 'greener' technology.
- A substantially increased demand for optically pure compounds, especially for pharmaceutical applications.
- Easier production of biocatalysts due to recombinant expression systems.
- Many enzymes are commercially available.
- Modern protein engineering methods allow the straightforward production of tailor-made biocatalysts for a given process.

The most important application of enzymes in organic chemistry is in the synthesis of optically active compounds. This is due to the excellent enantio- and stereoselectivity shown by many enzymes, which makes them attractive alternatives to asymmetric chemical synthesis or reactions starting from the chiral pool. Enzymes are also used for the synthesis of chemicals lacking a chiral center. Prominent examples are the production of acrylamide on the $>100\,000$ t y^{-1} scale and of nicotinamide, both using a nitrile hydratase, and the synthesis of simple esters, such as myristyl myristate for cosmetic application, catalyzed by a lipase.

Often, the targets for organic synthesis are identical when comparing an enzymatic approach with the use of transition metal catalysis; for instance a chiral alcohol can be made by stereoselective hydrogenation using a Pd-catalyst or by using a ketoreductase in the presence of a cofactor such as NAD(P)H. The decision between biocatalysis or chemocatalysis has to be made case by case and major determinants are the optical purity of the product, the efficiency of each process, the possibility to scale it up, and the overall process costs, which must include, for instance, the need to remove transition metal traces from the product.

The development of a biocatalytic route includes many steps which are depicted in Figure 7.1. This includes the discovery of the enzyme (commercial biocatalysts, screening in microorganisms, identification of the protein encoding sequence in databases followed by cloning of the gene, and expression of the enzyme, etc.), the decision to use either whole cells or isolated enzymes, biochemical characterization, possibly optimization by protein engineering, immobilization, process design (reactor system), and downstream processing to isolate the product.

Selected examples of biocatalysts useful for organic synthesis are given below. These are organized based on the classification suggested by the Enzyme Commission (EC). This commission has subdivided all enzymes into six distinct classes organized by the type of reaction catalyzed (EC 1: Oxidoreductases, EC 2: Transferases, EC 3: Hydrolases, EC 4: Lyases, EC 5: Isomerases, and EC 6: Ligases). Each enzyme is then given a four-digit number to classify the specific reaction,

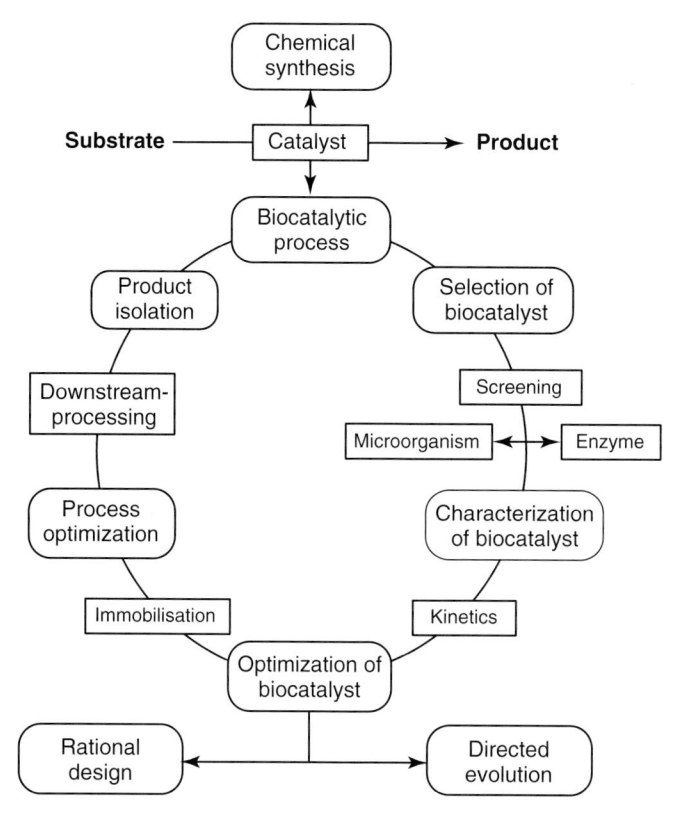

Figure 7.1 The biocatalysis cycle includes various aspects to be considered when establishing an enzymatic route as a counterpart to a chemical method.

including, for instance, cofactor requirement. All enzymes described so far are listed in databases [1].

Enzymes are polypeptides assembled from the 20 proteinogenic amino acids. Their primary structure is determined by the exact order of the amino acids, the secondary structure includes three-dimensional elements such as α-helices and β-sheets.

Functionally active enzymes must be in their tertiary structure: the correct three-dimensional folding of the polypeptide chain. Many enzymes also contain disulfide bonds to improve stability and correct folding and can contain sugar residues (glycosylation). Only then is it certain that the active site and binding pockets are correctly formed and efficient catalysis can occur.

Many enzymes also have a quaternary structure, that is, the assembly of several monomers to multimers or the coordination of two or more polypeptides to form an active heteromer (one example is the nitrile hydratases composed of an α- and a β-subunit).

In contrast to chemical catalysts, enzymes are rather sensitive toward experimental conditions because of the polyamide backbone and the necessity for a correct and stable three-dimensional structure. Enzyme deactivation can be promoted by temperature and pH-values outside their stability and activity range, hydrolysis of peptide bonds (also via degradation by proteases), disruption of disulfide bonds, oxidation or other chemical reactions at functional groups (i.e., formation of a Schiff's base between the ε-amino group of lysine and aldehydes), loss of metal ions necessary for catalytic activity, and so on.

Furthermore, many enzymes undergo substrate and/or production inhibition. For applications in biocatalysis, this is often a major issue, as the substrate concentrations should be as high as possible for high productivity in a process, but could be above the inhibition threshold [2]. All enzymes contain an active site in which the conversion of substrate to products takes places. The active site is usually composed of amino acids with functional residues in their side chains, asparagine (Asp), serine (Ser), histidine (His), and lysine (Lys) being the most prominent ones. In addition, metals ions such as zinc, iron, or magnesium are often essential for the catalytic step. The first proposal of an enzymatic mechanism was put forward by Emil Fischer in 1894 [3]. He assumed that an enzyme and its substrate interact mechanistically like a lock and a key. Although this assumption was quite sophisticated at that time, it means that the enzymes have a very rigid structure. This model fails to explain the stabilization of the transition state, and the Fischer model was later replaced by the induced-fit mechanism developed by Koshland [4]. He proposed that upon binding of the substrate, the enzyme can change its conformation under the influence of the substrate similarly to the interaction of a hand (substrate) with a glove (enzyme).

Not all enzyme classes are equally important for biocatalysis, and hence only the most useful ones are treated in this chapter. For each enzyme class, a brief introduction, the reaction principle, and some examples of further applications of enzymes are given. Note that with a few exceptions only reactions using isolated enzymes are included.

Various types of enzymes and numerous examples of their use in chemo-enzymatic synthesis have been described, especially in the past three decades. It is impossible to provide a sufficient coverage of all developments in this chapter, and hence only a brief introduction to the field can be given. Readers are encouraged to consult the broad range of excellent books [5–12] and reviews [13–15], which provide an in-depth coverage of the application of enzymes in organic synthesis.

7.1.1
Choice of Reaction Strategy: Kinetic Resolution or Asymmetric Synthesis

Enzymatic syntheses of optically active compounds can be by kinetic resolution of racemic mixtures or can be performed via asymmetric synthesis (Scheme 7.1). Kinetic resolution will only lead to a maximum yield of 50% unless the unwanted enantiomer is racemized. This can be achieved using a racemase or by chemical racemization. If kinetic resolution and racemization are performed simultaneously,

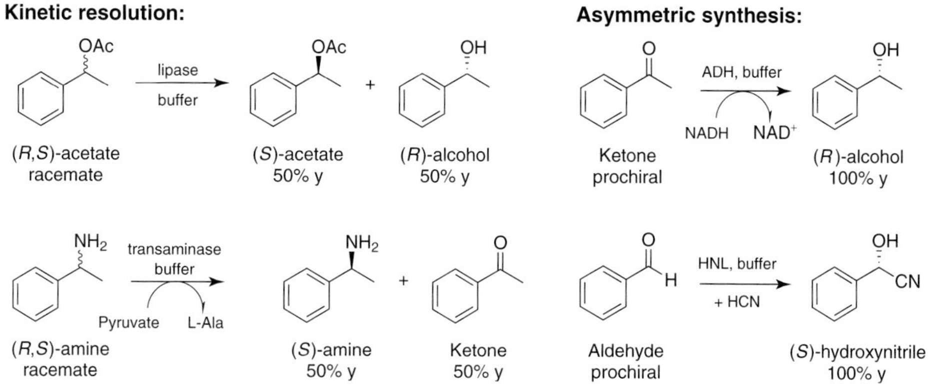

Scheme 7.1 Enzymatic reactions can be performed as kinetic resolutions yielding at maximum 50% product as shown for a lipase-catalyzed kinetic resolution or a transaminase-catalyzed reaction. In contrast, asymmetric synthesis gives 100% of the desired optically pure product exemplified for a reduction using an alcohol dehydrogenase (ADH/ketoreductase) and the cofactor NADH or by a C–C coupling using a hydroxy nitrile lyase (HNL).

the process is named dynamic kinetic resolution (DKR) [16]. In contrast, asymmetric synthesis allows the production of one enantiomer at 100% yield. Examples include the desymmetrization of prostereogenic compounds using, for example, an alcohol dehydrogenase (ADH) in the reduction of a ketone or the formation of a chiral compound by, for example, C–C-bond formation using an HNL.

The performance of an enzyme in a kinetic resolution can easily be judged by the E-value (enantioselectivity, enantiomeric ratio). This value is the ratio of the reaction rates (V_{max}/K_m) of one enantiomer compared with the other. Chen and co-workers developed simple equations to calculate the E-value from the optical purity (%ee, enantiomeric excess) of the substrate or product and the conversion [17, 18]. A non-selective enzyme has an E-value of 1; preparatively useful E-values are >50 and desired are E-values >100, as only then can optically pure product and substrate be isolated at 50% conversion. In asymmetric synthesis, the E-value is independent of the conversion.

7.1.2
Choice of Reaction Systems

Although enzymatic reactions usually take place in aqueous buffered systems and organic solvents generally denature enzymes quickly, it was already shown in the 1930s that lipase-catalyzed esterifications can take place in organic solvents containing approximately 10% water [19, 20], but most of this work was forgotten. Klibanov's group discovered many other examples of enzyme-catalyzed reactions in organic solvents and further demonstrated that enzymes require only traces of water (for reviews see Refs. [21, 22]). Since then, numerous examples of biocatalytic reactions in organic solvents have been described, and the majority of examples include the use of lipases. Biocatalysis in the presence of organic solvents can range from the addition of water-miscible solvents (e.g., lower alcohols, dimethyl sulfoxide (DMSO), acetone, and tetrahydrofurane (THF)) via the use of water-immiscible solvents (e.g., hexane, toluene, and ethers) to the use of pure organic solvents. Water-miscible solvents are often used to facilitate dissolution of substrate or product in the aqueous reaction system, but most enzymes tolerate only up to 10% of these solvents. Water-immiscible solvents have the advantage of much higher substrate (and/or product) solubility, the option that either substrate or product are in the organic phase to facilitate the reaction and to enable a simple work-up using phase separation. In both cases, the enzyme is still dissolved in its native state in the aqueous phase.

Reactions in pure organic solvents – containing a very small amount of water, usually $\leq$5% (v/v) – have the advantage that also certain synthesis reaction such as transesterifications with lipases are favored. As a disadvantage, the enzyme is not dissolved, and hence mass transfer can be rate-limiting. To avoid inactivation, immobilized enzymes are usually used. Recycling and reuse of the biocatalyst is facilitated by a simple filtration or centrifugation step.

7.2
Examples

7.2.1
Oxidoreductases (EC 1)

It is estimated that about 25% of the presently known enzymes are oxidoreductases. The most useful enzymes for preparative applications are dehydrogenases or reductases. Also, mono- and dioxygenases, oxidases, and peroxidases belong to this class. All of them require NADH or NADPH as a cofactor. Due to this cofactor-dependency, recycling is necessary for cost-effective processes unless the reaction is performed in a whole-cell system. Most oxidoreductases are not used as isolated biocatalysts and are therefore not covered extensively in the following sections.

7.2.1.1 Dehydrogenases (EC 1.1.1.-, EC 1.2.1.-, EC 1.4.1.-)

Synthesis of Alcohols The most important application of oxidoreductases in organic chemistry is in the reduction mode, as this yields chiral compounds such as alcohols, hydroxy acids, or amino acids. In case of ADHs, one hydrogen and two electrons are transferred from the reduced nicotinamide moiety to an acceptor molecule such as a ketone or an α-keto acid. In many cases, this reaction is highly stereoselective and the hydride is delivered by the dehydrogenase either from the *re*- or the *si*-face of the carbonyl, yielding the corresponding (*R*)- or (*S*)-products (Scheme 7.2). Many ADHs were found to obey the Prelog rule, which is based on

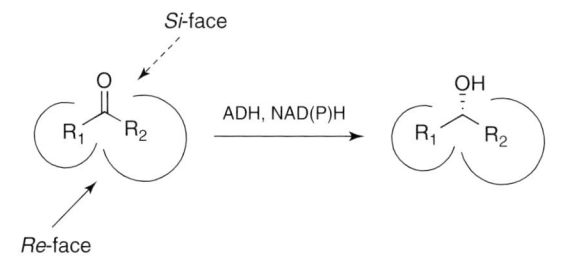

Scheme 7.2 According to the Prelog-rule, the size of the substituents R_1 and R_2 (here: $R_1 < R_2$) determines whether the carbonyl of a ketone is attacked by the hydride either from the re- or the si-face. Here the (S)-product is formed, as a sequence rule of $R_2 > R_1$ is assumed.

the size of substituents and allows us to predict which enantiomer will be produced; however, exceptions have also been described (Anti-Prelog).

In the literature, most examples of the use of ADHs deal with whole-cell systems, for example, bakers yeast. More recently, ADHs have also been used as isolated enzymes coupled with efficient cofactor regeneration systems which do not require a separate biocatalyst. Instead NAD(P)H is directly recycled by the ADH used to produce the optically pure alcohol in the presence of isopropanol. Key to success was the high stability of the ADH at high concentrations of isopropanol and acetone.

Synthesis of Amino Acids Amino acids (α-amino carboxylic acids) are widely used in nutrition, medical applications, and organic synthesis. They can be produced by one of the following routes:

- Cultivation of microorganisms, especially overproducers of certain amino acids
- Extraction from protein hydrolyzates
- Enzymatic synthesis from prochiral precursors or via kinetic resolution of racemates
- Chemical synthesis.

The first two methods provide access only to natural L-amino acids, whereas the last two approaches allow for the synthesis of non-natural amino acids and the D-enantiomers. L-glutamate (>1 000 000 tonnes) and L-lysine (>800 000 tonnes) are amino acids obtained by cultivation with *Corynebacterium glutamicum* as the main producer. Protein hydrolyzates are used for the isolation of L-cysteine, L-leucine, L-asparagine, L-arginine, and L-tyrosine.

The enzymatic synthesis can start from racemic amino acids obtained by chemical synthesis (i.e., the Strecker method) followed by kinetic resolution using esterases (i.e., enantioselective hydrolysis of amino acid carboxylic acid esters) or acylases/amidases (i.e., enantioselective hydrolysis of amides) or using hydantoinases. Alternatively, prochiral precursors can be subjected to reductive amination using amino acid dehydrogenases (AADHs). With a few exceptions, only L-amino

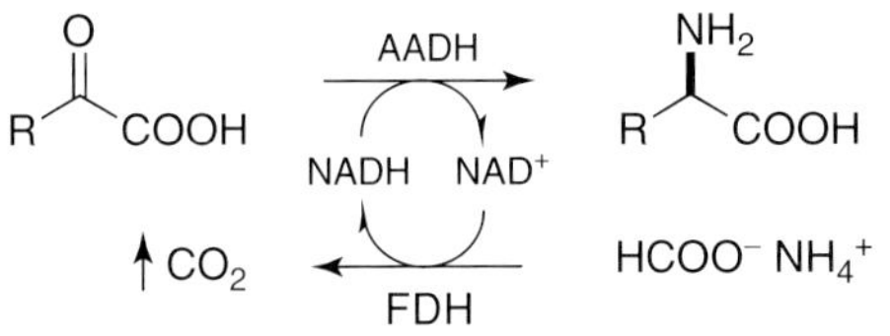

Scheme 7.3 Example of the synthesis of optically active L-amino acids using an amino acid dehydrogenase (AADH). The cofactor NADH is regenerated using a formate dehydrogenase (FDH) from *Candida boidinii*. The equilibrium is shifted toward NADH by using ammonium formate yielding carbon dioxide as by-product.

acids are accessible. Various AADHs have been described in the literature [23]; however, only a limited number are currently used. Numerous genes and protein sequences of AADHs have been identified, and several 3D-crystal structures have been solved. Although sequences vary considerably between AADHs, the residues involved in catalysis and nicotinamide cofactor binding are highly conserved.

Leucine and phenylalanine dehydrogenases (Leu-DH, Phe-DH) are the most important AADHs, especially as they are not restricted to their natural substrates, accept a broad range of other precursors and efficiently allow the synthesis of non-proteinogenic α-amino acids from the corresponding prostereogenic α-keto acids (Scheme 7.3).

Cofactor regeneration is required for cost-effective application of dehydrogenases. An elegant solution for the required recycling of the cofactor was developed by Kula and Wandrey based on the use of formate dehydrogenase (FDH), for example, from the yeast *Candida boidinii* using ammonium formate as co-substrate (Scheme 7.3). Carbon dioxide formed as by-product is highly volatile, leading to a favorable shift of the equilibrium. This allows for total turnover numbers (moles product per moles cofactor) of up to 600 000, as demonstrated for the continuous synthesis of phenylalanine [24]. The overall reaction is best performed in an enzyme membrane reactor, in which the AADH and the FDH are retained in the reactor compartment using ultrafiltration membranes. Covalent coupling of the cofactor NADH to polyethylene glycol (PEG) can be used to avoid leakage through the membrane. Further examples and more details can be found in reviews [23, 25].

One example of an amino acid produced by this approach is L-*tert*-leucine (L-^{t}Leu), which is an important building block for a range of pharmaceuticals, as peptide bonds involving this amino acid are only slowly hydrolyzed by peptidases. Reductive amination using an LeuDH in combination with the FDH for cofactor recycling was significantly better than other enzymatic strategies and was used for large-scale production [26]. Other non-natural amino acids accessible by this route are L-neopentylglycine, L-β-hydroxyvaline, and 6-hydroxy- L-norleucine.

7.2.1.2 Oxygenases

Oxygenases are enzymes which introduce one (monooxygenases) or two (dioxygenases) oxygen atoms into their substrates. Typically NADH or NADPH serve as reduction equivalents via electron-transfer proteins such as reductases. The major interest in these enzymes for organic synthesis is due to their high regio- and stereoselectivity. Moreover, many of these reactions are difficult to perform by chemical methods, especially if non-activated hydrocarbon moieties need to be transformed. Despite the fact that numerous oxygenases are known, their application in organic synthesis is limited due to a number of problems. These include limited availability of sufficient amounts of enzyme, insufficient stability and often very low specific activity, requirement of costly cofactors, and the need for a reductase. Many enzymes are also membrane-bound, which further limits their application. Some of these problems were overcome by the use of whole-cell systems, preferentially with overexpression of the oxygenase.

P450 Mono-Oxygenases (EC 1.14.13.-) P450-Monooxygenases (also named CYP, Cytochrome P450 enzymes) are widely distributed in nature and play a key role in various steps of primary and secondary metabolism as well as in detoxification of xenobiotic compounds. A range of reactions are catalyzed by these enzymes (Scheme 7.4), which all include the transfer of molecular oxygen to non-activated aliphatic or aromatic X–H bonds (X: –C, –N, –S) [27]. Furthermore, a remarkable number of P450 enzymes are able to epoxidize –C=C– double bonds [28]. For these oxygenation reactions P450 enzymes require cofactors like NADPH or NADH as reduction equivalents. P450 enzymes show characteristic spectral properties: the maximum absorption at 450 nm in the differential spectra of carbon monoxide, which gave them their name [29].

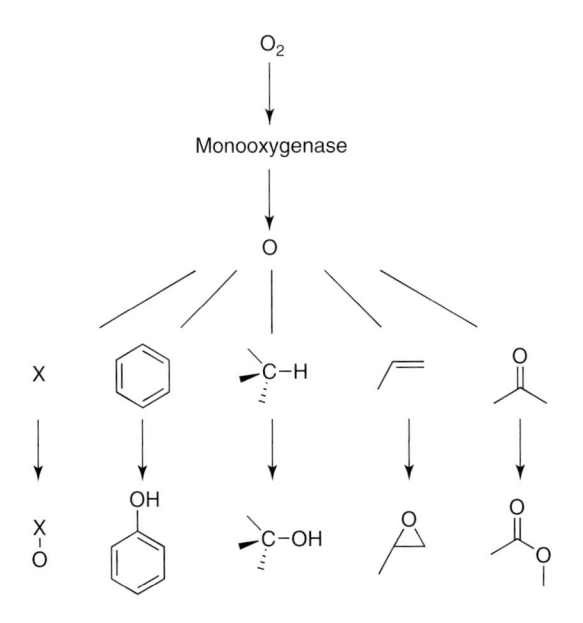

Scheme 7.4 Overview of reactions catalyzed by P450- and Baeyer–Villiger (right hand example) monooxygenases.

The P450 superfamily is one of the largest and oldest gene families [30]. In 2004, the number of P450 encoding sequences was estimated at over 1000 [31]. For example, the genome project of the plant *Arabidopsis thaliana* has led to the identification of more than 270 putative P450 genes [32]. The classification of P450 genes is based on primary sequence homologies. P450 genes are identified by the abbreviation CYP, (cytochrome P) followed by a number denoting the family, a letter designating the subfamily (when two or more exist), and a numeral representing the individual gene within the subfamily. For example, CYP4A1 represents the first gene in the P450 subfamily A of the P450 gene family 4.

Depending on the mechanism of the electron transfer system (reductase system), P450 enzymes are divided into four classes [33, 34]:

- **Class I:** these mainly occur in mitochondrial systems and most bacteria. The electron systems consist of a FAD-domain as reductase and a further iron-sulfur protein
- **Class II:** are often located in the endoplasmic reticulum and require only a single protein for the electron transfer, a FAD/FMN-reductase
- **Class III:** do not require reduction equivalents. The P450 enzymes directly convert peroxygenated substrates which have already 'incorporated' the oxygen
- **Class IV:** only one enzyme is known, which receives its electrons directly from NADH.

Many enzyme structures of CYP have been resolved, and, despite their low sequence homology, they share a close structural similarity. As the activated intermediates are not covalently bound, only three structures in complex with their natural substrates (for CYP101, $P450_{cam}$, and CYP102) have been determined.

Several proposals to bypass the cofactor regeneration of NADPH have been suggested, such as electrochemical methods or the use of cobalt(III)sepulchrate, which can be reduced by cheap zinc dust [35]. Most convenient is the shunt pathway using hydrogen peroxide instead of NADH or NADPH, which has been described for $P450_{cam}$ (Joo *et al.* [36] 1999) and P450 BM-3 [37].

Although P450 monooxygenases catalyze a broad range of interesting reactions which are difficult to perform by chemical means – for instance regioselective hydroxylation of a fatty acid – their application is hampered by their low specific activity, narrow substrate range, usually low stability, and the need for a complex electron transfer system. In addition, most P450 enzymes are only applicable in whole-cell systems. Nevertheless, the recent progress made in the protein engineering of several P450 enzymes may allow for their application in biocatalysis in the future. Further information on P450s can be found in a number of books and reviews [38–42].

Baeyer-Villiger Mono-Oxygenases (EC 1.14.13.16, 1.14.13.22) Baeyer-Villiger Mono-oxygenases (BVMOs) catalyze the biocatalytic counterpart of the chemical Baeyer-Villiger oxidation (BVO) [43], which was first described by Adolf Baeyer and Victor Villiger more than a century ago [44]. The chemical oxidation usually uses

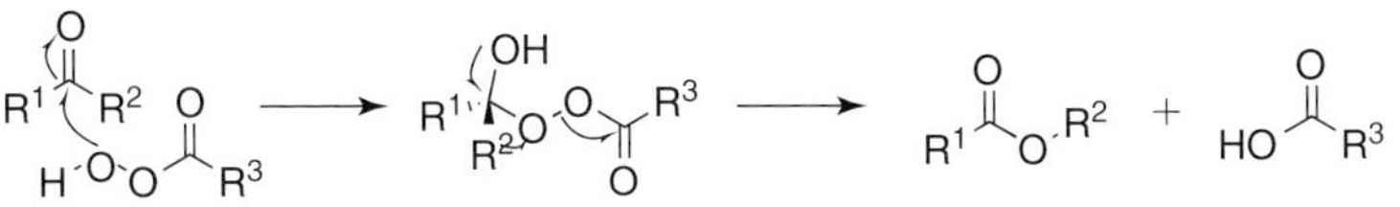

Scheme 7.5 Rearrangement occurring in the chemical Baeyer–Villiger oxidation as proposed by Criegee.

peracids. The mechanism is generally accepted to proceed by a two-step process, which was initially proposed by Criegee [45], Scheme 7.5.

However, the chemical BVO is not stereoselective unless organometallic reagents are employed. Even then, the optical purity of the products is unsatisfactory in most cases [46–49].

The first example of an enzymatic BVO dates back to 1948 for a fermentation of cholestanone with *Proactinomyces erythropolis* [50]. A few years later, a double BVO was suggested for the side-chain degradation of progesterone [51].

In the early 1980s, a range of publications about biocatalysis with newly discovered BVMOs appeared. One enzyme originates from *Pseudomonas putida* [52–55]: the other is produced by the S2-organism *Acinetobacter calcoaceticus* [56]. For both enzymes, it could be demonstrated, that they accept a relatively broad range of mono- and bicyclic ketones. In the case of racemic substrates, the oxidation often proceeded with good to high enantioselectivity. In the 1990s, the BVMO from *Acinetobacter* sp. was cloned and functionally expressed in yeast [57, 58] and later also in *E. coli* [59]. Moreover, it could be shown that a whole-cell biotransformation is possible, although side-reactions – such as a reduction of the ketone substrate to an alcohol by an endogenous ADH present in the whole-cell system – can occur. During the enzymatic BVO one oxygen atom is introduced into a ketone precursor molecule yielding either the corresponding lactone or an ester (depending on whether a cyclic or a acyclic ketone is used). The second oxygen atom is converted into water. For their catalytic action, BVMOs require a cofactor (NADH or NADPH) and they are flavin-dependent (FAD or FMN) (Scheme 7.6).

Recently, a fusion protein between a BVMO covalently linked to a soluble $NADP^+$-dependent phosphite dehydrogenase was created [61], enabling the usage of phosphite as a cheap and sacrificial electron donor for efficient cofactor recycling. Interestingly, kinetic parameters, substrate specificity, and stereoselectivity were not negatively influenced as the fusion enzymes displayed very similar biocatalytic behavior to that of wild-type enzymes.

By far the best-studied BVMO is the enzyme produced by *Acinetobacter calcoaceticus* NCIMB 9871, which is also available as recombinant enzyme, as mentioned above. Some examples exist of the application of this biocatalyst to the conversion of mono- and bicyclic ketones. A quite similar substrate spectrum was reported for BVMO from camphor-induced *Pseudomonas putida*, which produces three different BVMOs: two almost identical enzymes require NADH (each type is generated by induction with either (−) or (+)-camphor), and the third BVMO utilizes NADPH.

Also, acetophenone-converting BVMOs were described originating from, for example, *Arthrobacter* [62], *Nocardia* [63], and *Pseudomonas* species [64, 65]. These

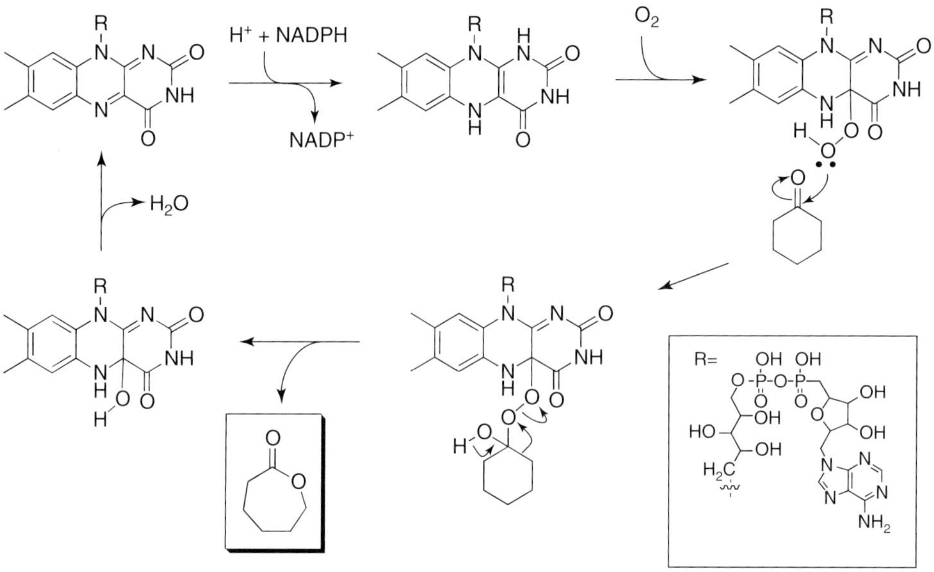

Scheme 7.6 Proposed mechanism for the enzyme-catalyzed Baeyer–Villiger oxidation.

enzymes accept a broad range of ring-substituted acetophenones, which were converted into the corresponding achiral acetate esters.

Conversion of aliphatic straight-chain ketones was described for monooxygenases from *Mycobacterium* sp. [66], *Pseudomonas* [67], and *Nocardia* sp. [68], and only very recently the formation of optically active products and remaining substrates was also shown for the BVMO-catalyzed kinetic resolution of β-hydroxyketones [69]. Although the wild-type enzyme from *Pseudomonas fluorescens* already showed good enantioselectivity, this could be further enhanced by protein engineering using directed evolution [70].

Extensive coverage of the state-of-the-art research regarding BVMOs can be found in recent reviews [71, 72].

7.2.2
Hydrolases (EC 3.1)

7.2.2.1 Lipases (EC 3.1.1.3) and Esterases (EC 3.1.1.1)

Lipases (triacylglycerol hydrolases) are probably the most frequently used hydrolases in organic synthesis [73, 74]. They are widely found in nature (animal, man, bacteria, yeast, fungi, and plants) and a considerable number of enzymes are commercially available. Their natural function is the hydrolysis and re-esterification of triglycerides, that is, natural fats and oils. The reaction is catalyzed by a catalytic triad composed of Ser, His, and Asp (sometimes Glu) similar to serine peptidases and carboxyl esterases (EC 3.1.1.1). The mechanism for ester hydrolysis or formation is

7.2 Examples

essentially the same for lipases and esterases and is composed of four steps. First, the substrate reacts with the active-site serine yielding a tetrahedral intermediate stabilized by the catalytic His- and Asp-residues. Next, the alcohol is released and a covalent acyl-enzyme complex is formed. Attack of a nucleophile (water in hydrolysis, alcohol in (*trans*)-esterification) forms again a tetrahedral intermediate, which collapses to yield the product (an acid or an ester) and free enzyme (Scheme 7.7).

Besides their use in organic synthesis for the production of optically active compounds, lipases are also used in various other areas such as laundry detergents, cheese-making, modification of natural fats and oils, synthesis of sugar esters, and simple esters used in personal care (e.g., myristyl myristate, decyl cocoate).

Lipases are distinguished from esterases by their substrate specificity: lipases accept long-chain fatty acids (in triglycerides) as substrates, esterases prefer short-chain fatty acids. More generally it can be stated that lipases readily accept water-insoluble substrates, whereas esterases prefer water-soluble compounds. A further difference was found in the 3D structures of these enzymes: lipases contain a hydrophobic oligopeptide (often called a '*lid*' or '*flap*') – which is not present in esterases – covering the entrance to the active site. Lipases preferentially act at a water-organic solvent (or oil) interface, which presumably accounts for a movement of the lid, making the active site accessible for the substrate. This phenomenon is referred to as '*interfacial activation*.' Further characteristic structural features of lipases are the α,β-hydrolase fold [75] and a consensus sequence around the active site serine (Gly-X-Ser-X-Gly, where X denotes any amino acid). It could be shown

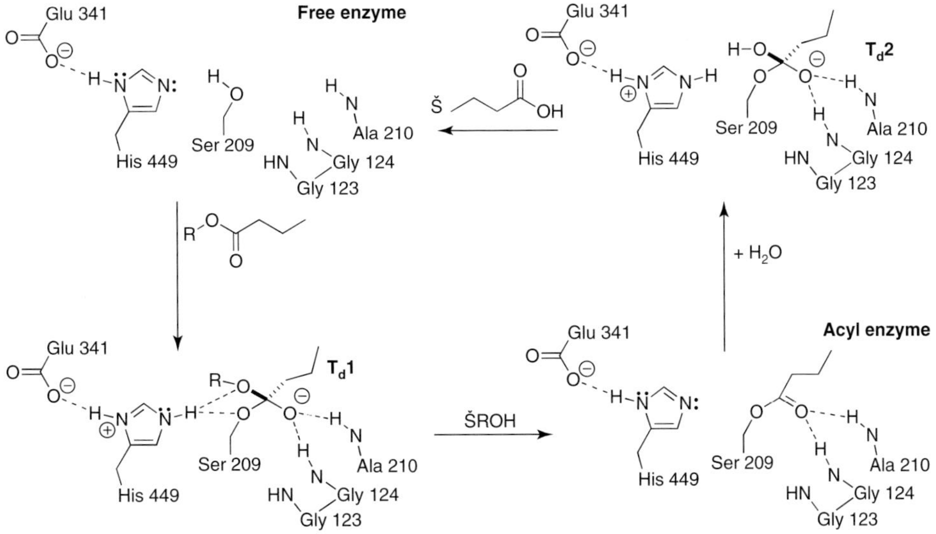

Scheme 7.7 Mechanism of lipase or esterase-catalyzed hydrolysis of an ester. The amino acid numbering corresponds to the active site of lipase from *Candida rugosa*, CRL.

that after removal of the lid by genetic engineering, the activity of a lipase was improved in solution, mainly for applications in the laundry/detergent area.

The main interest in the application of lipases – and to a lesser extent esterases – in organic chemistry is for the following reasons:

- Lipases are highly active in a broad range of non-aqueous solvents.
- They often exhibit excellent stereoselectivity.
- They accept a broad range of esters other than triglycerides.
- They accept nucleophiles other than water (i.e., alcohols, amines).

As lipases are very stable in a range of organic solvents, acylation is often the preferred reaction. In order to drive the kinetic resolution to completion, either one substrate is used in excess or activated acyl donors are used. Using enol esters such as vinyl acetate or isopropenyl acetate, the reaction is practically irreversible, as the vinyl alcohol generated undergoes keto-enol tautomerization to carbonyl compounds (i.e., acetaldehyde from vinyl esters, acetone from isopropenyl esters). However, acetaldehyde could inactivate the enzyme via formation of a Schiff's base with lysine residues. Anhydrides such as acetic acid anhydride are cheap, but the free acid generated causes a drop in pH and could also participate in a slower background acylation.

In the literature, >2000 examples (!) for the resolution with lipases can be found [76]. In general, reactions proceed with good to excellent enantioselectivity for secondary alcohols, but only moderate selectivity is found for primary alcohols. Researchers have tried to predict the outcome of a lipase-catalyzed resolution based on the substrate structure and the type of lipase. Kazlauskas and co-workers developed empirical rules which allow prediction of which enantiomer of a primary or a secondary alcohol reacts faster than the other [77, 78]. According to this rule, high enantioselectivities can be achieved with substrates bearing a medium-sized (e.g., methyl-) and large-sized (e.g., phenyl-) substituent such as α-phenylethanol.

From the considerable number of available lipases, only a few have been shown to be broadly applicable, exhibiting high enantioselectivity and having sufficient stability. These are porcine pancreatic lipase (PPL), lipase B from *Candida antarctica* (CAL-B, tradenames: Novozyme 435®), lipases from *Burkholderia cepacia* (BCL, former names: *Pseudomonas cepacia*, *Ps. fluorescens*), *Candida rugosa* (CRL, former name: *Candida cylindracea*), and *Rhizomucor miehei* (RML, tradename Lipozyme RM IM®).

In contrast to secondary and primary alcohols, fewer examples of the application of lipases can be found for the kinetic resolution of carboxylic acids. Interestingly, lipases which show high selectivity for alcohols (e.g., PCL (*Pseudomonas cepacia* lipase), CAL-B), are much less selective in the resolution of carboxylic acids, and vice versa (e.g., CRL).

Until recently, only a few lipases (e.g., CRL and lipase A from *Candida antarctica*) were known to accept tertiary alcohols. Due to their bulky structure, it was believed that these compounds do not fit into the active site of the enzyme. It was later discovered that a certain amino acid motif (GlyGlyGly(Ala)X-motif, where X denotes any amino acid) located in the oxyanion binding pocket of lipases and esterases determines activity toward tertiary alcohols [79]. All enzymes bearing this motif (e.g., PLE, several acetyl choline esterases, and an esterase from *Bacillus subtilis*) were active toward several acetates of tertiary alcohols, while enzymes bearing the more common GlyX-motif did not hydrolyze the model compounds. The low-to-modest enantioselectivity of these enzymes could be overcome by rational protein design and directed evolution [80–83]. A recent review covers the resolution of tertiary alcohols, including examples of protein engineering of esterases for efficient resolution of these compounds [84].

PLE is probably the most useful carboxyl esterase for organic synthesis. It accepts a broad range of substrates, which are often converted with excellent stereoselectivity. Besides the kinetic resolution of various racemates, PLE was shown to be efficient in the desymmetrization of prostereogenic and *meso*-compounds [85–87]. PLE was traditionally isolated from pig liver by extraction and consists of several isoenzymes with the α-, β-, and γ-subunits as the most dominant ones. Unfortunately, the isoenzymes can differ substantially in their substrate specificity and enantioselectivity. This was overcome in the last decade by the discovery and functional expression of various isoenzymes in the yeast *P. pastoris* or in *E. coli* [88, 89]. This now allows the production of recombinant PLE as a stable product without the interfering influence of other isoenzymes and hydrolases. Properties and applications of PLE have recently been reviewed [90].

7.2.2.2 Peptidases, Acylases, and Amidases

Peptidases (EC 3.4.21.x) and amidases (EC 3.5.1.x) catalyze both formation and hydrolysis of amide links. Although their natural role is hydrolysis, they are also used to form amide bonds. Two different strategies have been applied for this: a thermodynamic or a kinetic control. In thermodynamically controlled syntheses, reaction conditions are changed to shift the equilibrium toward synthesis instead of hydrolysis. Hydrolysis of peptides is favored by -5 kcal mol^{-1}. It is driven mainly by the favorable solvation of the carboxylate and ammonium ions and also depends on pH. One common way to shift the equilibrium toward synthesis is to replace water with an organic solvent. The organic solvent suppresses the ionization of the starting materials and also reduces the concentration of water. Other common ways to shift the equilibrium are to increase the concentrations of the starting materials or to choose protective groups that promote precipitation of the product.

In kinetically controlled syntheses, an activated carboxyl component, usually an ester or amide, is used. This reacts with the enzyme to form an acyl enzyme intermediate, which then reacts either with an amine to form the desired amide or with water to form a carboxylic acid. Because the starting material is an activated carboxyl component, reactions are faster in the kinetically-controlled approach than in the thermodynamically-controlled approach. Because the kinetically-controlled approach requires an acyl enzyme intermediate, only serine hydrolases (e.g., the peptidase subtilisin or the amidase penicillin G amidase) are suitable. Metallo-peptidases such as thermolysin work only in thermodynamically controlled syntheses. Peptidase-catalyzed peptide synthesis was first reported in 1901 [91]. Since the late 1970s peptidases have been more often used in peptide synthesis. The advantages of an enzyme-catalyzed peptide synthesis are mild conditions, no racemization, minimal need for protective groups, and high regio- and enantioselectivity. An example of a kinetically controlled peptide synthesis is the α-chymotrypsin-catalyzed production of kyotorphin (Tyr-Arg), an analgesic dipeptide [92]. To minimize the hydrolysis of the acyl enzyme intermediate, a high concentration of the nucleophile was used, which was possible only because the charged maleyl protective group increased the solubility of the carboxyl component. Subtilisin accepts a broader range of substrates than other peptidases, so it is also used for amide couplings involving non-natural substrates [93]. When coupling a D-amino acid, it is best to use it as the nucleophile, not as the carboxyl donor, because subtilisin is more tolerant of changes in the nucleophile than in the carboxyl group. Acylases and amidases also catalyze the formation of peptide bonds. An important application here is in the hydrolysis and synthesis of the β-lactam peptide antibiotics (penicillins and cephalosporins). Further information can be found in reviews [94–97].

7.2.2.3 Nitrilases (EC 3.5.5.1) and Nitrile Hydratases (EC 4.2.1.84)

Nitriles are important precursors for the synthesis of carboxylic acid amides and carboxylic acids. Chemical hydrolysis of nitriles requires a strong acid or base at high temperatures. Nitrile-hydrolyzing enzymes have the advantage that they react under mild conditions and do not produce large amounts of by-products. In addition, the enzyme can be regio- and stereoselective. Two different enzymatic pathways can be used to hydrolyze nitriles). Nitrilases (EC 3.5.5.1) directly catalyze the conversion of a nitrile into the corresponding acid plus ammonia. In the other pathway, a nitrile hydratase (NHase, EC 4.2.1.84; a lyase) catalyzes the hydration of a nitrile to the amide, which can be converted to the carboxylic acid and ammonia by an amidase (EC 3.5.1.4).

Pure nitrilases and nitrile hydratases are usually unstable, and thus the biocatalyst is often used as whole-cell preparation. The nitrile-hydrolyzing activity must be induced first. Common inducers are benzonitrile, isovaleronitrile, crotononitrile, and acetonitrile, but the inexpensive inducer urea also works. In addition, inducing with ibuprofen or ketoprofen nitriles can yield enantioselective enzymes [98]. After induction, preparative conversion is usually performed by adding the nitriles either during cultivation or by employing resting cells. Most commonly used strains are from *Rhodococcus* sp., the most important ones being subspecies of *Rh. rhodochrous*.

Nitrilases are cysteine hydrolases, which act via an enzyme-bound imine intermediate using a Glu-Lys-Cys catalytic triad. All nitrilases are inactivated by thiol reagents (e.g., 5,5′-dithiobis(2-nitrobenzoic acid)), indicating that they are sulfhydryl enzymes.

Two different groups of nitrile hydratases have been described, which require Fe(III) or Co(III) ions [99]. The *Rh. rhodochrous* J1 strain produces two kinds of NHases differing in their molecular weight (520 and 130 kDa). The high-molecular-weight NHase acts preferentially on aliphatic nitriles, whereas the smaller enzyme also has high affinity toward aromatic nitriles. Most nitrile hydratases accept aliphatic nitriles only (e.g., from *Arthrobacter* sp. J-1, *Brevibacterium*

R312, *Pseudomonas chlororaphis* B23); however strains have also been described which can act on arylalkylnitriles, arylacetonitriles, and heterocyclic nitriles.

The synthesis of acrylamide and nicotinic acid using nitrile hydratase from *Rhodococcus rhodochrous* [100] has been commercialized. Until recently, only approximately 15 nitrilases were described. Researchers at Diversa (San Diego, USA), now named Verenium Inc., discovered more than 200 unique nitrilases from genomic libraries obtained from DNA extracted from environmental samples [101]. Twenty-seven enzymes afforded mandelic acid in >90% *ee* under conditions of DKR. Out of these biocatalysts, one nitrilase afforded (*R*)-mandelic acid in 86% yield and 98% *ee*. Further studies showed that this nitrilase also converted a broad range of mandelic acid derivatives and analogs with high activity and similar stereoselectivity. Another nitrilase exhibited high activity and broad substrate tolerance toward aryllactic acid derivatives, which were also converted in a DKR.

A few groups investigated the hydrolysis of prochiral dinitriles. Hydrolysis of 3-hydroxyglutaronitriles revealed that Bn- or Bz-protecting groups were required to achieve acceptable enantiomeric excess for the monocarboxylic acid [102]. For the nitrilases discovered by Diversa Inc., it could be shown that a range of enzymes show high conversion (>95%) and selectivity (>90% *ee*). The best enzyme gave 98% yield and 95% *ee* for the (*R*)-product [101]. In addition, 22 enzymes that afford the opposite enantiomer with 90–98% *ee* were discovered. In a later study, the most effective (*R*)-nitrilase was optimized by protein engineering to withstand high substrate concentrations while maintaining high enantioselectivity. The best variant obtained by the 'gene-site saturation mutagenesis' technique contained a single mutation (Ala190His) and allowed the production of the (*R*)-acid at 3 M substrate concentration with 96% yield at 98.5% *ee* [103]. More examples of the characterization and application of nitrilases and nitrile hydratases/amidases can be found in recent reviews [76, 104, 105].

7.2.2.4 Hydantoinases (EC 3.5.2.-)

Hydantoinases are valuable enzymes for the production of optically pure D- and L-amino acids. They catalyze the reversible hydrolytic cleavage of hydantoins and 5′-monosubstituted hydantoins. In combination with carbamoylases (EC 3.5.1.-), the reaction yields L- or D-amino acids depending on the stereoselectivity of the enzymes (Scheme 7.8). To shift the yield above 50%, a DKR starting from racemic hydantoins is also possible by either working at slightly alkaline pH values (pH > 8) or using hydantoin-specific racemases. Since racemic hydantoins are readily available by chemical synthesis (e.g., Strecker synthesis), the synthesis of non-natural amino acids is also feasible if the enzyme has an appropriate substrate specificity.

Especially since the 1970s, a broad range of hydantoinases and carbamoylases were discovered by screening owing to their importance in the synthesis of optically pure amino acids. Nowadays, hydantoinases from, for example, *Arthrobacter* sp., *Nocardia* sp., *Bacillus* sp., and *Pseudomonas* sp. have been described, and a range of D-amino acids are accessible by using them. Crystal structures have been elucidated, for instance, for hydantoinases from *Arthrobacter aurescens* and *Thermus* sp. and the

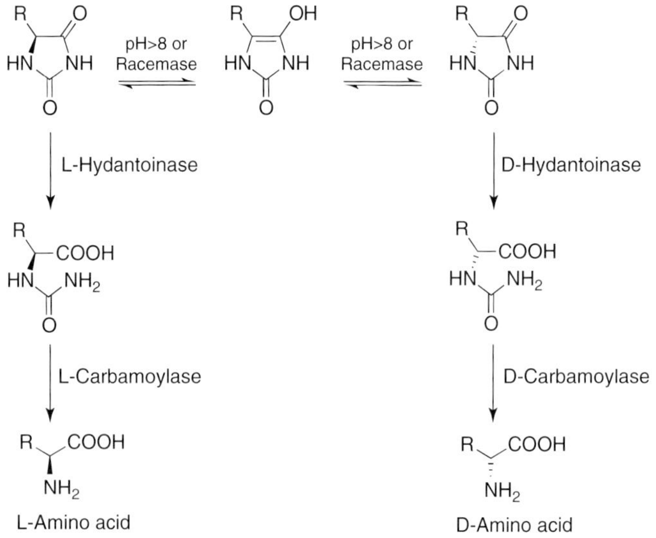

Scheme 7.8 Synthesis of L- or D-amino acids using a combination of hydantoinase and carbamoylase. Complete conversion of racemic D,L-hydantoin can be achieved in a dynamic kinetic resolution by racemization at alkaline pH or with specific racemases.

D-carbamoylase from *Agrobacterium*. For this D-carbamoylase, the catalytic center was found to consist of a glutamine, a lysine, and a cysteine residue [106, 107].

In the hydantoinase process, a hydantoinase, a carbamoylase, and a racemase are combined to afford 100% yield of the desired amino acid. Whereas isolated enzymes were used initially, current processes use a genetically engineered *E. coli* expressing all three enzymes under the control of a rhamnose-inducible promotor. With this system, L-tryptophan was produced with an efficiency six times that achieved by the wild-type strain *Arthrobacter aurescens*.

7.2.3
Lyases (EC 4)

7.2.3.1 Hydroxynitrile Lyases (EC 4.1.2.-)

HNLs, often also named oxynitrilases, catalyze the reversible stereoselective addition of HCN to aldehydes and ketones (Scheme 7.1). Thus, HNLs allow the synthesis of optically pure compounds from prostereogenic substrates at (theoretically) quantitative yield. In addition, many HNLs show excellent stereoselectivity (for reviews see Refs. [108–111]). The resulting α-hydroxynitriles are versatile intermediates for a broad variety of chiral synthons, which can be obtained by subsequent chemical synthesis.

One of the earliest reports on biocatalysis involved the use of an HNL, when Rosenthaler used an HNL-containing extract (emulsion) for the preparation of

(R)-mandelonitrile from benzaldehyde and HCN [112]. However, little attention was paid to this discovery until the 1960s, when the corresponding enzyme was isolated, characterized, and used for the production of enantiomerically enriched (R)-cyanohydrins.

More than 3000 plant species are known to release HCN from their tissues, and for approximately 300 plants the HCN is released from a cyanogenic glycoside or lipid. These cyanide donors can be cleaved either spontaneously or by the action of an enzyme such as HNL.

Until about a decade ago, all HNLs were isolated from different plant sources, and about 11 enzymes with either (S)- or (R)-stereoselectivity were described. The oxynitrilase from *Prunus* sp. contains the cofactor FAD, but this is not involved in redox reactions. Instead, it seems to have a structure-stabilizing effect. Some of the HNLs are glycosylated, and most of them are composed of subunits. As only small amounts of HNLs were available by extraction from plant sources, the enzymes from *Hevea brasiliensis* (rubber tree), *Manihot esculenta* (cassava), and *Linum usitatissimum* (flax) have been cloned and overexpressed in *E. coli* or *P. pastoris*. For instance, the gene encoding the enzyme from *Hevea brasiliensis* was expressed at very high levels in the yeast *Pichia pastoris* under the control of the AOX1 (alcohol oxidase) promoter. On a laboratory scale, a production of 23 g L^{-1} of pure HNL was achieved in a high-cell-density fermentation, and the protein could be recovered by one-step ion-exchange chromatography [109]. The crystal structures have been elucidated for enzymes from *Hevea brasiliensis*, *Sorghum bicolor*, *Manihot esculenta*, and *Prunus amygdalus* (almond). Interestingly, the enzymes share an α/β-hydrolase fold and also contain an active-site serine usually embedded in a GXSXG-motif, similarly to lipases and esterases. This might indicate an evolutionary relationship between the two enzyme classes hydrolases and lyases.

For the HNL from *Sorghum bicolor* the following mechanism was proposed (Scheme 7.9): In contrast to a histidine residue serving as a general base in serine peptidases and lipases, the carboxylate group of a C-terminal tryptophan (Trp$_{270}$ in *Sorghum bicolor*) abstracts a proton from the cyanohydrin hydroxyl group. A water molecule bound to the active site appears to be involved in proton transfer. Thus, the entering cyanohydrin is hydrogen-bonded to Ser$_{158}$ and Trp$_{270}$, which abstracts a proton from the OH-group of the substrate. Next, a proton is transferred from the tryptophan via the active site water to the nitrile leaving group. Protonation of the cyanide ion results in the products, 4-hydroxybenzaldehyde and HCN. This model can also explain the (S)-stereoselectivity observed for the reverse reaction [60].

Beside the availability of HNLs, two further problems had to be solved to allow an efficient application of HNLs in organic synthesis: (i) performing the reactions in water-immiscible organic solvents and (ii) reaction at low pH, typically in the range of pH 4–5.

Both methods suppress the competing chemical reaction which results in lower optical purities of the product. In contrast to the enzymes from *H. brasiliensis* and *M. esculenta*, where aliphatic and aromatic aldehydes function as substrates, the HNL from *Sorghum bicolor* only catalyzes the formation and cleavage of aromatic

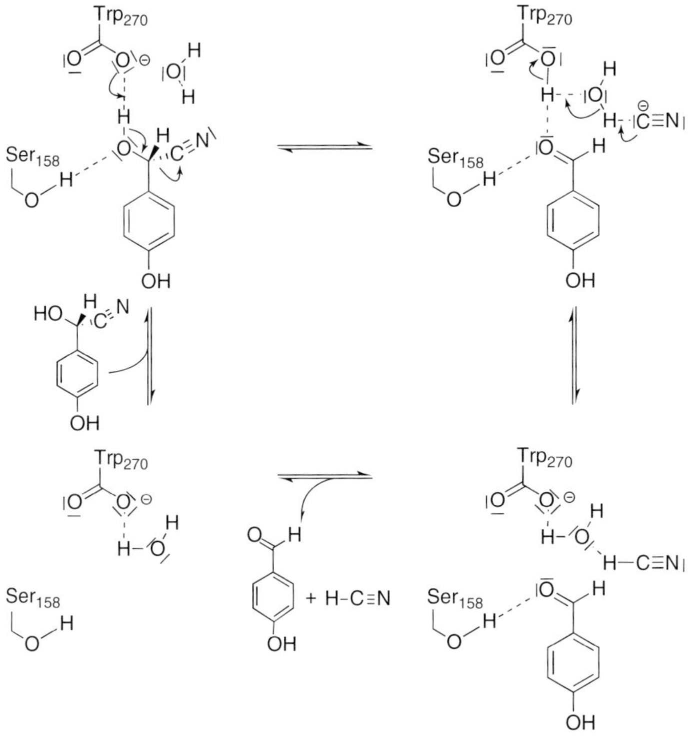

Scheme 7.9 Suggested reaction mechanism for HNL from *Sorghum bicolor* [60].

(S)-cyanohydrins. As mentioned above, the use of water-immiscible solvents is recommended for organic synthesis with HNLs to avoid a non-stereoselective chemical reaction. In literature reports, higher yields and excellent optical purities were only possible using PaHNL immobilized on Avicel (a cellulose membrane) in diisopropylether.

In contrast, reactions in water/ethanol mixtures gave products with inferior optical purity, for example, only 11% *ee* for the *meta*-substituted phenyl derivative. This effect was less pronounced for reactions using ketones. More recently, processes for the production of (S)-cyanohydrins catalyzed by HNLs from *Hevea brasiliensis* and *Manihot esculenta* and the subsequent chemical hydrolysis to (S)-hydroxycarboxylic acids have been developed. For instance, the synthesis of (S)-*m*-phenoxybenzaldehyde cyanohydrin – an intermediate in the synthesis of pyrethroids – in a biphasic system has been commercialized by DSM Chemie Linz and Nippon Shokubai.

7.2.3.2 Aldolases (EC 4.1.2.-; 4.1.3.-)

Aldolases catalyze the biological equivalent to the chemical Aldol-reaction, the formation of carbon-carbon bonds by (reversible) stereocontrolled addition of a nucleophilic ketone to an electrophilic aldehyde acceptor. Aldolases are usually classified according to the nature of the nucleophilic component into (i) pyruvate (and phosphoenolpyruvate)-, (ii) dihydroxyacetone phosphate (DHAP)-, (iii) acetaldehyde-, and (iv) glycine-dependent enzymes.

The most important aldolases use DHAP as they allow the formation of two new stereocenters in a single reaction. Also, a range of corresponding enzymes were identified. The ability of aldolases to accept a variety of non-natural acceptor substrates and to generate new stereocenters of known absolute and relative stereochemistry makes them powerful tools for asymmetric synthesis.

Depending on their mechanism, aldolases are classified into Type I and Type II enzymes. Type I enzymes are predominantly found in higher plants and animals and are metal-cofactor independent. The free amino group of a lysine residue in the active site reacts with DHAP with formation of a Schiff's base intermediate. An enamine is formed after deprotonation, and this then attacks the aldehyde. Finally the Schiff's base intermediate product decomposes after reaction with water, and the aldol and enzyme are released. Type II aldolases occur mostly in bacteria and fungi and are Zn^{2+}-dependent. The zinc ion acts as a Lewis acid, which polarizes the carbonyl group in DHAP.

As pointed out above, aldolases are highly specific, and the stereochemical outcome of an aldol reaction can be usually predicted independently of the substrate structure. Consequently, the synthesis of all four diastereomers accessible from DHAP and an aldehyde is possible by using four different aldolases (Scheme 7.10).

Numerous examples of the application of aldolases can be found in a number of excellent reviews and book chapters [113–115]. In the following paragraphs only two types of aldolases are discussed.

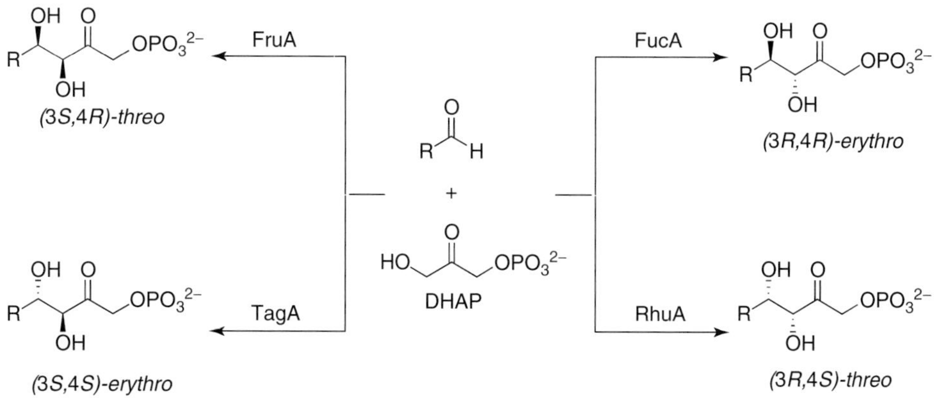

Scheme 7.10 Aldolase reactions catalyzed by the four stereocomplementary aldolases FruA (fructose-1, 6-diphosphate aldolase), FucA (Fuculose-1-phosphate aldolase), TagA (Tagatose-1,6-diphosphate aldolase), and RhuA (Rhamnulose-1-phosphate aldolase).

DHAP-Dependent Aldolases The most often used DHAP-dependent enzymes are Fru-aldolases (FruA, often also abbreviated D-fructose-diphosphate (FDP) aldolases, EC 4.1.2.13), which catalyze the reaction between DHAP and D-glyceraldehydes-3-phosphate to form FDP. The equilibrium constant for this reaction is $\sim 10^4$ M^{-1} in favor of FDP formation. The enzyme has been isolated from various eukaryotic and prokaryotic sources and both type I and type II biocatalysts were described. Type I enzymes are usually tetramers (160 kDa); type II FDP aldolases are dimers ($\sim$80 kDa). Sequence homologies are usually very low between type I and II aldolases and especially in-between type II enzymes. The crystal structures of the FDP-aldolase from rabbit muscle aldolase (RAMA) and others have been determined. RAMA and several type II aldolases from microbial sources have been cloned and overexpressed, which substantially facilitates access to the enzymes.

The phosphorylated ketone DHAP must be available for these aldolase-catalyzed reactions. Although DHAP is in principle available via the reverse reaction, that is, from a retro-aldol reaction with FDP aldolase from FDP, this approach requires triosephosphate isomerase and can hardly be used if a different aldolase is used in the forward aldol reaction, which then leads to a complex product mixture.

The kinase-catalyzed phosphorylation of dihydroxyacetone is hampered by the requirement of ATP regeneration. A chemical alternative is a multi-step synthesis, which usually provides a DHAP dimer, which is stable and can easily be converted into DHAP by acid hydrolysis. Alternatively, arsenate derivatives of DHAP can be used albeit with lowered aldolase activity. One approach for DHAP synthesis was based on phosphorylation of glycerol using a phytase – a cheap and readily available enzyme – and inorganic pyrophosphate. Its feasibility was demonstrated for the synthesis of 5-deoxy-5-ethyl-D-xylulose in a one-pot reaction combining four enzymatic steps with three different enzymes starting from glycerol.

First, phosphorylation of glycerol by reaction with pyrophosphate in the presence of phytase at pH 4.0 in 95% glycerol afforded racemic glycerol-3-phosphate quantitatively. The L-enantiomer was then oxidized with glycerolphosphate oxidase (GPO) to DHAP under aerobic conditions at pH 7.5. Hydrogen peroxide is removed with the aid of catalase.

In-situ generated DHAP reacts with butanal using FruA followed by dephosphorylation of the aldol adduct, again using phytase at pH 4. Overall, 5-deoxy-5-ethyl-D-xylulose was obtained in 57% yield from L-glycerol-3-phosphate. The phytase 'on/off-switch' by changing the pH value was the key to controlling phosphorylation and dephosphorylation [116]. More recently, it was shown that dihydroxyacetone borate is also accepted by DHAP-dependent aldolases. The major advantage is that dihydroxy acetone is simply added to borate buffer and DHAP is no longer required, making this approach extremely useful for organic synthesis with aldolases [117].

Pyruvate/Phosphoenolpyruvate-Dependent Aldolases The best studied enzyme utilizing pyruvate is *N*-acetylneuraminate (NeuAc) aldolase (EC 4.1.3.3). This catalyzes the reversible condensation of pyruvate with *N*-acetylmannosamine

(ManNAc) to form sialic acid. The initial products of aldol cleavage are α-ManNAc and pyruvate. Although *in vivo* the equilibrium favors the retro-aldol reaction ($K_{eq} \approx 12.7$ M^{-1}), the aldol reaction for organic synthesis can be achieved using excess pyruvate, which yields the β-anomer of NeuAc. NeuAc aldolase is a Schiff's base-type I aldolase and has been isolated from bacteria and animals. The enzymes from *Clostridium* and *E. coli* are commercially available and the *E. coli* enzyme has been cloned and overexpressed. Optimum pH is 7.5, but activity is retained between pH 6 and pH 9. It can be used in solution, in immobilized form, or enclosed in a membrane system. With this aldolase, glycoconjugates, which play important roles in cell-cell interactions and cell adhesion, were synthesized on the multigram scale.

7.2.4
Transaminases

Transaminases (TAs, EC 2.6.1.18) are highly versatile biocatalysts for the synthesis of optically active amines or α-amino acids [118] by transfer of an amino group from a donor substrate to an acceptor compound utilizing the cofactor pyridoxyl-5-phosphate (PLP) [97, 119]. This reaction can either be performed as a kinetic resolution of a racemic amine or as an asymmetric synthesis starting from a prostereogenic ketone (Scheme 7.1). α-TAs require the presence of a carboxylic acid function in the α-position to the keto or amine functionality and hence only allow the formation of α-amino acids; ω-TAs are much more useful as they in principle accept any ketone (or amine). The first groundbreaking work in this field was done in the late 1980s by the US-American company, Celgene [120]. In the last decade, ω-TAs, which have been extensively studied and identified in a mere dozen organisms [121–129], were biochemically characterized and also overexpressed in microbial hosts such as *E. coli*, with the enzyme from *Vibrio fluvialis* being probably the most intensively studied ω-TA. Most ω-TAs exhibit (S)-selectivity, but also a few examples of (R)-selective enzymes were discovered [123, 126]. Some of the ω-TAs are also capable of converting β-amino acids [130–133]. Because most ω-TAs show excellent stereoselectivity, they presently offer the unique possibility to synthesize optically active amines or β-amino acids directly from the prostereogenic ketone with a theoretically quantitative yield. Although the great potential of ω-TAs – especially for the asymmetric synthesis of chiral amines from ketones – was recognized many years ago, only more recently developed strategies allowed their efficient use.

A major limitation in the asymmetric synthesis starting from prostereogenic ketones is the unfavorable equilibrium. Kim and coworkers reported [134] that only a 0.5% yield of product amine is formed from acetophenone even if a 10-fold excess of alanine, serving as the amine donor, was used.

Hence, a powerful method is needed to shift the reaction equilibrium. Several approaches have therefore been developed to shift the equilibrium toward product formation in asymmetric synthesis.

The first strategy was the removal of pyruvate with lactate dehydrogenase (LDH), and this enabled up to 90% conversion in a system using dried whole cells containing the TA. Very recently, Kroutil's group reinvestigated the equilibrium shift using LDH more systematically, that is, in combination with a glucose dehydrogenase for cofactor recycling [126]. A range of ketones (50 mM) were efficiently converted into the respective amines at high-to-quantitative conversions and with excellent enantiomeric purities (>98% ee).

An elegant method is to recycle pyruvate with ammonia, NADH, and amino acid dehydrogenase to alanine [125]. The formed NAD^+ can be regenerated using the well-established FDH cofactor recycling system. In the overall reaction, a ketone is converted with ammonium formate yielding optically pure amine, water, and carbon dioxide – which essentially resembles an asymmetric reductive amination catalyzed by an amine dehydrogenase. In most cases, high yields of 90–99% were obtained.

Alternatively, pyruvate decarboxylase (PDC) can be used, yielding acetaldehyde and CO_2 [135, 136]. This is a very simple process since only one additional enzyme and no cofactor recycling of NADH is needed and the reaction equilibrium is irreversibly shifted due to carbon dioxide formation. Furthermore, PDCs are commercially available, or a crude extract from *Zymomonas mobilis* can be used. Indeed, conversions similar or slightly higher compared to processes using LDH were achieved [137].

The current status of the use of TAs can be found in a recent review [138].

7.3
Summary/Conclusions

In this chapter, various examples of the use of enzymes in organic synthesis have been summarized to provide an overview and to act as an 'appetizer' to spark an interest in biocatalysis. It has been shown that a very broad range of biocatalysts have been discovered, which can be employed efficiently in the synthesis of numerous organic compounds. Nature provides chemists with enzymes for many types of well-known and useful reactions. The major achievements made in enzyme discovery, molecular biology, enzyme production, and protein engineering have turned the application of biocatalyts into a mature field, which is already being applied successfully in numerous industrial syntheses and can also be readily used by researchers facing challenging synthesis problems.

References

1. Enzyme Nomenclature Database, http://www.expasy.ch/enzyme/.
2. Fersht, A. (1985) *Enzyme Structure and Mechanism*, vol. 2, Freeman, New York.
3. Fischer, E. (1894) Einfluss der configuration auf die wirkung der enzyme. *Ber. Dtsch. Chem. Ges.*, **27** (3), 2985–2993.

4. Koshland, D.E. (1958) Application of a theory of enzyme specificity to protein synthesis. *Proc. Natl. Acad. Sci. U.S.A.*, **44**, 98–104.
5. Grunwald, P. (2009) *Biocatalysis: Biochemical Fundamentals and Applications*, Imperial College Press, London.
6. Bornscheuer, U.T. and Kazlauskas, R.J. (2006) *Hydrolases in Organic Synthesis – Regio- and Stereoselective Biotransformations*, 2nd edn, Wiley-VCH Verlag GmbH, Weinheim.
7. Buchholz, K., Kasche, V., and Bornscheuer, U.T. (2005) *Biocatalysts and Enzyme Technology*, Wiley-VCH Verlag GmbH, Weinheim.
8. Faber, K. (2004) *Biotransformations in Organic Chemistry: A Textbook*, 5th edn, Springer, Berlin.
9. Liese, A., Seelbach, K., and Wandrey, C. (2006) *Industrial Biotransformations*, 2nd edn, Wiley-VCH Verlag GmbH, Weinheim.
10. Bommarius, A.S. and Riebel, B.R. (2004) *Biocatalysis*, vol. 1, Wiley-VCH Verlag GmbH, Weinheim, p. 611.
11. Patel, R.N. (2000) *Stereoselective Biocatalysis*, Marcel Dekker, New York.
12. Patel, R.N. (2006) *Biocatalysis in the Pharmaceutical and Biotechnology Industries*, CRC Press, Boca Raton, FL.
13. Schoemaker, H.E., Mink, D., and Wubbolts, M.G. (2003) Dispelling the myths – biocatalysis in industrial synthesis. *Science*, **299**, 1694–1697.
14. Schmid, A. et al. (2001) Industrial biocatalysis today and tomorrow. *Nature*, **409**, 258–268.
15. Faber, K. and Kroutil, W. (2005) New enzymes for biotransformations. *Curr. Opin. Chem. Biol.*, **9**, 181–187.
16. Martín-Matute, B. and Bäckvall, J.-E. (2007) Dynamic kinetic resolution catalyzed by enzymes and metals. *Curr. Opin. Chem. Biol.*, **11**, 226–232.
17. Chen, C.S. et al. (1982) Quantitative analyses of biochemical kinetic resolutions of enantiomers. *J. Am. Chem. Soc.*, **104**, 7294–7299.
18. Chen, C.S. et al. (1987) Quantitative analyses of biochemical kinetic resolution of enantiomers. 2. Enzyme-catalyzed esterifications in water-organic solvent biphasic systems. *J. Am. Chem. Soc.*, **109**, 2812–2817.
19. Sperry, W.M. and Brand, F.C. (1941) A study of cholesterol esterase in liver and brain. *J. Biol. Chem.*, **137**, 377–387.
20. Sym, E.A. (1936) Action of esterase in the presence of organic solvents. *Biochem. J.*, **30**, 609–617.
21. Klibanov, A.M. (1990) Asymmetric transformations catalyzed by enzymes in organic solvents. *Acc. Chem. Res.*, **23**, 114–120.
22. Klibanov, A.M. (1989) Enzymatic catalysis in anhydrous organic solvents. *Trends Biochem. Sci.*, **14**, 141–144.
23. Bommarius, A.S. (2002) in *Enzyme Catalysis in Organic Synthesis* (eds K. Drauz and H. Waldmann), Wiley-VCH Verlag GmbH, Weinheim, pp. 1047–1063.
24. Hummel, W. et al. (1987) Isolation of L-phenylalanine dehydrogenase from Rhodococcus sp. M4 and its application for the production of L-phenylalanine. *Appl. Microbiol. Biotechnol.*, **26**, 409–416.
25. Ohshima, T. and Soda, K. (2000) in *Stereoselective Biocatalysis* (ed. R.N. Patel), Marcel Dekker, New York, pp. 877–877.
26. Bommarius, A.S. et al. (1992) in *Chirality in Industry* (eds A.N. Collins, G.N. Sheldrake, and J. Crosby), John Wiley & Sons, Ltd, London, pp. 371–397.
27. Goldstein, J.A. and Faletto, M.B. (1993) Advances in mechanisms of activation and deactivation of environmental chemicals. *Environ. Health Perspect.*, **100**, 169–176.
28. Lewis, D.F.V. (1996) *Cytochromes P450: Structure, Function and Mechanism*, Taylor & Francis, London.
29. Omura, T. and Sato, R.J. (1964) *J. Biol. Chem.*, **239**, 2370–2378.
30. Nelson, D.R. et al. (1993) The P450 superfamily: update on new sequences, gene mapping, accession numbers, early trivial names of enzymes, and nomenclature. *DNA Cell Biol.*, **12** (1), 1–51.
31. P450 Encoding Sequences, http://www.icgeb.trieste.it/~p450srv/new/p450.html.

32. Nelson, D.R. (1999) Cytochrome P450 and the individuality of species. *Arch. Biochem. Biophys.*, **369** (1), 1–10.
33. Degtyarenko, K.N. (1995) Structural domains of P450-containing monooxygenase systems. *Protein Eng.*, **8**, 737–747.
34. Peterson, J.A. and Graham, S.E. (1998) A close family resemblance: the importance of structure in understanding cytochromes P450. *Structure*, **6** (9), 1079–1085.
35. Schwaneberg, U. *et al.* (2000) P450 in biotechnology: zinc driven ω-hydroxylation of p-nitrophenoxydodecanoic acid using P450 BM-3 F87A as a catalyst. *J. Biotechnol.*, **84**, 249–257.
36. Joo, H., Lin, Z., and Arnold, F. H. (1999) Laboratory evolution of peroxide-mediated cytochrome P450 hydroxylation. *Nature*, **399**, 670–673.
37. Farinas, E.T. *et al.* (2001) Directed evolution of a cytochrome P450 monooxygenase for alkane oxidation. *Adv. Synth. Catal.*, **343**, 601–606.
38. Urlacher, V.P., Lutz-Wahl, S., and Schmid, R.D. (2004) Microbial P450 enzymes in biotechnology. *Appl. Microbiol. Biotechnol.*, **64**, 317–325.
39. Urlacher, V. and Schmid, R.D. (2002) Biotransformations using prokaryotic P450 monooxygenases. *Curr. Opin. Biotechnol.*, **13**, 557–564.
40. Cirino, P.C. and Arnold, F.H. (2002) Protein engineering of oxygenases for biocatalysis. *Curr. Opin. Chem. Biol.*, **6**, 130–135.
41. Li, Z. *et al.* (2002) Oxidative biotransformations using oxygenases. *Curr. Opin. Chem. Biol.*, **6**, 136–144.
42. and Urlacher, V. (eds) Biooxidation, Wiley-VCH, Weinheim.
43. d Chen, Y.C.J. (1988) Baeyer-Villiger-oxidation hängige monooxyge- *Chem. Int. Ed. Engl.*, **27**,
44. Villiger, V. (1899) Einaro'schen reagens auf *Ber.*, **32**, 3625–3633.
45. Criegee, R. (1948) Die umlagerung der dekalin-peroxyester als folge von kationischem sauerstoff. *Justus Liebigs Ann. Chem.*, **560**, 127–141.
46. Strukul, G. (1998) Transition metal catalysis in the Baeyer-Villiger oxidation of ketones. *Angew. Chem. Int. Ed. Engl.*, **37** (9), 1198–1209.
47. Bolm, C., Schlingloff, G., and Weickhardt, K. (1994) Optically active lactones from a Baeyer-Villiger-type metal-catalyzed oxidation with molecular oxygen. *Angew. Chem. Int. Ed. Engl.*, **33** (18), 1848–1849.
48. Bolm, C. (1997) Catalyzed Baeyer–Villiger reactions. *Adv. Catal. Processes*, **2** (Asymmetric Catalysis), 43–68.
49. Del Todesco Frisone, M., Pinna, F., and Strukul, G. (1993) Baeyer-Villiger oxidation of cyclic ketones with hydrogen peroxide catalyzed by cationic complexes of Platinum(II): Selectivity properties and mechanistic studies. *Organometallics*, **12**, 148–156.
50. Turfitt, G.E. (1948) The microbiological degradation of steroids: 4. Fission of the steroid molecule. *Biochemistry*, **42**, 376–383.
51. Fried, J., Thoma, R.W., and Klingsberg, A. (1953) Oxidation of steroids by microorganisms. III. Side chain degradation, ring D-cleavage and dehydrogenation in ring A. *J. Am. Chem. Soc.*, **75**, 5764–5765.
52. Grogan, G., Roberts, S., and Willetts, A. (1992) Biotransformations by microbial Baeyer–Villiger monooxygenases stereoselective lactone formation in vitro by coupled enzyme systems. *Biotechnol. Lett.*, **14** (12), 1125–1130.
53. Grogan, G. *et al.* (1993) Camphor-grown Pseudomonas putida, a multifunctional biocatalyst for undertaking Baeyer–Villiger monooxygenase-dependent biotransformations. *Biotechnol. Lett.*, **15** (9), 913–918.
54. Jones, K.H., Roy, T.S., and Trudgill, P.W. (1993) Diketocamphane enantiomer-specific 'Baeyer-Villiger' monooxygenases from camphor-grown Pseudomonas putida ATCC 17453. *J. Gen. Microbiol.*, **139**, 797–805.
55. Taylor, D.G. and Trudgill, P.W. (1986) Camphor revisited: studies of

2,5-diketocamphane 1,2-monooxygenase from Pseudomonas putida ATCC 17453. *J. Bacteriol.*, **165** (2), 489–497.

56. Gagnon, R. et al. (1994) Biological Baeyer–Villiger oxidation of some monocyclic and bicyclic ketones using monooxygenases from Acinetobacter calcoaceticus NCIMB 9871 and Pseudomonas putida NCIMB 10007. *J. Chem. Soc., Perkin Trans. 1* (18), 2537–2543.
57. Stewart, J.D. et al. (1996) A 'Designer Yeast' that catalyzes the kinetic resolutions of 2-alkyl-substituted cyclohexanones by enantioselective Baeyer–Villiger oxidations. *J. Org. Chem.*, **61** (22), 7652–7653.
58. Kayser, M., Chen, G., and Stewart, J. (1999) Designer yeast: an enantioselective oxidizing reagent for organic synthesis. *Synlett*, **1**, 153–158.
59. Mihovilovic, M.D. et al. (2001) Baeyer–Villiger oxidations of representative heterocyclic ketones by whole cells of engineered Escherichia coli expressing cyclohexanone monooxygenase. *J. Mol. Catal., B Enzym.*, **11**, 349–353.
60. Lauble, H. et al. (2002) Crystal structure of hydroyxnitrile lyase from Sorghum bicolor in complex with the inhibitor benzoic acid: a novel cyanogenic enzyme. *Biochemistry*, **41**, 12043–12050.
61. Torres Pazmiño, D.E. et al. (2008) Self-sufficient Baeyer-Villiger monooxygenases – effective coenzyme regeneration for biooxygenations by fusion engineering. *Angew. Chem. Int. Ed.*, **47**, 2275–2278.
62. Cripps, R.E. (1975) The microbial metabolism of acetophenone. *Biochem. J.*, **152**, 233–241.
63. Cripps, R.E., Trudgill, P.W., and Whateley, J.G. (1978) The metabolism of 1-phenylethanol and acetophenone by nocardia T5 and an arthrobacter species. *Eur. J. Biochem.*, **86**, 175–186.
64. Tanner, A. and Hopper, D.J. (2000) Conversion of 4-hydroxyacetophenone into 4-phenyl acetate by a flavin adenine dinucleotide-containing Baeyer-Villiger- type monooxygenase. *J. Bacteriol.*, **182** (23), 6565–6569.
65. Kamerbeek, N.M. et al. (2001) 4-Hydroxyacetophenone monooxygenase from Pseudomonas fluorescens ACB. *Eur. J. Biochem.*, **268**, 2547–2557.
66. Hartmans, S. and de Bont, J.A.M. (1986) Acetol monooxygenase from mycobacterium py1 cleaves acetol into acetate and formaldehyde *FEMS Microbiol. Lett.*, **36**, 155–158.
67. Forney, F.W., Markovetz, A.J., and Kallio, R.E. (1967) Bacterial oxidation of 2-tridecanone to 1-undecanol. *J. Bacteriol.*, **93**, pp. 649–655.
68. Britton, L.N., Brand, J.M., and Markovetz, A.J. (1974) Source of oxygen in the conversion of 2-tri-decanone to undecyl acetate by *Pseudomonas cepacia* and *Nocardia* species. *Biochim. Biophys. Acta*, **369**, 45–49.
69. Kirschner, A. and Bornscheuer, U.T. (2006) Kinetic resolution of 4-hydroxy-2-ketones catalyzed by a Baeyer-Villiger monooxygenase. *Angew. Chem. Int. Ed.*, **45**, 7004–7006.
70. Kirschner, A. and Bornscheuer, U.T. (2008) Directed evolution of a Baeyer-Villiger monooxygenase to enhance enantioselectivity. *Appl. Microbiol. Biotechnol.*, **81** (3), 465–472.
71. Mihovilovic, M.D., Müller, B., and Stanetty, P. (2002) Monooxygenase-mediated Baeyer–Villiger oxidations. *Eur. J. Org. Chem.*, 3711–3730.
72. Rehdorf, J. and Bornscheuer, U.T. (2010) Monooxygenases, Baeyer-Villiger oxidations in organic synthesis, in *Encyclopedia of Industrial Biotechnology, Bioprocess, Bioseparation and Cell Technology* (ed. M.C. Flickinger), John Wiley & Sons, Inc., Hoboken, 1–35, DOI: 10.1002/9780470054581.eib451.
73. Schmid, R.D. and Verger, R. (1998) Lipases – interfacial enzymes with attractive applications. *Angew. Chem. Int. Ed. Engl.*, **37**, 1608–1633.
74. Kazlauskas, R.J. and Bornscheuer, U.T. (1998) in *Biotechnology* (eds H.J. Rehm et al.), Wiley-VCH Verlag GmbH, Weinheim, pp. 37–191.
75. Ollis, D.L. et al. (1992) The a/b hydrolase fold. *Protein Eng.*, **5**, 197–211.
76. Bornscheuer, U.T. and Kazlauskas, R.J. (1999) *Hydrolases in Organic*

Synthesis – Regio- and Stereoselective Biotransformations, Wiley-VCH Verlag GmbH, Weinheim.

77. Kazlauskas, R.J. et al. (1991) A rule to predict which enantiomer of a secondary alcohol reacts faster in reactions catalyzed by cholesterol esterase, lipase from Pseudomonas cepacia, and lipase from Candida rugosa. *J. Org. Chem.*, **56** (8), 2656–2665.

78. Weissfloch, A.N.E. and Kazlauskas, R.J. (1995) Enantiopreference of lipase from Pseudomonas cepacia toward primary alcohols. *J. Org. Chem.*, **60**, 6959–6969.

79. Henke, E., Pleiss, J., and Bornscheuer, U.T. (2002) Activity of lipases and esterases towards tertiary alcohols: insights into structure-function relationships. *Angew. Chem. Int. Ed.*, **41** (17), 3211–3213.

80. Henke, E. et al. (2003) A molecular mechanism of enantiorecognition of tertiary alcohols by carboxylesterases. *ChemBioChem*, **4**, 485–493.

81. Bartsch, S., Kourist, R., and Bornscheuer, U.T. (2008) Complete inversion of enantioselectivity towards acetylated tertiary alcohols by a double mutant of a Bacillus subtilis esterase. *Angew. Chem. Int. Ed.*, **47**, 1508–1511.

82. Heinze, B. et al. (2007) Highly enantioselective kinetic resolution of two tertiary alcohols using mutants of an esterase from Bacillus subtilis. *Protein Eng. Des. Sel.*, **20**, 125–131.

83. Kourist, R., Bartsch, S., and Bornscheuer, U.T. (2007) Highly enantioselective synthesis of arylaliphatic tertiary alcohols using mutants of an esterase from Bacillus subtilis. *Adv. Synth. Catal.*, **349**, 1393–1398.

84. Kourist, R., Dominguez de Maria, P., and Bornscheuer, U.T. (2008) Enzymatic synthesis of optically active tertiary alcohols: expanding the biocatalysis toolbox. *Chembiochem*, **9** (4), 491–498.

85. Jones, J.B., Hinks, R.S., and Hultin, P.G. (1985) Enzymes in organic synthesis. 33. Stereoselective pig liver esterase-catalyzed hydrolyses of meso cyclopentyl-, tetrahydrofuranyl-, and tetrahydrothiophenyl-1,3-diesters. *Can. J. Chem.*, **63**, 452–456.

86. Jones, J.B. (1990) Esterases in organic synthesis: present and future. *Pure Appl. Chem.*, **62** (7), 1445–1448.

87. Lam, L.K.P., Hui, R.A.H.F., and Jones, J.B. (1986) Enzymes in organic synthesis. 35. Stereoselective pig liver esterase catalyzed hydrolyses of 3-substituted glutarate diesters. Optimization of enantiomeric excess via reaction conditions control. *J. Org. Chem.*, **51**, 2047–2050.

88. Lange, S. et al. (2001) Cloning, functional expression, and characterization of recombinant pig liver esterase. *Chembiochem*, **2**, 576–582.

89. Musidlowska, A., Lange, S., and Bornscheuer, U.T. (2001) By overexpression in the yeast pichia pastoris to enhanced enantioselectivity: new aspects in the application of pig liver esterase. *Angew. Chem. Int. Ed.*, **40** (15), 2851–2853.

90. Dominguez de Maria, P. et al. (2007) Pig Liver Esterase (PLE) as biocatalyst in organic synthesis: from nature to cloning and to practical applications. *Synthesis*, **10**, 1439–1452.

91. Savjalov, W.W. (1901) Zur theorie der eiweissverdauung. *Pflügers Arch. Ges. Physiol.*, **85**, 171.

92. Fischer, A. et al. (1994) A novel approach to enzymic peptide synthesis using highly solubilizing N^{α}-protecting groups of amino acids. *Biocatalysis*, **8** (4), 289–307.

93. Moree, W.J. et al. (1997) Exploitation of subtilisin BPN' as catalyst for the synthesis of peptides containing non-coded amino acids, peptide mimetics and peptide conjugates. *J. Am. Chem. Soc.*, **119**, 3942–3947.

94. Bordusa, F. (2002) Proteases in organic synthesis. *Chem. Rev.*, **102**, 4817–4867.

95. Kasche, V. (2001) in *Proteolytic Enzymes: A Practical Approach* (eds M. Beynon and J. Bond), Oxford University Press, Oxford, pp. 265–292.

96. Schellenberger, V. and Jakubke, H.D. (1991) Protease-catalyzed kinetically controlled peptide synthesis. *Angew. Chem. Int. Ed. Engl.*, **30**, 1437–1449.

97. Drauz, K. and Waldmann, H. (2002) *Enzyme Catalysis in Organic Synthesis*,

vols. 1–3, 2nd edn, Wiley-VCH Verlag GmbH, Weinheim.
98. Layh, N. et al. (1997) Enrichment strategies for nitrile-hydrolysing bacteria. *Appl. Microbiol. Biotechnol.*, **47**, 668–674.
99. Nagasawa, T., Takeuchi, K., and Yamada, H. (1991) Characterization of a new cobalt-containing nitrile hydratase purified from urea-induced cells of Rhodococcus rhodochrous J1. *Eur. J. Biochem.*, **196**, 581–589.
100. Nagasawa, T. and Yamada, H. (1995) Microbial production of commodity chemicals. *Pure Appl. Chem.*, **67**, 1241–1256.
101. DeSantis, G. et al. (2002) An enzyme library approach to biocatalysis: development of nitrilases for enantioselective production of carboxylic acid derivatives. *J. Am. Chem. Soc.*, **124**, 9024–9025.
102. Crosby, J.A., Parratt, J.S., and Turner, N.J. (1992) Enzymic hydrolysis of prochiral dinitriles. *Tetrahedron: Asymmetry*, **3**, 1547–1550.
103. DeSantis, G. et al. (2003) Creation of a productive, highly enantioselective nitrilase through Gene Site Saturation Mutagenesis (GSSM). *J. Am. Chem. Soc.*, **125**, 11476–11477.
104. Kobayashi, M. and Shimizu, S. (1994) Versatile nitrilases: nitrile-hydrolysing enzymes. *FEMS Microbiol. Lett.*, **120**, 217–224.
105. Wieser, M. and Nagasawa, T. (1999) in *Stereoselective Biocatalysis* (ed. R. Patel), Marcel Dekker, New York, pp. 461–486.
106. Syldatk, C. et al. (1999) Microbial hydantoinases – industrial enzymes from the origin of life? *Appl. Microbiol. Biotechnol.*, **51**, 293–309.
107. Altenbuchner, J., Siemann-Herzberg, M., and Syldatk, C. (2001) Hydantoinases and related enzymes as biocatalysts for the synthesis of unnatural chiral amino acids. *Curr. Opin. Biotechnol.*, **12**, 559–563.
108. Effenberger, F. (2000) in *Stereoselective Biocatalysis* (ed. R.N. Patel), Marcel Dekker, New York, pp. 321–342.
109. Griengl, H., Schwab, H., and Fechter, M. (2000) The synthesis of chiral cyanohydrins by oxynitrilases. *Trends Biotechnol.*, **18**, 252–256.
110. Effenberger, F., Förster, S., and Wajant, H. (2000) Hydroxynitrile lyases in stereoselective catalysis. *Curr. Opin. Biotechnol.*, **11**, 532–539.
111. Schmidt, M. and Griengl, H. (1999) in *Topics in Current Chemistry* (ed. W.-D. Fessner), Springer, Berlin, pp. 193–226.
112. Rosenthaler, L. (1908) Durch enzyme bewirkte asymmetrische synthese. *Biochem. Z.*, **14**, 238–253.
113. Wong, C.H. (2002) in *Enzyme Catalysis in Organic Synthesis* (eds K. Drauz and H. Waldmann), Wiley-VCH Verlag GmbH, Weinheim, pp. 931–974.
114. Fessner, W.-D. and Helaine, V. (2001) Biocatalytic synthesis of hydroxylated natural products using aldolases and related enzymes. *Curr. Opin. Biotechnol.*, **12**, 574–586.
115. Fessner, W.-D. (2000) in *Stereoselective Biocatalysis* (ed. R.N. Patel), Marcel Dekker, New York, pp. 239–265.
116. Schoevaart, R., van Rantwijk, F., and Sheldon, R.A. (2000) A four-step cascade for the one-pot synthesis of non-natural carbohydrates from glycerol. *J. Org. Chem.*, **65**, 6940–6943.
117. Sugiyama, M. et al. (2006) Borate as a phosphate ester mimic in aldolase-catalyzed reactions: practical synthesis of L-fructose and L-iminocyclitols. *Adv. Synth. Catal.*, **348**, 2555–2559.
118. Stewart, J.D. (2001) Dehydrogenases and transaminases in asymmetric synthesis. *Curr. Opin. Chem. Biol.*, **5** (2), 120–129.
119. Frey, P.A. and Hegemann, P. (2007) *Enzymatic Reaction Mechanism*, Oxford University Press, New York.
120. Matcham, G.W. and Bowen, A.R.S. (1996) Biocatalysis for chiral intermediates: meeting commercial and technical challenges. *Chim. Oggi-Chem. Today*, **14** (6), 20–24.
121. Chen, D. et al. (2008) An amine: hydroxyacetone aminotransferase from Moraxella lacunata WZ34 for alaninol synthesis. *Bioprocess Biosyst. Eng.*, **31**, 283–289.

122. Hanson, R.L. et al. (2008) Preparation of (R)-amines from racemic amines with an (S)-amine transaminase from Bacillus megaterium. *Adv. Synth. Catal.*, **350**, 1367–1375.
123. Iwasaki, A. et al. (2003) Microbial synthesis of (R)- and (S)-3,4-dimethoxyamphetamines through stereoselective transamination. *Biotechnol. Lett.*, **25**, 1843–1846.
124. Kaulmann, U. et al. (2007) Substrate spectrum of w-transaminase from Chromobacterium violaceum DSM30191 and its potential for biocatalysis. *Enzyme Microb. Technol.*, **41**, 628–637.
125. Koszelewski, D. et al. (2008) Formal asymmetric biocatalytic reductive amination. *Angew. Chem.*, **120**, 9477–9480; (2008) *Angew. Chem. Int. Ed.*, **47**, 9337–9340.
126. Koszelewski, D. et al. (2008) Asymmetric synthesis of optically pure pharmacologically relevant amines employing w-transaminases. *Adv. Synth. Catal.*, **350**, 2761–2766.
127. Shin, J.-S. and Kim, B.-G. (2001) Comparison of the w-transaminases from different microorganisms and application to production of chiral amines. *Biosci. Biotechnol. Biochem.*, **65** (8), 1782–1788.
128. Shin, J.-S. and Kim, B.-G. (2002) Exploring the active site of amine:pyruvate aminotransferase on the basis of the substrate structure-reactivity relationship: how the enzyme controls substrate specificity and stereoselectivity. *J. Org. Chem.*, **67**, 2848–2853.
129. Yonaha, K. et al. (1976) Purification and crystallization of bacterial omega-amino acid-pyruvate aminotransferase. *FEBS Lett.*, **71** (1), 21–24.
130. Banerjee, A. et al. (2005) Methods for the Stereoselective Synthesis and Enantiomeric Enrichment of b-amino Acids.
131. Bum-Yeol, H. et al. (2008) Identification of ω-aminotransferase from caulobacter crescentus and sitedirected mutagenesis to broaden substrate specificity. *J. Microbiol. Biotechnol.*, **18**, 48–54.
132. Kim, J. et al. (2007) Cloning and characterization of a novel b-transaminase from Mesorhizobium sp. strain LUK: a new biocatalyst for the synthesis of enantiomerically pure b-amino acids. *Appl. Environ. Microbiol.*, **73**, 1772–1782.
133. Yun, H. et al. (2004) w-amino acid:pyruvate transaminase from Alcaligenes denitrificans Y2k-2: a new catalyst for kinetic resolution of ß-amino acids and amines. *Appl. Environ. Microbiol.*, **70** (4), 2529–2534.
134. Shin, J.-S. and Kim, B.-G. (1999) Asymmetric synthesis of chiral amines with w-transaminase. *Biotechnol. Bioeng.*, **65** (2), 206–211.
135. Raj, K.C. et al. (2002) Cloning and characterization of the Zymobacter palmae pyruvate decarboxylase gene (pdc) and comparison to bacterial homologues. *Appl. Environ. Microbiol.*, **68** (6), 2869–2876.
136. Goetz, G. et al. (2001) Continuous production of (R)-phenylacetylcarbinol in an enzyme-membrane reactor using a potent mutant of pyruvate decarboxylase from Zymomonas mobilis. *Biotechnol. Bioeng.*, **74** (4), 317–325.
137. Höhne, M. et al. (2008) Efficient asymmetric synthesis of chiral amines by combining transaminase and pyruvate decarboxylase. *Chembiochem*, **9** (3), 363–365.
138. Höhne, M. and Bornscheuer, U.T. (2009) Biocatalytic routes to optically active amines. *ChemCatChem*, **1**, 42–51.

8
Electrocatalysis

Timo Jacob

8.1
Introduction

In recent years there has been an increasing interest in electrochemistry, in particular fuel cells, for their ecologically desirable and economically efficient use of energy resources. An understanding of the processes occurring in electrochemical systems or during electrocatalytic reactions must include knowledge of the structure of the electrochemical double layer at metal/solution interfaces, which can be considered the site for electrochemical reactions. However, these interfaces are rather complex, involving different aspects such as surface dipoles, a potential drop, specific and non-specific adsorption (e.g., oxygen, water, impurities, etc.), or modified surface structures. Due to this complexity our knowledge about the ongoing processes is mostly limited to the macroscopic regime. However, nowadays theoretical methods are able to provide deeper insights into structures and processes at the atomistic level, which together with experiments could lead to a better understanding.

Regarding the electrode/electrolyte-interface, it is important to distinguish between two types of electrochemical systems: thermodynamically closed (and in equilibrium) and open systems. While the former can be understood by knowing the equilibrium atomic structure of the interface and the electrochemical potentials of all components, open systems require more information, since the electrochemical potentials within the interface are not necessarily constant. Variations could be caused by electrocatalytic reactions locally changing the concentration of the various species. In this chapter we will focus on the former situation, that is, interfaces in equilibrium with a bulk-electrode and a multi-component bulk-electrolyte, which are both influenced by temperature and pressures/activities, and are constrained by a finite voltage between electrode and electrolyte.

Experimentally, different structure- and surface-sensitive techniques such as *in-situ* scanning tunneling microscopy (STM), *in-situ* X-ray diffraction (XRD), transition electron microscopy (TEM), or *in-situ* infrared spectroscopy (IR) have been

---

Catalysis: From Principles to Applications, First Edition. Edited by Matthias Beller, Albert Renken, and Rutger van Santen.
© 2012 Wiley-VCH Verlag GmbH & Co. KGaA. Published 2012 by Wiley-VCH Verlag GmbH & Co. KGaA.

developed to study electrochemical interfaces (see [1–4] and references therein). Despite these developments, understanding the structure of the electric double layer is still an area of intense research [5, 6]. This is particularly true for the solution side of the double layer, the knowledge of which stems mostly from thermodynamic data [7]. As well as XRD and infrared absorption methods [8–10], STM is capable of yielding valuable structural information normal to the surface, which otherwise is difficult to obtain [11–13]. However, while the imaging of electrode surfaces with STM in an electrochemical cell under operating conditions (under potential control) is by now a well-established technique [14, 15], tunneling spectroscopy of the electrochemical interface is still in its infant stage [16, 17].

This situation is not the least caused by experimental problems, arising from the rather limited potential window set by the decomposition potential of aqueous solutions and by the fact that the tunnel voltage between tip and sample governs not only the tunneling process, but also the electrochemistry at the tip and sample surface. This problem is particularly relevant for so-called current–voltage spectroscopy [18], in which the potential of the sample or tip (or of both) is varied. Such problems are less severe for distance tunneling spectroscopy, in which the tunneling current is recorded as a function of tip distance, and the potentials of the sample and tip (and hence tunnel voltage) are held constant.

Regarding theoretical studies, different models and approaches have been developed in order to understand the structure and properties of systems under electrochemical conditions. An overview can be found in reviews [19–22], papers [23–32], and references therein.

In this work, mainly experimental input, semi-empirical approaches or still quite simplified models are used. The presence of the electrode potential is either neglected or introduced by charging the electrode surface or applying an external electric field. While most of the theoretical studies disregard the presence of an electrode potential, some try to consider its influence on catalytic reactions by shifting the Fermi energies by the value of the electrode potential. So far this approach has been applied to the hydrogen evolution reaction (HER) and oxygen reduction reaction (ORR) on different electrodes [28, 33]. Focusing more on the atomistic structure of the interface, *ab-initio* molecular dynamics simulations on charged electrodes have been performed [30], where a corresponding counter charge had been located at a certain distance from the electrode surface. This method tries to reproduce the potential profile within the interfacial region by inducing an external potential drop.

In this chapter we discuss the concepts determining electrochemical interfaces. First we will characterize each species by discussing the electrochemical potential, which is followed by a detailed description of the electrochemical double layer. Finally, the deduced concepts will be applied to the ORR at a Pt electrode, which can be considered as one of the most fundamental reactions in electrochemistry.

8.2 Theory

8.2.1 Electrochemical Potentials

Before we discuss the electrochemical system, it is important to define the properties and characteristics of each component, especially the electrolyte. In the following we assume macroscopic amounts of an electrolyte containing various ionic and non-ionic components, which might be solvated. If the bulk-electrolyte thus defined is in thermodynamic equilibrium, each of the species present is characterized by its electrochemical potential, which is defined as the free energy change with respect to the particle number N_i of species i:

$$\tilde{\mu}_i(T, c_i, \phi_S) = \left(\frac{\partial G}{\partial N_i}\right)_{T, c_i, N_j} + q_i \cdot \phi_s = \mu_i(T, c_i) + q_i \cdot \phi_s \quad (8.1)$$

Here G is the Gibbs free energy of the system without external electrostatic potential, and $q_i \cdot \phi_s$ refers to the energy contribution coming from the interaction of an applied constant electrostatic potential ϕ_s with the charge q_i of the species. The first term on the right side of Eq. (8.1) is the usual chemical potential $\mu_i(T, c_i)$, which for an ideal solution is given by

$$\mu_i(T, c_i) = \overline{\mu}_i(T, c^0) + k_B T \ln\left(\frac{c_i}{c^0}\right) \quad (8.2)$$

where $\overline{\mu}_i$ is the chemical potential of species i at standard conditions, which for the solvent and the solute (electrolyte/ions) are defined as follows:

Solvent: standard temperature and pure solvent, which means the absence of electrolyte ions, solute (Henry's law standard state): standard temperature and concentration c^0 in the hypothetical state of an ideal solution (infinite dilution), which is reached at $\lim_{c_i \to 0} f_i = 1$.

All deviations from an ideal behavior can be related to direct (electrostatic) or indirect (through the solvent) interactions between the electrolyte ions. Keeping with Eq. (8.2), these effects are usually treated by scaling the concentration with a so-called activity coefficient f_i, which then leads to the activity $a_i = f_i \cdot c_i \cdot$ [34]. With this definition Eq. (8.2) becomes:

$$\mu_i(T, a_i) = \overline{\mu}_i(T, c^0) + k_B T \ln\left(\frac{c_i}{c^0}\right) + k_B T \ln(f_i) \quad (8.3)$$

As long as the ion concentration is below 1 M, which is fulfilled by most electrochemical experiments, the last term of Eq. (8.3) contributes to the chemical potential by only a few percent. However, for concentrations above 5 M, effects of up to 10% could occur [35].

Using Eq. (8.3), one should in principle be able to evaluate all deviations from the ideal behavior, that is, the activity coefficient, for each ionic species within the electrolyte separately (anions: f_a; cations: f_c) [35–37]. However, only the mean activity coefficient, which can be approximated by $f_\pm = \sqrt{f_a \cdot f_c}$, is accessible experimentally. This is caused by the fact that when adding only one of the ionic species to the electrolyte the change in Gibbs free energy would result in an additional energy contribution coming from the interaction of the ionic species with an overall charged (electrified) solution. As the two energy contributions cannot be separated unambiguously, experimentally an entire salt molecule is added to the electrolyte at a time.

The second term on the right side of Eq. (8.1) is the energy required to transfer the charge q_i, associated with each particle of the ith-species, from a reference potential ϕ_{ref}, which can be chosen freely, to the electrostatic potential of the bulk-electrolyte (solution) ϕ_s.

8.2.2
Electric Double Layer

Compared to systems related to surface science, the electrode/electrolyte-interface is a multi-component system, which besides temperature and partial pressures/concentrations is additionally influenced by the electrode potential, meaning the electrostatic potential between the electrode and the bulk-electrolyte.

For the following description, we will assume a charged (plate) capacitor as the initial system and will discuss the changes induced by filling the space between the two electrodes with a liquid electrolyte. Due to its experimental relevance we will consider a single electrode/electrolyte interface only (i.e., electrochemical half-cell), where the electrostatic potential of the electrode will simply be designated as ϕ_e (see Figure 8.1).

As described above, the electrolyte usually contains anions and cations, which are partially or fully solvated, water molecules and various species being involved in an electrocatalytic reaction. The excess charge on the electrode surface is compensated by an accumulation of corresponding electrolyte counter ions, leading to overall charge neutrality.

Since a (plate) capacitor without electrolyte in principle shows a linear behavior of the electrostatic potential between the electrodes $\phi(x)$, at the hypothetical moment exactly when the space between both electrodes is filled with a liquid electrolyte ($t = 0$), the electrochemical potentials of the ions,

$$\tilde{\mu}_a = \mu_a + q_a \cdot \phi(x) \quad \text{and} \quad \tilde{\mu}_c = \mu_c + q_c \cdot \phi(x) \tag{8.4}$$

also show linear behavior as a function of distance to the electrode x. At $t > 0$ the system tries to reach thermodynamic equilibrium by redistributing the anions and cations within the electrolyte such that counter ions in the electrolyte are accumulated at or near the electrode, which will finally compensate the electrode excess charge. Some of these ions might even lose parts of their solvation shell and adsorb at the electrode surface (*specific adsorption*) or just weakly interact as solvated ions (*non-specific adsorption*). When thermodynamic equilibrium is reached, the electrochemical potentials of anions and cations are constant throughout the electrolyte. Furthermore, the electrolyte ions will exactly compensate the electrode excess charges, leading to a potential drop within the interfacial region.

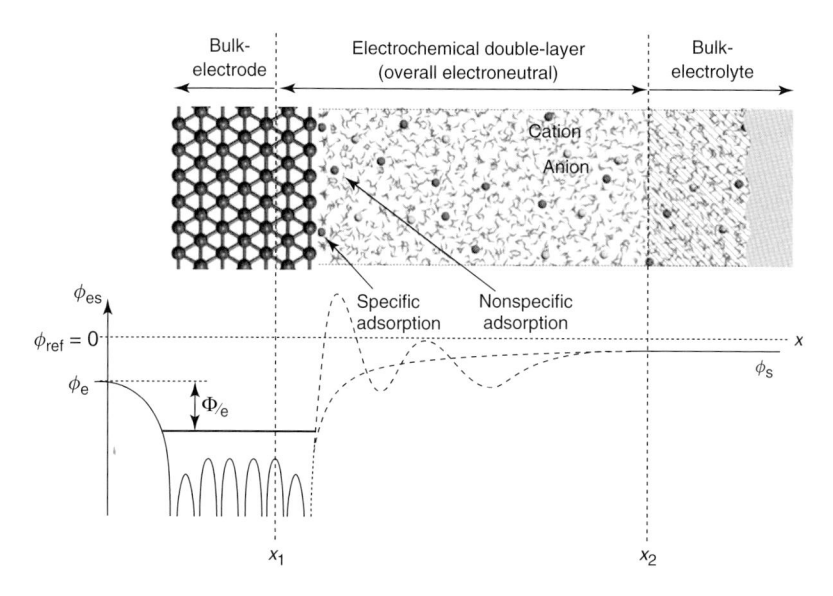

Figure 8.1 Atomistic model of the electrochemical half-cell, showing the electrode/electrolyte interface ($x_1 < x < x_2$), which is connected to the bulk-electrode and -electrolyte (reservoirs). The lower panel indicates the electrostatic potential of the electrodes and the bulk-electrolyte (solid lines), and possible shapes for the potential drop in-between (dashed lines).

Figure 8.1 shows schematically the electrochemical half-cell and the electrostatic potential after the system has reached equilibrium. It combines a single electrode/electrolyte interface, being in contact with two reservoirs: bulk-electrode ($x < x_1$) and bulk-electrolyte ($x > x_2$). Since by definition the excess charges on the electrode are fully shielded within the interface, far away from the electrode the electrostatic potential assumes the constant value ϕ_s with respect to the reference potential $\phi(x)$. The width of such an interface might range from a few to several hundreds of Ångströms, which is rather small compared to the size of the bulk-electrolyte region in realistic systems. Therefore, the latter can be considered as an *electrolyte-reservoir*.

Within the interface, the explicit profile of the electrostatic potential $\phi(x)$ as a function of electrode distance is unknown (in Figure 8.1 possible shapes are indicated as dashed lines), and the only well-defined values are the asymptotic limits that are the potential values of the electrode ϕ_e, the bulk-electrolyte ϕ_s, and their difference (electrode potential):

$$\Delta\phi = \phi_e - \phi_s \tag{8.5}$$

A number of classical descriptions have been formulated to describe the shapes of the potential drops across the electrochemical interface [35, 36], including the Helmholtz model, the Gouy–Chapman model and the Gouy–Chapman–Stern model (see Figure 8.2). The earliest model formulated by Helmholtz in 1879 treats the interface mathematically as a simple capacitor formed by the electrode and a

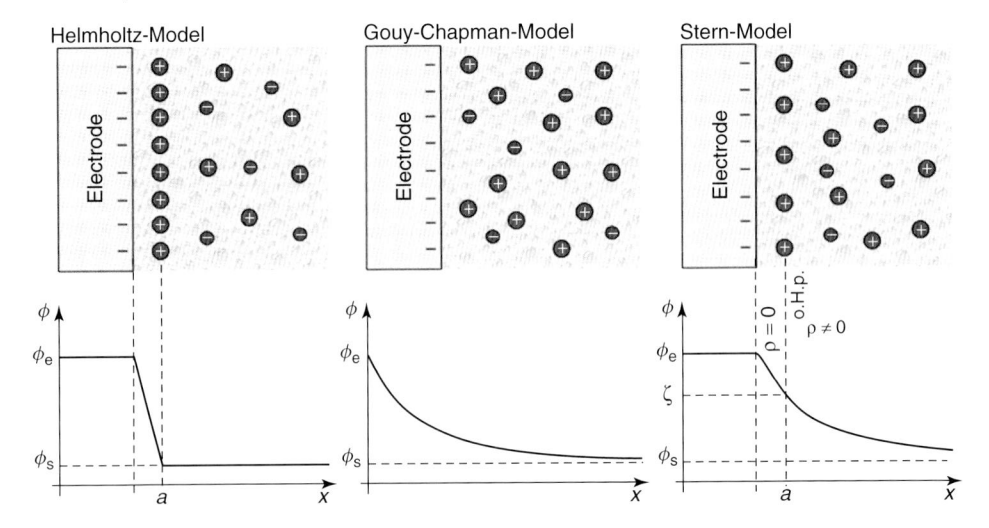

Figure 8.2 Illustration of different classical double-layer models (top) and the corresponding potential profiles originating from the surface excess charges and the distribution of counter ions within the electrolytes.

single layer of non-specifically adsorbed (solvated) ions, leading to the so-called *inner layer*. Later, Gouy and Chapman (1910–1913) made significant improvements by introducing the diffuse character of the electrolyte to the electric double-layer. In their model, the electrolyte ions, which are treated as point charges, reach their bulk concentrations only at larger distances from the electrode, leading to an exponential shape of the potential drop and the so-called *diffuse double-layer*. One of the current classical and widely applied descriptions is known as Gouy–Chapman–Stern model (often referred to as the Stern model), which combines the Helmholtz inner layer with the Gouy–Chapman diffuse layer. It states that ions do have finite size, so that they cannot approach the surface closer than their radius. The non-specifically adsorbed ions of the Gouy–Chapman diffuse double layer are not at the surface, but at some distance (larger than their radius) away from the surface. Combining these two models, the potential drop within the electrochemical double layer becomes (distances are in correspondence to Figure 8.2):

$$x < a: \quad \phi(x) = \phi_e - \frac{\phi_e - \zeta}{a} \cdot x$$

$$x \geq a: \quad \phi(x) = \zeta \exp\left[-\kappa(x-a)\right] + \phi_s \quad \text{with} \quad \kappa^{-1} \approx \sqrt{\frac{T}{\sum_i n_i^0 z_i}} \quad (8.6)$$

Here κ^{-1} is the Debye length, which can be attributed to the center of charge within the diffuse layer ($x \geq a$). The above equations show that up to the so-called outer Helmholtz plane (ohp) the potential decreases linearly from ϕ_e to the potential value at $x = a$, which is usually referred to as the ζ potential. Beyond the ohp the potential drops exponentially to the potential of the bulk-electrolyte ϕ_s. Here the exponential shape is a result of the Boltzmann-type ion distributions assumed for the diffuse part of the double-layer.

Although these classical models have been very successful in describing the overall behavior of the electrochemical interfaces, they neglect various aspects of the interface, such as image charges, surface polarization, or interactions between the excess charges and the water dipoles. Therefore, the width of the electrode/electrolyte interfaces is usually underestimated. In addition, the ion distribution within the interfaces is not fixed, which for short times could lead to much stronger electric fields near the electrodes.

8.3 Application to the Oxygen Reduction Reaction (ORR) on Pt(111)

The ORR is a key process in combustion, corrosion, cellular respiration, and energy technology. Under electrochemical conditions, with a supply of hydrogen, the electrocatalytic ORR [38–41] is also the reaction that is the basis of the polymer-electrolyte (or proton-exchange) membrane fuel cell (PEMFC). Economic and environmental factors around the globe are driving research to develop practical and environmentally sustainable energy sources as well as long-lived heterogeneous catalysts that conserve precious metal supplies. The ORR is expected to play a central role in these technologies, but a more fundamental understanding of the reaction is needed.

Atomic-level understanding of the ORR mechanism is still in its early stages due to the high complexity of ORR kinetics. What is known, however, is that the complete electrochemical ORR involves four net coupled proton and electron transfers (CPETs) to molecular oxygen at the cathode. While the idealized electrochemical reaction generates 1.23 V per electron (Scheme 8.1), the standard operating potential for the electrocatalytic ORR on Pt(111) is below 0.9 V. Determining the cause of the overpotential of ~0.3 V, reducing it, and improving the overall activity are the keys to unlocking ORR catalyst design and better harnessing the ORR as a practical means for energy conversion. Some ORR processes are believed to lead to surface oxides and/or strongly binding intermediates, which in turn are expected to be hindrances of the ORR.

$$2\,H_{2(g)} \rightarrow 4\,H^{+}_{(aq)} + 4\,e^{-} \quad E^{\circ} = 0\,V$$

$$O_{2(g)} + 4\,H^{+}_{(aq)} + 4\,e^{-} \rightarrow 2\,H_{2}O_{(l)} \quad E^{\circ} = 1.229\,V$$

$$\text{Net: } 2\,H_{2(g)} + O_{2(g)} \rightarrow 2\,H_{2}O_{(l)} \quad U = 1.229\,V$$

Scheme 8.1 Hydrogen oxidation (at the anode) and oxygen reduction (at the cathode) reactions at the (fuel cell) electrodes.

Although possible ORR intermediates only consist of H and O atoms, the overall mechanism is still elusive, even when considering highly studied Pt(111) electrodes. Over the past decades there have been substantial efforts to use quantum mechanics (QM) to determine binding energies (BEs) of oxygen on Pt and other transition metal surfaces and extend those data to investigate aspects of the ORR mechanism [42–46], and even to investigate the role of an applied electrode potential in a reaction mechanism [47, 48]. Most relevantly, QM simulations can provide accurate descriptions of chemical bonding energies, which can be used to predict ORR rate constants. Calculating these values is a first step toward understanding the complete electro-catalytic ORR from first principles, something that will likely require a multi-scale analysis explicitly addressing the electrochemical double-layer, electron dynamics, surface coverage effects, and transport issues. Although such a complete simulation is not yet feasible, rate constants calculated from QM can be used in a kinetic model and then compared to experimental observables.

Investigating the exact ORR mechanism requires explicitly determining the BEs of a list of possible intermediates: O^*, H^*, O_2^*, OH^*, OOH^*, $H_2O_2^*$, and H_2O^* (as well as their dissociation behavior) that link the intermediates to one another. Since these mechanisms occur under electrochemical conditions, both Langmuir–Hinshelwood (LH) and Eley–Rideal (ER) mechanisms must be considered. While the former mechanism assumes all reacting intermediates to be at the surface, the ER mechanisms allow for species from the electrolyte to directly react with a surface intermediate (e.g., H_3O^+).

As possible reaction mechanisms, we considered three possible pathways that are based on adsorbed intermediates (see Figure 8.3):

1) **The O_2 Pathway:** adsorption of O_2 is followed by its dissociation into $2O^*$. The O^* intermediates then undergo four CPETs leading to one water (in the two-electron pathway) or two waters (in the four-electron pathway). Note that

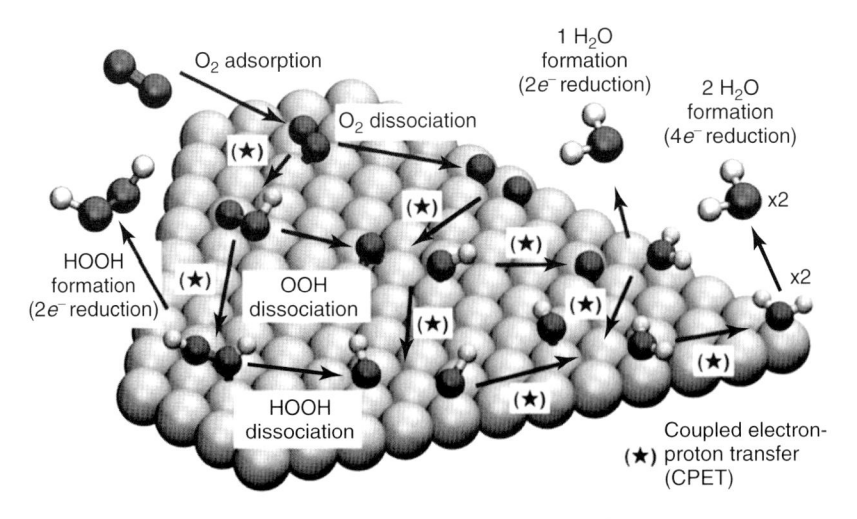

Figure 8.3 Schematic model of possible reaction mechanisms of the ORR on Pt(111).

due to strongly adsorbed O* species, the O_2 dissociation would appear in mechanisms as a one-electron reductive process [49].

2) **The OOH Pathway:** adsorption of O_2 is followed by one CPET and then O*−OH* dissociation to form O* and OH*. These intermediates may then undergo three CPETs to from one or two water molecules, as was the case in the O_2 pathway.

3) **The HOOH Pathway:** adsorption of O_2 is followed by two CPETs to form H_2O_2*. When H_2O_2* is released from the surface, this is a two-electron pathway. When H_2O_2* is allowed to dissociate on the surface, the two OH* intermediates can then undergo two CPETs to form two waters via a four-electron pathway.

The results at zero potential ($U = 0$ V vs the RHE) (reversible hydrogen electrode) are summarized in Figure 8.4a, whereby all free energies were evaluated according to Eq. (8.7):

$$\Delta G_{298} = \Delta E_{SCF} + \Delta E_{solv} + \Delta E_{ZPE} + \Delta H_{vib}(T) - T \cdot \Delta S_{vib}(T) \qquad (8.7)$$

Here the different terms on the right side represent the total energy as obtained by standard density functional theory (DFT) calculations (ΔE_{SCF}), contributions arising from the surrounding solution (ΔE_{solv}, here assumed to be water), zero-point energy contributions (ΔE_{ZPE}), and temperature-dependent contributions from the vibrational enthalpy (ΔH_{vib}) and entropy (ΔS_{vib}). Furthermore, in order to undertake first-principles investigations on the ORR, which involve various charge transfer steps, we use the general proposal from Damjanovic and Brusic [50] and assume that fast electronic relaxation times are in play within the outer Helmholtz plane of the electrode. Therefore, we treat each one-electron reduction as an overall CPET process, which means that every electron transfer is coupled with a proton transfer occurring simultaneously. This assumption permits us to use a common approach, namely treating an applied potential by explicitly shifting the electrode's Fermi level (see Ref. [35] for more details). Here, energies for intermediates undergoing a net CPET will be shifted by $+eU$, where e is the elementary charge and U is the electrode potential (referenced to the RHE) applied within the simulation. Regarding the overall protonation barriers, we have developed a scheme based on high-level *ab initio* CCSD(T) calculations; details may be found in Ref. [51].

Each of the three processes described above starts with the adsorption of O_2. When O_2 adsorbs on the Pt(111) surface, three stable binding geometries are found: bridge, fcc, and tilted [49, 52]. The most stable configuration is O_2* bound at a bridge position (BE = −1.35 eV), where both oxygens use a doubly-occupied π-orbital to form donor–acceptor bonds to the surface (peroxide-like O_2). The second stable structure corresponds to dioxygen above an fcc surface site, such that one oxygen binds on top of a Pt atom and the other oxygen at a bridge position (superoxide-like O_2). Finally, the last structure is somewhat comparable to bridge-bound O_2*, except that the molecule is tilted toward the surface such that the O=O π-bond can form a donor–acceptor bond to an adjacent Pt atom. Reaction energy diagrams corresponding to the subsequent reactions are shown in Figure 8.4.

After adsorption, O_2* may either dissociate, which would result in the O_2 pathway, or attract one or two hydrogen atoms from the surface or protons from

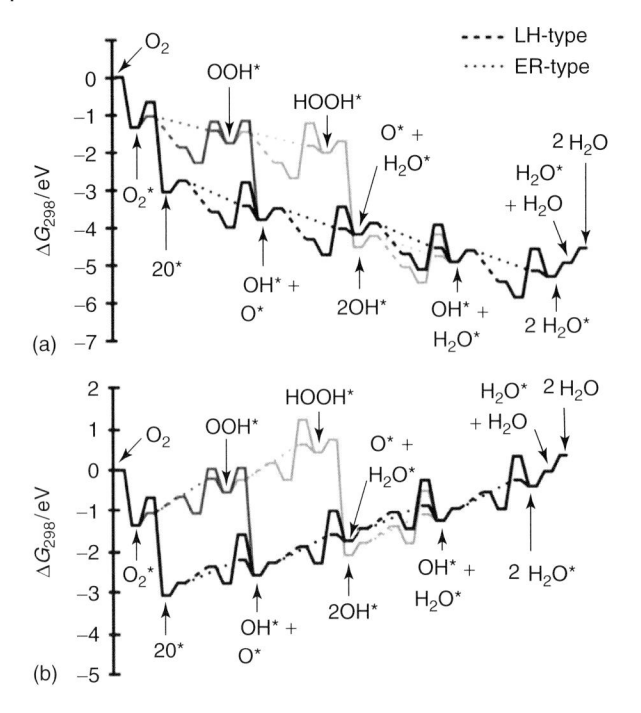

Figure 8.4 Electrochemical ORR mechanisms: (a) at U = 0V and (b) at the ideal Nernst-potential of U = 1.23V (vs RHE – reversible hydrogen electrode).

the electrolyte. The ΔG_{298} values of $O_2{}^*$, OOH^*, and $HOOH^*$ are -1.35, -1.77, and -2.03 eV, respectively. Following the different mechanisms, the next step has to involve dissociation of the adsorbed intermediate. The dissociation barriers for each of these adsorbates: O^*-O^*, O^*-OH^*, and HO^*-OH^* are 0.68, 0.59, and 0.31 eV, respectively, with barriers decreasing corresponding to the number of H atoms bound to O_2^*. While each of the dissociation steps involves some interesting features, in the following we will exemplarily concentrate on the O^*-O^* process. Details on the dissociations of all the other intermediates can be found in Refs. [49, 52].

As already described, O_2^* adsorbs on the surface in three possible configurations, each of which is associated with a different binding strength. However, each of these structures has also its own O^*-O^* dissociation energy. Thus, O_2^* may dissociate with one of three dissociation barriers corresponding to those from the bridge, the fcc, or the tilted configuration. Although, tilted-O_2^* forms the weakest surface bond, this adsorbate structure has the lowest dissociation barrier. Binding energy alone suggests that O_2^* resides at a bridge surface site. However, in the tilted configuration the dissociation barrier is very much lowered. Thus, bridge-bound O_2^* may change its structure to the tilted configuration (which does not require much reorganization energy) and then dissociate via the O_2^*-tilted route. By doing so, the overall dissociation barrier reduces to only 0.68 eV, which is agreement with the experimental values of 0.38–0.50 eV [53–55].

The products of O_2^* dissociation are two O^* atoms located in threefold sites: fcc/fcc, hcp/hcp, or fcc/hcp. Interestingly, we found that dissociation results in two O^* atoms at non-adjacent threefold positions (separated by two lattice constants), also in excellent agreement with the STM experiments by Ho et al. [56, 57]. When multiple O_2^* dissociations are considered to occur across the entire surface, the final structure would correspond to a $p(2 \times 2)$ overlayer, a result that was also obtained by evaluating the surface phase diagram of various adlayer configurations [58] and which is also in correspondence to the experiment [56]. Since the O–O interactions are already rather small, for coverages ≤ 0.25 ML it is justified to simulate O^* atoms as independent adsorbates. Therefore, coverage effects seem to be of minor importance for the remaining discussion. The barrier to hop from an hcp to an fcc site via a bridge position is only 0.24 eV, so we expect that most atomic oxygen will eventually migrate into the most thermodynamically stable fcc sites.

The above description has already shown how complex the theoretical modeling of an apparently simple reaction step such as the adsorption and subsequent dissociation of O_2 can be. However, after similarly elaborate considerations, one can obtain accurate dissociation barriers for the intermediates that are on the surface after the first reaction steps.

The dissociation barriers for each of these adsorbates: O^*-O^*, O^*-OH^*, and HO^*-OH^* are 0.68, 0.59, and 0.31 eV, respectively, with barriers decreasing corresponding to the number of H atoms bound to O_2^*.

After these dissociations, multiple processes may lead to either $2O^*$, $O^* + OH^*$, $2OH^*$, or $OH^* + H_2O^*$. The overall ΔG_{298} values for each set of adsorbates also decreases corresponding to the number of H atoms on the O^* intermediates: -3.06, -3.80, -4.54, and -4.93 eV. The most strongly bound intermediate at $U = 0$ V are two adsorbed water molecules ($2H_2O^*$), which bind by 0.76 eV with respect to $2H_2O_{(l)}$ (or a relative free energy of -5.33 eV with respect to $H_{2(g)} + O_{2(g)}$). Therefore, this intermediate represents the energetically lowest value in the energy diagram of Figure 8.4a. As the final removal of water requires 0.38 eV per water molecule, the final state, which is two water molecules in liquid phase, is 0.76 eV higher. Consequently, the overall ORR process denoted in Scheme 8.1 is 4.57 eV downhill in energy at $U = 0$ V with respect to the RHE.

Although trends are clearly seen from these energies, LH-type reaction barriers showing hydrogenations that involve adsorbed H^* atoms are more complicated. LH barriers leading to OH^*, H_2O^*, OOH^*, and $HOOH^*$ are quite high: 1.19, 1.28, 1.09, and 1.46 eV, respectively. These are in stark contrast to the (de)protonation steps encountered in ER-type reactions, which can all be approximated to 0.3 eV. Thus, we expect ER-type mechanisms to dominate over LH mechanisms without (or at low) electrode potentials.

Based on these calculations alone, one would deduce that, at low potentials, ER-type mechanisms will primarily operate through the HOOH pathway since this process has the lowest energy barriers. Interestingly, the $H_2O_2^*$ intermediate cannot be characterized with periodic DFT calculations on a clean Pt(111) surface

[43]. Our own periodic DFT calculations on (3 × 3) unit cells indicate that its dissociation barrier is negligibly low; however this indicates that the presence of other co-adsorbates, such as H atoms, possibly due to underpotential deposition, may prevent the HO–OH bond from breaking, and may thereby stabilize $H_2O_2^*$ intermediates [59].

A substantially different picture of the reaction profiles is seen when applying the ideal electrode potential of 1.23 V (i.e., the Nernst potential of the ORR). Here, energy levels are shifted for each net-CPET step, and the total reaction is no longer highly exothermic, but essentially thermodynamically neutral. The overall energies of O_2^* remain unchanged, though the adsorbates OOH* and HOOH* are destabilized to $\Delta G_{298} = -0.54$ and $+0.43$ eV. The positive ΔG_{298} value for $H_2O_2^*$ alone suggests that the HOOH pathway would be shut off entirely at higher potentials. The overall ΔG_{298} value for 2O* remains unchanged, and the ΔG_{298} values for the adsorbates $O^* + OH^*$, 2OH*, and $OH^* + H_2O^*$ are now -2.57, -2.08, and -1.24 eV. Note that these energies now *increase* based on the number of H atoms bound to O*.

Under the influence of an electrode potential (see Figure 8.4b), the individual hydrogenation barriers for LH-type mechanisms as well as O*–O*, O*–OH*, and HO*–OH* dissociations remain unchanged. An increased electrode potential, however, increases ER reaction barriers significantly. The ER reaction barriers for O_2^*–H, HOO*–H, O*–H, and HO*–H are each no longer ~0.3 eV as they were at $U = 0$ V, but rather 1.15, 1.16, 0.87, and 0.99 eV at $U = 1.23$ V. The quite high barriers in the first two cases support a prediction that the OOH and HOOH pathways are less favorable than the O_2 pathway at high electrode potentials. Furthermore, the increased reaction barriers at higher electrode potentials indicate that both ER- *and* LH-type reactions may even be in play at certain ambient conditions. This already shows that the ORR mechanisms combine electrochemical reaction steps, which are usually considered to be Eley–Rideal-type steps, with reaction processes usually thought to be related to surface science exclusively (i.e., LH-type steps).

8.4
Summary

In this chapter we have described the basic concepts of electrochemical systems, focusing on electrode/electrolyte-interfaces and their thermodynamics. After defining the electrochemical potential, which is one of the important quantities for these kinds of systems, the structure of the interface has been discussed. Based on the description of the nature of the potential drop within the interfacial region, we afterwards discussed some classical models allowing for a quantitative understanding of the electric double-layer.

Finally we have described the oxygen reduction reaction on Pt electrodes, which is one of the most fundamental reactions in electrochemistry, and which served as a case study for applying the concepts of electrochemistry to an electrocatalytic reaction. Distinguishing between different reaction mechanisms, we have shown how BEs and dissociation barriers obtained from first principles calculations can be used to reveal the exact mechanism of electrocatalytic reactions as complex as the ORR.

References

1. Abruna, H.D. (1991) Modern techniques for in-situ interface characterization, *Electrochemical Interfaces* Wiley-VCH Verlag GmbH, Weinheim.
2. Kolb, D.M. (1996) Reconstruction phenomena at metal-electrolyte interfaces. *Prog. Surf. Sci.*, **51** (2), 109–173.
3. Kolb, D.M. (2002) An atomistic view of electrochemistry. *Surf. Sci.*, **500** (1–3), 722–740.
4. Magnussen, O.M. (2002) Ordered anion adlayers on metal electrode surfaces. *Chem. Rev.*, **102** (3), 679–726.
5. Guidelli, R. and Schmickler, W. (2000) Recent developments in models for the interface between a metal and an aqueous solution. *Electrochim. Acta*, **45** (15–16), 2317–2338.
6. Wernersson, E. and Kjellander, R. (2006) On the effect of image charges and ion-wall dispersion forces on electric double layer interactions. *J. Chem. Phys.*, **125** (15), 154702–154709.
7. Parsons, R. (1980) in *Comprehensive Treatise of Electrochemistry*, vol. 1 (ed. J.O.M. Bockris, B.E. Conway, and E.B. Yeager), Plenum Press, New York. Chapter 1, 1.
8. Shi, Z. et al. (1994) Investigations of SO_4^{2-} adsorption at the Au(111) electrode by chronocoulometry and radiochemistry. *J. Electroanal. Chem.*, **366** (1–2), 317–326.
9. Ataka, K.-I. and Osawa, M. (1998) In situ infrared study of water-sulfate coadsorption on gold(111) in sulfuric acid solutions. *Langmuir*, **14** (4), 951–959.
10. Ito, M. and Yamazaki, M. (2006) A new structure of water layer on Cu(111) electrode surface during hydrogen evolution. *Phys. Chem. Chem. Phys.*, **8** (31), 3623–3626.
11. Nagy, G. (1995) Structure of platinum/water interface as reflected by STM measurements. *Electrochim. Acta*, **40** (10), 1417–1420.
12. Hugelmann, M. and Schindler, W. (2004) In situ distance Tunneling Spectroscopy at Au-(111)/0.02 M HClO4 – From Faradaic regime to quantized conductance channels. *J. Electrochem. Soc.*, **151** (3), E97–E101.
13. Nagy, G. and Wandlowski, T. (2003) Double layer properties of Au(111)/H_2SO_4 (Cl) + Cu2+ from distance tunneling spectroscopy. *Langmuir*, **19** (24), 10271–10280.
14. Itaya, K. (1998) In situ scanning tunneling microscopy in electrolyte solutions. *Prog. Surf. Sci.*, **58** (3), 121–247.
15. Kolb, D.M. (2000) Structure studies of metal electrodes by in-situ scanning tunneling microscopy. *Electrochim. Acta*, **45** (15–16), 2387–2402.
16. Hiesgen, R., Eberhardt, D., and Meissner, D. (2005) Direct investigation of the electrochemical double layer using the STM. *Surf. Sci.*, **597** (1–3), 80–92.
17. Halbritter, J. et al. (1995) Tunneling mechanisms in electrochemical STM --distance and voltage tunneling spectroscopy. *Electrochim. Acta*, **40** (10), 1385–1394.
18. Hugelmann, P. and Schindler, W. (2005) In-situ voltage tunneling spectroscopy at electrochemical interfaces. *J. Phys. Chem. B*, **109** (13), 6262–6267.
19. Schmickler, W. (1996a) Electronic effects in the electric double layer. *Chem. Rev.*, **96** (8), 3177–3200.
20. Schmickler, W. (1999) Recent progress in theoretical electrochemistry. *Annu. Rep. C Phys. Chem.*, **95**, 117–162.
21. Koper, M.T.M., van Santen, R.A., and Neurock, M. (2003) in *Catalysis and Electrocatalysis at Nanoparticle Surfaces* (eds E.R. Savinova, C.G. Vayenas, and A. Wieckowski), Marcel Dekker, New York. Chapter 1, 1.

22. Koper, M.T.M. et al. (2004) *Modern Aspects of Electrochemistry*, vol. 36, Springer, pp. 51–130.
23. Nazmutdinov, R.R. and Shapnik, M.S. (1996) Contemporary quantum chemical modelling of electrified interfaces. *Electrochim. Acta*, **41** (14), 2253–2265.
24. Halley, J.W., Schelling, P., and Duan, Y. (2000) Simulation methods for chemically specific modeling of electrochemical interfaces. *Electrochim. Acta*, **46** (2–3), 239–245.
25. Vassilev, P. et al. (2001) Ab initio molecular dynamics simulation of liquid water and water--vapor interface. *J. Chem. Phys.*, **115** (21), 9815–9820.
26. Haftel, M.I. and Rosen, M. (2003) Surface-embedded-atom model of the potential-induced lifting of the reconstruction of Au(1 0 0). *Surf. Sci.*, **523** (1, 2), 118–124.
27. Feng, Y.J., Bohnen, K.P., and Chan, C.T. (2005) First-principles studies of Au(100)-hex reconstruction in an electrochemical environment. *Phys. Rev. B*, **72** (12), 125401.
28. Kitchin, J.R. et al. (2004) Modification of the surface electronic and chemical properties of Pt(111) by subsurface 3d transition metals. *J. Chem. Phys.*, **120** (21), 10240–10246.
29. Gunnarsson, M. et al. (2004) Corrected Debye-Hückel analysis of surface complexation: III. Spherical particle charging including ion condensation. *J. Colloid. Interface Sci.*, **274** (2), 563–578.
30. Taylor, C.D. et al. (2006) First principles reaction modeling of the electrochemical interface: consideration and calculation of a tunable surface potential from atomic and electronic structure. *Phys. Rev., B*, **73** (16), 165402.
31. Jacob, T. (2007a) Potential-induced lifting of the Au(100)-surface reconstruction studied with DFT. *Electrochim. Acta*, **52** (6), 2229–2235.
32. Jacob, T. (2007b) Theoretical investigations on the potential-induced formation of Pt-oxide surfaces. *J. Electroanal. Chem.*, **607** (1–2), 158–166.
33. Rossmeisl, J. et al. (2006) Calculated phase diagrams for the electrochemical oxidation and reduction of water over Pt(111). *J. Phys. Chem. B*, **110** (43), 21833–21839.
34. Atkins, P.W. (1990) *Physikalische Chemie* Wiley-VCH Verlag GmbH, Weinheim.
35. Bockris, J.O.M., Reddy, A.K.N., and Gamboa-Aldeco, M. (2000) *Modern Electrochemistry 2A*, 2nd edn, Kluwer Academic/Plenum Publishers, New York.
36. Schmickler, W. (1996b) *Interfacial Electrochemistry* Oxford University Press, New York.
37. Abbas, Z. et al. (2002) Corrected Debye-Hückel theory of salt solutions: size asymmetry and effective diameters. *J. Phys. Chem. B*, **106** (6), 1403–1420.
38. Adzic, R. (1998) in *Electrocatalysis* (eds J. Lipkowski and P.N. Ross), Wiley-VCH Verlag GmbH, New York. 197–242.
39. Marković, N.M. and Ross, P.N. (1999) in *Interfacial Electrochemistry: Theory, Experiments and Applications* (ed. A. Wieckowski), Marcel Dekker, New York. 821–842.
40. Marković, N.M. et al. (2001) Oxygen reduction reaction on Pt and Pt bimetallic surfaces: a selective review. *Fuel Cells*, **1** (2), 105–116.
41. Ross, P.N. (2003) in *Handbook of Fuel Cells: Fundamentals, Technology, Applications* (eds W. Vielstich, A. Lamm, and H.A. Gasteiger), Wiley-VCH Verlag GmbH, Weinheim. 465–480.
42. Sidik, R.A. and Anderson, A.B. (2002) Density functional theory study of O2 electroreduction when bonded to a Pt dual site. *J. Electroanal. Chem.*, **528** (1–2), 69–76.
43. Panchenko, A. et al. (2004) Ab initio calculations of intermediates of oxygen reduction on low-index platinum surfaces. *J. Electrochem. Soc.*, **151** (12), A2016–A2027.
44. Sha, Y. et al. (2010) Theoretical study of solvent effects on the platinum-catalyzed oxygen reduction reaction. *J. Phys. Chem. Lett.*, **1** (5), 856–861.
45. Ma, Y. and Balbuena, P.B. (2007) OOH dissociation on Pt clusters. *Chem. Phys. Lett.*, **447** (4–6), 289–294.
46. Qi, L., Yu, J., and Li, J. (2006) Coverage dependence and hydroperoxyl-mediated pathway of catalytic water formation on

Pt (111) surface. *J. Chem. Phys.*, **125** (5), 054701–054708.

47. Anderson, A.B. and Albu, T.V. (2000) Catalytic effect of platinum on oxygen reduction an Ab initio model including electrode potential dependence. *J. Electrochem. Soc.*, **147** (11), 4229–4238.
48. Nørskov, J.K. *et al.* (2004) Origin of the overpotential for oxygen reduction at a fuel-cell cathode. *J. Phys. Chem. B*, **108** (46), 17886–17892.
49. Jacob, T. (2006a) The mechanism of forming H_2O from H_2 and O_2 over a Pt catalyst via direct oxygen reduction. *Fuel Cells*, **6** (3–4), 159–181.
50. Damjanovic, A. and Brusic, V. (1967) Electrode kinetics of oxygen reduction on oxide-free platinum electrodes. *Electrochim. Acta*, **12** (6), 615–628.
51. Keith, J.A., Jerkiewicz, G., and Jacob, T. (2010) Theoretical investigations of the oxygen reduction reaction on Pt(111). *Chemphyschem*, **11** (13), 2779–2794.
52. Jacob, T. and Goddard, W.A. (2006b) Water formation on Pt and Pt-based alloys: a theoretical description of a catalytic reaction. *Chemphyschem*, **7** (5), 992–1005.
53. Gland, J.L. (1980) Molecular and atomic adsorption of oxygen on the Pt(111) and Pt(S)-12(111) × (111) surfaces. *Surf. Sci.*, **93** (2–3), 487–514.
54. Steininger, H., Lehwald, S., and Ibach, H. (1982) Adsorption of oxygen on Pt(111). *Surf. Sci.*, **123** (1), 1–17.
55. Parker, D.H., Bartram, M.E., and Koel, B.E. (1989) Study of high coverages of atomic oxygen on the Pt(111) surface. *Surf. Sci.*, **217** (3), 489–510.
56. Stipe, B.C., Rezaei, M.A., and Ho, W. (1998) Inducing and viewing the rotational motion of a single molecule. *Science*, **279** (5358), 1907–1909.
57. Ho, W. (1998) Inducing and viewing bond selected chemistry with tunneling electrons. *Acc. Chem. Res.*, **31** (9), 567–573.
58. Venkatachalam, S. and Jacob, T. (2009) in *Handbook of Fuel Cells: Advances in Electrocatalysis, Materials, Diagnostics and Durability* (eds H.A.G.W. Vielstich and H. Yokokawa), John Wiley & Sons, Ltd, Chichester. 133–151.
59. Markovic, N.M. *et al.* (1994) Structural effects in electrocatalysis: oxygen reduction on platinum low index single-crystal surfaces in perchloric acid solutions. *J. Electroanal. Chem.*, **377** (1–2), 249–259.

9
Heterogeneous Photocatalysis: from Surface Chemistry to Reactor Design
Guido Mul

9.1
Introduction

The science of heterogeneous photocatalysis covers a wide range of disciplines including (i) band gap engineering and defect chemistry [1], (ii) material science to synthesize (nano)-structured materials [2], (iii) (ultra) fast spectroscopies to study the lifetime of photo-excited states [3, 4], (iv) studies of surface composition and catalytic events [5–7], and finally (v) photoreactor design [8]. It is impossible to cover the recent achievements and discuss fundamental aspects of all these disciplines in a 10 page text on photocatalysis. The focus here is on a general description of the concept of photocatalysis, the surface catalytic events, and finally some reactor design considerations.

9.1.1
What Is Photocatalysis?

It is important to realize what photocatalysis actually is. In principle, light of high enough energy can be used without any catalyst to activate chemical bonds, which is the field of photochemistry, also referred to as *photolysis*. Photocatalysis is the use of a photon-excited catalyst to accelerate a *thermal* reaction, in which the catalyst should not undergo a permanent transition, but be restored to its initial configuration [9]. The wavelength of the light should be of high enough energy to excite the catalyst, but not necessarily of enough energy for photo-activation of the reactant. This can in fact in many cases lead to undesired chemistry/selectivity, as illustrated in the following example of selective oxidation of cyclohexane to cyclohexanone:

$$C_6H_{12} + O_2 \rightarrow C_6H_{10}O + H_2O \tag{9.1}$$

In Figure 9.1 product formation in the photon-induced oxidation of cyclohexane over TiO_2 is shown, using different reaction conditions [11]. Product formation in the absence of catalyst but induced by absorption of deep UV radiation by cyclohexane (this occurs only below $\sim$270 nm) is shown in Figure 9.1a. Clearly

---

Catalysis: From Principles to Applications, First Edition. Edited by Matthias Beller, Albert Renken, and Rutger van Santen.
© 2012 Wiley-VCH Verlag GmbH & Co. KGaA. Published 2012 by Wiley-VCH Verlag GmbH & Co. KGaA.

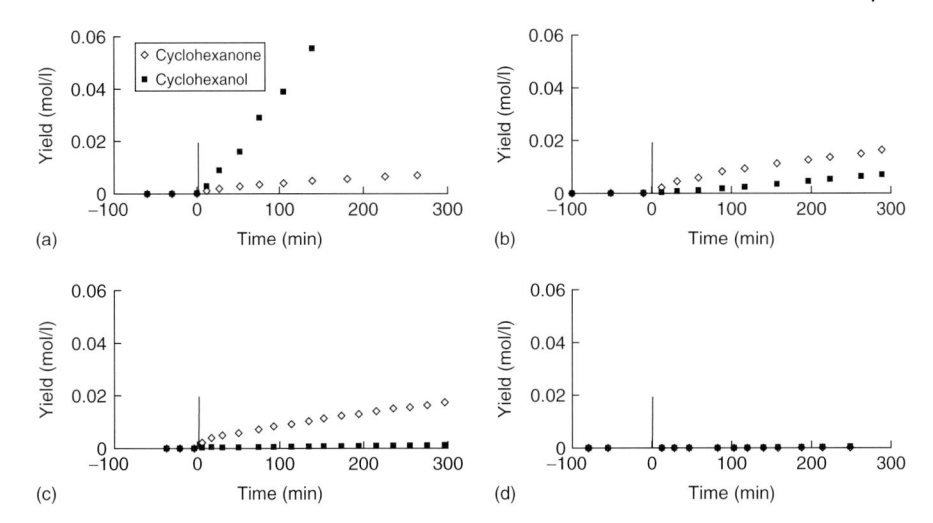

Figure 9.1 Effect of the experimental conditions on the cyclohexanone and cyclohexanol amounts produced in the absence or presence of Hombikat TiO_2 in photocatalytic oxidation of cyclohexane. (a) Quartz reactor, no catalyst (pure photolysis), (b) Quartz reactor, with catalyst (1 g L^{-1} of TiO_2, combination of photolysis and photocatalysis), (c) Pyrex reactor with catalyst (1 g L^{-1} of TiO_2, excluding photolysis), and (d) Pyrex reactor, no catalyst. Light was switched on at $t = 0$ (min) [10].

cyclohexanol, rather than cyclohexanone, which is the preferred product, as it has application in the production of Nylon. The effect of the addition of a TiO_2 catalyst on product selectivity, still in the presence of deep UV radiation (below ~270 nm), is evident by comparison of Figure 9.1a with Figure 9.1b. Although the catalyst decreases the overall productivity (by non-reactive light absorption), the selectivity in favor of cyclohexanone is greatly enhanced. Figure 9.1c illustrates the selectivity if photolysis is excluded (confirmed by Figure 9.1d), which is possible by using a pyrex reactor to filter the UV light. This ensures that only the catalyst is photoactivated, the 366 nm emission line of the Hg lamp being predominantly effective. A selectivity of >90% in favor of cyclohexanone can be achieved in this way. The role of the catalyst in photocatalysis at 366 nm is in principle similar to that in thermocatalysis: the rate is enhanced (or the catalyst allows less extreme conditions, in this case higher wavelengths), as is the selectivity for a desired product.

9.1.2
What Is the Principle of Photocatalysis?

When a photocatalyst absorbs UV/Vis light energy, a transition in the electronic state occurs, yielding the photo-excited state. For molecular catalysts, the light absorption process involves electron excitation from the HOMO (Highest Occupied Molecular Orbital) to the LUMO (Lowest Unoccupied Molecular Orbital) of a chromophore (light absorbing entity) [12]. For isolated Ti-sites, for example, in a mesoporous material, this can be described by the process: $Ti^{4+}\text{-}O^{2-} \rightarrow Ti^{3+}\text{-}O^{-}$ [13]. After light absorption by molecular catalysts, the energy of the excited state may be lost in various ways, often predominantly by thermal degradation (i.e., heat is produced), or by photon emission (radiative decay by fluorescence (immediate conversion of absorbed light into re-emitted energy), or phosphorescence (light energy is stored in a reservoir from which it slowly leaks)) [12].

As well as molecular catalysts, quite often crystalline materials are applied in photocatalysis, the most abundantly investigated being TiO_2. One could in a simplistic view state that the orbitals of oxygen constitute the so-called valence band (occupied in the ground state), and the orbitals of Ti the conduction band (contains electrons in the excited state). The energy difference between the highest level of the valence band and the lowest level of the conduction band is referred to as the 'band gap.' The nature of the active sites thus created by excitation of the crystalline catalysts is typically described by the terminology 'holes,' which are capable of oxidizing a substrate by accepting an electron, and 'electrons,' which are able to reduce a (second) substrate [9]. One can thus state that two active sites are created upon absorption of one photon.

The usual illustration that is shown in textbooks to describe heterogeneous photocatalysis is presented in Figure 9.2 [9]. As well as by the above-described redox reactions, relaxation of the excited state occurs by volume recombination or surface

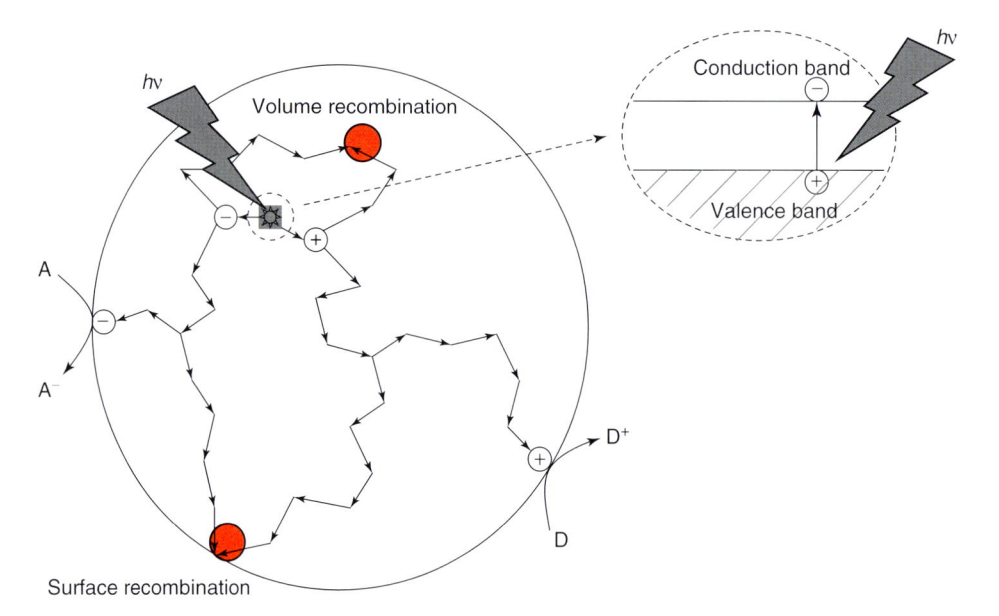

Figure 9.2 Representation of the processes occurring upon light absorption by a crystalline semi-conductor particle. An excited state is created which can in principle lead to the formation of two active surface sites, one capable of reduction, and the other of oxidation of surface-adsorbed substrates. Volume or surface recombination (indicated by the red circles) decreases the efficiency of the photocatalytic process [9].

recombination, the energy being dissipated in the form of heat or luminescence. In fact these processes are much more likely than the employment of the excited state in redox reactions, explaining the typically observed low rates ($mol_{product}\ g^{-1}_{cat}\ s^{-1}$) in photocatalysis.

Determination and manipulation of the quantity and lifetime of surface excited states (i.e., the catalytically active sites) is important for improving photocatalytic processes. The analysis of these features of the applied materials is in the domain of spectroscopists and physicists, and the theory behind these methodologies will not be extensively discussed here. Manipulation of the band gap energies to improve light absorption properties is, for example, described in the textbook of Cox [1].

So far we have not chemically identified the photo-excited active sites. Since the lifetime of these sites is extremely short, advanced spectroscopies have to be used to identify the nature of the (primary) catalytic sites formed by light absorption. For TiO_2, the most investigated photocatalytic material, particularly in photo-oxidation reactions, the chemical nature of the surface actives sites has been demonstrated to include surface hydroxyl radicals (generated by reaction of a surface hydroxyl group with a 'hole') in combination with a surface Ti^{3+} site, which transfers an electron to surface-adsorbed oxygen in oxidation processes, forming superoxide anions, O_2^-. In Case Study 1 (below) we will discuss in more detail how the TiO_2 morphology affects the efficiency of hydroxyl radical generation and thus the photocatalytic rate.

9.2
Applications of Photocatalysis

Photocatalysis is applicable for essentially any type of reaction. So why is photocatalysis so little applied in practice? Besides the challenge of engineering reactors to allow efficient introduction of light, the main reason is the real time concentration of the photoactivated sites, which is extremely small. As mentioned previously, this is due to the extremely short lifetimes of these photo-excited states. Research is ongoing to manipulate this, with emphasis on two types of applications, namely the photodegradation of small amounts (parts per million levels) of organic compounds in either air or water (converting contaminants to CO_2 and H_2O) and (at room temperature) thermodynamically uphill reactions such as water splitting and CO_2 activation (artificial photosynthesis). In the first application one can get away with the low concentration of active sites since the turnover frequency needed is small (contaminants are present in low concentrations), and in fact some commercial applications exist. For the second type of reaction there is no thermal catalytic alternative, and in fact photocatalysis is used to store solar energy in the form of chemical energy (solar-to-fuel conversion). To efficiently harvest solar energy and store this as chemical energy in a product such as hydrogen or methanol, creation of visible light-sensitive systems is required [14]. One can find a significant amount of literature focusing on this goal, as will be discussed in Case Study 2. Besides these two major areas of research, selective oxidation can also be achieved in photocatalysis, as already illustrated in Figure 9.1.

9.3
Case Studies

We continue the discussion on photocatalysis on the basis of three case studies. First, the current understanding of the structure-activity correlation for a model reaction representing the decontamination of water is illustrated. Second, a case study with further detail on the composition and role of a photocatalysts in thermodynamically uphill reactions is described. We conclude with a description of current trends in the development of potentially improved photoreactors.

9.3.1
Water Purification: the Quest for the Structure–Activity Relationship of TiO_2

An enormously large number of papers have appeared in the literature describing the activity of TiO_2 of various morphologies, suppliers, and chemical modification for the decomposition of aqueous-phase pollutants. It is often not easy to determine why a specific TiO_2 specimen shows improved performance, for example, compared to Degussa P25 TiO_2, particularly as the efficacy seems also to depend to a large extent on the dye or substrate chosen to be catalytically converted [15]. Research is ongoing to derive design rules for the improvement of TiO_2-based catalysts for water decontamination applications. Certainly the composition of the TiO_2 is important (e.g., the ratio of rutile to anatase), although the reasons for a regularly observed positive effect of the presence of a small fraction of rutile are still not completely understood. In the present case study we will focus on pure anatase, which is typically considered the more active of the TiO_2 phases. From a catalysis perspective one would expect *a priori* that the higher the surface area (i.e., the smaller the crystal size), the higher would be the activity of a given weight of TiO_2.

Specifically we focus here on the effect of the particle size of a TiO_2 catalyst known as 'Hombikat UV 100' (from Sachtleben Chemie), modified by calcination at different temperatures, on the aqueous-phase methylene blue decomposition activity. Attention is paid to the associated changes in the surface composition, in particular the number of available OH groups, as well as to the amount and lifetime of electron–hole pairs formed upon photo-excitation. The latter can be determined by a technique called Time-Resolved Microwave Conductivity (TRMC). A description of this technique can be found in Refs. [16–18] and references therein.

The effect of particle size on the performance in methylene blue decomposition is illustrated in Figure 9.3. First a decrease in rate constant, which was deduced from the exponential decay of the methylene blue concentration as a function of time, is observed by increasing the particle size from 5 to 15 nm. Subsequently a strong enhancement in performance can be noticed, maximizing at around 25 nm. In the same graph (Figure 9.3a), the qualitative trends in available hole concentration, deduced from the TRMC measurements, and available surface OH concentration, deduced from Infrared and NH_3-TPD measurements, are shown.

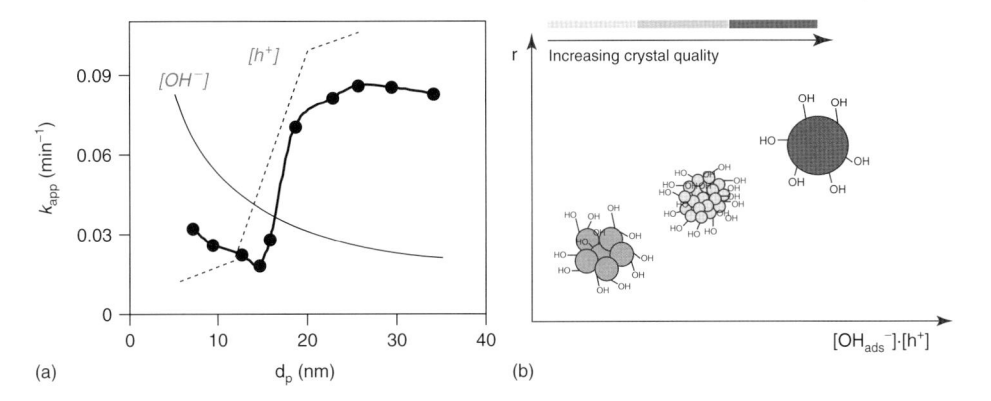

Figure 9.3 (a) Quantitative trend in the k_{app} of methylene blue degradation as a function of anatase crystallite size. A decrease is followed by a rapid increase and shallow maximum. Also shown are the qualitative trends in [h$^+$], as determined by TRMC measurements, and [OH$^-$], determined by ammonia TPD experiments and infrared measurements. (b) Schematic representation of the effect of crystallite size on hole availability (a dark color indicates a high availability) and number of OH groups present in a similar particle agglomerate volume. (After Ref. 16.)

The trend in reactivity can be described by the simple relation: $k_{app} = k[h^+][OH^-]$. This equation is in agreement with the general assumption that hydroxyl radicals are considered to be the primary oxidizing species in photocatalytic oxidation reactions. Principally these are formed by reaction of holes (h$^+$) with surface OH$^-$ groups. Apparently, the product [h$^+$][OH$^-$] decreases for particles increasing in size range from 5 to 15 nm. While the total amount of OH groups further reduces for larger crystallites, it is concluded that this reducing amount is over-compensated by a rapid increase in the availability of holes at the surface. This is proposed to be the result of the crystal quality, that is, a small amount of grain boundaries and bulk defects in larger crystallites. A threshold that categorizes the materials into two groups seems to exist: (i) the 5–15 nm range, where a relatively small amount of 'holes' reach the surface, resulting in a strong dependence of the rate on the OH group concentration (a function of surface area) and (ii) the 15–35 nm range, in which the increasing crystal quality leads to a significantly larger surface hole (h$^+$) concentration, and consequently a lower dependence of the rate on the concentration of surface OH groups. These findings are summarized in Figure 9.3b.

In Figure 9.3b, in which the darker color indicates a higher surface hole concentration, the number of OH groups drawn is representative (qualitatively) of the actual situation. As shown in the figure, a certain degree of agglomeration of the smaller particles in the suspension can be expected. As a final note, the concentration of other sites besides hydroxyl groups, such as surface defects, can also change by calcination and contribute to changes in k_{app}. Still, the results of the present study suggest that a combination of a high OH-group concentration (available for small crystallites) in combination with a high crystallinity (low amount of defects) should provide for highly active catalysts. Perhaps TiO$_2$ nanotubes and other recently reported nano-structures with high activity will conform to these requirements [19].

9.3.2
Energy Conversion: Advanced Materials to Go Thermodynamically Uphill!

9.3.2.1 Design of Crystalline Catalysts

In principle, many interesting conversions can be imagined where one would be able to drive a reaction thermodynamically uphill. The most exciting one is artificial photosynthesis, in which CO_2 and H_2O are converted to, for example, methanol or higher hydrocarbons. Two approaches are feasible to achieve this: (i) to focus on the 'everything in one pot' option, that is, CO_2 and H_2O are converted over a multifunctional system simultaneously or (ii) to focus on the development of water-splitting catalysts and convert the H_2 with CO_2 so produced in a conventional catalytic process [20]. First let us evaluate the options for water splitting. Various ways to harvest solar energy to produce hydrogen, including a description of molecular approaches (typically inspired by nature) and approaches based on irradiated semiconductor/liquid interfaces, have been reported [14]. Studies on the latter option are related to heterogeneous catalysis and mostly inspired by the data reported by Fujishima and Honda in 1972 [21]. They demonstrated that hydrogen and oxygen evolution from water splitting induced by UV light was possible in an electrochemical cell, using a TiO_2 photoanode and a Pt counterelectrode. The advantage of this configuration is that H_2 and O_2 are produced in separate compartments, eliminating the possibility of the back reaction:

$$2H_2 + O_2 \rightarrow 2H_2O \tag{9.2}$$

There are many reports in the literature showing that the electrochemical cell is not necessary in principle. Light-harvesting units based on the principle of colloidal suspensions of platinized TiO_2 in one simple reactor are described [22–26]. The function of a noble metal promoter (e.g., Pt) on TiO_2 or other semiconductors is twofold: it enhances the lifetime of the photo-excited state (i.e., alters physical properties), and it catalyzes the formation of H_2. Often a functionality for O_2 formation is also added to the TiO_2 surface, such as RuO_2. The disadvantage of noble metal promotors is that these are also able to catalyze the back reaction (Eq. (9.2)). This reaction can be prevented by poisoning the Pt sites with, for example, CO, which has indeed been found to inhibit the undesired back reaction [9]. Also, Na poisoning induced by performing the photocatalytic water decomposition in an Na_2CO_3 solution has been demonstrated to positively affect the overall efficiency of the system for the stoichiometric photodecomposition of water [9]. Finally, the back reaction can be prevented by creation of a protective oxide layer on the surface of the noble metal particles. This has been achieved for Rh-promoted photocatalysts by deposition of a Cr_2O_3 layer. This layer allows the hydrogen produced at the Rh surface to escape, but prevents the diffusion of oxygen to the metal surface necessary for the reaction of H_2 and O_2 [27–29].

The photochemical steps in the conversion of H_2O are not exactly known. Mainly for TiO_2 surfaces, detailed proposals for the reaction pathway have been provided. An early proposal involves an [OH-Ti^{4+}-O-Ti^{4+}-OH] surface site consisting of two OH groups and two Ti^{4+} centers. Upon photo-excitation (two photons are needed) the OH groups are converted to OH radicals, followed by recombination to H_2O_2 and finally decomposition to oxygen and H_2O [9]. The two Ti^{4+} sites form two Ti^{3+} sites on which, by a two-electron transfer to adsorbed H_2O and subsequently breaking of the H_2-O bond, H_2 is formed. It should be noted that this pathway is currently being further evaluated and refined [3, 4]. DFT calculations are being conducted, and new evidence shows that rather than acceptance of the hole by surface OH groups, the bridged oxygen in a [-Ti^{4+}-O-Ti^{4+}−] site could be involved, at least for rutile surfaces. Future studies will be focused on refining the pathway of photocatalytic water splitting on TiO_2 and other semi-conducting materials.

9.3.2.2 The Quest for Visible Light-Sensitive Systems

Development of improved visible light-activated systems is necessary to enhance the solar hydrogen production rate (reducing the light exposed area required), which is typically in the domain of inorganic chemists and 'band gap engineers.' For example, 'N' and 'C' doping have been investigated extensively to shift the band gap energy of TiO_2 toward the visible [30]. Factors that determine the applicability of other catalysts than TiO_2 in photocatalytic water splitting are and chemical stability (the catalyst should not change state in the process), the price, and the toxicity [31].

An interesting alternative design route to create a visible light-sensitive water decomposition system is based on the so-called Z-scheme [32]. This system is based on the combination of two metal oxides with visible light absorbance, as shown in Figure 9.4, and was in particular investigated by Kudo and co-workers [33, 34]. The system includes a visible light-active O_2-Photocatalyst catalyzing O_2 evolution, but the energy level of the photo-excited electrons is not high enough to induce H_2 formation. The second catalyst is also excited by visible light, leading to hydrogen generation, but now the hole is not of sufficient energy to oxidize water. However, the energy levels of the conduction band electrons of the O_2-Photocatalyst and valence band holes of the H_2-Photocatalyst are such that recombination is feasible through the mediation of an Fe^{2+}/Fe^{3+} couple. Further improvement of these systems is currently under investigation, including systems that would not require the electron mediator [33].

9.3.2.3 Supported Chromophores

Rather than focusing on crystalline materials, there is a further option to synthesize chromophores on readily accessible inert support materials. Just as in thermocatalysis, the photocatalysis community has embraced the advent of well-structured mesoporous materials, such as ETS-10, MCM-41, SBA-15, TUD-1, and so on, giving the opportunity to structure photocatalytically active sites and to enhance performance. These systems have mainly been investigated in the gas phase conversion of CO_2 and $H_2O(g)$, research endeavors having been initiated in the group

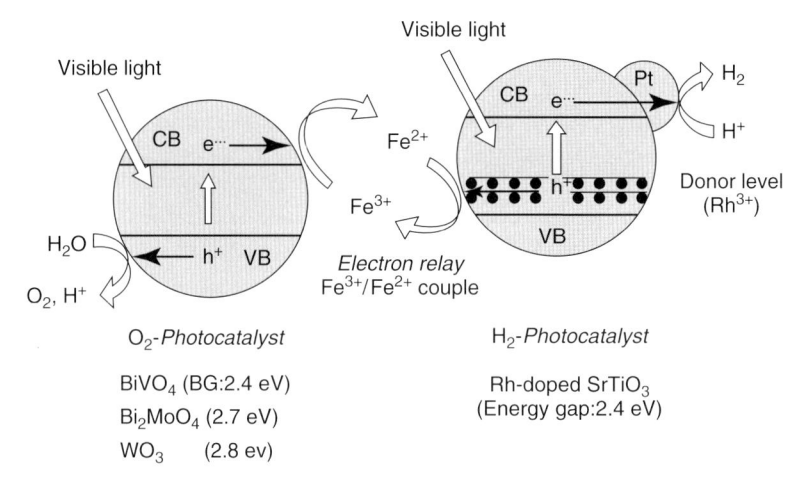

Figure 9.4 Representation of a Z-scheme process, mediated by the Fe^{2+}/Fe^{3+} couple. Oxides such as $BiVO_4$, Bi_2MoO_4, and WO_3 are active in O_2 evolution, while Rh^{3+}-doped, Pt-promoted $SrTiO_3$ is a very active visible-light catalyst for H_2 evolution. (Copied from Ref. 32.)

of Anpo [13, 35, 36]. Using supported TiO_2-based systems, CH_4 and CH_3OH (in a reported ratio of 4 : 1) were the main products formed in significantly larger quantities than those achieved with crystalline TiO_2. Large differences in performance were observed, depending on the mesoporous material used, with SBA-15 yielding the most active catalyst. The pathways to the observed products have been proposed to be related to surface 'C' formation and subsequent hydrogenation. In a recent advanced IR study on Ti supported on MCM-41, however, only CO could be demonstrated as the product of the reaction [5]. An energy scheme, indicated in Figure 9.5, was proposed.

Excitation of the Ti–O ligand-to-metal charge transfer transition of Ti centers leads to transient Ti^{+III} and a hole on a framework oxygen (O^{-1}). Electron transfer from Ti^{+III} to CO_2 yields CO_2^-, and transfer of the hole to water generates a surface OH radical, and H^+, CO_2^-, and H^+ recombine to CO and a second OH radical. The OH radicals either combine to yield H_2O_2 or directly dismutate to give O_2 and H_2O. Figure 9.5 indicates the free energies associated with the formation of the stable products. The only speculative energy is that of the surface OH radicals. Research is currently directed at combining the Ti sites with other (transition) metals to achieve visible light sensitivity. Examples of successful molecular design in mesoporous materials have been reported in the literature [37, 38]. As a final note, to evaluate the reactivity it is extremely important to use pure mesoporous materials, free from carbonaceous residues, which is often not easy to achieve. In view of the long lifetime of OH radicals in a room temperature molecular sieve due to random walk (hundreds of microseconds) it is likely that, even at a very low concentration, OH radicals react with these carbonaceous residues, at least partly

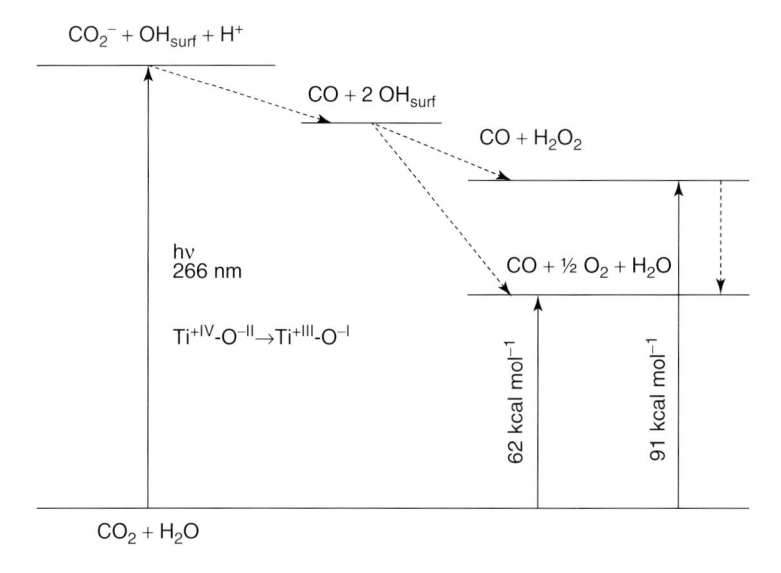

Figure 9.5 Scheme for CO_2 reduction in the presence of H_2O over isolated Ti sites in the MCM-41 framework. The conversion is induced by a 266 nm laser pulse. (Copied from Ref. [5].)

contributing to the product distribution of CO_2 reduction. Careful evaluation of the literature data is recommended.

9.3.3
Photocatalysis in Practice: Some Reactor Considerations

Currently photo-reactors for liquid phase oxidation are usually based on slurry systems; that is, the solid phase is dispersed within the liquid in the reactor. Usually the slurry is illuminated from the top (a Top Illumination Reactor) or by the use of an immersed light source, referred to as an *Annular Reactor* [8, 39, 40]. Although these designs offer ease of construction and high catalyst loading, they clearly have drawbacks, such as the difficulty of separation of catalyst particles from the reaction mixture and low light utilization efficiencies due to the scattering and shielding of light by the reaction medium and catalyst particles. This is illustrated for the Top Illumination Reactor in Figure 9.6 [8]. Clearly the light intensity at the catalyst particle surface, I_s, decreases as a function of the position of the catalyst particle in the reactor with respect to the incoming light (I_0). For given realistic catalyst concentrations (1 g L^{-1}) and density, the light intensity I_s drops down to half of its maximum when the penetration depth in the reaction medium is only 0.015 m. At the bottom of the reactor (0.054 m), the light reaching the catalyst surface is almost zero. This implies that effectively only the catalyst particles in the top of the vessel contribute to the conversion of the reactant! This translates into

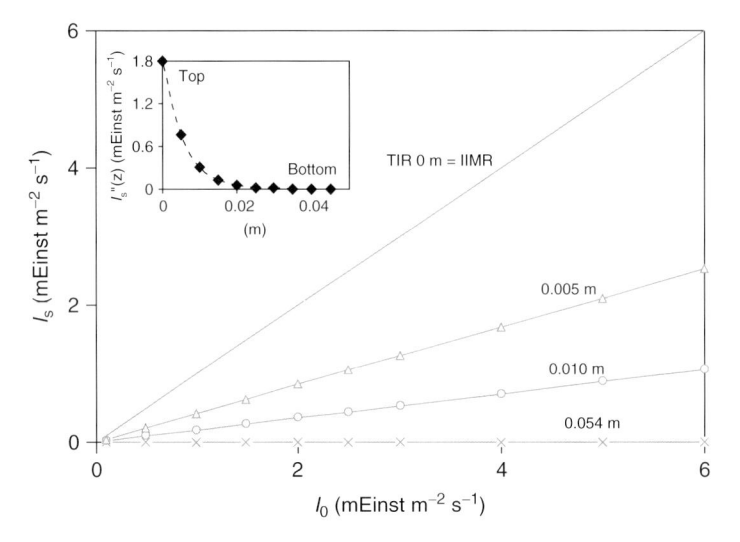

Figure 9.6 Light intensity that reaches the catalyst surface (I_s) for a top illumination reactor, for a given light intensity entering the reactor (I_0); this dependence is different for a certain distance from the light source. The inset shows the light intensity decay in the TIR (Top Illumination Reactor) through the reaction mixture when I_0 is 1.8×10^{-3} Einst m^{-2} s^{-1}. IIMR stands for Internally Illuminated Monolith Reactor. In this reactor concept the catalyst is immobilized and assumed to be completely exposed (i.e., $I_0 = I_s$) [8].

the rapidly decreasing slope in the relationship between I_0 and I_s at increasing liquid depth.

Various attempts have been made to amend the aforementioned light distribution problem by immobilizing photocatalytic systems. One approach was to employ optical fibers as a light distributing guide and support for photocatalysts. Light propagates through the fiber core, while a certain amount of photons are refracted into the coated titania layer. By this means, the optical fibers enable the remote delivery of photon energy to the reactive sites of the photocatalyst [41, 42]. Various groups have reported on the successful application of titania-coated optical fibers in photocatalytic reactions. However, a coating of a TiO_2 catalyst layer on quartz fibers has several intrinsic drawbacks, particularly in liquid phase applications. Firstly, the adhesion strength and layer thickness of catalyst coating on the fibers strongly affect its durability and performance. As the adhesion of TiO_2 particles on quartz fibers is primarily due to electrostatic interaction, it is unlikely that the coating layer will withstand severe liquid flow conditions in large-scale continuous operation modes. To enhance the durability of the titania coatings, fibers were often roughened before the immobilization of catalysts. However this will inevitably result in an uneven distribution of light along the axial direction of the fibers. Other significant problems are the short light-propagating length (<10 cm) and the heat build-up in the bundled array, which could lead to local deactivation of the catalyst.

An alternative to coating the catalyst onto the fibers is a combination of side-light-emitting fibers and ceramic monoliths. The side-light fibers are evenly distributed inside a ceramic monolith structure, on the inner walls of which, for example, a titania photocatalyst is coated [8]. The reaction system is so constructed that flows in various hydrodynamic regimes, such as Taylor flow (alternating gas bubbles and liquid slugs) and film flow can be realized. Because no catalyst is coated on the fibers, the emitted light can reach the catalyst-reactant interface without being strongly attenuated by the solid particles. This means that, by approximation, $I_s = I_0$, as indicated in Figure 9.6. Indeed, promising results have been achieved with this concept. A scheme of the Internally Illuminated Monolith Reactor (IIMR) is presented in Figure 9.7.

9.3.3.1 Microreactors

Another innovative photoreactor design is based on the microreactor concept. LED light sources are available to illuminate microreactors with well-defined wavelengths including in the UV [40, 43–46]. The distance between the light source and the catalyst is small, with the catalyst immobilized on the walls of the microchannels. Furthermore, in gas/liquid configurations microreactors allow

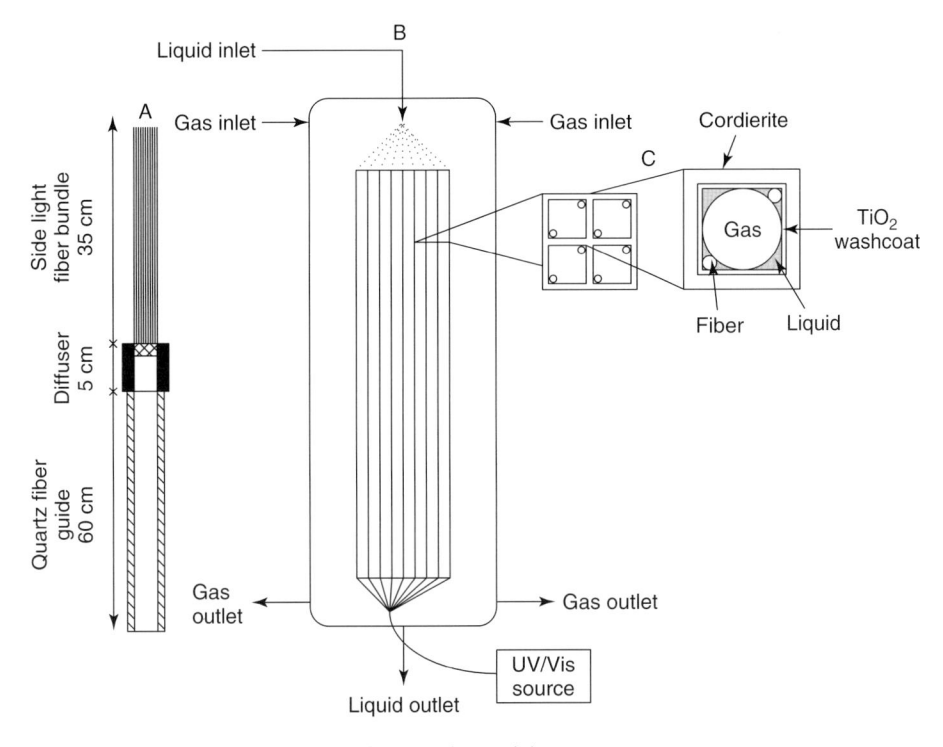

Figure 9.7 Scheme of the internally illuminated monolith reactor set-up, consisting of a ceramic monolith with square channels, on the walls of which the catalyst (TiO$_2$) is coated and in which two optical fibers are inserted and located in the corners of the channels. The detail shows the construction of the optical fiber bundle. (Copied from Ref. [8].)

good control of the gas–liquid distribution. The feasibility of microreactors in photocatalysis was demonstrated in the literature. Positive effects on selectivity have been observed, both as a result of the use of monochromatic light (the advantage of LEDs vs, e.g., mercury vapor lamps) and excellent control over the residence times of the reactants [45, 46].

9.4
Concluding Remarks

More and more research groups active in the field of homo- and heterogeneous catalysis realize that light can be an attractive alternative to heat for performing a catalytic conversion. With the advent of highly efficient LED light sources and novel (micro) reactor concepts, the practical application of light-induced catalysis might become feasible. Although the field of photocatalysis has significant history, analysis of the surface composition and surface chemistry in reaction conditions by vibrational spectroscopies is rather rare, in particular in the liquid phase, and much insight is still needed. The field would certainly benefit from an integrated approach rather than individual studies by band gap engineers, material scientists, spectroscopists, catalysis groups, and recently theoreticians. This will propel the science into the field, and may open up rules enabling rational design of novel catalyst formulations. Photocatalysis no doubt has a bright future.

References

1. Cox, P.A. (1987) *The Electronic Structure and Chemistry of Solids*, Oxford University Press, Oxford.
2. Frei, H. (2006) Selective hydrocarbon oxidation in zeolites. *Science*, **313** (5785), 309–310.
3. Nakamura, R. and Nakato, Y. (2004) Primary intermediates of oxygen photoevolution reaction on TiO_2 (rutile) particles, revealed by in situ FTIR absorption and photoluminescence measurements. *J. Am. Chem. Soc.*, **126** (4), 1290–1298.
4. Nakamura, R. et al. (2005) Molecular mechanisms of photoinduced oxygen evolution, PL emission, and surface roughening at atomically smooth (110) and (100) n-TiO_2 (rutile) surfaces in aqueous acidic solutions. *J. Am. Chem. Soc.*, **127** (37), 12975–12983.
5. Lin, W.Y., Han, H.X., and Frei, H. (2004) CO_2 splitting by H_2O to CO and O_2 under UV light in TiMCM-41 silicate sieve. *J. Phys. Chem. B*, **108** (47), 18269–18273.
6. Mul, G. et al. (2008) Cyclohexene photo-oxidation over vanadia catalyst analyzed by time resolved ATR-FT-IR spectroscopy. *Phys. Chem. Chem. Phys.*, **10** (21), 3131–3137.
7. Yeom, Y.H. and Frei, H. (2001) Photoactivation of CO in Ti silicalite molecular sieve. *J. Phys. Chem. A*, **105** (22), 5334–5339.
8. Carneiro, J.T. et al. (2009) An internally illuminated monolith reactor: Pros and cons relative to a slurry reactor. *Catal. Today*, **147**, S324–S329.
9. Serpone, N. and Pelizzetti, E. (1989) *Photocatalysis-Fundamentals and Applications*, Wiley-Interscience.
10. Hub, S., Hilaire, L., and Touroude, R. (1988) Hydrogenation of But-1-yne and But-1-ene on palladium catalysts: particle size effect. *Appl. Catal.*, **36**, 307–322.

11. Du, P., Moulijn, J.A., and Mul, G. (2006) Selective photo(catalytic)-oxidation of cyclohexane: effect of wavelength and TiO_2 structure on product yields. *J. Catal.*, **238** (2), 342–352.
12. Atkins, P.W. (1986) *Physical Chemistry*, Oxford University Press, Oxford.
13. Anpo, M. *et al.* (1998) Photocatalytic reduction of CO_2 with H_2O on Ti-MCM-41 and Ti-MCM-48 mesoporous zeolite catalysts. *Catal. Today*, **44** (1–4), 327–332.
14. Rajeshwar, K., McConnel, R., and Licht, S. (2008) *Solar Hydrogen Generation-Toward a Renewable Energy Future*, Springer Science and Business Media.
15. Hoffmann, M.R. *et al.* (1995) Environmental applications of semiconductor photocatalysis. *Chem. Rev.*, **95** (1), 69–96.
16. Carneiro, J.T. *et al.* (2010) Toward a physically sound structure-activity relationship of TiO_2-based photocatalysts. *J. Phys. Chem. C*, **114** (1), 327–332.
17. Carneiro, J.T. *et al.* (2011) How phase composition influences optoelectronic and photocatalytic properties of TiO_2. *J. Phys. Chem. C*, **115** (5), 2211–2217.
18. Carneiro, J.T., Savenije, T.J., and Mul, G. (2009) Experimental evidence for electron localization on Au upon photo-activation of Au/anatase catalysts. *Phys. Chem. Chem. Phys.*, **11** (15), 2708–2714.
19. Roy, S.C. *et al.* (2010) Toward solar fuels: photocatalytic conversion of carbon dioxide to hydrocarbons. *Acs Nano*, **4** (3), 1259–1278.
20. Haije, W., and Geerlings, H. (2011) Efficient production of solar fuel using existing large scale production technologies. *Environ. Sci. Technol*, **45**, 8609–8610.
21. Fujishima, A. and Honda, K. (1972) Electrochemical photolysis of water at a semiconductor electrode. *Nature*, **238** (5358), 37–38.
22. Borgarello, E. *et al.* (1982) Visible-light induced water cleavage in colloidal solutions of chromium-doped titanium-dioxide particles. *J. Am. Chem. Soc.*, **104** (11), 2996–3002.
23. Borgarello, E. *et al.* (1981) Sustained water cleavage by visible-light. *J. Am. Chem. Soc.*, **103** (21), 6324–6329.
24. Borgarello, E. *et al.* (1981) Photochemical cleavage of water by photocatalysis. *Nature*, **289** (5794), 158–160.
25. Kalyanasundaram, K., Borgarello, E., and Gratzel, M. (1981) Visible-light induced water cleavage in cds dispersions loaded with Pt and RuO_2, hole scavenging by RuO_2. *Helv. Chim. Acta*, **64** (1), 362–366.
26. Kiwi, J. *et al.* (1980) Cyclic water cleavage by visible-light–drastic improvement of yield of H_2 and O_2 with bifunctional redox catalysts. *Angew. Chem. Int. Ed. Eng.*, **19** (8), 646–648.
27. Maeda, K. *et al.* (2010) Simultaneous photodeposition of rhodium-chromium nanoparticles on a semiconductor powder: structural characterization and application to photocatalytic overall water splitting. *Energy Environ. Sci.*, **3** (4), 471–478.
28. Maeda, K. *et al.* (2010) Photocatalytic overall water splitting promoted by two different cocatalysts for hydrogen and oxygen evolution under visible light. *Angew. Chem.Int. Ed.*, **49** (24), 4096–4099.
29. Sakamoto, N. *et al.* (2009) Highly dispersed noble-metal/chromia (core/shell) nanoparticles as efficient hydrogen evolution promoters for photocatalytic overall water splitting under visible light. *Nanoscale*, **1** (1), 106–109.
30. Mohamed, A.E. and Rohani, S. (2011) Modified TiO_2 nanotube arrays (TNTAs): progressive strategies towards visible light responsive photoanode, a review. *Energy Environ. Sci.*, **4** (4), 1065–1086.
31. Carp, O., Huisman, C.L., and Reller, A. (2004) Photoinduced reactivity of titanium dioxide. *Prog. Solid State Chem.*, **32** (1–2), 33–177.
32. Kudo, A. and Miseki, Y. (2009) Heterogeneous photocatalyst materials for water splitting. *Chem. Soc. Rev.*, **38** (1), 253–278.
33. Sasaki, Y. *et al.* (2009) Solar water splitting using powdered photocatalysts driven by Z-schematic interparticle electron transfer without an electron

mediator. *J. Phys. Chem. C*, **113** (40), 17536–17542.
34. Kato, H. *et al.* (2004) Construction of Z-scheme type heterogeneous photocatalysis systems for water splitting into H_2 and O_2 under visible light irradiation. *Chem. Lett.*, **33** (10), 1348–1349.
35. Yamashita, H. *et al.* (1998) Selective formation of CH_3OH in the photocatalytic reduction of CO_2 with H_2O on titanium oxides highly dispersed within zeolites and mesoporous molecular sieves. *Catal. Today*, **45** (1–4), 221–227.
36. Anpo, M. *et al.* (1997) Photocatalytic reduction of CO_2 with H_2O on titanium oxides anchored within micropores of zeolites: effects of the structure of the active sites and the addition of Pt. *J. Phys. Chem. B*, **101** (14), 2632–2636.
37. Cuk, T., Weare, W.W., and Frei, H. (2010) Unusually long lifetime of excited charge-transfer state of all-inorganic binuclear TiOMnII unit anchored on silica nanopore surface. *J. Phys. Chem. C*, **114** (19), 9167–9172.
38. Han, H.X. and Frei, H. (2008) In situ spectroscopy of water oxidation at Ir oxide nanocluster driven by visible TiOCr charge-transfer chromophore in mesoporous silica. *J. Phys. Chem. C*, **112** (41), 16156–16159.
39. Du, P. *et al.* (2008) A novel photocatalytic monolith reactor for multiphase heterogeneous photocatalysis. *Appl. Catal. A Gen.*, **334** (1–2), 119–128.
40. Van Gerven, T. *et al.* (2007) A review of intensification of photocatalytic processes. *Chem. Eng. Process.*, **46** (9), 781–789.
41. Xu, J.J. *et al.* (2008) Photocatalytic activity on TiO_2-coated side-glowing optical fiber reactor under solar light. *J. Photochem. Photobiol. A Chem.*, **199** (2–3), 165–169.
42. Wu, J.C.S. *et al.* (2008) Application of optical-fiber photoreactor for CO_2 photocatalytic reduction. *Top. Catal.*, **47** (3–4), 131–136.
43. Vesborg, P.C.K. *et al.* (2010) Gas-phase photocatalysis in mu-reactors. *Chem. Eng. J.*, **160** (2), 738–741.
44. Lapkin, A.A. *et al.* (2008) Photo-oxidation by singlet oxygen generated on nanoporous silicon in a LED-powered reactor. *Chem. Eng. J.*, **136** (2-3), 331–336.
45. Coyle, E.E. and Oelgemoller, M. (2008) Micro-photochemistry: photochemistry in microstructured reactors. The new photochemistry of the future? *Photochem. Photobiol. Sci.*, **7** (11), 1313–1322.
46. Matsushita, Y. *et al.* (2007) Recent progress on photoreactions in microreactors. *Pure Appl. Chem.*, **79** (11), 1959–1968.

Part III
Industrial Catalytic Conversions

This part contains a number of chapters that describe the chemistry of several important industrial catalytic processes. The selection of topics has been made to complement the catalytic systems discussed in previous chapters.

The initial three chapters relate to molecular catalysis. Carbonylation catalysis is an important route toward the production of oxygenates other than through oxidation. Biocatalytic processes are especially important for the production of enantiomeric fine chemicals, and modern polyolefin processes depend on the use of early transition metal homogeneous catalytic systems.

The next two topics relate to heterogeneous transition metal catalysis. Two very important and classical processes are discussed: ammonia synthesis and the Fischer–Tropsch reaction.

The final three chapters deal with applications of inorganic materials as catalysts. Zeolite catalytic systems are discussed for acid-catalyzed hydrocarbon conversion reactions. Later chapters deal with catalytic oxidation by reducible oxides and oxidation at high temperatures. This part concludes with a chapter discussing the use of sulfides in desulfurization catalysis.

10
Carbonylation Reactions
Matthias Beller

10.1
General Aspects

From the very beginning homogeneous catalysis has been inspired by and has benefited significantly from advances in the area of organometallic chemistry. Clearly, the mechanistic understanding of elementary steps and the synthesis of new organometallic compounds provide a valuable source for stimulation for the development of novel catalytic reactions and improved catalyst systems. On a molecular basis, especially metal hydrides and carbonyl complexes have been proven to be important catalysts and intermediates in industrially relevant catalytic reactions.

The term carbonylation [1, 2] is used for a large number of closely related catalytic reactions that have in common the incorporation of carbon monoxide into a substrate by addition of CO to unsaturated compounds such as alkynes or alkenes in the presence of nucleophiles (NuHs) (Scheme 10.1).

These reactions are closely related to the hydroformylation process (oxo-synthesis) by which a formyl group and a hydrogen atom are attached to an olefinic double bond (see Section 10.2). In addition to carbonylations of unsaturated compounds, reactions of carbon monoxide with activated C–X compounds (X = OH, halide) are of practical interest. Here, CO is inserted into an existing C–X bond leading to aliphatic or aromatic carboxylic acid derivatives. The most important industrial example of this type of chemistry is the carbonylation of methanol to give acetic acid (Monsanto and Cativa processes).

Because a number of late and middle transition metal complexes are capable of undergoing the required elementary steps for carbonylation, for example, coordination and insertion of CO and olefins into M–C or M–H bonds, there exists a plethora of different catalysts based on Co, Rh, Ir, Ni, Pd, Pt, Fe, Mo, W, and others. However, rates of ligand exchange reactions, insertion into C–X bonds, and addition of CO to unsaturated bonds vary by orders of magnitude. Hence, catalyst performance of the different metals is quite diverse.

Catalysis: From Principles to Applications, First Edition. Edited by Matthias Beller, Albert Renken, and Rutger van Santen.
© 2012 Wiley-VCH Verlag GmbH & Co. KGaA. Published 2012 by Wiley-VCH Verlag GmbH & Co. KGaA.

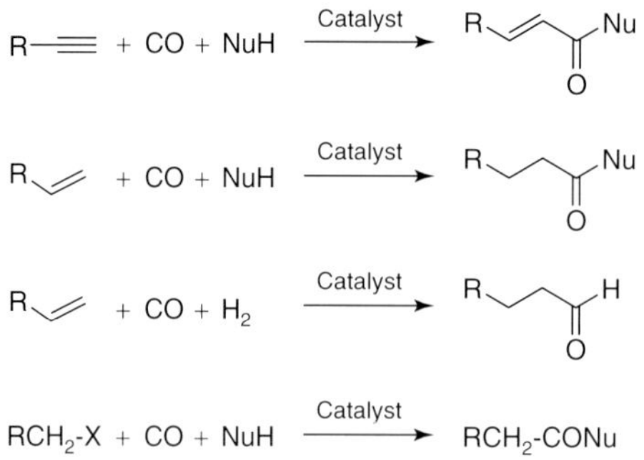

Scheme 10.1 Different types of industrially important carbonylation reactions.

10.2
Hydroformylation

Today, from an industrial point of view, one of the most important homogeneously catalyzed reactions is the hydroformylation of olefins (Scheme 10.2) [3–5]. In this reaction, discovered in 1938 by Otto Roelen at Ruhrchemie/Oberhausen, Germany, olefins and synthesis gas (hydrogen and carbon monoxide) react to form aldehydes.

As shown in Scheme 10.2, different regioisomers can be formed, the linear (n-) aldehyde and the branched (iso-) aldehyde. The regioselectivity is usually expressed as n/iso or l/b ratio.

Despite investigations for more than 40 years some mechanistic issues still remain open, for example, the hydrogenolysis of the metal acyl complex at different conditions.

Although various transition metal-based complexes (e.g., of Ru, Ir, Co, Rh, and Fe) are known to catalyze hydroformylation reactions, mainly rhodium- and (less) cobalt-based systems are used because of their superior activity [6, 7]. However, most recently, cationic palladium and iridium phosphine complexes have also shown promising activity for this methodology.

The generally accepted mechanism of the reaction using a triphenylphosphine rhodium-based catalyst is depicted in Scheme 10.3 [8–11]. The major share of the produced aldehydes is applied as starting material for the production of alcohols, for example, 2-ethylhexanol (made by aldol reaction of butyraldehyde and subsequent reduction), which in turn is used for the production of the corresponding terephthalic esters. These esters are applied as plasticizers for polyvinylchloride (PVC), and are therefore produced on a multi-million ton scale.

Scheme 10.2 Linear and branched aldehydes formed by hydroformylation of terminal olefins.

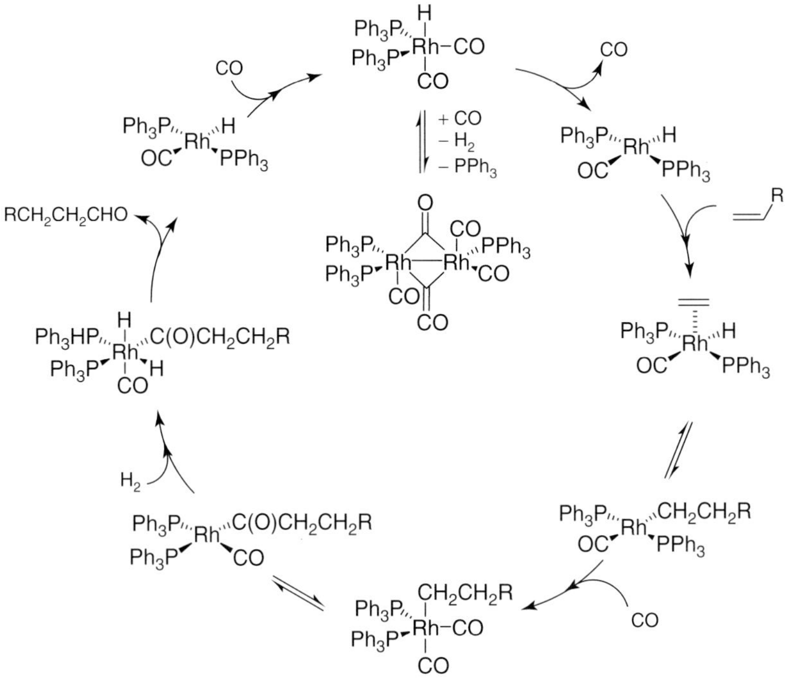

Scheme 10.3 General mechanism of the rhodium-catalyzed hydroformylation.

Although the volume of these established products still increases each year, in the last decades novel plasticizer alcohols based on either hydroformylation of C4-dimers or hydroformylation of butene mixtures and subsequent aldolization and reduction are becoming more important.

Nowadays, commercial hydroformylation plants are run exclusively with catalysts based on either rhodium or cobalt as the central metal. The first generation of hydroformylation processes (BASF, ICI, Ruhrchemie) using cobalt carbonyl complexes were run at relatively high temperatures (150–180 °C) and pressures (200–350 bar).

Later, Shell introduced a phosphine-modified cobalt carbonyl process for the synthesis of detergent alcohols, which is still in use today.

Since the technical and economical success of the homogeneous low-pressure oxo processes (LPO) by Union Carbide and Celanese in the mid seventies, substitution of cobalt catalysts by rhodium catalysts took place. Nevertheless, a significant amount of oxo products (>2.5 Mio tons y^{-1}) are still produced using cobalt catalysts, especially $HCo(CO)_4$ and $HCo(CO)_3PR_3$ complexes. Looking at the hydroformylation of propene, which accounts for the major share of hydroformylation capacity, the technology of the cobalt-based processes has remained unchanged over the last decades, whereas in the case of rhodium improved hydroformylation processes have been introduced.

An interesting industrial development in this area was the Ruhrchemie/Rhône-Poulenc process [12–14], whereby a water-soluble rhodium catalyst is employed in a two-phase hydroformylation process. Based on the original idea of Kuntz [15, 16], an Rh/TPPTS complex (TPPTS = tris sodium salt of *meta*-trisulfonated triphenylphosphine) is used as the catalyst for this process, which reaches a production level of more than 500 000 tons y^{-1} nowadays.

The economical competitiveness of the Ruhrchemie/Rhône-Poulenc process is based on the simple principle of catalyst recycling and low-cost catalyst/product separation.

Despite its advantages, the Rh/TPPTS catalyst also has limitations. For example, it is not possible to hydroformylate internal olefins or long-chain olefins in water with sufficient catalyst activity.

Current trends in hydroformylation reactions are concerned with the substitution of more expensive terminal olefin feedstocks by cheaper olefinic mixtures. Additional impetus for the development of alternative hydroformylation processes is provided by the discussion about new and more environmentally friendly plasticizer products [17]. Thus, hydroformylation of low-price internal olefins to longer-chain linear aldehydes represents an important task in industry. In fact P. W. N. M. van Leeuwen stated in 2000 in his excellent recent book '*The selective formation of linear aldehydes starting from internal alkenes is still one of the greater challenges in hydroformylation chemistry*' [3]. The recent realization of the hydroformylation of butene mixtures to *n*-valeraldehyde on an industrial scale is an illustrative example. In order to obtain linear aldehydes from internal olefins the catalyst has to perform a fast isomerization between the internal and terminal olefin (Scheme 10.4, reaction a).

Unfortunately the thermodynamic equilibrium mixture of butenes contains less than 5% of the terminal olefin (for this reason isomerization should be avoided if terminal olefins are used as starting material). In addition, the hydroformylation of the terminal olefin (Scheme 10.4, reaction c) must occur many times faster (and with high *n*-selectivity) than the reaction of the internal olefin (Scheme 10.4, reaction b).

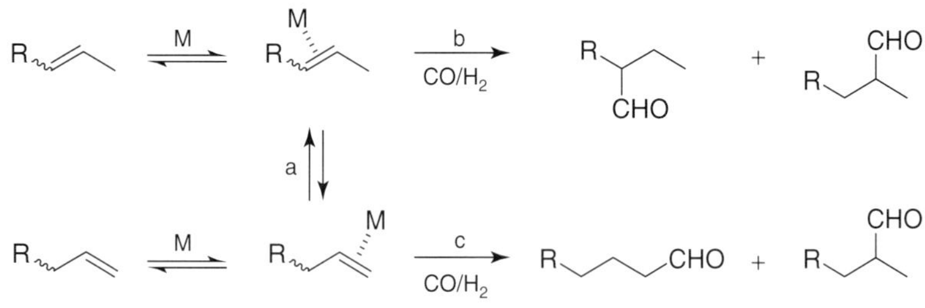

Scheme 10.4 Selective hydroformylation of internal olefins to give linear aldehydes: (a) isomerizaton; (b) hydroformylation of internal olefin; and (c) hydroformylation of terminal olefin.

To date different homogeneous catalysts have been tested for use in the hydroformylation of internal olefins. In general cobalt-based homogenous systems show the same hydroformylation activity for terminal and internal olefins. However, these catalysts need high temperatures and pressures (up to 190 °C and 250 bar), which is not desirable for industry. The selectivity in favor of the linear products, which is reported as *n/iso* ratio, is typically in between 1 : 1 and 7 : 1 even if phosphine-modified systems are used [18–20]. In the presence of cobalt catalysts it is often observed that the selectivity does not depend on whether the reacted olefins are terminal or internal isomers.

Interestingly, by using cobalt catalysts the rate of olefin isomerization is not only faster than hydroformylation but also the CO insertion step is faster than olefin dissociation.

Thus, internal olefins are converted to aldehydes without leaving the metal. In contrast to cobalt catalysts, rhodium phosphine systems show a much higher activity for terminal olefins, and internal olefins are converted only very slowly to branched aldehydes with little isomerization. Similarly to cobalt catalysts, coördinatively unsaturated rhodium species exhibit activity toward isomerization of the olefinic substrate. Such unsaturated rhodium species are formed either due to a lack of CO in the liquid phase or in the presence of sterically demanding ligands which prevent the coordination of more than one phosphorus-containing ligand [21, 22].

In this respect Billig *et al.* reported excellent selectivities for the rhodium-catalyzed hydroformylation of internal olefins using chelating bulky phosphites as ligands. Here, *n/iso* ratios up to 96 : 4 have been obtained for the hydroformylation of 2-butene by using the chelating, but sterically demanding, phosphites [23–25]. Since notable advantages, such as the induction of high reaction rates and easy synthesis, this class of ligands has been extensively studied [26–33]. In the last decade, phosphonites, phosphinites, and even phosphines have also been developed [34–41]. More recently, even tetraphosphorus ligands have been successfully applied [42–46]. This area clearly demonstrates that catalyst improvement in homogeneous catalysis is still based to a large extent on the synthesis of new organic ligands. Despite progress in the area of molecular modeling, successful ligand synthesis relies on experience, intuition, and trial and error.

Obviously, most of the ligands shown in Scheme 10.5 have a rather complicated structure, which makes them difficult to synthesize. Hence, further improvements are desired in this area.

An environmentally benign (atom efficient, one-pot) synthesis of amines from olefins using an *in situ* hydroformylation is the so-called hydroaminomethylation reaction [47]. After initial hydroformylation of the olefin, subsequent formation of an enamine (or imine) followed by hydrogenation takes place. Since its discovery by Reppe at BASF, the hydroaminomethylation reaction has hardly been studied.

In recent years especially, Eilbracht and co-workers have developed new methodologies based on tandem sequences using the hydroaminomethylation reaction. Interestingly, the synthesis of linear amines from internal olefins via hydroaminomethylation has also been described [48].

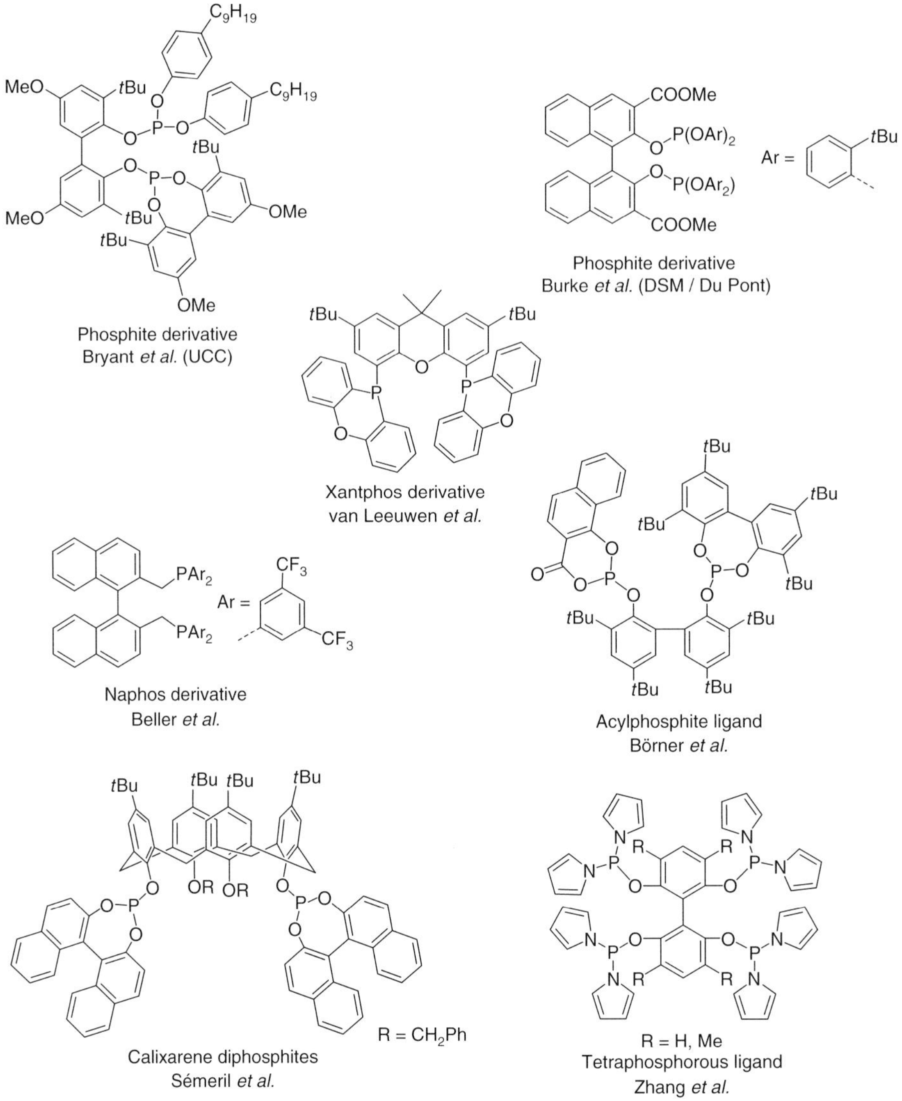

Scheme 10.5 Ligands for hydroformylation of internal olefins to linear aldehydes [23–46].

10.3
Other Carbonylations of Olefins and Alkynes

Pioneering work in the field of carbonylations was done by W. Reppe at BASF in the 1930s and 1940s [49]. He coined the term 'carbonylation' for the addition of carbon monoxide to unsaturated compounds in the presence of NuHs. In Scheme 10.6, typical carbonylation reactions of olefins are depicted. In these reactions, similarly

$$R-CH=CH_2 + CO + H_2O \longrightarrow R-CH_2-CH_2-COOH$$

$$R-CH=CH_2 + CO + ROH \longrightarrow R-CH_2-CH_2-COOR$$

$$R-CH=CH_2 + CO + RSH \longrightarrow R-CH_2-CH_2-COSR$$

$$R-CH=CH_2 + CO + HNR^1R^2 \longrightarrow R-CH_2-CH_2-CONR^1R^2$$

$$R-CH=CH_2 + CO + HOOCR^1 \longrightarrow R-CH_2-CH_2-COOCOR^1$$

Scheme 10.6 Typical carbonylations of alkenes in presence of various nucleophiles.

to hydroformylations, regioselectivity of linear versus branched products is an important issue.

Often mixtures of isomeric carboxylic acids are obtained, owing not only to the occurrence of both Markovnikov and *anti*-Markovnikov addition of the alkene to the metal hydride, but also to the metal-catalyzed alkene isomerization. If the NuH is water or an alcohol, the reaction is usually called *hydrocarboxylation* or *hydroesterification*. Thiols, amines, acids, and even CH-acidic compounds have also been applied as NuHs in these reactions.

In addition to olefins, alkynes are converted into α,β-unsaturated compounds. Formally, a hydrogen and a CONu group are attached to the C–C triple bond. Hence, the formation of E and Z double-bond isomers can be observed. In contrast to the carbonylations of alkynes, the carbonylation of alkenes is slower, and therefore higher pressures and temperatures are often required to reach acceptable rates.

Among the different carbonylation reactions, hydrocarboxylation has attracted most industrial interest. Importantly, this method provides a route to monocarboxylic acids, for example, ethylene to propanoic acid, acetylene to acrylic acid, or 1-olefins (readily available from oligomerization of ethylene) to higher carboxylic acids. In addition, the alkoxycarbonylation of 1,3-butadiene to give methyl 3-pentenoate (intermediate for ε-caprolactam) [50] and propyne to give methyl methacrylate (MMA) [51] were investigated intensively by industry in the last decade.

Catalysts for the hydrocarboxylation reaction are complexes of the transition metals Ni, Co, Fe, Rh, Ru, Pd, Pt, and Ir. Under reaction conditions, the corresponding metal carbonyls or hydridocarbonyls are formed from various catalyst precursors, which can be metal salts (halides preferred), complex salts, oxides, or in some special cases even fine metal powders. In the case of metal halides, the nature of the anion plays an important role. Traditionally, industrially important catalyst metals were Ni and Co. However, these catalysts required harsh conditions, as can be seen from Table 10.1.

The $Co_2(CO)_8$-catalyzed hydrocarboxylation of linear α-olefins usually gives 50–60% linear carboxylic acids, the total carboxylic acid yield being 80–90%. Typical conditions are: 150–200 °C, 150–250 bar.

A special feature of cobalt catalysts is the predominant formation of linear products starting from internal olefins. This effect is similar to what is observed in

Table 10.1 Hydrocarboxylations of olefins with various catalysts.

Catalyst	$Co_2(CO)_8$	$Ni(CO)_4$	PdX_2L_2	$PtX_2L_2 + SnX_2$	RhX_3
Temperature (°C)	150–200	200–320	70–120	80–100	100–130
Pressure (bar)	150–250	150–300	1–150	1–200	1–100

Co-catalyzed hydroformylations. In the case of Co catalysts, addition of hydrogen (5–10 vol%) is beneficial and accelerates the reaction. An increase in the reaction rate was also observed in the presence of 3–8 mol equiv. pyridine as ligand. The cobalt carbonyl/pyridine catalyst system is applied industrially for the synthesis of higher alkanoic acids, for example, the hydrocarboxylation of isomers of undecene yields dodecanoic acid with approximately 80% selectivity. The cobalt catalyst can be recovered on distilling off the products of the reaction.

It is commonly assumed for cobalt-catalyzed carbonylations that $Co_2(CO)_8$ reacts with hydrogen or an NuH with an acidic proton to form the catalytically active species $HCo(CO)_4$ [52].

After replacement of one CO ligand by the olefin, which occurs either by an associative or a dissociative mechanism, olefin insertion into the Co–H bond takes place (Scheme 10.7). Subsequent coordination and insertion of CO into the metal–alkyl bond leads to the corresponding acyl complex. Finally, hydrolysis of the acyl complex with the NuH gives the corresponding carboxylic acid or carboxylic acid derivative and completes the catalytic cycle. Presumably, the acyl cleavage takes place by a nucleophilic attack on the carbonyl carbon of the acyl group.

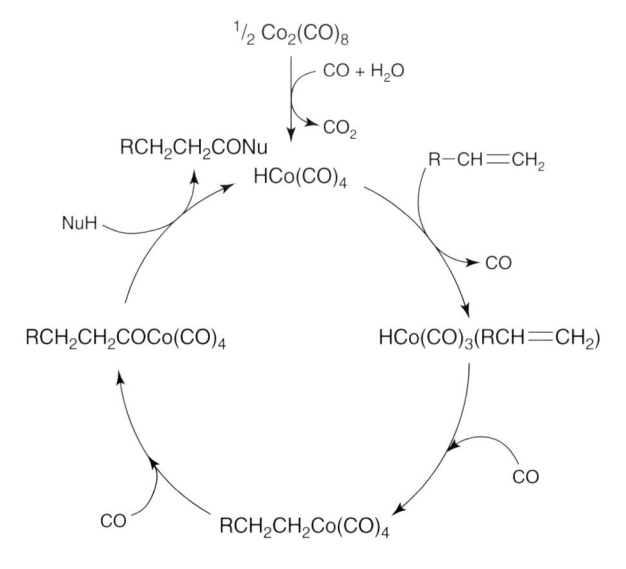

Scheme 10.7 Mechanism of the Co-catalyzed carbonylation.

Nickel catalysts are more suitable for the carbonylation of alkynes, whereas, for olefins, Co, Rh, Pd, Pt, and Ru are equally active if not better. It is characteristic of nickel catalysts in the hydrocarboxylation of α-olefins that the branched carboxylic acid is formed as the main product (60–70%). With internal olefins, branched products are formed exclusively.

More recently, Pd, Pt, Rh, and Ru found widespread use due to their improved catalytic performance under milder conditions [53]. Platinum catalysts are superior with respect to the regioselectivity, especially with tin compounds as co-catalysts.

However, the rates remain quite low even at high CO pressure. Similarly to hydroformylations, the catalysts may or may not be ligand-modified. Co and Ni catalysts are usually used in an unmodified way, whereas ligands are used with Pd and Pt. Due to the milder reaction conditions, substrates like butadiene and styrene can be efficiently carbonylated with Pd catalysts. Otherwise polymerization or other side reactions are problematic.

The regioselectivity of hydroesterification of alkyl acrylates or aromatic olefins catalyzed by $L_m PdX_n$ can be largely controlled by variation of the ligands. Triphenylphosphine promotes preferential carboxylation to the branched isomer, whereas with bidentate bisphoshines the linear product is produced overwhelmingly [54]. The generally proposed mechanisms for catalytic hydroesterification of alkenes are shown in Scheme 10.8. Cycle A, the hydride pathway, is the most accepted.

In the case of Pd-catalyzed carbonylations, there is support for the involvement of $HPdCl(PPh_3)_2$ as the active species under acidic conditions. Evidence for this comes from the isolation of trans-Pd-(COPr)Cl(PPh$_3$)$_2$ from propene hydroformylation [55, 56], while $Pd(CO)(PPh_3)_3$ is inactive as a catalyst in the absence of HCl. Furthermore, in the presence of the alcohol the acyl complex formed the desired

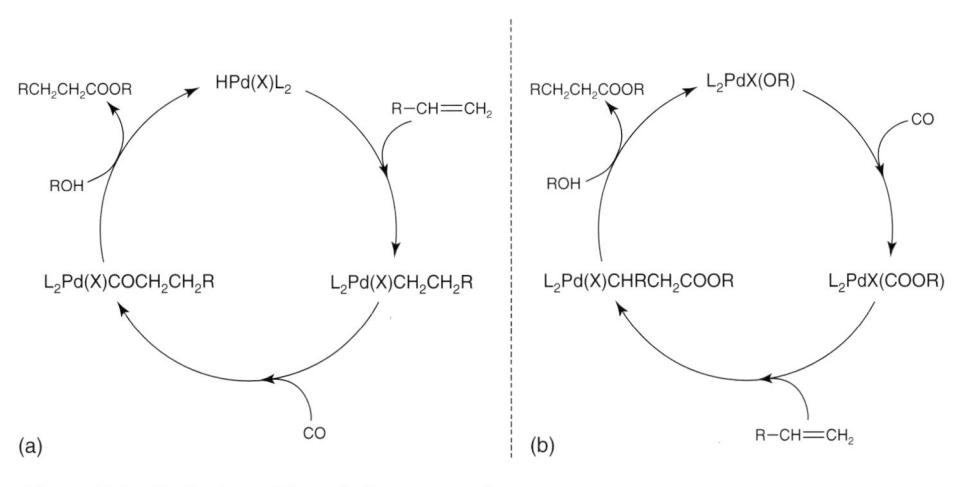

Scheme 10.8 Mechanism of the palladium-catalyzed carbonylation: (a) Hydride pathway and (b) alkoxycarbonyl pathway.

ester. However, under neutral or basic conditions it cannot be excluded that an alkoxycarbonyl complex could operate as the catalytically active species. In the case of PdX_2L_2/SnX_2 catalyst systems, olefins seem to be the hydrogen source for the formation of the active Pd-H species.

Future challenges for carbonylation reactions of olefins and alkynes remain the improvement of catalyst activity and general applicability of these reactions. Still, a number of functionalized olefins (interesting for fine chemicals) cannot be efficiently (and selectively) carbonylated.

Other challenging topics include carbonylations in water [57] or 'novel' solvents, for example, supercritical fluids and ionic liquids, as well as asymmetric carbonylations. While progress in controlling the regiochemistry of hydrocarboxylations has been made, stereoselective carboxylations, which are of interest for intermediates for pharmaceuticals and agrochemicals, are clearly underdeveloped. Valuable representatives of higher-value acids are the commercially important 2-arylpropionic acids. Despite reasonable research efforts, which have led to progress in this area [58–60], better catalyst systems which fulfill technical needs have yet to be developed.

Interestingly, little information is available regarding the recycling of olefin or alkyne carbonylation catalysts. Similarly to hydroformylation, heterogeneous or heterogenized catalysts always face the problem of leaching via the formation of volatile metal carbonyl complexes or clusters.

Of industrial interest is the methoxycarbonylation of 1,3-butadiene as a potential reaction step for ε-caprolactam synthesis. Therefore, carbonylation reactions of butadiene will be discussed in more detail. As shown in Scheme 10.9 1,3-butadiene can be functionalized by carbonylation reactions in several ways. Depending on reaction conditions, monocarboxylate, dicarboxylate, or telomerized products can be obtained.

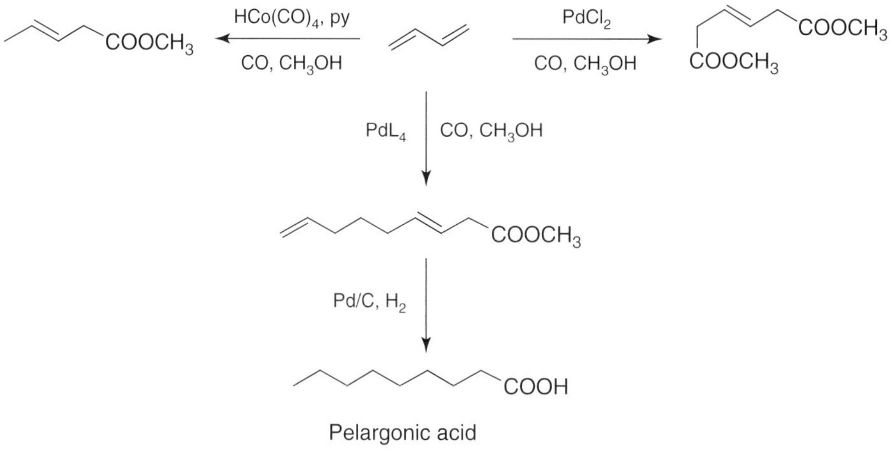

Scheme 10.9 Carbonylation reactions of 1,3-butadiene.

Telomerization of 1,3-butadiene occurs in the presence of methanol and CO using halide-free Pd as the catalyst to give 3,8-nonadienoate ester [61]. Hydrogenation of the telomerization product provides pelargonic acid in good yields.

Halide ions seem to inhibit dimerization by occupying the coordination site on the palladium center and consequently blocking the coordination site of a second diene group required for the dimerization. Thus, palladium acetate solubilized by tertiary amines gives the best yield of dimerization products.

In contrast, oxidative carbonylation of 1,3-butadiene in methanol catalyzed by $PdCl_2$ gives dimethyl-2-butene-1,4-dicarboxylate, while palladium complexes operating under non-oxidative conditions catalyze the hydrocarboxylation to yield preliminary 3-pentenoates. This latter reaction has been optimized considerably by DSM. Earlier on, BASF developed a three-stage process for the synthesis of adipic acid from the butadiene-containing C_4 cut. However, in this old process cobalt was the catalyst metal of choice (Scheme 10.10).

The reaction took place in two steps; the first of these, which involved a lower temperature (100–140 °C), used a fairly high concentration of $HCo(CO)_4$ and pyridine as the catalyst system to ensure rapid carbonylation of butadiene to give methyl pent-3-enoate in 90% selectivity, thus avoiding typical side reactions such as dimerization and oligomerization.

In the second step, the concentration of pyridine as a ligand must be low because it has an inhibitory effect on the isomerization reaction. *In-situ* isomerization to the 4-pentenoic acid ester is a prerequisite for the subsequent carbonylation which provides dimethyl adipate. To ensure internal double-bond rearrangement, the temperature of the reaction was increased to 160–200 °C to give dimethyl adipate with 80% selectivity. After hydrolysis of the ester, adipic acid was obtained with an overall selectivity of about 70%.

Finally, a special type of carbonylation reaction of unsaturated compounds should be mentioned: the so-called Pausen-Khand reaction. Here, an olefin and an alkyne are reacted with CO to give directly cyclopentenones. Apart from cobalt- also rhodium-based catalysts are known for this carbonylation reaction.

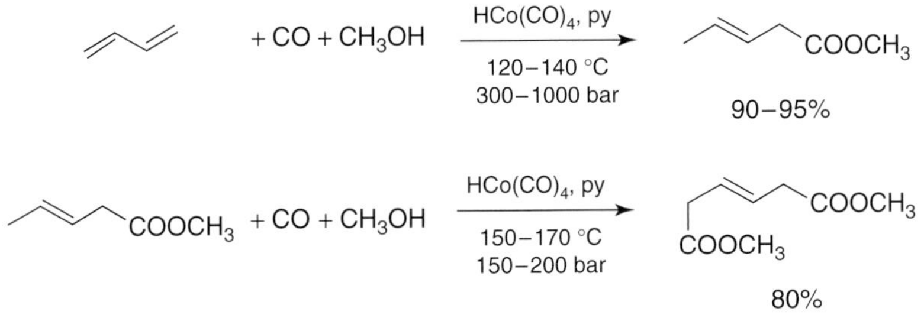

Scheme 10.10 BASF's pilot-plant process for carbonylation of 1,3-butadiene.

10.4
Carbonylations of Alcohols and Aryl Halides

Alcohols, amines, ethers, carboxylic acids, and alkyl as well as aryl halides can be carbonylated to give the corresponding acids, amides, esters, anhydrides, and acid halides, for instance. Typically these reactions are catalyzed by either Pd, Rh, Co, and Ir complexes under relatively drastic conditions (presence of acid, > 100 °C). While the carbonylation of aryl and benzyl halides attracted considerable interest from the organic synthesis community, the reaction of methanol to acetic acid constitutes the most important industrial example. Traditionally, rhodium or cobalt have been used as catalysts for the carbonylation of methanol.

The mechanism of the rhodium-catalyzed carbonylation of methanol (generally referred as the *Montsanto process*), is depicted in Scheme 10.11 [62].

The active species $[Rh^I(CO)_2I_2]^-$ can be easily formed from various Rh compounds. The rate-determining step is the oxidative addition of methyl iodide (formed from methanol and HI) to $[Rh^I(CO)_2I_2]^-$, which explains the independence of the reaction rate from the CO pressure. After the oxidative addition, CO insertion into the alkyl-Rh bond takes place. CO uptake gives an 18-electron complex that decomposes with reductive elimination of acetic acid iodide into $[Rh^I(CO)_2I_2]^-$.

In a second half-cycle, the iodide reacts with methanol, producing acetic acid and methyl iodide. Although the acetic acid selectivity in this process is 99% at low CO pressures (as low as 1 bar), since the 1990s an improved methanol carbonylation process emerged based on Ir as the catalyst metal (Cativa™ process) [63–67].

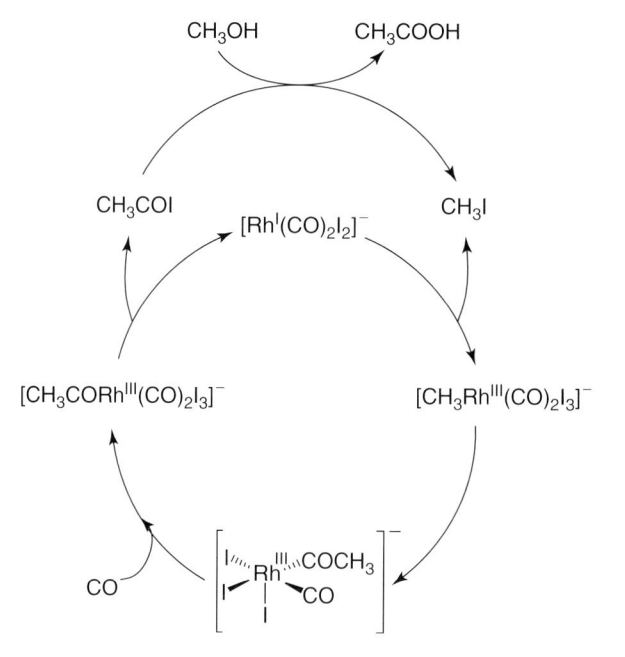

Scheme 10.11 Rhodium-catalyzed carbonylation of methanol.

Starting in the early 1990s, researchers at BP discovered that Ir catalysts provide high carbonylation activity even at low water content [68–70]. This is a significant advantage compared to the older Rh-based process because the catalyst is more stable and fewer by-products are formed. In addition, the CO efficiency is advantageous compared to the Montsanto process. The active catalyst in the Cativa™ process is $[Ir^I(CO)_2I_2]^-$.

The catalytic cycle is similar to the Rh process, but a different rate-determining step is involved. Interestingly, the performance of the Ir catalyst is improved by adding promoters such as iodides of Zn, Cd, and In, or carbonyl complexes of Re, Ru, Os, and W. Apparently these promoters abstract iodide from $[IrCH_3(CO)_2I_3]^-$, thereby facilitating the formation of the corresponding acyl complex. Several acetic acid plants of BP use this new technology.

Apart from bulk production, carbonylation reactions of aryl and benzyl halides have attracted the interest of the fine chemical industry. Basically, these three component coupling reactions offer numerous possibilities for the selective synthesis of aromatic carbonyl compounds, especially carboxylic acid derivatives (acids, esters, and amides). Additionally, aldehydes and ketones are accessible in a similar way (Scheme 10.12) [71].

An example of a carbonylation reaction applied for fine chemical production is Hoechst–Celanese's synthesis of Ibuprofen, which is one of the important non-steroidal anti-inflammatory agents.

The reaction sequence consists of the p-acetylation of iso-butylbenzene in liquid HF, reduction of the resulting acetophenone to the corresponding benzylic alcohol in the presence of a Pd/C catalyst, and subsequent palladium-catalyzed carbonylation in conc. HCl to yield the target molecule (Scheme 10.13) [72].

This catalytic route is nowadays the main industrial production process for Ibuprofen due to the reduced number of steps and the lower amount of by-products compared to the original Boots process.

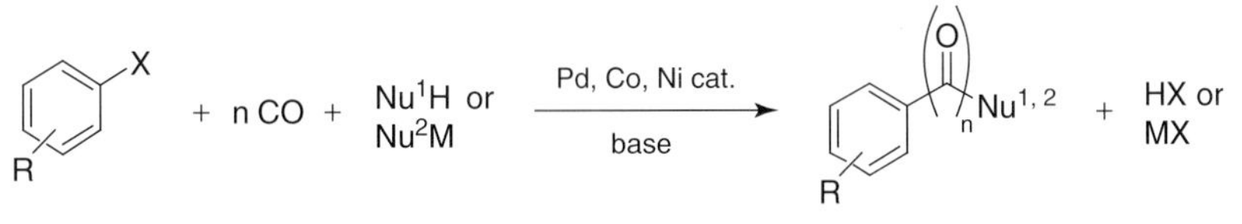

n = 1, 2
X = I, Br, Cl, N_2^+, OSO_2R, ArI^+, IO_2, SO_2Cl
M = Na, K, BR_2, AlR_2, SnR_3, SiR_3
Nu^1 = OH, OR, NR_2, F, Cl; Nu^2 = H, alkyl, aryl, alkenyl, alkynyl, CN, RCO_2

Scheme 10.12 Carbonylation of aryl–X derivatives.

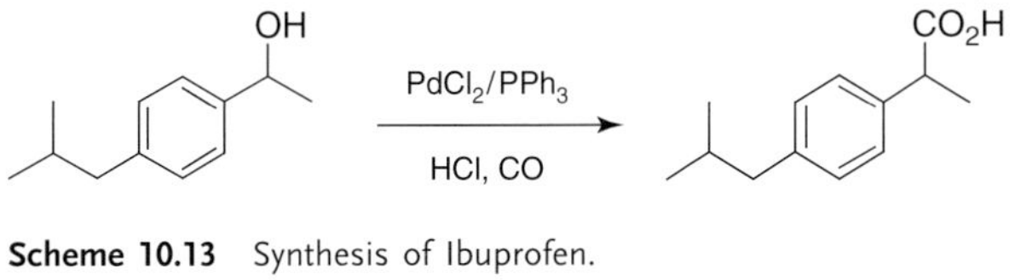

Scheme 10.13 Synthesis of Ibuprofen.

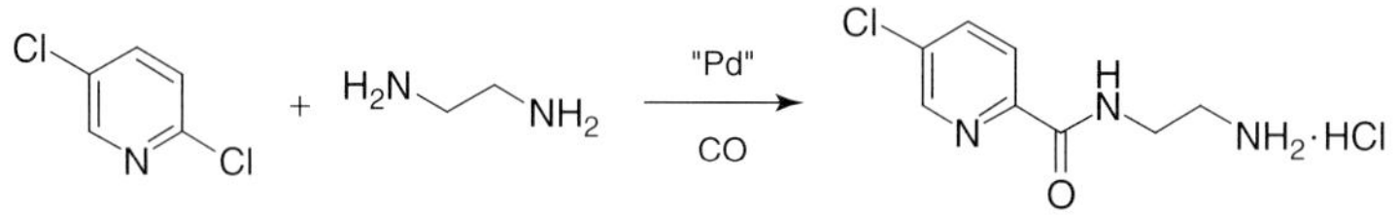

Scheme 10.14 Synthesis of lazabemide.

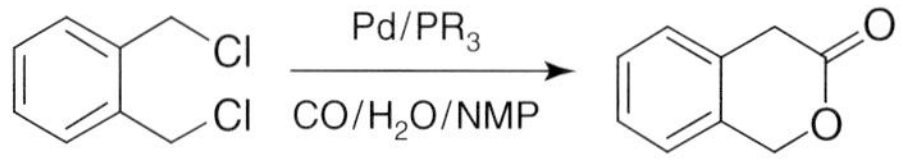

Scheme 10.15 Synthesis of isochroman-3-one by Clariant.

Among the carbonylation reactions of aryl halides, those of heteroaryl halides were of special interest to industrial research groups. This methodology provides easy access to valuable intermediates for the manufacture of herbicides and pharmaceuticals.

One example of the industrial application of a palladium-catalyzed carbonylation reaction is the synthesis of Lazabemide, a monoamine oxidase B inhibitor by Hoffmann-La Roche. The original eight-step laboratory synthesis of Lazabemide could be replaced by a one-step protocol starting from 2,5-dichloropyridine (Scheme 10.14) [73, 74]. The product was isolated in 65% yield. Traces of palladium in the product could be removed by appropriate work-up, because only small amounts of catalyst have to be used (TON = 3000).

In addition to aryl–X and vinyl–X starting materials, palladium-catalyzed carbonylations take place smoothly with allyl–X and benzyl–X (X = Cl, Br, I, OAc, OC(O)R, etc.) compounds. Apart from the synthesis of Ibuprofen, the industrially interesting carbonylations of benzyl chloride to give phenyl acetic acid [75] or N-acetyl phenyl alanine [76] have been studied.

As an example, the former Clariant process for the production of isochroman-3-one starting from o-xylylene dichloride is shown (Scheme 10.15). This process operated for some time in the last decade. The key step was the palladium/phosphine-catalyzed carbonylation and subsequent lactonization in the presence of water in NMP as the solvent [77, 78].

Future challenges for C–X carbonylations include, for example, direct reactions of alcohols. Even better would be the selective carbonylation of C–H bonds. Although some special examples demonstrate that such reactions are indeed possible, general catalytic methodologies have still to be developed.

References

1. Mullen, A. (1980) in *New Syntheses with Carbon Monoxide* (ed. J. Falbe), Springer, Berlin.
2. Bertleff, W. (1986) *Ullmann's Encyclopedia of Industrial Chemistry*, 5th edn, Wiley-VCH Verlag GmbH, Weinheim.
3. van Leeuwen, P.W.N.M. (2000) in *Rhodium Catalyzed Hydroformylation* (eds P.W.N.M. van Leeuwen and C. Claver), Kluwer, Dordrecht.
4. Frohning, C.D., Kohlpaintner, C.W., and Bohnen, H.-W. (2002) *Applied Homogeneous Catalysis with Organometallic Compounds*, Wiley-VCH Verlag GmbH, pp. 29–194.

5. Lenarda, M., Storaro, L., and Ganzerla, R. (1996) Hydroformylation of simple olefins catalyzed by metals and clusters supported on unfunctionalized inorganic carriers. *J. Mol. Catal. A: Chem.*, **111** (3), 203–237.
6. Alvila, L. et al. (1992) Comparative study of homogeneous catalytic activity of group 8-9 metal compounds in hydroformylation of 1-hexene. *J. Mol. Catal.*, **73** (3), 325–334.
7. Protzmann, G. and Wiese, K.-D. (2001) Status und Zukunftsaspekte der industriellen Hydroformylierung. *Erdöl Erdgas Kohle*, **117**, 235–240.
8. Heck, R.F. and Breslow, D.S. (1961) The reaction of cobalt hydrotetracarbonyl with olefins. *J. Am. Chem. Soc.*, **83** (19), 4023–4027.
9. Evans, D., Osborn, J.A., and Wilkinson, G. (1968) Hydroformylation of alkenes by use of rhodium complex catalysts. *J. Chem. Soc. A*, **33** (21), 3133–3142.
10. Evans, D., Yagupsky, G., and Wilkinson, G. (1968) The reaction of hydridocarbonyltris(triphenylphosphine)rhodium with carbon monoxide, and of the reaction products, hydridodicarbonylbis(triphenylphosphine)rhodium and dimeric species, with hydrogen. *J. Chem. Soc. A*, 2660–2665.
11. Brown, C.K. and Wilkinson, G. (1970) Homogeneous hydroformylation of alkenes with hydridocarbonyltris-(triphenylphosphine)rhodium(I) as catalyst. *J. Chem. Soc. A*, 2753–2764.
12. Cornils, B. (1998) *In d. Catal. News*, (2), 7.
13. Cornils, B. and Wiebus, E. (1995) *CHEMTECH*, **25**, 33–38.
14. Cornils, B. and Wiebus, E. (1996) Virtually no environmental impact: the biphasic oxo process. *Recl. Trav. Chim. Pays-Bas*, **115** (4), 211–215.
15. Kuntz, E.G. (1987) *CHEMTECH*, **17**, 570–575.
16. Cornils, B. and Kuntz, E.G. (1995) Introducing TPPTS and related ligands for industrial biphasic processes. *J. Organomet. Chem.*, **502** (1–2), 177–186.
17. (1998) *Chem. Ind.*, 379.
18. Macaluso, A. and Rigdon, O.W. (1976) Synthesis of linear primary alcohols from internal olefins. US Patent US3907909.
19. Slaugh, L.H. and Mullineaux, R.D. (1966) Hydroformylation of olefins, S.O. Co, Editor. US3239569.
20. Beller, M. and Krauter, J.G.E. (1999) Cobalt-catalyzed biphasic hydroformylation of internal short chain olefins. *J. Mol. Catal. A: Chem.*, **143** (1–3), 31–39.
21. van Leeuwen, P.W.N.M. and Roobeek, C.F. (1983) Hydroformylation of less reactive olefins with modified rhodium catalysts. *J. Organomet. Chem.*, **258** (3), 343–350.
22. van Rooy, A. et al. (1991) *J. Chem. Soc., Chem. Commun.*, 1096–1097.
23. Billig, E., Abatjoglou, A.G., and Bryant, D.R. (1988) Transition metal complex catalyzed processes. US 4769498.
24. Billig, E., Abatjoglou, A.G., and Bryant, D.R. (1991) Bis-phosphite compounds. EP 0213639.
25. Billig, E., Abatjoglou, A.G., and Bryant, D.R. (1992) Transition metal complex catalyzed processes. EP 0214622.
26. Burke, P.M. et al. (1997) Process to prepare a terminal aldehyde D.D. Pont, Editor. WO 9733854.
27. Kreutzer, K.A. et al. (2001) D. Pont, Editor. WO 0121567.
28. Bunel, E.E. et al. (2001) D. Pont, Editor. WO 0121627.
29. Selent, D., Hess, D., Wiese, K.-D., Röttger, D., Kunze, C., and Börner, A. (2001) Rhodiumkatalysierte isomerisierende Hydroformylierung interner Olefine mit einer Klasse neuartiger Phosphorliganden. *Angew. Chem. Int. Ed.*, **113**, 1739–1741.
30. Sémeril, D. et al. (2006) Regioselectivity with hemispherical chelators: increasing the catalytic efficiency of complexes of diphosphanes with large bite angles. *Angew. Chem. Int. Ed.*, **45** (35), 5810–5814.
31. Sémeril, D., Matt, D., and Toupet, L. (2008) Highly regioselective hydroformylation with hemispherical chelators. *Chem. – Eur. J.*, **14** (24), 7144–7155.
32. Monnereau, L. et al. (2009) Micellar effects in olefin hydroformylation catalysed by neutral, calix[4]arene-diphosphite rhodium complexes. *Adv. Synth. Catal.*, **351** (10), 1629–1636.

33. Monnereau, L., Sémeril, D., and Matt, D. (2010) Solvent-free olefin hydroformylation using hemispherical diphosphites. *Eur. J. Org. Chem.*, **16**, 3068–3073.
34. Beller, M., Zimmermann, B., and Geissler, H. (1999) Dual catalytic systems for consecutive isomerization–hydroformylation reactions. *Chem. – Eur. J.*, **5** (4), 1301–1305.
35. van der Veen, L.A., Kamer, P.C.J., and van Leeuwen, P.W.N.M. (1999) Neuartige Rhodiumkatalysatoren für die Hydroformylierung interner Olefine zu linearen Aldehyden. *Angew. Chem.*, **111**, 349–351.
36. van der Veen, L.A., Kamer, P.C.J., and van Leeuwen, P.W.N.M. (1999) Hydroformylation of internal olefins to linear aldehydes with novel rhodium catalysts. *Angew. Chem. Int. Ed.*, **38** (3), 336–338.
37. Selent, D. et al. (2000) Neuartige oxyfunktionalisierte Phosphonitliganden für die hydroformylierung isomerer n-olefine. *Angew. Chem. Int. Ed.*, **112** (9), 1694–1696.
38. Selent, D. et al. (2000) Novel oxyfunctionalized phosphonite ligands for the hydroformylation of isomeric n-olefins. *Angew. Chem. Int. Ed.*, **39** (9), 1639–1641.
39. Breit, B. and Seiche, W. (2001) Recent advances on chemo-, regio- and stereoselective hydroformylation. *Synthesis*, **2001** (1), 0001–0036.
40. Klein, H. et al. (2001) Hoch selektive katalysatoren für die hydroformylierung interner olefine zu linearen aldehyden. *Angew. Chem. Int. Ed.*, **113** (18), 3505–3508.
41. Klein, H. et al. (2001) Highly selective catalyst systems for the hydroformylation of internal olefins to linear aldehydes. *Angew. Chem. Int. Ed.*, **40** (18), 3408–3411.
42. Yan, Y., Zhang, X., and Zhang, X. (2006) A tetraphosphorus ligand for highly regioselective isomerization-hydroformylation of internal olefins. *J. Am. Chem. Soc.*, **128** (50), 16058–16061.
43. Yu, S. et al. (2008) Highly regioselective isomerization-hydroformylation of internal olefins to linear aldehyde using Rh complexes with tetraphosphorus ligands. *Org. Lett.*, **10** (16), 3469–3472.
44. Yu, S., Chie, Y.-M., and Zhang, X. (2009) Highly regioselective and rapid hydroformylation of alkyl acrylates catalyzed by a rhodium complex with a tetraphosphorus ligand. *Adv. Synth. Catal.*, **351** (4), 537–540.
45. Yu, S. et al. (2009) Highly regioselective hydroformylation of styrene and its derivatives catalyzed by Rh complex with tetraphosphorus ligands. *Org. Lett.*, **11** (1), 241–244.
46. Yu, S. et al. (2009) Highly regioselective hydroformylation of 1,5-hexadiene to linear dialdehyde catalyzed by rhodium complexes with tetraphosphorus ligands. *Tetrahedron Lett.*, **50** (40), 5575–5577.
47. Eilbracht, P. et al. (1999) Tandem reaction sequences under hydroformylation conditions? New synthetic applications of transition metal catalysis. *Chem. Rev.*, **99** (11), 3329–3366.
48. Seayad, A. et al. (2002) Internal olefins to linear amines. *Science*, **297** (5587), 1676–1678.
49. Reppe, W. (1953) Carbonylierung I. über die umsetzung von acetylen mit kohlenoxyd und verbindungen mit reaktionsfähigen wasserstoffatomen synthesen α,β-ungesättigter carbonsäuren und ihrer derivate. *Justus Liebigs Ann. Chem.*, **582** (1), 1–37.
50. Beller, M., Krotz, A., and Baumann, W. (2002) Palladium-catalyzed methoxycarbonylation of 1,3-butadiene: catalysis and mechanistic studies. *Adv. Synth. Catal.*, **344** (5), 517–524.
51. Drent, E., Arnoldy, P., and Budzelaar, P.H.M. (1994) Homogeneous catalysis by cationic palladium complexes. Precision catalysis in the carbonylation of alkynes. *J. Organomet. Chem.*, **475** (1–2), 57–63.
52. Cornils, B. (1980) in *New Syntheses with Carbon Monoxide* (ed. J. Falbe), Springer, Berlin.
53. Brennführer, A., Neumann, H., and Beller, M. (2009) Palladium-catalyzed carbonylation reactions of alkenes and alkynes. *ChemCatChem*, **1** (1), 28–41.
54. Kiss, G. (2001) Palladium-catalyzed Reppe carbonylation. *Chem. Rev.*, **101** (11), 3435–3456.

55. Bardi, R. et al. (1979) Metals in organic syntheses. III. Highly regioselective propene hydrocarboxylation promoted by a PdCl2(PPh3)2-PPh3 catalyst precursor: trans-Pd(COPr-n)Cl(PPh3)2 as an active catalytic species. *Inorg. Chim. Acta*, **35**, L345–L346.
56. Cavinato, G. and Toniolo, L. (1990) On the mechanism of the hydrocarbalkoxylation of olefins catalyzed by palladium complexes. *J. Organomet. Chem.*, **398** (1–2), 187–195.
57. Cornils, B. and Herrmann, W.A. (1998) *Aqueous-Phase Organometallic Catalysis*, Wiley-VCH Verlag GmbH, Weinheim.
58. Consiglio, G. and Roncetti, L. (1991) Enantioselective carbonylation reactions of para-substituted 2-phenylpropenes. *Chirality*, **3** (4), 341–344.
59. Hiyama, T., Wakasa, N., and Kusumoto, T. (1991) Hydroesterification of 6-Methoxy-2-naphthylethene. *Synlett*, **1991** (08), 569,570.
60. Chelucci, G. et al. (1994) (-)-(4S,5R)-4-(2-pyridyl)-5-(diphenylphosphino)methyl-2,2-dimethyl-1,3-dioxolane a new chiral ligand for enantioselective catalysis. *Tetrahedron: Asymmetry*, **5** (3), 299–302.
61. Neibecker, D., Poirier, J., and Tkatchenko, I. (1989) Carbonylation of organic allyl moieties. Part 1. Synthesis of pelargonic (nonanoic) and margaric (heptadecanoic) acid methyl esters from 1,3-butadiene. *J. Org. Chem.*, **54** (10), 2459–2462.
62. Forster, D. (1976) On the mechanism of a rhodium-complex-catalyzed carbonylation of methanol to acetic acid. *J. Am. Chem. Soc.*, **98** (3), 846–848.
63. Garland, C.S., Giles, M.F., and Sunley, J.G. (1995) Process for the production of acetic acid, B. Chemicals, Editor. EP 0643034.
64. Baker, M.J. et al. (1996) Process for the carbonylation of alkyl alcohols and/or reactive derivatives thereof, B. Chemicals, Editor. EP 0749948.
65. Garland, C.S. et al. (1996) Process for the production of a carboxylic acid, B. Chemicals, Editor. EP 0728726.
66. Baker, M.J. et al. (1997) Process for the production of acetic acid by carbonylation, B. Chemicals, Editor. EP 0752406.
67. Sunley, G.J. and Watson, D.J. (2000) High productivity methanol carbonylation catalysis using iridium: the Cativa(TM) process for the manufacture of acetic acid. *Catal. Today*, **58** (4), 293–307.
68. Haynes, A. and M. Beller (2006) *Acetic Acid Synthesis by Catalytic Carbonylation of Methanol Catalytic Carbonylation Reactions*, Springer, Berlin, Heidelberg, pp. 179–205.
69. Haynes, A. et al. (2004) Promotion of iridium-catalyzed methanol carbonylation: mechanistic studies of the cativa process. *J. Am. Chem. Soc.*, **126** (9), 2847–2861.
70. Jones, J.H. (2000) The Cativa™ process for the manufacture of acetic acid. Iridium catalyst improves productivity in an established industrial process. *Platinum Met. Rev.*, **44** (3), 94–105.
71. Beller, M. et al. (1995) Progress in hydroformylation and carbonylation. *J. Mol. Catal. A Chem.*, **104** (1), 17–85.
72. Beller, M. (1996) in *Applied Homogeneous Catalysis with Organometallic Compounds* (eds B. Cornils and W. Herrmann), Wiley-VCH Verlag GmbH.
73. Schmid, R. (1996) Homogeneous Catalysis with Metal Complexes in a Pharmaceuticals' and Vitamins' Company: Why, What for, and Where to Go?. *Chimia*, **50**, 110.
74. Scalone, M. and Vogt, P., and Roche, H.-L. (1990) EP 385210.
75. Kohlpaintner, C.W. and Beller, M. (1997) Palladium-catalyzed carbonylation of benzyl chlorides to phenylacetic acids – a new two-phase process. *J. Mol. Catal.: A: Chem.*, **116** (1–2), 259–267.
76. de Vries, J.G. et al. (1996) Preparation of d,l-phenylalanine by amidocarbonylation of benzyl chloride. *J. Org. Chem.*, **61** (5), 1842–1846.
77. Geissler, H. and Pfirmann, R. (1998) Production of isochroman-3-one, Clariant, Editor. DE 19815323.
78. Geissler, H. and Pfirmann, R. (2000) Process for the preparation of isochroman-3-ones, Clariant, Editor. EP 1086949.

11
Biocatalytic Processes
Uwe Bornscheuer

11.1
Introduction

As already outlined in Chapter 7, biocatalysis refers to the use of enzymes in industrial applications. Indeed, numerous processes have been developed using a broad range of biocatalysts or a key enzymatic reaction in a microorganism. As early as 1908, Rosenthaler used a hydroxynitrile lyase-containing extract for the preparation of (R)-mandelonitrile from benzaldehyde and hydrogen cyanide (HCN). Since then, more and more enzymes have been identified, and, in parallel, their use in organic chemistry has steadily increased. Especially since the mid-1970s, the number of publications utilizing enzymes as well as the number of industrialized processes has increased substantially. Hundreds of processes have been established, and excellent overviews and detailed examples can be found in books [1–3] and reviews [4–6].

In most applications, the excellent enantio- and stereoselectivity shown by many enzymes is the key feature, which makes them competitive to the alternative chemical reaction (i.e., by asymmetric synthesis or starting from the chiral pool). Moreover, the optically pure chiral products are rather expensive compounds, and hence the cost of the biocatalyst plays a less decisive role. However, in recent years the use of biocatalysts for relatively cheap commodities was also commercialized on industrial scale. The most prominent examples are the production of acrylamide (on the $>100\,000$ t y^{-1} scale) and nicotinamide, both using a nitrile hydratase (NHase), and the synthesis of simple esters, such as myristyl myristate for cosmetic application, catalyzed by a lipase [7].

11.1.1
How to Choose the Best Route?

The decision whether a certain synthesis is to be performed by biocatalytical or chemical means is governed by various considerations, as shown in a simplified manner in Figure 7.1. Most important are, of course, the availability of the enzyme in sufficient amounts at low prices, its properties (activity, selectivity),

---

Catalysis: From Principles to Applications, First Edition. Edited by Matthias Beller, Albert Renken, and Rutger van Santen.
© 2012 Wiley-VCH Verlag GmbH & Co. KGaA. Published 2012 by Wiley-VCH Verlag GmbH & Co. KGaA.

and especially its stability under process conditions. For all intended applications, careful decisions have to be made case-by-case, which should also include such aspects as solvent system used, process design with respect to reactor set-up, integration into an existing process or production plant, up- and downstream processing, and so on. Furthermore, it has often been observed that an upscaling of an enzymatic process is hampered by the low substrate concentrations tolerated by the enzyme. In contrast to chemical catalysts, many enzyme also seriously suffer from substrate and/or product inhibition, which needs to be solved by process design (e.g., substrate feeding strategies or *in-situ* product removal) or protein engineering to enable reactions at molarities and reaction rates sufficiently high for industrial purposes.

Hence, it can happen that, for the same target, various biocatalytic routes can be considered, which may use the same enzyme but different substrates or reaction systems, or enzymes from different classes. This is illustrated in Figure 11.1 for the α-hydroxy acid mandelic acid serving as model compound to exemplify this challenge. Some of the routes have been described in the literature, while others are purely theoretical. As well as the availability of the enzyme with desired properties, the availability of a suitable starting material can be of equal importance to the decision in favor of a specific synthesis route. Furthermore, problems encountered with product isolation (i.e., substrate and product cannot be separated efficiently or cost effectively) can mean that some alternative enzymatic route has to be chosen.

Straightforward access to a single enantiomer of mandelic acid in theoretically 100% yield is possible by the use of a ketoreductase, starting from the corresponding ketone (Route I, Figure 11.1). However, this reaction requires the cofactor nicotinamide adenine dinucleotide (phosphate)(NAD(P)H) efficient recycling is therefore needed, which adds to the overall costs of the process.

The alternative, C–C-bond formation from benzaldehyde and HCN catalyzed by a hydroxynitrile lyase (Route II), has the advantages of readily available starting materials and various enzymes to choose from, but there are disadvantages, namely that the chemical background formation of racemic mandelic acid at certain pH-values must be suppressed and safe handling of the highly toxic HCN in the factory must be guaranteed.

Good atom efficiency is also feasible by using either a nitrilase or a combination of NHase with an amidase starting from racemic hydroxynitrile (Route III). Although this is at first glance a kinetic resolution with only 50% yield of the desired enantiomer, the disproportion of hydroxynitrile to benzaldehyde and HCN – the same chemical background reaction which affects the hydroxynitrile lyase-catalyzed route negatively – here enables a dynamic kinetic resolution.

Disadvantages are the often low stereoselectivities of the nitrile-converting enzymes, which however could be improved by protein engineering (see below for a recent example in the synthesis of a Lipitor building block). The use of hydrolases such as lipases, proteases, or amidases is a reasonably good route despite the fact that it is a kinetic resolution (Route IV). The strength of this approach lies in the low prices and high stabilities and activities of the enzymes (especially lipases),

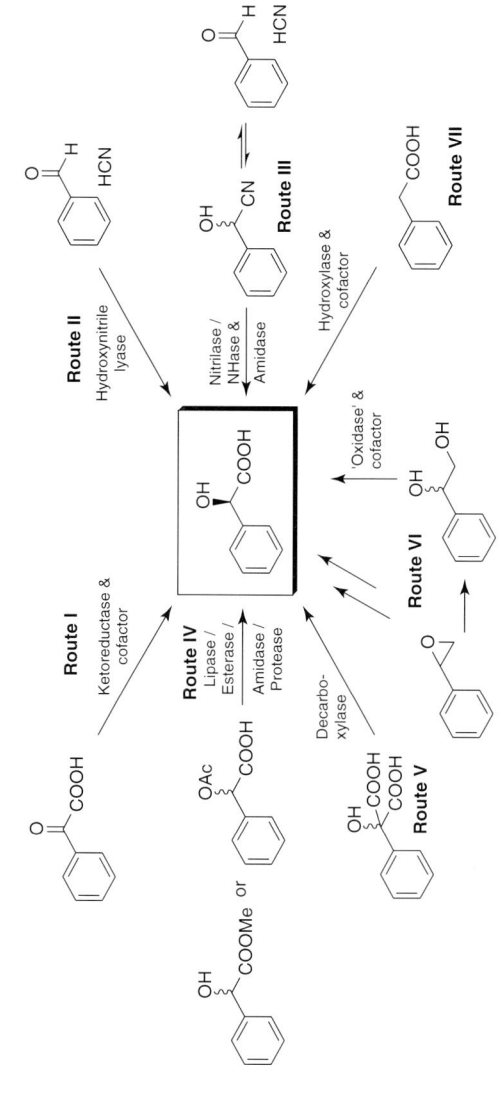

Figure 11.1 Possible enzymatic routes to optically pure (R)-mandelic acid. Routes I–V have been described, but routes VI and VII are theoretical.

which reduce the costs of the biocatalysts and hence can make such a process economical, although one enantiomer needs to be recycled or even discarded.

The rather elegant application of decarboxylases (Route V) is hampered by the lack of availability of the dicarboxylic acid starting material.

The route starting from styrene oxide or the corresponding diol suffers from the lack of availability of suitable enzymes and the requirement for a cofactor (Route VI).

Finally, the direct and stereoselective hydroxylation of phenylacetic acid (Route VII) would be very useful, but the P450 monooxygenases required for this approach show very low turnover numbers and stereoselectivity. Furthermore, their stability is a major problem, and cofactor recycling and oxygen supply have to be ensured as well.

This simple example compound demonstrates that where various enzymatic routes can be envisaged a careful evaluation of the entire process is needed before a decision on a specific route can be made.

11.2
Examples

11.2.1
General Applications

Amino acids are very important compounds used mainly as flavor enhancers or in animal nutrition. Proteinogenic α-amino acids in their natural L-form are commonly produced by fermentation using, for instance, *Corynebacterium glutamicum*, which is industrially used for the production of L-glutamate ($>1\,500\,000$ t y^{-1}) and L-lysine ($>800\,000$ t y^{-1}). Biocatalytic routes are rather used for the production of the D-enantiomer and/or non-natural amino acids using kinetic resolution with esterases (i.e., enantioselective hydrolysis of amino acid carboxylic acid esters) or acylases/amidases (i.e., enantioselective hydrolysis of amides). The hydantoinase process (see also Chapter 7) for the production of L-methionine was improved by directed evolution of the hydantoinase from *Arthrobacter* sp. DSM9771 expressed in *E. coli*. A combination of error-prone polymerase chain reaction (PCR) and saturation mutagenesis led to the identification of a variant with higher L-selectivity and fivefold increased specific activity. Recombinant technology was also used to clone and express together a D-hydantoinase from *Bacillus stearothermophilus* SD1 and a D-N-carbamoylase from *Agrobacter tumefaciens* in *E. coli*. Both enzymes form approximately 20% of the total cell protein, and both proteins were expressed at a comparable level (ratio 1 : 1.2). Thus, D,L-p-hydroxyphenylhydantoin (30 g L^{-1}) was efficiently converted by this recombinant whole-cell catalyst to D-p-hydroxyphenylglycine – an important precursor in the synthesis of semisynthetic β-lactam antibiotics – within 15 h at 96% yield [8].

Alternatively, prochiral precursors can be subjected to reductive amination using amino acid dehydrogenases (AADHs) [9]. Leucine dehydrogenase (Leu-DH) and phenyl alanine dehydrogenase (Phe-DH) are the most important AADHs, especially as they are not restricted to their natural substrates, accept a broad range of other precursors, and efficiently allow the synthesis of non-proteinogenic α-amino acids from the corresponding prostereogenic α-keto acids. One example of an amino acid produced by this approach is L-*tert*-leucine (L-^{t}Leu), which is an important building block for a range of pharmaceuticals, as peptide bonds involving this amino acid are only slowly hydrolyzed by peptidases. Reductive amination using a LeuDH in combination with the formate dehydrogenase (FDH) for cofactor recycling was significantly better than other enzymatic strategies and was used for large-scale production [10]. Other non-natural amino acids accessible by this route are L-neopentylglycine, L-β-hydroxyvaline, and 6-hydroxy-L-norleucine.

Lipases have been used for the industrial production of various compounds. This includes modification of triglycerides for human nutrition and synthesis of simple esters used in cosmetics or as detergents, but the major interest is in the synthesis of optically pure compounds. Overviews can be found in a book [11] and a review [6]. A few representative compounds are given in Figure 11.2. Researchers at DSM developed a lipase-catalyzed process for the production of a Captopril intermediate, (R)-3-chloro-2-methyl propionate. All lipases preferentially hydrolyzed the (S)-enantiomer, up to 98% ee at 64% conversion being observed with a lipase from *Candida cylindracea*.

As lipases naturally cleave ester bonds, it is not surprising that the majority of compounds resolved have been chiral alcohols or carboxylic acids. However, lipases were also found to be most efficient in the kinetic resolution of amines by kinetic resolution. One example is the highly enantioselective acylation of (R,S)-1-phenethyl amine in a process established at BASF (Figure 11.3). The (R)-amide is separated from the (S)-amine by distillation or extraction and the free (R)-amine is released through basic hydrolysis. As the lipase from *Burkholderia plantarii* shows broad substrate specificity, a wide variety of aryl alkyl amines, alkyl amines, and amino

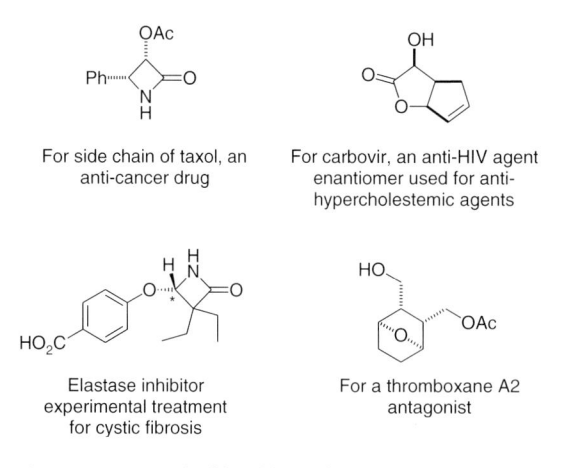

For side chain of taxol, an anti-cancer drug

For carbovir, an anti-HIV agent enantiomer used for anti-hypercholestemic agents

Elastase inhibitor experimental treatment for cystic fibrosis

For a thromboxane A2 antagonist

Figure 11.2 Four building blocks for pharmaceuticals obtained using lipase-catalyzed reactions [11].

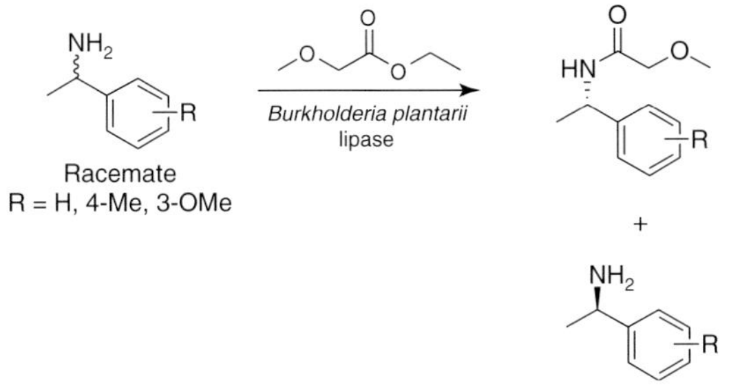

Figure 11.3 A lipase is used by the BASF AG for the kinetic resolution of chiral amines [6, 12].

alcohols could be resolved, in several cases on the multiton scale. The undesired enantiomers can be racemized, and the acylating agent (such as methoxyacetic acid ethylester) can be recovered. BASF can produce more than 4000 t y^{-1} with this process [12].

An example of the use of an esterase is in the synthesis of (+)-*trans*-(1R,3R) chrysanthemic acid, an important precursor of pyrethrin insecticides. The enzyme from *Arthrobacter globiformis* catalyzed the sole formation of the desired enantiomer (>99% *ee*, at 38% conversion). The enzyme was purified and the gene was cloned in *E. coli* [13]. In a 160 g scale process, hydrolysis was performed at pH 9.5 at 50 °C. Acid produced was separated through a hollow-fiber membrane module and the esterase was very stable over four cycles of 48 h.

Proteases are mainly used in peptide synthesis (for reviews see Refs. [14–17]). The advantages of an enzyme-catalyzed peptide synthesis are mild conditions, no racemization, minimal need for protective groups, and high regio- and enantioselectivity. The largest-scale application (hundreds to thousands of tons) of peptidase-catalyzed peptide synthesis is the thermolysin-catalyzed synthesis of aspartame (Figure 11.4), a low-calorie sweetener [18]. Precipitation of the product drives this thermodynamically controlled synthesis. The high regioselectivity of thermolysin ensures that only the α-carboxyl group in aspartate reacts. Thus, there

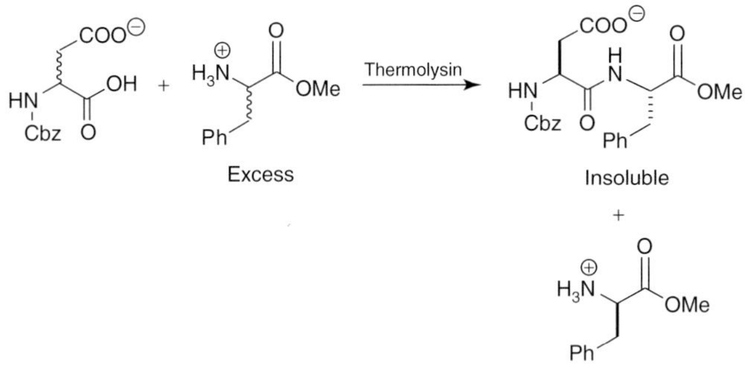

Figure 11.4 Enzymatic synthesis of the sweetener Aspartame is achieved by using the peptidase thermolysin.

is no need to protect the β-carboxylate. The high enantioselectivity allows the use of racemic amino acids, as only the L-enantiomer reacts.

Acylases and amidases also catalyze the formation of peptide bonds. An important application here is in the hydrolysis and synthesis of β-lactam peptide antibiotics (i.e., Penicillins and Cephalosporins). About 40 000 tons of penicillin are hydrolyzed per year in an equilibrium-controlled process catalyzed by penicillin amidases to produce 6-aminopenicillanic acid (6-APA). New processes for the synthesis of a range of semi-synthetic β-lactam antibiotics starting from 6-APA or 7-aminocephalosporanic acid (7-ACA) have been developed [19]. For the synthesis, penicillin amidase is used in a kinetically-controlled approach, and the antibiotic can be obtained in yields >95% (of 6-APA or 7-ACA) in aqueous solution when the synthesis is performed with a greater than twofold excess of activated substrate (Figure 11.5).

Two applications based on nitrile hydratase from *Rhodococcus rhodochrous* [20] have been commercialized (Figure 11.6). The large-scale production of the commodity chemical acrylamide is performed by Nitto Chemical (Yokohama, Japan) on a >100 000 metric tons per year scale. Initially, strains from *Rhodococcus* sp. N-774 or *Pseudomonas chlororaphis* B23 were used. However, the current process uses the 10-fold more productive strain *Rhodococcus rhodochrous* J1. The productivity is >7 kg acrylamide per g cells at a conversion of acrylonitrile of 99.97%. Formation of acrylic acid is barely detectable at the reaction temperature

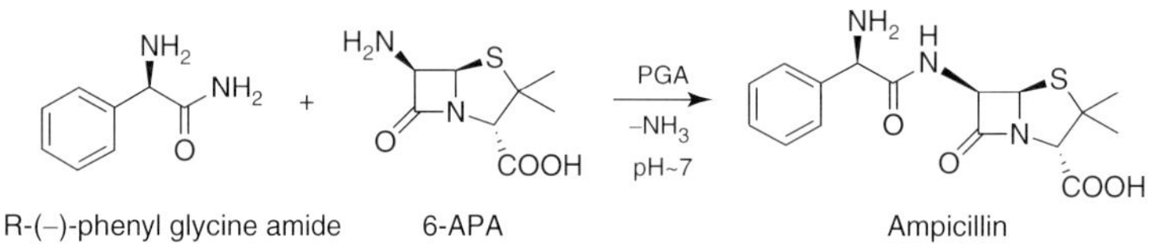

Figure 11.5 Synthesis of the antibiotic Ampicillin is achieved using Penicillin G amidase (PGA) under mild conditions.

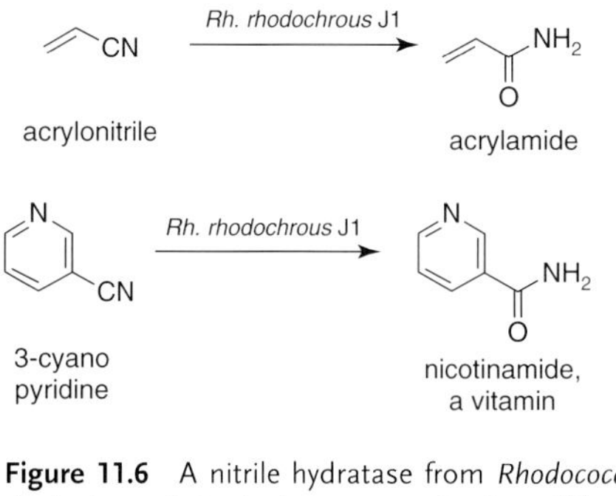

Figure 11.6 A nitrile hydratase from *Rhodococcus rhodochrous* J1 is the key enzyme in the mild and selective formation of acrylamide from acrylonitrile and nicotinamide from 3-cyano pyridine.

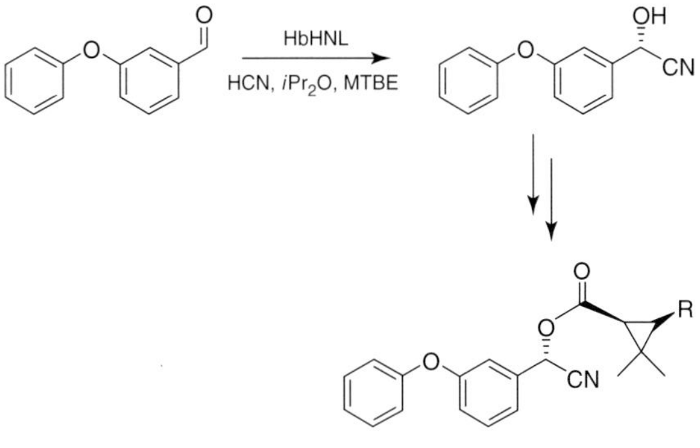

Figure 11.7 Synthesis of a chiral hydroxynitrile using a hydroxynitrile lyase from *Hevea brasiliensis*.

of 2–4 °C. In laboratory-scale experiments with resting cells up to 656 g acrylamide/L (!) reaction mixtures were achieved. Besides acrylamide, a wide range of other amides can also be produced, for example, acetamide (150 g L^{-1}, isobutyramide (100 g L^{-1}), methacrylamide (200 g L^{-1}), propionamide (560 g L^{-1}), and crotonamide (200 g L^{-1}).

Despite the requirement to handle HCN on a large scale, processes for the production of (S)-cyanohydrins catalyzed by hydroxynitrile lyase (HNLs) from *Hevea brasiliensis* (HbHNL) and *Manihot esculenta* and the subsequent chemical hydrolysis to (S)-hydroxycarboxylic acids have been developed. The synthesis of (S)-*m*-phenoxybenzaldehyde cyanohydrin – an intermediate in the synthesis of pyrethroids – in a biphasic system (Figure 11.7) has been commercialized by DSM Chemie Linz and Nippon Shokubai [6].

11.3
Case Study: Synthesis of Lipitor Building Blocks

Lipitor (trade name Atorvastatin, Figure 11.8) is a cholesterol-lowering drug acting as HMG-CoA reductase inhibitors and achieved sales of 12 billion US$ in 2004. Due to this extremely high commercial significance, several biocatalytic routes have been developed to synthesize key building blocks with desired absolute configuration and high optical purity.

Researchers at the company Diversa Inc. (San Diego, USA; now Verenium Inc.) developed a nitrilase-based route for the enzymatic synthesis of the side-chain precursor (R)-4-cyano-3-hydroxybutyric acid [21]. The best wild-type enzyme identified in metagenomic libraries gave 98% yield and 95% *ee* for the (R)-product [22]. As the (R)-specific nitrilase gave lower optical purity at 3 M substrate concentration, it was optimized by protein engineering.

The best variant contained a single mutation (Ala190His) and allowed the production of the (R)-enantiomer at 96% yield having 98.5% *ee* [21]. A process using this engineered enzyme was then scaled up in collaboration with researchers

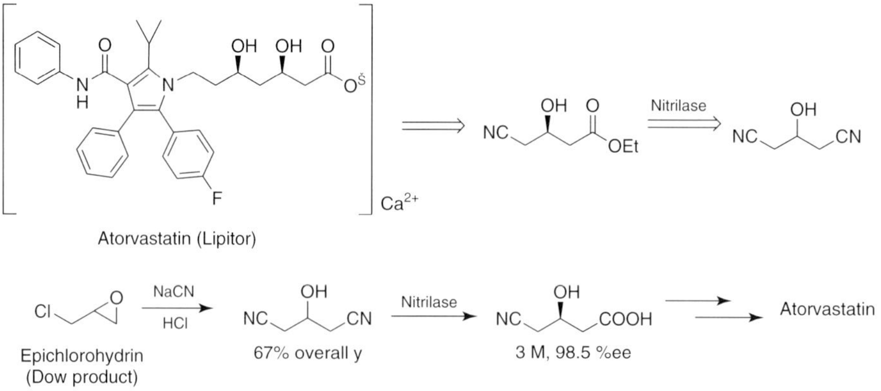

Figure 11.8 The total synthesis of Atorvastatin can include (R)-4-cyano-3-hydroxybutyric acid as key chiral intermediate. This can be produced starting from epichlorohydrin using a nitrilase-catalyzed desymmetrization.

from Dow Pharma [23]. Under optimized conditions, 6 wt% of enzyme led to complete conversion within 13 h reaction time at 3 M substrate concentration leading to 82% isolated yield of the desired building block. Figure 11.8 shows also how the biocatalytic step fits into an overall synthesis scheme starting from readily available epichlorohydrin.

A very elegant route for the synthesis of the side chain of Lipitor is based on the use of an aldolase [24]. The enzyme (DERA, 2-deoxy-D-ribose 5-phosphate aldolase) catalyzed the highly enantioselective tandem aldol reaction between chloroacetaldehyde and two equivalents of acetaldehyde (Figure 11.9). Key to success was again the protein engineering of the enzyme with respect to stability against acetaldehyde [25], which could be improved 10-fold by mutagenesis [26].

A third alternative was the use of a halohydrin dehalogenase (HHDH) from *Agrobacterium radiobacter* (Figure 11.10) leading to the same precursor as does the nitrilase route described above. Again, the wild-type enzyme needed to be

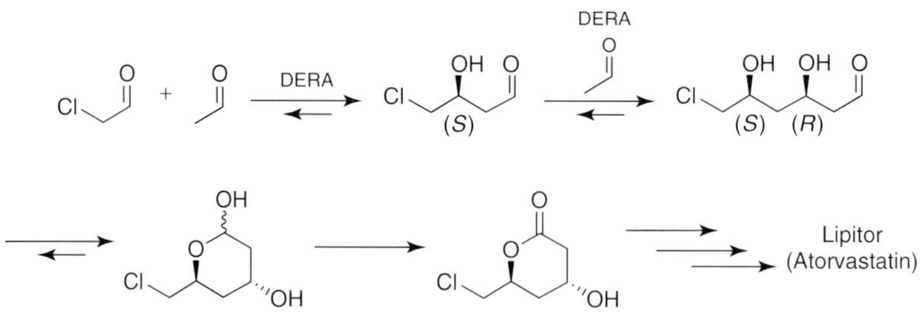

Figure 11.9 Synthesis of a key precursor of Lipitor using 2-deoxy-D-ribose-5-phosphate aldolase (DERA) in a tandem aldol reaction with chloroacetaldehyde and two equivalents of acetaldehyde as starting materials [26].

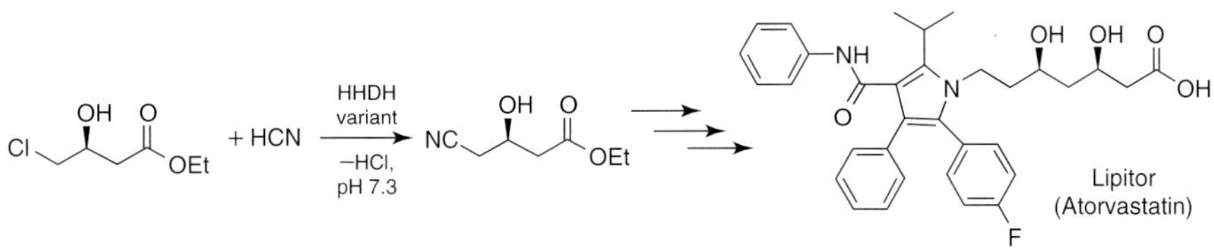

Figure 11.10 Synthesis of a key precursor of Lipitor using a halohydrin dehalogenase (HHDH). The best variant contained 35 mutations.

improved by protein engineering [27]. The new enzyme had a 4000-fold volumetric productivity in the cyanation process compared to the wild type.

11.4
Conclusions

The examples in this chapter show that biocatalysis is already a mature technology for the large-scale production of various (chiral) products for the chemical and pharmaceutical industries. The plethora of diverse enzymes and different activities available from nature and advanced methods of protein engineering have thus enabled the development of highly useful industrial processes.

References

1. Tao, J., Lin, G.-Q., and Liese, A. (2009) *Biocatalysis for the Pharmaceutical Industry*, Wiley-VCH Verlag GmbH, Weinheim.
2. Liese, A., Seelbach, K., and Wandrey, C. (2006) *Industrial Biotransformations*, 2nd edn, Wiley-VCH Verlag GmbH, Weinheim.
3. Aehle, W. (ed.) (2007) *Enzymes in Industry*, Wiley-VCH Verlag GmbH, Weinheim.
4. Schoemaker, H.E., Mink, D., and Wubbolts, M.G. (2003) Dispelling the myths – biocatalysis in industrial synthesis. *Science*, **299**, 1694–1697.
5. Schmid, A. et al. (2001) Industrial biocatalysis today and tomorrow. *Nature*, **409**, 258–268.
6. Breuer, M. et al. (2004) Industrial methods for the production of optically active intermediates. *Angew. Chem. Int. Ed.*, **43**, 788–824.
7. Thum, O. (2004) Enzymatic production of care specialties based on fatty acid esters. *Tenside Surf. Det.*, **41** (6), 287–290.
8. Park, J.H., Kim, G.J., and Kim, H.S. (2000) Production of D-amino acid using whole cells of recombinant Escherichia coli with separately and coexpressed D-hydantoinase and N-carbamoylase. *Biotechnol. Prog.*, **16**, 564–570.
9. Bommarius, A.S. (2002) in *Enzyme Catalysis in Organic Synthesis* (eds K. Drauz and H. Waldmann), Wiley-VCH Verlag GmbH, Weinheim, pp. 1047–1063.
10. Bommarius, A.S. et al. (1992) in *Chirality in Industry* (eds A.N. Collins, G.N. Sheldrake, and J. Crosby), John Wiley & Sons, Inc., London, pp. 371–397.
11. Bornscheuer, U.T. and Kazlauskas, R.J. (2006) *Hydrolases in Organic Synthesis – Regio- and Stereoselective Biotransformations*, 2nd edn, Wiley-VCH Verlag GmbH, Weinheim.
12. Balkenhohl, F. et al. (1997) Optically active amines via lipase-catalyzed

methoxyacetylation. *J. Prakt. Chem.*, **339**, 381–384.

13. Nishizawa, M. et al. (1995) Stereoselective production of (+)-trans-chrysanthemic acid by a microbial esterase: cloning, nucleotide sequence, and overexpression of the esterase gene of Arthrobacter globiformis in Escherichia coli. *Appl. Environ. Microbiol.*, **61** (9), 3208–3215.

14. Bordusa, F. (2002) Proteases in organic synthesis. *Chem. Rev.*, **102**, 4817–4867.

15. Kasche, V. (2001) in *Proteolytic Enzymes: A Practical Approach* (eds M. Beynon and J. Bond), Oxford University Press, Oxford, pp. 265–292.

16. Schellenberger, V. and Jakubke, H.D. (1991) Protease-catalyzed kinetically controlled peptide synthesis. *Angew. Chem. Int. Ed. Engl.*, **30**, 1437–1449.

17. Drauz, K. and Waldmann, H. (2002) *Enzyme Catalysis in Organic Synthesis*, 2nd edn, vols. **1–3**, Wiley-VCH Verlag GmbH, Weinheim.

18. Oyama, K. (1992) in *Chirality in Industry* (eds A.N. Collins, G.N. Sheldrake, and J. Crosby), John Wiley & Sons, Ltd, Chichester, pp. 237–247.

19. Liese, A., Seelbach, K., and Wandrey, C. (2000) *Industrial Biotransformations*, Wiley-VCH Verlag GmbH, Weinheim.

20. Nagasawa, T. and Yamada, H. (1995) Microbial production of commodity chemicals. *Pure Appl. Chem.*, **67**, 1241–1256.

21. DeSantis, G. et al. (2003) Creation of a productive, highly enantioselective nitrilase through Gene Site Saturation Mutagenesis (GSSM). *J. Am. Chem. Soc.*, **125**, 11476–11477.

22. DeSantis, G. et al. (2002) An Enzyme library approach to biocatalysis: development of nitrilases for enantioselective production of carboxylic acid derivatives. *J. Am. Chem. Soc.*, **124**, 9024–9025.

23. Bergeron, S. et al. (2006) Nitrilase-catalysed desymmetrisation of 3-hydroxyglutaronitrile: preparation of a Statin side-chain intermediate. *Org. Proc. Res. Dev.*, **10**, 661–665.

24. Liu, J., Hsu, C.C., and Wong, C.H. (2004) Sequential aldol condensation catalyzed by DERA mutant Ser238Asp and a formal total synthesis of atorvastatin. *Tetrahedron Lett.*, **45**, 2439–2441.

25. Greenberg, W.A. et al. (2004) Development of an efficient, scalable, aldolase-catalyzed process for enantioselective synthesis of statin intermediates. *Proc. Natl. Acad. Sci. U.S.A.*, **101** (16), 5788–5793.

26. Jennewein, S. et al. (2006) Directed evolution of an industrial biocatalyst: 2-deoxy-D-ribose 5-phosphate aldolase. *Biotechnol. J.*, **1**, 537–548.

27. Fox, R.J. et al. (2007) Improving catalytic function by ProSAR-driven enzyme evolution. *Nat. Biotechnol.*, **25** (3), 338–344.

12
Polymerization

Vincenzo Busico

12.1
Introduction

Polymers together represent roughly one fourth by value of the global chemical industry output. With a world market of 80 Mt per year (2008 production data [1]), polyethylene (PE) is one of the 10 largest-volume chemical products, and other polymers – in particular, polypropylene (PP) and polyvinylchloride (PVC) – are not too far behind [2].

This chapter, on the other hand, does not need to cover a correspondingly large fraction of the book. In fact, most polymers of industrial relevance (e.g., PVC, polystyrene, polyesters, polyamides, polycarbonates, polyacrylates, polyurethanes, and in part synthetic rubbers) are made by mature polymerization processes involving comparatively unsophisticated catalysts/initiators (such as inorganic or organic acids, bases, or radicals), or even not involving catalysts at all. The chemistry and technology of such processes has been comprehensively and thoroughly covered in dozens of excellent books and encyclopedias [3]. Moreover, the catalysts – when used – have limited impact on polymer structure design. For some monomers, such as methylmethacrylate [4], lactide [5, 6], aliphatic epoxides, and carbon dioxide [7, 8], to mention just a few, recent results achieved with the mediation of coordination catalysts suggest that the situation may change to some extent in the future, but until now the impact on production has been negligible. Other monomers (e.g., vinylchloride) instead have so far resisted all attempts at innovation [9, 10].

In Ziegler–Natta-type olefin polymerizations, on the other hand, science and technology are still progressing at a pace that fully justifies an update [11]. This is mainly because the catalysts control practically all aspects of polymer architecture from molecular to macro scale (that is, from microstructure to morphology), thus providing access to tailor-made products [12]. It is on the fascinating and inter-related aspects of polyolefin catalyst, product and process design that the various sections of the present chapter will focus.

Catalysis: From Principles to Applications, First Edition. Edited by Matthias Beller, Albert Renken, and Rutger van Santen.
© 2012 Wiley-VCH Verlag GmbH & Co. KGaA. Published 2012 by Wiley-VCH Verlag GmbH & Co. KGaA.

12.2
Polyolefins in Brief

The growth of the polyolefin market has been exponential since the discovery of Ziegler–Natta catalysis in 1953/1954, and in spite of the financial crisis of 2008/2009 the positive trend is expected to continue in the foreseeable future. In 2012, with the announced start-up of several large production lines (mostly concentrated in the Persian Gulf area, next to the oil fields), the capacity of the polyolefin industry will approach 150 Mt per year [1]. All this originates from two seemingly humble by-products of oil refining, namely the two simple and small molecules of ethene and propene. In fact, one possible way to look at the polyolefin industry is one which lends part of the lightest fraction of refined oil, solidifies it into products with smart applications, and after a life cycle which may last many years and include one or more recyclings, gives them back for safe incineration, thus ultimately releasing their energy potential anyway.

If one adds that the atom economy of olefin polymerization is practically 100%, one can only conclude that polyolefins are the greenest of all non-green chemicals [12].

Another important reason for the extraordinary success of polyolefins on the market is that out of two basic monomers it is possible, with a proper choice of catalyst and process, to produce dozens of different materials ranging from commodities to specialties, and from ultra-rigid thermoplastics to super-soft elastomers [12, 13]. As is well known, ethene is homopolymerized to a linear, highly crystalline polymer (High Density PolyEthylene, HDPE) with transition metal catalysts [12] (Figure 12.1a), and to a branched, practically amorphous polymer (Low Density PolyEthylene, LDPE) with radical initiators [14, 15] (Figure 12.1b). Propene, in turn, is a non-symmetrical prochiral monomer which can be homopolymerized to an endless variety of regio- and stereostructures, approaching more or less closely one in a set of ideal cases [13, 16] (Figure 12.2). Last but not least, ethene

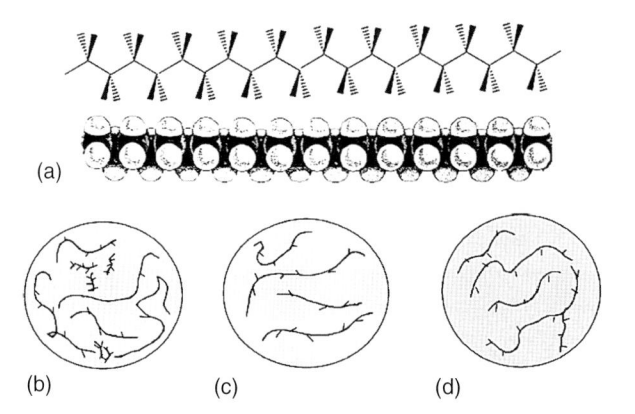

Figure 12.1 The microstructure of HDPE (a), LDPE (b), LLDPE (c), and long-chain-branched LLDPE (d).

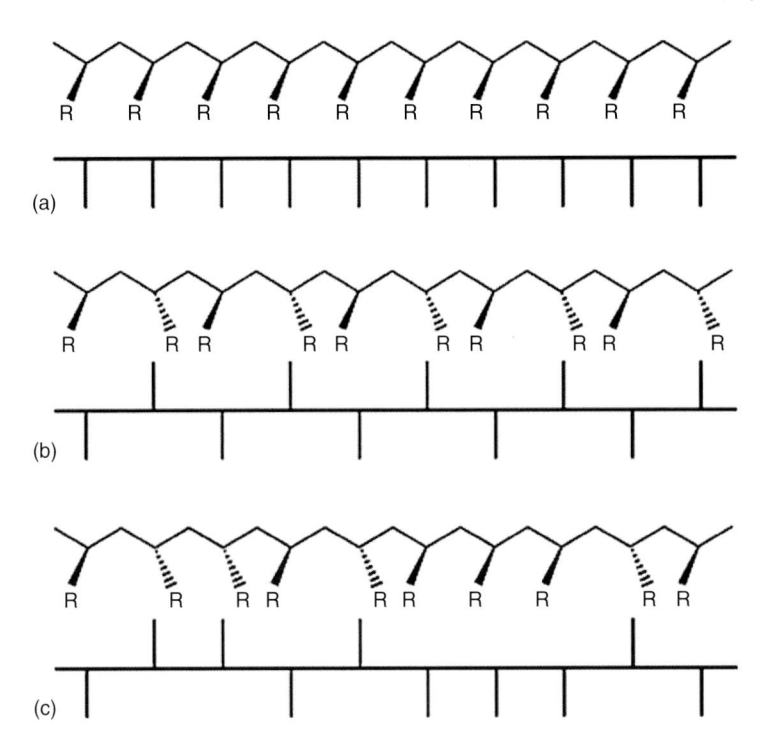

Figure 12.2 The microstructure of isotactic (a), syndiotactic (b), and atactic (c) polypropylene (R=CH$_3$).

and propene can copolymerize with each other and/or with other monomers, among which higher 1-alkenes like 1-butene, 1-hexene or 1-octene are of special importance for the production of the so-called Linear Low Density Poly Ethylene [17] (LLDPE; Figure 12.1c). In most cases, the chain microstructure – that is the sequence distribution of the (co)monomeric units – can be controlled with high precision almost at will. The key point is that the physical and application properties of a polyolefin-based material are exquisitely sensitive to the microstructure of the polymer chains of which it is made, as well as to their molecular mass distribution, which can also be tuned so as to span from oligomers to ultra-high molecular weights, and from monodisperse to very broad. Physical or reactor blending, usually with thorough control of phase morphology, complete the options of material design [12].

The following are simple examples. A regioregular homopolymer of propene can resemble more or less closely the ideal isotactic stereostructure (Figure 12.2a), depending on the catalyst used to prepare it. Thermoplastic materials made of highly isotactic PP chains can develop high degrees of crystallinity with a melting temperature (T_m) above 165 °C. Such materials are strong and rigid, but become brittle once below 0 °C due to a comparatively high glass transition temperature ($T_g \sim 0 - 5$ °C) [13]. The deliberate introduction of point irregularities in the PP chains in the form of either stereodefects (with a suitable catalyst) or ethylene units via copolymerization, provides a way to tune (lower) the degree of crystallinity, and the values of T_m and T_g. In particular, with increasing ethylene unit content the chains become more flexible due to reduced interference between the methyl side groups, and are ultimately unable to crystallize. With the proper combination of catalyst and process, one can produce in one reactor single-phase propylene-rich plastomers or, with a reactor cascade, heterophasic materials featuring a rigid and high-melting ($T_m > 160\ °C$) homo-PP matrix and a finely dispersed, largely amorphous ethylene/propylene copolymer phase with low T_g ($< -20\ °C$) [13]. We will come back to this in more detail later (Section 12.4.1).

For linear PE, of course, stereochemistry is not an option, and copolymerization is the only way to tune the microstructure, but it is a powerful and highly versatile one.

Random copolymers of ethene and propene, with or without a third monomer (usually a diene in view of a subsequent chemical crosslinking), have long been marketed as elastomers ('ethylene/propylene rubber,' EPR) [18]. More recently, catalysts have become available for the copolymerization of ethene with long-chain 1-alkenes such as 1-hexene or 1-octene; the resulting materials, comprehensively known as LLDPE, as already noted, have properties intermediate between those of HDPE and LDPE (Figure 12.1) and represent a fast-growing segment of the polyolefin market [17]. The most recent breakthroughs have occurred in the area of ethylene *block* copolymers, as we shall see in Section 12.5.

12.3
Olefin Polymerization Catalysts

12.3.1
The Catalytic Species: Structure and Reactivity

Olefin polymerization mediated by transition metal (M) catalysts is a poly-insertion reaction of the unsaturated monomer(s) into the M-Polymeryl σ-bond. The commonly accepted mechanism is shown in Scheme 12.1. The monomer π-coordinates to a vacant site of M, is activated due to a strong polarization of the double bond induced by a complementary polarization of the M-Polymeryl bond (more than to back-donation from the metal to the olefin π^* orbital), and undergoes chain-migratory insertion [12, 13, 16, 19].

Typical catalyst formulations include a transition metal precursor L_yMX_z (where L_y denotes the ancillary ligand framework, and X is a monodentate ligand such as

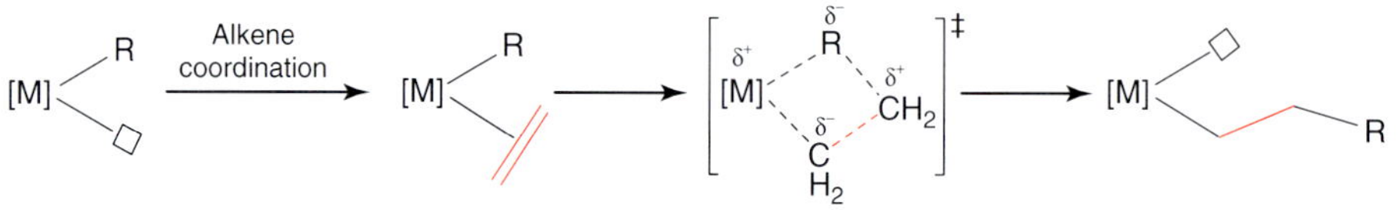

Scheme 12.1

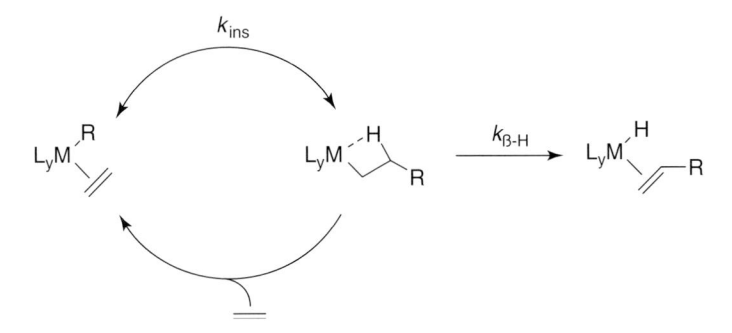

Scheme 12.2

alkyl, aryl, or halide), and a suitable activator which in most cases is a main-group metal alkyl M′R$_n$. The function of the latter is to generate the active site, that is, the initial M-R σ bond, and/or to create the necessary coordination vacancy (□) on M by abstracting one X [20]:

$$L_yMX\,(\square) + M'R_n \rightarrow L_yMR\,(\square) + M'R_{n-1}X \tag{12.1}$$

$$L_yMX_2 + M'R_n \rightarrow \left[L_yMR\,(\square)\right]^+ + \left[M'R_{n-1}X_2\right]^- \tag{12.2}$$

Although a variety of transition metals (M = Ti, Zr, Hf, V, Cr, Fe, Co, Ni, Pd) have been reported to be catalytically active in olefin polymerizations [12, 13, 16–19, 21], the vast majority of the industrially important catalysts are based on early transition metals, and in particular Ti, Zr, Hf, V, and Cr [12, 13]. One common feature of early compared with late transition metals is a lower propensity of M-alkyl fragments to undergo β-H elimination [22], which represents a possible exit from the catalytic cycle and therefore a limitation to chain growth (Scheme 12.2).

The choice of the main-group metal alkyl is more limited, Al-alkyls representing by far the main option, followed far behind by Zn-alkyls [12, 13, 18, 20]. If the active species is a cation (Eq. (12.2)), the counter-anion must be poorly coordinating, so as not to obstruct the coordination vacancy on M and thus hinder access to the monomer [20]. A paradigmatic example is that of bis (cyclopentadienyl) Column 4 metal catalysts (Figure 12.3), known since the late 1950s but considered to be of no practical interest for more than 20 years, until the serendipitous discovery of methylalumoxane (MAO; Figure 12.4) and its ill-defined anion with highly delocalized charge revealed that the poor catalytic activity of [Cp$_2$MR]$^+$ cations (Cp = cyclopentadienyl) generated by the activation of Cp$_2$MX$_2$ precursors with AlR$_3$ or AlR$_2$Cl is due to their tight ion-pairing with [AlR$_2$X$_2$]$^-$ or [AlRClX$_2$]$^-$ anions [20].

As is the case in general with coordination catalysts, the ligand framework of the transition metal defines an active pocket which, due to electronic and

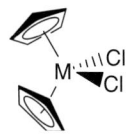

Figure 12.3 The basic structure of Column 4 metallocenes.

Figure 12.4 Model structure of MAO.

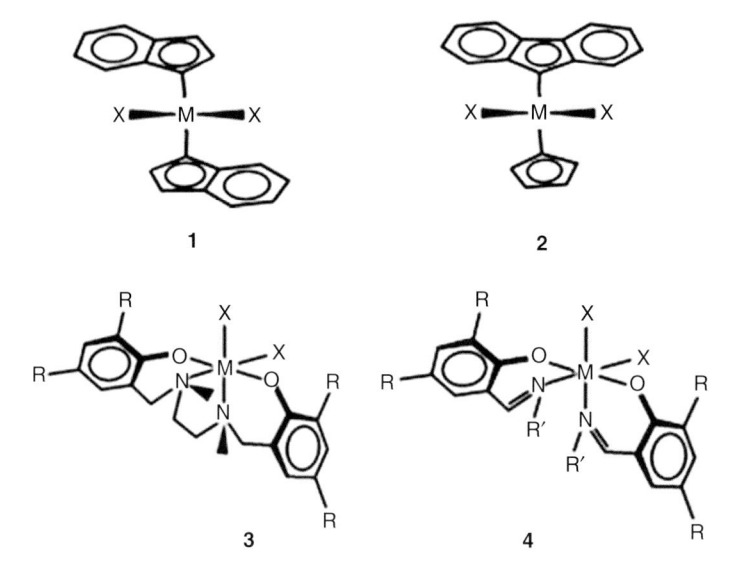

Figure 12.5 Precursors of stereoselective propene polymerization catalysts. Upon activation, the stereorigid C2-symmetric ansa-metallocene 1 and the Cs-symmetric ansa-metallocene 2 (M = Zr or Hf) are isotactic- and syndiotactic-selective, respectively [16, 19] (see also text and Figures 12.7 and 12.8). The stereorigid C2-symmetric bis(phenoxyamine) complex 3 (M = Zr or Hf, R bulky) is isotactic-selective [25], whereas the deceptively similar C2-symmetric bis(phenoxyimine) complex 4 (M = Ti, R bulky) is syndiotactic-selective [24] as a result of the 2,1 insertion regiochemistry of propene and of the systematic inversion of configuration at the metal at each insertion step allowed by the fluxional ligand framework.

steric effects, determines practically all aspects of the catalytic behavior. In olefin polymerization, this holds for molecular catalysts, like metallocene [19, 23] and 'non-metallocene' [21, 24] ones deriving from well-defined precursor complexes (Figure 12.5), as well as for the classical heterogeneous Ziegler–Natta systems whose elusive active species are generated by the alkylation of chiral Ti atoms exposed on certain edges of $TiCl_3$ crystals [13, 16, 18] or assembled on an $MgCl_2$ support by the reaction of a Ti precursor (usually $TiCl_4$) with an AlR_3 cocatalyst and one or more organic electron donors added as selective surface modifiers [12, 13, 16] (Figure 12.6).

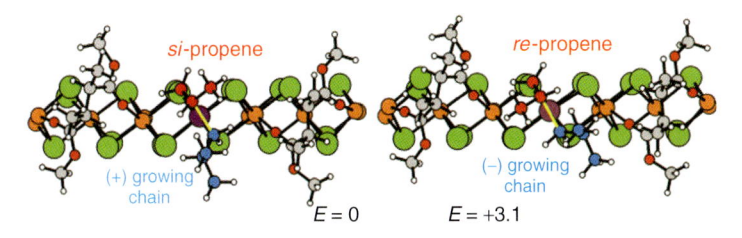

Figure 12.6 Model of isotactic-selective Ti active species on the (110) edge of an MgCl$_2$ crystal [13, 16]. The two-electron donor (succinate) molecules adsorbed at adjacent positions create a quasi-C2-symmetric pocket around the Ti center (colored brown), in which the 'growing chain orientation' mechanism of stereocontrol [26] can be highly effective (see also text and Figure 12.7). (Courtesy of L. Cavallo (University of Salerno).)

A unique feature of polymerization catalysts, on the other hand, is the possibility to trace the mechanistic origin of the selectivity from the elucidation of polymer microstructure, even in cases when the (detailed) structure of the active species is unknown. This active site 'fingerprinting' is of special importance with polyolefins, whose microstructure can be determined in great detail. Elegant applications of ^{13}C NMR to the statistical analysis of PP configuration, in particular, highlighted the fine details of the enantioselectivity of propene insertion for most catalysts, including the aforementioned heterogeneous Ziegler–Natta ones [16]. It has been demonstrated that in the vast majority of cases the chiral recognition of propene in the active pocket entails a subtle transfer of information from the chirotopic environment of the active site to the first C–C bond of the growing polymeryl, constrained in a chiral orientation by very specific non-bonded contacts, and then from this to the incoming monomer, which can find a smooth insertion path in the favored regiochemistry only with the enantioface pointing the methyl substituent *anti* to the said C–C bond [13, 16, 19, 26] (Figure 12.7). A compelling demonstration is that the first propene insertion into an M–CH$_3$ bond is usually non-enantioselective even when the subsequent ones are fully so. Along with the chain migratory insertion path, this 'growing chain orientation' mechanism provides a simple explanation for the observed selectivities of a variety of catalysts, the best-known example certainly being that of *ansa*-metallocenes of different symmetries [13, 16, 19, 26] (Figure 12.8).

On the other hand, the simplicity of the explanation does not imply that the fact is obvious. As a matter of fact, due to the possible occurrence of site epimerization [27–29] or chain epimerization [30, 31] processes in competition with monomer insertion, the relationship between the enantioselectivity of monomer insertion and the stereoselectivity of the polymerization is not necessarily straightforward.

A stereoregular polymer, indeed, is the outcome of a long string of enantioselective insertions in the right sequence, and the sequence can be disordered whenever the active sites are not homotopic and/or a catalyst has the ability to racemize a monomeric unit ex-post.

Many features of the polymerization other than the enantioselectivity in the insertion of a prochiral monomer can be tuned with a proper design of the active

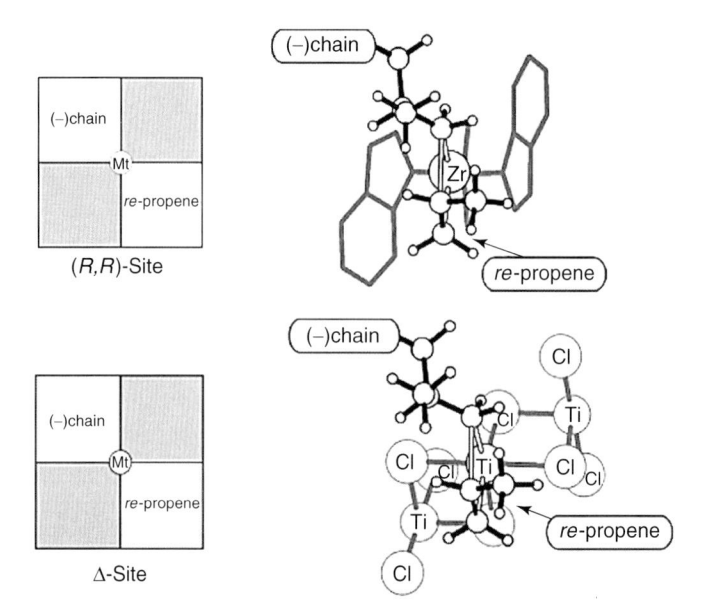

Figure 12.7 The 'growing chain orientation' mechanism of stereocontrol [26] at C2-symmetric ansa-zirconocene (a) and heterogeneous Ti-based Ziegler–Natta (b) model active species. The similarity is immediately evident [16, 19]. (Reproduced from Ref. [26].)

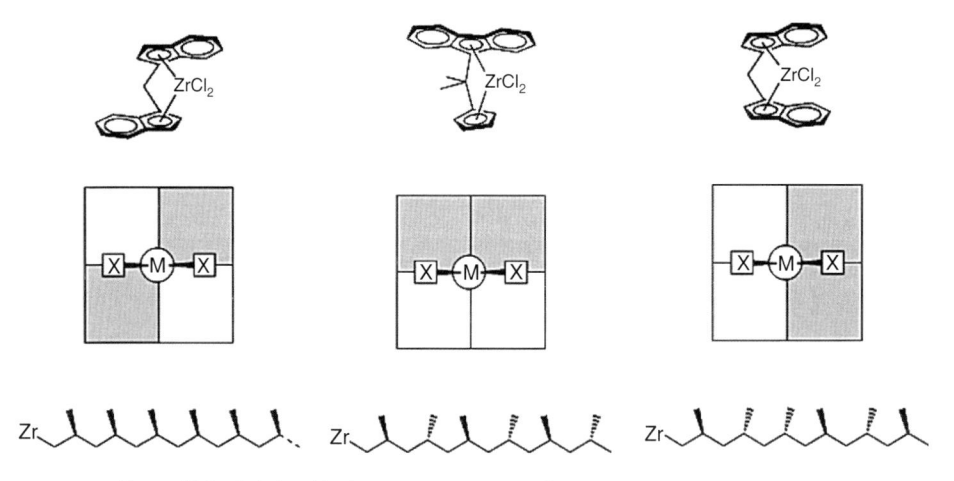

Figure 12.8 Relationship between symmetry and stereoselectivity in propene polymerization at C2-symmetric (left), Cs-symmetric (center), and meso-Cs-symmetric (right) ansa-zirconocene active species, according to the 'growing chain orientation' mechanism of stereocontrol [16, 19, 26].

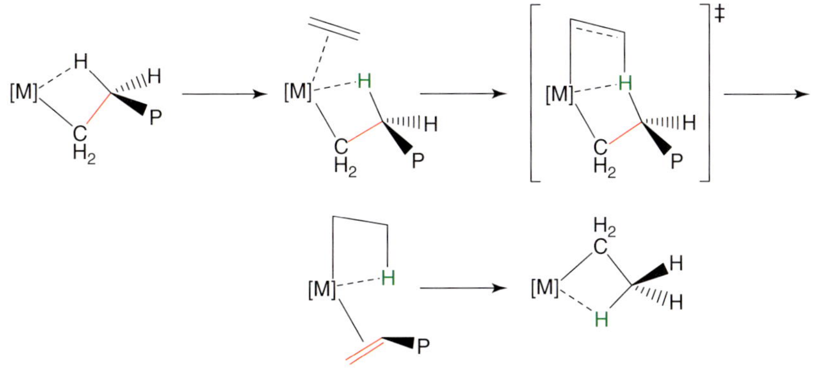

Scheme 12.3

pocket. For instance, the relative rates of chain propagation and transfer, that is, the average polymer molecular mass, can be altered by orders of magnitude, typically by slowing the dominant chain transfer pathway. When this is β-H transfer to the monomer, in particular, for a number of different catalysts – including late transition metal ones [32] – ways have been found to de-stabilize the space-demanding six-centre transition state (Scheme 12.3) by means of suitable substituents on the ancillary ligand (Figure 12.9), with truly outstanding results [33]. It is important here to recall that the ability to produce polyolefins with very high average molecular weight (well above 10^5 Da) is extremely important, because due to the lack of functional groups the only strong interactions between polyolefin chains are via physical entanglements, and high to very high molecular weights are needed to develop the physical and mechanical material properties required for useful applications [12, 13, 34].

Recently, several catalysts have been implemented for which *all* chain transfer pathways, including β-H transfer to the metal (Scheme 12.2) and to the monomer (Scheme 12.3), as well as trans-alkylation with the main-group metal-alkyl cocatalyst (Scheme 12.4), are negligible in the time scale of a practical polymerization [35]. The result is a chain propagation under a 'controlled' kinetic regime, which in the initial phase mimics a 'living' process, and can open the door to important applications like block polymerizations (see Section 12.5).

12.3.2
Polymerization Kinetics: Active, 'Dormant' and 'Triggered' (?) Sites

The rate law of catalytic olefin polymerization is deceptively simple [36]:

$$v_p = -d[C_nH_{2n}]/dt = k_p[M^*][C_nH_{2n}]^a \qquad (12.3)$$

The main problem in the application of Eq. (12.3) is that in the vast majority of cases the fraction of transition metal actually involved in catalysis ([M*]) is unknown [37], and there are reasons to believe that it can be well below 100%. This holds not only

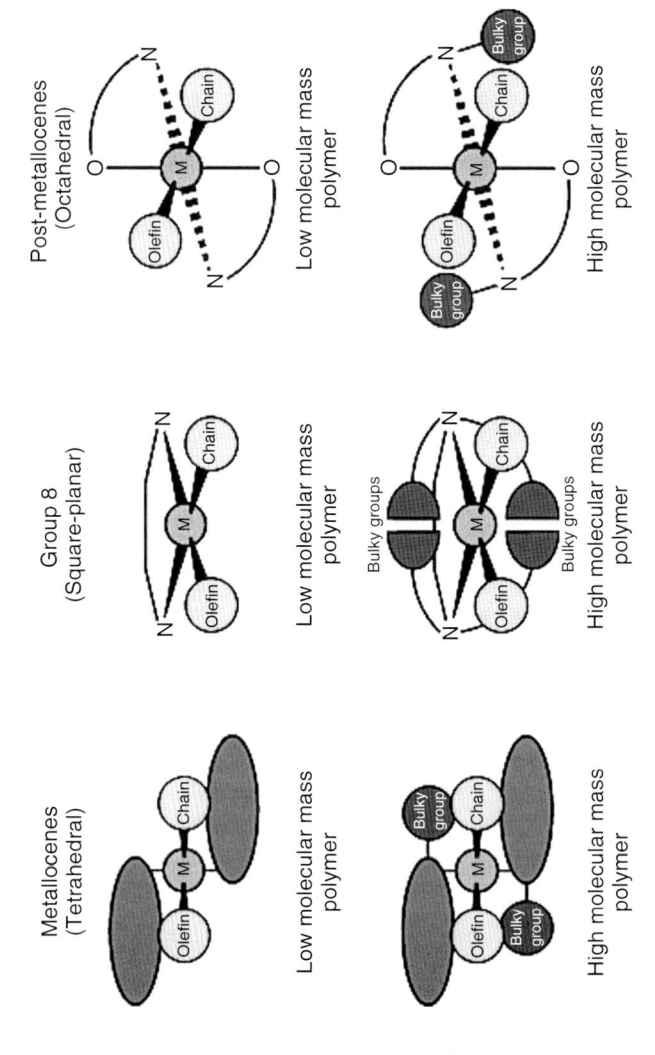

Figure 12.9 Modifications of ligand framework in *ansa*-zirconocene (left), (α-diimine)Ni/Pd (centre) and bis(phenoxyimine)Ti (right) model active species resulting in an increased molecular weight capability. (Reproduced from Ref. [33].)

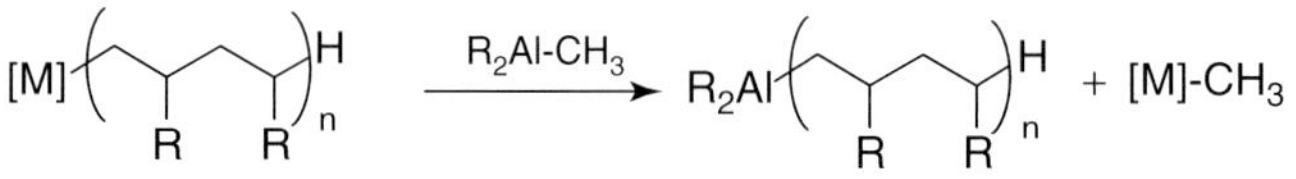

Scheme 12.4

for heterogeneous catalysts [38], but also for molecular ones in homogeneous phase [39, 40]. As an example, metallocene cations are known to give rise to a variety of aggregation processes involving the counterion and/or the main-group metal-alkyl cocatalyst, which remove part of the transition metal to the catalytic cycle [40, 41] (Figure 12.10).

Unfortunately, measuring [M*] is very difficult [37–40], and it is often impossible in practice to disaggregate the k_p[M*] product in Eq. (12.3). Even if [M*] were known, however, the use of k_p for molecular kinetic applications would be questionable. In fact, with the only possible exception of ethene homopolymerization in the presence of a molecular catalyst derived from a well-defined precursor, the value of k_p is an average over a population of different active species. With heterogeneous catalysts this is fairly obvious, in view of the multiplicity of surface structures that is typical of such systems. The problem with molecular catalysts, on the other hand, is more subtle, and is mainly related to the possible occurrence of so-called 'dormant' species [40–43].

The best-known example is the formation of secondary M-Polymeryl bonds in propene polymerizations with predominant 1,2 regiochemistry (Scheme 12.5a) as a result of occasional regioirregular 2,1 monomer insertions (Scheme 12.5b) [42–46]. For many industrially important catalysts (like heterogeneous Ziegler–Natta ones) considerable experimental evidence exists that M-CH(CH$_3$)-CH$_2$-P moieties are poorly reactive toward propene insertion (Scheme 12.5c,d; $k_{sp} < 10^{-2}\ k_{pp}$; $k_{ss} \sim 0$), to the point that even in cases when the fraction of 2,1 insertions is very low ($k_{ps} < 10^{-2}\ k_{pp}$) the majority of the M-Polymeryls are in a stand-by state [44–46]. This is revealed by the strong activating effect of H$_2$, which is often added in these polymerizations as a chain transfer agent to tune polymer molecular weight via σ-bond metathesis [12, 13, 47] (Scheme 12.5e,f), and at odds with propene is small enough to react with primary and secondary M-Polymeryls at similar rates ($k_{pH} \sim k_{sH}$), thus removing the steric blockage in the latter case [42–46]. The unsubstituted ethene molecule can also play a similar wake-up action on the 'dormant' sites [48] (Scheme 12.5g,h; $k_{pE} \sim k_{sE}$).

In view of this complicated mechanistic picture, one should not be surprised that the reaction order with respect to the monomer, that is, the exponent 'a' in Eq. (12.3), can be a fractional number between 0 and 2 [36]. Olefin polymerization is evidently not an elementary process, and k_p and [M*] are apparent quantities which depend on the reaction conditions (including chemical and physical variables). This notwithstanding, in the vast majority of cases the value of a is not far from 1 [36]. Sporadic cases where $a \approx 0$ have been reported, for instance in 'living' polymerizations at very low temperature, and unambiguously explained in terms of catalyst saturation [49]. Cases where $a \approx 2$, on the other hand, have been observed in heterogeneous Ziegler–Natta catalysis [50] and also, more recently,

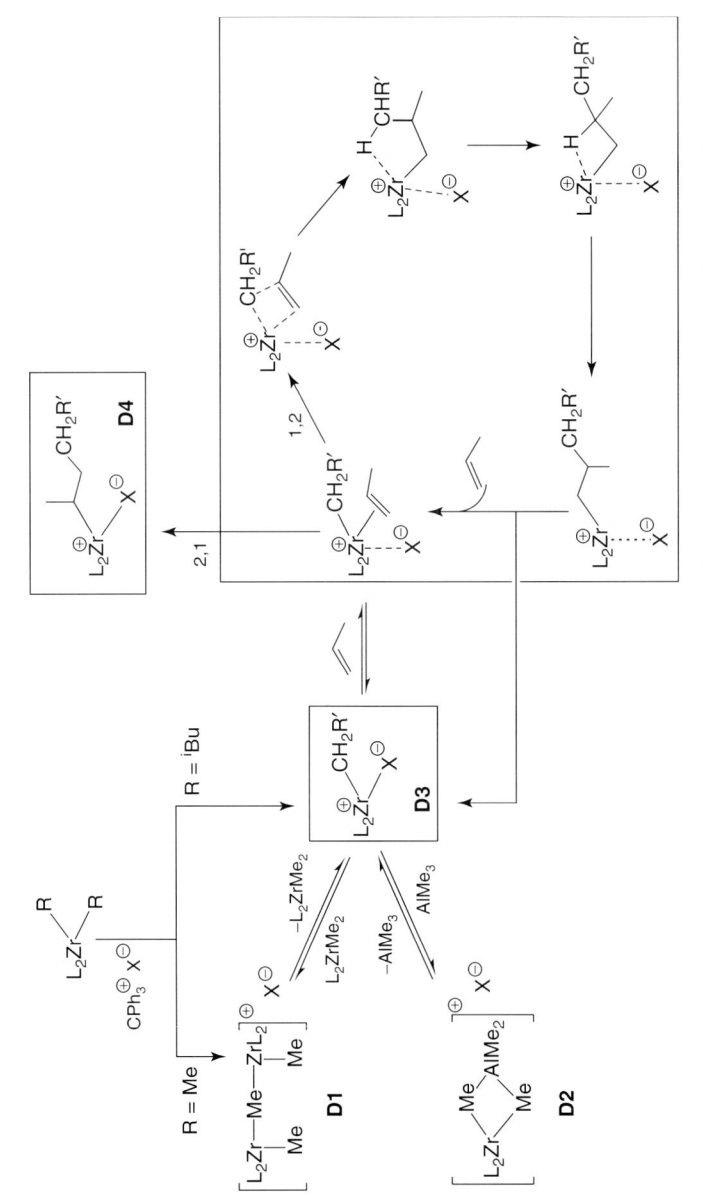

Figure 12.10 Proposed ion pairing and dinuclear aggregation equilibria in ansa-zirconocene-based olefin polymerization catalysis. (Adapted from Ref. [40].)

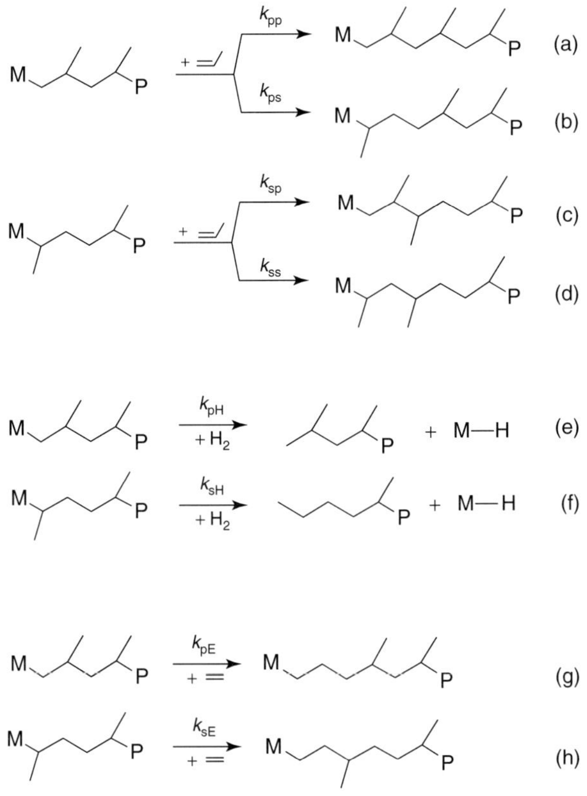

Scheme 12.5

with some metallocene catalysts [51]; quite surprisingly, this has been taken by some as an indication of a bimolecular insertion mechanism ('trigger' hypothesis) [52]. This rather naïve interpretation has been subsequently disproved with more careful kinetic analyses, demonstrating that the fraction of 'dormant' sites can be a function – *inter alia* – of monomer concentration, and that such a dependence can result in reaction orders intermediate between 1 and 2 [53, 54].

12.4
Olefin Polymerization Process Technology

12.4.1
Heterogeneous Catalysis

Like all large-volume products, polyolefins in general are low-value-added materials [12, 13]. The market price of a commodity polyolefin like HDPE is less than twice that of the monomer, which means that for an adequate profit catalysts and processes need to be close to perfection. Typical industrial olefin polymerization catalysts are exceedingly active, with TONs and TOFs well above 10^6 and $10^3\,s^{-1}$ respectively (referred to *monomer insertion*), moreover under rather mild conditions ($T < 100$ °C, $P < 50$ bar with few exceptions). The stereoselectivity in the production of isotactic PP can also be close to 100%, and in fact the separation of less-tactic byproducts (which was a major drawback of first-generation Ziegler–Natta catalysts based on $TiCl_3$ [18, 36]) is now just a memory of the past [13]. Along with these outstanding features a low cost is mandatory, and a number of otherwise excellent catalysts never made their way to the market 'simply' because they are too expensive. The most representative case is the evolution of *ansa*-zirconocenes for isotactic PP [19] (Figure 12.11), ultimately becoming able to compete with $MgCl_2$-supported Ziegler–Natta catalysts with respect to productivity and stereoselectivity, but definitely not to synthesis and application costs [11].

The production units, in turn, must be as large as technically feasible for economy of scale. Modern polyolefin plants can produce up to 400 kT per year in one line, usually with liquid monomer or gas phase technologies which avoid the need for a solvent with all the related problems of recovery and purification, and of course require heterogeneous catalysts [12, 13]. This means that, whenever the catalyst precursor is a molecular entity, a protocol for its immobilization on a suitable support without sacrificing activity and selectivity must be implemented; the technological challenges and the economic impact of this task should never be underestimated [12, 55].

In gas phase processes, of course, the polymer particles grow around the catalyst ones, but also when produced in liquid phase polymers like HDPE, isotactic PP (including the heterophasic grades with impact resistance) and even some LLDPE grades are insoluble in the reaction medium. In all cases, for smooth reactor operation a highly controlled morphology of the polymer particles is a must [12, 55]. In fact, the formation of 'fines' (that is, of small particles sensitive to the effects of static charge accumulation) triggers phenomena of reactor fouling (Figure 12.12) ultimately hindering heat transfer, which is potentially catastrophic, with a fast reaction liberating ~ 20 kcal mol^{-1} of converted monomer. Particle agglomeration, to the other extreme, can produce plugs and also decrease heat exchange, block reactor discharge, and heavily perturb flowability in fluidized-bed units [12, 55].

In general, it has been demonstrated that in well-behaved processes the morphology of the polymer particles mirrors that of the catalyst particles in practically all respects (shape, porosity, average size, size distribution, etc.), due to what is known as the *'replication'* phenomenon [12, 56]. Mastering this replication is one of the most challenging and fascinating aspects of modern polyolefin production, and is key to the tremendous success of polyolefin-based materials. Here we can only describe the main elements in brief.

The starting point is the support to be used for the immobilization of the catalytic species. In classical Ziegler–Natta systems, $MgCl_2$ is by far the most used carrier [12, 13, 56]. It has been suggested that its crystal structure acts as a template for the build-up of the active Ti species via epitactic chemisorption of a convenient precursor compound (usually $TiCl_4$) on certain surfaces, followed by their selective modification by means of one or more electron donors [13, 16] (Section 12.3.1).

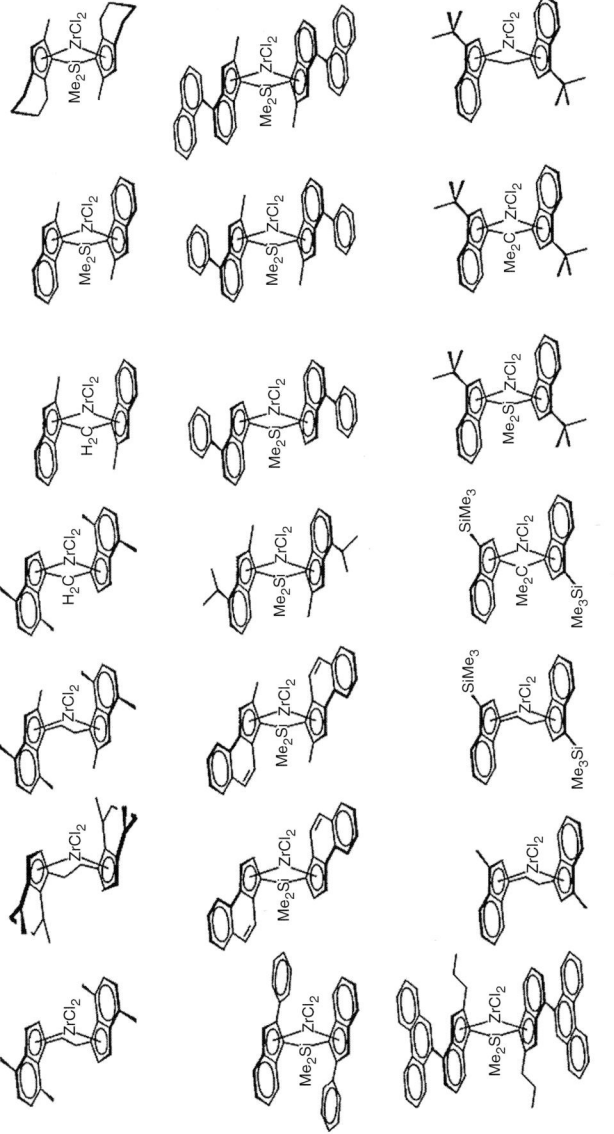

Figure 12.11 Representative structures of C2-symmetric *ansa*-zirconocenes for isotactic PP.

Figure 12.12 Example of reactor fouling. (Courtesy of J. Pater (Lyondell Basell)).

Irrespective of the true mechanism, it is a fact that $MgCl_2$-supported Ti-based catalysts, with a productivity of up to 5 t of polymer per gram of Ti and <2% by weight of less-tactic byproducts, feature an unbeatable cost/performance balance for the production of isotactic PP materials [13], and are highly competitive for that of HDPE [12, 47]. The multiplicity of active species typical of such catalysts, on the other hand, is a serious drawback in copolymerization reactions, and it is in the LLDPE segment of the market, in particular, that molecular catalysts are superior in terms of comonomer and molecular mass distributions [12, 17]. Here the dominant support is porous amorphous silica, typically impregnated with MAO for the immobilization of a suitable precatalyst (usually but not always a metallocene) via formation of an ion pair [12, 55] (Eq. (12.2)). Both for $MgCl_2$ and silica, sophisticated methods are available for the preparation of spherical particles (average diameter typically between 20 and 80 μm, depending on the application) with narrow size distribution, very high specific surface, and controlled porosity [12, 13, 56] (Figure 12.13).

It is important to note that in the case of $MgCl_2$ such particles are secondary aggregates of millions of primary nano-particles (with an average size below 50 nm) kept together by relatively weak interactions; therefore, it is common practice to improve the mechanical resistance of these aggregates prior to their introduction

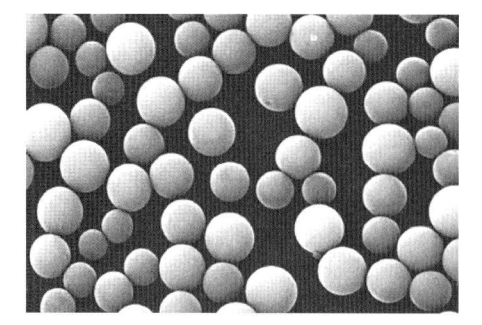

Figure 12.13 Particles of porous amorphous silica with controlled morphology (20 μm average diameter).

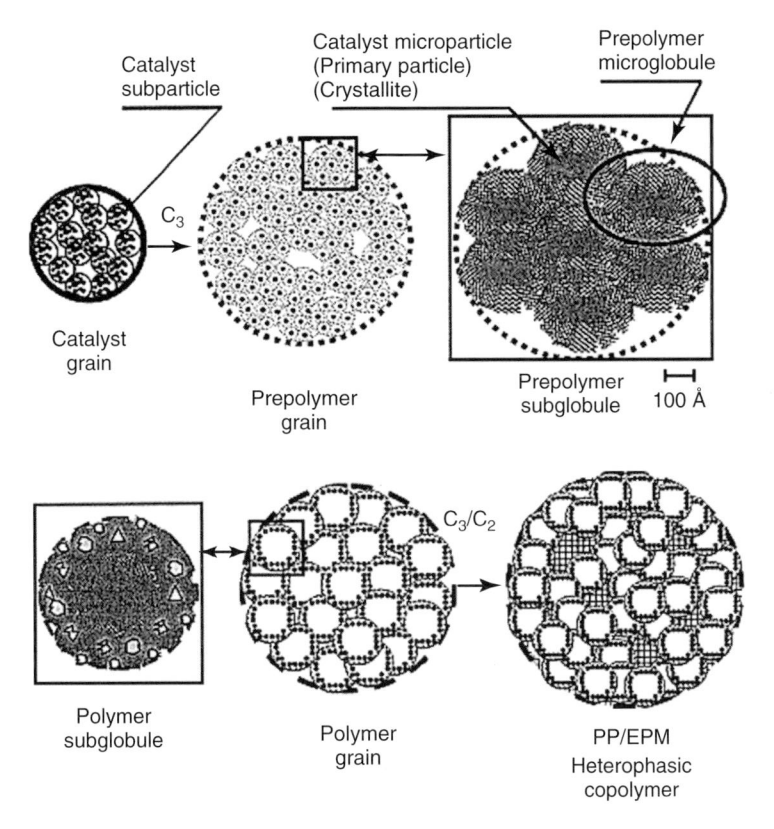

Figure 12.14 Model of replication of catalyst particle morphology into polymer particle morphology for MgCl$_2$-supported Ti-based Ziegler–Natta catalysts [13], as exploited for the production of heterophasic PP (Section 12.2) with the so-called 'reactor granule technology' [56]. (Reproduced from Ref. [56].)

into the polymerization reactor by means of a mild pre-polymerization treatment which 'glues' the primary particles together [13, 56].

Polymer growth on a catalyst particle is a complicated process which has been studied in depth at micro-, meso- and macroscale without reaching unequivocal mechanistic conclusions [12, 56]. What is certain is that with increasing polymer yield mechanical stress mounts inside the particle, which soon breaks up. A controlled morphology of the polymer granule via replication requires a smooth and uniform fragmentation, so that – in an over-simplified picture – each fragment acts as a radial source of polymer ending up with a spherical domain of globular sub-units (Figure 12.14-top). Although good replication can be achieved both with silica- and MgCl$_2$-supported catalysts, the results are usually superior for the latter, because the expansion of the primary particles does not require the breakage of strong chemical bonds [12, 13, 56].

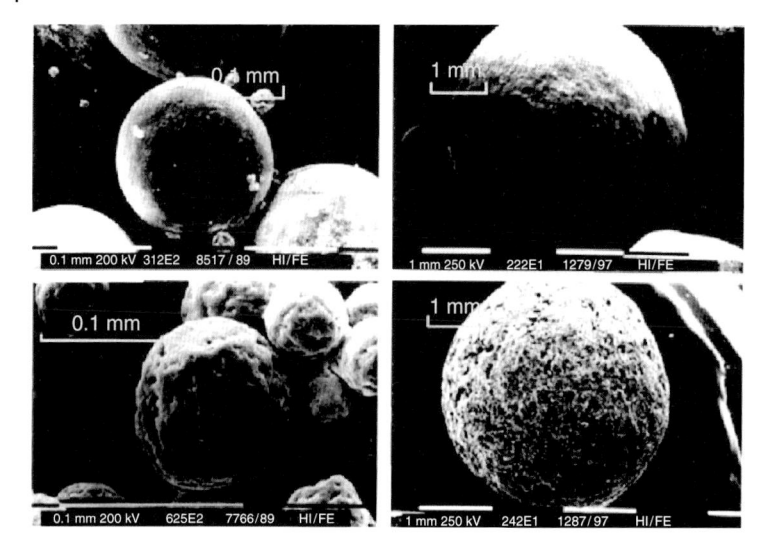

Figure 12.15 Replication of MgCl$_2$-supported Ti-based Ziegler–Natta [13] catalyst particles of different porosity in propene polymerization. SEM. (Courtesy of J. Pater (Lyondell Basell).)

As noted above, the resulting polymer granules are porous like the catalyst particles from which they originated (Figure 12.15). This porosity is very important for reactor blending applications ('reactor granule technology') [13, 56]. In particular, an EPR phase can be produced in a second stage inside the pores of isotactic PP particles. These can contain up to 70% by weight of rubber without becoming sticky as long as the rubber phase is confined inside the PP matrix. The fine dispersion of the rubber domains results in materials with remarkable impact resistance, usable for a variety of sophisticated applications [13, 56] (Figure 12.14-bottom).

12.4.2
Homogeneous Catalysis

Homogeneous olefin polymerization processes can be looked at as niche applications, but in a market of 150 Mt per year a niche can easily represent some megatonnes. An important reason for the rather limited diffusion of solution process technologies in catalytic olefin polymerization is that they are very demanding in terms of catalyst performance. For the process to be truly homogeneous, the polymer produced must be soluble in the reaction medium. In case of semicrystalline polyolefins like HDPE or isotactic PP, this condition can only be attained at temperatures above 110–130 °C, at which most (but not all!) catalysts undergo fast deactivation and/or have an inadequate polymer molecular weight capability (that is, a low ratio between the rates of chain propagation and transfer). A lower reaction temperature could be used, in principle, for polymers with little or no ability to

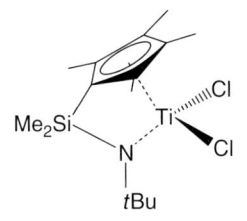

Figure 12.16 Prototypical structure of a Constrained Geometry Catalyst.

crystallize, and most typically for copolymers like LLDPE or EPR; however, even in such cases it would be advantageous to operate the process above 100 °C so as to avoid the problems arising from an excessive viscosity of the reaction medium as soon as monomer conversion reaches practical levels.

A limited number of molecular catalyst classes are especially suited to olefin polymerization in solution. In terms of industrial significance, the Constrained Geometry Catalysts of Figure 12.16 are leading for the production of LLDPE due to a high ability to incorporate higher 1-alkenes in copolymerizations with ethene; this holds not only for typical comonomers like 1-hexene and 1-octene, but also for allyl-terminated macromonomers that form when growing PE chains undergo β-H elimination [17].

It has been found that at high temperature and comparatively low monomer concentration the probability that such macromonomers are incorporated in the growing chains is non-negligible, with the resulting formation of long-chain branches (Scheme 12.6). Long-chain-branched LLDPE (Figure 12.1d) has a very peculiar rheological behavior (Figure 12.17) that makes it of interest for a number of applications [17, 34].

Other classes of half-metallocene catalysts [57] (Figure 12.18) are also known as good incorporators of higher olefins, including branched ones, and are attracting considerable attention for application in in ethene/propene/diene terpolymerization (dominated for decades by V-based Ziegler–Natta catalysts dating back to the 1960s, such as VCl_4 or $V(acac)_3$ (acac = acetylacetonate) in combination with $AlEt_2Cl$ and a Lewis base, e.g., anisole) [18].

Ethene homopolymerization in the presence of Ni and Pd catalysts with sterically hindered α-diimine ancillary ligands [32] (Figure 12.19) must also be mentioned in this section, although until now no practical application has been reported. As noted in Section 12.3.1, β-H elimination at late transition metal alkyls in general is fast, which makes such metals more suitable for use in olefin oligomerizations. However, as first noted by Brookhart and co-workers in a series of pioneering papers [32], this propensity can be contrasted with a proper design of the ligand framework, at least as far as β-H transfer to the monomer is concerned (Figures 12.9 and 12.20). When the latter process is inhibited by sterically hindering the approach of the monomer, intramolecular β-H transfer to the metal – which does occur – tends to be followed by the re-insertion of the resulting macro-olefin, which has a lower barrier than its displacement. Due to the possibility of the double bond to rotate, the said re-insertion can lead to chain isomerization with the generation of a side branch. Iteration of the process, described as *'chain walking'* [32], can ultimately

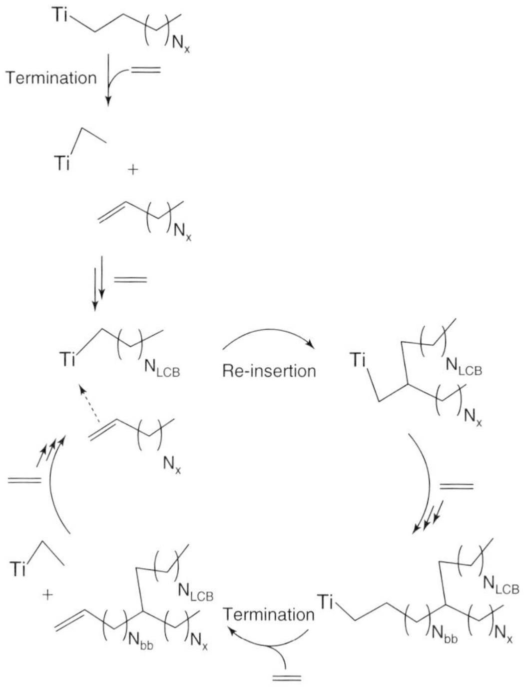

Scheme 12.6

lead to a highly branched PE mimicking LDPE (Scheme 12.7). More than as an alternative to the radical route to LDPE, however, late transition metal catalysts are appealing because, due to a much lower oxophilicity compared with early transition metal ones, they could be usable for the polymerization of functional olefins like, for example, acrylates. This has been shown to be feasible, particularly in copolymerizations with ethene, but up to now with inadequate productivity and molecular weight capability [58].

12.5
The Latest Breakthroughs

As noted in the Introduction, olefin polymerization is still developing fast [11]. In the area of commodity and commodity-plus products, the progress is mainly in technology, a typical example being the development of the new multi-zone reactors, which enable a much finer control in polymer particle morphology and phase behavior [12, 59]. For specialty polyolefins, on the other hand, advances in catalysis are still the dominant factor. In spite of a more and more thorough mechanistic understanding, the approach to catalyst discovery is still largely empirical, and the recent massive introduction of High Throughput Experimentation (HTE) tools and methods [60] has contributed quite significantly to speeding up the search [11, 61].

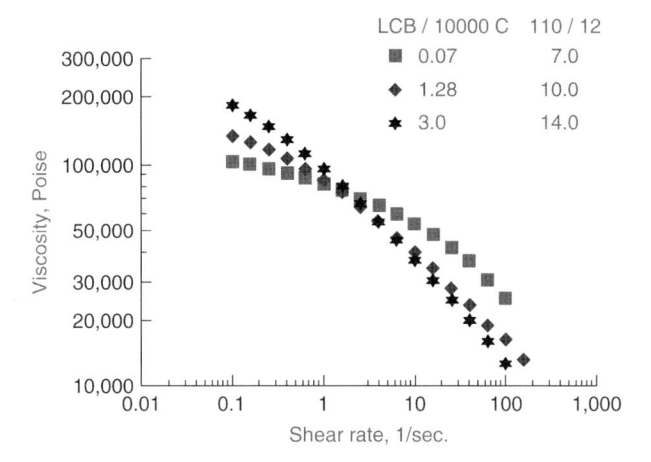

Figure 12.17 Effect of long-chain branching on the shear thinning behavior of LLDPE. (Reproduced from Refs. [30, 31].)

Figure 12.18 Representative half-titanocene precatalyst structures.

The main novelty in the first decade of the new millennium is the introduction of catalysts with controlled kinetic behavior, that is, with fast chain initiation relative to propagation and negligible chain transfer/termination in the time scale of a practical experiment [35]. The term '*living*' is often used [35] – improperly [62] – to describe such properties. For a long time, the only known controlled olefin polymerization catalysts have been the classical V-based Ziegler–Natta ones, but only at very low temperature (typically, −78 °C), and with moderate selectivity in syndiotactic sense when polymerizing propene [18, 49]. Nowadays, several catalyst classes exhibit controlled kinetics combined with a high stereoselectivity for PP at room temperature or even above [35], as already noted in Section 12.3.1.

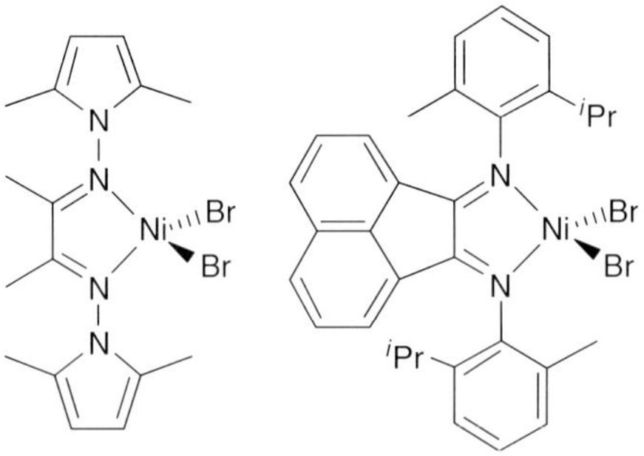

Figure 12.19 Representative α-diimine-Ni(II) precatalyst structures.

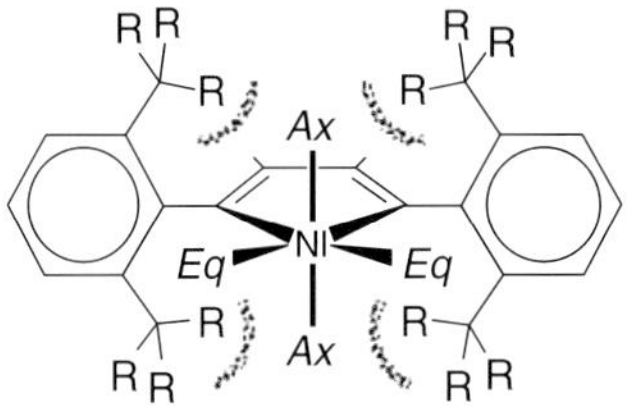

Figure 12.20 Steric hindrance at α-diimine-Ni(II)-based catalytic species, inhibiting β-H transfer to the monomer.

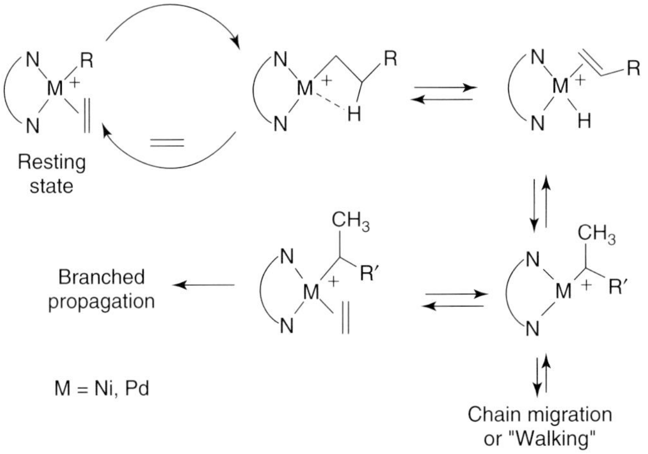

Scheme 12.7

Some are claimed to have been specifically 'designed' for a slow β-H elimination to the monomer (normally the dominant chain transfer pathway) [25, 35, 63]. Interestingly, there can be disagreement on the real mechanism which operates in some cases (see, e.g., the so-called *ortho*-F effect in bis(phenoxy-imine)Ti(IV) catalysis, described by the inventors as electronic in nature [25] and subsequently found by other authors to be mainly steric [33]); this suggests that sometimes the term '*design*' may have been used too optimistically.

An elective application for this catalysis is block copolymerization, typically via sequential comonomer feeding. A number of well-defined polyolefin architectures have been prepared and characterized also with respect to possible end-use properties [35]. The growing polymeryls can also be functionalized so as to make them capable to initiate a new block with a different chain growth mechanism, thus combining Ziegler–Natta with, for example, radical or ionic chemistry and ending up with novel 'hybrid' materials which feature polyolefin and non-polyolefin parts [64]. However, an important drawback for industrial application is the inherent limitation of this approach that only one polymer chain (at most) is produced per transition metal center; this inevitably translates into very low productivities and very high costs.

A possible solution is controlled polymerization with degenerative chain transfer [65, 66]. Of special importance are processes in which each polymer chain experiences in sequence periods of growth at a transition metal center, and 'dormancy' at a main-group metal center. Under proper conditions, the vast majority of the chains at any given time can be parked in the 'dormant' pool on the main-group metal. An outstanding example entails tandem catalysis in the presence of ZnR_2 as a 'chain shuttling' agent, recently reported by a team of Dow Chemical scientists [67]. These authors managed to implement a chain growth process in which each individual chain alternates periods of lengthening on one or the other in a couple of transition metal catalyst (generically denoted as 'Cat1' and 'Cat2') with periods of dormancy on Zn (Scheme 12.8).

Compared with classical block copolymerization with controlled catalysts, the chain shuttling route has the great advantage that the process is catalytic with respect to the usually expensive transition metal species (hundreds of chains being

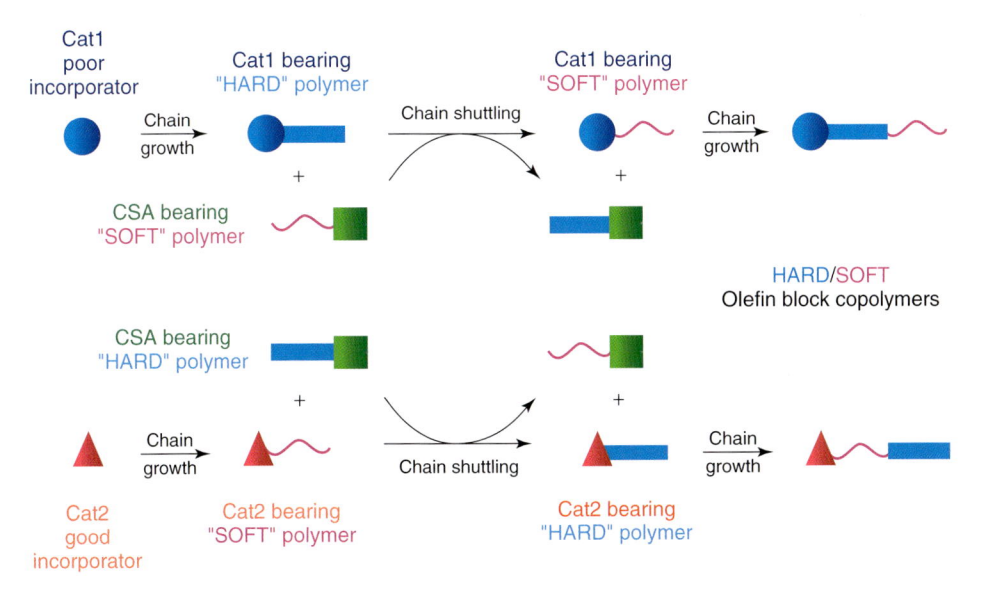

Scheme 12.8

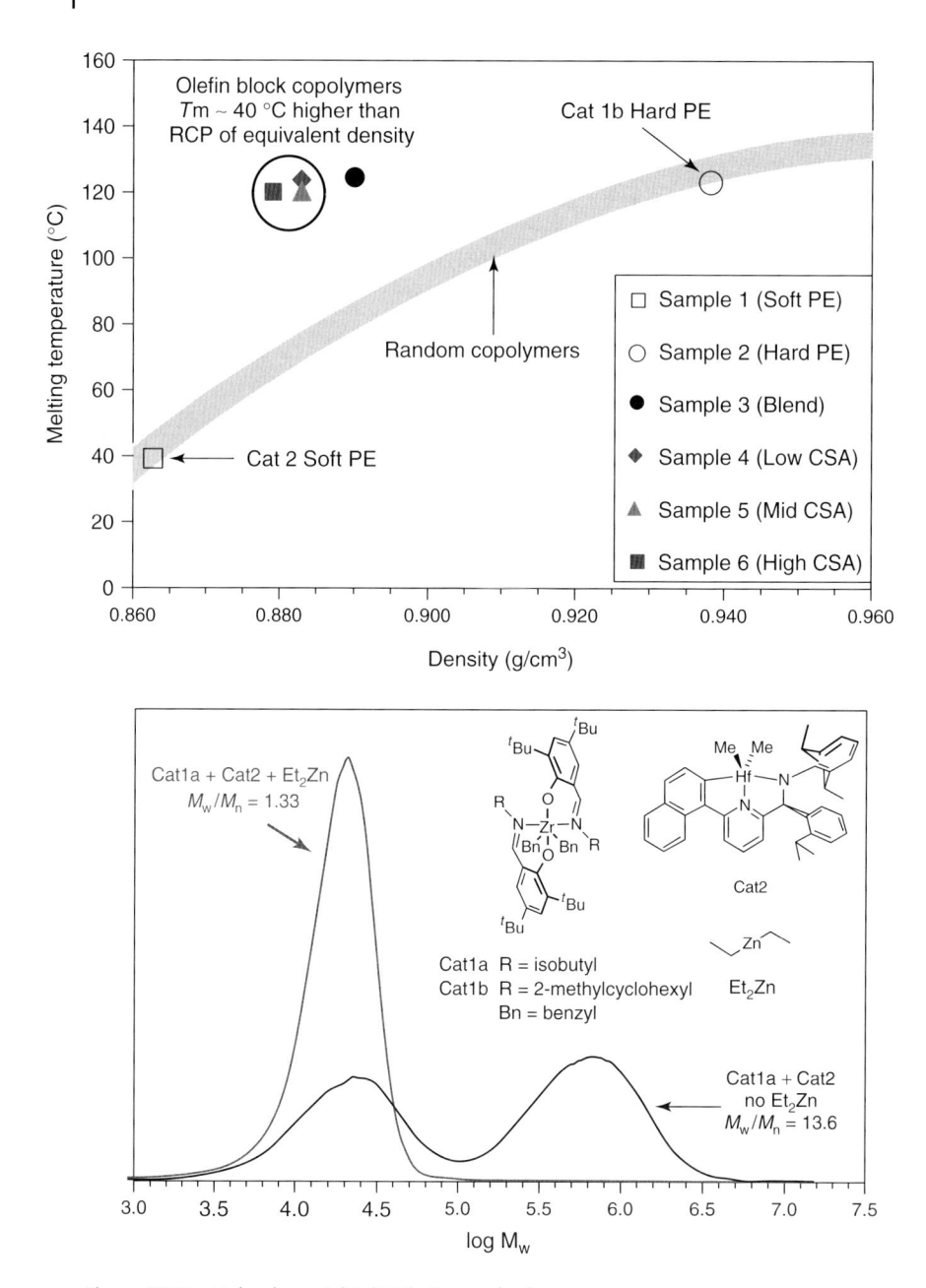

Figure 12.21 Molecular weight distribution and relationship between density and melting temperature for ethene/1-octene block copolymers obtained via tandem catalysis under 'chain shuttling' conditions. (Reproduced from Ref. [67].)

produced per transition metal center), and does not require repeated switches between different (co)monomer feeds (a technically demanding operation). In the case where Cat1 and Cat2 are, respectively, a good and a poor comonomer incorporator in ethene/1-octene copolymerization, the reaction products are statistical multi-block copolymers with sequential crystallizable (low octene unit content) and non-crystallizable (high octene unit content) random copolymer segments. These novel materials, whose microstructure can be tuned by adjusting the Cat1/Cat2 mole ratio, the comonomer feed and the concentration of ZnR_2, have outstanding thermoplastic-elastomeric properties, and broke a long-standing paradigm in structure/properties relationships, because the possibility to adjust independently the average lengths of the crystallizable and non-crystallizable blocks makes T_m and density two uncorrelated parameters [67, 68] (Figure 12.21). Therefore, not surprisingly, large-scale industrial application has already been announced [17].

It is important to note that of the two catalysts used in the process – namely, a bis(phenoxy-imine)Zr(IV) species [24] as the poor 1-octene incorporator, and a (pyridylamide)Hf(IV) one [69, 70] as the good 1-octene incorporator (Figure 12.21) – the latter belongs in a class of compounds discovered by means of HTE [69, 70]. The combination of the two with the chain shuttling agent was also identified by HTE [67]. All this is an impressive demonstration of the dramatic impact of HTE on the innovation process in the polyolefin field (probably the one with the highest industrial investments in new discovery tools outside pharmaceutical chemistry) [11, 61]. On the other hand, as soon as a better understanding of the mechanism of trans-alkylation and of the factors making it reversible is achieved, possibly also with the aid of HTE tools [11, 61], it is plausible to anticipate that new systems and applications will be identified on a more rational basis.

References

1. Terano, M. (2009) in *Next Generation Polyolefins*, vol. 3 (ed. M. Terano), Japan Chemical Management Think Tank, Tokyo pp. 197–253.
2. (2007) *Chem. Eng. News*, **85** (27), 55–64.
3. Kirk-Othmer (2007) *Kirk-Othmer Concise Encyclopedia of Chemical Technology*, 5th edn, John Wiley & Sons, Inc.
4. Chen, E.Y.X. (2009) Coordination polymerization of polar vinyl monomers by single-site metal catalysts. *Chem. Rev.*, **109** (11), 5157–5214.
5. Chamberlain, B.M. *et al.* (2001) Polymerization of lactide with zinc and magnesium β-diiminate complexes: stereocontrol and mechanism. *J. Am. Chem. Soc.*, **123** (14), 3229–3238.
6. Ovitt, T.M. and Coates, G.W. (2002) Stereochemistry of lactide polymerization with chiral catalysts: new opportunities for stereocontrol using polymer exchange mechanisms. *J. Am. Chem. Soc.*, **124** (7), 1316–1326.
7. Moore, D.R. *et al.* (2003) Mechanism of the alternating copolymerization of epoxides and CO_2 using β-diiminate zinc catalysts: evidence for a bimetallic epoxide enchainment. *J. Am. Chem. Soc.*, **125** (39), 11911–11924.
8. Cohen, C.T., Chu, T., and Coates, G.W. (2005) Cobalt catalysts for the alternating copolymerization of propylene oxide and carbon dioxide: combining high activity and selectivity. *J. Am. Chem. Soc.*, **127** (31), 10869–10878.

9. Stockland, R.A. and Jordan, R.F. (2000) Reaction of vinyl chloride with a prototypical metallocene catalyst: stoichiometric insertion and β-Cl elimination reactions with rac-(EBI)ZrMe+ and catalytic dechlorination/oligomerization to oligopropylene by rac-(EBI)ZrMe$_2$/MAO. *J. Am. Chem. Soc.*, **122** (26), 6315–6316.
10. Foley, S.R. et al. (2003) Reaction of vinyl chloride with late transition metal olefin polymerization catalysts. *J. Am. Chem. Soc.*, **125** (14), 4350–4361.
11. Busico, V. (2009) Metal-catalysed olefin polymerisation into the new millennium: a perspective outlook. *Dalton Trans.*, (41), 8794–8802.
12. Severn, J.R. and Chadwick, J.C. (2008) in *Tailor-Made Polymers via Immobilization of Alpha-Olefin Polymerization Catalysts* (eds J.R. Severn and J.C. Chadwick), Wiley-VCH Verlag GmbH, Weinheim.
13. Moore, E.P. Jr. (1996) in *Polypropylene Handbook: Polymerization, Characterization, Properties, Applications* (eds E.P. Moore), Hanser Publishers, Munich.
14. Bersted, B.H., Slee, J.D., and Richter, C.A. (1981) Prediction of rheological behavior of branched polyethylene from molecular structure. *J. Appl. Polym. Sci.*, **26** (3), 1001–1014.
15. Kiparissides, C. et al. (2005) Mathematical modeling of free-radical ethylene copolymerization in high-pressure tubular reactors. *Ind. Eng. Chem. Res.*, **44** (8), 2592–2605.
16. Busico, V. and Cipullo, R. (2001) Microstructure of polypropylene. *Prog. Polym. Sci.*, **26** (3), 443–533.
17. Chum, P.S. and Swogger, K.W. (2008) Olefin polymer technologies–history and recent progress at The Dow Chemical Company. *Prog. Polym. Sci.*, **33** (8), 797–819.
18. Boor, J. (1979) *Ziegler-Natta Catalysts and Polymerizations*, Academic Press, New York.
19. Resconi, L. et al. (2000) Selectivity in propene polymerization with metallocene catalysts. *Chem. Rev.*, **100** (4), 1253–1346.
20. Chen, E.Y.-X. and Marks, T.J. (2000) Cocatalysts for metal-catalyzed olefin polymerization: activators, activation processes, and structure-activity relationships. *Chem. Rev.*, **100** (4), 1391–1434.
21. Gibson, V.C. and Spitzmesser, S.K. (2003) Advances in non-metallocene olefin polymerization catalysis. *Chem. Rev.*, **103** (1), 283–316.
22. Ziegler, T. et al. (1988) Theoretical study on the difference in the relative strengths of metal-hydrogen and metal-methyl bonds in complexes of early transition metals and complexes of middle to late transition metals. *Inorg. Chem.*, **27** (20), 3458–3464.
23. Alt, H.G. and Köppl, A. (2000) Effect of the nature of metallocene complexes of Group IV metals on their performance in catalytic ethylene and propylene polymerization. *Chem. Rev.*, **100** (4), 1205–1222.
24. Makio, H. and Fujita, T. (2009) Development and application of FI catalysts for olefin polymerization: unique catalysis and distinctive polymer formation. *Acc. Chem. Res.*, **42** (10), 1532–1544.
25. Busico, V. et al. (2006) Design of stereoselective Ziegler-Natta propene polymerization catalysts. *Proc. Natl. Acad. Sci. U.S.A.*, **103** (42), 15321–15326.
26. Corradini, P., Guerra, G., and Cavallo, L. (2004) Do new century catalysts unravel the mechanism of stereocontrol of old Ziegler–Natta catalysts? *Acc. Chem. Res.*, **37** (4), 231–241.
27. Nele, M. et al. (2001) Two-state models for propylene polymerization using metallocene catalysts. 2. Application to ansa-metallocene catalyst systems. *Macromolecules*, **34** (12), 3830–3841.
28. Busico, V. et al. (2003) Metallocene-catalyzed propene polymerization: from microstructure to kinetics. C2-symmetric ansa-zirconocenes. *Macromolecules*, **36** (12), 4258–4261.
29. Chen, M.-C. and Marks, T.J. (2001) Strong ion pairing effects on single-site olefin polymerization: mechanistic insights in syndiospecific propylene enchainment. *J. Am. Chem. Soc.*, **123** (47), 11803–11804.
30. Busico, V. and Cipullo, R. (1994) Influence of monomer concentration on the stereospecificity of 1-alkene polymerization promoted by C2-symmetric

ansa-metallocene catalysts. *J. Am. Chem. Soc.*, **116** (20), 9329–9330.
31. Busico, V. et al. (1998) C2-symmetric ansa-metallocene catalysts for propene polymerization: Stereoselectivity and enantioselectivity. *J. Mol. Catal. A: Chem.*, **128** (1–3), 53–64.
32. Ittel, S.D., Johnson, L.K., and Brookhart, M. (2000) Late-metal catalysts for ethylene homo- and copolymerization. *Chem. Rev.*, **100** (4), 1169–1204.
33. Talarico, G., Busico, V., and Cavallo, L. (2004) 'Living' propene polymerization with bis(phenoxyimine) Group 4 metal catalysts: new strategies and old concepts. *Organometallics*, **23** (25), 5989–5993.
34. Gahleitner, M. (2001) Melt rheology of polyolefins. *Prog. Polym. Sci.*, **26** (6), 895–944.
35. Domski, G.J. et al. (2007) Living alkene polymerization: new methods for the precision synthesis of polyolefins. *Prog. Polym. Sci.*, **32** (1), 30–92.
36. Kissin, Y.V. (1985) *Isospecific Polymerization of Olefins*, Springer-Verlag, New York.
37. Liu, Z. et al. (2001) Kinetics of initiation, propagation, and termination for the [rac-(C_2H_4(1-indenyl)$_2$)ZrMe][MeB(C_6F_5)$_3$]-catalyzed polymerization of 1-hexene. *J. Am. Chem. Soc.*, **123** (45), 11193–11207.
38. Mori, H. and Terano, M. (1997) Stopped-flow techniques in olefin polymerization *Trends Polym. Sci.*, **5**, 314–321.
39. Busico, V., Cipullo, R., and Esposito, V. (1999) Stopped-flow polymerizations of ethene and propene in the presence of the catalyst system rac-Me$_2$Si(2-methyl-4-phenyl-1-indenyl)2ZrCl2/methylaluminoxane. *Macromol. Rapid Commun.*, **20** (3), 116–121.
40. Song, F., Cannon, R.D., and Bochmann, M. (2003) Zirconocene-catalyzed propene polymerization: a quenched-flow kinetic study. *J. Am. Chem. Soc.*, **125** (25), 7641–7653.
41. Busico, V. et al. (2009) Hafnocenes and MAO: beware of trimethylaluminum! *Macromolecules*, **42** (6), 1789–1791.
42. Busico, V. et al. (2005) Reactivity of secondary metal-alkyls in catalytic propene polymerization: how dormant are 'Dormant Chains'? *J. Am. Chem. Soc.*, **127** (6), 1608–1609.
43. Landis, C.R., Sillars, D.R., and Batterton, J.M. (2004) Reactivity of secondary metallocene alkyls and the question of dormant sites in catalytic alkene polymerization. *J. Am. Chem. Soc.*, **126** (29), 8890–8891.
44. Tsutsui, T., Kashiwa, N., and Mizuno, A. (1990) Effect of hydrogen on propene polymerization with ethylenebis(1-indenyl)zirconium dichloride and methylalumoxane catalyst system. *Makromol. Chem. Rapid Commun.*, **11** (11), 565–570.
45. Busico, V., Cipullo, R., and Corradini, P. (1993) Ziegler-Natta oligomerization of 1-alkenes: a catalyst's 'fingerprint', 1. Hydrooligomerization of propene in the presence of a highly isospecific MgCl$_2$-supported catalyst. *Makromol. Chem.*, **194** (4), 1079–1093.
46. Chadwick, J.C., van Kessel, G.M.M., and Sudmeijer, O. (1995) Regio- and stereospecificity in propene polymerization with MgCl$_2$-supported Ziegler-Natta catalysts: effects of hydrogen and the external donor. *Macromol. Chem. Phys.*, **196** (5), 1431–1437.
47. Boehm, L.L., Enderle, H.-F., and Fleissner, M. (1994) in *Catalyst Design for Tailor-Made Polyolefins* (eds K. Soga and M. Terano), Kodansha, Tokyo, pp. 351–363.
48. Busico, V. et al. (2004) Propene/Ethene-[1-^{13}C] copolymerization as a tool for investigating catalyst regioselectivity. MgCl$_2$/Internal Donor/TiCl$_4$ ˆ 'External Donor/AlR3 systems. *Macromolecules*, **37** (20), 7437–7443.
49. Doi, Y., Ueki, S., and Keii, T. (1979) 'Living' coordination polymerization of propene initiated by the soluble V(acac)3-Al(C_2H_5)2Cl system. *Macromolecules*, **12** (5), 814–819.
50. Pino, P., Rotzinger, B., and von Achenbach, E. (1985) The role of some bases in the stereospecific polymerization of propylene with titanium catalysts supported on magnesium chloride. *Makromol. Chem.*, **13** (S19851), 105–122.

51. Stehling, U. et al. (1994) Ansa-zirconocene polymerization catalysts with anelated ring ligands – effects on catalytic activity and polymer chain length. *Organometallics*, **13** (3), 964–970.
52. Ystenes, M. (1991) The trigger mechanism for polymerization of ±-olefins with Ziegler-Natta catalysts: a new model based on interaction of two monomers at the transition state and monomer activation of the catalytic centers. *J. Catal.*, **129** (2), 383–401.
53. Fait, A. et al. (1999) A possible interpretation of the nonlinear propagation rate laws for insertion polymerizations: a kinetic model based on a single-center, two-state catalyst. *Macromolecules*, **32** (7), 2104–2109.
54. Busico, V. et al. (2002) Metallocene-catalyzed propene polymerization: from microstructure to kinetics. 1. C2-symmetric ansa-metallocenes and the 'Trigger' Hypothesis. *Macromolecules*, **35** (2), 349–354.
55. Severn, J.R. et al. (2005) Bound but not gagged immobilizing single-site alpha-olefin polymerization catalysts. *Chem. Rev.*, **105** (11), 4073–4147.
56. Cecchin, G., Morini, G., and Pelliconi, A. (2001) Polypropene product innovation by reactor granule technology. *Macromol. Symp.*, **173** (1), 195–210.
57. Nomura, K. (2009) Half-titanocenes containing anionic ancillary donor ligands as promising new catalysts for precise olefin polymerisation. *Dalton Trans.*, (41), 8811–8823.
58. Nakamura, A., Ito, S., and Nozaki, K. (2009) Coordination–insertion copolymerization of fundamental polar monomers. *Chem. Rev.*, **109** (11), 5215–5244.
59. Mei, G. et al. (2006) The spherizone process: a new PP manufacturing platform. *Macromol. Symp.*, **245–246** (1), 677–680.
60. Hagemeyer, A., Strasser, P., and Volpe, A.F. (2004) in *High Throughput Screening in Catalysis* (eds A. Hagemeyer, P. Strasser, and A.F. Volpe), Wiley-VCH Verlag GmbH, Weinheim.
61. Busico, V. et al. (2009) High-throughput screening in olefin-polymerization catalysis: from serendipitous discovery towards rational understanding. *Macromol. Rapid Commun.*, **30** (20), 1697–1708.
62. Busico, V., Talarico, G., and Cipullo, R. (2005) Living Ziegler-Natta polymerizations: true or false? *Macromol. Symp.*, **226** (1), 1–16.
63. Mitani, M., Nakano, T., and Fujita, T. (2003) Unprecedented living olefin polymerization derived from an attractive interaction between a ligand and a growing polymer chain. *Chem. Eur. J.*, **9** (11), 2396–2403.
64. Kawahara, N. et al. (2008) *Polymer Hybrids Based on Polyolefins – Syntheses, Structures, and Properties New Frontiers in Polymer Synthesis*, Springer, Berlin, Heidelberg, pp. 79–119.
65. Mueller, A.H.E. et al. (1995) Kinetic analysis of 'Living' polymerization processes exhibiting slow equilibria. 1. Degenerative transfer (Direct Activity Exchange between Active and 'Dormant' Species). Application to group transfer polymerization. *Macromolecules*, **28** (12), 4326–4333.
66. Kempe, R. (2007) How to polymerize ethylene in a highly controlled fashion? *Chem. Eur. J.*, **13** (10), 2764–2773.
67. Arriola, D.J. et al. (2006) Catalytic production of olefin block copolymers via chain shuttling polymerization. *Science*, **312** (5774), 714–719.
68. Hustad, P.D. et al. (2008) An exploration of the effects of reversibility in chain transfer to metal in olefin polymerization. *Macromolecules*, **41** (12), 4081–4089.
69. Frazier, K.A. et al. (2005) High activity olefin polymerization catalyst and process, D.G.T. Inc, Editor. US Patent 6,953,764.
70. Boussie, T.R. et al. (2006) Nonconventional catalysts for isotactic propene polymerization in solution developed by using high-throughput-screening technologies. *Angew. Chem. Int. Ed.*, **45** (20), 3278–3283.

13
Ammonia Synthesis

Jens Rostrup-Nielsen

13.1
Ammonia Plant

Ammonia is a commodity chemical with an annual production of 120 million tons. It is used mostly for fertilizers, and production rate normally increases with population growth as long as the efficiency of nitrogen uptake by plants is not improved.

Ammonia is produced by synthesis, the raw material being 'synthesis gas,' which is produced from any carbon source, usually natural gas, water, and air.

$$\text{Natural gas} + \text{water} + \text{air} \longrightarrow \text{syngas} \ (H_2 : N_2 = 3 : 1)$$
$$\longrightarrow \text{ammonia} + \text{heat} + CO_2 \quad (13.1)$$

For commodities, costs are the most important parameter. The variable costs are linked to feed costs, energy efficiency, selectivity, and environmental costs. As a result of globalization of the economy, ammonia plants are situated at locations where natural gas is cheap (Middle East, Trinidad, Nigeria ...), although ammonia production is hardly feasible where there is a big market for natural gas as fuel.

Plants have become larger to take advantage of economy of scale. Typical plants have a capacity of 1000–1500 tons NH_3/day, but single ammonia synthesis loops of 3500 tons/day are under construction.

Plants have become more integrated to minimize energy consumption, which, for ammonia made from natural gas has decreased over the last 50 years from 40 to 29 GJ/t corresponding to a thermal efficiency (LHV) of 65%, or 73% of the theoretical minimum [1, 2]. The process scheme for ammonia involves eight catalytic steps, as shown in Figure 13.1 for a natural gas-based plant [2].

The natural gas is desulfurized by absorption of H_2S and lower mercaptans over zinc oxide. If the feed contains higher mercaptans and thiophene a hydrogenation step is installed prior to the ZnO vessel. The hydrogen is made by tubular steam reforming of natural gas followed by a secondary autothermal reforming step in which nitrogen is added with air, the oxygen reacting with the unconverted methane from the primary reformer. If the feed contains large amounts of higher hydrocarbons, these are converted in a low-temperature adiabatic prereformer.

Catalysis: From Principles to Applications, First Edition. Edited by Matthias Beller, Albert Renken, and Rutger van Santen.
© 2012 Wiley-VCH Verlag GmbH & Co. KGaA. Published 2012 by Wiley-VCH Verlag GmbH & Co. KGaA.

13 Ammonia Synthesis

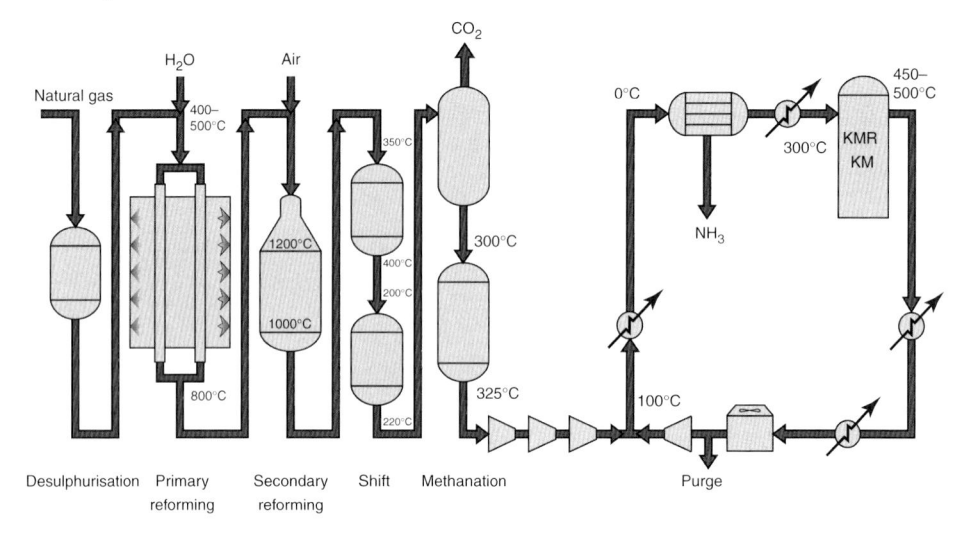

Figure 13.1 Process scheme for ammonia synthesis based on natural gas.

The CO-rich gas from the secondary reformer is converted by the water gas shift reaction in a high-temperature step (HTS) and a low-temperature shift (LTS). CO_2 is removed in a wash, and residual amounts of CO_2 are converted in a low-temperature methanation step. The purified synthesis gas ($H_2/N_2 = 3$) is compressed and sent to the ammonia synthesis loop. The ammonia produced is separated from the unconverted synthesis gas by cooling in a chiller system operating with ammonia as the working fluid. Typical catalysts and process conditions for catalytic steps are summarized in Table 13.1. Heat produced by the combustion of natural gas as well as the heat recovered from the various process steps is used to run the plant.

An industrial breakthrough resulted from the introduction of high-pressure tubular steam reforming in the 1960s and the integration of the steam turbine cycle and process cycle, which meant that less work was required for compression of the synthesis gas. This was followed by steady improvement and sophistication of the technology, and not least by application of better catalysts. Even small improvements in the process scheme can show short pay-back times. On the other hand, uncertainties associated with new technology can easily nullify an economic advantage. The high degree of integration means that improvement of one step could harm the performance of another step, which means that the weakest part of the plant could determine the performance of the entire plant. The reliability of the plant, reflected in the on-stream factor, is crucial for the plant economy. Therefore, long catalyst life and control of secondary phenomena such as catalyst poisoning and carbon formation are important issues [1].

This chapter describes the impact of basic research on the design of two of the catalytic steps: the ammonia synthesis and the primary reformer. The catalyst activity is the dominating parameter for the synthesis, the catalyst being well protected at the back end of the plant; at the same time, control of secondary phenomena is crucial for operation of the tubular reformer.

Table 13.1 Typical catalysts and process parameters for a natural gas-based ammonia plant.

Reaction	Catalyst	P (bar)	T (°C)	Gas
Desulfurization				
$RHS + H_2 = RH + H_2S$	$CoMo/Al_2O_3$	45	365	10 ppb S
$H_2S + ZnO = ZnS + H_2O$	ZnO	45	365	
Steam reforming				
Prereforming				
$C_nH_m + nH_2 = 0 =$	Ni, X/MgO,	40	450/500	$H_2O/C = 3.0$
$nCO + 1/2(n+m)H_2$	Al_2O_3			
Tubular reforming				
$CH_4 + H_2O = CO + 3H_2$	$Ni/MgAl_2O_4$	37	520/817	–
Secondary reforming				
$CH_4 + \frac{1}{2}O_2 = CO + 2H_2$	$Ni/MgAl_2O_4$	37	550/1000	0.3% CH_4, 13% CO
Water gas shift				
$CO + H_2O = CO_2 + H_2$	–	–	–	–
HTS	Fe_3O_4 (Cr), Al	6	360/430	3% CO
LTS	Cu, ZnO, Al_2O_3	34	205/225	0.3% CO
Methanation				
$CO + 3H_2 = CH_4 + H_2O$	Ni/Al_2O_3	32	300/323	10 ppm (CO + CO_2)
Synthesis				
$N_2 + 3H_2 = 2NH_3$	Fe (K,Ca,Al)	190	233/456	20% NH_3

13.2 Synthesis

13.2.1 Technology Development

The Haber–Bosch synthesis was introduced almost 100 years ago. The iron catalyst developed for the process by Mittasch [3] by tedious trial and error work is not very different from the industrial catalyst used today, but the know-how gained since then about the reaction and the catalyst structure has led to further optimization of modern ammonia synthesis.

The Temkin–Pyzhev kinetic equation and laboratory experiments under well-defined conditions by A. Nielsen *et al.* [4] formed the basis for a one-dimensional heterogeneous reactor model by Kjær [5] including diffusion effects. This showed that the effectiveness factor was less than one. Smaller catalyst particles would, however, result in a high pressure drop in the synthesis reactor, requiring more compression of recycle gas and less energy efficiency. A decrease in the mass flow by increasing the diameter of the high pressure reactor would be expensive. The solution was the radial flow reactor [6], allowing a lower pressure drop when using

Figure 13.2 Topsøe radial flow converter (capacity 1630 tons/day).

small catalyst particles without increasing the reactor diameter. Figure 13.2 shows a photograph of a radial converter for a 1630 tons/day ammonia plant. The feedback from industrial operation has provided a large amount of data with which to check the computer model [7]. The reactor typically consists of two or three catalyst beds with inter-bed heat exchangers [2, 8].

13.2.2
The Catalysis

Ammonia synthesis has served as a model reaction for fundamental studies in catalysis. It has been the 'bellwether' [9] because of its industrial importance and its simplicity, with no side reactions. Hence, an understanding of the mechanism of this synthesis has been closely linked to the progress of fundamental catalysis theory.

The development of ultra-high-vacuum techniques for the study of catalysts gave important information on adsorption and desorption rates in well-defined systems, not the least in the work by Ertl [10], although at pressures far from industrial conditions. It was, however, possible to 'bridge the gap' by microkinetic analysis [11], as illustrated in Figure 13.3 and later by DFT calculations [12, 13].

The Temkin Pyzhev kinetics assumed the dissociation of nitrogen to be the rate-determining step. This was confirmed by first principles calculations using the DFT method, leading to a full description of the energetics of the elementary steps involved in the synthesis on iron catalysts [13, 14] and later on ruthenium catalysts [14] with a mechanism involving dissociation of nitrogen followed by successive hydrogenation (Table 13.2):

The full energy diagram for the synthesis on ruthenium is shown in Figure 13.4.

The DFT analysis also identified step sites playing a key role, as shown in Figure 13.4. This was demonstrated by experiments on Ru bimetallic catalysts [15]. Blocking the steps on the ruthenium surface with gold atoms resulted in a drastic decrease in rate.

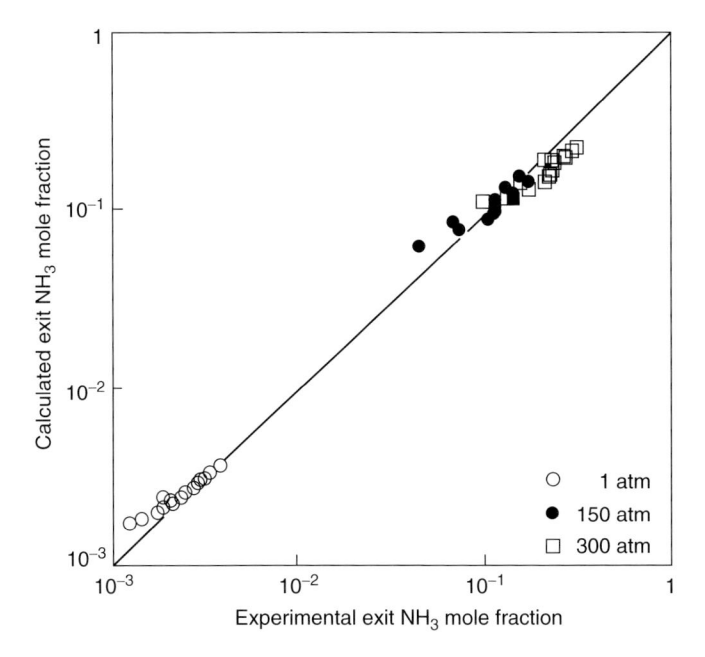

Figure 13.3 NH$_3$ synthesis. Calculated and measured conversions [11].

Table 13.2 The elementary surface reactions of the ammonia synthesis reaction [2].

N$_2$ + *	= 2N − *
H$_2$ + *	= 2H*
N* + H*	= NH − * + *
NH − * + H*	= NH$_2$ − * + *
NH$_2$ − * + H − *	= NH$_3$ + 2*

Activity trends for various metals were obtained by the Nørskov–Topsøe team by combining results from a micro-kinetic model for the synthesis with DFT calculations, showing a linear relationship between activation energy and nitrogen-surface bond energy [16]. This results in a volcano-shaped dependence of the catalyst activity on nitrogen adsorption energy [16]. This also led to a prediction of the optimum catalyst filling, depending on reaction conditions [17]. For each set of reaction conditions, nitrogen binding energy corresponding to the optimum in the volcano plot was calculated [18] as shown in Figure 13.5. Iron is the best catalyst for low concentrations of ammonia (inlet conditions), whereas ruthenium is the preferred choice at high concentration, that is, close to equilibrium. Ruthenium has been applied in industry in the final catalyst bed [2], but with limited success mainly due to its high price.

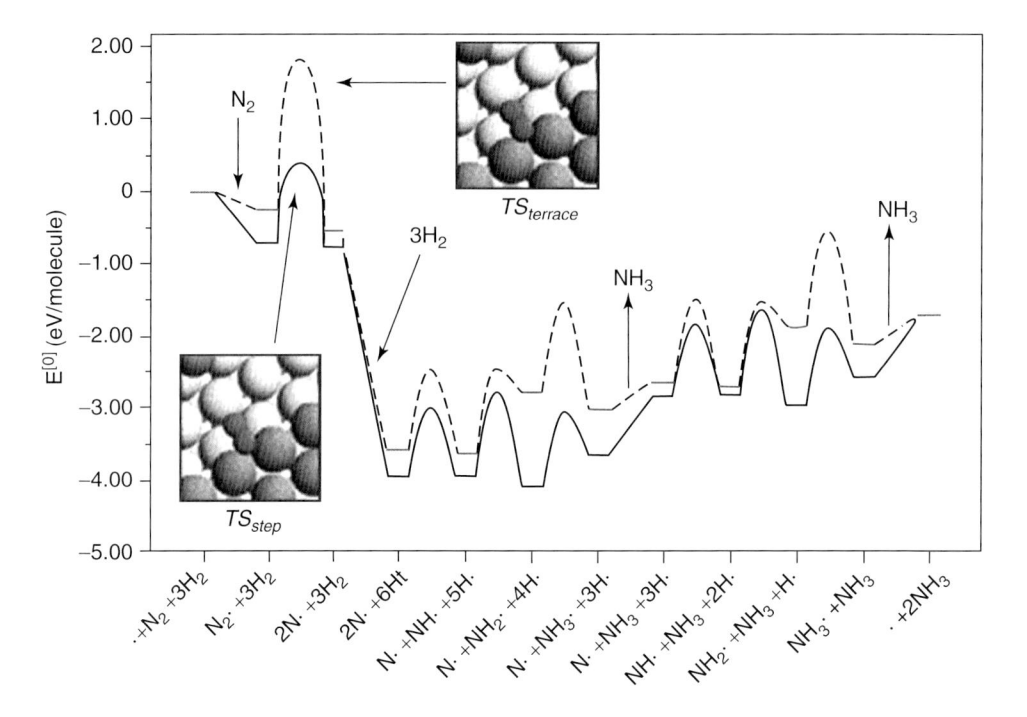

Figure 13.4 Energy diagram for elementary steps of the ammonia synthesis on ruthenium (Honkala et al. [13]).

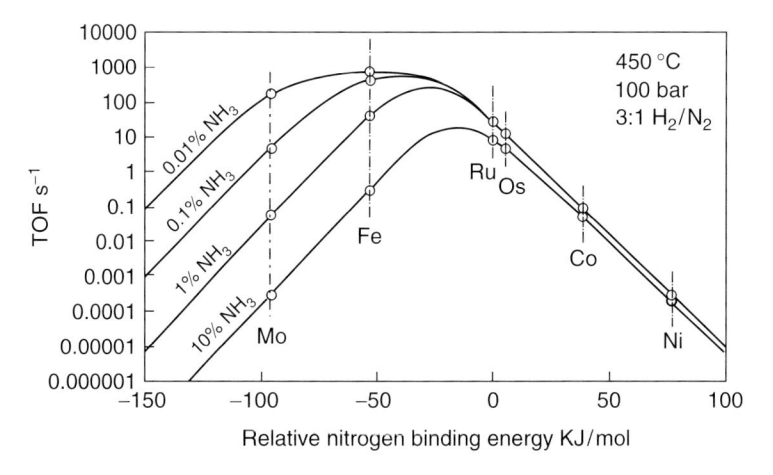

Figure 13.5 Activity (TOF) of promoted ammonia synthesis catalyst for different ammonia conversion levels. 450 °C, 100 bar, $H_2/N_2 = 3$ [18].

It is a challenge to find catalysts closer to the optimum [18, 19], but a high TOF is not sufficient. It is essential to achieve a high surface area, high filling density, and stability as well.

13.2.3
Process Optimization

The ammonia synthesis is well understood. The industrial performance can be predicted from fundamentals. Basic research has given a strong tool for a deductive approach for scale-up and optimization. Nevertheless, the process scheme introduces constraints in the choice of process parameters.

The optimum synthesis pressure is determined by the operation of the three compressors in the synthesis loop, the syngas compressor, the recycle compressor, and the ammonia compressor in the chiller loop. High loop pressure means higher conversion per pass and hence less recycle. Low loop pressure means more recycle and more energy for the ammonia compressor in the chiller because of lower ammonia partial pressure. This results in an optimum loop pressure in the range 80–220 bar [2].

In principle, a more active catalyst should allow the loop to operate at lower temperature with higher equilibrium conversion, but low temperature means less value of the steam raised from the heat recovery in the loop.

Of course, higher catalyst activity for a given pressure and temperature means smaller reactor volume, but long catalyst life may be more important for the economy of the plant. In spite of the strong input from fundamental research including a Nobel Prize it appears difficult to beat the iron catalyst discovered 100 years ago.

13.3
Steam Reforming

13.3.1
Technology

Steam reforming of hydrocarbons is a well-established process for the manufacture of synthesis gas and hydrogen [1, 20]. The reaction between the two stable molecules, steam and methane, is strongly endothermic. Industrial steam reforming is carried out in a fired reactor, the tubular reformer [1, 21], in which catalyst tubes are placed in a fired furnace supplying the heat for the reaction and the heat to provide the exit temperature required for the desired conversion (Figure 13.6).

The reformer tubes are filled with catalyst, which must show high mechanical stability because of exposure to high temperatures and steam partial pressures [1, 21]. With a typical activity of a nickel catalyst, the gas composition in the reformer tube quickly arrives close to equilibrium [22]. It can be shown [22] that there is a huge surplus of catalyst activity in a tubular reformer. This because

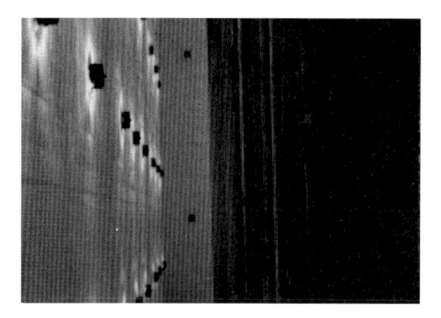

Figure 13.6 Tubular reformer. Steam reforming is a complex coupling of catalysis, heat transfer, and mechanical design.

the tube dimensions and the number of tubes are determined by the maximum allowable heat flux, other mechanical considerations leaving the catalyst volume as a dependent variable [21]. Still, there is a need for high catalyst activity because this means that the heat transfer for a given conversion can take place at a lower tube wall temperature.

13.3.2
The Catalysis

The development of the steam reforming process was not based on an initial understanding of the catalysis. In contrast to the ammonia synthesis, progress was driven mainly by an 'inductive approach,' with feedback from industrial pilot tests [23]. The initial vague ideas about the mechanism were mainly formulated by industrial groups in the 1970s [24, 25], and this was followed by input from surface science and theoretical methods [26].

The mechanism of steam reforming of methane is well described on the basis of recent fundamental studies [22], DFT calculations [27], and adsorption studies [22, 27]. Also, observations by *in-situ* high-resolution transmission electron microscopy (HRTEM) [28] have shown that step sites play a key role in methane activation. This is illustrated by the energy diagram for the individual reaction steps [27] shown in Figure 13.7 on the dense Ni(111) surface and the stepped Ni(2) surface.

The stepped surface has a lower barrier for the initial dissociation of methane. For both surfaces the highest barrier is the surface reaction to carbon monoxide, which may indicate a two-step mechanism [21, 26]:

$$CH_4 + * \longrightarrow CH_x * + \frac{1}{2}(4-x)H_2 \longrightarrow C* + \gamma H_2$$

$$C* + OH* \longrightarrow CO + \frac{1}{2}H_2 + *^*$$

$$H_2O = OH* + \frac{1}{2}H_2 \qquad (13.1)$$

This was supported by an expanded analysis [26] using the free energy (not the total energy, as used in Figure 13.7) and scaling principles [29] for activation

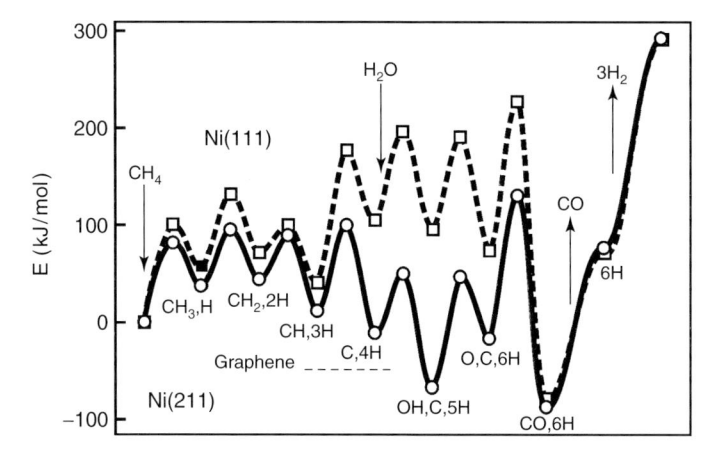

Figure 13.7 Energy diagram for steam reforming of methane. DFT calculations [27].

of methane and carbon monoxide. It was shown for nickel that the barrier for formation of CO decreases with temperature, indicating that a one-step mechanism may dominate at high temperatures. This trend is reflected by kinetic studies [1, 20, 26].

In general, the kinetics of the steam reforming reaction is found to be first order with respect to methane partial pressure [19, 20, 30]. However, at lower temperatures (<550 °C) a denominator term may result in an overall zero dependence of total pressure [20, 21].

The group VIII metals are active for the steam reforming reaction, with rhodium and ruthenium being the most active and nickel the most preferred because of price. Most studies [1, 21, 26] find a sequence of activity (turn over frequency) Ru, Rh > Ni Ir > Pt Pd. This sequence has also been confirmed by DFT calculations [26].

The higher hydrocarbons are more reactive (per carbon atom) than methane, except for aromatic molecules [21]. This means that liquid hydrocarbons in principle can be easily converted by steam reforming, but in practice this is limited by the higher potential for sulfur poisoning and by the higher risk of carbon formation.

13.3.3
Secondary Phenomena

The group VIII metals are subject to sulfur poisoning, with nickel being the most sensitive [1, 21]. The H_2S/Ni system is well described in terms of a two-dimensional sulfide [20].

Carbon is formed on nickel as fibers ('whiskers') from carbon monoxide, methane, and higher hydrocarbons [1, 22]. They grow from a nickel crystal with a diameter close to that of the nickel crystal. The fibers are strong and may result in a breakdown of the catalyst pellet. This may have a serious impact on the operation

(a)

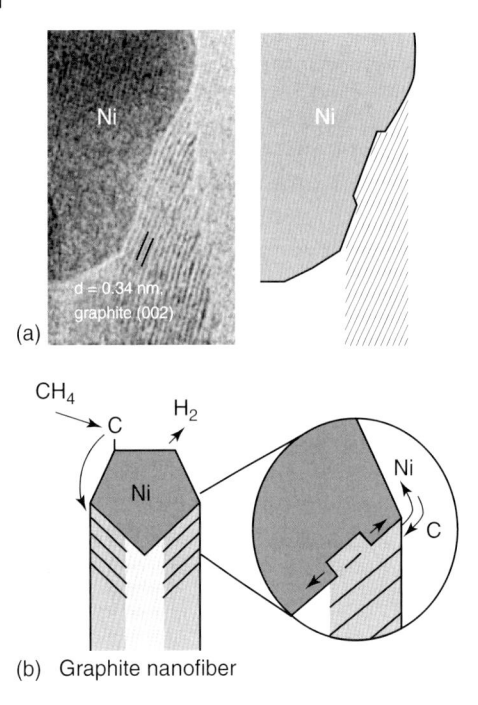

CH₄ → C
H₂ ↗
Ni

(b) Graphite nanofiber

Figure 13.8 Snapshot of a whisker growth. HRTEM [28]. The step site moves as the graphite plane grows.

of the tubular reactor, as the carbon and the broken catalyst pellets may result in mal-distribution of feed and overheating of the reformer tubes [21].

Early TGA studies [31] and *in-situ* electron microscopy work [32] resulted in an understanding of parameters governing the nucleation and growth of the carbon fibers, which led to design models for carbon-free operation [21]. Recent work by the Nørskov/Topsøe team has given more insight into the mechanism at atomic level. As shown in Figure 13.7, adsorbed carbon atoms are most stable on the Ni(211) surface [22, 27]. This was confirmed by HRTEM studies [28] and DFT calculations [33] showing that the nucleation of carbon takes place at step sites, as illustrated in Figure 13.8.

The blockage of step sites has been one method to eliminate carbon formation [1, 22]. This was demonstrated on a model Ni, Au catalyst [34], and it may explain the function of promoters (alkali, Sn, ...) [1] as well as the role of sulfur passivation [1, 35] in inhibiting nucleation of carbon.

The catalysis of steam reforming is an example of important secondary phenomena that could be explained in terms of basic principles. Catalyst activity is less critical for conventional tubular reforming, but an understanding of the mechanism has provided input for the design of advanced reformer concepts [1], taking advantage of the huge activity accessible by using catalyzed hardware [36, 37] or small monolithic catalyst beds [38]. This is applied both for fuel cells and for chemical recuperation in heat exchange reforming [1, 39].

13.4 Conclusions

Ammonia technology has benefited from a close interaction between industrial catalysis and basic research. The basic principles give a framework for analyzing the empirical data, and the feedback from industrial operation provides a strong basis for optimization, trouble shooting, and scale-up of new process concepts.

Abbreviations

DFT	density function theory
HRTEM	high resolution transmission electron microscopy
LHV	lower heating value
TGA	thermogravimetric analysis
TOF	turnover frequency

References

1. Rostrup-Nielsen, J.R. and Christiansen, L.J. (2010) *Concepts in Syngas Preparation*, Imperial College Press, London, (2011).
2. Dybkjær, I. (1995) in *Ammonia* (ed. A. Nielsen), Springer-Verlag, Berlin, Chapter 6, 199–328.
3. Mittasch, A. (1951) *Geschichte der Ammoniaksynthese*, Verlag Chemie, Weinheim.
4. Nielsen, A., Kjaer, J., and Hansen, B. (1964) Rate equation and mechanism of ammonia synthesis at industrial conditions. *J. Catal.*, **3** (1), 68–79.
5. Kjær, J. (1972) *Computer Methods in Catalytic Reactor Calculations*, Haldor Topsøe, Vedbæk.
6. Nielsen, A. (1972) Radial flow ammonia converter. *Ammonia Plant Saf.*, **14**, 46.
7. Jarvan, J.E. (1970) A general purpose program for calculation and analysis of fixed-bed catalytic reactors. *Ber. Bunsen-Ges. Phys. Chem.*, **74** (2), 142–149.
8. Nielsen, S.E. (2007) Proceedings FAI International Conference Fertilizer Technology, New Delhi.
9. Boudart, M. (1994) Ammonia synthesis: the bellwether reaction in heterogeneous catalysis. *Top. Catal.*, **1** (3), 405–414.
10. Ertl, G. (1983) Primary steps in catalytic synthesis of ammonia. *J. Vac. Sci. Technol., A*, **1**, 1247–1253.
11. Stoltze, P. and Nørskov, J.K. (1985) Bridging the 'Pressure Gap' between ultrahigh-vacuum surface physics and high-pressure catalysis. *Phys. Rev. Lett.*, **55** (22), 2502.
12. Nørskov, J.K. et al. (2002) Universality in heterogeneous catalysis. *J. Catal.*, **209** (2), 275–278.
13. Honkala, K. et al. (2005) Ammonia synthesis from first-principles calculations. *Science*, **307** (5709), 555–558.
14. Hellman, A. et al. (2009) Ammonia synthesis and decomposition on a Ru-based catalyst modeled by first-principles. *Surf. Sci.*, **603** (10–12), 1731–1739.
15. Dahl, S. et al. (1999) Role of steps in N2 activation on Ru(0001). *Phys. Rev. Lett.*, **83** (9), 1814.

16. Nørskov, J.K. et al. (2002) Universality in heterogeneous catalysis. *J. Catal.*, **209** (2), 275–278.
17. Jacobsen, C.J.H. et al. (2001) Catalyst design by interpolation in the periodic table: bimetallic ammonia synthesis catalysts. *J. Am. Chem. Soc.*, **123** (34), 8404–8405.
18. Jacobsen, C.J.H. and Nielsen, S.E. (2002) Comparative evaluation of ammonia synthesis catalyst features. *Ammonia Plant Saf.*, 212.
19. Nørskov, J.K. et al. (2009) Towards the computational design of solid catalysts. *Nat. Chem.*, **1** (1), 37–46.
20. Rostrup-Nielsen, J.R. (2008) in *Handbook of Heterogeneous Catalysis* (eds G. Ertl et al.), Wiley-VCH Verlag GmbH & Co. KGaA, Weinheim, pp. 1–24.
21. Rostrup-Nielsen, J.R. (1984) in *Catalysis, Science and Technology* (eds J.R. Anderson and M. Boudart), Springer-Verlag, Berlin, Vol. 5, Chapter 1, pp. 1–118.
22. Rostrup-Nielsen, J.R., Sehested, J., and Nørskov, J.K. (2002) *Advances in Catalysis*, Academic Press, pp. 65–139.
23. Rostrup-Nielsen, J., Bao, X., and Xu, Y. (2004) *Studies in Surface Science and Catalysis*, Elsevier, pp. 121–126.
24. Andrew, S.P.S. (1969) Catalysts and catalytic processes in the steam reforming of naphtha, I&EC. *Product R&D*, **8** (3), 321–324.
25. Rostrup-Nielsen, J.R. (1975) *Steam Reforming Catalysts*, Danish Technical Press, Copenhagen.
26. Jones, G. et al. (2008) First principles calculations and experimental insight into methane steam reforming over transition metal catalysts. *J. Catal.*, **259** (1), 147–160.
27. Bengaard, H.S. et al. (2002) Steam reforming and graphite formation on Ni catalysts. *J. Catal.*, **209** (2), 365–384.
28. Helveg, S. et al. (2004) Atomic-scale imaging of carbon nanofibre growth. *Nature*, **427** (6973), 426–429.
29. Abild-Pedersen, F. et al. (2007) Scaling properties of adsorption energies for hydrogen-containing molecules on transition-metal surfaces. *Phys. Rev. Lett.*, **99** (1), 016105.
30. Wei, J. and Iglesia, E. (2004) Mechanism and site requirements for activation and chemical conversion of methane on supported Pt clusters and turnover rate comparisons among noble metals. *J. Phys. Chem. B*, **108** (13), 4094–4103.
31. Rostrup-Nielsen, J. and Trimm, D.L. (1977) Mechanisms of carbon formation on nickel-containing catalysts. *J. Catal.*, **48** (1–3), 155–165.
32. Baker, R.T.K. et al. (1972) Nucleation and growth of carbon deposits from the nickel catalyzed decomposition of acetylene. *J. Catal.*, **26** (1), 51–62.
33. Abild-Pedersen, F. et al. (2006) Mechanisms for catalytic carbon nanofiber growth studied by ab initio density functional theory calculations. *Phys. Rev. B*, **73** (11), 115419.
34. Besenbacher, F. et al. (1998) Design of a surface alloy catalyst for steam reforming. *Science*, **279** (5358), 1913–1915.
35. Rostrup-Nielsen, J.R. (1984) Sulfur-passivated nickel catalysts for carbon-free steam reforming of methane. *J. Catal.*, **85** (1), 31–43.
36. Rostrup-Nielsen, J.R. (1993) Production of synthesis gas. *Catal. Today*, **18** (4), 305–324.
37. Farrauto, R.J. et al. (2007) Precious metal catalysts supported on ceramic and metal monolithic structures for the hydrogen economy. *Catal. Rev.*, **49** (2), 141–196.
38. Seris, E.L.C. et al. (2005) Demonstration plant for distributed production of hydrogen from steam reforming of methane. *Chem. Eng. Res. Des.*, **83** (6), 619–625.
39. Rostrup-Nielsen, J.R. (2009) Steam reforming and chemical recuperation. *Catal. Today*, **145** (1–2), 72–75.

14
Fischer–Tropsch Synthesis in a Modern Perspective
Hans Schulz

14.1
Introduction

Excitingly, Franz Fischer and Hans Tropsch observed the formation of liquid hydrocarbons when they passed a mixture of hydrogen and carbon monoxide (then known as 'water gas' from coke gasification) at normal pressure over a cobalt catalyst, and called the new reaction 'gasoline synthesis' [1]. It seemed like a miracle that aliphatic hydrocarbons (and H_2O) were formed from simple inorganic molecules. But miracle actually means complexity of unknown order (Figure 14.1). In those days (~1926), oil reserves were actually thought to be limited, and the production of automotive fuel from coal appeared attractive [2, 3].

Today, Fischer–Tropsch (FT) synthesis is of particular interest to provide clean diesel fuel not only from coal (CTL: coal to liquid) but also from natural gas (GTL: gas to liquid) and maybe in the future from bio-mass (BTL: bio-mass to liquid) (Figure 14.2). Even with $H_2 + CO_2$ as synthesis gas, FT synthesis would be possible [4] if hydrogen were readily available. Low-molecular-weight olefins can also be the aimed-at products of FT synthesis [5, 6]. Present commercial FT production includes plants on a coal basis in South Africa and plants on a natural gas basis in Qatar, Malaysia, and South Africa. FT reactors must deal with exothermicity of a reaction requiring 'isothermal' conditions. Big reactors for slurry phase, fluid-bed, and fixed-bed synthesis have been developed [5, 7, 8].

The FT product consists of many compounds, distributed over gas phase, liquid organic phase, liquid aqueous phase, and an organic material solid at room temperature. The complexity of the product composition raises problems in analysis. Figure 14.3 shows the gas chromatogram of the volatile fraction (at 200 °C) of a product from FT synthesis on a cobalt catalyst [10, 11]. In Figure 14.4 the fractions C_6 and C_7 are presented as original and as hydrogenated in a GC pre-column, showing only paraffins – a major n-paraffin peak and small peaks of the methyl-branched isomers at each carbon number. This chromatogram presents information about the branching reaction during growth of aliphatic chains. FT synthesis for diesel and jet fuel on cobalt catalysts is combined with

Catalysis: From Principles to Applications, First Edition. Edited by Matthias Beller, Albert Renken, and Rutger van Santen.
© 2012 Wiley-VCH Verlag GmbH & Co. KGaA. Published 2012 by Wiley-VCH Verlag GmbH & Co. KGaA.

14 Fischer–Tropsch Synthesis in a Modern Perspective

CO + H₂ ⟹ [Black box of catalysis] ⟹ **Gasoline** $CH_3-(CH_2)_x-CH_3$ $(+H_2O)$

High complexity of unknown order

- Various adsorbed intermediates not easily identified
- Complex kinetic scheme
- Kinetic constants not directly accessible:
- *Models needed*, assumptions necessary
- Steady state develops slowly (self-organization)
- The *true catalyst is built during reaction* (reconstruction)
- Difficulties with product analysis:

Figure 14.1 The miracle of Fischer–Tropsch synthesis.

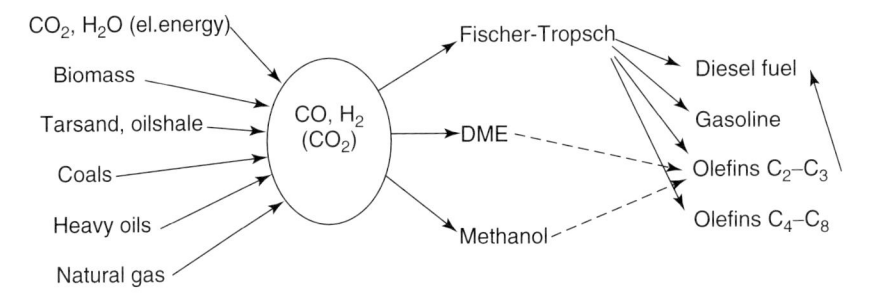

Inputs: CO_2, H_2O (el.energy), Biomass, Tarsand, oilshale, Coals, Heavy oils, Natural gas → CO, H₂ (CO_2) → Fischer-Tropsch → Diesel fuel, Gasoline, Olefins C_2-C_3, Olefins C_4-C_8; also →DME, →Methanol

Figure 14.2 Synthesis gas options for fuels and chemicals. Competing with Fischer–Tropsch synthesis, olefins (and other hydrocarbons) can also be obtained via methanol (or dimethyl ether) followed by shape-selective conversion on zeolite ZSM-5 or SAPO-34 [9].

mild hydrocracking of the heavier product fractions. Olefins C_2 to C_8 can be obtained with iron catalysts at relatively high temperature [5].

The order of product composition points in the direction of a polymerization mechanism, a C_1-monomer being added to the growing chain. Among the principles ruling selectivity, suppression of distinct elemental reactions and spatial constraints on growth sites appear important. Complicating FT catalysis, a steady state develops slowly. Self-organization of the FT regime with changing selectivity and rate and restructuring of the catalyst are pertinent [12, 13]. Iron behaves differently compared with cobalt, ruthenium, or nickel as catalysts.

Catalyst development addresses high activity and high stability at low temperature and iron carbide formation at high temperature.

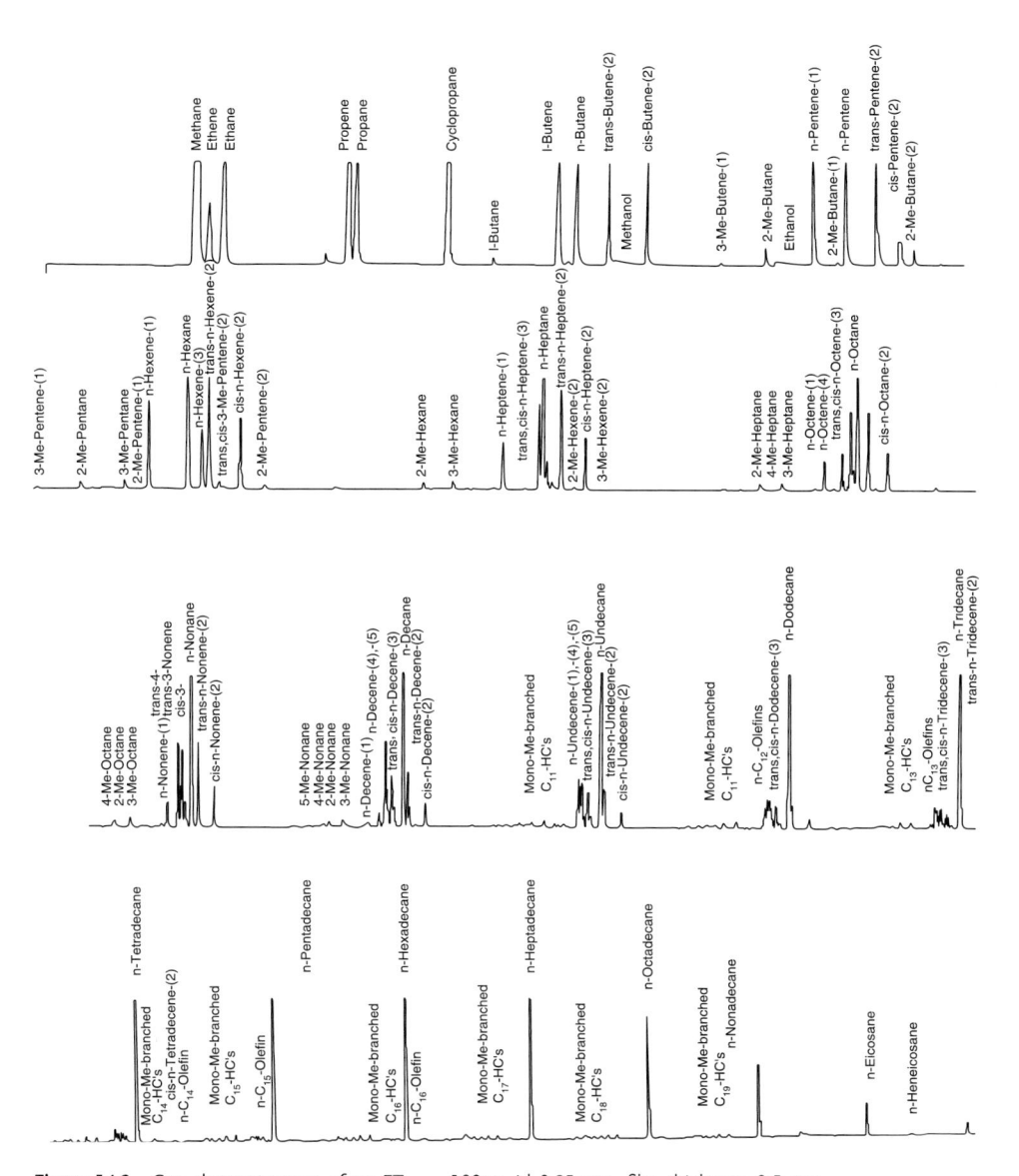

Figure 14.3 Gas chromatogram of an FT product: FT synthesis on cobalt. Ampoule sample (sampling time less than 1 s, samples taken from the volatile products flow at about 200 °C), GC-conditions: capillary 100 m, id 0.25 mm, film thickness 0.5 μm, methyl silicone cross-linked, temperature: −80 to 275 °C, carrier gas H_2, introducing gas N_2, FID, and duration of chromatogram 75 min.

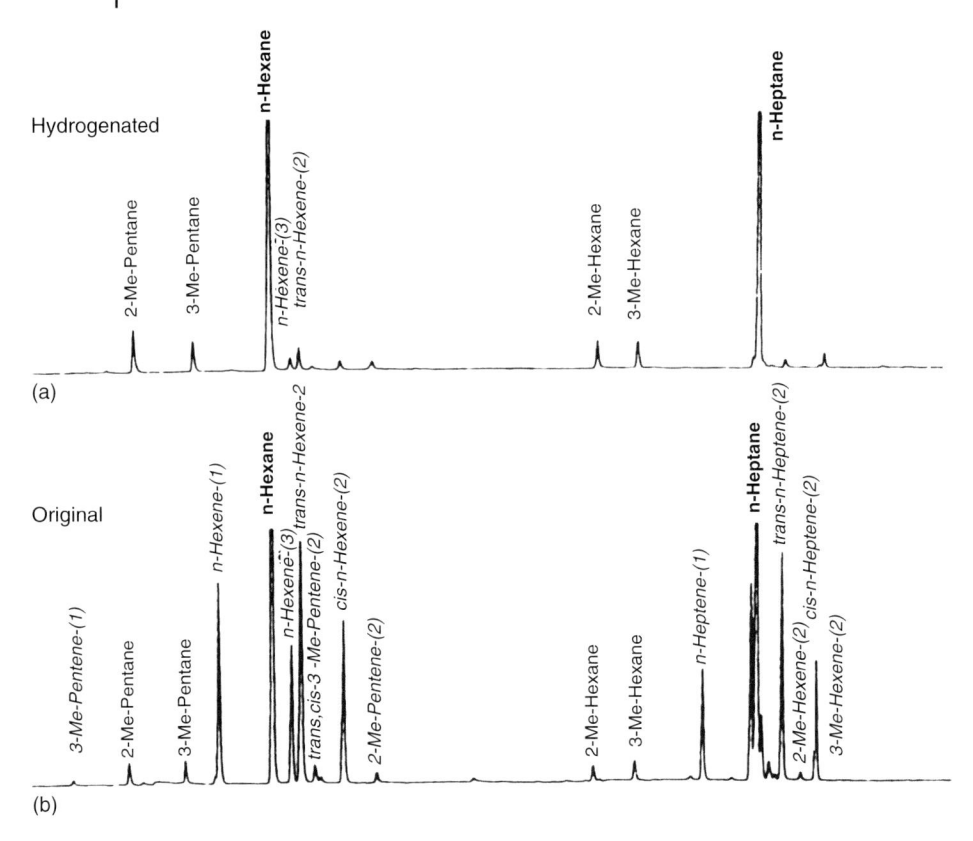

Figure 14.4 Section C_6/C_7 of FT product chromatograms, original and pre-column hydrogenated. GC-conditions as in Figure 14.3, GC-hydrogenation at 250 °C on 0.35 wt% Pt on quartz particles particle diameter (dp) = 0.2–0.25 mm [11].

14.2
Stoichiometry and Thermodynamic Aspects

14.2.1
Stoichiometry

Stoichiometry of FT synthesis is flexible, depending on catalyst properties and reaction conditions. The simplest equation is

$$CO + 2\,H_2 = (CH_2) + H_2O \quad \Delta H_R = -165\,\text{kJ (at 227 °C)} \quad (14.1)$$

(CH_2) means a CH_2-group of an aliphatic chain. The molar H_2-to-CO consumption ratio is 2 : 1. Consumption ratio refers to the molar ratio in which the gases are consumed from the gas which is fed to the reactor. In a fixed bed FT reactor the H_2/CO ratio changes over the reactor length as the degree of conversion increases if the H_2/CO-ratio of the feed gas is not the same as the H_2/CO-consumption ratio.

14.2 Stoichiometry and Thermodynamic Aspects

For paraffins (C_nH_{2n+2}) as products, the H_2/CO consumption ratio will be a little higher than that for olefins (C_nH_{2n}). For methane as the product compound, the H_2/CO consumption-ratio is as high, at 3 : 1, (Eq. (14.2)).

$$CO + 3H_2 = CH_4 + H_2O \tag{14.2}$$

With iron catalysts, the H_2/CO consumption ratio is influenced by the simultaneous water gas shift reaction (Eq. (14.3)). At low-temperature FT synthesis (~230 °C) the water gas shift equilibrium is on the right hand side of Eq. (14.3).

$$CO + H_2O = CO_2 + H_2 \quad \Delta H_R = -39.8 \text{ kJ (at 227 °C)} \tag{14.3}$$

Then the FT stoichiometry including water gas shift can be

$$2\,CO + H_2 = (CH_2) + CO_2 \quad \Delta H_R = -204 \text{ kJ (at 227 °C)} \tag{14.4}$$

corresponding to an H_2/CO consumption ratio of only 0.5.

At a high reaction temperature (about 350 °C), the reverse water gas shift reaction is favored, Eq. (14.5), where CO_2 supplies carbon for FT synthesis

$$CO_2 + 3H_2 = (CH_2) + 2\,H_2O \tag{14.5}$$

In practice, feed gas composition must be adjusted for the actual FT process, and the FT process to the available gas composition. The composition of as-prepared syngas depends on the properties of the raw material and the gasification process. Entrained-flow coal gasification (operating at high temperature, above 1000 °C) produces a CO-rich syngas. Here the use of iron catalysts with simultaneous water gas shift activity could be favorable. Syngas from methane (natural gas) – being rich in hydrogen – is preferably used in low-temperature FT synthesis on cobalt with no water gas shift activity.

FT process performance depends on synthesis gas composition. For example, with cobalt as catalyst, too high a H_2/CO-ratio would favor (undesirable) methane formation. With iron as catalyst at high temperature (depending on the degree of catalyst alkalization) a low H_2/CO-ratio could cause carbon formation and respective catalyst deactivation.

14.2.1.1 Thermodynamic Aspects

Efficient heat recovery from FT reactors is important [14, 15]. Free enthalpies at assumed FT conditions (200 °C, 5 bar) and inversion temperatures (temperature at which formally the reaction changes its direction, $K_p = 1$; $\Delta G_R = 0$) are significant in FT synthesis. Formation of n-hexane according to Eq. (14.6)

$$6\,CO + 13\,H_2 \rightarrow C_6H_{14} + 6\,H_2O$$
$$\Delta G_R \,(200\,°C,\, 5\,\text{bar}) = -377 \text{ kJ/mol}; \; T_{inv} = 487\,°C \tag{14.6}$$

is thermodynamically favored up to 487 °C. Hydrogenation of n-hexene-(1), according to Eq. (14.7),

$$\text{n-hexene-1} + H_2 \rightarrow \text{n-hexane};$$
$$\Delta G_R \,(200\,°C,\, 5\,\text{bar}) = -71 \text{ kJ/mol} \tag{14.7}$$

appears well possible at FT conditions; however, olefins are the main primary products of FT synthesis. Suppression of olefin hydrogenation under FT conditions can be assumed.

Methane as the FT product has an interesting aspect (Eq. (14.8)).

$$CO + 3H_2 \rightarrow CH_4 + H_2O \tag{14.8}$$
$$\Delta G_R (200\,°C,\, 5\,\text{bar}) = -115\,\text{kJ/mol};\ T_{inv} = -725\,°C$$

The formation of methane is thermodynamically much favored compared with that of any higher hydrocarbon. It follows that there must be a definite kinetic barrier against the formation of methane in the FT regime. Otherwise no higher hydrocarbons would be obtained.

The direction of the water gas shift reaction:

$$CO + H_2O \rightarrow CO_2 + H_2 \quad \Delta H_0^R (250\,°C) = 39.6\,\text{kJ/mol} \tag{14.9}$$

can change within the range of FT conditions. At low FT temperature, it is on the side of CO_2 and H_2, but at high FT temperature, CO may be formed from CO_2 and H_2. The water gas shift reaction can be useful in FT synthesis with CO-rich syngases. With gases rich in CO_2 but lean in CO the reverse reaction can proceed. Remarkably, the water gas shift reaction is catalyzed only by iron catalysts, not with cobalt as catalyst.

14.2.1.2 Rate Equations and Operation Ranges

Rate Equations A number of rate equations for the FT synthesis have been presented, and their fit with experimental data has been discussed, but the complexity of the synthesis can merely be adequately mirrored or theoretically deduced.

Principle relations as published by Anderson [16] are

$$\text{For iron catalysts: } -r_{FT} = \frac{k_{FT}\, P_{CO}\, P_{H_2}}{P_{CO} + b\, P_{H_2O}} \tag{14.10}$$

$$\text{For cobalt catalysts: } -r_{FT} = \frac{k_{FT}\, P_{CO}\, P^2_{H_2}}{1 + b\, P_{CO}\, P^2_{H_2}} \tag{14.11}$$

The most important practical difference is inhibition by H_2O with iron but not with cobalt. For reactor design purposes the kinetics must be determined individually for the actual catalyst.

14.2.1.3 Operating Ranges (Pichler)

Only a few metals – cobalt, iron, ruthenium, and nickel – are catalysts for FT synthesis. In an important diagram presented by Pichler, operating ranges of temperature and pressure for synthesis gas-converting catalysts are indicated (Figure 14.5) [2].

Cobalt, nickel, and ruthenium show a similarity in behavior: any increase in the reaction temperature leads to excessive methane formation. Hence, as methane is

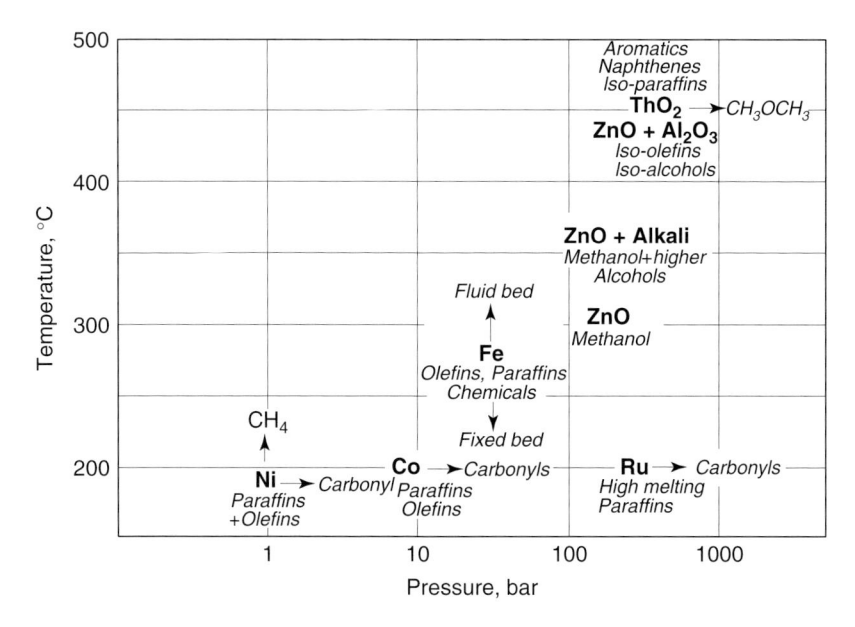

14.2 Stoichiometry and Thermodynamic Aspects | 307

Figure 14.5 Synthesis gas conversion on different catalysts as indicated by Pichler.

thermodynamically much favored (as previously noted), it has been concluded that the kinetic barrier to CH_4 formation that normally exists in the FT synthesis no longer does so at increasing temperature [13].

But there is also a pressure limitation. Too high pressure can lead to reaction of the metal with CO with formation of the metal carbonyl, because this then becomes thermodynamically feasible. Formation of nickel carbonyl may occur above 1 bar. With cobalt, reaction pressures up to, for example, 40 bar are possible [13]. At pressures higher than 100 bar, cobalt hydrocarbonyl complexes become stable, catalyzing the hydroformylation of olefins to aldehydes (alcohols).

It has been suggested that using these metals near the conditions of carbonyl formation can lead to a thermodynamically controlled surface segregation, an effect of self-organization of the FT regime [17].

With ruthenium, very high pressures (up to 1000 bar) can be applied, and high-molecular-weight paraffins are formed at low reaction temperatures on highly dispersed catalysts [18–20].

Iron as an FT catalyst behaves differently. The reaction temperature can be increased up to 350 °C without excessive methane formation. Metallic iron is converted to iron carbides. At increasing CO partial pressure, formation of carbon on the catalyst proceeds. Mainly 1-olefins are obtained as products, indicating the absence of (secondary) olefin hydrogenation and isomerization [5, 13].

Olefin hydrogenation is typically catalyzed on metallic surfaces. It may be assumed that the consumption of the metallic iron phase by formation of iron carbide is the reason for high olefin selectivity in the FT synthesis.

Similarly, the low methanation trend in FT synthesis on iron may be explained. The further reactions of synthesis gas in Figure 14.5 refer to oxidic catalysts.

14.3
Processes and Product Composition

14.3.1
Commercial FT-Synthesis

Today's major industrial operators of the FT synthesis are listed in Table 14.1 [5, 7, 8, 21–25]. Two kinds of application – low-temperature and high-temperature synthesis – may be distinguished. Low-temperature synthesis (about 230 °C) aims at long-chain hydrocarbons. Automotive fuel (particularly clean diesel) is presently the main desired product. High-value waxes, excellent jet fuels, lubricating oils, and feedstocks for detergent manufacture can also be produced.

Low-temperature FT synthesis is today combined with mild hydrocracking (and isomerization) of the heavy part of the FT product for maximizing the yield of diesel fuel. Chain branching leads to improved low-temperature fuel behavior and improved viscosity properties. The high-temperature application (about 350 °C) with iron catalysts aims at a low-molecular-weight olefinic product. The olefins C_2–C_8 can be isolated for use as chemicals or oligomerized to maximize the overall gasoline and diesel yield.

Table 14.1 Commercial FT synthesis.

Location	Sasolburg	Secunda	Mossel Bay	Bintulu		Las Raffan
Country		South Africa		Malaysia		Qatar
Name	Arge	Synthol	Synthol	SMDS	Oryx	Pearl
Company	Sasol	Sasol	Petro SA	Shell	Sasol/QP	Shell
Start of Operation	1956	1959	1992	1993	2006	2010
Capacity						
t/a	0.5×10^6	8.8×10^6	2.0×10^6	0.8×10^6	1.9×10^6	4×10^6
Bpd	8000	160 000	36.000	14 7000	34 000	70 000
Feed	Coal	Coal	Natural gas	Natural gas	Natural gas	Natural gas
Reactor	Fixed bed	Fluid bed	Entrained catalyst	Fixed bed	Slurry	Fixed bed
Catalyst	Iron	Iron	Iron	Cobalt	Cobalt	Cobalt
Temperature (°C)	220–240	330–350	330–350	230	230	230
Pressure (bar)	25	25	25	35	25	35
Main products	Wax, diesel	Gasoline, olefins	Gasoline, diesel	Diesel	Diesel	Diesel
References	[7]	[5]	[25]	[8, 21–23]	[7, 24]	[8]

14.3.1.1 Low-Temperature Synthesis

FT synthesis at 'low temperature' needs effective reactor cooling in order to approach isothermal operation. Multi-tubular fixed-bed reactors and slurry-phase bubble column reactors are used commercially.

The multi-tubular reactor with iron as catalyst (Arge process, Lurgi and Ruhrchemie company) was developed on the basis of the "iron middle pressure synthesis" invention by Fischer and Pichler and has been in use by Sasol at Sasolburg since 1956 [7]. The catalyst tubes of about 5 cm diameter and 12 m length are surrounded by boiling water, the heat of reaction being efficiently recovered through evaporation. At Sasol, hard paraffinic materials are the most valuable products [15]. Each reactor contains about 2000 parallel tubes. A tail-gas-to-fresh-gas recycle ratio of about 2 is used.

Cobalt as catalyst in fixed-bed reactors was used by Ruhrchemie at normal pressure, according to the Fischer and Tropsch invention of 'normal pressure FT synthesis' [1]. The catalyst was held in a box with the cooling medium flowing in tubes through the catalyst bed. Today, cobalt catalysts are used in a pressure range of about 20–40 bar in multi-tubular reactors in the Shell middle distillate synthesis, this referring to the Fischer–Pichler cobalt middle pressure synthesis [2]. High activity of the cobalt catalyst restricts the catalyst tube diameter even more than with iron as catalyst. A plant with a capacity of about 800 000 t/a capacity is operated by Shell at Bintulu, Malaysia. A Shell plant with a capacity of about 4 million tons per year (named '*Pearl*') is being brought on stream at Las Raffan, Qatar. In order to enhance the overall diesel yield, mild hydrocracking of the heavy part of the FT product is applied to attain a diesel selectivity of about 60 C% (of the liquid FT product).

Hydrocracking and hydroisomerization of the heavy FT product fraction differ from the petroleum application. The feed (after simple hydrogenation of olefins and alcohols) consists of paraffin molecules only, without any sulfur compounds and aromatics, allowing 'ideal bifunctional hydrocracking' [26, 27]. Ideal hydrocracking is attained with a highly active hydrogenation metal (like platinum) on an acidic support (alumina–silica, H-zeolite). Then 'no' secondary cracking (of the olefinic product molecules from primary cracking, these being immediately hydrogenated) will be noticed. Typical performance of *n*-paraffin ideal hydrocracking and hydroisomerization is explained in Table 14.2 for *n*-dodecane conversion on platinum on an acidic zeolite-Y at four temperatures in the range 265–300 °C. At 275 °C, isomerization conversion is 48% with only 17% hydrocracking.

Isomerization is faster than cracking. Branched paraffins crack preferentially (due to the branched carbenium ions involved). At higher temperatures, the cracking yield is higher (e.g., 56%) than that of isomerization (e.g., 35%), but still only primary cracking (200 mol of crack products per 100 mol of C_{12} cracked) occurs. Correspondingly, in hydrocracking of FT paraffins the distribution on carbon number fractions will be gradually shifted to lower carbon number products by just cracking the biggest molecules preferentially (see also results at 300 °C in Table 14.2).

Table 14.2 'Ideal' bifunctional hydrocracking of paraffins, n-dodecane model feed (0.5 wt% Pt on zeolite CaY, 40 bar, $nH_2/nHC = 20$).

Temperature (°C)	265	275	285	300
Conversion (%)				
Cracking	5	17	56	99.5
Isomerization	34	48	35	0.47
	Moles of products per 100 mol dodecane cracked			
Methanes	–	–	–	0.1
Ethanes	–	–	–	0.1
Propanes	6.7	6.7	7.0	9.0
Butanes	29.7	30.4	31.8	38.9
Pentanes	42.3	41.9	41.9	46.3
Hexanes	43.5	43.5	42.9	44.2
Heptanes	42.3	41.2	41.0	40.3 (−1.4%)
Octanes	29.5	30.7	30.6	25.9 (−15%)
Nonanes	6.3	6.0	5.9	3.4 (−42%)
Decanes	–	–	–	–
Undecanes	–	–	–	–
Sum	200	200	201	208

14.3.1.2 Slurry Reactors

In a bubble column reactor, small catalyst particles are suspended in paraffinic oil (from FT synthesis) in a column to which the synthesis gas is introduced from below. The advantages of the FT slurry bubble column reactor are simpler heat removal and reactor up-scaling. In the 'ORYX' plant at Qatar, two parallel slurry reactors, with a capacity of about 1 million tons per year each, are in use (see Table 14.1). The original work on FT synthesis with iron catalysts in slurry bubble column reactors was performed by Koelbel and co-workers [28].

14.3.1.3 High-Temperature Fischer–Tropsch Synthesis

FT synthesis on iron at 'high temperature' (330–350 °C) is associated with catalyst phase changes and carbon formation. Small hydrocarbon molecules are formed at the relatively high reaction temperature. The mixture of FT products is all volatile at reaction conditions. Fluidized-bed reactors have been developed for high-temperature FT synthesis. This technology allows scaling up to big reactor units. Reactors with a capacity of 20 000 bpd are operated at Secunda, having replaced the former circulating catalyst reactors [5], which are still in use by Petro SA at Mossel Bay (SA) [25]. There the process is optimized for automotive fuel production. Small olefins are converted by oligomerization on an HZSM-5 zeolite catalyst to liquid hydrocarbons, mainly in the diesel boiling range. High-temperature FT synthesis is practiced successfully by Sasol at Secunda for producing olefins in

Table 14.3 Potential of olefin production through FT synthesis (assumed plant capacity 4×10^6 t/a, iron catalyst, fluid bed).

Product fraction[a]		Olefin-1 content[a] (C%)		Potential of olefin-1 production[a]
Carbon number	Amount (C%)	In fraction	In product	t/a
C_2	7	30	2.1	85 000
C_3	13	80	10.4	415 000
C_4	13	75	9.8	390 000
C_5	12	65	7.8	310 000
C_6	10	55	5.5	220 000
C_7	8	50	4.0	160 000
C_8	6	45	2.7	120 000
	69	–	42	1.7×10^6

[a] Estimated values from various sources.

addition to fuels. This process may well become more widespread in the chemical industry [6].

The potential for olefin-(1) production through FT synthesis is tentatively visualized in Table 14.3, assuming a plant capacity of 4 million tons per year. The earlier development of an FT fluidized-bed process in the US (Hydrocoal plant at Brownsville) has been well documented in the literature [29].

14.3.1.4 Synthesis Gas

Industrial FT synthesis today relies on coal (South Africa) and natural gas (Malaysia, Qatar, South Africa). At Secunda (SA), coal of high ash content (e.g., 25 wt%) is gasified in numerous parallel Lurgi fixed-bed units. The process is about thermally neutral, using both oxygen and water in exothermic and endothermic reactions. Coal-derived synthesis gas needs exhaustive purification, particularly from sulfur compounds such as H_2S and COS.

Most of the CO_2 is also removed. The purified gas contains about 15 vol% of methane. Together with the FT methane it can be converted to synthesis feed gas by steam reforming or be used as fuel gas (e.g., town gas at Johannesburg).

FT synthesis gas from natural gas is obtained by catalytic auto-thermal reforming [24] or partial oxidation [21, 23].

14.4
Catalysts, General

As discussed above, the metals cobalt, iron, ruthenium, and nickel are active in FT synthesis, each in its range of temperature and pressure (Figure 14.5).

Ruthenium works at the lowest temperature (150 °C and below) without any promoter or support. At high pressures – up to 1000 bar – high-molecular-weight paraffins similar to polyethylene are formed [18–20]. Ruthenium appears suited for basic FT studies, but high costs and limited reserves are restrictive for industrial application.

Nickel catalysts for FT synthesis at about 190 °C have been prepared by precipitation in a similar procedure to that used for cobalt catalysts, using thoria or zirconia as promoters and kieselguhr or aerosil as support [17, 30]. However, methane formation becomes excessive at increasing temperature, and at pressures higher than 1 bar nickel may react with CO to form the volatile and very poisonous nickel carbonyl [2].

14.4.1
Cobalt

Cobalt as FT catalyst is used industrially today for producing mainly diesel fuel and jet fuel at about 230 °C in fixed-bed or slurry reactors.

The original cobalt FT catalyst – developed by Fischer and co-workers – had the weight composition 100 Co 18 ThO_2 100 Kieselguhr, and was obtained by fast co-precipitation from a hot solution of the metal nitrates with the slurried kieselguhr particles suspended in a potassium solution. This procedure leads to a very 'homogeneous' distribution of Co, Th, and Kieselguhr. The Kieselguhr particles provide a very open porosity and the non-reducible thoria stabilizes the metal dispersion.

It was observed early that performance of a cobalt catalyst depended on a 'formation' step ('Formierung') at the beginning of a run with the before reduced catalyst. Slowly increasing the reaction temperature up to the steady-state conditions was favorable (H. Pichler, personal communication). In modern terms, this is known as catalyst restructuring. With surface-tunneling electron microscopy, roughening of the cobalt crystal planes under FT conditions has been observed [31]. This surface segregation has been interpreted as creating additional sites which simultaneously disproportionate into higher and lower coordination sites of different catalytic nature, being active for different elementary reactions of the complex mechanism [13, 17].

Small amounts of Pt or Ru are added to ease catalyst reduction [32].

14.4.2
Iron

Iron as a catalyst in FT synthesis can be linked to 'iron middle pressure synthesis' invention by Fischer and Pichler [2]. Iron catalysts differ considerably from the other FT catalyst metals (see Figure 14.5). Iron catalysts must be promoted with alkali-commonly potassium. Iron catalysts are also active for the water gas shift reaction during FT synthesis. In general, there is a tendency of iron to react with CO to form iron carbide and elemental carbon, depending on alkalization [33].

For synthesis of a waxy product at low temperature, very active iron catalysts are prepared by precipitation from nitrate solutions. The catalyst particles are then obtained by extruding.

For synthesis of a low-molecular-weight olefinic product at high temperature in a fluidized-bed reactor, the catalyst is prepared by fusion of iron oxide (and promoters) in an arc furnace. Remarkably, the actual working iron FT catalyst is not the iron metal but iron carbide, which is formed by reaction of iron with CO [4, 5, 7].

14.5 Reaction Fundamentals

14.5.1 Ideal Polymerization Model

FT synthesis exhibits the feature of linear polymerization, the probability of chain prolongation being independent of the length of the chain. The relevant ideal polymerization model is presented in Figure 14.6.

In the kinetic scheme (Figure 14.6a) there is only one kind of intermediate Sp_N. The product distribution can be represented in diagrams of moles of products over carbon number (Figure 14.6b), log moles over carbon number (Figure 14.6c) and growth probability pg over carbon number (Figure 14.6d). The logarithmic diagram, called the *Anderson–Schulz–Flory curve* in the literature, is commonly

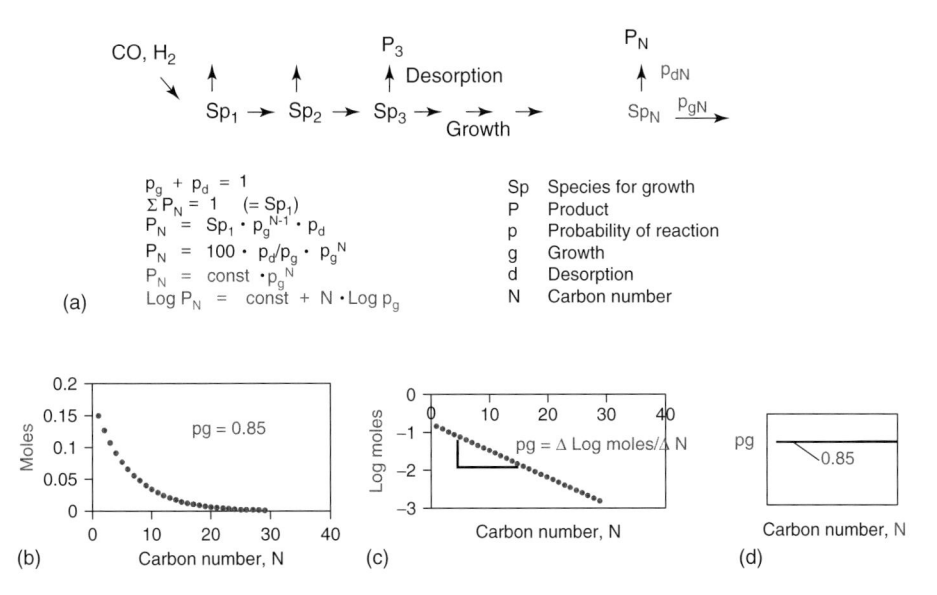

$p_g + p_d = 1$
$\Sigma P_N = 1 \; (= Sp_1)$
$P_N = Sp_1 \cdot p_g^{N-1} \cdot p_d$
$P_N = 100 \cdot p_d/p_g \cdot p_g^N$
$P_N = \text{const} \cdot p_g^N$
$\text{Log } P_N = \text{const} + N \cdot \text{Log } p_g$

Sp Species for growth
P   Product
p   Probability of reaction
g   Growth
d   Desorption
N   Carbon number

Figure 14.6 Basic kinetic scheme of FT hydrocarbon formation.

Product distribution being defined by only one number, the value of growth-probability p_g

used. But the diagram pg versus N is best suited for discussing the real mechanism as it directly shows deviations from ideal polymerization.

14.5.1.1 Chain Growth

Probability of chain growth as a function of carbon number has been calculated from observed FT product distributions. Typical results for iron and cobalt as catalysts are shown in Figure 14.7 [4, 17].

Looking at the results for FT on cobalt at steady state (full circles, Figure 14.7b), several deviations are noticed: a low value at $N = 1$ indicates much higher methane selectivity than expected from the ideal model. This 'extra methane' will be discussed below.

A relatively high value at C_2 is due to high ethylene reactivity.

From C_3 to higher values of carbon number, increasing values of pg are seen.

As shown by reactions of olefins added in steady-state experiments with a gradientless reactor [34], the increase is due to increasing reactor residence times because of increasing solubility in the reactor liquid phase, supporting secondary reactions such as olefin re-adsorption on growth sites.

The growth probability in general is surprisingly low in the early time of the experiment with cobalt. This result is linked to initial cobalt catalyst restructuring.

With iron the results are different (Figure 14.7a):

- Methane is not formed with higher selectivity than expected from the ideal model.
- Deviations from a horizontal line (increasing values in the range C_2-C_{10}) can be related to re-adsorption of olefins on growth sites. This occurs most at steady state when liquid product has accumulated on the catalyst and the reactor residence time for volatile product molecules has increased (full circles).
- There is less selectivity change during the formation time with iron, compared with using cobalt as catalyst, indicating re-assembling of catalyst to be of a different nature.

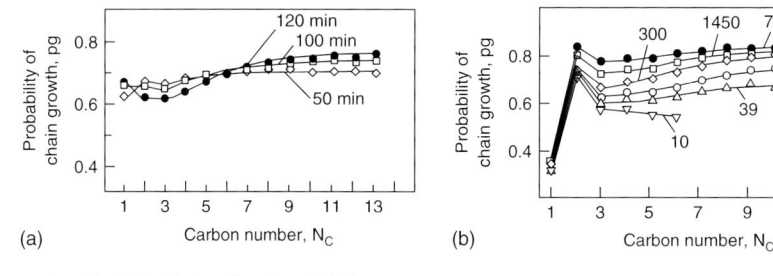

(a) Iron (Fe-Al-Cu-K); 250 °C, 10 bar, H_2/CO = 2.3:1

Schulz, Schaub, Claeys, Riedel

(b) Cobalt (Co-Zr-Pt-Aerosil); 190 °C, 5 bar, H_2/CO = 2:1

Thesis Zh. Nie

Figure 14.7 Chain growth probability in dependence on chain length at different reaction times; (a) iron catalyst and (b) cobalt catalyst; fixed bed reactor.

It appears that with iron in the present case, the nature of the growth sites does not change significantly during the catalyst formation time.

14.5.1.2 Alternative Reactions on Growth Site

The ideal FT polymerization model distinguishes only between growth and desorption of a chain Sp_N. Referring to real FT selectivity, further reactions (of Sp_N) must be acknowledged, as shown in Figure 14.8: (i) chain branching (reaction 2 to $Sp_{N+1\,br}$), (ii) desorption for two kinds of products – olefins (reaction 3 to $P_{N,olefin-1}$) and paraffins (reaction 4 to $P_{N,paraffin}$), (iii) reversibility of chain growth (reaction 1), and (iv) reversibility of olefin desorption (reaction 3). Additionally, secondary olefin reactions of hydrogenation (reaction 6) and double-bond shift (reaction 7) – on sites not capable for growth via species Sp^*_{N+1} – are taken into account.

14.5.1.3 Branching

Mono-methyl branched chains are formed in addition to straight chains, as can be seen in the C_6/C_7-chromatogram range of a hydrogenated FT product Figure 14.4 (reaction 2 in the kinetic scheme Figure 14.8). Probability of branching as a function of carbon number (of growing species) is shown in Figure 14.9 (calculated from the detailed product composition data).

Branching probability is generally low (1–10%) and is strongly carbon number dependent. At steady state (full circles Figure 14.9) branching probability declines strongly from $N_C = 3$ toward higher values of N_C. This is related to increasing spatial constraints on the site as the length of the chain increases. (A two-molecular branching reaction, for example, of a methyl species with an alkyl-carbene species could be visualized as spatially demanding.)

The first value of branching probability at $N_C = 3$ is specifically low. This 'irregular' behavior was explained by an argument of spatial constraints by Wojciechowski, being due to particular spatial demands on desorption after a branching step [35].

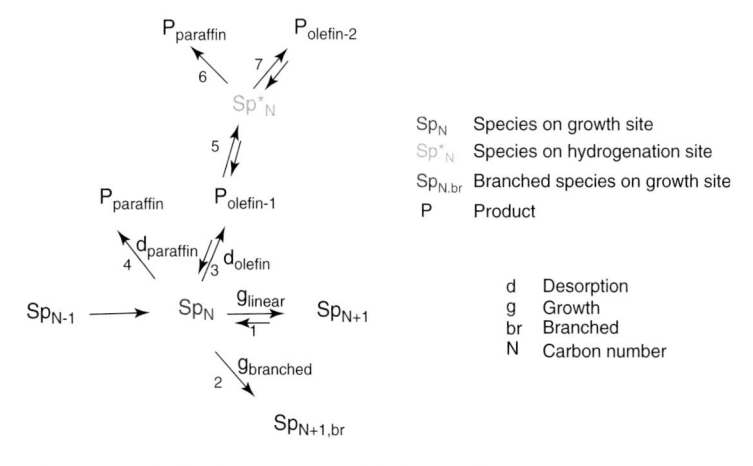

Figure 14.8 Ideal polymerization model of FT synthesis.

Sp_N Species on growth site
Sp^*_N Species on hydrogenation site
$Sp_{N,br}$ Branched species on growth site
P   Product

d   Desorption
g   Growth
br   Branched
N   Carbon number

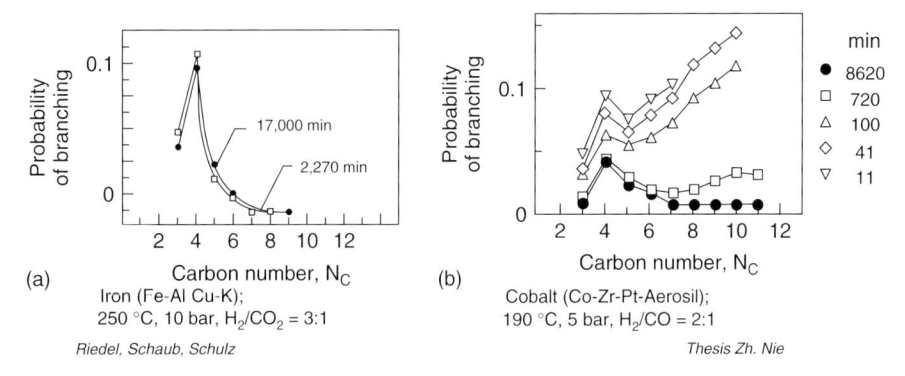

Figure 14.9 Chain branching with cobalt and iron as catalyst.

With iron, merely a change of branching probability during the catalyst formation time has been noticed (Figure 14.9a), indicating no change in the nature of the growth sites.

With cobalt (Figure 14.9b) initially (at 41 and 100 min time on stream) the branching probability has been found much higher than at steady state, and the pattern of the curves is different initially, indicating a different further mechanism. Assuming less stringent spatial constraints on the sites initially – before segregation of the cobalt surface – re-adsorption of olefins also with their internal C-atoms on the growth sites could be possible, and this can be expected to be favored by the size of the species.

14.5.1.4 Alcohols in FT-Synthesis

In the kinetic scheme of reactions on growth sites (Figure 14.8), the minor formation of alcohols has not been included; however, it is essential for the system, namely in relation to the reaction of olefin hydroformylation, commonly performing as homogeneous catalysis (e. g. wth cobalt hydrocarbonyl complexes as catalyst). On adding 1 vol% propene to the feed gas during FT synthesis on cobalt (178 °C, 30 bar, $H_2/CO = 2$, CO-conversion = 40%), the selectivity for n-butanol increased [36]. This is attributed to CO insertion, assuming the species desorbs as aldehyde – as further growth is not possible. The aldehyde is easily hydrogenated to alcohol. The kinetic scheme is shown in Figure 14.10a [13]. Regarding the increasing contents of alcohols in carbon number fractions at increasing reaction pressure (Figure 14.10b) [13, 36], this is explained to reflect an increasing probability of CO-insertion (as competing with CH_2-insertion) relating to a shift toward hydroformylation. This is linked to cobalt surface segregation, increasing the specific number of growth sites [17] and changing their nature in the direction of that of the metal atom in Co-carbonyl complexes in a thermodynamically controlled manner. The specifically high ethanol selectivity would relate to the particularly high reactivity of the CH_3 species with CO.

14.5.1.5 Desorption (Olefins/Paraffins)

Product desorption from the FT growth sites mainly concerns olefins, but also paraffins to some extent (kinetic scheme in Figure 14.11a). Evidently, the olefins of FT chain growth possess terminal double bonds. Desorption of olefins is

14.5 Reaction Fundamentals

(a)

$$CH_3-Co \xrightarrow{+CO} CH_3-CO-Co \xrightarrow{+H} CH_3-CHO \xrightarrow{+H_2} CH_3-CH_2-OH$$

$$CH_3-Co \xrightarrow{+CH_2} CH_3-CH_2-Co$$

$$\xrightarrow{-H} CH_2=CH_2$$
$$\xrightarrow{+H} CH_3-CH_3$$
$$\xrightarrow{+CH_2}$$

(b)

Moles in NC-fractions, %

33 bar, 17 bar, 5 bar, 1.2 bar

Carbon number, N_C

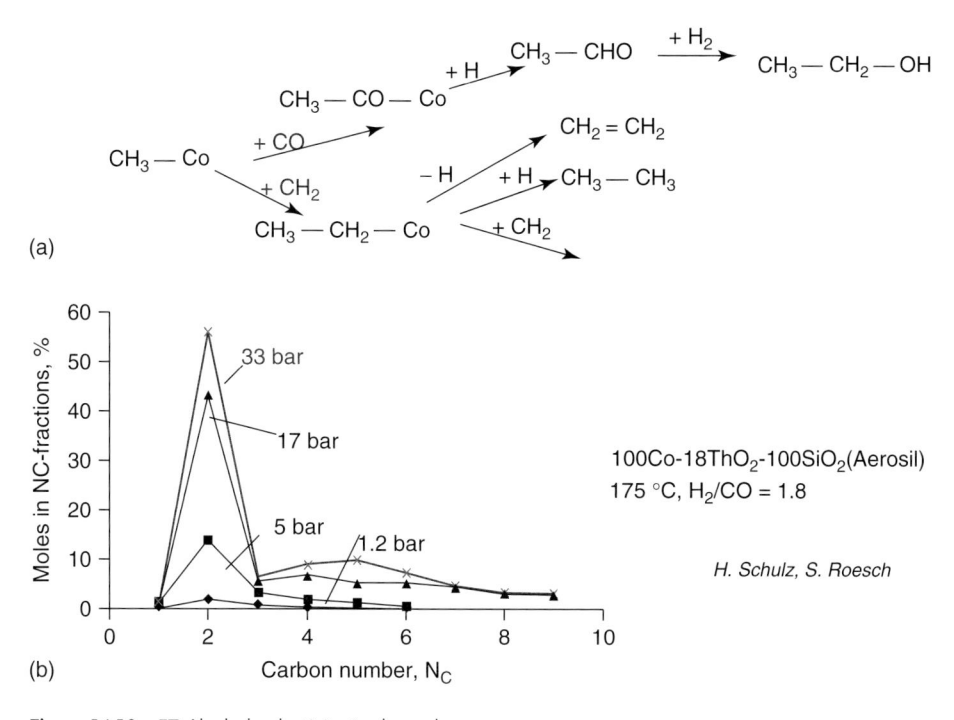

100Co-18ThO$_2$-100SiO$_2$(Aerosil)
175 °C, H$_2$/CO = 1.8

H. Schulz, S. Roesch

Figure 14.10 FT Alcohol selectivity in dependence on pressure: cobalt catalyst (b), and kinetic scheme (a).

Olefin- and paraffin formation on FT-sites

(a)

$$R-CH_2-CH_2- \xrightarrow{-H} R-CH=CH_2 \quad 70–80\%$$
$$\xrightarrow{+H} R-CH_2-CH_3 \quad 20–30\%$$

Olefin-reactions on "Non-FT-sites"

(b)

$$R-CH_2-CH=CH_2 \xrightleftharpoons{+H} R-CH_2-CH_2-CH_2- \xrightarrow{+H} R-CH_2-CH_2-CH_3$$

$$R-CH=CH-CH_3 \xrightleftharpoons{+H}{-H} R-CH_2-CH-CH_3 \xrightarrow{+H}$$

Figure 14.11 Kinetic schemes of primary and secondary olefin and paraffin formation ((a) primary reactions and (b) secondary reactions).

reversible to some extent, as shown by added ^{14}C-labeled olefins [37]. It is evident from experiments with added olefins that olefin double-bond shift and olefin hydrogenation proceed on sites different from sites for growth (kinetic scheme in Figure 14.11b and species Sp^*_N in Figure 14.8).

Typical FT product compositions with reference to olefins and paraffins are represented in Figures 14.12 and 14.13 [4, 17]. With iron as catalyst, the olefin content of 70–80 C% is fairly independent of carbon number, and (>95%) olefins with a terminal double bond are obtained very exclusively. This characterizes product formation from linear alkyl intermediates. H-elimination in the beta-position (about 75%) and H-addition (about 25%) are both possible.

Evidently, alkyl species on the FT growth sites are terminally bonded (otherwise beta-olefins would be primarily obtained also). These characteristics are considered to be caused by spatial constraints on the FT sites.

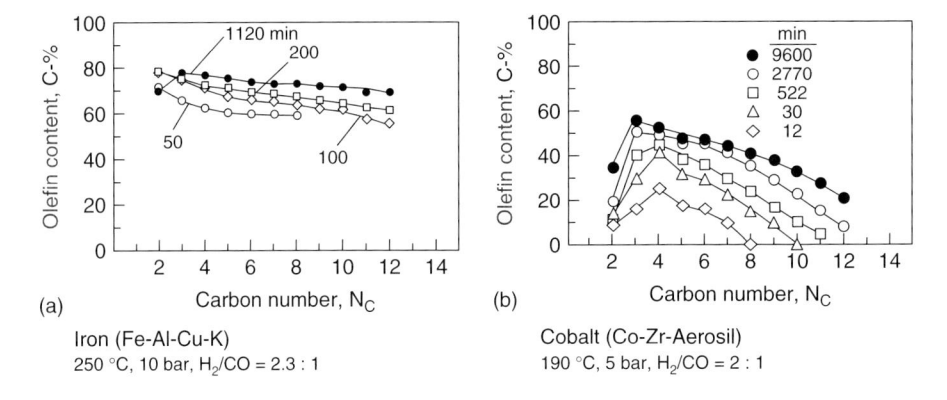

(a) Iron (Fe-Al-Cu-K)
250 °C, 10 bar, $H_2/CO = 2.3 : 1$

Riedel, Schaub, Schulz

(b) Cobalt (Co-Zr-Aerosil)
190 °C, 5 bar, $H_2/CO = 2 : 1$

Thesis Zh. Nie

Figure 14.12 Olefin content in carbon number fractions; (a) iron and (b) cobalt catalyst.

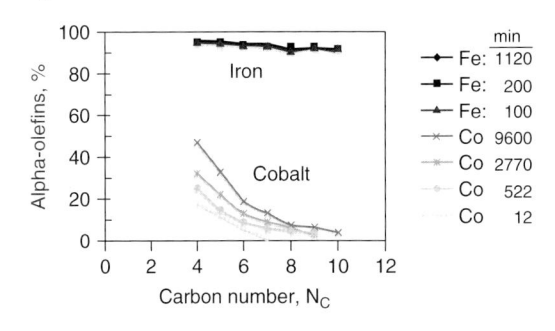

Iron (Fe-Al-Cu-K); 250 °C, 10 bar, $H_2/CO = 2.3:1$
Cobalt (Co-Zr-Aerosil); 190 °C, 5 bar, $H_2/CO = 2:1$

Figure 14.13 Alpha olefins among n-olefins in FT products, depending on carbon number, reaction time, and catalyst (iron or cobalt).

The pattern of olefin content over carbon number with the cobalt catalyst is indicative of secondary olefin reactions on non-FT hydrogenation sites (Figure 14.12b). The extent of hydrogenation and the pattern of the curves initially change with time, indicating increasing suppression of secondary hydrogenation. Seeing catalyst restructuring as surface segregation, this relates to the disappearance of cobalt atoms on plane sites (capable of olefin hydrogenation and associated double bond shift but not capable of chain growth).

The pattern of the curves in Figure 14.12b initially changes with time because of double bond shift accompanying hydrogenation. The curve's maximum is at C_3 at steady state, but at C_4 initially (12 and 30 min on stream). This is explained by the changing olefin composition due to isomerization in the direction of olefins with internal double bonds, which are less reactive for hydrogenation.

Generally, the content of olefins declines with carbon number from C_4 onwards because the higher olefins react preferentially. Initially the olefin content in general increases with time because of catalyst restructuring. However, C_3 is a special case. There is only one olefin isomer possible, respectively, two identical ones with terminal double bonds. Then initially the C_3-olefin hydrogenation is favored (as compared with C_4-olefins) because of lack of olefins with internal double bonds. The maximum in the curves in Figure 14.12b shifts from C_4 to C_3.

14.5.1.6 Catalyst Formation *in situ*

Self-Organization of the FT Regime During the pioneering work of Fischer and co-workers it was realized that the steady state of synthesis only developed under reaction conditions. Catalyst 'Formierung' (formation) was recognized, but no deeper insight was possible. As for iron, it was connected with the formation of iron carbides. But it was imagined that some remaining metallic iron would be the active phase in synthesis. Regarding the catalyst 'formation' phenomenon with cobalt, which did not react to form a carbide but also developed activity only with time, 'cobalt surface complexes' were postulated by Pichler. Today, surface restructuring of catalysts has become visible on an atomic scale, and the idea by Pichler is thus supported in principal.

Iron Catalyst Formation Time-resolved measurements of activity and selectivity together with time-resolved analysis of phases of the catalyst composition have been performed with a precipitated (alkalized) iron catalyst, using a syngas consisting of H_2 and CO_2 ($H_2/CO_2 = 3/1$) [4].

Looking at the results from Mössbauer spectroscopy (Figure 14.14a), we see the consumption of the alpha iron phase and the formation of an iron carbide phase. Also, formation of a tentatively identified (amorphous) 'FeOx' phase is noticed. Looking at the X-ray spectra in Figure 14.14b, it is noticed that the metallic iron phase disappears completely.

The corresponding yield of products as a function of time is shown in Figure 14.15. Activity for FT synthesis only begins in episode IV, and attains

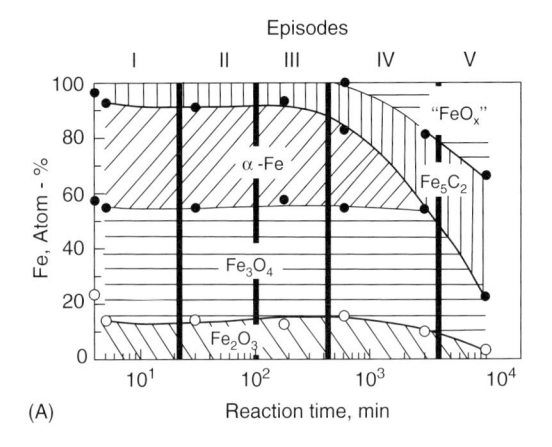

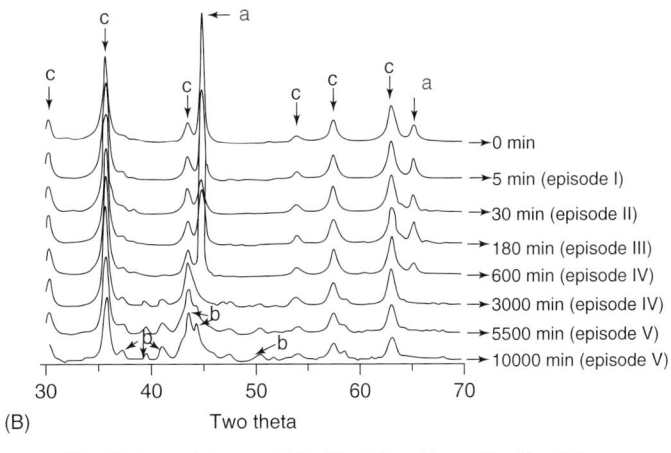

Figure 14.14 'True iron catalyst formation' in FT synthesis. (A) Phase composition by Moessbauer spectroscopy and (B) X-ray spectra for iron phases.

steady state after about four days in episode V, when all metallic iron has disappeared. The important conclusions from these results – together with the data in Figures 14.7, 14.9, 14.12, and 14.13 are that no metallic iron was involved in FT synthesis, iron carbide being the active phase. Under these conditions, no olefin hydrogenation and isomerization occurs – the only olefins found in the product are alpha olefins. Otherwise it is suspected, as there are no secondary olefin reactions when no metallic iron is present, that these reactions would need a metallic iron phase to proceed. With the other FT catalysts (Co, Ni, and Ru), the metal phase is generally dominant and olefin secondary reactions can merely be suppressed completely.

The activity for the water gas shift reaction, a common reaction on alkalized iron oxides, develops earlier than FT activity (Figure 14.15). Water gas shift activity includes fast reversible hydrogen activation. It is speculated that this activated hydrogen can be supplied to the iron carbide phase to assist FT synthesis on the carbide phase.

The kinetic model of FT synthesis in Figure 14.8 then simplifies, no second kind of active site and attached species (SpN*) being needed, and the only secondary olefin reaction being re-adsorption on growth sites with their terminal C-atom. This could proceed more extensively, as alpha olefins are not consumed by hydrogenation and double bond shift.

Cobalt Catalyst Formation Former work on the FT mechanism suffered from lack of quantitative data about detailed product composition, this being more problematic as composition and activity changes in course of self-organization. Efficient methods of gas chromatography and special sampling for time resolution have provided time-resolved composition data of high accuracy as a basis for mechanistic conclusions [13, 38, 39].

Self-organization of the FT regime with cobalt as catalyst has already been addressed above with regard to the range of conditions for cobalt operation. The phenomenon of catalyst 'Formierung' and the selectivity change during the initial time of a synthesis run, together with cobalt surface roughening visualized as surface segregation – all have to be recognized.

Surface segregation is interpreted as surface-plane-sites disproportionation and cosequent multiplication of exposed sites. Disproportionation for higher and lower

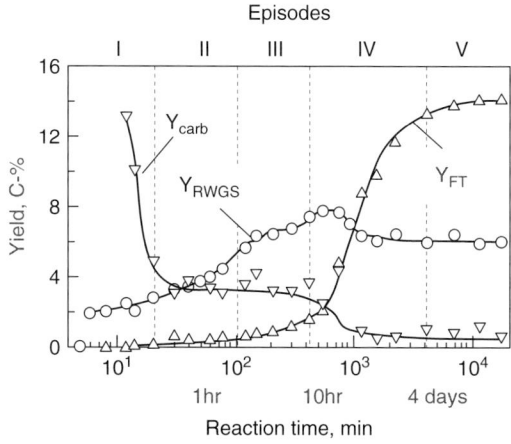

Yields:
Y_{FT} = Hydrocarbons, Y_{RWGS} = Reverse water gas shift,
Y_{carb} = Carbide

Iron (Fe-Al-Cu-K); 250 °C, 10 bar, H_2/CO_2 = 3:1

Figure 14.15 'Formation' of the FT regime with iron as catalyst.

surface-atom coordination is explained as creating sites of different nature, active for different elemental steps in the FT reaction mechanism. In Figures 14.16 and 14.17 the observed increase in initial rate is presented together with the thermodynamic view of segregation [17], as pictured qualitatively in Figure 14.16 with a free enthalpy profile. There will be a need for additional surface energy with decreasing particle diameter and even more energy to form the state of segregated surface with more exposed Co-atoms.

At the FT reaction conditions the energy for higher dispersion and segregation is compensated by the energy of adsorption of additional CO molecules. This process of self-organization will attain an equilibrium state at steady state of FT synthesis, meaning also that the surface structure is dynamic, depending on the conditions of reaction (CO partial pressure, temperature).

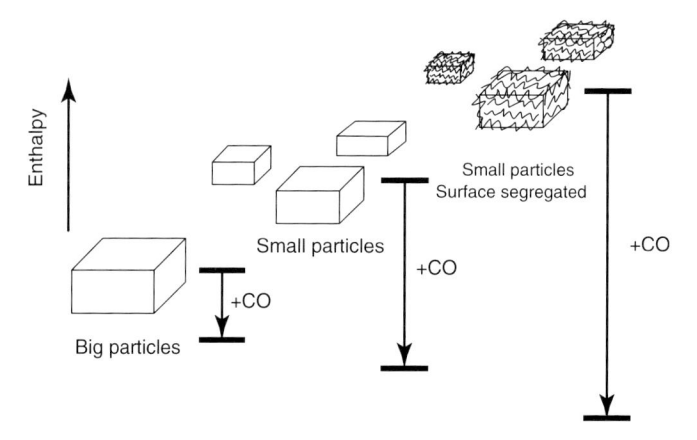

Figure 14.16 Thermodynamic view of 'true cobalt catalyst formation' through sites disproportionation.

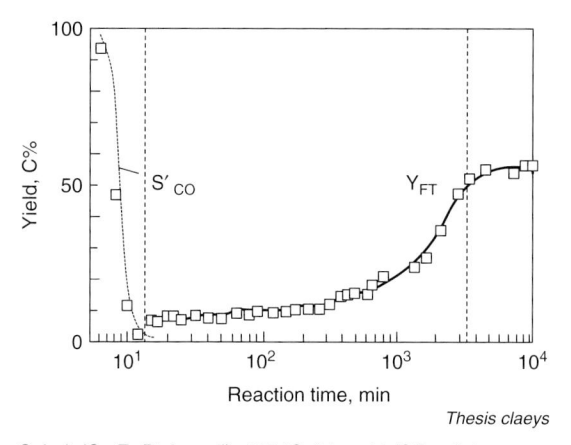

Cobalt (Co-Zr-Pt-Aerosil); 190 °C, 5 bar, $H_2/CO = 2:1$

Figure 14.17 FT reaction rate during 'true cobalt catalyst formation'.

14.6
Concluding Remarks

Recent advances in FT synthesis concern the development of large FT reactor units (fixed bed, fluid bed, and slurry phase) and the building of big commercial plants. Development also addresses the use of bio-mass for FT synthesis.

Our understanding of FT catalysis is showing great progress. The complexity of the reactions involved is being revealed down to the level of basic reactions and catalytic sites. The FT regime is observed to be developed in a process of self-organization. The considerable complexity of the composition of the product is no longer a hindrance to research, but is rather a source of information about fundamental reactions and the nature of the catalytic sites. Self-organization is brought about by constraints, including spatial demands, which govern the network of reactions. There is much to learn about heterogeneous catalysis from FT synthesis.

References

1. Fischer, F. and Tropsch, H. (1926) Die Erdölsynthese bei gewöhnlichem Druck aus den Vergasungsproducten der Kohlen. *Brennstoff. Chem.*, **7**, 97.
2. Pichler, H. (1952) Twenty-five years of synthesis of gasoline by catalytic conversion of carbon monoxide and hydrogen in *Advances in Catalysis* (eds W.G. Frankenburg, V.I. Komarewsky, and E.K. Rideal), Academic Press, pp. 271–341.
3. Schulz, H. (1999) Short history and present trends of Fischer-Tropsch synthesis. *Appl. Catal., A: Gen.*, **186** (1–2), 3–12.
4. Schulz, H. *et al.* (1999) Transient initial kinetic regimes of Fischer-Tropsch synthesis. *Appl. Catal., A: Gen.*, **186** (1–2), 215–227.
5. Steynberg, A.P. *et al.* (1999) High temperature Fischer-Tropsch synthesis in commercial practice. *Appl. Catal., A: Gen.*, **186** (1–2), 41–54.
6. Schwab, E. *et al.* (2009) Syngas to lower olefins. *DGMK-Tagungsbericht*, **2**, 25.
7. Espinoza, R.L. *et al.* (1999) Low temperature Fischer-Tropsch synthesis from a Sasol perspective. *Appl. Catal., A: Gen.*, **186** (1–2), 13–26.
8. Overtoom, R. *et al.* (2009) *Proceedings of the 1st Annual Gas Processing Symposium*, Elsevier, Amsterdam, pp. 378–386.
9. Schulz, H. (2010) 'Coking' of zeolites during methanol conversion: basic reactions of the MTO-, MTP- and MTG processes. *Catal. Today*, **154** (3–4), 183–194.
10. Claeys, M. (1997) Selektivität, Elementarschritte und kinetische Modellierung bei der Fischer-Trosch-Synthese, Dissertation University of Karlsruhe, Karlsruhe, Germany.
11. van Steen, E. (1993) Elementarschritte der Fischer-Tropsch CO-Hydrierung mit Eisen- und Kobaltkatalysatoren Dissertation University of Karlsruhe, Karlsruhe, Germany.
12. Schulz, H., Davis, B.H., and Occelli, M.L. (2007) Comparing Fischer-Tropsch synthesis on iron and cobalt catalysts: The dynamics of structure and function. *Stud. Surf. Sci. Catal. and Catalysis*, **163**, 177–199.
13. Schulz, H. (2010) Fischer-Tropsch Synthesis and Hydroformylation on Cobalt Catalysts: The Thermodynamic Control in *Advances in Fischer-Tropsch Synthesis, Catalysts, and Catalysis* (eds B.H. Davis and M.L. Occelli), CRC Press, Taylor and Francis Group, Boca Raton, FL, London, New York, pp. 165–183.
14. Anderson, R.B. (1984) *The Fischer-Tropsch Synthesis*, Academic Press, Orlando, FL.

15. Schulz, H. and Cronje, J.H. (1977) Kohle, Fischer-Tropsch Synthese in *Ullmanns Enzyklopädie der Technischen Chemie*, 4th Edition, Vol. 14, Verlag Chemie, Weinheim, pp. 329–355.
16. Anderson, R.B. (1956) Hydrocarbon synthesis, hydrogenation and cyclization, in *Catalysis*, Vol. 4, (ed. P.H. Emmet), Reinhold, New York, p. 1.
17. Schulz, H., Nie, Z., and Ousmanov, F. (2002) Construction of the Fischer-Tropsch regime with cobalt catalysts. *Catal. Today*, **71** (3–4), 351–360.
18. Pichler, H. and Buffleb, H. (1938) Über das besondere Verhalten von Rutheniumkatalysatoren bei der Synthese hochmolekularer Paraffinkohlenwasserstoffe, *Brennsoff-Chem.*, **19**, 226.
19. Pichler, H. and Buffleb, H. (1940) Über das besondere Verhalten von Rutheniumkatalysatoren bei der Synthese hochmolekularer Paraffinkohlenwasserstoffe, *Brennsoff-Chem.*, **21**, 273, 257, 285.
20. Bellstedt, F. (1970) Untersuchungen auf dem Gebiet der Polymethylen-Hochdrucksynthese an Rutheniumkatalysatoren, Dissertation University of Karlsruhe, Karlsruhe, Germany.
21. van Wechem, V.M.H. et al. (1994) *Studies in Surface Science and Catalysis*, Elsevier, pp. 43–71.
22. Geerlings, J.J.C. et al. (1999) Fischer-Tropsch technology – from active site to commercial process. *Appl. Catal., A: Gen.*, **186** (1–2), 27–40.
23. Sie, S.T. and Krishna, R. (1999) Fundamentals and selection of advanced Fischer-Tropsch reactors. *Appl. Catal., A: Gen.*, **186** (1–2), 55–70.
24. Vogela, A., Steynberg, A., and Breman, B. (2007) *Studies in Surface Science and Catalysis*, Elsevier, pp. 61–66.
25. PetroSA (2011) Petroleum, Oil and Gas Corporation of South Africa. http://www.petrosa.co.za/.
26. Schulz, H.F. and Weitkamp, J.H. (1972) Zeolite catalysts: hydrocracking and hydroisomerization of n-dodecane. *Int. Eng. Chem. Prod. Res. Dev.*, **11** (1), 46–53.
27. Pichler, H., Schulz, H., and Weitkamp, J. (1972), Ueber das Hydrocracken gesaettigter Kohlenwasserstoffe, *Erdoel und Kohle* **25**, 494.
28. Kölbel, H. and Ackermann, P. (1956) Großtechnische versuche zur Fischer-Tropsch-synthese im flüssigen medium. *Chem. Ing. Techn.*, **28** (6), 381–388.
29. Weitkamp, A.W. (1953) *Ind. Eng. Chem.*, **45** (2), 343, 350, 539, 363.
30. Fischer, F. and Meyer, K. (1931) Ueber die Verwendbarkeit von Nickelkatalysatoren fuer die Benzinsynthsse. *Brennstoff-Chem.*, **12**, 225.
31. Wilson, J. and de Groot, C. (1995) Atomic-scale restructuring in high-pressure catalysis. *J. Phys. Chem.*, **99** (20), 7860–7866.
32. Oukaci, R., Singleton, A.H., and Goodwin, J.G. Jr. (1999) Comparison of patented Co F-T catalysts using fixed-bed and slurry bubble column reactors. *Appl. Catal., A: Gen.*, **186** (1–2), 129–144.
33. Dry, M. (1981) in *Catalysis* (eds J. Anderson and M. Boudart), Springer-Verlag, Berlin, p. 159.
34. Schulz, H. and Claeys, M. (1999) Reactions of alpha-olefins of different chain length added during Fischer-Tropsch synthesis on a cobalt catalyst in a slurry reactor. *Appl. Catal., A: Gen.*, **186** (1–2), 71–90.
35. Wojciechowski, B.W. (1988) The Kinetics of the Fischer-Tropsch synthesis. *Catal. Rev.*, **30** (4), 629–702.
36. Roesch, S. (1980) Selektivitaetsbeziehungen bei der Fischer-Tropsch Kohlenoxidhydrierung an Kobaltkatalysatoren, Dissertation. University of Karlsruhe, Karlsruhe.
37. Schulz, H. (2003) Major and minor reactions in Fischer-Tropsch synthesis on cobalt catalysts: Fischer-Tropsch catalysis-science and practice. *Top. Catal.*, **26** (1–4), 73–85.
38. Schulz, H. et al. (1984) Entwicklung und Anwendung der Kapillar-GC-Gesamtprobentechnik für Gas/Dampf-Vielstoffgemische. DGMK-Forschungsbericht 320, DGMK Hamburg.
39. Schulz, H. (2011) Time resolved selectivity for unsteady regimes in catalytic petroleum chemistry, Catalysis Today, **178** (151–156).

15
Zeolite Catalysis
Rutger van Santen

15.1
Introduction

In this chapter we discuss reactivity aspects of zeolite catalysis, focusing on three different zeolitic systems: solid acids of the Brønsted type, Lewis acidic, and redox systems.

As an example of solid acid catalysis we discuss hydrocracking catalysis, which produces branched hydrocarbons in the diesel or gasoline range from heavier feedstocks. This provides an opportunity to analyze shape-selective selectivity aspects.

Then we consider the activation of hydrocarbons by Lewis acids such as cations of Ga or Zn. These are usually part of oxycationic clusters that selectively dehydrogenate alkanes. The important difference from the protonic zeolites is that in contrast to the latter they selectively produce alkenes without competitive cracking reactions.

True redox catalysis occurs in selective oxidation reactions. We discuss the need for extra-framework single-center catalysis in the Panov reaction, which produces phenol by oxidation of benzene with N_2O. Single-center catalysis by Ti, substituting for framework Si, is discussed for the epoxidation of propylene.

Alkane oxidation activated by framework and non-framework redox cations in $AlPO_4$ frameworks isostructural with the zeolites is discussed in a final subsection.

15.2
The Hydrocracking Reaction; Acid Catalysis

The hydrocracking reaction proceeds at relatively mild temperatures. Zeolites in the protonic state are activated with particles of transition metal, reducible oxide, or sulfide. These catalytically active components convert alkanes to their corresponding olefins. High hydrogen pressures are used to suppress formation of deactivating carbonaceous residue. Noble metals such as Pt or Pd can be used

Catalysis: From Principles to Applications, First Edition. Edited by Matthias Beller, Albert Renken, and Rutger van Santen.
© 2012 Wiley-VCH Verlag GmbH & Co. KGaA. Published 2012 by Wiley-VCH Verlag GmbH & Co. KGaA.

when the reaction is to be executed at relatively low temperatures. At higher temperatures or with some sulfur impurity present, MoS_2 or WS_2 can be used. These can be present as small particles in the micropores of the zeolite or as large particles external to the zeolite particles. Under the hydrocracking reaction conditions of relatively high temperature and high ratio of hydrogen to alkane, equilibrium between alkane and alkene leads to a very low concentration of alkenes in the reactive system and also reduces the formation of carbonaceous residues.

The relatively low concentrations of olefins generated in the micropores of the zeolite react with the zeolite protons, leading to isomerization and cracking reactions. Subsequent hydrogenation of the olefinic product molecules leads to formation of product alkanes. However, at higher temperatures aromatics may also be formed by consecutive conversion reactions.

In hydroisomerization reactions where the number of carbon atoms in the product and reactant are the same, cracking reactions are suppressed. Such ideal cracking will only occur when a sufficient amount of the olefin-forming promoters is added, so that alkane–alkene equilibrium is maintained.

Especially when large hydrocarbons are converted, micropore occupation will be high, so that product desorption will be hindered by diffusion. Then the assumption of product equilibrium within the micropores is often justified. The product distribution then reflects desorption equilibria of dominant reaction intermediates. Adsorption equilibria of adsorbed intermediates sensitively depend on zeolite structure, which explains the large differences in product distribution of the different zeolites. In such systems, catalysis by very small particles is preferred. In practice, when crystals are larger, mesopores through the crystal are highly beneficial to overall catalyst performance. The wide pore mesoporous channels assist diffusion of the heavier molecules through the zeolite.

Two fundamental principles are operational, as explained below.

15.2.1
The Dependence of Cracking Selectivity and Activity on Hydrocarbon Chain Length

In Chapter 5 a short mechanistic presentation of zeolite catalysis was given. We noted the relationship between the apparent activation energy and the hydrocarbon chain length for acid cracking of intermediate-size hydrocarbons. Catalyzed by the narrow-pore ZSM-5 zeolite the apparent activation energy decreases linearly with hydrocarbon chain length. The corresponding reaction is catalyzed by high-energy carbonium ion type transition states. At the high temperatures applied, zeolite micropores have a low concentration of hydrocarbons, and hence their concentration increases linearly with the hydrocarbon adsorption coefficient. As long as proton activation of the hydrocarbon is independent of hydrocarbon chain length, protons are statistically distributed through the zeolite and proton activation is independent of proton position. The apparent activation energy will decrease linearly with hydrocarbon chain length.

Hydroisomerization and hydrocracking reactions take place at lower temperatures because now olefins instead of alkanes become activated by the zeolite protons. The intermediates of the reaction react through low-activation-energy transition-state carbonium ions. Reaction occurs at high hydrogen-to-hydrocarbon ratios in order to suppress carbonaceous residue-forming reactions.

It is interesting to compare experimental data for hydroisomerization of hexane on different zeolites. Crystal particle sizes have to be selected to suppress diffusion effects. A low Al/Si ratio has to be used to suppress deactivation of protons by Al concentration effects.

The Langmuir-Hinshelwood kinetics expression has been used to analyze experimental data, summarized in Table 15.1:

Since the coverage with hydrocarbon is now substantially higher than for catalytic cracking, changes in the respective Henry coefficients of adsorption for the different zeolites now have a much smaller effect on overall measured rate differences. The rate data have been normalized per proton.

We also note that variation in the elementary rate constants of proton activation are small compared to the observed variation in overall rate constants. Again the overall rate differences correlate with the differences in hydrocarbon adsorption energies.

There is an interesting anomaly in the observed trends. We note that the zeolites with narrower pores (TON versus MOR) have comparatively lower rates than the wider-pore seolites. This opposes the trend of increasing adsorption energy with decreasing pore size dimension. This trend with decreasing channel and microcavity dimensions is due to the decreased average distance between hydrocarbon and zeolite oxygen atom walls. It stops when the microcavity is so small that the molecule cannot enter it. The decreased mobility of the hydrocarbon with increasing interaction with the micropore channel decreases its entropy. At hydroisomerization temperature the two opposing dependences start to interfere. As a consequence the concentration of molecules in the wider micropore is the larger. Thus, the differences in rates of conversion of the different zeolites relate to the differences in the adsorption entropy of reacting molecules (see also Chapter 2).

These considerations assist interpretation of Weitkamp's data on hydroisomerization and hydrocracking as a function of hydrocarbon chain length (Figure 15.1).

Table 15.1 Hydroisomerization data for four zeolites ($T = 240°C$, $p_{nC6} = 779$ Pa) [1].

	TOF (s^{-1})	$K_{ads,nC6}$ (Pa^{-1})	K_{isom} (s^{-1})
H-Beta	5.0×10^{-3}	6.4×10^{-5}	1.7×10^{-2}
H-Mor	1.1×10^{-2}	3.3×10^{-4}	2.7×10^{-2}
H-ZSM-5	4.1×10^{-3}	3.3×10^{-5}	2.8×10^{-2}
H-ZSM-22	1.6×10^{-3}	1.4×10^{-5}	1.7×10^{-2}

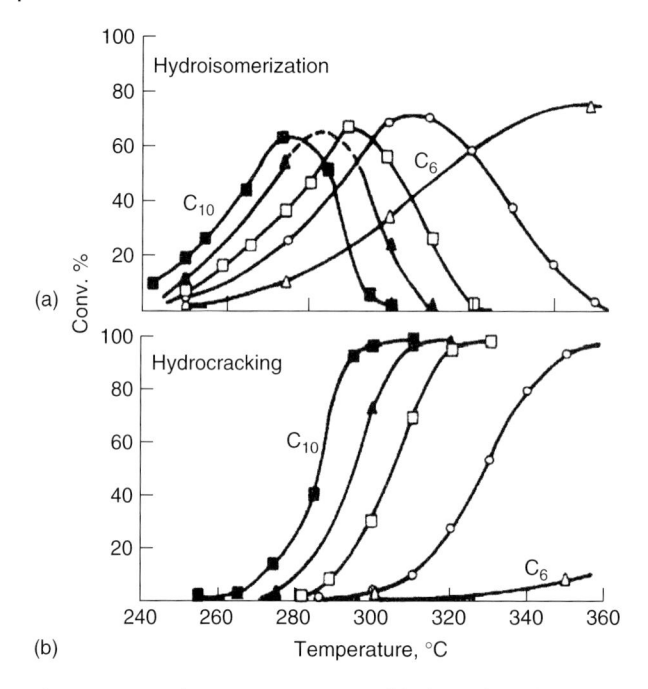

Figure 15.1 Hydroisomerization (a) and hydrocracking (b) production plots for the conversion of linear alkanes with different chain lengths. The catalyst used is an acidic bifunctional Pt/Ca-zeolite Y material [2].

In Figure 15.1, temperatures of conversion and product formation are compared for a wide-pore zeolite.

One notes the decreasing reaction temperature with increasing chain length. The lower temperature is due to the reduction of the apparent activation energy when chain length increases. We also note the more favorable hydroisomerization selectivity with increasing hydrocarbon chain length, which is due to the lower temperature of the reaction. The activation energies for β C–C cracking are higher than those for isomerization.

15.2.2
Symmetric versus Asymmetric Cracking Patterns. Stereoselectivity, Pore Size, and Topology Dependence

Cracking is suppressed as long as temperatures are low and olefin concentrations remain low. Long hydrocarbon chain molecules have a lower selectivity for isomerization, because continued methyl type isomerization and methyl shift reactions increase the relative concentration of tertiary hydrocarbons. These readily crack through the β C–C bond cleavage reaction.

Isomerization and cracking may also proceed through four- or six-ring hydrocarbon structures, as illustrated in Figure 15.2.

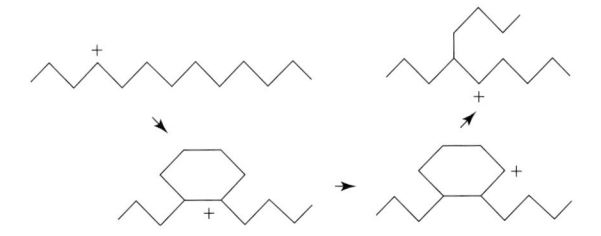

Figure 15.2 Butylbranching mechanism of 4-tridecyl cation via substituted cornerprotonated cyclohexanes (the formal representation of C–H bonds is omitted) [3].

Figure 15.3 Mechanism of side-chain elongation through ring contraction in methylcyclohexyl cation [3].

Interestingly, six ring-cations can rearrange to five-ring cations, similarly to the pairing reaction discussed in Chapter 5. It leads to C–C bond formation (see Figure 15.3).

Cracking of longer linear chain molecules through intermediate five-ring cation formation may lead to preferential formation of isobutenes due to continued methyl shift isomerization.

The less active sulfides are preferred activating promoters for hydrocracking. Most reactions now do not proceed at alkane–alkene equilibrium.

When steric constraints are absent a statistical cracking pattern is expected [3]:

$$C_3 = C_{n-3} \ll C_4 = C_{n-4} < \ldots < C_{n/2} \tag{15.1}$$

Equation (15.1) is found for even-chain-number hydrocarbons, Eq. (15.2) for uneven hydrocarbons [3]:

$$C_3 = C_{n-3} \ll C_4 = C_{n-4} < \ldots < C_{(n-1)/2} = C_{(n+1)/2} \tag{15.2}$$

In Figure 15.4, measured cracking distributions are shown for conversion by wide-pore zeolite Y, that causes no steric constraints to the reaction, and ZSM-5, having narrow pores, that cause steric selectivity effects.

The symmetric cracking patterns are readily recognized for the wide-pore zeolite Y system, but deviations due to steric effects to be discussed next are seen for the narrower-pore system. Suppression of the medium-range products is striking.

Under hydrocracking reaction conditions the micropores of zeolites are highly occupied by hydrocarbons. This implies that diffusion becomes inhibited. Now it is justified to assume that the relative concentrations of reaction intermediates are equilibrated within the zeolite. There is no equilibrium between these intermediates

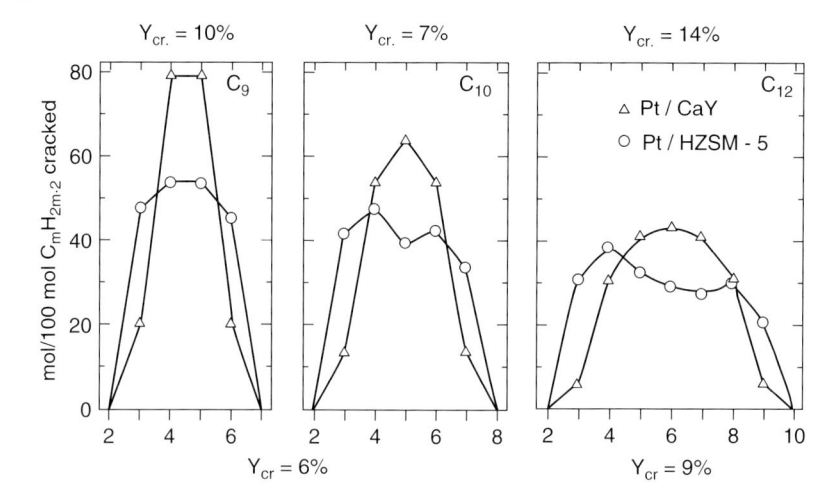

Figure 15.4 Carbon number of the cracked products from n-alkanes over Pt-ZSM-5 and Pt-CaY, representing the regular pattern [3].

and the gas phase. This physical picture has been used to predict the effect of differences in channel dimensions and topology on the selectivity of hydrocracking reactions.

This can be illustrated by the different selectivities of n-C10 cracking [4] by the two related zeolite structures MFI and MEL. MFI has the same structure as ZSM-5. (see Figure 15.5).

The cracking product distribution for the two zeolites can be seen to be very different, notwithstanding the very small differences in channel structure. Within the MFI zeolite, preferential formation of 4,4-dimethyloctane occurs because it is commensurate with the zig-zag channel. The MEL structure prefers formation of 2,4-dimethyloctane, which fits in the larger intersection. 2,4-MeC$_8$ and 4,4-MeC$_8$ are cracking precursors that yield branched and linear butane, respectively.

This example illustrates reaction intermediate shape selectivity, which is to be distinguished from kinetic stereoselectivity that is due to differences in the diffusion rates of reactant or product molecules.

Stereo-selective behavior can also be strongly influenced by reactant pressure. This enhances the contribution to selectivity due to diffusional blocking, which decouples the internal equilibrium between reaction intermediates from the gas phase. An example is provided by a comparison of the 2,3-dimethylbutane to n-hexane ratio of the C_6 hydrocracking products formed during n-C_{16} hydrocracking. Figure 15.6 illustrates the changes in this selectivity as a function of micropore size.

When the pores are small, as with the MTW type structure, repulsive adsorbate–zeolite wall interactions disfavor adsorption of the bulky molecule in favor of adsorption of the linear molecule. When the pores increase in size these unfavorable interactions disappear. Because of the high loading of the micropores the additional

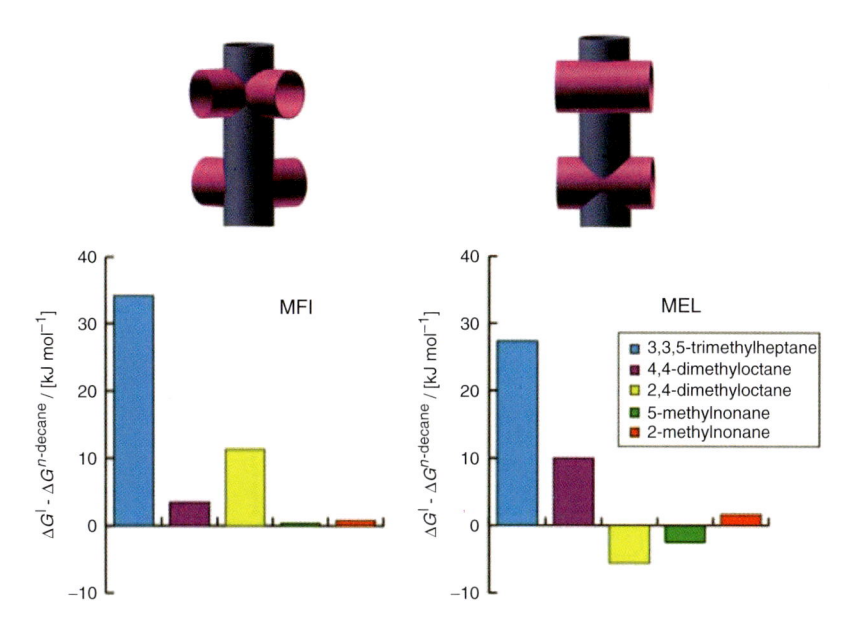

Figure 15.5 Comparison of MFI- and MEL-types of pores: top, an artist's impression of the structures and, bottom, free energy of formation of selected reaction intermediates relative to that of adsorbed n-decane. (Figure reproduced with permission from Ref. [4].)

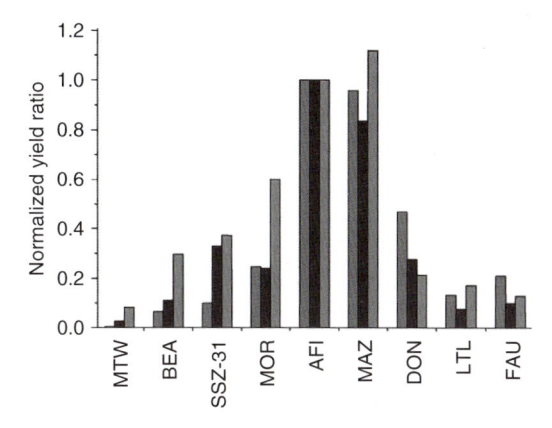

Figure 15.6 Effect of pore size on the hydroisomerization selectivity (2,3-dimethylbutane-to-n-hexane ratio) of the C$_6$ hydrocracking products formed during n-c15 hydroconversion (increasing pore size from MTW to FAU). When the pores are small (as with MTW type zeolites), repulsive adsorbent-adsorbate Van der Waals interactions impede dimethylbutane (DMB) formation; when the pores increase in size, those impeding interactions disappear and interadsorbent interactions favor formation of the better packing DMB; when the pore size increases above 0.74 nm, differences in packing efficiencies disappear because the adsorbents no longer have to line up head-to-tail but can pack in an increasingly more random, liquid-like fashion [4].

interactions with the other adsorbed molecules now become important. The bulky molecules occupy less space than the linear molecules, which experience more of a spatial constraint in the still narrow channels and have to align along the channel direction. The effective volume of the bulky more compact molecules is less. Favorable entropy gives a lower free energy and hence a substantial increase in selectivity. When the pores become very large, the linear molecules can freely rotate, and the entropy gain of the bulky molecules disappears. This selectivity advantage will only be found at high pressures, which maintain high hydrocarbon coverage of the micropores.

15.3
Lewis Acid–Lewis Base Catalysis; Hydrocarbon Activation

Ion exchange of soft Lewis acid cations such as Zn^{2+} and Ga^+ leads to zeolitic systems that can activate CH bonds of alkanes. The advantage of catalysis by these cations is that there is no C–C bond cleavage and systems can operate at low hydrogen pressure. The Cyclar process, which that converts short alkanes into olefins and aromatics, is based on the activation of Brønsted acidic zeolites by soft Lewis acids.

These cations can be present in the micropores as single-center cations, as dimers, or as higher oxycationic complexes. Usually the dimer oxycationic complex is the most reactive. We will illustrate this by the example of ethane activation by extra-framework Ga^+, which we compare with activation by the dimer oxycation. Activation energies for some elementary steps and different Ga complexes in the micropores of Mordenite are illustrated in Figure 15.7.

Activation of a CH bond by the Ga^+ cation can occur homolytically over the single center with formation of $(CH_3GaH)^+$ or heterolytically with formation of $GaCH_3$ and a proton attached to an oxygen atom of the zeolite framework. The oxygen atom that bridges an Al and Si framework cation is Lewis basic and accepts such a proton.

The homolytic dissociation process is the analog of the oxidative addition reaction known from homogenous complexes of transition metals such as Pd. In such a process, two electrons are transferred to the reagent.

On Ga^+ the activation energy of this reaction, in which Ga^+ is converted into Ga^{3+}, is nearly 400 kJ mol^{-1}. This is because of the very high ionization potential of the Ga valence d-atomic orbitals. On the other hand, the heterolytic process only requires a barrier slightly higher than 200 kJ mol^{-1}, and therefore is the preferred activation mechanism.

In Figure 15.7 one can observe that the $(GaO)^+$ intermediate activates the CH bond with a lower barrier, which is less than half that of the heterolytic activation barrier of Ga^+. This is due to the high reactivity of the rather weakly bonded oxygen to the Ga^{3+} cation. The corresponding O atom is a strong Lewis base. Also, in the dimer complexes the bridging O atom is highly reactive, so that a low barrier

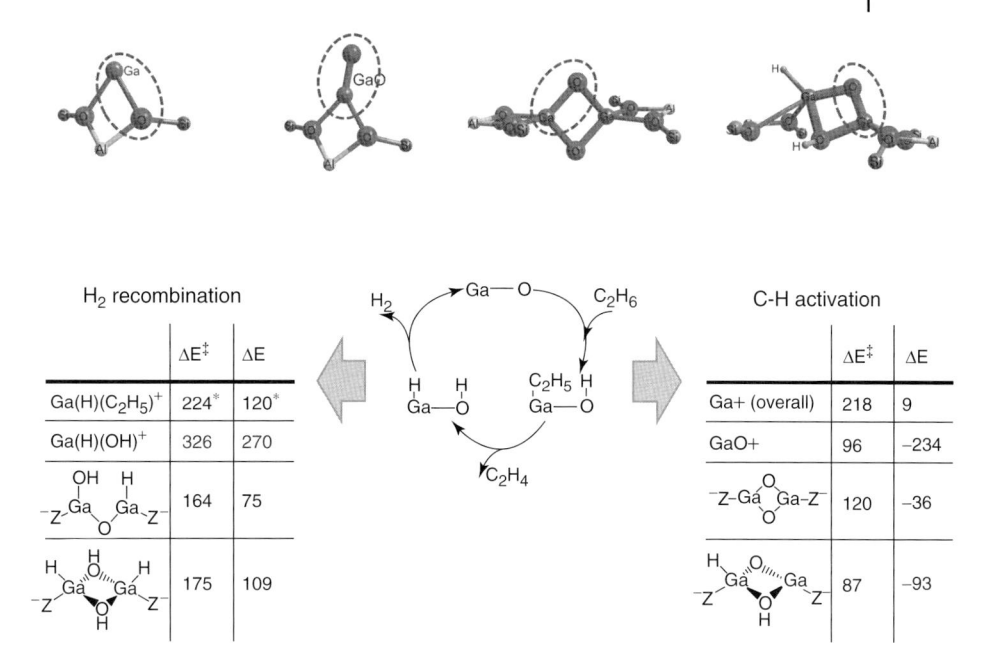

15.4 Selective Oxidation; Redox Catalysis

Figure 15.7 The reactivity of mononuclear versus binuclear gallium cations in Mordenite zeolite [5].

H$_2$ recombination

	$\Delta E^\ddagger$	ΔE
Ga(H)(C$_2$H$_5$)$^+$	224*	120*
Ga(H)(OH)$^+$	326	270
$^-$Z–Ga(OH)–O–Ga–Z$^-$ (with H)	164	75
$^-$Z–Ga(H)–O(H)–Ga(H)–Z$^-$	175	109

C-H activation

	$\Delta E^\ddagger$	ΔE
Ga+ (overall)	218	9
GaO+	96	−234
$^-$Z–Ga(O)(O)Ga–Z$^-$	120	−36
$^-$Z–Ga(H)–O(H)–Ga–Z$^-$	87	−93

* one-step H$_2$ and C$_2$H$_4$ desorption for Ga$^+$

for CH activation is found. CH activation can be considered a Lewis base-assisted Lewis acid activation process.

Also in the same Figure 15.7 the activation energies for hydrogen recombination are presented. One now observes a very high activation energy on the GaO$^+$ complex. It actually competes with desorption of water. In contrast the dimer, oxycationic complexes show low activation energies because of the lower reactivity of the complex oxygen atoms. Clearly, overall catalysis with favorable activation energies is only possible for the oxycationic complexes.

This example illustrates the high Lewis basic reactivity of O that is part of extra-framework oxycationic complexes. The increased reactivity of zeolites containing extra-framework Al containing oxycation complexes may also relate to the presence of extra highly reactive Lewis basic oxygen sites that can accommodate cationic intermediates.

15.4
Selective Oxidation; Redox Catalysis

As in the previous case, reducible cations can be located in extra-framework sites as single-center cations or oxydicationic complexes. Such is the situation for Fe cations in the Panov reaction that converts benzene to phenol with N_2O, to be discussed first.

We also discuss the reactivity of reducible cations located in the framework of zeolites. Two systems are discussed. The first is Ti-catalyzed epoxidation of propylene with hydrogen peroxide. Whereas Ti is reducible, we will see that it actually acts as a Lewis acid.

The other examples are from the Thomas systems (see Ref. [8]), with Co or Mn substituted on Al positions of the zeolite polymorphic framework of the $AlPO_4$ systems.

15.4.1
The Reactivity of Extra-Framework Single-Site versus Two-Center Fe Oxycations

In the Panov reaction, N_2O and benzene are selectively converted over Fe-activated ZSM-5. In active catalysts, a mixture of Fe oxycationic complexes is present. The activity of the catalyst has been found to correlate with the concentration of Fe^{2+}. Catalyst activity and stability strongly correlate. Systems with a bimodal pore size have increased activity. The wider pores reduce catalyst deactivation, because access to the catalytically reactive is less hindered. Catalyst activity also depends on catalyst composition. Addition of alumina has a beneficial effect. The inorganic chemistry of Fe located in the micropores is complex. As illustrated in Figure 15.8, the different intermediates can interconvert. Autoreduction and hydrolysis reactions occur, and pH plays an important role. High acidity favors formation of Fe^{2+}.

The reactivities of single-center Fe^{2+} and different dimeric oxycationic complexes have been compared computationally (see Figure 15.9).

The essential conclusion one can draw from these reaction energy diagrams is that there is very little difference between the reactivity of the single-center complex and that of the dicationic complex. N_2O decomposition is slightly more difficult on the dimeric complex than on the single-center Fe^{2+} complex. Phenol desorption has the highest activation energies, with a lower desorption energy on the dicationic oxycation.

Whereas the reactivity toward phenol formation is similar, it is different when one also considers a possible deactivating reaction. As expected, the extra-framework oxygen that is part of the oxycationic complex has a strong Lewis basicity. Adsorbed phenol will readily donate its proton to such basic oxygen atoms with formation of phenolate. Phenolates will deactivate the catalyst through oligomerization processes.

Figure 15.10 illustrates the difference in reactivity of transfer of proton to basic zeolite wall O versus acceptance by an oxygen atom of the dicationic oxycation complex.

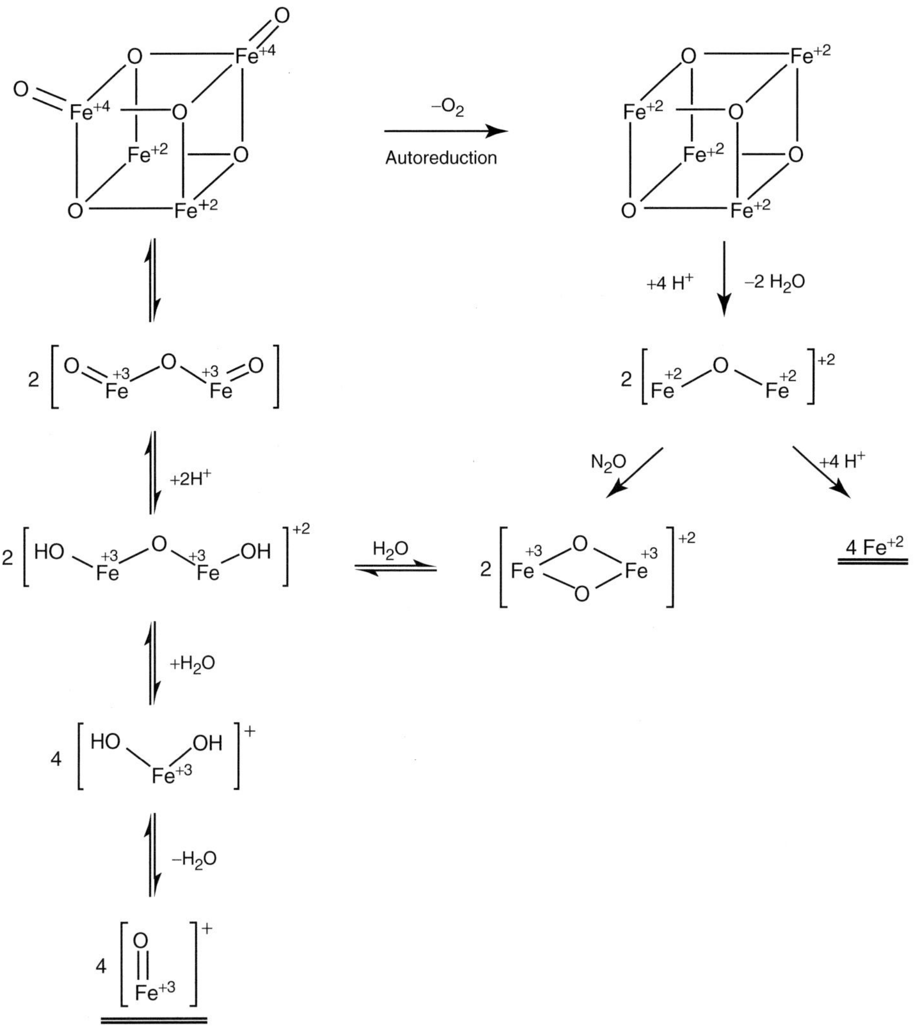

Figure 15.8 Self-organization of oxide clusters in N_2O oxidation (interplay of protons, water, and N_2O).

15.5 Framework-Substituted Redox Ions

15.5.1 Ti-Catalyzed Epoxidation

Ti substituted into the framework of siliceous MFI is a selective catalyst for the epoxidation of propylene by hydrogen peroxide. It has recently been implemented in a process by Solvay/BASF. This is different from the reactivity of single-center

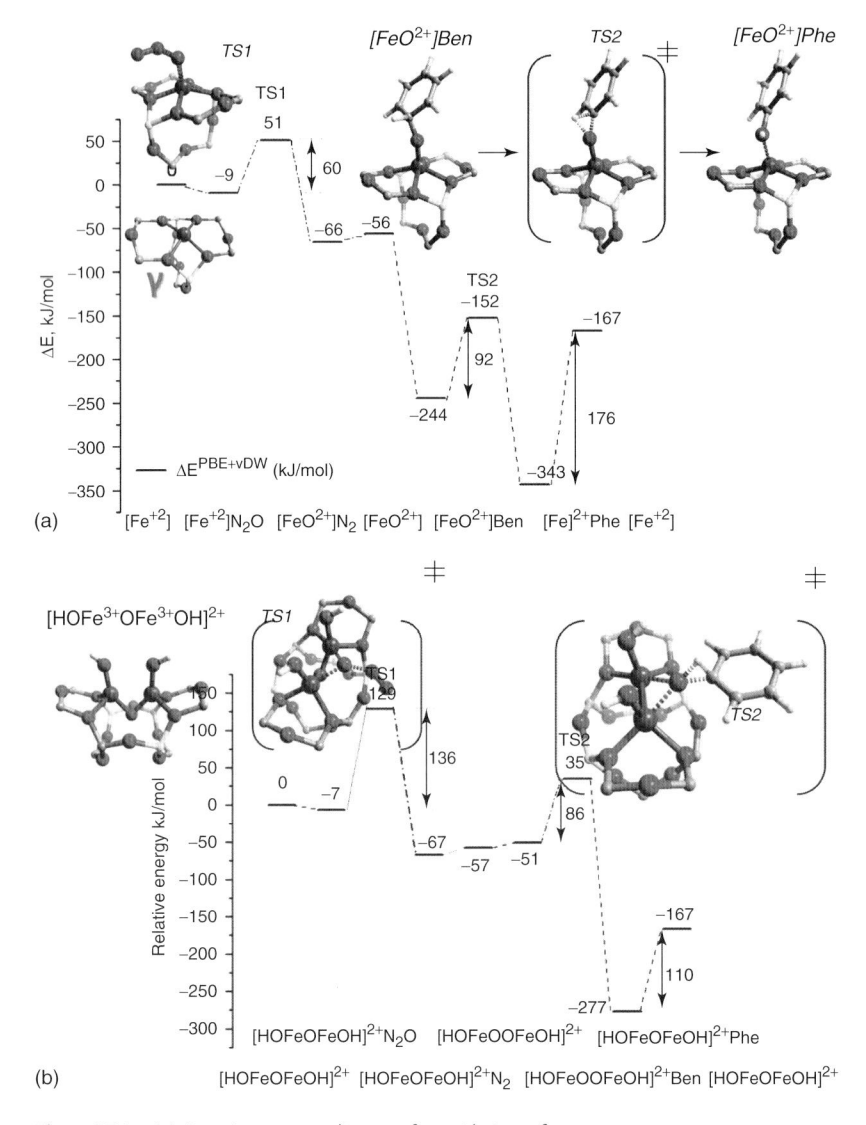

Figure 15.9 (a) Reaction energy diagram for oxidation of benzene by N_2O catalyzed by Fe^{2+} cation in ZSM-5. (b) Reaction energy diagram for oxidation of benzene by N_2O catalyzed by $[HOFe^{3+}OF^{3+}OH]^{2+}$ oxycation in ZSM-5 [6].

Ti immobilized on silica. This is used in the Shell SMPO process, and can catalyze propylene epoxide formation only from a hydroperoxide. When hydrogen peroxide is used with this catalyst, Ti hydrolyzes. The difference in stability relates to the rigidity of the zeolite lattice and its high hydrophobicity.

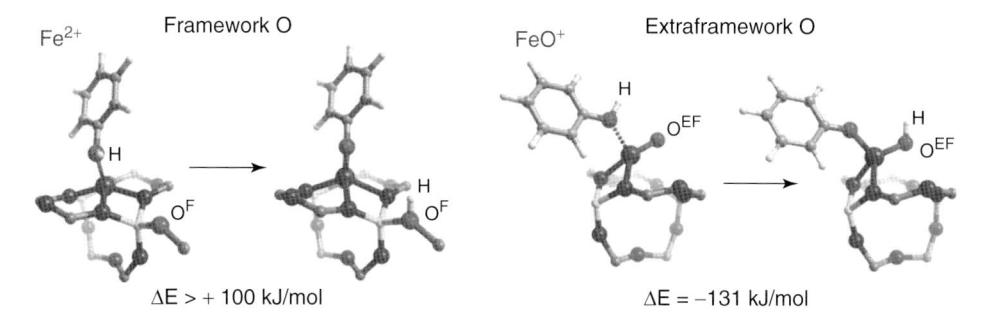

Figure 15.10 The presence of basic extra-framework oxygen species favors decomposition of the adsorbed phenol via proton transfer [6].

Figure 15.11 Epoxidation mechanism of propene by Ti in zeolite ZSM-5 (schematic) [7].

Figure 15.11 illustrates the mechanism of epoxidation with hydrogen peroxide by the Ti single-center site. Reaction is initiated by hydrolysis of one of the TiOSi bonds around Ti. Ti now becomes three-coordinated with other lattice oxygen atoms, and an OH attaches. This can be readily converted with hydrogen peroxide to a TiOOH group. This peroxy group is now activated by Ti and will epoxidize propylene. The solvent will increase the reactivity of TiOOH.

In this case the reducible redox cation in essence functions as a Lewis acid. Hydrolysis and formation of TiOH activates the system. When larger cations, such as Sn or Zr, are substituted in the zeolite framework, they behave like Lewis acidic. They cannot usefully be applied as oxidation catalysts, but activate keto groups in reactions that give C–C bond formation. As for Ti hydrolysis, the metal-oxide bond has to be hydrolyzed, and the cation is active as an XOH species.

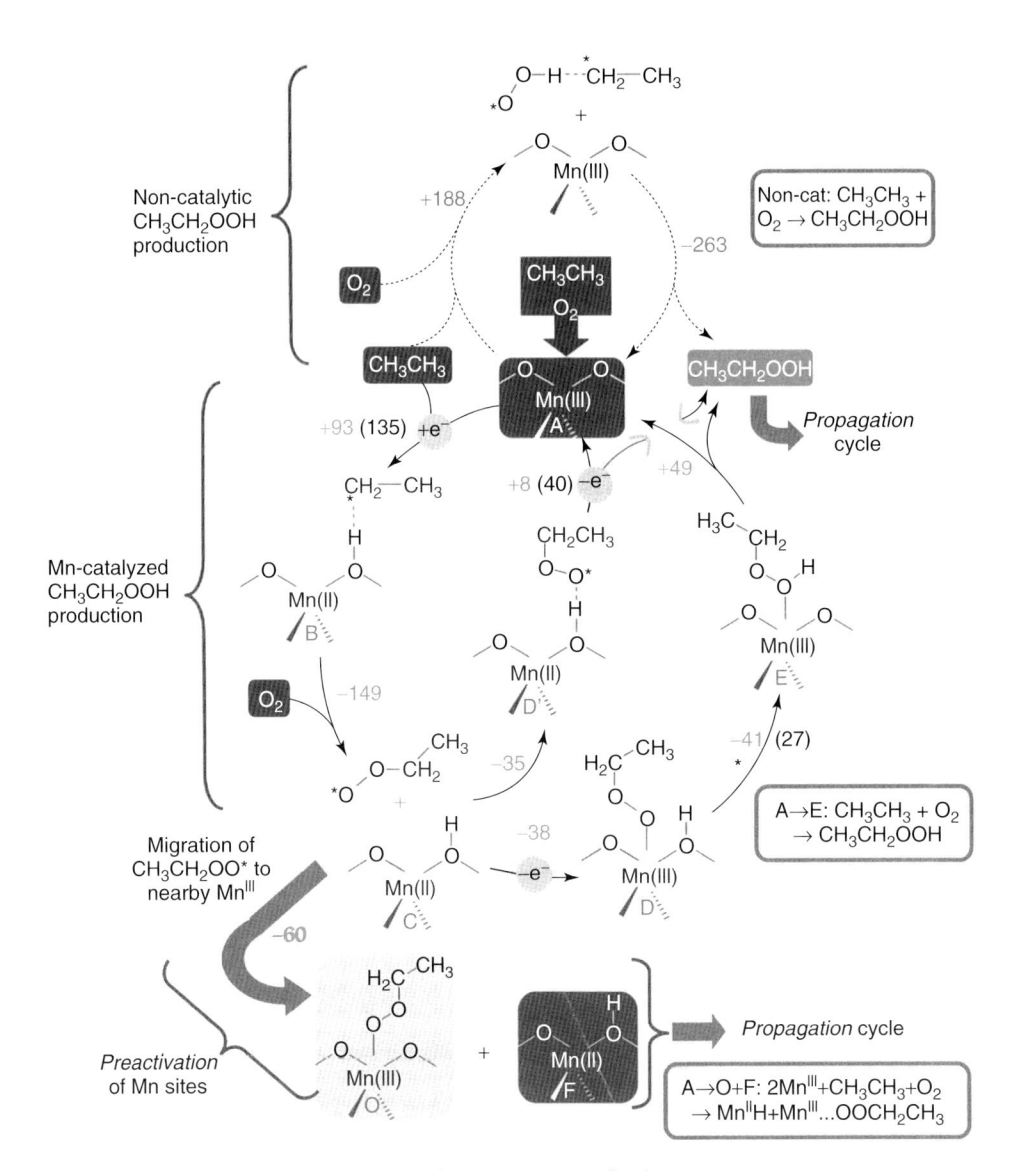

Figure 15.12 Preactivation mechanism (A → E) for the production of CH_3CH_2OOH from RH and O_2 without (top) and with the assistance of Mn(III) (middle). A dark background indicates initial catalyst and reactant molecules while a light background indicates reaction intermediate or product molecule [8, 9].

15.5.2
Thomas Chemistry; Redox Cations in the AlPO$_4$ Framework

Cr^{3+}, Mn^{3+}, or Fe^{3+}, when substituted for Al^{3+} in the framework of AlPO$_4$, can be used to oxidize alkanes directly with oxygen. Shape selectivity can be exploited to increase the selectivity of the radical reactions involved. These systems have been explored for the selective oxidation of *n*-hexane to adipic acid, an important intermediate for nylon production.

In zeolites with small dimensions, initial reaction will occur with the terminal CH$_3$ hydrogen atoms that are most reactive. This is assisted by shape selectivity. Those AlPO$_4$ structures that have such small cavity apertures that alkanes can only enter by an end-on approach give dominance of primary CH activation. Chain termination reactions through recombination of two peroxy radicals that would lead to alcohol or aldehyde will be suppressed, and instead the peroxy intermediates will form hydroperoxides and additional primary radicals. The hydroperoxides will decompose by reaction with the redox cations. The primary decomposition product of oxidation with Mn^{2+} is l-hexanal. Mn^{2+} can be produced by reduction with alkane, as illustrated in Figure 15.12.

The overall mechanism of selective oxidation of this radical reaction to give the acids has been elucidated for Mn and is shown in Figure 15.12.

The hydroperoxide reacts with two Mn^{3+} framework cations to peroxide radical adsorbed to Mn^{3+}, a bridging framework proton, and a reduced Mn^{2+} cation. In a consecutive step the peroxide radical reacts with Mn^{2+} and the proton to give acetic acid. This reaction competes with decomposition of peroxide to give the hexanal and Mn^{3+}.

For exclusive oxidation on the two ends of the hexane molecule, each AlPO$_4$ cavity should contain two M^{3+} ions at opposite positions. The optimum cavity structure is that of chabasite. It was established that for an M^{3+}/P ratio of 0.08 or higher two cations would be located in the chabasite cavity of the AlPO$_4$-18 system.

References

1. van Santen, R.A. and Neurock, M. (2006) *Molecular Heterogeneous Catalysis*, Wiley-VCH Verlag GmbH, Weinheim, p. 199.
2. Weitkamp, J. (1982) Isomerization of long-chain n-alkanes on a Pt/CaY zeolite catalyst. *Ind. Eng. Chem. Prod. Res. Dev.*, **21** (4), 550–558.
3. Jacobs, P.A. et al. (1991) *Studies in Surface Science and Catalysis*, Chapter 12, Elsevier, pp. 445–496.
4. Smit, B. and Maesen, T.L.M. (2008) Molecular simulations of zeolites: adsorption, diffusion, and shape selectivity. *Chem. Rev.*, **108** (10), 4125–4184.
5. Pidko, E.A. and van Santen, R.A. (2010) in *Zeolites and Catalysis* (eds J. Cejka, A. Corma, and S. Zones), Wiley-VCH Verlag GmbH & Co. KGaA, pp. 301–334.
6. Li, G. et al. (2011) Stability and reactivity of active sites for direct benzene oxidation to phenol in Fe/ZSM-5: A comprehensive periodic DFT study. *J. Catal.*, in press.
7. Jackson, S.D. and Hargreaves, J.S.J. (1979) in *Metal Oxide Catalysis* (eds S.D. Jackson and J.S.J. Hargreaves), Wiley-VCH Verlag GmbH & Co. KGaA, p. 721.

8. Thomas, J.M. (2008) Heterogeneous catalysis: enigmas, illusions, challenges, realities, and emergent strategies of design. *J. Chem. Phys.*, **128**, 182502.
9. Gómez-Hortigüela, L. *et al.* (2010) Catalytic reaction mechanism of Mn-doped nanoporous aluminophosphates for the aerobic oxidation of hydrocarbons. *Chem. – Eur. J.*, **16** (46), 13638–13645.

16
Catalytic Selective Oxidation – Fundamentals, Consolidated Technologies, and Directions for Innovation

Fabrizio Cavani

16.1
Catalytic Selective Oxidation: Main Features

Catalytic selective oxidation accounts for the production of several organic and inorganic chemicals in the petrochemical, fine, and specialty chemicals and pharmaceutical industries [1–5]. Table 16.1 lists the main industrial oxidation processes used in the petrochemical industry for the synthesis of organic compounds [2, 6–10]. Many of the chemicals listed are intermediates for the polymer industry, and are currently produced in huge quantities – several million tons per year.

The reactor technologies used include gas–solid, gas–liquid, and gas–liquid–solid reactors. Catalysts are either dissolved in the liquid reaction medium, take the form of solid particles suspended in the fluid phase (stirred slurry reactors, fluidized-bed reactors, and bubbling-slurry columns), or are shaped into pellets in single-tube or multitubular fixed-bed reactors. Reaction conditions range from mild temperatures, in liquid-phase reactions, to >400 °C for some gas-phase transformations.

Such a wide variety of conditions and technologies is unique in the petrochemical industry: for instance, most hydrogenation reactions are carried out in the liquid phase. This peculiarity is due to the characteristics of oxidation processes:

1) All selective oxidation reactions are exoenthalpic, and almost all oxidations are substantially irreversible throughout the entire range of temperatures used in the chemical industry.
2) Although present, deactivation phenomena – very often a major problem for other classes of reactions – do not represent the main drawback for oxidation reactions; this permits the use of operation conditions that, if carried out in the absence of an oxidant, would lead to rapid deactivation of the catalyst.
3) The types of substrate that can be functionalized by means of catalytic oxidation range from very reactive molecules, such as olefins and alkylaromatics, to less reactive or even apparently inert compounds, such as aromatics and alkanes. On the other hand, various types of oxidants can be chosen for liquid-phase and even gas-phase oxidations; the nature of the substrate to be transformed

Catalysis: From Principles to Applications, First Edition. Edited by Matthias Beller, Albert Renken, and Rutger van Santen.
© 2012 Wiley-VCH Verlag GmbH & Co. KGaA. Published 2012 by Wiley-VCH Verlag GmbH & Co. KGaA.

Table 16.1 Main oxidation processes for the synthesis of organic compounds in the petrochemical industry.

Oxidation process PHASE reactor	Catalyst Reactant conversion selectivity	Reaction conditions	Remarks on the technology	Main by-products
Methane + NH_3 to hydrogen cyanide (ammoxidation) $CH_4 + NH_3 + 1/2 O_2$ $\rightarrow HCN + H_2O$ (HCN also produced by ammonolysis, without oxygen, and as a by-product in acrylonitrile synthesis) GAS PHASE Converter chamber	Pt/Rh gauzes 100%, 80–85%	Oxidant air. Feed: 15% O_2, 15% NH_3, 15% CH_4. T 1100–1200 °C, P 2–4 bar. It includes the endothermic reaction between ammonia and methane to HCN and H_2, and the exothermic combustion of hydrogen and methane. Overall the process is autothermal.	A rapid quenching of products is done after the gauze to avoid the degradation of HCN. Unreacted ammonia is recovered and recycled. Ammonia must be removed from the off-gas before HCN recovery because it promotes HCN polymerization.	H_2 (co-product of ammonolysis), CO_2, CO, traces of cyanogen, acrylonitrile, acetonitrile, and propionitrile.
Methanol to formaldehyde (oxidative dehydrogenation, ODH) $CH_3OH + 1/2 O_2$ $\rightarrow CH_2O + H_2O$ GAS PHASE Fixed-bed multitubular reactor	$Fe_2(MoO_4)_3$-MoO_3 + dopants 97–99%, 94–95%	Oxidant air Feed: 9.5–10% O_2, 7–10% methanol, rest N_2, and other inerts. T 260–310 °C, P 1.3–1.7 bar, or P 1.6–2 bar (high P plants). Tailgas recirculation.	Once-through methanol conversion 97–99%, to avoid energy-intensive methanol separation by distillation (that favors by-products formation).	Dimethylether, methylformate, CO, CO_2. Formic acid has to be less than 100 ppm.

Reaction	Catalyst	Conv%, Sel%	Process conditions	Notes	Byproducts
Methanol to formaldehyde (dehydrogenation + ODH + hydrogen combustion) GAS PHASE Fixed-bed multitubular reactor	Ag crystals (shallow catalyst bed), or Ag gauze	98–99%, 98–99%	Oxidant air Feed: 10–12% O_2, 40–55% methanol, rest N_2 and other inerts. T 600–650 °C. P 1.1–1.2 bar. Ballast steam. Eventual tailgas recirculation (tailgas also contains H_2).	High methanol conversion despite the low concentration of O_2, because of methanol dehydrogenation and hydrogen combustion.	as above
Ethene to ethene oxide (epoxidation) $= + 1/2\, O_2 \rightarrow \triangle\!\!\!\!^{O}$ GAS PHASE Fixed-bed multitubular reactor	Ag-αAl$_2$O$_3$ + dopants	15–20% pp^a (overall >99%), 76–82%	Oxidant O_2 Feed: 7–9 % O_2, 20–40% ethene, rest ballast (methane). T 250–270 °C, P 12–15 bar Tail gas recirculation; purge and ethene loss are minimal.	A limited ethene per-pass conversion disfavors consecutive transformations of ethene oxide and irreversible ethene combustion.	Acetaldehyde, ethene glycol, hydrocarbons, and CO_2.
		96–98%, 65–70%	Oxidant air Feed: 5–8 % O_2, 3–5% ethene, rest inerts (N_2, CO_2). T 250–280 °C. P 10–20 bar.	Alkylhalides are added in parts per million amount as gas phase promoters. Air-based process: two or three in-series reactors. First reactor: conversion <40%, select to EO (ethene oxide) >70%. After EO abatement, a part of the exit stream is recycled, a part is sent to the second reactor.	

(continued overleaf)

Table 16.1 (continued)

Oxidation process PHASE reactor	Catalyst Reactant conversion selectivity	Reaction conditions	Remarks on the technology	Main by-products
Ethene + HCl to 1,2-dichloroethane (oxychlorination) $== + 2\,HCl + 1/2\,O_2 \rightarrow Cl\diagdown\diagup Cl + H_2O$ GAS PHASE Fixed-bed multitubular reactor or fluidized-bed reactor	$CuCl_2\text{-}\gamma Al_2O_3 +$ dopants (K, Mg) 97–99%, 95–98% <40% pp[a], 96–99% 97–99%, 94–97%	Oxidant O_2. Feed: 24–28% ethene, 47–55% HCl, rest O_2. Recycle of tailgas to minimize losses. Oxidant O_2 Feed: 30–50% ethene, 20–30% HCl, rest O_2 plus ballast. Recycle of unconverted ethene. Oxidant air. Feed: 12–15% ethene, 24–30% HCl, 8–10% O_2, rest N_2 and other inerts. Fluidized-bed: T 220–250 °C, P 3–5 bar. Fixed-bed (three in-series reactors, with differentiated oxidant feed): T 230–300 °C, P 3–15 bar.	HCl conversion has to be total, to avoid downstream corrosion problems. With the O_2-based process, ethene per-pass conversion can be either partial (<40%) or almost total. Oxychlorination is integrated with ethene chlorination, to close the balance upon Cl. 1,2-dichloroethane is then cracked to vinyl chloride monomer.	Ethylchloride, CO_2, other chlorinated compounds (polychlorinated ethane and ethene, chloromethanes, heavies).

16.1 Catalytic Selective Oxidation: Main Features

Reaction	Catalyst, conversion, selectivity	Process conditions	Notes	Byproducts
Ethene to acetaldehyde (less used than in the past, because of the more economical methanol carbonylation for acetic acid production). $\ce{=} + 1/2\, O_2 \longrightarrow \overset{O}{\underset{H}{\|}}$ LIQUID/VAPOR PHASE Titanium column with perforated trays.	$PdCl_2$-$CuCl_2$ in HCl medium. 25–30% pp[a], 93–96%. 97–98%, 95–97%.	Single-step process: oxidant O_2 T 120–130 °C, P 3–5 bar. Recycle of ethane. Diluent steam. Two-step process: oxidant air (for the oxidation of the reduced catalyst) T 110 °C, P 8–9 bar. No recycle of ethane.	Vaporization of water removes the heat of reaction. In the O_2-based process, high purity ethene is needed in make-up, to limit the purge stream. O_2 has to be less than 8%. The catalyst solution is continuously withdrawn, regenerated separately and recycled.	Acetic acid, oxalic acid, crotonaldehyde, and chlorinated organics.
Ethene + acetic acid to vinyl acetate (acetoxylation) $\ce{=} + \overset{O}{\underset{OH}{\|}} + 1/2\, O_2 \longrightarrow \overset{O}{\underset{O}{\|}}\!\!\!\diagup\!\!\diagdown\!\!= + H_2O$ VAPOR PHASE Fixed-bed reactor (recently, fluidized-bed technologies developed).	Pd/Pt-support + dopants or Pd/Au-support Ethene 8–12% pp[a] (total 88–90%), acetic acid 20–25% pp[a], select. 92–93%.	Oxidant O_2 Feed: 50% ethene, 15% acetic acid, 6% O_2, 29% inerts T 160–180 °C, P 5–8 bar.	O_2 is preferred to avoid inert and limit ethene losses in purge stream. Low O_2 concentration is used, to avoid formation of flammable mixtures. The process is sometimes integrated with the acetic acid production.	CO_2, ethylacetate, and acetaldehyde.
Acetaldehyde to acetic acid (radicalic autoxidation) $\overset{O}{\underset{H}{\|}} + 1/2\, O_2 \longrightarrow \overset{O}{\underset{OH}{\|}}$ LIQUID/VAPOR PHASE Bubble column.	Mn or Co acetate 96–98%, 96–98%. 91–92%, 93–94%.	Oxidant O_2. T 70–80 °C, P 1.5–3 bar. Oxidant air. T 55–65 °C, P 2–10 bar.	Despite the high once-through conversion achieved in the O_2-based process, acetaldehyde is separated and recycled.	Formaldehyde, formic acid, alkylacetates, acetone, and CO_2.

(continued overleaf)

Table 16.1 (continued)

Oxidation process PHASE reactor	Catalyst Reactant conversion selectivity	Reaction conditions	Remarks on the technology	Main by-products
Two-step propene to acrylic acid (via intermediate acrolein) (allylic oxidation) $\diagup\!\!\diagup + 3/2\,O_2 \longrightarrow \diagup\!\!\diagup\!\!\overset{OH}{\underset{O}{\diagdown}} + H_2O$ GAS PHASE Two in-series fixed-bed reactors without intermediate separation of acrolein.	First step: Fe/Co/Bi/Mo/O (crystalline multimetal molybdate)-SiO_2 + dopants 92–95%, 90–95% Second step: V/Mo/W/O 95–97%, 92–94%	Oxidant air. First step: Feed 4–5% propene, 30–40% steam, and 55–65% air T 300–350 °C, P 1–3 bar Also steam is added. Second step T 250–280 °C, P 1–2 bar. Also steam is added.	High once-through conversion in both steps. Rapid quenching of exit gas is necessary to prevent consecutive reactions.	Acetaldehyde, acetic acid, CO, and CO_2.
Propene + NH_3 to acrylonitrile (ammoxidation) $\diagup\!\!\diagup + NH_3 + 3/2\,O_2 \longrightarrow \diagup\!\!\diagup\!\!CN + 3\,H_2O$ GAS PHASE Fluidized-bed reactor	Co/Fe/Ni/Bi/P/K/Mo/O (crystalline multimetal molybdate) -SiO_2 97–100%, 75–82%.	Oxidant air or O_2-enriched air. Feed: 7% propene, 8% NH_3, rest air. T 420–480 °C, P 3–4 bar.	Once-through complete propene conversion and almost complete NH_3 conversion. Rapid neutralization of unconverted NH_3 is required	Acetonitrile, hydrogen cyanide, CO, CO_2, N_2. Acetonitrile and HCN are valuable by-products.

16.1 Catalytic Selective Oxidation: Main Features

Reaction	Catalyst	Conversion / Selectivity	Process	Notes
Propene to propene oxide (epoxidation) $\diagup\!\!\!\diagdown$ + Cl$_2$ + Ca(OH)$_2$ → $\underset{O}{\triangle}\!\!\diagdown$ + CaCl$_2$ + H$_2$O LIQUID PHASE Unpacked in-series columns	No catalyst	93–95%, 94–96%	CHPO (ChloroHydrin to Propene Oxide) technology. Oxidant Cl$_2$. Two steps. T 40 °C. First step: synthesis of chlorohydrin. Second step: dehydrochlorination to PO, at basic conditions.	Process integrated with Cl$_2$ production. Fresh propene also contains propane. Chlorine conversion is complete. Chlorohydrin per-pass conversion is almost total, with 96% selectivity to PO.	Co-product NaCl or CaCl$_2$. Main by-products: 1,2-dichloropropane, dichloropropanol, chlorinated diisopropylether.
$\diagup\!\!\!\diagdown$ + $\underset{\text{Ph}}{\overset{\text{OOH}}{\underset{|}{\text{CH}}}}$ → $\underset{O}{\triangle}\!\!\diagdown$ + $\underset{\text{Ph}}{\overset{\text{OH}}{\underset{|}{\text{CH}}}}$ LIQUID PHASE Series of staged reactors, with intermediate cooling	TiO$_2$-SiO$_2$ or Mo naphthenate	15% pp^a (overall 95–97%), 95–97%.	PO/SM (styrene monomer) technology. Oxidant: ethylbenzene hydroperoxide (EBHP). T 100–130 °C, P 35 bar. Propene is fed in excess (propene/hydroperoxide 10/1–2/1).	EBHP conversion >97%. PO selectivity with respect to EBHP >70%. Acetophenone (by-product of hydroperoxide synthesis) and residual hydroperoxide are hydrogenated to phenyl-1-ethanol.	Co-product phenyl-1-ethanol (dehydrated to styrene). Main by-products: acetaldehyde, propionaldehyde, acetone, heavies from polymerization of styrene.
$\diagup\!\!\!\diagdown$ + $\overset{\text{OOH}}{\underset{|}{\text{t-Bu}}}$ → $\underset{O}{\triangle}\!\!\diagdown$ + $\overset{\text{OH}}{\underset{|}{\text{t-Bu}}}$ LIQUID PHASE Series of staged reactors	Mo naphthenate	15% pp^a (overall 95–97%), 95–97%.	PO/TBA (t-butylalcohol) technology. Oxidant: t-butylhydroperoxide (t-BuOOH). Feed: 20% t-BuOOH, 20% t-BuOH (formed in the synthesis of t-BuOOH), 60% propene. T 80–100 °C, P 30–40 bar.	t-BuOOH per-pass conversion 90–95%. Propene is injected at each reactor. A large excess of propene is used, that is recycled to the first stage.	Co-product t-butanol. Main by-products: acetone, aldehydes, propene glycol, and methylformate.

(continued overleaf)

348 | *16 Catalytic Selective Oxidation*

Table 16.1 (continued)

Oxidation process PHASE reactor	Catalyst Reactant conversion selectivity	Reaction conditions	Remarks on the technology	Main by-products
[propene + CHP → PO + cumyl alcohol] LIQUID PHASE Fixed-bed reactor structured into multiple catalyst layers, with heat exchangers in between the layers	TiO_2-SiO_2 (mesoporous Ti-silicate). 10–15% pp[a] (hypothetical), 99%. Reuse of the co-product cumyl alcohol, reoxidized to CHP, and recycled.	Oxidant cumylhydroperoxide (CHP). Large excess of propene in feed (propene/CHP 10/1). T 60 °C.	CHP conversion >95%; selectivity to PO on CHP basis 95%. The per-pass propene conversion is low, to limit consecutive reactions. Unconverted propene is recycled.	Co-product cumyl alcohol, that however is oxidized to CHP and recycled. Main by-products: acetaldehyde, propionaldehyde, methanol, propene glycol, ketones, and esters.
[propene + H_2O_2 → PO + H_2O] LIQUID PHASE Probably two in-series fixed-bed reactors with differentiated propene feed	TiO_2-SiO_2 (microporous Ti-silicalite: TS-1). Propene conversion unknown. Yield on propene basis 95%.	HPPO (Hydrogen Peroxide for Propene Oxide) technology. Oxidant: H_2O_2 (HP): dilute solution in water/methanol. T 40 °C, P 20 atm.	HP conversion >99%; selectivity to PO on an HP basis 95–96%. In each reactor, the high propene-to-HP ratio leads to a low propene conversion, but the selectivity to PO with respect to HP is improved.	1,2-Propandiol, 1-methoxy-2-propanol, 2-methoxy-1-propanol, dipropeneglycol monomethylethers, propanol hydroperoxides.
n-Butane to maleic anhydride (MA also produced by benzene oxidation) [n-butane + 7/2 O_2 → MA + 4 H_2O] GAS PHASE Fluidized-bed or fixed-bed reactor.	V/P/O (vanadyl pyrophosphate) 80–90%, 65–73%.	Oxidant air. Fluidized-bed: feed 3–5% n-butane, rest air. Fixed-bed: inlet <2.4% n-butane, rest air. Steam is also added.	Alkylphosphates/phosphites are added as gas-phase promoters. Conversion is not total to limit consecutive MA degradation. In fluidized-bed technology, unconverted n-butane is burnt to generate additional high-P steam.	Acrylic acid, acetic acid, CO, CO_2, phthalic anhydride (from benzene).

16.1 Catalytic Selective Oxidation: Main Features

Reaction	Catalyst	Conditions	Notes	By-products
Isobutane to t-butylhydroperoxide (autoxidation) LIQUID PHASE Series of agitated reactors. ![isobutane + O2 → t-BuOOH] 35–40% pp, 50–60%.	No catalyst.	Oxidant O_2. O_2 inlet concentration < 8%. T 110–130 °C, P 30–35 bar.	Injection of citric acid prevents excessive formation of by-products. Unconverted O_2 and isobutane are recycled. Vaporization of a part of the reaction medium is used to remove the heat generated.	t-Butyl alcohol.
Two-step t-butanol (or isobutene) to methacrylic acid (via intermediate methacrolein) ![t-BuOH + 3/2 O2 → methacrolein + 2 H2O]	First step: Co/Fe/Ni/Bi//Mo/O-SiO$_2$ + dopants (crystalline multimetal molybdate) 100%, 85–90%. Second step: P/Mo/V/O (Keggin-type polyoxometalate)-support VAPOR PHASE In-series fixed-bed reactors. 90–95%, 90–95%.	Oxidant air. First step: T 330–350 °C, P 3–4 bar. Steam is also added. Second step: T 300–320 °C, P 2–4 bar. Steam is also added.	Unconverted methacrolein is separated after the second step and recycled. This process is used only in Japan for methylmethacrylate production.	Acetic acid, acetone, acetic acid, CO, CO_2, and heavies.
Cyclohexane to cyclohexanol/cyclohexanone (radicalic autoxidation) ![cyclohexane + O2 → cyclohexanone + H2O] LIQUID/VAPOR PHASE Two or three in-series bubble columns.	Co (or Mn) naphthenate + boric acid 10–15% pp^a, 90–91%.	Oxidant air or O_2. T 150–160 °C, P 8–10 bar. Air feed is differentiated to each reactor. Boric acid makes an ester with cyclohexanol and hinders consecutive oxidations.	Low once-through conversion, to limit consecutive reactions. Water is continuously removed as azeotrope with cyclohexane. The true oxidation product is cyclohexylhydroperoxide, which then undergoes deperoxidation to cyclohexanol and cyclohexanone.	6-Hydroxyhexanoic, n-butyric, n-valeric, succinic, glutaric, and adipic acids.

(continued overleaf)

Table 16.1 (continued)

Oxidation process PHASE reactor	Catalyst Reactant conversion selectivity	Reaction conditions	Remarks on the technology	Main by-products
Cyclohexanol/one to adipic acid O=⟨cyclohexanone⟩ + 3/2 HNO$_3$ → HOOC–(CH$_2$)$_4$–COOH + 3/4 N$_2$O + 3/4 H$_2$O LIQUID PHASE Two (or more) in-series reactors.	Cu/V salts. 100%, 95%.	Oxidant HNO$_3$. First reactor T 60–80 °C. Second reactor 90–100 °C, P 1–4 bar.	A large excess of 50–65% HNO$_3$ is used. The oxidant is reduced to NO, NO$_2$ (both recycled) and N$_2$O (that is either abated or used in a downstream process).	Glutaric acid, succinic acid.
Cyclohexanone + NH$_3$ to cyclohexanone oxime (ammoximation) O=⟨cyclohexanone⟩ + NH$_3$ + H$_2$O$_2$ → NOH=⟨cyclohexanone oxime⟩ + 2 H$_2$O LIQUID PHASE	TiO$_2$-SiO$_2$ (microporous Ti-silicalite: TS-1). 100%, >99% (selectivity on H$_2$O$_2$ basis, 98%).	Oxidant H$_2$O$_2$. T 70–90 °C. Solvent water/methanol.	The reaction between ammonia and hydrogen peroxide generates NH$_2$OH in situ.	

Reaction	Catalyst	Conditions	Notes	Byproducts/Comments
Toluene to benzoic acid (radicalic autoxidation) $\displaystyle\text{C}_6\text{H}_5\text{CH}_3 + 3/2\, O_2 \longrightarrow \text{C}_6\text{H}_5\text{COOH} + H_2O$ **LIQUID/GAS PHASE** In-series continuous stirred reactors.	Co benzoate + NaBr (initiator). 20–40% pp^a, 91–93%.	Oxidant air. T 160–180 °C, P 2–10 bar. Part of vent gas recycled.	Partial vaporization of toluene maintains the temperature.	Benzyl alcohol, benzaldehyde, benzylbenzoate, diphenyl, formic acid, acetic acid, and CO_2.
p-Xylene to terephthalic acid (radicalic autoxidation) $\text{p-}(CH_3)_2C_6H_4 + 3\,O_2 \longrightarrow \text{HOOC-C}_6H_4\text{-COOH} + 2\,H_2O$ **LIQUID/GAS PHASE** Bubble column.	Co/Mn acetates + NaBr (initiator). 98–99%, 93–97%.	Oxidant air. T 180–220 °C, P 15–30 bar. Solvent acetic acid.	The heat of reaction is removed by partial vaporization, condensation, and reflux. Unconverted p-xylene and solvent are recycled.	p-Toluic acid, 4-carboxybenzaldehyde (highly undesired impurity; to eliminate it, the acid suspension in water is hydrogenated).
o-Xylene to phthalic anhydride (PA also produced by naphthalene oxidation) $\text{o-}(CH_3)_2C_6H_4 + 3\,O_2 \longrightarrow \text{phthalic anhydride} + 3\,H_2O$ **GAS PHASE** Fixed-bed multitubular reactor.	$V_2O_5\text{-}TiO_2$ + dopants (Cs, Sb, P). Three–four layers of catalysts with different composition. >99.9%, 78–83%.	Oxidant air. Inlet: 1.5–2.0% o-xylene, rest air. T 350–370 °C (in hot spots, 390–420 °C).	Total conversion is required, to limit the o-xylene concentration in vent gas. S-containing compounds can also be fed as gas phase promoters.	Maleic anhydride, benzoic acid o-toluic acid, phthalide, CO and CO_2.

(continued overleaf)

Table 16.1 (continued)

Oxidation process PHASE reactor	Catalyst Reactant conversion selectivity	Reaction conditions	Remarks on the technology	Main by-products
Cumene to cumylhydroperoxide (radicalic autoxidation) $+ O_2 \longrightarrow $-OOH LIQUID/GAS PHASE Series of multistage reactors (bubble columns).	Mn or Co salts. 25–35% pp[a], 95–97%.	Oxidant O_2 or O_2-enriched air. T 115–130 °C, P 1 bar. Reaction carried out in alkaline aqueous medium.	Cumene must be free of impurities. Once-through conversion is low to limit decomposition of CHP to by-products.	Acetic acid, formic acid, acetophenone, and 2-phenyl-2-propanol.

[a] pp = per passage.

dictates the choice of both the most convenient oxidant and the type of catalyst to be used. This also implies that very different mechanisms may indeed be involved, and different reaction conditions have to be used.

Table 16.1 shows another significant and unusual aspect: with the exception of autoxidations, there are no generally used industrial oxidation catalysts: each reaction uses a specific catalyst which ensures the best performance exclusively in that specific application. Differences between catalysts include not only 'minor' components, such as dopants or types of support, but also the main active component. On the other hand, the discovery of new active phases has boosted both the development and the industrial implementation of new oxidation processes. Examples include:

(i) Ti-substituted silicalite (TS-1), the discovery of which allowed aqueous hydrogen peroxide to be used as an oxidant for compounds insoluble or slightly soluble in water, and is now used industrially for both propene epoxidation and the ammoximation of cyclohexanone to the oxime, the intermediate for caprolactam [11];

(ii) the so-called M1 phase, containing Mo, V, Nb, Te, and/or Sb as components of a crystalline mixed oxide with defined composition [12]. This phase has quite unique properties for the oxidative dehydrogenation (ODH) of ethane to ethene and oxidation to acetic acid and for the ammoxidation of propane to acrylonitrile and oxidation to acrylic acid, and is currently used commercially for the synthesis of acrylonitrile; and

(iii) vanadyl pyrophosphate, $(VO)_2P_2O_7$, shows outstanding catalytic properties for the oxidation of n-butane, which made possible the progressive replacement of benzene with the alkane as raw material for the synthesis of maleic anhydride, starting from the 1980s [4].

16.2
Catalytic Selective Oxidation: What Makes the Development of an Industrial Process More Challenging (and Troublesome) than Other Reactions

The development and industrial implementation of an oxidation process entail not only facing and solving problems related to the determination of the optimal catalyst composition and morphology and the best reaction conditions and reactor type for achieving the maximum space-time yield and selectivity (normal targets for any industrial chemical reaction), but also dealing with both hazards deriving from the possible formation of flammable mixtures and associated phenomena such as thermal run-away (Figure 16.1). Managing highly exothermic reactions is not an unusual problem in chemical productions, but in oxidation the efficient removal of the considerable amounts of heat released per unit mass and unit time cannot

16 Catalytic Selective Oxidation

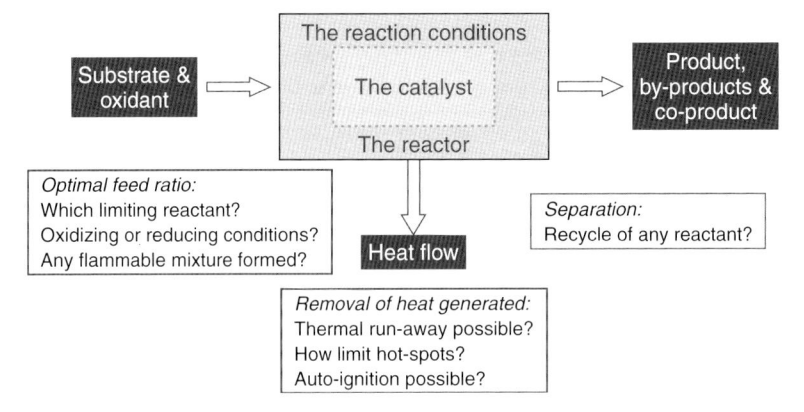

Figure 16.1 The complexity of catalytic selective oxidations, and the problems to be dealt with for the development and management of an oxidation process.

be separated from the control of the composition of the gas or vapor phase, in both gas/solid and gas(vapor)/liquid/(solid) reactors.

The control of the composition of the gas phase not only aims to avoid the formation of flammable mixtures in the reactor dome and in all unfilled up- and downstream process sections (e.g., in adiabatic compressors and in recycle streams), but may also have important implications on both the overall process performance and the management of the effluent streams [2].

This is seen, for instance, in Figure 16.2, comparing the vent composition in ethene oxychlorination for two different compositions of the combined feed to the reactor. The flow rate of the vent stream for the air-based process is much higher than that of the purge stream for the oxygen-based (with recycle of unconverted ethene) process. Moreover, in the latter case the higher concentration of combustible components permits a less costly incineration than with the diluted vent stream of the air-based process.

The overall process performance in the two cases is not very different – indeed, a slightly higher selectivity for 1,2-dichloroethane due to the reduced production of carbon dioxide is claimed with the oxygen-based process; moreover, a better ethene process conversion and thus a better process yield derive from the recycle of the ethene unconverted per pass. What makes the oxygen process definitely more convenient, and environmentally more acceptable, than the air process is the better economics deriving from the easier and less costly treatment of effluent streams.

It is worth mentioning that the catalysts used in the two processes are slightly different; the main component of the active phase is the same in the two cases ($CuCl_2$ supported over γ- or θ-Al_2O_3), but the type of dopant may differ, also depending on the type of reactor technology used (either fixed-bed or fluidized-bed).

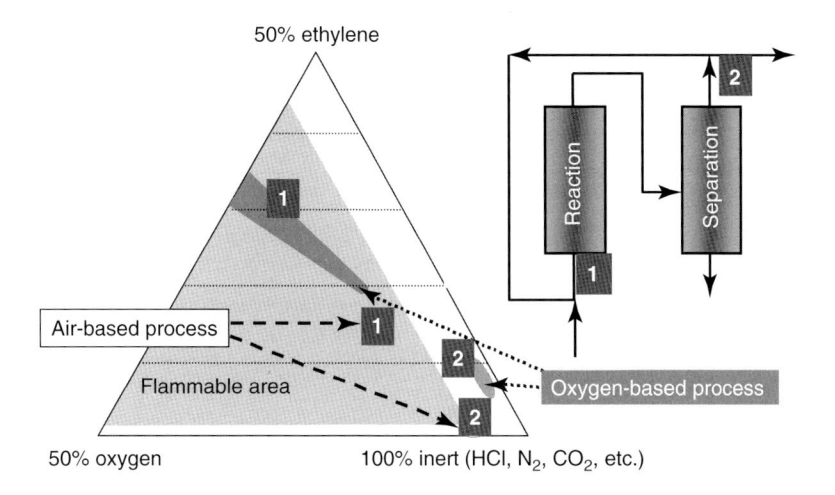

Figure 16.2 A comparison of the effluent stream composition (before incineration) in a typical industrial unit for ethene oxychlorination, either air- or oxygen-based.

Component	Air-based process content (vol%); flow-rate (m³ h⁻¹)	Oxygen-based process content (vol%); Flow-rate (m³ h⁻¹)
O_2 + Ar	4-8; 400–2400	0.1–2.5; <25
C_2H_4	0.1–0.8; 10–240	2–5; <50
CO_x	1–3; >100	15–30; <300
1,2-$C_2H_4Cl_2$ + other chlorinated compounds	0.02–0.2; 2–60	0.5–1; <10
Overall flow rate	10000–30000	<1000

16.3
Catalytic Selective Oxidation: the Forefront in the Continuous Development of More-Sustainable Industrial Technologies

Today it is clear that the challenge of creating a sustainable future has to be directly addressed by the chemical industry [3, 4, 13]. It is necessary to reconsider the types of chemicals, how they are produced, and how they are used; emphasis must be put on the need to combine social objectives with both the management of natural resources and the preservation of the natural bases for life. We need alternative paths for chemical processes that either require less energy input or even produce energy, while generating the desired compound with high selectivity and producing limited waste that is eventually used as a feedstock for other processes: these have always been the targets for better process economics, but must now be considered rational first choices.

Indeed, more sustainable processes often also have better economics than traditional ones; in fact, fewer by-products mean lower costs for raw materials,

> 1. Make reactions where the amount of co-product is minimal
>   - use reactants with maximum atomic incorporation into final products.
> 2. Search for reaction conditions at which the selectivity is the highest, rather than productivity or yield. Recycle of the reactant is often a feasible option.
> 3. Search for methods to immobilize catalysts (anchoring, grafting,...) over supports that can be filtered and easily recovered.
> 4. Search for solvents, when needed, that are non-toxic, and can be easily separated and re-utilized.
> 5. Consider that the reaction investigated may be a part of a more complex process. Process integration may lead to less expensive and more efficient technologies.

Scheme 16.1 Some 'Green Chemistry' rules for the development of more sustainable oxidation processes.

smaller investment for their separation, and lower costs for their disposal. Higher productivity means less investment because smaller reactors can be used, and increasing energy demand leads to investment in heat integration in order to save energy.

In this context, oxidation catalysis has been playing and will continue to play a leading role. The reason for this is that oxidation is the tool for the production of huge quantities of intermediates and monomers for the polymer industry. The impact of these production processes on the environment might have been much greater than it actually is if considerable efforts had not been devoted to the continuous improvement of the technologies used for the production of these chemicals, including: the replacement of toxic or dangerous reactants, better heat recovery and energy integration in the plant, recovery of waste streams and abatement of tail emissions, and downstream use of by-products. However, some processes still co-produce large amounts of wastes, or operate under conditions leading to non-optimal selectivity for the desired compound.

Innovation in oxidation catalysis should not only involve the adoption of advanced synthetic procedures and methods in compliance with Green Chemistry rules (Scheme 16.1), but should also follow the more general concepts that allow chemical production to meet the goal of better sustainability, for example, (i) integrate chemical production to minimize transport and storage, (ii) integrate processes (catalysis, reaction, and separation), (iii) design inherently safe processes, (iv) adopt the intensification strategy, and (v) develop direct synthesis, avoiding multistep processes. The strategies that can be adopted for better selectivity for the desired product – which is certainly still the main problem in catalytic oxidation – fit within the broader scope of a more sustainable chemistry.

16.4
The Main Issue in Catalytic Oxidation: the Control of Selectivity

For many catalytic oxidations, selectivity is still the major issue, either potentially providing considerable profit increases in the industrial processes already being

16.4 The Main Issue in Catalytic Oxidation: the Control of Selectivity

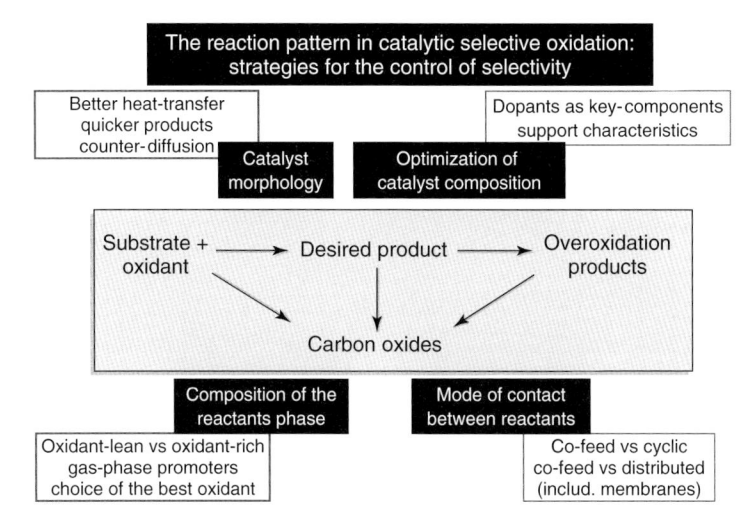

Figure 16.3 Strategies to be adopted for a better selectivity in oxidation processes, in relation to the reaction network.

used (for some reactions, 1% additional selectivity for the desired product could lead to huge economic returns) or being the main problem to be solved to obtain new economically viable processes. In fact, the achieved selectivity is often not optimal because of the following problems: (i) the desired product is more reactive toward the oxidant than the reactant itself under the conditions necessary for the reaction to occur at an acceptable rate; (ii) in complex reactions, intermediately formed compounds (such as products of partial oxidation other than the desired one or products of oxidative cracking or combustion) may also be a source of by-product formation; (iii) the reactant used may contain various functional groups where the attack by the oxidant may occur; and (iv) in heterogeneous, gas-phase reactions a contribution to by-product formation may derive from homogeneous radical reactions. Indeed, the reaction network in catalytic selective oxidations may include both parallel reactions (a selective pathway finally leading to the desired compound plus various unselective routes leading to by-products) and consecutive reactions involving both the reaction intermediates and the desired product. Therefore, careful control of the reaction conditions and an optimal mode for contacting the reactants are fundamental to achieve the highest selectivity (Figure 16.3).

It is not uncommon for the various reactions involved in the complex reaction network to have different activation energies; therefore, temperature may be a powerful tool for the control of the relevant rates for each of the reactions involved. On the other hand, the selectivity is obviously also affected by the conversion, and thus a per-pass moderate conversion (with recycle of the unconverted reactant) may be economically more convenient than a total conversion. This is illustrated in Figure 16.4, which shows a qualitative assessment of the profits generated as a

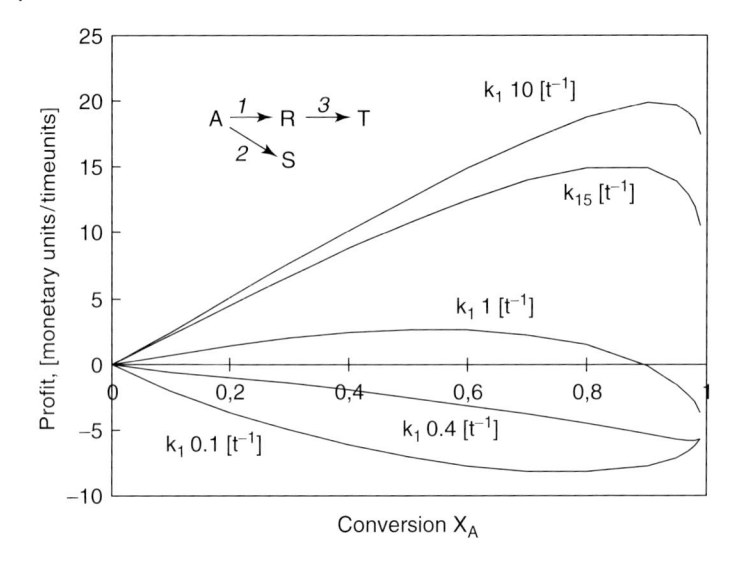

Figure 16.4 Effects of the conversion of reactant A, X_A, on profits for different values of the kinetic constant $k_1 [s^{-1}]$. Initial concentration of A, $C_{A0} = 1$ mol (volume unit)$^{-1}$. Reactor volume $V = 100$ volume unit (batch reactor). Value of the desired product R, $S_R = 1$ monetary unit (mole)$^{-1}$. Cost of the reactant A, $S_A = -0.4$ monetary unit (mole)$^{-1}$. Cost of by-products S and T (for separation and abatement or disposal, $S_S = -0.2$ and $S_T = -0.1$ monetary unit (mole)$^{-1}$, respectively). A simplified profit equation has been used, including only stream costs and an operative cost $S_{op} = 2$ monetary unit (batch time)$^{-1}$. The reaction network includes the following reactions (all being first order with respect to the reactant): A → R; A → S ($k_2 = 0.5$ (time unit)$^{-1}$); R → T ($k_3 = 0.3$ (time unit)$^{-1}$). The equation $X_A = 1 - e^{[-(k_1+k_2)t]}$ correlates t with X_A.

function of the conversion of reactant A (and hence of the reaction time t, or contact time τ) to yield the desired product R (reaction 1); the hypothetical reaction network also includes the parallel transformation of A into the by-product S (reaction 2), and the consecutive transformation of R into the by-product T (reaction 3) [14]. In oxidation catalysis, S and T are usually both represented by CO_2, and in this case the reaction network is known as a *'triangular scheme'*.

Needless to say, the catalyst composition and morphology also play a fundamental role (Figure 16.3). For instance, a proper morphology may be important to foster the counterdiffusion of the desired reactant toward the fluid phase, thus saving it from undesired consecutive reactions. In addition, the control of both mass- and heat-transfer phenomena – which may become limiting because of the considerable heat of reaction that may develop at the catalyst surface and the resulting surface temperature– is also an important tool. In fact, under these conditions, the surface reaction becomes so fast that interparticle transfer phenomena may become the limiting step.

16.5
Dream Reactions in Catalytic Selective Oxidation: a Few Examples (Some Sustainable, Some Not Sustainable)

Scheme 16.2 consists of a list of catalyzed 'dream oxidations' that have been the subject of study over the past 20 years. Some of these are now commercial processes, whereas others are not yet very promising in terms of obtaining yields or selectivities that could make the process commercially exploitable. Some of the reactions shown could be better alternatives to those currently used in industry because they would make use of greener or more economical oxidants, for example: the synthesis of propene oxide with oxygen instead of hydroperoxides, the synthesis of adipic acid by means of cyclohexanone oxidation with oxygen (instead of nitric acid, which coproduces N_2O), the ammoximation of cyclohexanone with hydrogen

---

Alternative oxidants
Methane to syngas with oxygen
Propene to propene oxide with air (oxygen)
Propene to propene oxide with N_2O
Propene to propene oxide with hydrogen peroxide (now commercial)
Propene to propene oxide with hydrogen and oxygen
Cyclohexanone to adipic acid with air (oxygen)
Benzene to phenol with hydrogen peroxide (eventually generated in-situ)
Benzene to phenol with N_2O (implemented in a demonstration unit)
Cyclohexanone to cyclohexanoneoxime with air (oxygen)
Cyclohexanone to cyclohexanoneoxime with hydrogen peroxide (now commercial)

Alternative raw materials or new routes
Methane to methanol (single step)
Methane to ethene
Methane to formaldehyde (single step)
Ethene to vinyl chloride (single step)
Ethene to vinyl chloride (single step)
Ethene to acetic acid (single step)
Ethane to acetic acid (single step) (implemented in demonstration unit)
Propane to acrylonitrile (single step) (now commercial)
Propane to acrylic acid (single step)
Propane to propene oxide (single step)
Alkanes (ethane, propane, n-butane) to olefins with air (exothermal)
Isobutane to methacrylic acid
n-Butane to 1,4-butanediol
n-Hexane to adipic acid
Cyclohexane to adipic acid (single step)
Cyclohexene to adipic acid
β-Picoline to nicotinic acid (niacin) (single step)

Scheme 16.2 Some 'Dream Catalytic Selective Oxidations' studied in recent years, mainly in the field of base chemicals and petrochemicals.

peroxide (now a commercial process) or even with air. Other reactions represent new alternative approaches to those currently used because of the potential benefits deriving from greener reactions (which produce less wastes and less co-products), from single-step rather than multi-step processes, or from using cheaper raw materials. It is worth noting that many of the reactions use an alkane as the starting reactant, and this obviously poses new challenges for its activation and selective transformation compared with conventional reactions that start from alkenes or other more reactive molecules.

One of the most attractive, but also most challenging, new synthetic routes is the direct oxidation of isobutane to methacrylic acid [15]. The interest in this reaction derives from the non-sustainability of the traditional process for obtaining methylmethacrylate, which uses toxic reactants and intermediates and also coproduces large amounts of ammonium sulfate. This one-step synthesis of an unsaturated acid starting from an alkane is certainly an interesting option; however, despite the considerable effort spent in seeking a catalyst offering better than 60% selectivity to methacrylic acid, even at medium-to-low isobutane conversion, and the patents and scientific papers published, the performance reported is inadequate for making this option economically viable. The best-performing catalysts are based on Keggin-type P/Mo/V polyoxometalates; these systems are able to perform the various successive steps in the multi-electron transformation of the alkane into the unsaturated acid: (i) the ODH of isobutane to isobutene (in fact, the olefin is the reaction intermediate), (ii) the allylic oxidation of the olefin into methacrolein, and (iii) the oxidation of the aldehyde into methacrylic acid. Apparently, the reaction is not very different from the one-step oxidation of *n*-butane to maleic anhydride, and in fact the same polyoxometalates also catalyze the latter reaction, but with a selectivity that is much lower than that obtained with vanadyl pyrophosphate (the industrial catalyst for *n*-butane oxidation to maleic anhydride).

On the other hand, vanadyl pyrophosphate catalyzes the oxidation of isobutane to methacrylic acid, but with lower selectivity than that achieved with the polyoxometalates. Indeed, the two reactions, although apparently similar, present substantial differences: (i) maleic anhydride is more stable under reaction conditions than methacrylic acid, and thus less prone to undergoing consecutive reactions of oxidative degradation (to acetic acid) or combustion and (ii) the mechanism of *n*-butane oxidation involves a first step of ODH of alkanes into butenes, but then the latter are oxidehydrogenated very quickly again to yield butadiene (a reaction that can obviously not occur in the case of isobutane) (Scheme 16.3) [1]. This means that the catalyst for *n*-butane oxidation must possess the ability to cause the electrophilic O insertion into the intermediately formed butadiene to yield 2,5-dihydrofuran – a property characteristic of vanadyl pyrophosphate, but not of polyoxometalates. On the other hand, vanadyl pyrophosphate is less efficient then polyoxometalates in performing allylic O-insertion into isobutene. This does not mean that some allylic oxidation may not occur on butene even with vanadyl pyrophosphate; in fact crotonaldehyde and methylvinylketone are by-products obtained in small amounts from *n*-butane. These compounds, however, undergo combustion quickly and do not contribute to the formation of maleic anhydride.

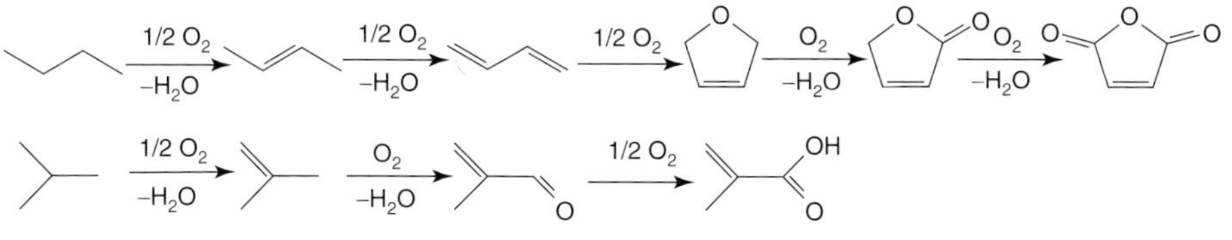

Scheme 16.3 Reaction network in n-butane oxidation to maleic anhydride and in isobutane oxidation to methacrylic acid.

For the reactions shown in Scheme 16.2, the following are obstacles that limit their industrial implementation:

1) The low selectivity or the low productivity (space-time yield) to the desired product, such as in (i) propane ODH to propene (low selectivity because of the consecutive combustion of the olefin) [16], (ii) benzene hydroxylation to phenol (low selectivity with respect to hydrogen peroxide), also in reactions which generate hydrogen peroxide *in situ* (low selectivity with respect to hydrogen), and (iii) in ethane or ethene direct oxychlorination to vinyl chloride (low selectivity because of the formation of other chlorinated compounds).
2) The excessive cost of the reactant, as in the case of cyclohexene direct oxidation to adipic acid with hydrogen peroxide, where the 4 mol of H_2O_2 needed per mole of product are already more expensive that the value of the adipic acid itself. Generally speaking, this problem is encountered when hydrogen peroxide is used as the oxidant. However, this can be overcome by integrating the hydrogen peroxide production using the classical anthraquinone route in a large dedicated plant with downstream utilization, for example, propene epoxidation [4]. Another example is the direct oxidation of n-hexane to adipic acid; the limited availability of this hydrocarbon would constitute a serious problem for large-scale production facilities.
3) The need for reaction conditions or solvents that may lead to harmful environments or pose problems in product separation or purification, such as in the case of the one-step oxidation of cyclohexane to adipic acid, in which the acetic acid solvent not only causes corrosion problems, but also remains in traces in the final product even after several purification steps.
4) The catalyst lifetime; this is the problem encountered, for instance, with polyoxometalates when they are used for gas-phase reactions.

16.6
A New Golden Age for Catalytic Selective Oxidation?

Over the past few years, several new liquid-phase oxidation processes have been successfully implemented at an industrial level, while others have been used in pilot or demonstration units [4, 6, 13]. Examples include, among others: (i) the epoxidation of propene with hydrogen peroxide (the BASF-Dow process and the Evonik-Uhde process); (ii) the epoxidation of propene with cumylhydroperoxide (CHP), developed by Sumitomo, the peculiarity of this process being the reuse of the co-product cumylalcohol, which is reoxidized to hydroperoxide; (iii) the non-catalyzed carboxidation of cyclododecatriene to cyclododecandienone with N_2O, intermediate for cyclododecanone synthesis, developed by BASF; (iv) the one-step oxidation of cyclohexane to adipic acid with oxygen, catalyzed by N-hydroxyphthalimide and Co/Mn complexes, currently used by Daicel in a pilot unit; (v) the ammoximation of cyclohexanone to cyclohexanoneoxime with ammonia and hydrogen peroxide, intermediate for caprolactam synthesis, formerly developed by eni SpA and now commercially applied by Sumitomo; and (vi) the ammoximation of cyclododecanone to the oxime, intermediate for laurolactam synthesis, the monomer for polyamide 12, developed by Evonik. Also worth mentioning is the oxidation of hydrogen to hydrogen peroxide with oxygen, which was implemented in a pilot unit by Evonik-Headwaters using a Pd nanocatalyst developed by Headwaters Technology.

These innovations have been possible not only because of the development of new catalytic systems capable of obtaining the desired product with high selectivity and space-time yield, but also because of the integration of the oxidant production with its use, as in propene epoxidation with hydrogen peroxide. Also, the development of a new process making use of N_2O, such as the carboxidation of cyclo-olefins and the catalyzed hydroxylation of benzene to phenol (the latter having been developed by Solutia and the Boreskov Institute of Catalysis and implemented up to a demonstration unit [17]) open up new prospects for the integration of processes that coproduce N_2O with new oxidation processes for olefin epoxidation, aromatics hydroxylation, and alkanes oxidation.

An emblematic example of how various approaches, starting from different organic reactants and using different oxidants, can be used in oxidation catalysis is the synthesis of adipic acid (Scheme 16.4) [6, 18]. The only industrial processes are those starting from benzene, which is either totally hydrogenated to cyclohexane, partially hydrogenated to cyclohexene, or oxidized to phenol (and acetone) via intermediate CHP. These intermediates are then transformed into either cyclohexanol

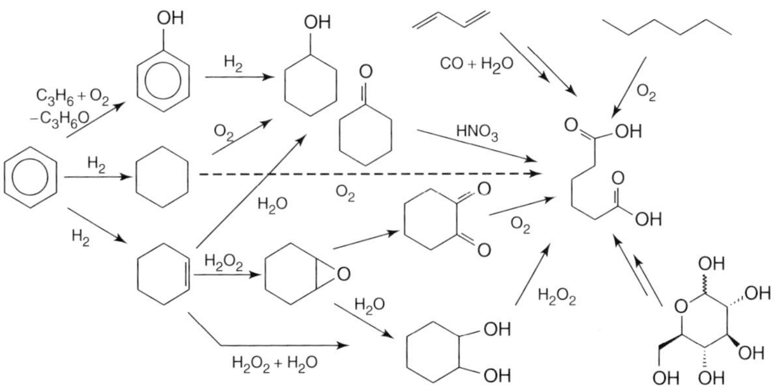

Scheme 16.4 Several possible synthetic routes for adipic acid production.

or cyclohexanol/cyclohexanone mixtures, which are lastly oxidized to adipic acid with nitric acid (and co-production of N_2O). However, several alternative pathways are possible, for example, from butadiene (a non-oxidative approach) or even from glucose or n-hexane. For each of these reactions, the key issues are the development of more selective catalytic systems and the decrease in the operating costs.

Also, in the case of gas-phase oxidations, new processes – such as the ammoxidation of propane to acrylonitrile, which was developed by Asahi with a 70 000 tons/year plant started up in Korea – have in recent years been implemented at an industrial level. Furthermore, various technologies have been scaled up to pilot or demonstration units, such as the epoxidation of propene with hydrogen peroxide vapor in a micro-structured unit, developed by a consortium including Evonik and Uhde, and the ethane oxidation to acetic acid, developed by Sabic. Various industrial processes have also been considerably improved, often with a significant reduction in the sustainability stress mark [19], as in the case of the Shell OMEGA process, integrating ethene oxide production with its hydrolysis to ethylene glycol, and the BP Leap and Celanese Vantage processes for vinylacetate production, integrated with upstream acetic acid production.

16.7
Conclusions: Several Opportunities for More Sustainable Oxidation Processes

There are several promising fields of development for a greener and more efficient chemical industry in the area of catalytic selective oxidation. These range from the use of alternative feedstock, for example, bioplatform molecules and alkanes in place of traditional building blocks derived from oil, to the use of new technologies like miniaturized devices, alternative approaches for contacting reactants (e.g., anaerobic oxidation), and new types of reactions and catalysts. It is also clear that a sizable effort is needed in order to develop alternative technologies that are more sustainable than traditional ones, while being economically sound at the same time. However, the recently developed oxidation processes show that there is still room not only for the improvement of the processes currently in use, but also for an innovative vision of chemical production.

References

1. Centi, G., Cavani, F., and Trifirò, F. (2001) *Selective Oxidation by Heterogeneous Catalysis. Recent Developments*, Plenum Publishing Corporation, London, New York.
2. Arpentinier, P., Cavani, F., and Trifirò, F. (2001) *The Technology of Catalytic Oxidations*, Editions Technip, Paris.
3. Cavani, F. and N. Ballarini (2009) in *Modern Heterogeneous Oxidation Catalysis*, Mizuno, N. (Ed), Wiley-VCH Verlag GmbH & Co. KGaA, pp. 289–331.
4. Cavani, F. and Teles, J.H. (2009) Sustainability in catalytic oxidation: an alternative approach or a structural evolution? *ChemSusChem*, **2** (6), 508–534.
5. Hodnett, B.K. (2000) *Heterogeneous Catalytic Oxidation*, John Wiley & Sons, Ltd, Chichester.

6. Cavani, F. et al. (2009) *Sustainable Industrial Chemistry*, Wiley-VCH Verlag GmbH & Co. KGaA.
7. Chauvel, A. and Lefebvre, G. (1989) *Petrochemical Processes*, E. Technip, Paris.
8. Green, M.M. and Wittcoff, H.A. (2003) *Organic Chemistry Principles and Industrial Practice*, Wiley-VCH Verlag GmbH, Weinheim.
9. Weissermel, K. and Arpe, H.-J. (2003) *Industrial Organic Chemistry*, Wiley-VCH Verlag GmbH.
10. Moulijn, J.A., Makkee, M., and van Diepen, A. (2001) *Chemical Process Technology*, John Wiley & Sons, Ltd, Chichester.
11. Clerici, M.G., Ricci, M., and Strukul, G. (2007) *Metal-Catalysis in Industrial Organic Processes*, Chapter 2, The Royal Society of Chemistry, pp. 23–78.
12. Cavani, F., Centi, G., and Marion, P. (2009) in *Metal Oxide Catalysis* (eds S.D. Jackson and J.S.J. Hargreaves), Wiley-VCH Verlag GmbH & Co. KGaA, pp. 771–818.
13. Cavani, F., Ballarini, N., and Luciani, S. (2009) Catalysis for society: towards improved process efficiency in catalytic selective oxidations. *Top. Catal.*, **52** (8), 935–947.
14. Cavani, F. (2004) *La Gestione e lo Sviluppo dei Processi Chimici Industriali (in Italian)*, Edizioni CLUEB, Bologna.
15. Ballarini, N. et al. (2007) in *Methods and Reagents for Green Chemistry* (eds P. Tundo, A.Perosa, and F. Zecchini), John Wiley & Sons, Inc., pp. 265–279.
16. Cavani, F., Ballarini, N., and Cericola, A. (2007) Oxidative dehydrogenation of ethane and propane: how far from commercial implementation? *Catal. Today*, **127** (1–4), 113–131.
17. Panov, G.I., Dubkov, K.A., and Starokon, E.V. (2006) Active oxygen in selective oxidation catalysis. *Catal. Today*, **117** (1–3), 148–155.
18. Cavani, F. (2010) Catalytic selective oxidation faces the sustainability challenge: turning points, objectives reached, old approaches revisited and solutions still requiring further investigation. *J. Chem. Technol. Biotechnol.*, **85** (9), 1175–1183.
19. Cavani, F. (2010) Catalytic selective oxidation: the forefront in the challenge for a more sustainable chemical industry. *Catal. Today*, **157** (1–4), 8–15.

17
High-Temperature Catalysis: Role of Heterogeneous, Homogeneous, and Radical Chemistry
Olaf Deutschmann

17.1
Introduction

This chapter focuses on heterogeneously catalyzed gas phase reactions and their interaction with the surrounding flow field in high-temperature catalysis. Understanding and optimization of heterogeneous reactive systems require knowledge of the physical and chemical processes on a molecular level. In particular, the short contact times and high temperatures at which reactions occur on the catalyst and in the gas phase mean that the interactions between transport and chemistry become important.

High-temperature catalysis is not a new concept; the Ostwald process for NO production by oxidation of ammonia over noble metal gauzes at temperatures above 1000 °C and residence times of less than a microsecond has been technically applied for decades; total oxidation of hydrogen and methane (catalytic combustion) over platinum catalysts were used even before Berzelius proposed the term '*catalysis*.' Recently, however, high-temperature catalysis has been extensively discussed again, in particular in the light of the synthesis of basic chemicals and hydrogen, and high-temperature fuel cells.

Catalytic partial oxidation (CPOX) of natural gas over noble metal catalysts at short contact times offers a promising alternative route for the production of synthesis gas [1, 2], olefins [3, 4], and hydrogen. For instance, synthesis gas, catalytically produced by steam and autothermal reforming, is needed in (gas-to-liquids) plants for synthetic fuels which are currently under development. CPOX of gasoline, diesel, or alcohols to synthesis gas or hydrogen may soon play a usually role in mobile applications for the reduction of pollutant emissions and in auxiliary power units.

For any fuel other than hydrogen, catalytic reactions are likely to occur in the anode of a solid oxide fuel cell (SOFC), leading to a complex chemical composition at the anode-electrolyte interface [5]. Primarily, the products of the electrochemical reactions, H_2O and CO_2, drive the catalytic chemistry in the anode. For the application of hydrocarbon and alcohol-containing fuels, an understanding of the catalytic kinetics is vital for the precise prediction of fuel utilization and

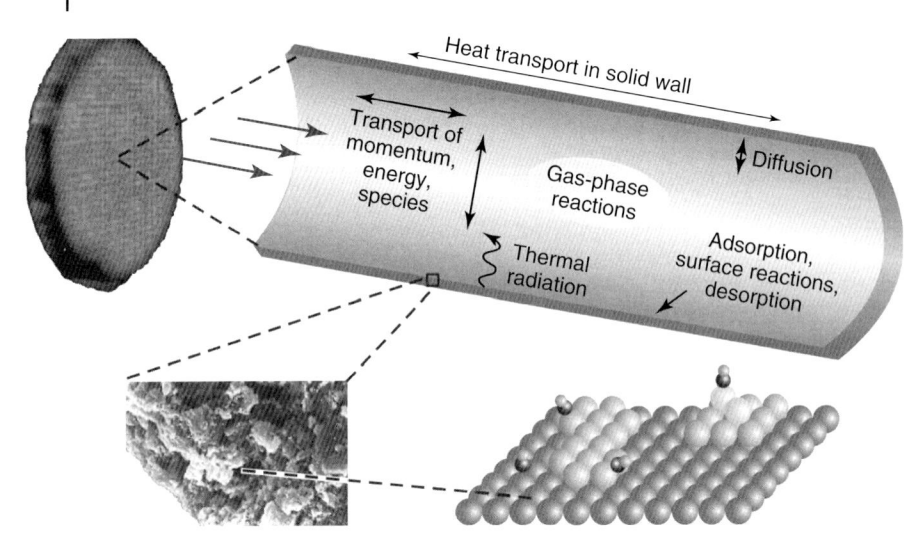

Figure 17.1 Catalytic combustion monolith and physical and chemical process occurring in the single monolith channel [6].

performance [7]. Coupling of the thermocatalytic reactions with the electrochemical processes and mass and heat transport in the cell will be discussed as an example of an anode-supported SOFC operated with methane-containing fuels and Ni/YSZ (yttria-stabilized zirconia) anode structure.

17.2
Fundamentals

Catalytic reactors are generally characterized by a complex interaction of various physical and chemical processes. Monolithic reactors can serve as an example in which partial oxidation and reforming of hydrocarbons, combustion of natural gas, and reduction of pollutant emissions from automobiles are frequently carried out. Figure 17.1 illustrates the physics and chemistry in a catalytic combustion monolith that glows at a temperature of about 1300 K due to the exothermic oxidation reactions. In each channel of the monolith, the transport of momentum, energy, and chemical species occurs not only in the flow (axial) direction, but also in the radial direction. The reactants diffuse to the inner channel wall, which is coated with the catalytic material, where the gaseous species adsorb and react on the surface. The products and intermediates desorb and diffuse back into the bulk flow. Due to the high temperatures, the chemical species may also react homogeneously in the gas phase. In catalytic reactors, the catalyst material is often dispersed in porous structures like washcoats or pellets. Mass transport in the fluid phase and chemical reactions are then superimposed by diffusion of the species to the active catalytic centers in the pores.

The temperature distribution depends on the interaction of heat convection and conduction in the fluid, heat release due to chemical reactions, heat transport in the solid material, and thermal radiation. If the feed conditions vary in time and space and/or heat transfer occurs between the reactor and the ambience, a non-uniform temperature distribution over the entire monolith will result, and the behavior will differ from channel to channel [8].

Today, the challenge in catalysis is not only to develop new catalysts to synthesize a desired product, but also to understand the interaction of the catalyst with the surrounding reactive flow field. Sometimes, exploitation of these interactions can lead to the desired product selectivity and yield. Detailed introductions to fluid dynamics and transport phenomena can be found in Refs. [9–13], and to the coupling with heterogeneous reactions in Refs. [13–14].

17.2.1
Heterogeneous Reaction Mechanisms

Following the concept of the mean-field approximation, simple rate equations can be used to model the surface reaction rate. However, first a reaction mechanism needs to be known. A tentative mechanism can often be proposed based on experimental surface science studies, on analogy to gas phase kinetics and organometallic compounds, and on theoretical studies, increasingly including density functional theory (DFT) calculations. This mechanism should include all possible paths for the formation of the chemical species under consideration in order to be 'elementary-like' and thus applicable over a wide range of conditions. The idea of the mechanism then needs to be evaluated by comparing numerous experimentally derived data with theoretical predictions based on the mechanism. Here, the simulations of the laboratory reactors require appropriate models for all significant processes in order to evaluate the intrinsic kinetics. Sensitivity analysis leads to the crucial steps in the mechanism for which refined kinetic experiments and data may be needed.

Since the early 1990s, many groups, following this concept, have developed surface reaction mechanisms for high-temperature catalysis which have been adapted from modeling homogeneous gas phase reactions, particularly in the fields of combustion [11] and pyrolysis [15] of hydrocarbons. Consequently, this concept is now often used when high-temperature processes in catalysis are considered, particularly in radical interactions between the solid phase (catalyst) and the surrounding gas phase (fluid flow).

In this concept, the surface reaction rate is related to the size of the computational cell in the flow field simulation assuming that the local state of the active surface can be represented by mean values for this cell. Hence, this model assumes randomly distributed adsorbates. The state of the catalytic surface is described by the temperature T and a set of surface coverages θ_i. The surface temperature and the coverages depend on time and the macroscopic position in the reactor, but are averaged over microscopic local fluctuations. Under those assumptions the molar net production rate $\dot{s}_i$ of a chemical species on the catalyst is given as

$$\dot{s}_i = \sum_{k=1}^{K_s} v_{ik} k_{f_k} \prod_{j=1}^{N_g+N_s+N_b} c_j^{v'_{jk}} \qquad (17.1)$$

Here, K_s is the number of surface reactions, c_j are the species concentrations, which are given, for example, in moles per square meter for the N_s adsorbed species and in, for example, moles per cubic meter for the N_g and N_b gaseous and bulk species.

According to the relation

$$\Theta_i = c_i \sigma_i \Gamma^{-1} \qquad (17.2)$$

where Γ is surface site density with a coordination number σ_i describing the number of surface sites which are covered by the adsorbed species, the variations of surface coverages follow the equation

$$\frac{\partial \Theta_i}{\partial t} = \frac{\dot{s}_i \sigma_i}{\Gamma} \qquad (17.3)$$

Since the reactor temperature and concentrations of gaseous species depend on the local position in the reactor, the set of surface coverages also varies with position. However, no lateral interaction of the surface species between different locations on the catalytic surface is modeled. This assumption is justified by the fact that the computational cells in reactor simulations are usually much larger than the range of lateral interactions of the surface processes.

Since the binding states of adsorption of all species vary with the surface coverage, the expression for the rate coefficient k_{f_k} is commonly extended by two additional parameters, μ_{i_k} and ε_{i_k} [13, 16]:

$$k_{f_k} = A_k T^{\beta_k} \exp\left[\frac{-E_{a_k}}{RT}\right] \prod_{i=1}^{N_s} \Theta_i^{\mu_{i_k}} \exp\left[\frac{\varepsilon_{i_k} \Theta_i}{RT}\right] \qquad (17.4)$$

A crucial issue with many of the surface mechanisms published is thermodynamic consistency [14]. Lately, optimization procedures enforcing overall thermodynamic consistency have been applied to overcome this problem [17, 18].

In particular, oxidation reactions, in which radical interactions play a very significant role, have been modeled extensively using this approach, for example, the oxidation of hydrogen [19–25], CO [26–28], methane [29–34], and ethane [4, 35–37] over Pt, and the formation of synthesis gas over Rh [34, 38]. Lately, mechanisms have been established for more complex reaction systems, for instance, three-way catalysts [39] or Chemical Vapor Deposition (CVD) reactors for the formation of diamond [40, 41], silica [42], and nanotubes [43]. In most of these reactions, adsorption and desorption of radicals is included and these steps are significant not only for the heterogeneous reaction but also for homogeneous conversion in the surrounding fluid. In most cases, the catalyst acts as sink for radicals produced in the gas phase, and hence radical adsorption inhibits gas phase reactions such as those discussed below for oxy-dehydrogenation of alkanes by high-temperature catalysis.

17.2.2
Homogeneous Reactions

In many catalytic reactors, the reactions do not only occur on the catalyst surface but also in the fluid flow. In some reactors even, the desired products are mainly produced in the gas phase, for instance in the oxidative dehydrogenation of paraffins to olefins over noble metals at short contact times and high temperature as discussed below [4, 37, 44–49]. Such cases are dominated by the interaction between gas phase and surface kinetics and transport. Therefore, any reactor simulation needs to include an appropriate model for the homogeneous kinetics along with the flow models. With v'_{ik}, v''_{ik} being the stoichiometric coefficients and an Arrhenius-like rate expression, the chemical source term of homogeneous reactions can be expressed by

$$R_i^{\text{hom}} = M_i \sum_{k=1}^{K_g} (v''_{ik} - v'_{ik}) A_k T^{\beta_k} \exp\left[\frac{-E_{a_k}}{RT}\right] \prod_{j=1}^{N_g} \left(\frac{Y_j \rho}{M_j}\right)^{a_{jk}} \quad (17.5)$$

Here, A_k is the pre-exponential factor, β_k is the temperature exponent, E_{a_k} is the activation energy, and a_{jk} is the order of reaction k related to the concentration of species j. Various reliable sets of elementary reactions are available for modeling homogeneous gas phase reactions, for instance, for total [11] and partial oxidation, and for pyrolysis of hydrocarbons. The advantage of the application of elementary reactions is that the reaction orders a_{jk} in Eq. (17.5) equal the stoichiometric coefficients v'_{jk}.

17.2.3
Coupling of Chemistry with Mass and Heat Transport

The chemical processes at the surface can be coupled with the surrounding flow field by boundary conditions for the species-continuity equations at the gas–surface interface [13, 16]:

$$\vec{n}(\vec{j}_i + \rho \vec{v}_{\text{Stef}} Y_i) = R_i^{\text{het}} \quad (17.6)$$

Here $\vec{n}$ is the outward-pointing unit vector normal to the surface, $\vec{j}_i$ is the diffusion mass flux of species i, and R_i^{het} is the heterogeneous surface reaction rate, which is given per unit geometric surface area, corresponding to the reactor geometry, in kilograms per meter per second. The Stefan velocity $\vec{v}_{\text{Stef}}$ occurs at the surface if there is a net mass flux between the surface and the gas phase [29, 50]. The calculation of R_i^{het} requires the knowledge of the amount of catalytically active surface area in relation to the geometric surface area, here denoted by $F_{\text{cat/geo}}$, at the gas–surface interface:

$$R_i^{\text{het}} = \eta F_{\text{cat/geo}} M_i \dot{s}_i \quad (17.7)$$

Here, $\dot{s}_i$ is the molar net production rate of gas phase species i, given in moles per square meter per second; the area now refers to the actual catalytically active surface area. $F_{\text{cat/geo}}$ can be determined experimentally, for example, by chemisorption measurements. The effect of internal mass transfer resistance for catalyst dispersed in a porous media is included by the effectiveness factor η [12, 51]. For more detailed models for transport in porous media, see Refs [52] and [53].

Modeling the flow field in laminar and turbulent flows is discussed in many textbooks and review articles [9, 11]. Even though the implementation of Eq. (17.5) in these fluid flow models is straightforward, an additional highly nonlinear coupling is introduced into the governing equations describing the flow field (leading to considerable computational efforts). The nonlinearity, the large number of chemical species, and the fact that chemical reactions exhibit a large range of time scales, in particular when radicals are involved, make the solution of these equation systems challenging. In particular for turbulent flows, but sometimes even for laminar flows, the solution of the system is too CPU time-consuming with current numerical algorithms and computer capacities. This calls for the application of reduction algorithms for large reaction mechanisms, for instance, by the extraction of the intrinsic low-dimensional manifolds of trajectories in chemical space [54], which can be applied for heterogeneous reactions [55]. Another approach is to use 'as little chemistry as necessary.' In these so-called adaptive chemistry methods, the construction of the reaction mechanism includes only steps relevant to the application studied [56].

17.2.4
Monolithic Catalysts

As an example of a modeling a high-temperature catalyst, catalytically coated monolithic structures such as that shown in Figure 17.1 are discussed. An efficient approach, which still includes all fundamental aspects, is often used for modeling catalytic monoliths, this being based on the combination of simulations of a representative number of channels with the simulation of the temperature profiles of the solid structure, treating the latter one as a continuum [57, 58]. This approach is the basis for the computer code DETCHEM^{MONOLITH} [59], which has been applied to model the transient behavior of catalytic monoliths. The code combines a transient three-dimensional simulation of a catalytic monolith with a 2D model of the single-channel flow field based on the boundary layer approximation. It uses detailed models for homogeneous gas phase chemistry and heterogeneous surface chemistry, and contains a model for the description of pore diffusion in washcoats.

The numerical procedure, as sketched in Figure 17.2, is based on the following ideas: The residence time of the reactive gas in the monolith channels is much smaller than the unsteadiness of the inlet conditions and the thermal response of the solid monolith structure. Under these assumptions, the time scales of the channel flow are decoupled from the temporal temperature variations of the solid, and the following procedure can be applied: A transient multi-dimensional heat balance is solved for the monolithic structure including the thermal insulation and reactor walls, which are treated as porous continuum. This simulation of the heat balance provides the temperature profiles along the channel walls. At each time step the reactive flow through a representative number of single channels is

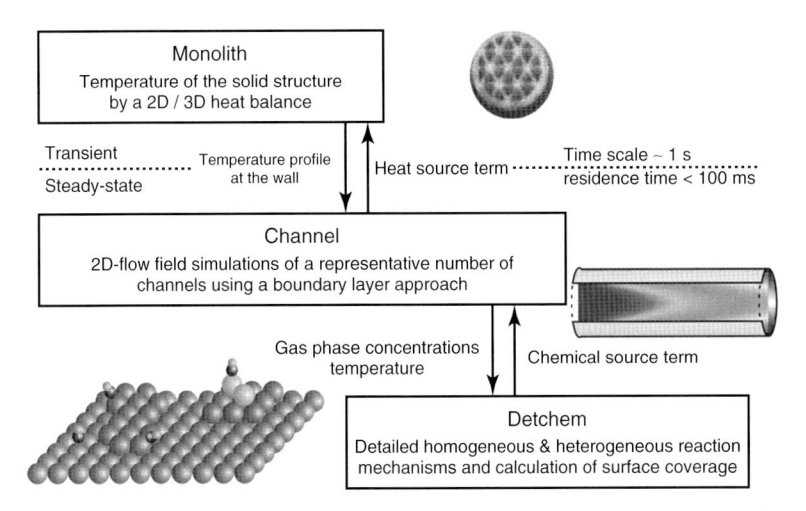

Figure 17.2 Structure of the code DETCHEM^{MONOLITH} [59].

simulated including detailed transport and chemistry models. These single-channel simulations also calculate the heat flux from the fluid flow to the channel wall due to convective and conductive heat transport in the gaseous flow and heat released by chemical reactions. Thus, at each time step, the single-channel simulations provide the source terms for the heat balance of the monolith structure while the simulation of the heat balance provides the boundary condition (wall temperature) for the single-channel simulations. At each time step, the inlet conditions may vary. This very efficient iterative procedure enables a transient simulation of the entire monolith without sacrificing the details of the transport and chemistry models, as long as the prerequisites for the time scales remain valid. Furthermore, reactors with alternating channel properties such as flow directions, catalyst materials, and loadings can be treated.

17.2.5
Experimental Evaluation of Models Describing Radical Interactions

The coupling of several complex models introduces a large number of parameters into the simulations. Hence, agreement between predicted and experimentally observed overall conversion and selectivity alone is not sufficient to evaluate individual sub models. Time and locally resolved profiles provide a more stringent test for model evaluation. Useful data arise from the experimental resolution of local velocity profiles by laser Doppler anemometry/laser Doppler velocimetry (LDA/LDV) [60, 61] and of spatial and temporal species profiles by *in-situ*, non-invasive methods such as Raman and laser-induced fluorescence (LIF) spectroscopy. For instance, an optically accessible catalytic channel reactor can be used to evaluate models for heterogeneous and homogeneous chemistry as well as transport by the simultaneous detection of stable species by Raman measurements and of OH radicals by Planar laser-induced fluorescence (PLIF) [62, 63].

17.2.6
Mathematical Optimization of Reactor Conditions and Catalyst Loading

In a chemical reactor, the initial and boundary conditions can be used to optimize the performance of the reactor, that is, maximize the conversion, the selectivity or the yield of certain product species. In particular, at the inlet of the catalytic monolith, the mass or molar fractions of the species, the initial velocity, or the initial temperature can be controlled to optimize one product composition. Furthermore, it may be possible to control the temperature profile $T_{wall}(z)$ at the channel wall, and vary the loading with catalyst along the channel, that is, $F_{cat/geo}(z)$. Moreover, the length of the catalytic monolith z_{max} can be optimized. Recently, algorithms have been established not only to optimize those control parameters but also to be applied to achieve a better understanding of the interactions between heterogeneous and homogeneous chemical reactions in catalytic reactors [64–66]. Radical interactions may play a decisive role, as shown in the example given below.

17.3
Applications

17.3.1
Turbulent Flow through Channels with Radical Interactions

Mantzaras et al. [67] applied the k-ε model, a presumed (Gaussian) probability density function for gaseous reactions, and a laminar-like closure for surface reactions to study turbulent catalytically stabilized combustion of lean hydrogen–air mixtures in plane platinum-coated channels. Here the two-dimensional OH profiles as shown in Figure 17.3 reveal the interaction of gas phase and surface chemistry; depending on the position in the reactor, the profiles as a function of distance from the catalyst vary significantly.

They also examined different low-Reynolds-number near-wall turbulence models and compared the numerically predicted results with data derived from PLIF measurements of OH radicals, Raman measurements of major species, and LDV

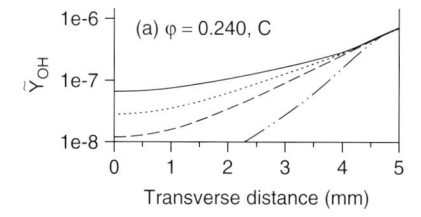

Figure 17.3 Transverse profiles of normalized Favre-averaged OH mass fractions in catalytically supported oxidation of hydrogen over a platinum wall. Streamwise distances: $x = 25$ mm (dash-double dotted), $x = 61$ mm (dashed), $x = 98$ mm (dotted), and $x = 150$ mm (solid) lines. The wall is located at $y = 5$ mm. (Taken from Mantzaras et al. [67].)

measurements of local velocities and turbulence [68]. They found that discrepancies between predictions and measurements are ascribed to the capacity of the various turbulence models to capture the strong flow laminarization induced by the heat transfer from the hot catalytic surfaces.

17.3.2
Synthesis Gas from Natural Gas by High-Temperature Catalysis

Another class of tube-like reactors is the monolith or honeycomb structure, which consists of numerous passageways with diameters varying between 0.1 mm and a few millimeters. The flow field in the thin channels of this reactor type is usually laminar. The catalytic material is mostly dispersed in a washcoat on the inner channel wall. Monolith channels are manufactured with various cross-sectional shapes, for example, circular, hexagonal, square, or sinusoidal. In the next sections, the configuration of catalytic monolithic reactors will be used to discuss the interaction of gas phase and surface chemistry in high-temperature catalysis.

The high-temperature CPOX of methane over Rh-based catalysts in short-contact-time (milliseconds) reactors has been intensively studied because it offers a promising route to convert natural gas into synthesis gas (syngas, H_2, and CO), which can subsequently be converted into higher alkanes or methanol or be used in fuel cells [1, 69, 70]. The indirect route for syngas formation has meanwhile been accepted; at the catalyst entrance total oxidation occurs to form steam as long as oxygen is available at the surface, and then methane is steam-reformed to hydrogen. Basically no dry reforming occurs, and the surface acts as sink for radicals, inhibiting significant gas phase reactions at pressures below 10 bar [71]. Also, the transient behavior during light-off of the reaction has been revealed. As an example, Figure 17.4 shows the time-resolved temperature and species profiles in a single channel of a catalytic monolith for partial oxidation of methane for the production of synthesis gas and the temperature distribution of the solid structure during light-off [38].

Natural gas contains higher alkanes and other minor components besides methane, and conversion and selectivity can be influenced by these other components. Consequently, conversion of methane in steam reforming (SR) of pure methane and in SR of natural gas (North Sea H) differ, as shown in Figure 17.5. Here, fuel/steam mixtures (molar steam-to-carbon ratios of 2.5 and 4, diluted by 75% Ar, 10 000 h^{-1} space velocity) were fed into a furnace containing an Rh-coated honeycomb catalyst [72].

17.3.3
Olefin Production by High-Temperature Oxidative Dehydrogenation of Alkanes

While gas phase reactions are not significant in CPOX of methane at atmospheric pressure, CPOX of ethane to ethylene over platinum-coated catalysts at short contact times [73, 74] is characterized by complex interactions of homogeneous gas phase and heterogeneous surface reactions [4, 46, 75].

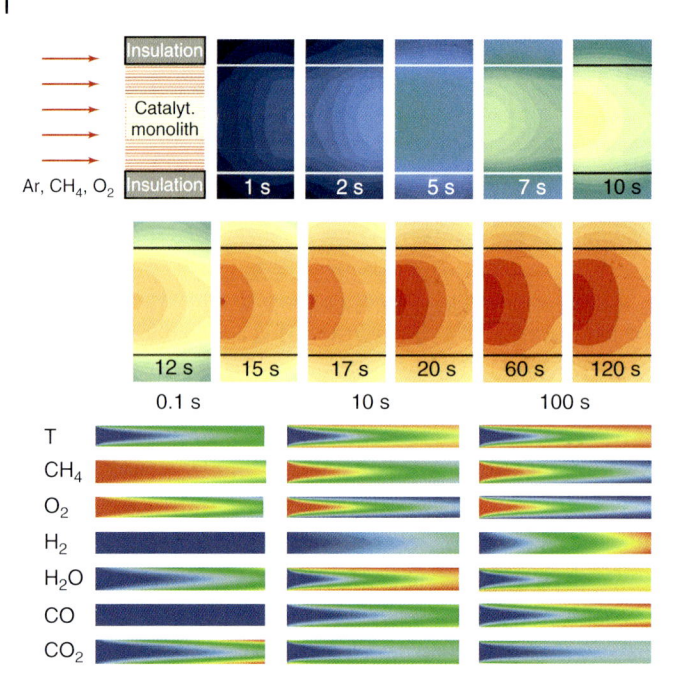

Figure 17.4 Simulation of light-off of a monolithic reactor coated with Rh for partial oxidation of methane to synthesis gas [38]. Temperature of the solid structure (top) and gas phase temperature and species mole fractions in a single channel in the center of the monolith (bottom), red = maximum, blue = minimum.

A picture of the complete reaction process is shown in Figure 17.6.

At the catalyst entrance oxygen is completely consumed at the surface within 1 mm, primarily producing CO_2 and H_2O. This total combustion of ethane leads to a rapid temperature increase from RT to 1000 °C. This high temperature drives the pyrolysis of ethane in the gas phase. After a decade of discussions on the reaction pathways, most studies today conclude that most of the ethylene (the desired product) is actually homogeneously produced in the gas phase. Further downstream, additionally, reforming and shift reactions occur.

Based on the molecular understanding of the interaction of gas phase and surface chemistry in oxy-dehydrogenation of ethane over Pt [4], Minh et al. [64] used a recently developed optimization code to find the optimal Pt catalyst loading in the along-the-flow direction in a Pt/Al_2O_3-coated honeycomb catalyst. The gas phase mechanism used consists of 25 reactive species (mainly C1 and C2 species) involved in 131 reversible reactions and one irreversible reaction, and the surface reaction mechanism consists of another 82 elementary-step-like reactions involving another 19 surface species [4]. This mechanism was later also used to study on-line catalyst addition effects [75]. The Pt-coated monolith had a diameter of 18 mm and a length of 10 mm. Each channel had a diameter of 0.5 mm. The monolith was fed

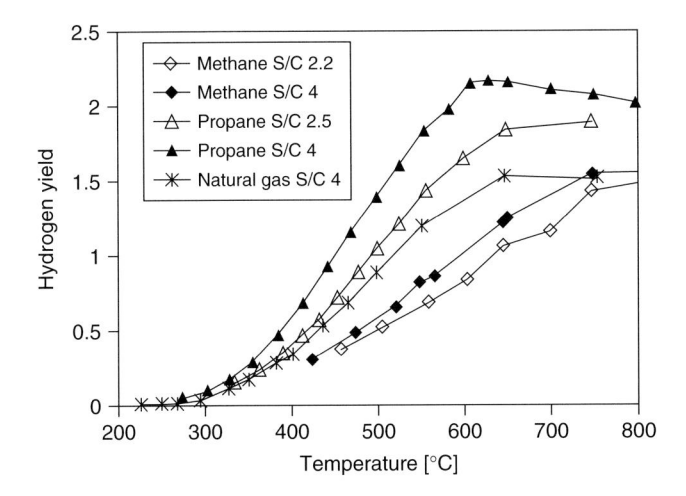

Figure 17.5 Comparison of experimentally determined hydrogen yields in steam-reforming of methane with S/C 2.5 (◊) and S/C 4 (♦), steam reforming of natural gas with S/C 4 (∗), steam reforming of propane with S/C 2.5 (△) and S/C 4 (▲); Rh catalyst. (Taken from Ref. [72].)

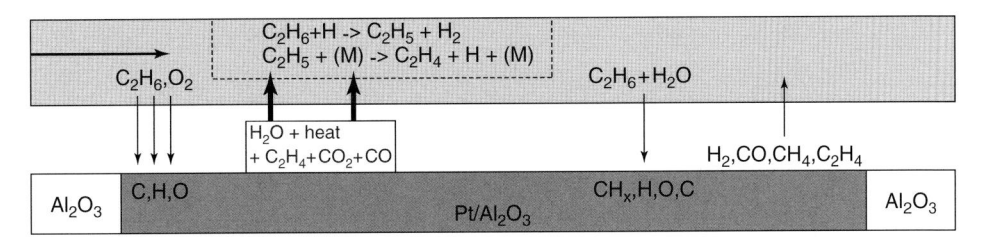

Figure 17.6 General picture of oxy-dehydrogenation of ethane over Pt at millisecond contact times and temperatures of ∼900 °C; inlet ethane/oxygen ratio ∼2. The upper and lower panels show gas phase and surface processes, respectively. (Adapted from Zerkle et al. [4].)

with an ethane/oxygen/nitrogen mixture of varying C/O ratio at five standard liters per minute leading to a residence time of few milliseconds.

17.3.3.1 Formulation of an Optimal Control Problem

The initial and boundary conditions can be used to optimize the performance of the reactor, that is, maximize the conversion, the selectivity, or the yield of certain product species. In particular, at the inlet of the catalytic monolith, the mass or molar fractions of the species, the initial velocity, or the initial temperature can be controlled to optimize the composition of one product. Furthermore, it may be possible to control the temperature profile $T_{wall}(z)$ along the channel wall, and vary the loading with catalyst along the channel, that is, $F_{cat/geo}(z)$. Moreover, the length of the catalytic monolith z_{max} can be optimized. Here, the objective function to

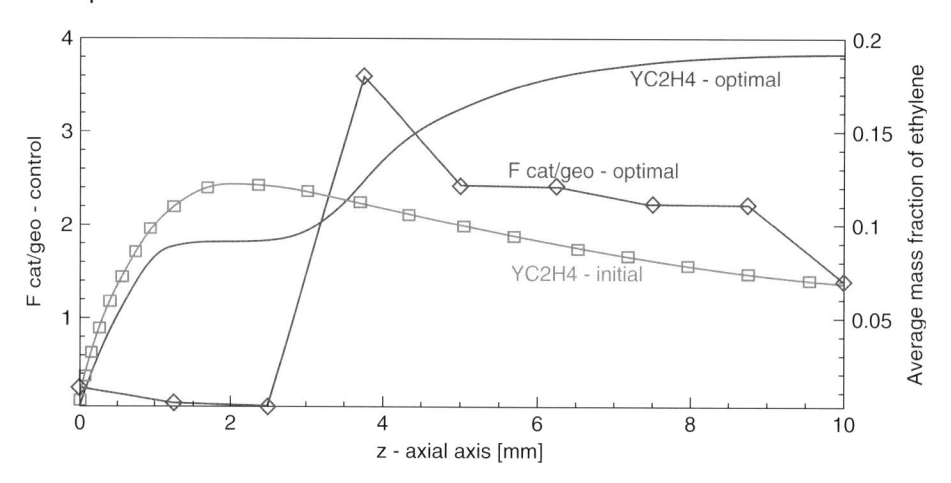

Figure 17.7 Catalyst loading expressed by $F_{cat/geo}(z)$ and the radial averaged mass fraction of ethylene at the initial and at optimal solutions in oxy-dehydrogenation of ethane over Pt-coated honeycomb monoliths. (Taken from Minh et al. [64].)

be maximized is the mass fraction of ethylene at the outlet of the channel. The control considered here is the catalyst loading, expressed, which is a function of the axial coordinate z. For practical reasons, there are often equality and inequality constraints on the control and state variables, such as upper and lower bounds for the catalyst loading and the (trivial) fact that the sum of all mass fractions must be one.

In the case considered, the inlet gas temperature is $T_{gas} = 600$ K, and the wall temperature $T_{wall}(z)$ is kept fixed at 1000 K. As a constraint the $F_{cat/geo}(z)$ is required to be between 0, that is, no catalyst, and 100, that is, highly loaded.

The optimization was started with a constant $F_{cat/geo}(z)$ profile of 20.0, leading to an objective value of 0.06. In the optimal solution the objective value is 0.19. Figure 17.7 shows the standard and optimal profiles of $F_{cat/geo}(z)$ and average mass fraction profiles of ethylene.

Figure 17.8 reveals the mass fraction profiles of ethane and ethylene with the optimal profile of $F_{cat/geo}(z)$. Platinum is a very efficient catalyst for the oxidation of ethane. In the first 2 mm of the catalyst, oxygen is almost completely consumed by surface reactions (catalytic oxidation of ethane), leading to the total oxidation products CO_2 and H_2O, ethylene, and some CO. Ethylene, however, is substantially produced in the gas phase as well. However, the conversion in the gas phase only occurs if a sufficiently large radical pool is build up, which takes a certain time/distance (so-called ignition delay time). Furthermore, some of the ethylene formed by surface reactions adsorbs on the surface as well, where in the region around 2 mm mainly reforming reactions occur though the total reaction rate is much smaller than that in the first millimeter of the catalyst where oxygen was still available. Since the production of ethylene by gas phase reactions really takes off

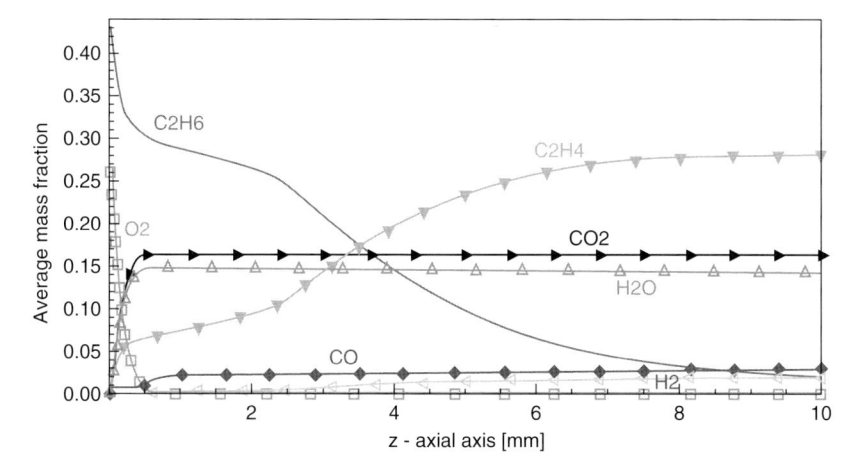

Figure 17.8 Average mass fractions of major species at the optimal solution in optimization of catalyst loading in oxy-dehydrogenation of ethane over Pt-coated honeycomb monoliths. (Taken from Minh et al. [64].)

further downstream (most of the ethylene is produced in the gas phase) due to the radicals available there and due to the fact that the surface is relatively inactive in the region around 2 mm, a plateau appears in that region around 2 mm (Figure 17.7) due to the competition between ethylene production in the gas phase and (partial) oxidation on the surface, both at relatively low rates. Optimization of the catalyst loading means a very low loading in this region, because here the catalyst not only oxidizes ethylene but also adsorbs radicals from the gas phase, and these are needed to initiate ethylene formation in the gas phase. The optimization means relatively low catalyst loading in the very active initial catalyst section, which can be understood as follows: Within the first millimeter of the catalyst, where oxygen is available, the process is limited by mass-transfer of ethane and even more oxygen to the surface. Here, primarily, oxidation of ethane occurs at a very high rate, the catalyst being very active. Consequently, catalyst is needed here but not at a high loading; this effect has also been observed in several experiments.

The hydrocarbon/oxygen (C/O) ratio can easily be used to tune the product selectivity in CPOX of alkanes. The group of L.D. Schmidt has studied a variety of fuels and basically found the same trend as that shown in Figure 17.9 for CPOX of n-decane at millisecond contact times over Rh catalysts [48].

At low C/O ratio, synthesis gas is the primary product; actually its maximum is reached close to the point where the reaction switches to total combustion, a flame often being formed in that transition region. With increasing C/O ratio, more and more olefins are formed, primarily α-olefins, but at higher C/O ratio a larger variety of other olefins are formed as well, and eventually acetylene, benzene, and polycyclic aromatic hydrocarbons (PAH) are formed at very low oxygen content. Coking of the catalyst and formation of soot in the gas phase quickly becomes a technical problem. The studies over noble metal-based catalysts and at millisecond

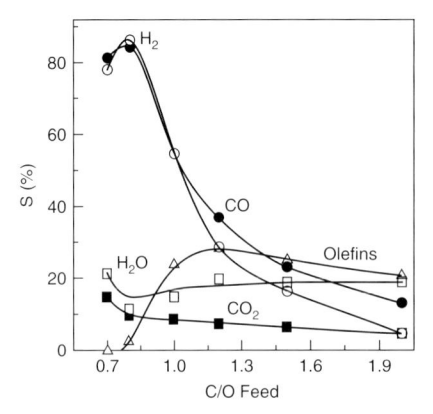

Figure 17.9 Effect of the *n*-decane/oxygen feed ratio on product selectivity in CPOX of *n*-decane at millisecond contact times over Rh catalysts [48].

contact times were recently extended to CPOX of biodiesel [76]. Also, autothermal reforming of ethanol was realized at short contact times and high temperatures in a two-stage reactor [77, 78]. An Rh–ceria catalyst on alumina foams/spheres served as the first stage for reforming, with some oxygen addition. The second reactor stage consists of Pt–ceria on alumina spheres to accelerate the water gas shift reaction for maximizing the hydrogen yield.

17.4
Hydrogen Production from Logistic Fuels by High-Temperature Catalysis

The production of hydrogen and synthesis gas (syngas, H_2, and CO) from logistic fuels such as gasoline, diesel, and kerosene by CPOX and SR is currently in the focus of both academic and industrial research. In contrast to the complex and costly supply of compressed and stored hydrogen for mobile fuel cell application, CPOX of liquid fuels allows the production and utilization of hydrogen through existing routes, which, in particular, is of interest for on-board applications. At operating temperatures around 1000 K and above, conversion of the fuel could occur not only on the solid catalyst but also in the gas phase.

Heterogeneous and homogeneous reactions in CPOX of hydrocarbons are linked, not only with respect to adsorption and desorption of fuel and oxygen molecules and the products, but also with respect to adsorption and desorption of intermediates and radicals. Therefore, mass transport of radicals and intermediates to and from the gaseous bulk phase and the catalytically active channel wall, mainly by radial diffusion in the small channels of the monolith (these being on the order of 0.25–1 mm) is crucial for the interaction of heterogeneous and homogeneous reactions in CPOX reactors. Hartmann *et al.*, for instance, studied the CPOX of *iso*-octane over rhodium/alumina-coated honeycomb monolith serving as gasoline

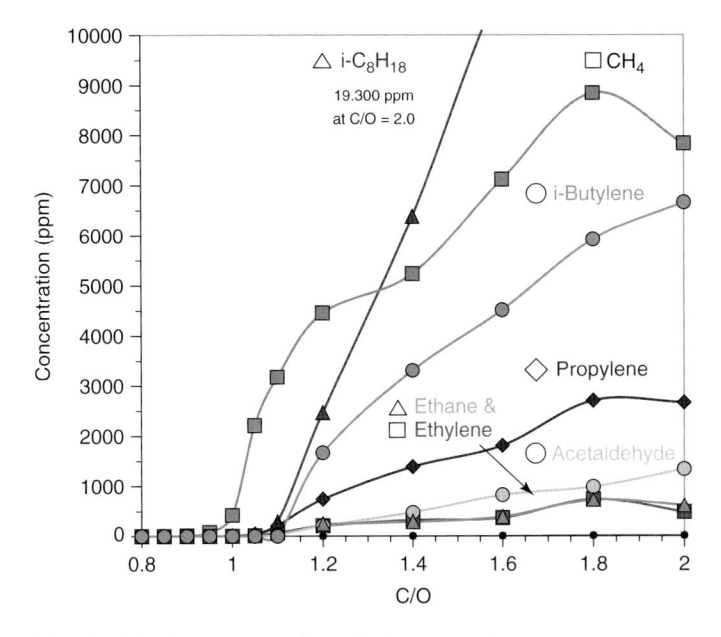

Figure 17.10 Experimentally determined concentrations of the side products and the fuel remaining in the outlet stream of CPOX of *iso*-octane over an Rh/alumina coated monolith as a function of C/O ratio. (Taken from Ref. [79].)

surrogate [79]. Very high hydrogen and carbon monoxide selectivities were found at stoichiometric conditions (C/O = 1), while at lean conditions more total oxidation occurs.

At rich conditions (C/O > 1), homogeneous chemical conversion in the gas phase is responsible for the formation of by-products such as olefins, shown in Figure 17.10, that also have the potential for coke formation, which was observed experimentally and numerically.

This study also revealed that the chemical models applied – even though the most detailed ones available (857 gas phase and 17 adsorbed species in over 7000 elementary reactions) were used – need further improvement.

Nevertheless, this combined modeling and experimental study revealed the role of surface, gas phase, and radical chemistry in high-temperature oxidative catalytic conversion of larger hydrocarbons.

From Figure 17.11, it can clearly be concluded that the major products (syngas) are produced in the entrance region of the catalyst on the catalytic surface; radial concentration profiles are caused by a mass transfer-limited process. As soon as the oxygen is consumed on the catalytic surface – similarly to CPOX of natural gas – hydrogen formation increases due to SR, and the major products are formed within a few millimeters. At rich conditions (C/O > 1.0), a second process, now in the gas phase, begins in the downstream part, as shown in Figure 17.12.

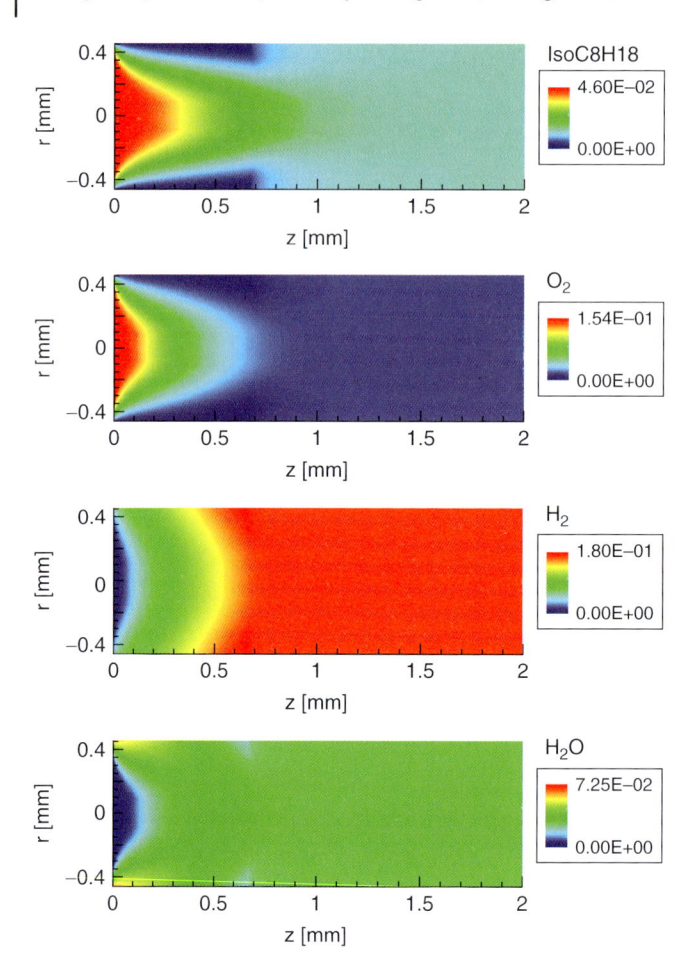

Figure 17.11 Numerically predicted profiles of molar fractions of reactants and major products in the entrance region of the catalyst at C/O = 1.2 in CPOX of *iso*-octane over an Rh/alumina-coated monolith. (Taken from Hartmann et al. [79].) Flow direction is from left to right.

The number of radicals in the gas phase is sufficiently large to initiate gas phase pyrolysis of the remaining fuel and formation of coke precursors such as ethylene and propylene. In the experiment, the downstream part of the catalyst is coked up, and here the Rh surface cannot act as a sink for radicals.

17.5
High-Temperature Catalysis in Solid Oxide Fuel Cells

In the recent past SOFCs have attracted considerable attention due to their potential to convert chemical energy into electrical energy at high efficiencies. SOFCs are

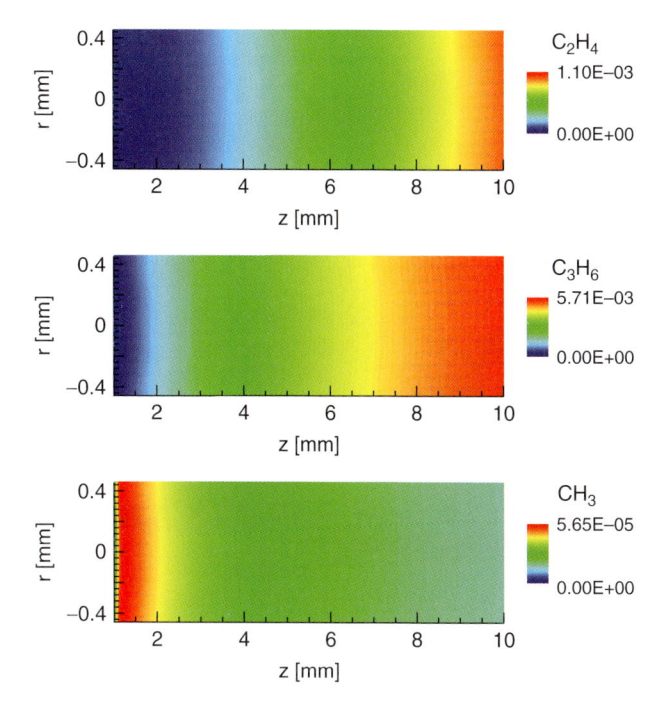

Figure 17.12 Numerically predicted profiles of molar fractions of minor products and radicals along the entire catalyst at C/O = 1.2 in CPOX of *iso*-octane over an Rh/alumina coated monolith. (Taken from Hartmann *et al.* [79].) Flow direction is from left to right.

also called *high-temperature fuel cells*, because they are operated at temperature of approximately 800 °C, and therefore (no CO poisoning) they have the potential for a direct use of hydrocarbons and alcohols as fuels [80]. A sketch of the major processes in an anode-supported SOFC operated with hydrocarbons is shown in Figure 17.13 [5].

The anode contains an electron conductor, for example, nickel, and an ion-conducting material, for example, YSZ. High-temperature catalytic reactions proceed on the Ni surface; steam as the major product of the electrochemical reaction drives the endothermic reforming of the incoming fuel, producing hydrogen, which eventually takes part in the charge transfer reaction.

Understanding the performance of an SOFC requires an understanding of the interaction of electrocatalysis, electric fields, current density, thermocatalytic reactions on the metal surface, potential homogeneous reactions in the gas phase (in particular for fuels containing liquid hydrocarbons), and mass and heat transport in the fuel and air channels and inside the porous membrane electrode assembly (MEA). Several groups [5, 81] have applied the following approach for their the analysis of temperature, species concentrations, and current density profiles, as well as the efficiency and power density of a planar SOFC at direct internal reforming

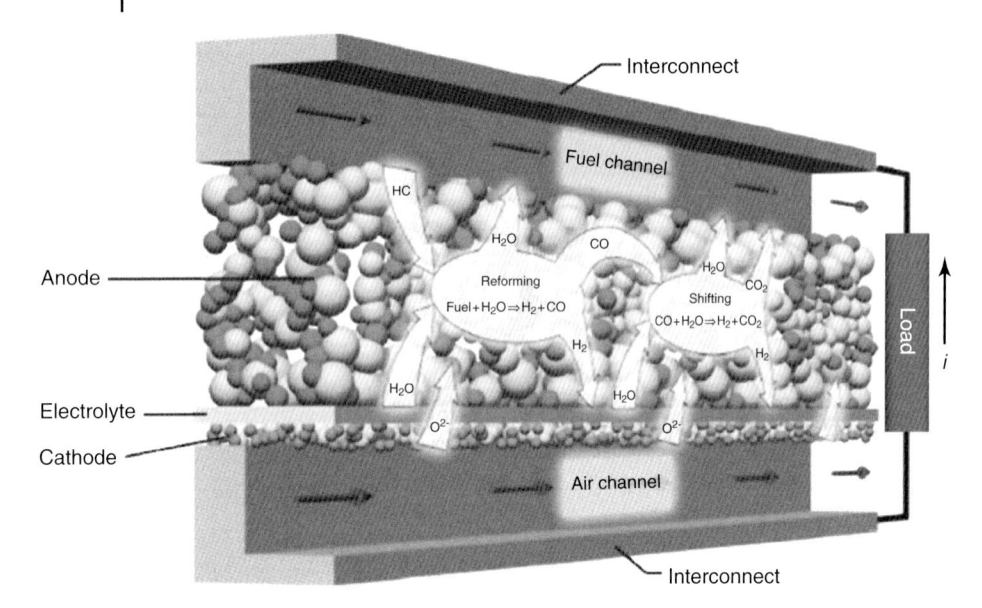

Figure 17.13 A schematic presentation of a cut-away view of a planar, anode-supported SOFC unit in co-flow configuration operated with hydrocarbon-containing fuels. (Adapted from Zhu et al. [5].)

conditions: the heterogeneous chemistry model of the catalytic reactions in the anode structure uses a multi-step reaction mechanism for the SR of methane on Ni-based catalysts [7]. The porous media transport is modeled using the Dusty Gas Model, and electrochemistry is modeled using a modified Butler–Volmer setting, assuming hydrogen to be the only electrochemically active species [82]. The channel flows are usually modeled by plug flow equations.

High-temperature catalysis in SOFCs will now be discussed using the example of an anode-supported SOFC operated with methane-containing fuels and an Ni/YSZ anode structure. The simulation reveals the species concentrations in the fuel channel and the porous anode structure, as shown in Figure 17.14 [83].

It is quite obvious from the figure that H_2 and H_2O have opposite fluxes within the anode, because the former is being consumed and the latter is being produced as a result of electrochemical reactions at the three-phase interface. The computed temperature distribution in a single planar cell operated with humidified methane (97% CH_4 and 3% H_2O) as fuel at adiabatic conditions assuming H_2 as the only electrochemically active species and the local variation of temperature in the fuel and air channels as well as in the interconnects are shown in Figure 17.15 [84].

At the entrance, the fuel stream loses heat to the comparatively cold air entering the air channel, and hence the temperature of the air stream increases. As reforming starts, the temperature of the fuel stream drops further, as does the temperature of the air stream. However, further downstream the temperature begins to increase as a result of exothermic thermocatalytic cell reactions such as

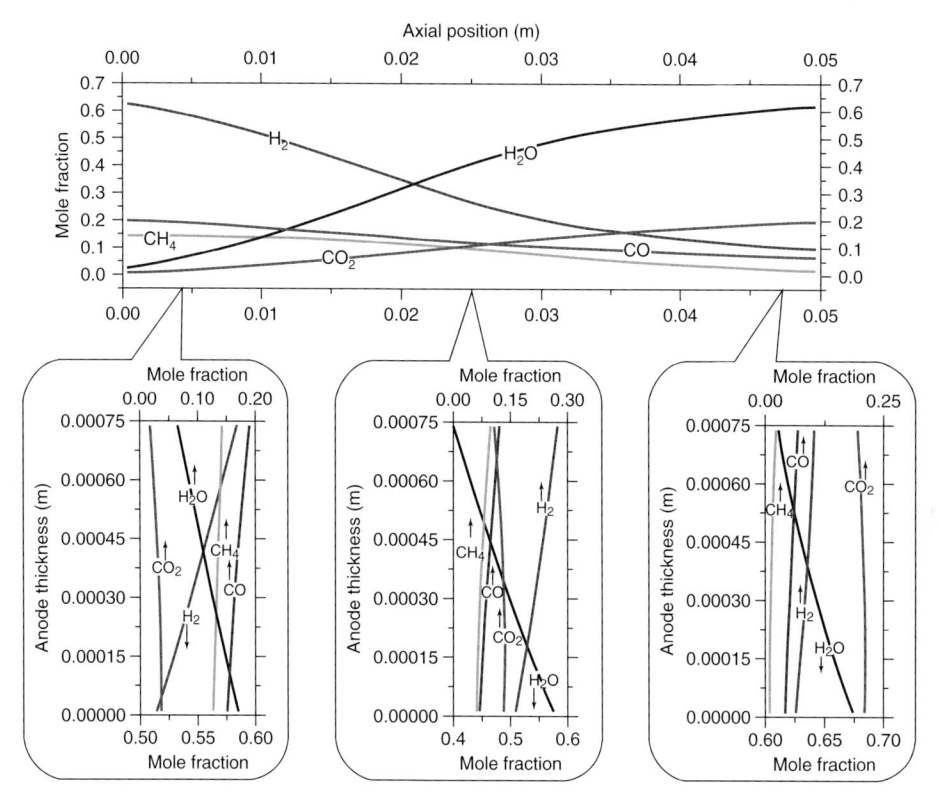

Figure 17.14 Numerically predicted species profiles within the fuel channel and within the anode of an SOFC. The inlet fuel at 800 °C is assumed to consist of 14% CH_4, 63% H_2, 2% H_2O, 20% CO, and traces of CO_2. Air is assumed to enter the cathode channel at 650 °C. The drop-down panels show the profiles across the thickness of the anode at selected axial positions taken from reference [83].

water gas shift, electrochemical oxidation, and ohmic and activation losses. There is no significant temperature difference between the anode side interconnect and the fuel channel in contrast to the behavior at the cathode side interconnect/air channel. Interestingly, the temperature of the anode side interconnect is lower than that of the fuel channel, while the cathode side interconnect has a higher temperature than the air channel, which is primarily caused by thermal radiation from the cathode electrode to the interconnect.

This model approach was also used to study the influence of air flow rate, anode thickness, catalyst loading, and pre-reforming on the performance of a planar anode-supported cell, which was 10 cm in length and operated with 40% CH_4 and 60% H_2O as fuel at 0.7 V. Since the cell voltage was fixed, fuel utilization is the most significant parameter for cell efficiency. For any particular operating condition, an optimal anode thickness exists for efficiency and power density (Figure 17.16) [85]; maxima were achieved for an anode thickness of approximately 0.5 mm under the conditions of this study.

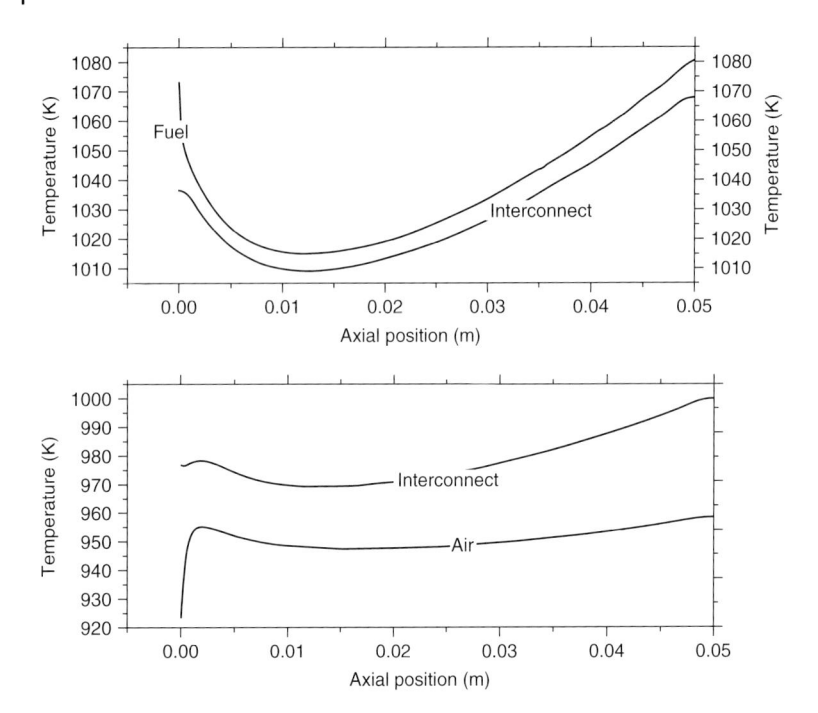

Figure 17.15 Numerical predictions of temperature profiles within the flow (a) and air (b) channels as well as in the interconnects in an SOFC cell operated with humidified methane as fuel. The inlet fuel and air streams are at 800 °C and 650 °C, respectively; operating cell potential is 0.7 V. (Taken from Ref. [84].)

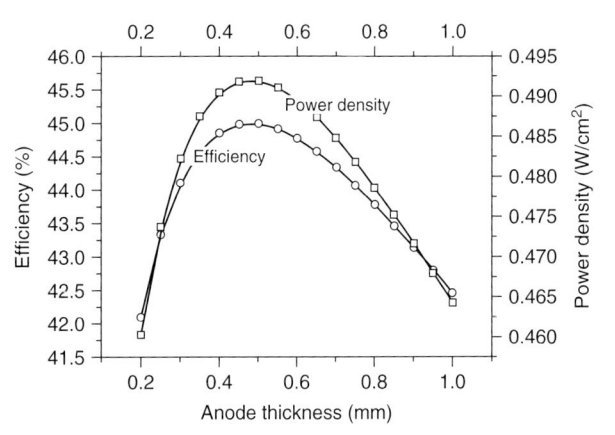

Figure 17.16 Effect of anode thickness on efficiency and power density for a cell operated under adiabatic conditions. The inlet fuel consists of 40% CH_4 and 60% H_2O at 800 °C. Cathode inlet is assumed to be air at 650 °C. (Figure taken from Janardhanan et al. [85].)

In the case of direct internal fuel reforming, the anode serves as catalyst for the production of H_2 and CO, which further participate in the charge transfer reactions at the three-phase boundary. In the case of thin anodes, the smaller amount of catalyst limits the amount of H_2 produced, leading to lower average current density, efficiency, and power density. For thick anodes, the H_2 and CO production by the (endothermic) fuel reforming is increased. However, this extended reforming rate results in larger temperature drop close to the channel inlet resulting in decreased performance.

References

1. Hickman, D.A. and Schmidt, L.D. (1992) Synthesis gas formation by direct oxidation of methane over monoliths. *J. Catal.*, **138** (1), 267–282.
2. Deutschmann, O. and Schmidt, L.D. (1998) Modeling the partial oxidation of methane in a short-contact-time reactor. *Am. Inst. Chem. Eng. J.*, **44** (11), 2465–2477.
3. Huff, M.C. and Schmidt, L.D. (1993) Production of olefins by oxidative dehydrogenation of propane and butane over monoliths at short contact times. *J. Catal.*, **149**, 127–141.
4. Zerkle, D.K. et al. (2000) Understanding homogeneous and heterogeneous contributions to the platinum-catalyzed partial oxidation of ethane in a short-contact-time reactor. *J. Catal.*, **196** (1), 18–39.
5. Zhu, H.Y. et al. (2005) Modeling elementary heterogeneous chemistry and electrochemistry in solid-oxide fuel cells. *J. Electrochem. Soc.*, **152** (12), A2427–A2440.
6. Janardhanan, V.M. and Deutschmann, O. (2011) in *Modeling and Simulation of Heterogeneous Catalytic Reactions: From the molecular process to the technical system.* (ed O., Deutschmann) Wiley-VCH, Weinheim,
7. Hecht, E.S. et al. (2005) Methane reforming kinetics within a Ni-YSZ SOFC anode support. *Appl. Catal., A: Gen.*, **295** (1), 40–51.
8. Windmann, J. et al. (2003) Impact of the Inlet Flow Distribution on the Light-off Behavior of a 3-way Catalytic Converter, SAE Technical Paper,2003-01-0937.
9. Bird, R.B., Stewart, W.E., and Lightfoot, E.N. (2001) *Transport Phenomena*, 2nd edn, John Wiley & Sons, Inc., New York.
10. Patankar, S.V. (1980) *Numerical Heat Transfer and Fluid Flow*, McGraw-Hill, New York.
11. Warnatz, J., Dibble, R.W., and Maas, U. (1996) *Combustion, Physical and Chemical Fundamentals, Modeling and Simulation, Experiments, Pollutant Formation*, Springer-Verlag, New York.
12. Hayes, R.E. and Kolaczkowski, S.T. (1997) *Introduction to Catalytic Combustion*, Gordon and Breach Science Publishers, Amsterdam.
13. Kee, R.J., Coltrin, M.E., and Glarborg, P. (2003) *Chemically Reacting Flow*, Wiley-Interscience.
14. Deutschmann, O. (2007) Computational fluid dynamics simulation of catalytic reactors, in *Handbook of Heterogeneous Catalysis*, 2nd edn, Chapter 1 (eds H.K.G. Ertl, F. Schüth, and J. Weitkamp), Wiley-VCH Verlag GmbH, p. 1811–1828.
15. Ranzi, E. et al. (1994) A new comprehensive reaction mechanism for combustion of hydrocarbon fuels. *Combust. Flame*, **99**, 201–211.
16. Coltrin, M.E., Kee, R.J., and Rupley, F.M. (1991) *Surface Chemkin (Version 4.0): A Fortran Package for Analyzing Heterogeneous Chemical Kinetics at a Solid-Surface–Gas phase Interface, SAND91-8003B*, Sandia National Laboratories.
17. Mhadeshwar, A.B., Wang, H., and Vlachos, D.G. (2003) Thermodynamic consistency in microkinetic development

of surface reaction mechanisms. *J. Phys. Chem. B*, **107** (46), 12721–12733.
18. Maier, L., Schädel, B., Herrera Delgado, K., Tischer, S., and Deutschmann, O. (2011) Steam Reforming of Methane over Nickel: Development of a Multi-Step Surface Reaction Mechanism. *Topics in Catalysis*, **54**, 845–858.
19. Williams, W.R., Marks, C.M., and Schmidt, L.D. (1992) Steps in the reaction $H2 + 1/2O2 = H2O$ on Pt–OH desorption at high-temperatures. *J. Phys. Chem.*, **96** (14), 5922–5931.
20. Hellsing, B., Kasemo, B., and Zhdanov, V.P. (1991) Kinetics of the hydrogen-oxygen reaction on platinum. *J. Catal.*, **132**, 210–228.
21. Warnatz, J. (1992) Resolution of gas phase and surface combustion chemistry into elementary reactions (Invited Lecture). *Proc. Combust. Inst.*, **24**, 553–579.
22. Rinnemo, M. et al. (1997) Experimental and numerical investigation of the catalytic ignition of mixtures of hydrogen and oxygen on platinum. *Combust. Flame*, **111** (4), 312–326.
23. Veser, G. (2001) Experimental and theoretical investigation of H_2 oxidation in a high-temperature catalytic microreactor. *Chem. Eng. Sci.*, **56**, 1265–1273.
24. Bui, P.-A., Vlachos, D.G., and Westmoreland, P.R. (1997) Modeling ignition of catalytic reactors with detailed surface kinetics and transport: oxidation of H_2/Air mixtures over platinum surfaces. *Ind. Eng. Chem. Res.*, **36** (7), 2558–2567.
25. Andrae, J.C.G. and Björnbom, P.H. (2000) Wall effects of laminar hydrogen flames over platinum and inert surfaces. *Am. Inst. Chem. Eng. J.*, **46** (7), 1454–1460.
26. Mai, J., von Niessen, W., and Blumen, A. (1990) The $CO+O_2$ reaction on metal surfaces: simulation and mean-field theory: the influence of diffusion. *J. Chem. Phys.*, **93**, 3685–3692.
27. Zhdanov, V.P. and Kasemo, B. (1994) Steady-state kinetics of CO oxidation on Pt: extrapolation from 10^{-10} to 1 bar. *Appl. Surf. Sci.*, **74** (2), 147–164.
28. Aghalayam, P., Park, Y.K., and Vlachos, D.G. (2000) A detailed surface reaction mechanism for CO oxidation on Pt. *Proc. Combust. Inst.*, **28**, 1331–1339.
29. Deutschmann, O. et al. (1996) Numerical modeling of catalytic ignition. *Proceed. Combust. Inst.*, **26**, 1747–1754.
30. Veser, G. et al. (1997) Catalytic ignition during methane oxidation on platinum: experiments and modeling. *Stud. Surf. Sci. Catal.*, **109**, 273–284.
31. Bui, P.-A., Vlachos, D.G., and Westmoreland, P.R. (1997) Catalytic ignition of methane/oxygen mixtures over platinum surfaces: comparison of detailed simulations and experiments. *Surf. Sci.*, **386** (2–3), L1029–L1034.
32. Dogwiler, U., Benz, P., and Mantzaras, J. (1999) Two-dimensional modelling for catalytically stabilized combustion of a lean methane-air mixture with elementary homogeneous and heterogeneous chemical reactions. *Combust. Flame*, **116** (1), 243.
33. Aghalayam, P. et al. (2003) A C1 mechanism for methane oxidation on platinum. *J. Catal.*, **213** (1), 23–38.
34. Hickman, D.A. and Schmidt, L.D. (1993) Steps in CH_4 oxidation on Pt and Rh surfaces: high-temperature reactor simulations. *Am. Inst. Chem. Eng. J.*, **39**, 1164–1176.
35. Huff, M. and Schmidt, L.D. (1993) Ethylene formation by oxidative dehydrogenation of ethane over monoliths at very short-contact times. *J. Phys. Chem.*, **97** (45), 11815–11822.
36. Huff, M.C. et al. (2000) The contribution of gas phase reactions in the Pt-catalyzed conversion of ethane-oxygen mixtures. *J. Catal.*, **191**, 46–45.
37. Donsi, F., Williams, K.A., and Schmidt, L.D. (2005) A multistep surface mechanism for ethane oxidative dehydrogenation on Pt- and Pt/Sn-coated monoliths. *Ind. Eng. Chem. Res.*, **44** (10), 3453–3470.
38. Schwiedernoch, R. et al. (2003) Experimental and numerical study on the transient behavior of partial oxidation of methane in a catalytic, monolith. *Chem. Eng. Sci.*, **58** (3–6), 633–642.
39. Chatterjee, D., Deutschmann, O., and Warnatz, J. (2001) Detailed surface reaction mechanism in a three-way catalyst. *Faraday Discuss.*, **119**, 371–384.

40. Ruf, B. et al. (1996) Simulation of homoepitaxial growth on the diamond (100) surface using detailed reaction mechanisms. *Surf. Sci.*, **352**, 602–606.
41. Harris, S.J. and Goodwin, D.G. (1993) Growth on the reconstructed diamond (100) surface. *J. Phys. Chem.*, **97** (1), 23–28.
42. Romet, S., Couturier, M.F., and Whidden, T.K. (2001) Modeling of silicon dioxide chemical vapor deposition from tetraethoxysilane and ozone. *J. Electrochem. Soc.*, **148** (2), G82–G90.
43. Scott, C.D. et al. (2003) Iron catalyst chemistry in modeling a high-pressure carbon monoxide nanotube reactor. *J. Nanosci. Nanotechnol.*, **3** (1–2), 63–73.
44. Beretta, A., Forzatti, P., and Ranzi, E. (1999) Production of olefins via oxidative dehydrogenation of propane in autothermal conditions. *J. Catal.*, **184** (2), 469–478.
45. Beretta, A. and Forzatti, P. (2001) High-temperature and short-contact-time oxidative dehydrogenation of ethane in the presence of Pt/Al2O3 and BaMnAl11O19 catalysts. *J. Catal.*, **200** (1), 45–58.
46. Beretta, A., Ranzi, E., and Forzatti, P. (2001) Oxidative dehydrogenation of light paraffins in novel short contact time reactors: experimental and theoretical investigation. *Chem. Eng. Sci.*, **56** (3), 779–787.
47. Subramanian, R. and Schmidt, L.D. (2005) Renewable olefins from biodiesel by autothermal reforming. *Angew. Chem. Int. Ed.*, **44** (2), 302–305.
48. Krummenacher, J.J. and Schmidt, L.D. (2004) High yields of olefins and hydrogen from decane in short contact time reactors: rhodium versus platinum. *J. Catal.*, **222** (2), 429–438.
49. Schmidt, L.D., Siddall, J., and Bearden, M. (2000) New ways to make old chemicals. *Am. Inst. Chem. Eng. J.*, **46**, 1492–1495.
50. Raja, L.L., Kee, R.J., and Petzold, L.R. (1998) Simulation of the transient, compressible, gas-dynamic, behavior of catalytic-combustion ignition in stagnation flows. *Proc. Combust. Inst.*, **27**, 2249–2257.
51. Papadias, D., Edsberg, L., and Björnbom, P.H. (2000) Simplified method of effectiveness factor calculations for irregular geometries of washcoats: a general case in a 3D concentration field. *Catal. Today*, **60** (1–2), 11–20.
52. Keil, F. (1999) *Diffusion und Chemische Reaktionen in der Gas-Feststoff-Katalyse*, Springer-Verlag, Berlin.
53. Keil, F.J. (2000) Diffusion and reaction in porous networks. *Catal. Today*, **53**, 245–258.
54. Maas, U. and Pope, S. (1992) Simplifying chemical kinetics: intrinsic low-dimensional manifolds in composition space. *Combust. Flame*, **88**, 239–264.
55. Yan, X. and Maas, U. (2000) Intrinsic low-dimensional manifolds of heterogeneous combustion processes. *Proc. Combust. Inst.*, **28**, 1615–1621.
56. Susnow, R.G. et al. (1997) Rate-based construction of kinetic models for complex systems. *J. Phys. Chem., A*, **101**, 3731–3740.
57. Tischer, S., Correa, C., and Deutschmann, O. (2001) Transient three-dimensional simulations of a catalytic combustion monolith using detailed models for heterogeneous and homogeneous reactions and transport phenomena. *Catal. Today*, **69** (1–4), 57–62.
58. Tischer, S. and Deutschmann, O. (2005) Recent advances in numerical modeling of catalytic monolith reactors. *Catal. Today*, **105** (3–4), 407–413.
59. Deutschmann, O. et al. (2011) DETCHEM Software Package. 2.3 ed., Karlsruhe, www.detchem.com.
60. Appel, C. et al. (2005) Turbulent catalytically stabilized combustion of hydrogen/air mixtures in entry channel flows. *Combust. Flame*, **140** (1–2), 70–92.
61. Calis, H.P.A. et al. (2001) CFD modelling and experimental validation of pressure drop and flow profile in a novel structured catalytic reactor packing. *Chem. Eng. Sci.*, **56** (4), 1713–1720.
62. Appel, C. et al. (2002) An experimental and numerical investigation of turbulent catalytically stabilized channel flow

combustion of hydrogen/air mixtures over platinum. *Proc. Combust. Inst.*, **29**, 1031–1038.
63. Appel, C. et al. (2002) An experimental and numerical investigation of homogeneous ignition in catalytically stabilized combustion of hydrogen/air mixtures over platinum. *Combust. Flame*, **128** (4), 340–368.
64. Minh, H.D. et al. (2008) Optimization of two-dimensional flows with homogeneous and heterogeneously catalyzed gas phase reactions. *AIChE J.*, **54** (9), 2432–2440.
65. Minh, H.D. et al. (2008) *Computational Science and its Applications−ICCSA 2008, Part 1, Proceedings*, LNCS, Vol. 5072, pp. 1121, 1126–1130.
66. von Schwerin, M., Deutschmann, O., and Schulz, V. (2000) Process optimization of reactives systems by partially reduced SQP methods. *Comput. Chem. Eng.*, **24** (1), 89–97.
67. Mantzaras, J. et al. (2000) Numerical modelling of turbulent catalytically stabilized channel flow combustion. *Catal. Today*, **59** (1–2), 3–17.
68. Appel, C. et al. (2003) An experimental and numerical investigation of turbulent catalytically stabilized channel flow combustion of hydrogen/air mixtures over platinum. *Proc. Combust. Inst.*, **29**, 1031–1038.
69. Schmidt, L.D., Deutschmann, O., and Goralski, C.T. (1998) *Natural Gas Conversion V*, Elsevier, Amsterdam, pp. 685–692.
70. Horn, R. et al. (2006) Syngas by catalytic partial oxidation of methane on rhodium: mechanistic conclusions from spatially resolved measurements and numerical simulations. *J. Catal.*, **242** (1), 92–102.
71. Quiceno, R. et al. (2006) Modeling the high-temperature catalytic partial oxidation of methane over platinum gauze: Detailed gas phase and surface chemistries coupled with 3D flow field simulations. *Appl. Catal., A: Gen.*, **303** (2), 166–176.
72. Schadel, B.T., Duisberg, M., and Deutschmann, O. (2009) Steam reforming of methane, ethane, propane, butane, and natural gas over a rhodium-based catalyst. *Catal. Today*, **142** (1–2), 42–51.
73. Huff, M.C. and Schmidt, L.D. (1993) Ethylene formation by oxidative dehydrogenation of ethane over monoliths at very short contact times. *J. Phys. Chem.*, **97**, 11815–11822.
74. Bodke, A.S. et al. (1999) High selectivities to ethylene by partial oxidation of ethane. *Science*, **285** (5428), 712–715.
75. Zerkle, D.K. et al. (2000) Modeling of on-line catalyst addition effects in a short contact time reactor. *Proc. Combust. Inst.*, **28**, 1365–1372.
76. Krummenacher, J.J., West, K.N., and Schmidt, L.D. (2003) Catalytic partial oxidation of higher hydrocarbons at millisecond contact times: decane, hexadecane, and diesel fuel. *J. Catal.*, **215** (2), 332–343.
77. Wanat, E.C., Venkataraman, K., and Schmidt, L.D. (2004) Steam reforming and water-gas shift of ethanol on Rh and Rh-Ce catalysts in a catalytic wall reactor. *Appl. Catal., A: Gen.*, **276** (1–2), 155–162.
78. Deluga, G.A. et al. (2004) Renewable hydrogen from ethanol by autothermal reforming. *Science*, **303** (5660), 993–997.
79. Hartmann, M., Maier, L., and Deutschmann, O. (2010) Catalytic partial oxidation of iso-octane over rhodium catalysts: an experimental, modeling, and simulation study. *Combust. Flame*, **157**, 1771–1782.
80. Park, S., Vohs, J.M., and Gorte, R.J. (2000) Direct oxidation of hydrocarbons in a solid-oxide fuel cell. *Nature (London)*, **404** (6775), 265–267.
81. Janardhanan, V.M. and Deutschmann, O. (2006) CFD analysis of a solid oxide fuel cell with internal reforming: coupled interactions of transport, heterogeneous catalysis and electrochemical processes. *J. Power Sources*, **162**, 1192–1202.
82. Zhu, H. and Kee, R.J. (2003) A general mathematical model for analyzing the performance of fuel-cell membrane-electrode assemblies. *J. Power Sources*, **117** (1–2), 61–74.
83. Janardhanan, V.M. and Deutschmann, O. (2007) Modeling of solid-oxide fuel

cells. *Z. Phys. Chem.-Int. J. Res. Phys. Chem. Chem. Phys.*, **221** (4), 443–478.

84. Janardhanan, V.M. and Deutschmann, O. (2007) Numerical study of mass and heat transport in solid-oxide fuel cells running on humidified methane. *Chem. Eng. Sci.*, **62** (18–20), 5473–5486.

85. Janardhanan, V.M., Heuveline, V., and Deutschmann, O. (2007) Performance analysis of a SOFC under direct internal reforming conditions. *J. Power Sources*, **172** (1), 296–307.

18
Hydrodesulfurization
Roel Prins

18.1
Introduction

The major industrial use of supported metal sulfides is in the removal of sulfur, nitrogen, oxygen, and metal atoms from oil fractions by reductive treatments in so-called hydrotreating processes. Such processes are of paramount importance, because oil products must be purified to diminish air-polluting emissions of sulfur and nitrogen oxides, which contribute to acid rain. Furthermore, most catalysts that are used for the upgrading of oil fractions to commercial fuels cannot tolerate sulfur and metals, and many oil streams in a refinery must therefore be hydrotreated. As a consequence, hydrotreating is the largest application of industrial catalysis on the basis of the amount of material (oil fractions) processed per year. Based on the amount of catalyst sold per year, hydrotreating catalysts constitute the third largest catalyst business after exhaust gas catalysts and fluid cracking catalysts.

Hydrotreating catalysts contain molybdenum and cobalt or nickel, supported on γ-Al_2O_3. When supported alone on alumina, Mo sulfide has a much higher activity for the removal of S, N, and O atoms than Co and Ni sulfide, but sulfided Co-Mo/Al_2O_3 and Ni-Mo/Al_2O_3 catalysts have a substantially higher catalytic activity than Mo/Al_2O_3. Cobalt and nickel are therefore considered as promoters of the Mo activity [1–4]. Cobalt is mainly used as a promoter for sulfided Mo/Al_2O_3 in hydrodesulfurization (HDS), while nickel is favored in hydrodenitrogenation (HDN). In addition to Mo and Co or Ni, hydrotreating catalysts often contain modifier elements such as P, B, F, or Cl, which are added to improve the catalytic as well as the mechanical properties of the catalyst.

In the future, biomass may develop into an alternative energy source. Biomass is rich in oxygen-containing compounds, and the oxygen atoms can be removed by similar processes and catalysts to thse used for sulfur- and nitrogen-containing molecules. Therefore, metal sulfides may find new applications in the treatment of biomass.

The most important hydrotreating process is HDS, the removal of sulfur from sulfur-containing molecules in oil fractions. Mechanisms to explain these reactions

Catalysis: From Principles to Applications, First Edition. Edited by Matthias Beller, Albert Renken, and Rutger van Santen.
© 2012 Wiley-VCH Verlag GmbH & Co. KGaA. Published 2012 by Wiley-VCH Verlag GmbH & Co. KGaA.

and the relationship between catalyst structure and reaction mechanisms are presented in this chapter.

18.2
Hydrodesulfurization

Organosulfur compounds are present in almost all crude oil fractions. The higher-boiling-point fractions of oil contain relatively more sulfur compounds and with a higher molecular weight. The low-boiling crude oil fraction contains mainly aliphatic organosulfur compounds, namely mercaptans, sulfides, and disulfides. These are very reactive, and the sulfur can easily be removed. The higher-boiling fractions, such as heavy straight-run naphtha, straight-run diesel, and light naphtha originating from the fluid catalytic cracking (FCC) process contain thiophenes, benzothiophenes, and their alkylated derivatives. These compounds are more difficult to convert via HDS than mercaptans and sulfides. The heaviest fractions blended to the gasoline and diesel pools (bottom FCC naphtha, coker naphtha, FCC, and coker diesel) contain mainly alkylbenzothiophenes, dibenzothiophenes, and alkyldibenzothiophenes.

Deep desulfurization of the fuels implies that more and more of the least reactive sulfur compounds must be converted. Aliphatic and aromatic thiols are intermediates in ring-opening reactions of cyclic sulfur-containing compounds. They have high reactivity, and the sulfur atom can be removed easily, which explains why they do not occur in oil. Aliphatic thiols can react through elimination and hydrogenation

$$\text{R–CH}_2\text{–CH}_2\text{–SH} \xrightarrow{-H_2S} \text{R–CH=CH}_2 \xrightarrow{H_2} \text{R–CH}_2\text{–CH}_3 \tag{18.1}$$

and through hydrogenolysis

$$\text{R–CH}_2\text{–CH}_2\text{–SH} + H_2 \longrightarrow \text{R–CH}_2\text{–CH}_3 + H_2S \tag{18.2}$$

Hofmann-type β-hydrogen elimination takes place via well-established acid–base catalysis via the surface of the metal sulfide [4]. Hydrogenation and hydrogenolysis take place at the metal sulfide surface, probably analogously to hydrogenolysis on metals, via C–S and H–H bond scissions and C–H and S–H bond formations [4]. Aliphatic thiols which contain a β-H atom undergo elimination faster than hydrogenolysis, while aliphatic thiols without β-H atoms, such as methyl mercaptan and thiophenol, undergo HDS through hydrogenolysis.

Thiophene, benzothiophene, dibenzothiophene, and their alkylated derivatives are the major sulfur-containing molecules in oil and coal-derived liquids. While the HDS mechanism of thiophene at atmospheric pressure is still under debate, at elevated H_2 pressures the major reaction path is via hydrogenation of thiophene to tetrahydrothiophene [5]. This intermediate can react to butadiene through two

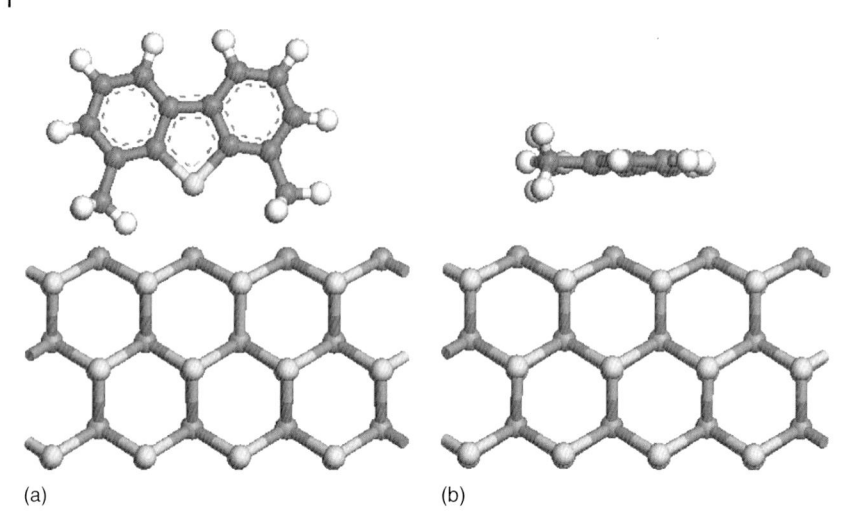

Figure 18.1 Adsorption of 4,6-dimethyldibenzothiophene in σ fashion (a) and in π fashion (b) on the metal edge of MoS_2.

successive β-H eliminations, to *n*-butane through two hydrogenolysis steps, or to butene by a mixture of these two reactions.

$$\tag{18.3}$$

In the HDS of dibenzothiophene, biphenyl is formed through a twofold hydrogenolysis, and this so-called direct desulfurization (DDS) is the major reaction pathway (80–90%) [1, 4, 6]. Alkyl groups adjacent to the sulfur atom, as in 4,6-dialkyldibenzothiophene, sterically hinder the adsorption of the molecule perpendicular to the catalyst surface (Figure 18.1a). Consequently, no strong bond between the S atom of 4,6-dialkyldibenzothiophene and a metal atom at the catalyst surface can be established, and the removal of the S atom through the DDS route is strongly suppressed. Also, H_2S adsorbs strongly on metal atoms on the catalyst surface and suppresses the DDS route.

Dibenzothiophene can also react to cyclohexylbenzene by partial hydrogenation of one of the phenyl rings followed by a double C–S bond scission. Another possibility is that one hydrogenolysis and one elimination take place, followed by hydrogenation of the resulting double bond. The route from di-alkyldibenzothiophene to di-alkylcyclohexylbenzene is called the *hydrogenation route*. It is about as fast for 4,6-dialkyldibenzothiophene as it is for dibenzothiophene, which can be explained by adsorption and reaction of the 4,6-dialkyldibenzothiophene parallel to the catalyst surface (Figure 18.1b). Because the DDS rate is fast for dibenzothiophene but slow for 4,6-dialkyldibenzothiophene, the hydrogenation route is the minority route (10–20%) for dibenzothiophene and the major (but equally slow) route for 4,6-dialkyldibenzothiophene. For these reasons, 4,6-dialkyldibenzothiophene molecules are among the molecules that are most difficult to desulfurize.

The HDS of benzothiophene occurs similarly to the hydrogenation route of dibenzothiophene [1, 4, 6], with first hydrogenation to dihydrobenzothiophene and then ring-opening hydrogenolysis and sulfur-removal hydrogenolysis.

18.3
The C-X Bond-Breaking Mechanism

Direct cleavage of the C–X bond in C_6H_5–X is the rule for C_6H_5–SH but not for C_6H_5–OH and C_6H_5–NH_2 [5]. Direct cleavage of the C–N bond in aniline and the C–O bond in phenol occurs to a limited extent but only under severe conditions, at high H_2/H_2S ratios, and above 400 °C. Under normal hydrotreating conditions, only aliphatic C–N and C–O bonds can be broken. Consequently, hydrogenation of the heterocycle is required in order to remove the nitrogen (HDN) or oxygen atom (hydrodeoxygenation, HDO). Because of the different hydrogenolysis behavior, the products of HDS are usually rich in aromatics, while the products of HDO and HDN are mainly aliphatic. This difference is due to the fact that the C_6H_5–SH bond is weaker than the C_6H_5–OH and C_6H_5–NH_2 bonds.

Removal of the heteroatom from an aliphatic C–SH fragment can, according to classical organic chemistry, take place either via elimination of H_2S or via substitution by H_2S followed by H_2S elimination or hydrogenolysis [4]. In the Hofmann elimination reaction, an acid helps in quaternizing the nitrogen atom, thereby creating a better leaving group, while a base promotes the elimination by removal of a β-H atom. Evidence for the existence of Brønsted acid sites in sulfided Mo, Co–Mo, and Ni–Mo catalysts has been provided [4], and the removal of a sulfur atom by elimination is clearly established.

$$\text{H-C-C-SH} \xrightarrow{HB} B^- + \text{H-C-C-SH}_2^+ \xrightarrow{-HB} \text{C=C} + H_2S \qquad (18.4)$$

18.4
Structure of the Sulfidic Catalyst

18.4.1
Structure of Mo

Oxidic catalyst precursors are made by impregnation, drying, and calcination steps, and are transformed into the actual hydrotreating catalysts by sulfidation in a mixture of H_2 and H_2S or a sulfur-containing molecule. The properties of the final sulfidic catalyst depend to a great extent on the calcination and sulfidation steps. Calcination at high temperature induces a strong interaction between the Al_2O_3 support and the Mo and Co or Ni ions in the oxidic state, which makes it difficult to transform the compounds formed into sulfides. The higher the calcination temperature, the higher the sulfidation temperature needed to reach the best hydrotreating activity. When the sulfidation temperature is too high, however, the metal oxides and metal sulfides will sinter. Optimum calcination and sulfidation temperatures for Al_2O_3 as a support are in the range 400–500 °C. A sulfur–oxygen exchange reaction takes place even at low temperatures, leading to an MoS_3-type intermediate structure, which at higher temperature reduces to the MoS_2 structure.

MoS_2 has a layer lattice, and the sulfur–sulfur interaction between successive MoS_2 sandwiches is weak. Crystals grow in the form of platelets with relatively large dimensions parallel to the basal sulfur planes and a small dimension perpendicular to the basal plane. High-resolution transmission electron microscopy of HDS model catalysts, consisting of MoS_2 crystallites on planar Al_2O_3, showed that the MoS_2 crystallites occurred in the form of platelets with a height-to-width ratio between 0.4 and 0.7. Some of these platelets were oriented with their basal plane parallel to the Al_2O_3 surface and some were oriented at a non-zero angle to the surface. This suggests that Mo–O–Al bonds bind the MoS_2 platelets to the Al_2O_3 surface. γ-Al_2O_3 is the preferred choice as the support of industrial hydrotreating catalysts because of its good textural and mechanical properties. Hydrotreating catalysts have to be regenerated several times during their existence, and this puts high demands on the catalyst support. Regeneration encompasses burning-off of the coke from the catalyst with air and resulfiding of the oxidized catalyst. In between, the catalyst must be purged to avoid explosive conditions.

18.4.2
Structure of the Promoter

Cobalt may be present in three forms after sulfidation, as Co_9S_8 crystallites on the support, as cobalt ions adsorbed on the edges of MoS_2 crystallites (the so-called Co-Mo-S phase, see below), and as cobalt ions in tetrahedral sites in the γ-Al_2O_3 lattice. Analogously, nickel can be present as segregated Ni_3S_2, as Ni-Mo-S, and as Ni ions in the support. The structure of the catalyst in the sulfided state is predetermined by the structure of the oxidic precursor [2]: Co_3O_4 transforms into Co_9S_8, cobalt ions in octahedral support sites transform into the Co-Mo-S phase, and cobalt ions in tetrahedral support sites remain in their positions. The promoter effect of cobalt is related to the cobalt ions in the Co-Mo-S phase and not to separate Co_9S_8.

An infrared study of the adsorption of NO molecules on sulfided Co-Mo/Al_2O_3 catalysts confirmed that the Co atoms are adsorbed on MoS_2 in positions along the edge [2]. At increasing cobalt loading, the spectrum of NO adsorbed on Co sites increased in intensity, while that of NO adsorbed on Mo sites decreased in intensity because Co ions at the edge decoration sites cover Mo ions and block adsorption of NO on these Mo ions. The observed behavior is, therefore, in accordance with the edge decoration location. Extended X-ray absorption fine structure (EXAFS) studies are also in accordance with this model. They showed that, in supported sulfided Ni-Mo catalysts, the Ni atoms are surrounded by five sulfur atoms at 2.22 Å, by one or two Mo atoms at 2.8 Å, and by one Ni atom at 3.2 Å [4]. These data are fully consistent with a model in which the Ni ions are located at the MoS_2 edges in the Mo plane in a square pyramidal coordination. The Ni ions are connected to the MoS_2 by four sulfur atoms, and the fifth sulfur atom is in the apical position in front of the Ni ion. A neighboring Ni atom is located at the next edge position at 3.2 Å.

At very low Co loadings and mild sulfiding temperatures all Co atoms can be positioned around the MoS_2 edges. Initially, the catalytic activity increases with increasing Co/Mo (or Ni/Mo) ratio, but a point will be reached where all edge positions are occupied and additional Co atoms will have to be put on top of Co atoms which are already present or have to form separate Co_9S_8 crystallites. High sulfiding temperatures also lead to large Co_9S_8 particles because of growth of the MoS_2 crystallites and, as a consequence, decrease in the MoS_2 edge area. Since the Co_9S_8 particles have a low catalytic activity and cover the MoS_2 particles, the HDS activity of Co-Mo catalysts decreases at high Co/Mo ratios. Maximum activity is usually observed at a Co/Mo ratio of 0.3–0.5.

Co-Mo/Al_2O_3 catalysts sulfided at high temperature are intrinsically more active per Co ion in the Co-Mo-S phase than those sulfided at 400–500 °C [2]. The more active phase was called *Co-Mo-S II* and the less active phase *Co-Mo-S I*. The type I structure was said to be bonded to the support via Mo–O–Al linkages, while type II may have few if any such linkages. Since the Mössbauer and EXAFS signals of the type I and II structures do not differ at all, the local Co and Ni environments must be exactly the same in both cases. This suggests that type I is less active than II for steric reasons. Catalyst–support linkages in type I probably hinder the approach of reactant molecules to the catalytic sites.

Daage and Chianelli proposed that the Mo atoms at the edges of the top layers of stacks of MoS_2 layers have a higher activity for hydrogenation of dibenzothiophene than the intermediate MoS_2 layers [2, 4]. Steric arguments were used to explain this situation. Adsorption in a σ mode ($\eta^1(S)$) should be possible at the edge Mo sites of all layers; this adsorption mode was assumed to be necessary for hydrogenolysis. On the other hand, π adsorption, with the aromatic ring parallel to the edge surface, results in hydrogenation; this geometry is unlikely on the Mo sites at the edges of the intermediate MoS_2 layers.

18.4.3
DFT Calculations

Density functional theory (DFT) calculations have furnished a wealth of information on the coordination of the metal atoms at the surface of unpromoted and promoted MoS_2 crystals. They have indicated that under the usual HDS conditions (10–50 kPa H_2S and 3–7 MPa H_2) the Mo catalyst contains 50% sulfided Mo edges, with sulfur atoms in bridge positions between Mo atoms [7]. The Mo atoms at the Mo edge

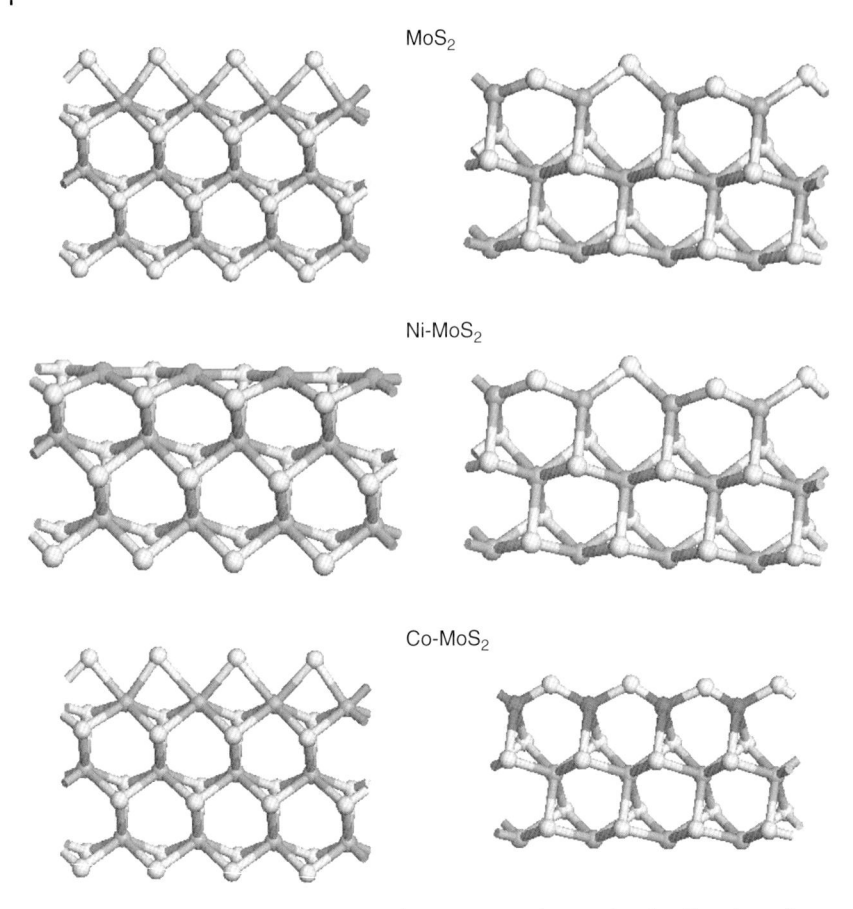

Figure 18.2 Structures of the Mo, Ni, and Co atoms at the metal and sulfur edges of MoS_2.

are thus fully coordinated by six sulfur atoms in a trigonal prismatic arrangement (Figure 18.2, top left). The Mo atoms at the S edges of MoS_2 are 50% sulfided under normal HDS conditions. The Mo atoms are surrounded by four sulfur atoms, the outer two in bridge positions being between the Mo atoms in a zigzag configuration, so that the Mo atoms are in a distorted tetrahedral sulfur coordination (Figure 18.2, top right).

The most stable position for the Co or Ni promoter atom is at the edges of the MoS_2 particles, substituting a molybdenum atom [7] and forming the Co-Mo-S and Ni-Mo-S structures. In the NiMo catalyst the Ni atoms are preferentially located at the metal edge, and are not covered by sulfur atoms. As a result, the Ni atoms have a square-planar sulfur coordination with open coordination positions (Figure 18.2, middle left). The Co atoms in the CoMo catalyst are preferentially situated at the S edges (Figure 18.2, bottom right). The DFT calculations indicate that, like the Mo atoms at the S edges of MoS_2, the Mo atoms at the S edges of Ni-MoS_2 as well as the Co atoms at the S edges of Co-MoS$_2$ are 50% sulfided under normal HDS conditions. The Mo and Co atoms are coordinated by four sulfur atoms, the outer two of which are in bridge positions between the metal atoms. The bridge positions of the sulfur atoms on the Co atoms are regular, leading to a tetrahedral sulfur coordination around the Co atoms, while the bridging sulfur atoms on the Mo atoms of the S edge of Ni-MoS$_2$ have the same zigzag configuration as the sulfur atoms at the S edge of MoS$_2$.

DFT calculations predict that the MoS$_2$, Co-MoS$_2$, and Ni-MoS$_2$ catalyst particles have a different morphology under HDS conditions. MoS$_2$ particles should have a deformed hexagonal shape, with molybdenum as well as sulfur edges [7]. Co- and Ni-promoted MoS$_2$ should have a more triangular shape, with the Co-promoted MoS$_2$ particles mainly terminated by Co-covered sulfur edges and Ni-promoted MoS$_2$ mainly by Ni-covered metal edges. Scanning tunneling microscopy (STM) of MoS$_2$ and Co-MoS$_2$ nanoclusters on a gold (111) substrate indeed showed that the morphology of the nanoclusters depends on the presence of Co and on the synthesis conditions [8]. At high H$_2$S/H$_2$ ratio the clusters had a triangular shape, and at a more normal ratio of 0.07 they were hexagonally truncated. The STM results indicate that the Co atoms are located at the S edge, confirming the DFT calculations. They also show one-dimensional metallic states at the Mo as well as at the S edges. The metallic character explains why MoS$_2$-type catalysts are reasonably good catalysts for hydrogenation and thus also for hydrotreating.

18.5
Hydrodenitrogenation

The question how a C–N bond is broken has caused much more discussion than the question how a C–S bond is broken. Originally, evidence for elimination as well as nucleophilic substitution was presented, but recent investigations have shown that nucleophilic substitution by H$_2$S is the prevailing reaction for all amines with the exception of amines with a tertiary carbon atom (e.g., *tert*-butylamine), which undergo elimination of ammonia [4]. The nucleophilic substitution of the amine group by a sulfhydryl group does not take place by a classic organic S$_N$2 reaction, but by dehydrogenation to an imine followed by H$_2$S addition, NH$_3$ removal, hydrogenation, and hydrogenolysis.

$$\underset{H}{\overset{|}{-C}}-N\underset{H}{\diagup} \xrightarrow{-H_2} \diagdown C=N\diagup \xrightarrow{H_2S} \underset{\underset{H}{S}}{\overset{|}{-C}}-N\underset{H}{\diagup} \longrightarrow \diagdown C=S + -N\diagup^H_H$$

$$\diagdown C=S \xrightarrow{H_2} \underset{H}{\overset{|}{-C}}-SH \xrightarrow{H_2} H_2S + \underset{H}{\overset{|}{-C}}-H$$

(18.5)

18.6
Determination of Surface Sites

Whereas there are several methods for the determination of the (specific) surface area of supported metal catalysts, no generally accepted method for supported metal sulfides exists. The catalyst is complex and consists of several components, and it is not sufficient to measure the average MoS_2 dispersion (surface-to-volume ratio), since only the edge surface and not the basal planes are catalytically active [2, 4]. O_2 chemisorption has limited applicability, because the oxygen chemisorption does not stop at oxidation of the exposed Mo ions; with time, subsurface and deeper-lying Mo ions are also oxidized. Since the oxidation process is topotactic, following the shape of the MoS_2 crystallites, a linear relationship between O_2 chemisorption and HDS activity may still be observed for a certain class of catalyst (like MoS_2) and another relationship for another class of catalysts such as promoted MoS_2. Since O_2 chemisorption reflects the general state of dispersion of the (un)promoted catalyst rather than the specific sites, O_2 chemisorption cannot be used for the quantitative determination of active sites. CO and NO have also been tried as adsorbents for the determination of the active surface area of sulfide catalysts. CO binds weakly to MoS_2, only slightly more strongly than to Al_2O_3, and CO adsorption is therefore normally performed at low temperature. Infrared and adsorption studies demonstrated a good correlation between NO chemisorption and HDS activity for MoS_2/Al_2O_3 catalysts but not for promoted catalysts. X-ray photoelectron spectroscopy (XPS) studies, on the other hand, indicated that NO adsorbs dissociatively above $-140\,°C$ on polycrystalline MoS_2 and that the resulting oxygen atoms lead to surface oxidation above $-73\,°C$. On the other hand, combined NO adsorption, IR, and magnetic susceptibility studies gave extensive information on the surface structure of $CoMo/Al_2O_3$ catalysts and indicated that the Co atoms are present as dinuclear clusters at the edges of the MoS_2 particles.

References

1. Girgis, M.J. and Gates, B.C. (1991) Reactivities, reaction networks, and kinetics in high-pressure catalytic hydroprocessing. *Ind. Eng. Chem. Res.*, **30** (9), 2021–2058.
2. Topsøe, H., Clausen, B.S., and Massoth, F.E. (1996) *Hydrotreating Catalysis: Science and Technology*, Springer, Berlin.
3. Kabe, T., Ishihara, A., and Qian, W. (1999) *Hydrodesulfurization and Hydrodenitrogenation*, Kodansha Ltd./Wiley-VCH, Tokyo/Weinheim.
4. Prins, R. (2008) in *Handbook of Heterogeneous Catalysis* (eds G. Ertl et al.), Wiley-VCH Verlag GmbH, Weinheim, p. 2695.
5. Schulz, H., Schon, M., and Rahman, N.M. (1986) in *Studies in Surface Science and Catalysis*, Chapter 6 (ed. L. Cerveny), Elsevier B.V, pp. 201–255.
6. Whitehurst, D.D., Isoda, T., and Mochida, I. (1998) in *Advances in Catalysis* (eds D.D. Eley, W.O.H.B. Gates, and K. Helmut), Academic Press, pp. 345–471.
7. Krebs, E., Silvi, B., and Raybaud, P. (2008) Mixed sites and promoter segregation: a DFT study of the manifestation of Le Chatelier's principle for the Co(Ni)MoS active phase in reaction conditions. *Catal. Today*, **130** (1), 160–169.
8. Lauritsen, J.V. et al. (2004) Atomic-scale insight into structure and morphology changes of MoS2 nanoclusters in hydrotreating catalysts. *J. Catal.*, **221** (2), 510–522.

Part IV
Catalyst Synthesis and Materials

This part dealing with the chemistry and practice of catalyst preparation is initiated with a chapter on the chemical reactivity of oxidic surfaces. This is discussed as a problem in molecular surface chemistry. Then a chapter follows on the preparation of supported catalysts. The third chapter deals with the inorganic chemistry of the preparation of porous oxidic materials. In the final chapter preparation of catalytic materials in relation to product development is addressed.

19
Molecularly Defined Systems in Heterogeneous Catalysis

Fernando Rascón and Christophe Copéret

19.1
Introduction

Heterogeneous catalysts play an essential part in about 90% of the world's industrial chemical transformations [1], which enable raw materials to be converted into valuable chemicals and fuels in an economical, efficient, and environmentally friendly manner. However, the development of heterogeneous catalysis is hindered by the difficulty of applying logic to what has always been essentially an empirical matter of trial and error [2]. In contrast, the development of homogeneous catalysts has benefited from an understanding of their structure–activity relationship, which has led to a myriad of new and diverse catalytic processes and a high level of efficiency [3, 4]. However, homogeneous catalysis has its own problems, principally recurrent deactivation of the catalysts, difficulties in recovering them after use, and purification of the final product. This makes it even more important to focus on heterogeneous catalysis as a possible way of circumventing these problems, since it could offer an answer to many of the economical and environmental issues faced by industry in the production of molecules of tremendous impact, such as ammonia, basic petrochemicals, pharmaceuticals, and polymers.

From a chemical point of view, heterogeneous catalysis can be understood at a molecular level, since the chemical reactions taking place at the surface of solid catalysts are essentially molecular in nature. Thus, the proper transfer of tools of molecular chemistry to surface science, along with a deeper knowledge of surface behavior, should permit us to improve chemical transformations taking place at a surface using methods similar to those used for homogeneous catalytic processes.

In this chapter, we first describe the preparation of surfaces – so-called single sites – as key to a molecular understanding of heterogeneous catalysis. Secondly, we show how this knowledge is used to immobilize catalysts and improve their output through a structure–activity relationship, feeding back to both homogeneous and heterogeneous processes. Later, we will explore how the single-site approach can help to understand and improve classical heterogeneous catalysis. Finally, we also describe how the immobilization of metal hydrides at single sites leads to the discovery of new reactions.

Catalysis: From Principles to Applications, First Edition. Edited by Matthias Beller, Albert Renken, and Rutger van Santen.
© 2012 Wiley-VCH Verlag GmbH & Co. KGaA. Published 2012 by Wiley-VCH Verlag GmbH & Co. KGaA.

19.2
Single Sites: On the Border between Homogeneous and Heterogeneous Catalysis

As in homogeneous catalysis, catalytic processes in the heterogeneous phase are considered an essentially molecular phenomenon, which implies that molecularly defined surface intermediates and/or transition states must be involved. Therefore, an appropriate study on heterogeneous catalysts must give access to the structure of a given active site, as well as accurately testing its catalytic performance. It is clear that the combination of these possibilities would establish the structure–activity relationship in heterogeneous catalysis, thus leading to the design of better catalysts or even new ones. The whole idea of this approach consists in showing that homogeneous and heterogeneous catalysis are indeed complementary, and that combining the advantages of both disciplines should lead in principle to optimal systems. One question is therefore: is it possible to implement a structure–reactivity approach to rationally improve these systems? In this chapter, we will show how the transfer of tools from molecular chemistry to surfaces can help to tackle and answer this question positively.

In order to develop heterogeneous catalysis through a structure–reactivity relationship, the first step is to be able to evaluate the structure of surface species at a molecular level, and it is therefore necessary to use combined characterization methods: chemical analysis (elemental analyses combined with chemical reactivity studies), soluble molecular models of supports (in particular for oxide surfaces), and spectroscopic methods adapted to amorphous solids (for more detailed description see Table 19.1), including Extended X-ray absorption fine spectroscopy (EXAFS) as a way to replace X-Ray crystallography, IR, solid-state NMR, UV-Vis, theoretical modeling (DFT calculations) as well as more specific techniques for solids (N_2 adsorption, XPS, and TEM). By this transfer of tools and techniques, it becomes possible to gain access to molecular control of heterogeneous catalysis.

The next step is to control and understand the surface chemistry to obtain well-defined surface species by understanding the reaction of molecules with a support – the heterogeneous phase – which can usually be considered as a rigid ligand (see below for further comments). Support is essential to the concept of immobilized molecularly defined systems, and therefore it is of prime importance to know its characteristics in detail in order to control the nature of the interactions with the incoming molecules that will take place as well as the potential structure of the supported species. It is also clear that the support should provide stability to the immobilized species.

Silica is one of the most widely used supports. Its surface is composed of siloxane bridges ($\equiv$Si–O–Si$\equiv$) and silanol groups ($\equiv$Si–OH). Thermal treatment at elevated temperatures leads to the condensation of adjacent silanols yielding water and a siloxane bridge. This process can be used to control the OH group density at the surface. For instance, at 700 °C the silica surface is mainly composed of isolated silanols, which are statistically distributed at an average distance of circa 1.3 nm. Treatment at higher temperatures further decreases the OH density and leads to the formation of reactive siloxane bridges, but this process is further

19.2 Single Sites: On the Border between Homogeneous and Heterogeneous Catalysis

accompanied by a loss of specific surface area due to sintering. Using this tunable property, it is possible to generate well-defined single sites on silica supports by controlling the OH density. For instance, the vacuum treatment at 200, 500, and 700 °C of silica with a surface area of 200 m^2 g^{-1} generates silica with 2.6, 1.2, and 0.7 ± 0.2 accessible silanol groups per nm^2, respectively (equal to a concentration of approximately 0.86, 0.42, and 0.23 mmol of OH per gram of silica, respectively) [5]. Starting from here, it is possible to perform stoichiometric reactions between the support and molecular organometallic compounds. This is done by grafting a given catalytically active complex or a molecular precursor with the appropriate ligand sets, in particular, one which is reactive, here symbolized

Table 19.1 Molecular vs surface chemistry: characterization tools.

Characterization tools for molecular chemistry	Is it possible to extrapolate them to surface studies on single sites?
IR (Raman) spectroscopy	Yes, useful for functional group identification and kinetic studies.
NMR spectroscopy	Yes, multinuclear and multidimensional solid-state NMR gives access to the full structure of a given catalyst, but it often requires labeling studies due to the low density of sites (the surface active sites correspond to much less than 10 wt% of the actual solid).
ESR spectroscopy	Yes, useful to determine electronic states at the metal centers and distinguish between different grafted species. It is, however, often very complex due to the high sensitivity of this method to distortion.
UV spectroscopy	Yes, identification of functional groups and concentration of active species can be studied.
Elemental analysis	Yes, mass balance of the reaction is straightforward.
Chemical characterization (reactivity)	Yes, reactions in batch/flow reactors in the solid–gas and/or solid–liquid phase can easily be performed. In particular, TPD can be used to characterize surface species.
Mass spectrometry	Not often used, but the appearance of TOF-SIMS methods could probably gain attention.

X-ray crystallography	EXAFS and XANES provide useful information to assess the structure, the coordination geometry, and oxidation state at the grafted metal centers.
	Specific techniques for solids:
	    N$_2$ adsorption: Surface characterization methods for solids allow the morphology of the surface and the interaction with substrates to be investigated.
	    Transmission electron microscopy: Microscopy permits observation of the surface at nanometer scale, where pores, channels as well as the formation of particles can be studied.
	    XPS spectroscopy: access to oxidation state.

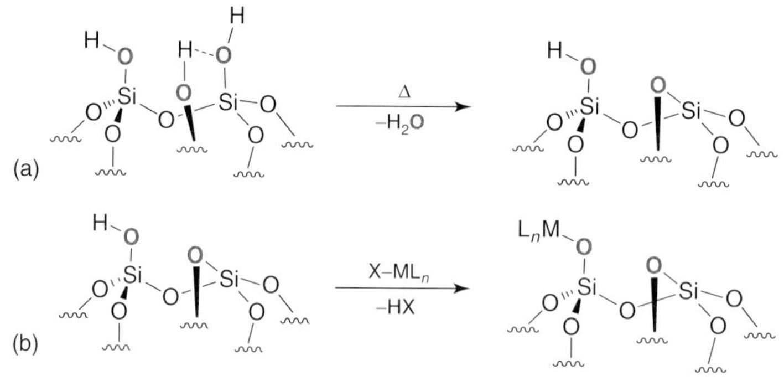

Scheme 19.1 (a) Dehydroxylation process upon thermal treatment. (b) Grafting on isolated silanols.

as X in [L_nMX]. On $SiO_{2-(700)}$, it will typically yield the corresponding monosiloxy complex [($\equiv$ SiO)ML_n], with formation of XH as by-product (Scheme 19.1) [6].

19.2.1
Taking Homogeneous Catalysis to the Heterogeneous Phase via a Molecular Approach: the Case of Single-Site Alkene Metathesis Catalysts

Alkene metathesis is a key petrochemical process: in particular, the cross-metathesis of ethene and 2-butenes, both formed in large amount in crackers, is massively used to increase the yield of propene, for which world demand has been increasing in the past few years. Very soon after the discovery of alkene metathesis, it was shown that this reaction probably involved metallocarbene intermediates [7, 8], and this led to the development of highly efficient homogeneous catalysts based on well-defined metallocarbenes [4, 9, 10]. Such progress was only possible thanks to the power of molecular organometallic chemistry in homogeneous catalysis. This has been translated into a series of applications from polymer science to medicinal chemistry, and it has more recently been seen as an efficient way to create the carbon–carbon bond skeleton of complex fine chemicals [11–13]. However, because of the nature of these chemicals, homogeneous catalysts suffer from several drawbacks such as decomposition, in particular via bimolecular deactivation processes, and problems with metal recovery (separation of products from the catalyst as well as metal contamination of the product). Therefore, immobilization of homogeneous catalysts emerges as an intuitive method of solving these issues, providing stability to the coordinative environment at the metal center and avoiding bimolecular deactivation, while metal recovery becomes an easy task.

As stated above, alkylidene-containing compounds are key intermediates in alkene metathesis. Taking this as a starting point, a large research effort has consisted in the preparation of surface-bound and well-defined metallocarbenes, which could in turn lead to a more rational development of heterogeneous catalysts.

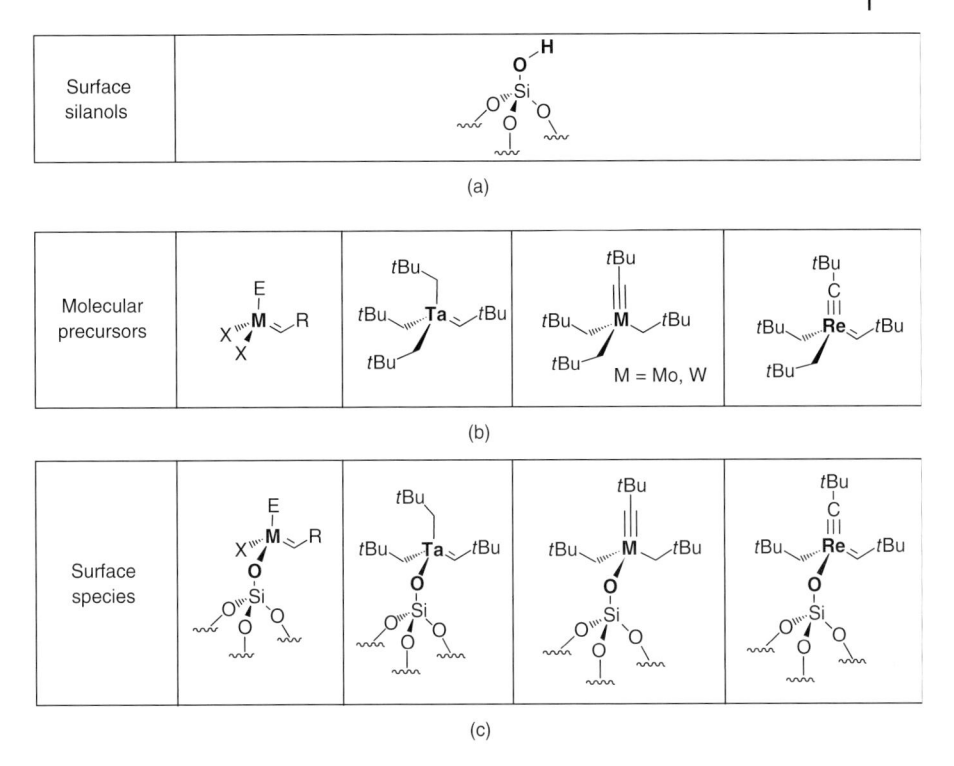

Scheme 19.2 (a–c) Grafting of an organometallic complex onto silica. Molecular alkylidene and alkylidyne complexes and their correspondent grafted alkylidene and alkylidyne complexes.

The grafting of alkylidene molecular precursors based on Ta, group 6 and Re transition metals yields overall surface species in which one alkyl is replaced by a surface siloxy ligand (see Scheme 19.2) [14]. However, the well-defined tantalum alkylidene surface complex only displays marginal activity in alkene metathesis, as previously observed for homologous molecular complexes [15]. Further access to the corresponding well-defined group 6 alkylidene surface complexes was investigated by grafting alkylidyne molecular complexes. While most of these are highly active in olefin metathesis, it has been shown that they are in fact alkylidyne surface complexes and that their corresponding alkylidene ligands are probably formed *in situ* in the presence of alkenes [16–18].

In the case of rhenium, the structure of the surface complex is noteworthy, as the coordination sphere of rhenium contains four different ligands (Schemes 19.2 and 19.3) [19, 20]. The activity in alkene metathesis of this well-defined system is also worth mentioning: it displays greater activity in alkene metathesis than the isoelectronic and isostructural rhenium homogeneous catalysts [21–23], [(X)(Y)(Re ≡ C^{t}Bu)(= CH$_2^t$Bu)], whether having two neopentyl (X = Y = CH$_2^t$Bu) or two alkoxy (X = Y = O^{t}Bu or OC(CH$_3$)(CF$_3$)$_2$) ligands in place of the neopentyl and siloxy

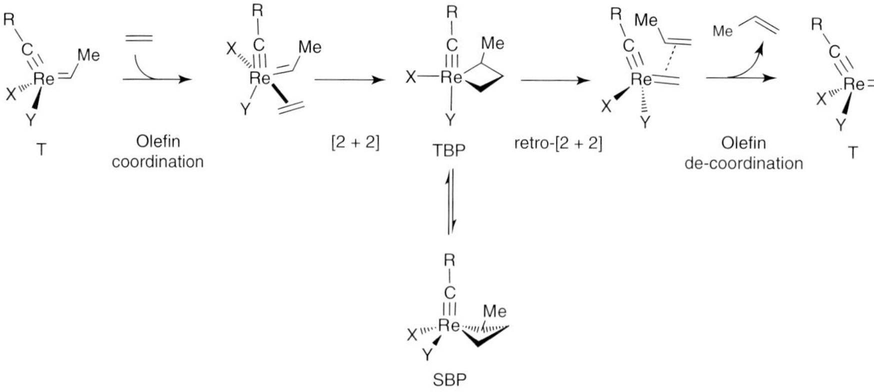

Scheme 19.3 Reaction pathway of alkene metathesis for tetrahedral (T) rhenium(0) complexes.

ligands. Interestingly, this catalyst is far more active than the classical heterogeneous catalysts Re_2O_7/Al_2O_3, and it does not require activating agents such as tin compounds to catalyze the metathesis of functionalized alkenes. This unexpected activity was studied in detail through computational studies, where it was found that having a catalyst structure dissymmetric at the metal center $(D)(A)Re(C^tBu)$ $(= CH^tBu)$ was key to its reactivity. Indeed, when D and A have different σ-bond donating abilities [24, 25], this translates into transition states of low energy and the formation of not-too-stable reaction intermediates. The overall reaction pathways correspond to a four-step pathway involving coordination and cyclo-addition, and the corresponding reverse elementary steps (cyclo-reversion and de-coordination, see Scheme 19.3).

The key step is the coordination (or de-coordination) of the alkene, which is associated with a distortion of the geometry of the metal complex from a tetrahedron into a trigonal prism in order to open a coordination site, and this is most favored when the most σ-donating ligand (D) is trans to the incoming alkenes and the less σ-donating ligand (A) shares the same basal plane of the trigonal bipyramidal structure. The overall energy span, which is related to the overall rate of the reaction [26], is also influenced by this dissymmetry because of the presence of a σ-donor ligand destabilizing the most stable intermediates, that is, the metallacyclobutanes. This simple concept can therefore be readily generalized to other iso-electronic and iso-structural complexes [27, 28], and the gained knowledge with the aforementioned well-defined alkylidene Re-based catalysts has turned out to be key to the development of the iso-electronic group-6 analogs (see Scheme 19.4), which are known for their better performance in solution. In fact, the silica-supported systems also turned out to display higher catalyst performances, in particular in terms of selectivities and stabilities, which greatly exceeds what is known in both homogeneous and heterogeneous catalysis.

Despite these clear improvements, it is worth mentioning that 1-butene is also observed in the metathesis of propene with these systems, albeit in a lesser amount

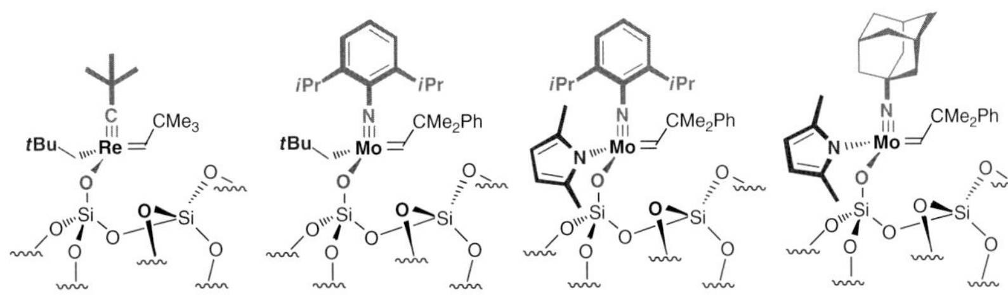

TOF (TON/min)	120	120	320	780
TON (1500 min)	6000	22000	101000	275000
Selectivity in 2-butenes	96%	99.4%	>99.9%	>99.9%

Scheme 19.4 Catalytic performance of some isoelectronic and isostructural molybdenum grafted species.

than for the Re-based catalyst. Note that this type of isomerization product is not observed in the bisalkoxy homogeneous catalysts for the metathesis of 1-alkenes, which indicates that their formation is related to the presence of the pendant alkyl ligand. As a result, further studies focused on the replacement of this ligand by amido and alkoxy ligands. This transformation led to greatly improved activity, selectivity, and stability, resulting in overall much higher turnover numbers (up to circa 300 000) and the absence of the previously observed and undesired 1-butene by-product. Noteworthy, these catalysts greatly out-performed the corresponding homogeneous catalysts.

These results and in particular the computational studies revealing that dissymmetry at the metal center was key to reactivity have now been used to develop improved homogeneous catalysts. Replacing the surface siloxy ligand of silica by bulkier aryloxy molecular ligands has led to the development of highly active, selective, and stable homogeneous catalysts, moreover having high enantio- and Z-selectivity (Scheme 19.5) [29, 30]. This breakthrough will likely lead to new

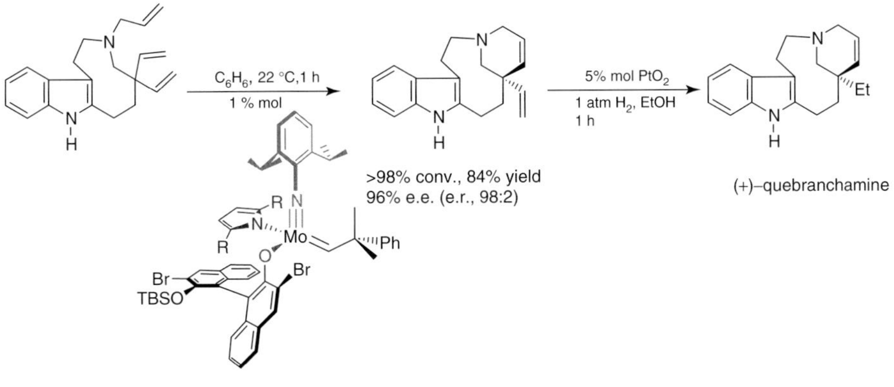

Scheme 19.5 Enantioselective ring-closing metathesis mediated by a molybdenum chiral complex (stereogenic-at-Mo). (Redrawn from Ref. [29].)

19.2.2
Bridging the Gap with Classical Heterogeneous Systems by a Molecular Approach: the Case of Re_2O_7/Al_2O_3 vs $MeReO_3/Al_2O_3$

Out of all the heterogeneous alkene metathesis technical catalysts, an interesting case is Re_2O_7 dispersed on γ-Al_2O_3, which is the only one known to work at ambient temperature and to be compatible with functionalized alkenes upon the addition of a co-catalyst [31, 32]. However, despite abundant studies it has not been used commercially, probably for several reasons: cost of Re, relatively fast deactivation, necessity for high regeneration temperatures (500 °C), chemical engineering considerations, and so on. Moreover, it has so far been difficult to improve the catalyst through the powerful structure–reactivity relationships because of the lack of understanding of this system, which is associated with a small number of active sites, complexity, and the absence of true molecular understanding of this system [33]. In contrast to silica-supported systems, those supported on alumina are more complex. In fact, while silica is constituted solely of tetrahedral silicon atoms surrounded by oxygens, aluminum atoms in γ-alumina are either in a tetrahedral (25%) or an octahedral (75%) geometry, leading to a wide variety of surface aluminum atoms that can adopt a wide range of coordination numbers ranging from 3 to 6. This leads to the presence of Lewis acid sites of different strengths and to a population of surface hydroxyl groups, from μ^1-OH on different types of Al to μ^2-OH and μ^3-OH groups (Scheme 19.6). These μ^n-OH sites are associated with different spectroscopic properties, acidities, and reactivities, a situation that seriously complicates any serious surface study [34–36].

While it is clear that Lewis-acidic sites have a distinct role in the overall performance of the Re_2O_7/Al_2O_3 system (initiation, stability, low tolerance to external poison, and deactivation) [33], the structure(s) of the active sites and their

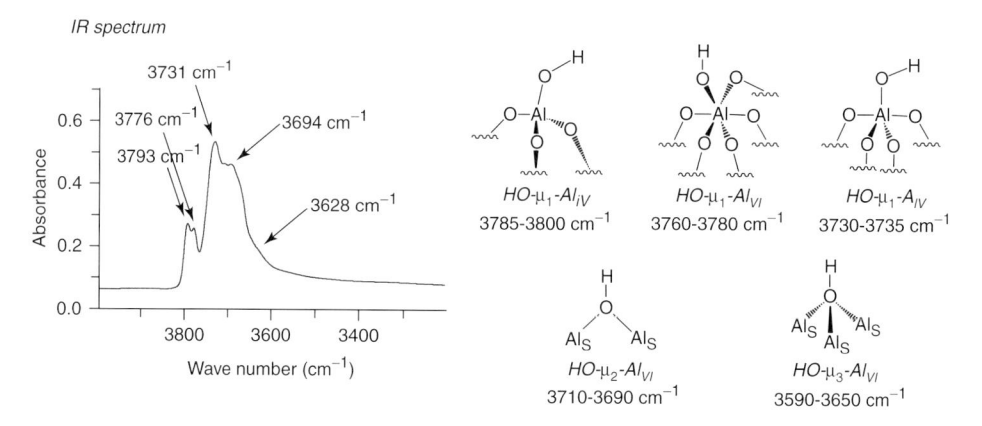

Scheme 19.6 Aluminum sites and their corresponding frequencies in IR spectroscopy.

mechanisms of initiation and deactivation are still not resolved. For instance, it is still not known how a carbene is formed from the starting oxides and why it deactivates after a few hundreds of cycles [37]. Similarly, the role of organotin agents as co-catalysts is not clearly understood. As a result, a molecular approach to the surface phenomena was needed.

Notably, the alumina-supported methyltrioxorhenium system (CH_3ReO_3/Al_2O_3) catalyzes alkene metathesis, and could provide some information about Re_2O_7/Al_2O_3. In fact, both display similar selectivity at low contact time ($Z/E = 0.4$), which suggests a very similar structure of active sites. Moreover, they both require an activation of the support at 500 °C (prior to grafting for CH_3ReO_3/Al_2O_3 or to catalytic testing for Re_2O_7/Al_2O_3). However, in contrast to the classical Re_2O_7 system, CH_3ReO_3/Al_2O_3 catalyzes the metathesis of functionalized alkenes in the absence of co-catalyst [38, 39]. A deeper study of CH_3ReO_3/Al_2O_3 reveals that the Re–C bond is not cleaved upon grafting, but that several surface species are formed: (i) the major surface species (1.0 nm^{-2}) corresponding to CH_3ReO_3 chemisorbed on aluminum Lewis-acid sites through its oxo ligands and (ii) the minor surface species (0.15 Re per nm^2) formed via the heterolytic cleavage of the C–H bond of CH_3ReO_3 on reactive surface Al_s–O_s sites [40], yielding $Al_sCH_2ReO_3$ surface species along with new hydroxyls (see Scheme 19.7). These reactive sites are in fact generated upon thermal treatment of alumina (calcinations and partial dehydroxylation typically performed at 500 °C) prior to the grafting of CH_3ReO_3, and correspond to Lewis aluminum sites. The site reactivity depends on coordination number, following the order: $(O_s)_3Al_s > (O_s)_4Al_s > (O_s)_5Al_s \gg (O_s)_6Al_s$ [40]. Of the various possible $Al_sCH_2ReO_3$ surface species, titration studies in combination with solid-state NMR data and calculations point to the formation of $Al_VCH_2ReO_3$ or $Al_{VI}CH_2ReO_3$ species [39, 41].

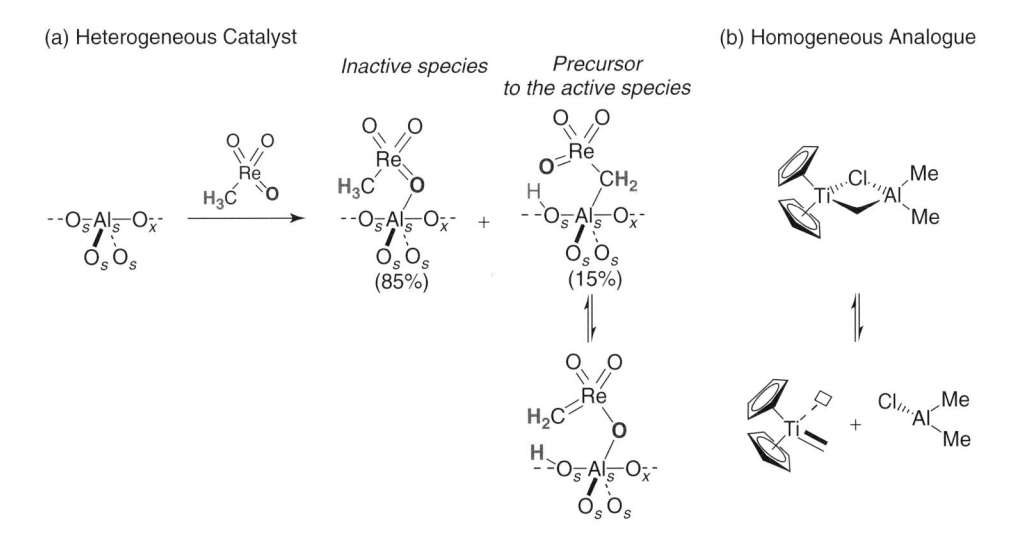

Scheme 19.7 (a) Grafting of CH_3ReO_3 over $Al_2O_{3\text{-}(500)}$. (b) Formation of the 'masked carbene' in Tebbe's catalyst.

Moreover, it was found that the only active species is in fact associated with the μ^2-CH$_2$ bridged compound, Al$_s$CH$_2$ReO$_3$, which in turn is changed into Al$_s$CH(R)ReO$_3$ through metathesis. While the actual carbenic species – necessary for an alkene metathesis transformation – has never been observed, it is probably generated *in situ* in the presence of alkenes.

Overall, CH$_3$ReO$_3$/Al$_2$O$_3$ corresponds to a heterogeneous equivalent of the Tebbe reagents, for which the carbene is also masked through coordination with a Lewis acid site [42, 43]. The existence of a 'masked carbenic' active site probably explains why this catalyst is compatible with functionalized alkenes, while the parent system based on Re$_2$O$_7$/Al$_2$O$_3$ requires organotin activators. Indeed, the formation of active sites for Re$_2$O$_7$/Al$_2$O$_3$ requires the presence of Lewis acid sites, which are probably poisoned by the functional group of the functionalized alkenes. Moreover, it is known that the reaction of SnMe$_4$ with Re$_2$O$_7$ generates CH$_3$ReO$_3$ [44], which is therefore probably formed *in situ* upon treatment of Re$_2$O$_7$/Al$_2$O$_3$ with SnMe$_4$ leading to the formation of active sites such as Al$_s$CH$_2$ReO$_3$.

While the investigation of CH$_3$ReO$_3$/γ-Al$_2$O$_3$ has shed new light on the parent rhenium oxide system in terms of the structure of the active sites, showing new paths to understanding its catalytic performance, reactivity, and stability, there is still much research to be done to reach a full understanding of Re$_2$O$_7$/Al$_2$O$_3$ system at a molecular level.

19.2.3
Toward New Reactivity: the Case of Supported Transition-Metal Hydrides

Upon treatment under H$_2$ at high temperatures (commonly at 150 °C and up to 500 °C), grafted hydrocarbyl-containing systems [EOMR$_x$] (E = Si for silica or Al for alumina, R = hydrocarbyl ligand) evolve into mononuclear metal hydride surface complexes through successive steps involving the formation of putative surface hydride intermediates ([EOMH$_y$]) and subsequent reaction with nearby surface functionalities such as siloxane and Al–O–Al bridges for silica and alumina, respectively.

This leads to the formation of new M–O bonds along with E–H surface species [45–47]. For example, [($\equiv$ SiO)Zr(CH$_2^t$Bu)$_3$] is transformed into a mixture of [($\equiv$ SiO)$_3$Zr – H] and [($\equiv$ SiO)$_2$Zr(–H)$_2$], **[Zr–H]** [48], whereas its tantalum analog [($\equiv$ SiO)Ta(= CH^{t}Bu)(CH$_2^t$Bu)$_2$] gives [($\equiv$ SiO)$_2$Ta – H$_x$], **[Ta–H]** ($x = 1$ and 3, Scheme 19.8) [49, 50]. Interestingly, for [(EO)W($\equiv$ C^{t}Bu)(CH$_2^t$Bu)$_2$], the silica-supported system (E = Si) leads mainly to de-grafting and formation of small clusters, while the alumina-supported systems (E = Al) afford the hydride species, the proposed species being [(Al$_s$O)W(–H)$_3$(=O)], **[W(O)–H]** [51].

The above-mentioned early transition-metal hydrides (group 4–6) possess a relatively unsaturated coordination sphere (formally 8 and 10 electron species) which translates into highly reactive species. In fact, they readily catalyze numerous reactions involving the low-temperature C–H and C–C bond activation of inactivated alkanes (Scheme 19.9). While unusual, these reactions proceed via the classical

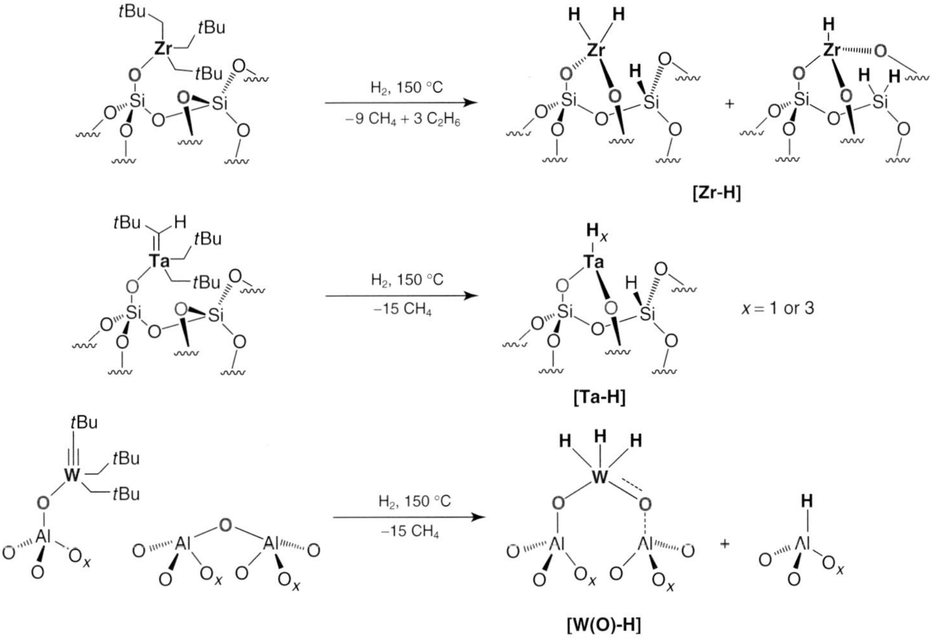

19.2 Single Sites: On the Border between Homogeneous and Heterogeneous Catalysis | 411

Scheme 19.8 Hydrogenolysis of grafted hydrocarbyl-containing systems.

elementary steps of organometallic mononuclear complexes. Over the years, the following reactions have been discovered:

1) **H/D reaction of D_2/H_2 or D_2/RH mixtures**: such reactions very likely take place by a σ-bond metathesis mechanism on high-oxidation-state metal hydride intermediates, for example, Zr^{IV}–H and Ta^V–H [52].
2) **Hydrogenolysis of alkanes**: while the C–C bond cleavage takes place via β-alkyl transfer for group 4 metal hydrides [M–H] (M = Ti, Zr, and Hf), it involves an α-alkyl transfer for [Ta–H], as evidenced by the relative reactivity toward substrate and the product selectivity [53, 54].
3) **Alkane metathesis**: [Ta–H] and [W(O)–H] transform an alkane into its lower and higher homologs. For instance, propane is converted into a mixture of mainly ethane and butane, and typical linear alkanes give selectively linear alkanes [51, 55, 56]. This unprecedented reaction involves metallacarbene and alkene intermediates, and the key C–C bond formation/cleavage step is in fact a π-bond metathesis [56–59]. More recently, it has been shown that [Zr–H] can also catalyze this reaction, but it leads to different products [60]. For instance, propane is converted into mainly methane, ethane, and 2-methylpropane. It has been proposed that this reaction involves a bis-hydride species and a different mechanism (β-alkyl transfer/insertion processes).
4) **Cross-metathesis of methane and higher alkanes**: [Ta–H] can catalyze the reaction of methane with a higher hydrocarbon, namely propane, and thereby incorporate methane into a higher alkane by cross-metathesis [61]. This

19 Molecularly Defined Systems in Heterogeneous Catalysis

(a) H/D exchange

$$H_2 + D_2 \xrightarrow{[M-H]} 2\,HD$$

$$CH_4 + D_2 \xrightarrow{[M-H]} CH_3D + HD$$

σ-Bond metathesis

M = Zr or Ta; X = H or CH$_3$

(b) Hydrogenolysis

$$C_3H_8 + H_2 \xrightarrow{[Zr-H]} CH_4 + C_2H_6$$

β-Alkyl transfer

$$C_3H_8 + H_2 \xrightarrow{[Ta-H]} CH_4 + C_2H_6 \xrightarrow[H_2]{[Ta-H]} 2\,CH_4$$

α-Alkyl transfer

(c) Alkane metathesis

$$3\,C_3H_8 \xrightarrow{[Zr-H]} CH_4 + 2\,C_2H_6 + \text{(isobutane)}$$

β-Alkyl transfer/
H-transfer/Insertion

$$2\,C_3H_8 \xrightarrow{[Ta-H]} C_2H_6 + C_4H_{10}$$

π-Bond metathesis/
α-H and β-H transfer

(d) Methane/alkane cross-metathesis

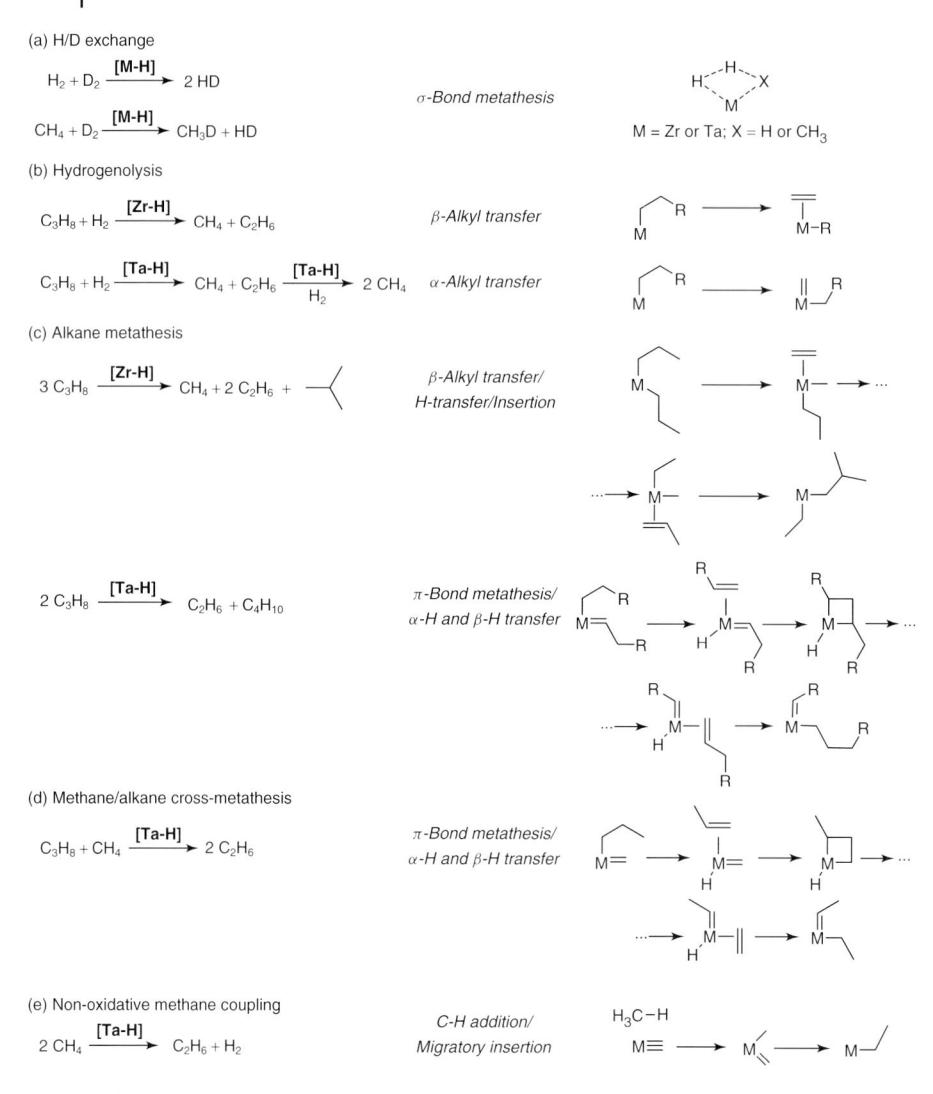

$$C_3H_8 + CH_4 \xrightarrow{[Ta-H]} 2\,C_2H_6$$

π-Bond metathesis/
α-H and β-H transfer

(e) Non-oxidative methane coupling

$$2\,CH_4 \xrightarrow{[Ta-H]} C_2H_6 + H_2$$

C-H addition/
Migratory insertion

Scheme 19.9 Reactivity of early transition-metal hydrides [M−H] (M = Zr, Ta) toward alkanes.

transformation can be seen as the inverse of the so-called alkane metathesis, and it is likely to proceed via carbene intermediates and olefin metathesis.

5) **Non-oxidative coupling of methane**: [Ta−H] was found to convert CH$_4$ into ethane, a reaction that presumably proceeds via C−H activation and leads to tantalum carbenic and carbynic species. The reaction is thus likely to involve a tantalum-methylmethylidene species as key intermediate, where the methyl ligand can migrate to the tantalum-methylidene, affording the tantalum-ethyl intermediate. Interestingly, such elementary steps coincide with

those proposed in reaction involving alkane formation such as Fischer–Tropsch synthesis [62].

Moreover, **[Ta–H]** was found to activate ammonia and dinitrogen [63]. While some of these transformations are still not fully understood, they bring new knowledge to analyze interesting processes on the surfaces. Besides, as activators of small molecules, these catalysts would in turn contribute to the transformation of natural gas and nitrogen sources into easily transportable carbon higher hydrocarbons and ammonia, respectively, both of these being urgent scientific challenges.

19.2.4
Beyond a Molecular Viewpoint: a Closer Look at the Role of the Surfaces

The rich catalytic diversity found in single-site systems such as well-defined alkene metathesis catalysts or the early transition metal hydrides presented above, is an eloquent example of how the support can exert dramatic changes in the stability and reactivity of a catalytic system. Now, we present some examples where the role of the support clearly influences the final grafted species or catalytic behavior.

In particular, silica and alumina offer very different environments, which influence the stability, structure, and thus catalytic performance of the grafted species. An interesting example is the grafting of the perhydrocarbyl complex $[Zr(CH_2{}^tBu)_4]$: on $SiO_{2\text{-}(700)}$, the resulting surface species is $[(\equiv SiO)Zr(CH_2^t Bu)_3]$, a neutral complex (see Scheme 19.8 above), while on $\gamma\text{-}Al_2O_{3\text{-}(500)}$, a cationic surface species is formed (Scheme 19.10) [64].

Here, the subsequent protonolysis of two Zr–C bonds by surface OH groups is observed, followed by a transfer of one of the remaining neopentyl ligands onto an adjacent Al_s site, thus affording a cationic species. Consequently, the two systems display very different reactivity: the alumina-supported system is a highly active polymerization catalyst, whereas the silica-supported system is totally inactive, consistent with the much higher reactivity of cationic species in alkene polymerization [65–68]. Besides the remarkable differences in catalytic performances due to the nature of the surface species, interaction of the substrate with the surface can translate into higher reaction rates because of an increased surface coverage ($v = k\theta;\ \theta = KP/(1 + KP)$). In fact, the sole interaction of an alkene such as ethene is by far more exothermic on alumina (circa 80 kJ mol^{-1}) than on silica (circa 10 kJ mol^{-1}) [69, 70]. This beneficial effect also applies to hydrogenation catalysts based on early transition metals, lanthanides, and actinides complexes supported

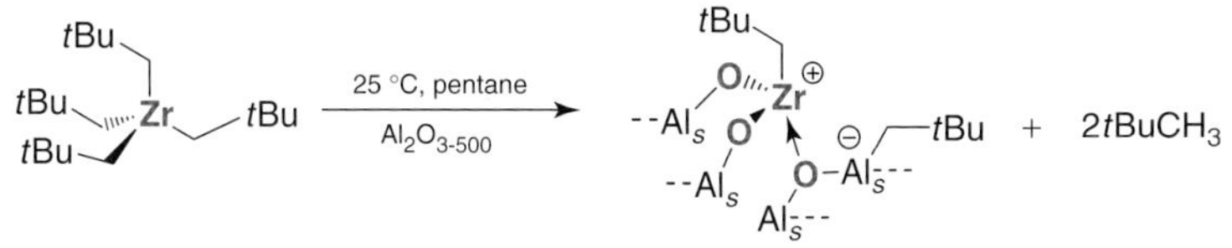

Scheme 19.10 Grafting of $Zr(CH_2tBu)_4$ over $Al_2O_{3\text{-}(500)}$.

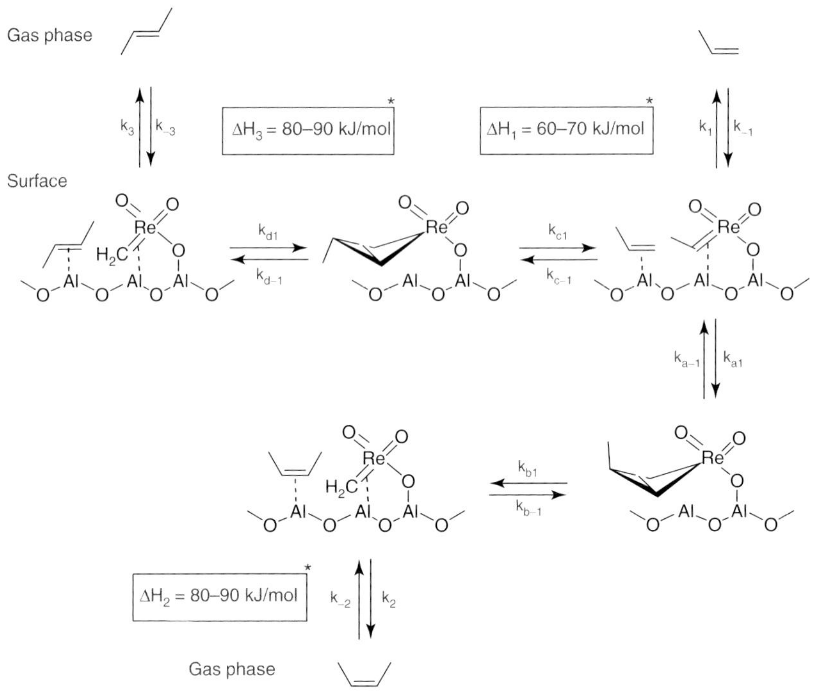

Scheme 19.11 Adsorption of C_3H_8 on CH_3ReO_3/Al_2O_3 and the corresponding equilibria upon the formation of (E) and (Z)-isomers.

on alumina [71]. This brings up a new question: Is it possible to tune the catalytic properties of a given system by modifying the surface properties of the support?

This can be illustrated with the grafting of $(CH_3)ReO_3$ onto an alumina whose surface has been previously modified with trimethylsilyl fragments (Scheme 19.11), which results into an improved activity (380 vs 150 min^{-1} for the parent $(CH_3)ReO_3/Al_2O_3$), overall performances (330 000 vs 170 000 TON), and selectivity (ratio of (E/Z)-2-butenes is 1/3 vs 3/1 for the parent $(CH_3)ReO_3/Al_2O_3$) [72]. In particular, the catalyst displays a very high selectivity in (Z)-2-butene even at high conversions (up to 10%), in contrast to most alumina-supported catalysts, including $(CH_3)ReO_3/Al_2O_3$ (see Section 19.2.2). The origin of these dramatic differences is not due to different active sites (they have the same NMR signatures), but can be explained by the difference in the adsorption properties of the surface of the two catalysts. For the parent system, $(CH_3)ReO_3/Al_2O_3$, the high adsorption properties of alumina (see above) requires the use of very high inverse space velocities (flow rates and low conversion) to reach kinetic selectivity (1/3 E/Z ratio). This shows that the support is more than a mere rigid and inert ligand, and these findings offer a new perspective to the study and development of catalytic systems, connecting the macroscopic phenomena with the molecular chemistry by an understanding of the interactions at molecular scale.

19.3
Conclusion and Perspectives

We have illustrated how the concept of single-site heterogeneous catalysis fits exactly between homogeneous and heterogeneous catalysis, helping to bridge the gap by providing molecular understanding of surface phenomena. Firstly, a 'single-site' approach is a way to immobilize homogeneous catalysts, where the dialog with theoretical studies permits us to establish structure–reactivity relationships and rational development of catalysts. This interplay has in fact fed back homogeneous catalysis on the development of highly active and stereo-selective catalysts. Moreover, this approach has also proven helpful to shed light on classical heterogeneous systems such as Re_2O_7/Al_2O_3 (e.g., structure of active site, activation mechanism, role of the co-catalysts ...), our understanding of which was mostly speculative. We have also shown how the single-site approach is a powerful method to generate highly active metal hydrides, which display unprecedented reactivity such as alkane metathesis, methane conversion processes, and dinitrogen activation.

The concept of single-site catalysts is not restricted to grafting molecular complexes onto oxide surfaces. First, it can be extended to the isolation of ions confined in the well-defined environment of a solid, such as zeolites. For instance, the active sites of titanium-substituted silicate-1 (TS-1) are often considered to be isolated titanium atoms (i.e., single sites, where titanium replaces only 2% of the silicon framework atoms). This leads to excellent catalytic properties in a wide range of selective oxidation reactions, in particular in the epoxidation of propene with aqueous hydrogen peroxide, an environmentally friendly oxidant [73]. This concept also encompasses the stabilization of highly reactive organic intermediates in the pores of inorganic materials like zeolites, such as in the methanol-to-olefin (MTO) process, where five- and six-membered cyclic carbenium ions stabilized in the cavities of the inorganic solid transform methanol into ethylene [74]. Moreover, site isolation at the surface of metallic nanoparticles by selective poisoning can also bring about single-site behavior. For instance, tin-modified nanoparticles of platinum can become totally selective for the dehydrogenation of isobutane into isobutene, while pure platinum surfaces also lead to hydrogenolysis products (cracking) [75].

The concept of single sites has diversified, now pointing to the rationalization of increasingly complex phenomena where a supramolecular point of view is necessary to understand the reactions taking place at the surface.

References

1. Farrauto, R.J. and Bartholomew, C.H. (1997) *Fundamentals of Industrial Catalytic Processes*, Chapman & Hall, London, p. 754.
2. Hutchings, G.J. (2009) Heterogeneous catalysts - discovery and design. *J. Mater. Chem.*, **19** (9), 1222–1235.
3. Wilkinson, G. (1993) in *Nobel Lecture in Nobel Lectures, Chemistry 1971–1980* (eds T. Frängsmyr and S. Forsén), World Scientific Publishing Co., Singapore, pp. 137–145.
4. Schrock, R.R. (2006) Multiple metal-carbon bonds for catalytic

metathesis reactions (Nobel Lecture). *Angew. Chem. Int. Ed.*, **45** (23), 3748–3759.
5. Legrand, A.P. (ed.) (1998) *The Surface Properties of Silica*, John Wiley & Sons, Inc., New York.
6. Copéret, C. *et al.* (2003) Homogeneous and heterogeneous catalysis: bridging the gap through surface organometallic chemistry. *Angew. Chem. Int. Ed.*, **42** (2), 156–181.
7. Hérisson, J.L., and Chauvin, Y. (1971) Catalysis of olefin transformations by tungsten complexes. II. Telomerization of cyclic olefins in the presence of acyclic olefins. *Makromol. Chem.*, **141**, 161–176.
8. Chauvin, Y. (2006) Olefin metathesis: the early days (Nobel Lecture). *Angew. Chem. Int. Ed.*, **45** (23), 3741–3747.
9. Schrock, R.R. (2002) High oxidation state multiple metal-carbon bonds. *Chem. Rev. (Washington, D.C.)*, **102** (1), 145–179.
10. Grubbs, R.H. (2006) Olefin-metathesis catalysts for the preparation of molecules and materials (Nobel Lecture). *Angew. Chem. Int. Ed.*, **45** (23), 3760–3765.
11. Schuster, M. and Blechert, S. (1997) Olefin metathesis in organic chemistry. *Angew. Chem. Int. Ed. Engl.*, **36** (19), 2037–2056.
12. Connon, S.J. and Blechert, S. (2003) Recent developments in olefin cross-metathesis. *Angew. Chem. Int. Ed.*, **42** (17), 1900–1923.
13. Hoveyda, A.H. and Zhugralin, A.R. (2007) The remarkable metal-catalysed olefin metathesis reaction. *Nature*, **450** (7167), 243–251.
14. Rendón, N., Blanc, F., and Copéret, C. (2009) Well-defined silica supported metallocarbenes: formation and reactivity. 38th International Conference on Coordination Chemistry, Jerusalem, ISRAEL, 2009.
15. Rocklage, S.M. *et al.* (1981) Multiple metal-carbon bonds. 19. How niobium and tantalum complexes of the type $M(CHCMe_3)(PR_3)_2Cl_3$ can be modified to give olefin metathesis catalysts. *J. Am. Chem. Soc.*, **103** (6), 1440–1447.
16. Petroff Saint-Arroman, R. *et al.* (2001) Characterization of surface organometallic complexes using high resolution 2D solid-state NMR spectroscopy. Application to the full characterization of a silica supported metal carbyne:. tplbond.SiO-Mo(.tplbond.C-Bu-t)$(CH_2$-Bu-t)$_2$. *J. Am. Chem. Soc.*, **123** (16), 3820–3821.
17. Le Roux, E. *et al.* (2005) Well-defined surface tungstenocarbyne complexes through the reaction of [W(sbCtBu)$(CH_2tBu)_3$] with silica. *Organometallics*, **24** (17), 4274–4279.
18. Blanc, F. *et al.* (2005) Reactivity differences between molecular and surface silanols in the preparation of homogeneous and heterogeneous olefin metathesis catalysts. *J. Organomet. Chem.*, **690** (23), 5014–5026.
19. Chabanas, M. *et al.* (2001) A highly active well-defined rhenium heterogeneous catalyst for olefin metathesis prepared via surface organometallic chemistry. *J. Am. Chem. Soc.*, **123** (9), 2062–2063.
20. Chabanas, M. *et al.* (2003) Perhydrocarbyl Re(VII) complexes: comparison of molecular and surface complexes. *J. Am. Chem. Soc.*, **125** (2), 492–504.
21. Toreki, R. and Schrock, R.R. (1990) A well-defined rhenium(VII) olefin metathesis catalyst. *J. Am. Chem. Soc.*, **112** (6), 2448–2449.
22. Toreki, R., Schrock, R.R., and Davis, W.M. (1992) Synthesis and characterization of rhenium(VII) alkylidene alkylidyne complexes of the type Re(CR')(CHR')(OR)$_2$ and related species. *J. Am. Chem. Soc.*, **114** (9), 3367–3380.
23. Toreki, R. *et al.* (1993) Metathetical reactions of rhenium(VII) alkylidene-alkylidyne complexes of the type Re(CR')(CHR')[OCMe$(CF_3)_2]_2$ (R' = CMe$_3$ or CMe$_2$Ph) with terminal and internal olefins. *J. Am. Chem. Soc.*, **115** (1), 127–137.
24. Solans-Monfort, X. *et al.* (2005) d0 Re-based olefin metathesis catalysts, Re(tpCR)(=CHR)(X)(Y): the key role of X and Y ligands for efficient active sites. *J. Am. Chem. Soc.*, **127** (40), 14015–14025.
25. Poater, A. *et al.* (2007) Understanding d_0-olefin metathesis catalysts: which

26. Kozuch, S. and Shaik, S. (2006) A combined kinetic-quantum mechanical model for assessment of catalytic cycles: application to cross-coupling and Heck reactions. *J. Am. Chem. Soc.*, **128** (10), 3355–3365.
27. Blanc, F. *et al.* (2007) Dramatic improvements of well-defined silica supported Mo-based olefin metathesis catalysts by tuning the N-containing ligands. *J. Am. Chem. Soc.*, **129** (27), 8434–8435.
28. Copéret, C. (2007) Design and understanding of heterogeneous alkene metathesis catalysts. *Dalton Trans.*, 5498–5504.
29. Malcolmson, S.J. *et al.* (2008) Highly efficient molybdenum-based catalysts for enantioselective alkene metathesis. *Nature*, **456** (7224), 933–937.
30. Flook, M.M. *et al.* (2009) Z-selective olefin metathesis processes catalyzed by a molybdenum hexaisopropylterphenoxide monopyrrolide complex. *J. Am. Chem. Soc.*, **131** (23), 7962–7963.
31. Mol, J.C. (2004) Catalytic metathesis of unsaturated fatty acid esters and oils. *Top. Catal.*, **27** (1–4), 97–104.
32. Mol, J.C. (2004) Industrial applications of olefin metathesis. *J. Mol. Catal. A: Chem.*, **213** (1), 39–45.
33. Mol, J.C. (1999) Olefin metathesis over supported rhenium oxide catalysts. *Catal. Today*, **51** (2), 289–299.
34. Digne, M. *et al.* (2002) Hydroxyl groups on γ-alumina surfaces: a DFT study. *J. Catal.*, **211** (1), 1–5.
35. Digne, M. *et al.* (2002) Structure and stability of aluminum hydroxides: a theoretical study. *J. Phys. Chem. B*, **106** (20), 5155–5162.
36. Digne, M. *et al.* (2004) Use of DFT to achieve a rational understanding of acid-basic properties of g-alumina surfaces. *J. Catal.*, **226** (1), 54–68.
37. Salameh, A. *et al.* (2007) Rhenium(VII) oxide/aluminum oxide: more experimental evidence for an oxametallacyclobutane intermediate and a pseudo-Wittig initiation step in olefin metathesis. *Adv. Synth. Catal.*, **349** (1–2), 238–242.
38. Herrmann, W.A. *et al.* (1991) Multiple bonding between main group elements and transition metals. 100. Part 1. Methyltrioxorhenium as catalyst for olefin metathesis. *Angew. Chem. Int. Ed.*, **30** (12), 1636–1638.
39. Salameh, A. *et al.* (2007) CH_3ReO_3 on γ-Al_2O_3: understanding its structure, initiation, and reactivity in olefin metathesis. *Angew. Chem. Int. Ed.*, **46** (21), 3870–3873.
40. Joubert, J. *et al.* (2006) Heterolytic splitting of H_2 and CH_4 on g-alumina as a structural probe for defect sites. *J. Phys. Chem. B*, **110** (47), 23944–23950.
41. Salameh, A. *et al.* (2008) CH_3-ReO_3 on γ-Al_2O_3: activity, selectivity, active site and deactivation in olefin metathesis. *J. Catal.*, **253** (1), 180–190.
42. Tebbe, F.N., Parshall, G.W., and Ovenall, D.W. (1979) Titanium-catalyzed olefin metathesis. *J. Am. Chem. Soc.*, **101** (17), 5074–5075.
43. Tebbe, F.N., Parshall, G.W., and Reddy, G.S. (1978) Olefin homologation with titanium methylene compounds. *J. Am. Chem. Soc.*, **100** (11), 3611–3613.
44. Herrmann, W.A., Fischer, R.W., and Correia, J.D.G. (1994) Multiple bonds between main-group elements and transition metals. Part 133. Methyltrioxorhenium as a catalyst of the Baeyer-Villiger oxidation. *J. Mol. Catal.*, **94** (2), 213–223.
45. Zakharov, V.A. *et al.* (1977) Formation of zirconium hydrides in supported organozirconium catalysts and their role in ethylene polymerization. *J. Mol. Catal.*, **2** (6), 421–435.
46. Yermakov, Y.I. *et al.* (1989) Hydride complexes of titanium and zirconium attached to silicon dioxide as hydrogenation catalysts. *J. Mol. Catal.*, **49** (2), 121–132.
47. Corker, J. *et al.* (1996) Catalytic cleavage of the C-H and C-C bonds of alkanes by surface organometallic chemistry: an EXAFS and IR characterization of a Zr-H catalyst. *Science*, **271** (5251), 966–969.
48. Rataboul, F. *et al.* (2004) Molecular understanding of the formation of surface zirconium hydrides upon thermal treatment under hydrogen of [(.tplbond.SiO)Zr(CH_2tBu)$_3$] by

49. Vidal, V. et al. (1996) Synthesis, characterization, and reactivity, in the C-H bond activation of cycloalkanes, of a silica-supported Tantalum(III) monohydride complex: (.tplbond.SiO)$_2$Ta(III)-H. *J. Am. Chem. Soc.*, **118** (19), 4595–4602.

50. Soignier, S. et al. (2006) Tantalum hydrides supported on MCM-41 mesoporous silica: activation of methane and thermal evolution of the tantalum-methyl species. *Organometallics*, **25** (7), 1569–1577.

51. Le Roux, E. et al. (2005) Development of tungsten-based heterogeneous alkane metathesis catalysts through a structure-activity relationship. *Angew. Chem. Int. Ed.*, **44** (41), 6755–6758.

52. Copéret, C. et al. (2003) Discrimination of σ-bond metathesis pathways in H/D exchange reactions on [(.tplbond.SiO)$_3$Zr-H]: a density functional theory study. *ChemPhysChem*, **4** (6), 608–611.

53. Lécuyer, C. et al. (1991) Organometallic chemistry on oxide surfaces. Selective, catalytic low-temperature hydrogenolysis of alkanes by a highly electrophilic, silica-gel-supported zirconium hydride. *Angew. Chem. Int. Ed. Engl.*, **30** (12), 1660–1661.

54. Chabanas, M. et al. (2000) Low-temperature hydrogenolysis of alkanes catalyzed by a silica-supported tantalum hydride complex, and evidence for a mechanistic switch from Group IV to Group V metal surface hydride complexes. *Angew. Chem. Int. Ed.*, **39** (11), 1962–1965.

55. Vidal, V. et al. (1997) Metathesis of alkanes catalyzed by silica-supported transition metal hydrides. *Science*, **276** (5309), 99–102.

56. Basset, J.-M. et al. (2006) From olefin to alkane metathesis: a historical point of view. *Angew. Chem. Int. Ed.*, **45** (37), 6082–6085.

57. Le Roux, E. et al. (2004) Detailed structural investigation of the grafting of [Ta(=CHtBu)(CH$_2$tBu)$_3$] and [CpTaMe$_4$] on silica partially dehydroxylated at 700 °C and the activity of the grafted complexes toward alkane metathesis. *J. Am. Chem. Soc.*, **126** (41), 13391–13399.

58. Basset, J. et al. (2005) Primary products and mechanistic considerations in alkane metathesis. *J. Am. Chem. Soc.*, **127**, 8604–8605.

59. Blanc, F. et al. (2006) Alkane metathesis catalyzed by a well-defined silica-supported Mo imido alkylidene complex: [(.tplbond.SiO)Mo(=NAr)(=CHtBu)(CH$_2$tBu)]. *Angew. Chem. Int. Ed.*, **45** (37), 6201–6203.

60. Thieuleux, C. et al. (2007) Homologation of propane catalyzed by oxide-supported zirconium dihydride and dialkyl complexes. *Angew. Chem. Int. Ed.*, **46** (13), 2288–2290.

61. Soulivong, D. et al. (2004) Cross-metathesis of propane and methane: a catalytic reaction of C-C bond cleavage of a higher alkane by methane. *Angew. Chem. Int. Ed.*, **43** (40), 5366–5369.

62. Soulivong, D. et al. (2008) Non-oxidative coupling reaction of methane to ethane and hydrogen catalyzed by the silica-supported tantalum hydride: ($\equiv$ SiO$_2$Ta-H. *J. Am. Chem. Soc.*, **130** (15), 5044–5045.

63. Avenier, P. et al. (2007) Dinitrogen dissociation on an isolated surface tantalum atom. *Science*, **317** (5841), 1056–1060.

64. Joubert, J. et al. (2006) Molecular understanding of alumina supported single-site catalysts by a combination of experiment and theory. *J. Am. Chem. Soc.*, **128** (28), 9157–9169.

65. Ballard, D.G.H. (1975) Transition metal alkyl as polymerization. *J. Polym. Sci. A*, **13**, 2191–2212.

66. Burwell, R.L. Jr. (1984) Organometallic complexes on alumina. *J. Catal.*, **86** (2), 301–314.

67. Tullock, C.W. et al. (1989) Polyethylene and elastomeric polypropylene using alumina-supported bis(arene) titanium, zirconium, and hafnium catalysts. *J. Polym. Sci., Part A: Polym. Chem.*, **27** (9), 3063–3081.

68. Tullock, C.W., Mulhaupt, R., and Ittel, S.D. (1989) Elastomeric polypropylene (ELPP) from alumina-supported bis(mesitylene)titanium catalysts. *Makromol. Chem., Rapid Commun.*, **10** (1), 19–23.
69. Grinev, V.E. *et al.* (1983) Heats of adsorption of olefins on supported molybdenum trioxide/aluminum oxide catalysts. *Kinet. Katal.*, **24** (2), 429–435.
70. Grinev, V.E. *et al.* (1981) Heats of adsorption of olefins on aluminum oxide. *Izv. Akad. Nauk SSSR, Ser. Khim.*, **30** (7), 1648–1651.
71. Marks, T.J. (1992) Surface-bound metal hydrocarbyls. Organometallic connections between heterogeneous and homogeneous catalysis. *Acc. Chem. Res.*, **25** (2), 57–65.
72. Salameh, A. *et al.* (2008) Tuning the selectivity of alumina-supported CH_3-ReO_3 by modifying the surface properties of the support. *Angew. Chem. Int. Ed.*, **47** (11), 2117–2120.
73. Bordiga, S. *et al.* (2007) Reactivity of Ti(IV) species hosted in TS-1 towards H_2O_2-H_2O Solutions investigated by ab initio cluster and periodic approaches combined with experimental XANES and EXAFS data: a review and new highlights. *Phys. Chem. Chem. Phys.*, **9** (35), 4854–4878.
74. Haw, J.F. *et al.* (2003) The mechanism of methanol to hydrocarbon catalysis. *Acc. Chem. Res.*, **36** (5), 317–326.
75. Basset, J.M. *et al.* (2004) Design, building, characterization and performance at the nanometric scale of heterogeneous catalysts. *Nanotechnol. Catal.*, **2**, 447–466.

20
Preparation of Supported Catalysts
Krijn P. de Jong

20.1
Introduction

Supported catalysts are used extensively in major industrial areas such as oil refining, gas-to-liquid conversions, chemicals manufacturing, and environmental processes. It is fair to say that supported catalysts are the most important class of catalysts in industry. Examples of these catalysts with their respective applications are summarized in Table 20.1. In most cases the active components comprise a metal (e.g., Ni, Co, Pt), metal oxide (Cr, V), or metal sulfide (Co, Mo) dispersed as nanoparticles on a high-surface-area support consisting of an oxide (e.g., Al_2O_3, SiO_2, or TiO_2) or carbon. Typically, the active component comprises 1–25 wt% of the total catalyst mass, and the nanoparticles have sizes of 1–10 nm.

The structure of a solid supported catalyst is schematically shown in Figure 20.1. Important roles of the support comprise, among others, prevention of sintering of the active phase, mechanical stability, and reduction of pressure drop in the reactor. Due to their high melting points and related thermal stability, metal oxides and carbon are almost exclusively the support materials of choice. The macroscopic support bodies are either millimeter-sized for fixed-bed or micrometer-sized for fluid-bed or slurry-bed reactor applications. At the mesoscopic length scale the support particles are important, on the one hand, to carry the nanoparticles and, on the other hand, to create empty spaces (pores) which are essential for mass transfer in catalysis. At the microscopic scale the active sites at the surface of the nanoparticles are most relevant for catalysis. The size and shape of the nanoparticles more often than not affect the density and nature of the active sites. The preparation of supported catalysts should take into account all these aspects at different length scales. For more than a century, many methods for the preparation of supported catalysts have been developed and used in laboratories and industry. Broadly speaking, we can discern three fundamentally different routes for the preparation, which are (i) selective removal, (ii) co-precipitation, and (iii) application of the active component on a separately produced support.

Selective removal is used to manufacture the ammonia synthesis catalysts (removal of oxygen from iron oxide to obtain porous metallic iron) and so-called Raney

Catalysis: From Principles to Applications, First Edition. Edited by Matthias Beller, Albert Renken, and Rutger van Santen.
© 2012 Wiley-VCH Verlag GmbH & Co. KGaA. Published 2012 by Wiley-VCH Verlag GmbH & Co. KGaA.

20.1 Introduction

Table 20.1 Survey of selected catalysts with their main applications.

Catalyst	Applications
Ni/SiO_2	Hydrogenation
$K_2O/Al_2O_3/Fe$	Ammonia synthesis
$Ag/\alpha\text{-}Al_2O_3$	Epoxidation
CrO_x/SiO_2	Polymerization
$CoMoS_2/\gamma\text{-}Al_2O_3$	Hydrotreating
Co/SiO_2	Fischer–Tropsch synthesis
$Cu/ZnO/Al_2O_3$	Methanol synthesis
Zeolite Y composite	Catalytic cracking
Pt/mordenite	Hydro-isomerization of light alkanes
V_2O_5/TiO_2	NO_x abatement
Pt/C	Hydrogenation; fuel cell

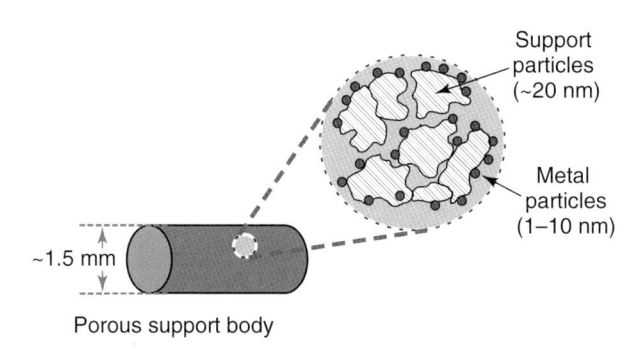

Figure 20.1 Structure at different length scales of cylindrical catalyst body suitable for fixed-bed reactor application [1].

nickel (removal of Al from NiAl alloy to obtain porous Ni). Co-precipitation is used to obtain, among others, Ni hydrogenation catalysts and Cu methanol synthesis catalysts via hydrotalcite-like precursors. These precursors contain elements of both the support and the active component and are together converted into supported catalysts, often with high loadings (>50 wt%) of the latter. In this chapter we will focus on the third and most widely used method, which is deposition of the active component on an existing support. This may be performed in many ways, including grafting of organometallic precursors, deposition of colloidal particles, sol–gel techniques, and chemical vapor deposition. For a full account of these, we refer the reader to the literature, in particular reviews and books [1–7].

In the present chapter we discuss deposition of precursors of the active components from aqueous solutions, more specifically, by ion adsorption, impregnation and drying, and deposition precipitation (DP). These methods all rely on the interaction of the precursor with the support surface in aqueous systems. We first discuss the nature of the support surface in water, and this is followed by an introduction to the three deposition methods mentioned. Finally, we discuss thermal treatment (calcination and reduction) of the supported precursors to give the active catalysts.

20.2
Support Surface Chemistry

Small metal oxide particles are the most used support materials, and these are considered in this section. Related to their synthesis as well as to surface stabilization, metal oxide surfaces are commonly terminated by hydroxyl groups. Depending on the metal cations, lattice structure, and detailed bonding of the OH groups, their nature may be considered as acidic, neutral, or basic in the context of aqueous acid–base chemistry. For example, at the surface of γ-Al_2O_3, infrared spectroscopy has shown that a variety of OH groups are present that can be considered as basic (terminal OH, bonded to one Al atoms), neutral (bridged OH, bonded to two Al atoms), and acidic (bridged OH, bonded to three Al atoms). Similar statements can be made concerning other oxide surfaces. Quoted densities of OH groups on oxide surfaces are typically 8 OH nm^2 for high-surface-area alumina and 5 OH nm^2 for silica. Although it is a simplification, it often suffices to consider the OH groups as a homogeneous population for which charge development in water is the most important phenomenon.

At low pH the hydroxyl groups will pick up a proton from solution and become positively charged, whereas at elevated pH deprotonation of the OH group will give rise to a negative charge (Figure 20.2). The pH at which the net charge of the surface is zero is referred to as the Point of Zero Charge (PZC). PZC values for common support materials are listed in Table 20.2. As indicated in Figure 20.2, at

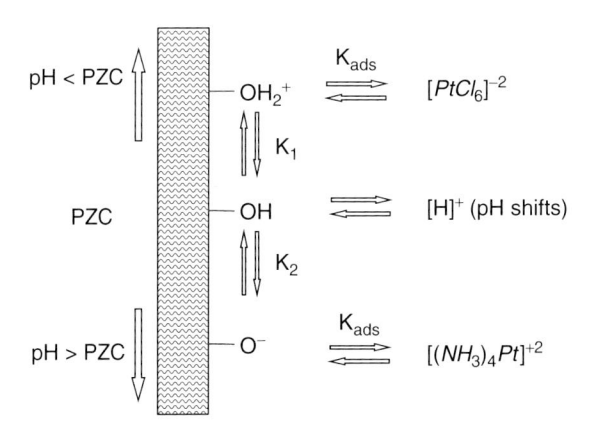

Figure 20.2 Surface charging of oxides in aqueous systems depending on pH and PZC (see text for explanation) and concomitant adsorption of Pt cations or anions. (Reproduced from Regalbutto [8].)

Table 20.2 PZC of common oxide and carbon supports.

Support	PZC
MoO_3	<1
Nb_2O_5	2–2.5
SiO_2	4
Oxidized activated carbon	2–4
Graphitic carbon	4–5
TiO_2	4–6
CeO_2	7
ZrO_2	8
Co_3O_4	7–9
Al_2O_3	8.5
Activated carbon	8–10

Reproduced from Ref. [8].

pH < PZC anion adsorption is dominant, whereas at pH > PZC cation adsorption is dominant.

The description of metal-ion adsorption as depicted in Figure 20.2 originates from colloidal chemistry (charged double layer of colloidal particles in water) and involves non-specific interactions of an electrostatic nature. Sometimes this non-specific interaction is referred to as 'outer-sphere complex formation.' Specific interactions, or 'inner-sphere complex formation,' are important in many cases, however [4]. For the latter type of interactions the oxygen of the oxide support may enter into the first coordination sphere of the metal ion. Several examples of this will be presented in the next sections of this chapter. Here it suffices to say that often discrimination between these adsorption models is very difficult because, among other reasons, local pH changes in the support pores may also bring about ligand exchange without direct involvement of support oxygen.

20.3
Ion Adsorption

This method for the preparation of supported metal catalysts is sometimes also referred to as 'ion exchange' or '(strong) electrostatic adsorption' [8–10]. The method is particularly attractive for deposition of noble metals, such as Ru, Rh, Pd, Ir, and Pt, at low loadings, since molecular dispersion of the precursor, and hence, in principle, high metal dispersions can be obtained.

In Section 20.6 we discuss the proper activation of these catalyst precursors, which is very important in order to realize their potential advantages of high dispersion.

Ion adsorption comprises contacting the support particles (powder or bodies) with a large excess of a dilute solution of the metal precursor. After proper

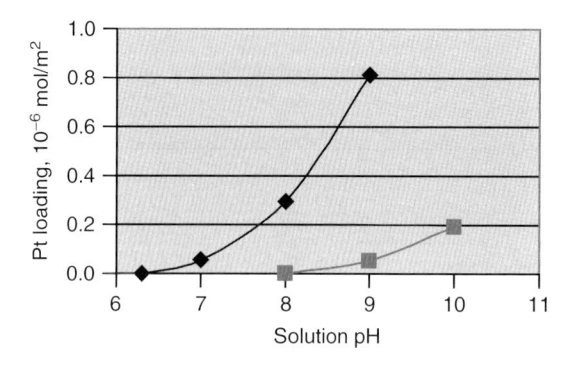

Figure 20.3 Extent of ion adsorption of $[Pt(NH_3)_4]^{2+}$ (aq) onto silica (diamonds) or alumina (squares) as a function of solution pH. (Data calculated from Ref. [11].)

equilibration, particularly important when using millimeter-sized support bodies, the loaded support is filtered off, washed to remove non-adsorbed species from the solution, and dried. Careful selection of the pH of the solution with respect to the PZC of the support in combination with the charge of the metal ions in solution is mandatory. The amount of Pt adsorbed onto silica or alumina as a function of pH is shown in Figure 20.3 [11]. On silica, the $Pt[NH_3]_{2+}^{4}$ ions adsorb well above the PZC of about four. The maximum uptake of the silica ($\sim$370 m^2 g^{-1}) amounts to $\sim$6 wt% Pt. Higher pH values will not bring about higher Pt uptake because of the effects of the ionic strength of the solution and, moreover, would bring about dissolution of the silica. The maximum uptake of platinum on silica, 0.8 μmol nm^{-2}, can be interpreted as an effective surface area of the Pt cation of 2 nm^2 complex^{-1}. From Figure 20.2 it follows that cation adsorption onto alumina is not very extensive, and anion adsorption is therefore used more often. Typically, $PtCl_6^{2-}$ ions are used, which display a smaller effective surface area, that is, 1 nm^2 complex^{-1}, which allows higher Pt loadings to be attained [8].

Proper attention should be paid to whether or not uniform metal distributions are obtained over the support particles, whether they are a powder or larger particles. This warning is particularly relevant when strong adsorption, low metal loadings, short equilibration times, and low solution volume during ion adsorption have been selected. If the adsorption can be considered as irreversible at the prevailing conditions, a so-called egg-shell catalyst often results, in which the metal is concentrated in an outer shell of the support particles or bodies. Additional ions may be added to the solution, for example, Cl$^-$ ions in the case of $PtCl_6^{2-}$ ions, so that they will compete with each other for the same adsorption sites, whereby the metal ions could be adsorbed more uniformly. Nevertheless, even with a competitive ion present, non-uniform metal distributions, for example, egg-white or egg-yolk, may still arise [12].

20.4
Impregnation and Drying

As a first step using this most important catalyst preparation technique, the porous support is contacted with a solution of a precursor of the active component. If the volume of the solution exceeds the pore volume of the support we refer to this as *'wet impregnation.'* If the volumes match with each other we call this *'dry impregnation'* or more often *'incipient wetness impregnation'* or *'pore volume impregnation (PVI).'* Since the latter technique is by far the most important we will restrict our discussion to PVI [12]. Usually, capillary forces suffice to draw the liquid into the pores of the support. In the case of an aqueous solution and a hydrophobic support, it is advised to add ethanol to the solution or to change to an organic solvent altogether. In the case of large pores, the capillary pressure may be too low to fill the pores, and evacuation prior to impregnation could be useful. The details of contacting the support and the solution become more important when working with large support bodies. It is preferred to add the solution in droplets whose volume is much smaller than that of the bodies, in order to prevent occlusion of air and breakage of the bodies.

After PVI, extensive interactions between precursor and support may take place. When an aqueous ammonium heptamolybdate (AHM) solution interacts with γ-alumina, support dissolution is promoted and Mo-Al hetero-polyanions are formed (Figure 20.4) [13]. This interaction may play a large role in the preparation of hydrotreating catalysts as the Mo-Al species may give rise to low dispersion of the resulting MoO_3, thereby lowering catalyst activity [14].

Strong interactions between precursor and support after PVI may give rise to gradients in large support bodies, as has been mentioned for ion adsorption in Section 20.3. In Figure 20.5 the time-resolved and space-resolved Mo concentrations in alumina pellets are shown during equilibration of the pellets after PVI [15]. In the case of the Mo–citrate complex, it took 3 h before equilibration had taken place

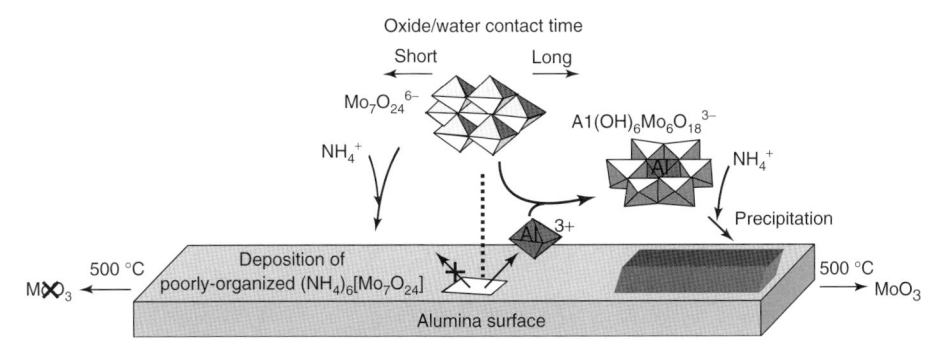

Figure 20.4 Schematic overview of interactions of ammonium heptamolydate dissolved in water and an alumina surface upon short and long contact time as well as after surface precipitation and calcination at 500 °C [12, 13].

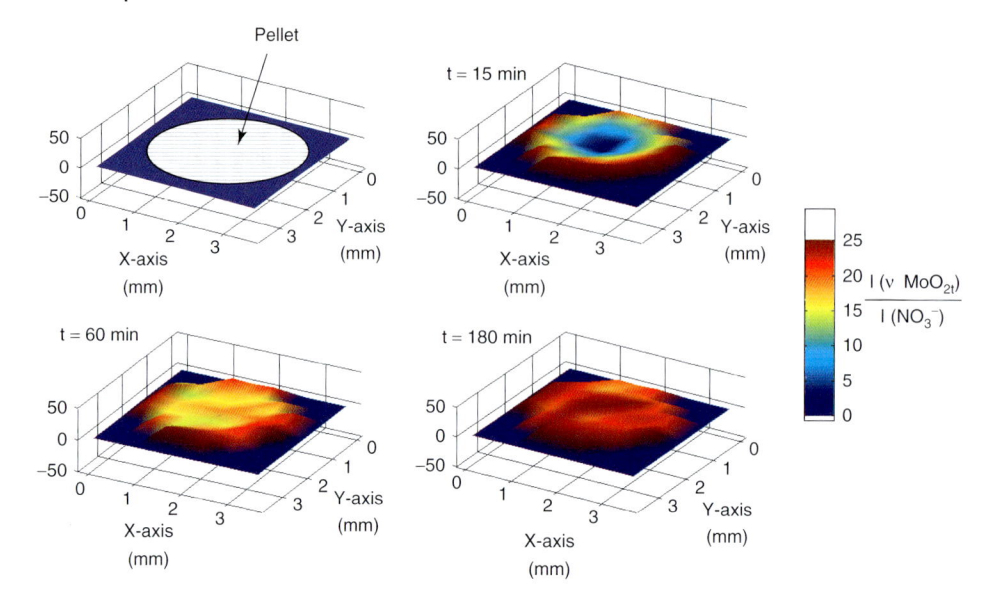

Figure 20.5 Raman microscopy data of Mo species after pore volume impregnation of 3 mm alumina pellets with an aqueous solution of an Mo(VI)-citrate complex with nitrate ions as internal standard. After 15 min an egg-shell distribution was present; 3 h were needed to obtain a homogeneous Mo distribution [15].

over the pellet. Aside from the development of concentration gradients in time, the detailed spectroscopic information also reveals differences between the species as a function of time and place.

For example, when alumina pellets were exposed to an aqueous solution of AHM, at the edge of the pellet AHM was detected, whereas in the center ammonium monomolybdate prevailed, pointing to the reaction

$$Mo_7O_{24}^{6-} + 8OH^- \rightarrow 7\,MoO_4^{2-} + 4H_2O \tag{20.1}$$

in the center of the pellet but not at the edge, probably due to the higher pH at the pellet center [15].

Elimination of the solvent is brought about in the drying step. It should be noted that, because of this approach, no waste water is produced, and this fact is a most important reason why impregnation and drying is the preferred industrial route. If weak interactions prevail, drying will cause deposition of an amorphous phase and/or crystallization of the precursors. A case in point involves impregnation and drying using a concentrated nickel nitrate solution in water (pH = 2) and silica supports (PZC = 4). In view of pH and PZC, little or no interaction between Ni species and the support is to be expected. After drying at 120 °C a basic nickel nitrate is formed whose crystal size nicely coincides with the silica support pore diameter (Figure 20.6). Clearly, crystallization in confinement has taken place [16].

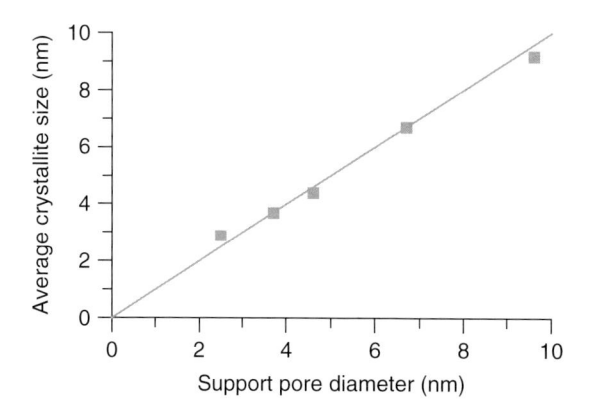

Figure 20.6 Average crystallite size from XRD line broadening for nickel hydroxy nitrate on silica supports with different pore diameters. Pore volume impregnation of supports with aqueous nickel nitrate solution produced a loading of about 24 wt% NiO. This was followed by drying at 120 °C [16].

Next to crystallization, drying may bring about redistribution of the precursor. At a macroscopic scale this phenomenon has been found and described extensively [6, 17]. More recently, also at the mesoscopic length scale, redistribution phenomena have been reported on the basis of TEM results [18]. If solute–support interactions are weak, redistribution could be extensive, and details of drying (gas flow and temperature ramp) are important parameters to control and to report. In order to limit redistribution during drying, several measures can be taken, such as solution viscosity enhancement and the use of chelated metal complexes [5].

20.5
Deposition Precipitation

DP enables one to deposit metals, metal hydroxides, or metal sulfides onto a support that is suspended in the solution of the metal precursor. Usually, the solution is present in great excess over the support pore volume and the support is suspended as a powder. In the case of DP, deposition is not brought about by evaporation of the solvent but rather by a chemical reaction. This chemical reaction may involve (i) increase of pH, (ii) reduction of metal ions, and (iii) ligand removal. For an extensive overview of the modes of DP we refer to the literature [1, 19]. Here we focus on the most widely studied method of pH increase by urea decomposition. Typically, urea is added to a diluted suspension of silica powder in an acidified (pH = 2) solution of nickel nitrate in water. The suspension is heated to 90 °C, whereupon slow hydrolysis of urea takes place according to the equation

$$CO(NH_2)_2 + 3H_2O \longrightarrow CO_2 + 2NH_4^+ + 2OH^- \tag{20.2}$$

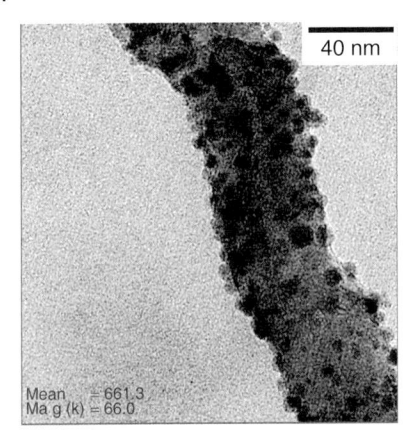

Figure 20.7 Carbon nanofiber-supported nickel catalysts obtained by deposition precipitation of Ni hydroxide [20]. After deposition, the catalyst was reduced in hydrogen at 450 °C. The Ni loading is 20 wt%.

Which, at pH ≈ 6, brings about the reaction

$$Ni^{2+} + 2OH^- \longrightarrow Ni(OH)_2 \qquad (20.3)$$

If there is sufficient interaction between support and precipitate, nucleation will take place exclusively at the support, and a highly dispersed nickel hydroxide phase will result. In the case of nickel hydroxide and silica, a new compound, nickel phyllosilicate, is in fact formed, and this provides the driving force for precipitation at the support rather than in the bulk of the liquid phase. Recent research has also enabled other support materials to be used, such as carbon nanofibers [20]. In Figure 20.7 a TEM image of Ni dispersed on a carbon nanofiber is shown.

DP is useful to obtain supported catalysts at high metal loadings and high dispersion. At the laboratory scale it is characterized by excellent reproducibility, since it is less sensitive to experimental conditions than are impregnation and drying.

20.6
Thermal Treatment

Following deposition of the precursor by one of the techniques described in Sections 20.3–20.5, one needs to convert this precursor, often in several steps, to the active phase. Thermal treatments may comprise calcination (heating in air unless specified otherwise), reduction, or sulfidation. For the sake of brevity we restrict ourselves to two examples of calcination to show the dramatic importance of the subsequent steps of catalyst preparation.

Highly dispersed Pt-on-zeolite catalyst can be obtained via ion-exchange using $Pt(NH_3)_4^{2+}$ ions in aqueous solution. A case in point is the preparation of Pt on ultra stable zeolite Y, 1 wt% Pt/USY. Following ion-exchange, washing, and drying at

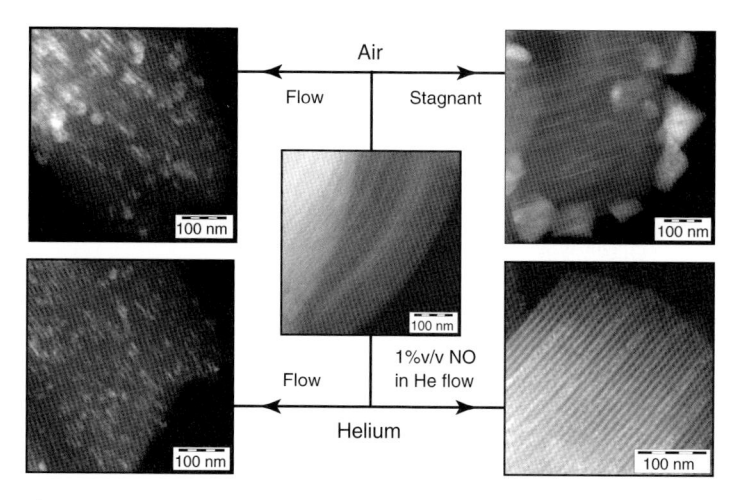

Figure 20.8 HAADF-STEM images of NiO supported on ordered mesoporous silica SBA-15 obtained by impregnation of nickel nitrate followed by drying (120 °C) and calcination (450 °C) in different gas atmospheres (air, He, or NO/He). The central image is of the dried catalyst displaying nickel nitrate nanowires of ∼9 nm diameter [22, 23].

120 °C, calcination in a flow of 20%O_2/Ar was carried out using different ramps to arrive at a final temperature of 350 °C. Using a ramp of 0.2 °C min^{-1} a mono-modal particle-size distribution was obtained of ∼1 nm Pt particles, whereas a ramp of 1 °C min^{-1} led to a bimodal distribution of 4–9 nm and ∼1 nm Pt particles [21].

Following nickel nitrate deposition via impregnation and drying onto ordered mesoporous silica (SBA-15), calcination in different gas atmospheres to obtain 24 wt% NiO/SiO_2 catalysts was carried out. After drying, the nickel nitrate was deposited as ∼9 nm nanowires in the mesopores of similar diameter (Figure 20.8, central image). Clearly, as also inferred from Figure 20.6, nickel nitrate had crystallized inside the mesopores in confinement. Calcination in flowing air and even more so in stagnant air gave rise to extensive redistribution and sintering of the nickel phase, leading to 10–100 nm NiO particles inside and outside of the mesopores (Figure 20.8, top images). Changing from air to an inert gas atmosphere (He) led to less extensive redistribution of the nickel phase, while calcination in NO/He atmosphere provided ∼4 nm NiO nanoparticles exclusively inside the mesopores [22, 23].

Aside from the examples presented here, other studies [24] have shown the prime importance of thermal treatments in the preparation of supported catalysts.

References

1. de Jong, K.P. (2009) *Synthesis of Solid Catalysts*, Wiley-VCH Verlag GmbH & Co. KGaA.

2. Regalbuto, J.R. (2007) *Catalyst Preparation, Science and Engineering*, CRC Press, Boca Raton, FL.

3. de Jong, K.P. (1999) Synthesis of supported catalysts. *Curr. Opin. Solid State Mater. Sci.*, **4** (1), 55–62.
4. Bourikas, K., Kordulis, C., and Lycourghiotis, A. (2006) The role of the liquid-solid interface in the preparation of supported catalysts. *Catal. Rev.*, **48** (4), 363–444.
5. van Dillen, A.J. *et al.* (2003) Synthesis of supported catalysts by impregnation and drying using aqueous chelated metal complexes. *J. Catal.*, **216** (1–2), 257–264.
6. Lekhal, A., Glasser, B.J., and Khinast, J.G. (2001) Impact of drying on the catalyst profile in supported impregnation catalysts. *Chem. Eng. Sci.*, **56** (15), 4473–4487.
7. Che, M. (1993) in *New Frontiers in Catalysis* (ed. L. Guczi), Akademiai Kiado, Budapest, p. 31–68.
8. Regalbuto, J.R. (2009) in *Synthesis of Solid Catalysts*, (ed. K. P. de Jong), Wiley-VCH Verlag GmbH & Co. KGaA, pp. 33–58.
9. Brunelle, J.P. (1978) Preparation of catalysts by metallic complex adsorption on mineral oxides. *Pure Appl. Chem.*, **50** (9–10), 1211–1229.
10. Hao, X., Spieker, W.A., and Regalbuto, J.R. (2003) A further simplification of the revised physical adsorption (RPA) model. *J. Colloid Interface Sci.*, **267** (2), 259–264.
11. Benesi, H.A., Curtis, R.M., and Studer, H.P. (1968) Preparation of highly dispersed catalytic metals: platinum supported on silica gel. *J. Catal.*, **10** (4), 328–335.
12. Marceau, E., Carrier, X., and Che, M. (2009) *Synthesis of Solid Catalysts*, Wiley-VCH Verlag GmbH & Co. KGaA, pp. 59–82.
13. Carrier, X. *et al.* (2003) Influence of ageing on MoO3 formation in the preparation of alumina-supported Mo catalysts. *J. Mol. Struct.*, **656** (1–3), 231–238.
14. Bergwerff, J.A. *et al.* (2006) Influence of the preparation method on the hydrotreating activity of MoS2/Al2O3 extrudates: a Raman microspectroscopy study on the genesis of the active phase. *J. Catal.*, **243** (2), 292–302.
15. Bergwerff, J.A. *et al.* (2004) Envisaging the physicochemical processes during the preparation of supported catalysts: Raman microscopy on the impregnation of Mo onto Al_2O_3 extrudates. *J. Am. Chem. Soc.*, **126** (44), 14548–14556.
16. Wolters, M., de Jongh, P.E., and de Jong, K.P. (2011) *Catal. Today*, **163**, 27–32.
17. Neimark, A.V., Kheifets, L.I., and Fenelonov, V.B. (1981) Theory of preparation of supported catalysts. *Ind. Eng. Chem. Prod. Res. Dev.*, **20** (3), 439–450.
18. Sietsma, J.R.A. *et al.* (2008) Ordered mesoporous silica to study the preparation of Ni/SiO_2 ex nitrate catalysts: impregnation, drying, and thermal treatments. *Chem. Mater.*, **20** (9), 2921–2931.
19. Louis, C. (2007) Deposition-precipitation syntheses of supported metal catalysts, in *Catalyst Preparation: Science and Engineering* (ed. J. Regalbuto), CRC Press, New York.
20. van der Lee, M.K. *et al.* (2005) Deposition precipitation for the preparation of carbon nanofiber supported nickel catalysts. *J. Am. Chem. Soc.*, **127** (39), 13573–13582.
21. de Graaf, J. *et al.* (2001) Preparation of highly dispersed Pt particles in zeolite Y with a narrow particle size distribution: characterization by hydrogen chemisorption, TEM, EXAFS spectroscopy, and particle modeling. *J. Catal.*, **203** (2), 307–321.
22. Sietsma, J.R.A. *et al.* (2007) The preparation of supported NiO and Co3O4 nanoparticles by the nitric oxide controlled thermal decomposition of nitrates. *Angew. Chem., Int. Ed.*, **46** (24), 4547–4549.
23. Sietsma, J.R.A. *et al.* (2008) How nitric oxide affects the decomposition of supported nickel nitrate to arrive at highly dispersed catalysts. *J. Catal.*, **260** (2), 227–235.
24. Toupance, T., Kermarec, M., and Louis, C. (2000) Metal particle size in silica-supported copper catalysts. Influence of the conditions of preparation and of thermal pretreatments. *J. Phys. Chem. B*, **104** (5), 965–972.

21
Porous Materials as Catalysts and Catalyst Supports
Petra de Jongh

Nanoporous materials play an important role in heterogeneous catalysis, either providing catalytic functionality themselves, or by acting as a support for the active phase. Most prominent examples of the first class of materials are zeolites – crystalline and microporous alumina-silicate materials that are used in fluid catalytic cracking and hydro-isomerization. The second class comprises mesoporous oxidic or carbon materials acting as supports for nanoparticulate metal or metal oxide catalysts. In a few more complex systems the nanoporous material is both a support and an active phase, or multi-phase systems are formed, for example, by co-precipitation or selective removal. The previous chapter gives an overview of how the active phase can be deposited onto a support. In this chapter we discuss the preparation and characteristics of the nanoporous materials that are most relevant for catalysis: oxidic materials with disordered porosity, carbon support materials, ordered mesoporous materials, zeolites, and molecular materials.

21.1
General Characteristics

In heterogeneous catalysis, the catalytically active sites are located at the surface of a solid phase and in contact with the vapor or liquid phase reactants. Hence porosity of the solid phase is essential, as it provides a high surface area for catalysis and good accessibility to the active sites. According to the IUPAC definition, pores can be classified as either micropores (<2.0 nm diameter), mesopores (2–50 nm diameter) or macropores (>50 nm diameter). Catalyst supports are typically mesoporous, as this ensures a high specific surface area for anchoring the active material while at the same time providing enough space for the active phase nanoparticles (typically 1–50 nm in diameter) and diffusion of the reactants and products. Zeolites are microporous, and it is not possible to prepare them with primary pores larger than roughly 1.5 nm diameter; however, this small pore size is also essential for size- and shape-selective catalysis. The microstructures of a few of the most relevant nanoporous catalyst materials are shown in Figure 21.1.

Catalysis: From Principles to Applications, First Edition. Edited by Matthias Beller, Albert Renken, and Rutger van Santen.
© 2012 Wiley-VCH Verlag GmbH & Co. KGaA. Published 2012 by Wiley-VCH Verlag GmbH & Co. KGaA.

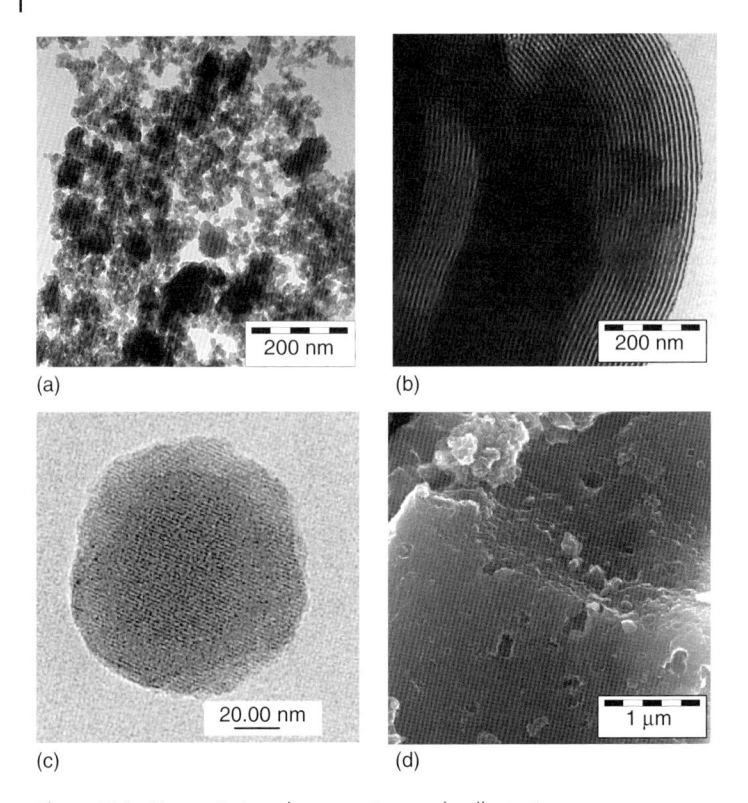

Figure 21.1 Transmission electron micrographs illustrating the microstructure of (a) disordered silica gel, (b) ordered mesoporous silica (SBA-15), (c) a nanocrystal of zeolite A, and (d) a scanning electron micrograph of activated carbon (Norit, ROZ3 extrudate). Note the differences in scale.

In addition to its structural characteristics, the chemical nature of the support also plays an important role. The active phase typically consists of metal nanoparticles. Due to their high surface-to-volume ratio, these particles are intrinsically less stable and more mobile than the corresponding bulk metal phase. A support surface should hence offer enough interaction to allow the preparation or deposition of these nanoparticles, and guard them against mobility and sintering under reaction conditions. On the other hand, the support material itself should be chemically stable under typical reaction conditions.

Although we refer to the support materials as being 'inert,' nevertheless the choice of support material does have a large impact on the catalytic functionality. The size and shape resulting from the active phase deposition depend on the morphology and surface properties of the support. Also after deposition the support can play a role, for instance, by electronic interaction with the metal nanoparticles or by providing adsorption or catalytic sites at or near the support–metal interface.

The combined requirements have led to the widespread use of mesoporous oxides (such as SiO_2, TiO_2, and Al_2O_3) as catalyst supports. In general they

Table 21.1 Overview of the properties of the porous materials most commonly used in catalysis.

Support	Prize	Surface area (m²g⁻¹)	Main porosity (other porosity)	Preparation
Silica gel based on alkali silicates	Low	100–800	Meso (micro, macro)	Condensation in aqueous solution
Fumed silica	Medium	50–400	Meso (micro)	Flame hydrolysis
γ-Al$_2$O$_3$	Low	200–300	Meso	Precipitation in aqueous solution followed by heating
η-Al$_2$O$_3$		100–200		
α-Al$_2$O$_3$	Low	<50	Macro	Treating (hydr)oxides >1100 °C
Activated carbon	Low	500–1500	Micro (meso, macro)	Pyrolysis of natural materials
Carbon black	Medium	10–400	–	Incomplete combustion of hydrocarbons
Zeolites	Medium	100–300	Micro	Hydrothermal growth

combine high stability (chemical inertness, high melting point, good mechanical strength, and attrition resistance) with low cost, wide availability, and flexibility in terms of porosity and shaping. For some specific applications carbon materials are used. An overview of the commonest catalytic materials and some of their properties is given in Table 21.1.

21.2
Sol-gel and Fumed Silica

Silica (SiO$_2$) is the most abundant compound in the earth's crust and is used in a wide range of applications such as glass, filler material, optical fiber, desiccant, chromatography stationary phase, and catalyst support. α-Quartz is the stable crystalline phase at ambient conditions. Silica is chemically rather inert, but can be dissolved in strongly alkaline and in HF-containing solutions. Relatively pure particulate silica is produced on a large scale. A first production route is via gelation in aqueous solution (Figure 21.2).

In this case the starting materials are typically natural sodium silicates, which are cheap and abundantly available. Sodium silicates are dissolved in alkaline solution. The chemistry of silicates in solution is very rich, and a wealth of oligomeric species such as $Si_2O_7^{6-}$, $Si_3O_{10}^{8-}$, and cyclic structures such as $Si_4O_8^{12-}$ can be formed, which are soluble at high pH. Upon lowering the pH of the solution (typically by adding

21 Porous Materials as Catalysts and Catalyst Supports

Table 21.2 Overview of the most common shaping techniques, and the associated feature size and reactor for which these shaped catalytic materials are most relevant.

Shaping technique	Typical size	Reactor
Extrusion	1–30 mm	Fixed-bed reactor
Pelletization	3–10 mm	Fixed-bed reactor
Granulation	1–20 mm	Fixed-bed reactor
Oil drop method	1–5 mm	Fixed-bed reactor
		Riser reactor
Spray drying	20–100 µm	Fluid-bed reactor
		Slurry reactor

Adapted from Ref. [5].

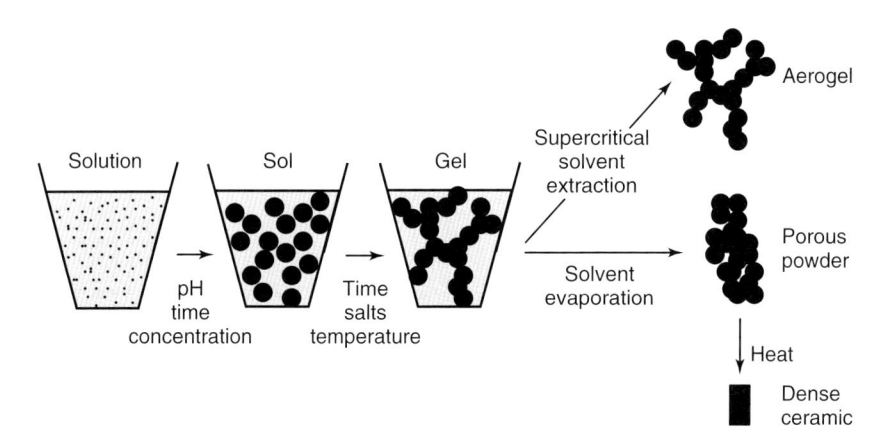

Figure 21.2 Steps in the preparation of a porous powder via the sol-gel method.

H_2SO_4 or alternatively by evaporating off the solvent), condensation of hydroxylated species occurs, leading to the formation of Si–O–Si bonds:

$$\equiv Si-OH + HO-Si \equiv \rightarrow \equiv Si-O-Si \equiv + H_2O \tag{21.1}$$

The rate and degree of condensation depend on the pH of the solution, temperature, and concentration. In neutral and alkaline solutions the silicate species are negatively charged and hence repel each other; this repulsion is effectively shielded by the presence of ions (from the starting material as well as from the added acid), making approach and condensation possible. With increasing degree of condensation, larger silica particles gradually appear in the suspension (a sol). At some point, an interconnected network of particles (a gel) is formed, which is evident from a large increase in the viscosity of the solution and in increase in light scattering [1, 2].

Alternatively, very pure (and more expensive) silica gels can be obtained starting from metal-organic precursors Si(OR)$_4$ such as tetra-ethoxy-silane, Si(OC$_2$H$_5$)$_4$. In this case the addition of water first leads to a hydrolysis step:

$$Si(OR)_4 + H_2O \rightarrow Si(OH)_4 + 4ROH \tag{21.2}$$

which is then followed by condensation reactions. The organic group is not incorporated into the final SiO$_2$ material, but is removed as an alcohol after hydrolysis. However, it is possible to include organic groups in the final material, for instance, by using a partly non-hydrolyzable precursor such as methyltrimethoxysilane (MTMS, CH$_3$Si(OCH$_3$)$_3$). This leads to the formation of hybrid (i.e. inorganic/organic) materials, which are more flexible and hydrophobic but less thermally stable than pure SiO$_2$. After condensation, the gel is washed to remove the salts. Drying is typically done at 150 °C, and leads to significant shrinkage of the material and a concomitant decrease in porosity. Typically, after drying, the gel has a specific surface of 100–800 m^2 g^{-1} with a primary particle size of 2–30 nm and a porosity of ∼65%. Shrinkage can be prevented by drying under supercritical conditions, in which case aerogels with porosities up to 98% are obtained.

In an alternative process, relatively high purity pyrogenic (fumed) silica is produced via a vapor phase process. The so-called Aerosil® process was invented in 1942 in Germany. The starting material is a macroscopic silica source such as sand, which is first converted into a chloride by carbochlorination:

$$SiO_2 + C + Cl_2 \rightarrow SiCl_4 + CO_2 \tag{21.3}$$

In the next step the tetrachloride is converted into an oxide at high temperature (∼1000 °C) by reaction with either a mixture of methane and oxygen or with water ('flame hydrolysis'):

$$SiCl_4 + H_2O \rightarrow SiO_2 + 4HCl \tag{21.4}$$

The water is formed *in situ* by co-feeding H$_2$ and O$_2$. The result is very fine primary clusters (∼1 nm), which are densely packed to form 10–30 nm particles. Specific surface areas can be tuned from 50 to 400 m^2 g^{-1} by varying the flame temperature, H$_2$/O$_2$ ratio, and the residence time/zone length in the reactor. Fumed silica is sold under brand names such as Aerosil®, Cabosil®, HDK®, and REOLOSIL®. Other oxides, such as TiO$_2$, Al$_2$O$_3$, ZrO$_2$, and even mixed oxides, can also be prepared in a similar manner by the flame hydrolysis process.

The surface chemistry of silica materials is determined by the presence of hydroxyl (silanol) groups. The silanol groups are either vicinal, geminal, or isolated (see Figure 21.3a). The density of silanol groups at room temperature is typically 4–6 OH nm^{-2} [2]. As a result, silica at room temperature is hydrophilic and physisorbs water that can be removed by heating to 100–200°C. Above 150–200 °C, especially when heating in vacuum, gradual condensation of the surface silanol groups takes place, leading to the formation of siloxane (bridged oxygen) groups. As a result the material becomes hydrophobic. The condensation of these surface groups, as well as the reverse reaction induced by water, is very slow. The concentration and nature

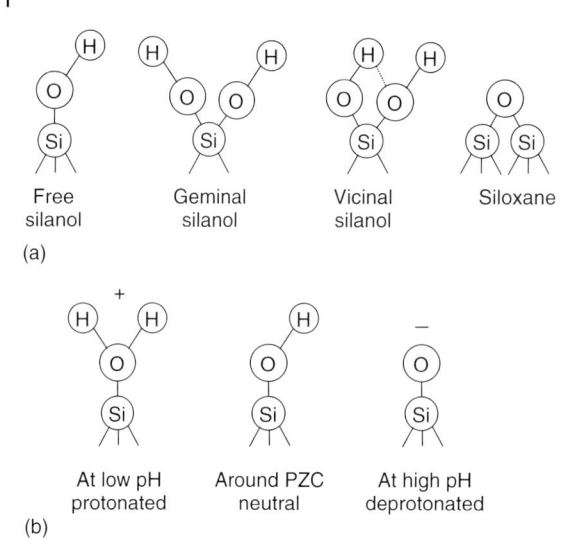

Figure 21.3 Silica surface groups: (a) the different types of silanol and siloxane groups and (b) an illustration of how the surface charge in aqueous solution is related to (de)protonation of the silanol groups.

of the surface groups determine the wetting properties, but also the interaction of SiO_2 with precursor phases.

In contact with a proton-containing liquid such as water, the surface can be charged, as illustrated in Figure 21.3b. At low pH the silanol groups are protonated, and the surface is positively charged. At high pH the silanol groups are deprotonated, and the surface is negatively charged. The pH at which the surface has no net charge is called the point of zero charge (PZC). For SiO_2 this is at pH 2–4. Each metal oxide has a different PZC, as the likelihood of (de)protonation of the hydroxyl groups is influenced by the electropositivity of, and distance to, the nearby metal atoms.

21.3
Alumina and Other Oxides

Alumina (Al_2O_3) exists in many different polymorphs, and is used in refractories and ceramics, as a filler material, and in catalysts, among other uses. Most relevant are the crystalline α-Al_2O_3 ('corundum'), and the amorphous γ-Al_2O_3 and η-Al_2O_3 phases, which are widely used in catalysis. Aluminum oxide is produced on a large scale in the so-called Bayer process from the mineral bauxite, in which it is separated from other oxides by selective precipitation.

Porous Al_2O_3 powders can be prepared via precipitation in aqueous solutions or flame hydrolysis. Al_2O_3 is an amphoteric oxide, that is, it dissolves in both

21.3 Alumina and Other Oxides

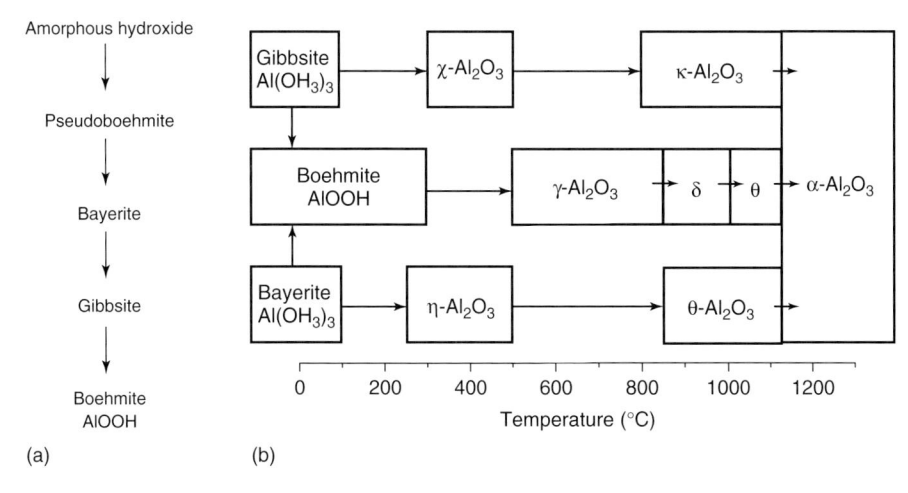

Figure 21.4 (a) The precipitates from a sodium aluminate solution when the pH is changed by the addition of a base (going through neutral to slightly alkaline pH), and (b) conversions to different crystalline polymorphs of Al_2O_3 upon heating. (Adapted from Ref. [3].)

acidic solutions as $Al(H_2O)_6^{3+}$ and in alkaline solutions as $Al(OH)_4^-$ or AlO_4^{5-}. Upon neutralization of the pH, the precipitation behavior in aqueous solutions is complex. The composition and structure of the precipitates depend not only on temperature, pH, and concentration, but also on time (aging) and history of the solution.

Figure 21.4a depicts a typical series of species precipitating from an aluminum salt solution [3]. Upon addition of a base to the acidic solution, first an amorphous hydroxide is formed, which is gradually converted into pseudoboehmite at neutral pH. The structure of pseudoboehmite resembles that of the crystalline boehmite (AlOOH), but a large amount of additional water is included in the structure. Upon further addition of a base, the pseudoboehmite is gradually converted into the aluminum hydroxide bayerite, and then into gibbsite. Finally crystalline boehmite (AlOOH) can be formed from the precipitate by heating to above 80 °C. Precipitation can be brought about by neutralizing either an acid or alkaline aluminum solution, by directly mixing the two, or by solvent evaporation. Precipitation from acidic solutions generally yields less ordered precipitates than that from alkaline solutions.

Different polymorphs of Al_2O_3 powder are formed by drying and heating the precipitates, as illustrated in Figure 21.4b [3]. γ-Al_2O_3 has a surface area of 200–300 m² g^{-1} and can be formed by heating boehmite to above 500 °C. Another relevant high-surface-area polymorph is η-Al_2O_3, which is formed from bayerite by heating above 300 °C. The high-surface-area compounds γ-Al_2O_3 and η-Al_2O_3 can be described as having a defective spinel-type structure. The oxygen atoms form a regular face-centered cubic lattice. However, there is some disorder in the position of the cations, which are partly located in the octahedral and partly in the tetrahedral positions, leaving many positions empty. Furthermore, water is included in the material, of which a major part is present at the surface as either linear, or two- or three-fold bridged OH groups. The PZC of Al_2O_3 is at a pH of around 8, with the linear groups being more basic and the bridged groups more acidic. Upon heating and dehydroxylation, the exposed Al sites represent Lewis acidity. In general, for γ-Al_2O_3, the [110] plane is predominantly exposed, while for η-Al_2O_3, the [111] plane is the typically exposed surface [4]. This has implications for catalysis, as the η-Al_2O_3 surface has stronger acid sites (with the surface OH-groups mainly associated with Al in tetrahedral positions). As a result γ-Al_2O_3 is more suitable, for instance, for hydrodesulfurization catalysts, whereas the η-Al_2O_3 is better suited for reforming catalysts [5].

Porous Al_2O_3 can alternatively be formed by flame hydrolysis. In this case γ-Al_2O_3 and δ-Al_2O_3 are formed around 1000 °C (for instance ALU-C® with a typical surface area of 100 $m^2 g^{-1}$). Crystalline α-Al_2O_3 is the most stable polymorph and has a low surface area. It is formed by heating any aluminum (hydr)oxide material to above 1100 °C. Despite its rather low specific surface (<50 $m^2 g^{-1}$), α-Al_2O_3 also finds application as a catalyst support for conditions in which mechanical and/or thermal stability and inertness are highly important, such as in steam reforming, and ethene epoxidation.

The choice of either a silica or an alumina support depends on the specific application. In general, alumina has higher mechanical robustness as well as higher thermal stability, especially when heated in the presence of steam. However, compared to silica, γ-Al_2O_3 shows higher reactivity with certain metals, as its structure allows easy incorporation of metal cations into empty octahedral positions. On the other hand, the rich surface chemistry of alumina could make it the preferred choice for the deposition of specific active phases.

Although the vast majority of catalyst supports used are either silica or alumina, other oxides, such as TiO_2, MgO, ZrO_2, ZnO, and mixed oxides, are also relevant. These oxides can also be prepared via either sol-gel reaction or via the flame hydrolysis process. In general, they possess smaller specific surface areas than can be obtained for either silica or alumina, but are less reactive and mechanically and thermally more stable. For some reactions, specific support interactions can play an important role, an example being the selective catalytic reduction of nitrogen oxides using V_2O_5/TiO_2.

21.4
Carbon Materials

At ambient pressure, the most abundant polymorph of carbon is crystalline graphite. However, porous carbons materials are available in a wide range of morphologies, purities, and cost. Most widely applied are activated carbon materials produced from cheap natural resources such as wood, coal, peat, or coconut shells. After heating the organic material to 800–1500 °C, a carbon with modest porosity remains. The porosity is enhanced (the material is 'activated') by a further treatment at high temperature under oxidizing conditions, or by leaching. Most pores are slit-like micropores, but also meso and macro porosity is present. Surface areas as high as 1000–1500 m^2 g^{-1} can be reached. Carbon black is formed by the burning of hydrocarbons in the presence of a sub-stoichiometric amount of oxygen. It consists of almost spherical nanometer-sized particles and derives its porosity from the interparticle space. High-purity carbon can be prepared by using well-defined organic precursors.

Graphite is hydrophobic, as there are virtually no functional groups at the surface. However, activated carbon typically contains a large amount of foreign elements, both inorganic (e.g., S, Fe, and Si) as well as up to 10–30 at% H and O [6]. The surface properties of activated carbon are determined by the presence of the oxygen-containing surface groups (Figure 21.5), which mostly terminate the edges of the graphene sheets [7].

Functional groups can also be added intentionally, for instance, by oxidation with aqueous HNO_3 solution [8]. Depending on the predominant type of functional groups (carboxylic, phenolic, lactonic, etc.) the PZC of carbon materials varies widely from acidic to basic pH.

Activated carbon supports offer low cost and high specific surface area. They are generally not reactive to metals, although in some specific cases intercalation or carbide formation can occur. An advantage compared to oxide supports is the possibility to reclaim expensive metal catalysts by combustion of the carbon. However, in general, carbon materials are less suitable for high-temperature gas phase reactions. Carbon is inherently unstable both under oxidizing conditions (forming CO_2) and under hydrogen (forming CH_4). However, it is an excellent support for liquid-phase reactions, such as hydrogenation reactions and fine-chemical synthesis, especially involving non-polar liquids.

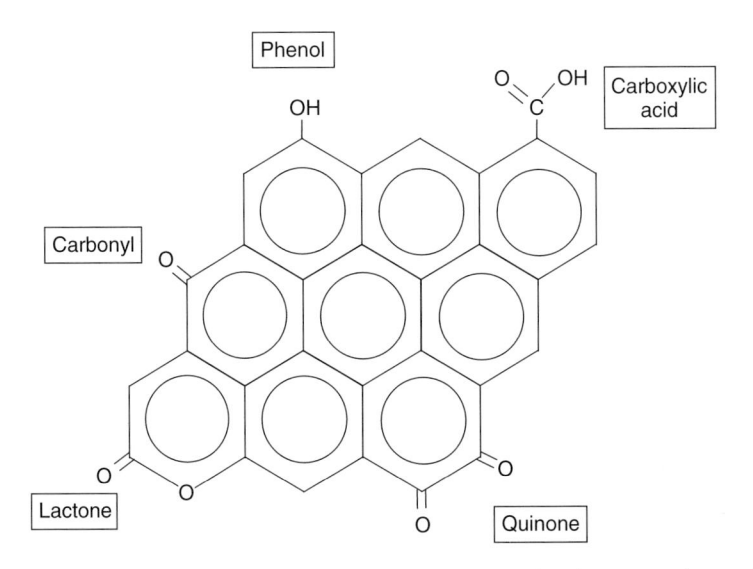

Figure 21.5 Surface groups frequently found in activated carbons. (Based on Ref. [7].)

Recently, carbon materials with well-defined and specific geometry have attracted much attention. Parallel single- or multi-wall carbon nanotubes can be grown from CH_4 or CO and H_2 in the gas phase using supported metal nanoparticulate growth catalysts. Also herringbone-type carbon nanofibers, spherical C_{60} (buckyballs), and single graphene sheets are produced in small quantities in gas phase processes. Although these model supports are very useful for fundamental research into the preparation of supported catalysts and into the specific influence of the support on catalytic activity, in general they are up to now too expensive and difficult to handle to be of practical relevance for applications.

21.5
Zeolites

Zeolites are crystalline aluminosilicates with well-defined micropores. They are widely used as sorbents, ion exchangers, and in catalysis. Zeolites can be mined (over 40 natural zeolite structures are known), but are usually grown from alkaline aluminosilicate solutions. Templates (for instance, cationic, such as alkyl ammonium cations, or small molecules or ion pairs) are added to the solution to induce the microporosity and a specific crystalline structure. The solution is first equilibrated at lower temperature, and then the zeolite crystals are grown slowly at high temperatures (100–150 °C) in an autoclave under high autogeneous water pressure. After washing and filtering, the solid is dried and calcined to remove the template [9].

The zeolite structure is based on TO_4 tetrahedra, where T represents metal atoms, mostly Si which is partly replaced by Al atoms. The tetrahedra (primary building units) can be linked together, sharing faces, edges, or corners, leading to different 3D structures containing micropores in the shape of channels, cages, and/or cavities, as illustrated for zeolite Y and mordenite (MOR) in Figure 21.7. At present, almost 200 different zeolite structures are reported, with microporosities in the range of 0.1–0.5 cm^3 g^{-1} [10] (Figure 21.6). As part of the Si^{4+} in the framework is replaced by Al^{3+}, charge compensating cations are present. In natural or as-synthesized zeolites, these are typically alkali metal ions such as Na^+ or Li^+.

However, these cations are mobile and can be easily exchanged by other large cations. If the cations are exchanged by NH_4^+, after subsequent heat treatment H^+ remains in the channels as charge compensation. These protons, located in the micropores of the zeolite, provide strong acid catalytic sites. Therefore zeolites in catalysis are sometimes referred to as '*solid acids*'.

Zeolites are widely used as active and selective catalysts, for instance for hydrocracking, isomerization, and alkylation reactions. The size of the micropores is comparable to that of organic molecules, and hence their specific size and shape can induce reactant or product selectivity. The density of acid sites is high, and correlates with the Al content. With increasing Si/Al ratio the density of acid sites decreases, but the acid strength (per site) increases as the Si in the lattice is more

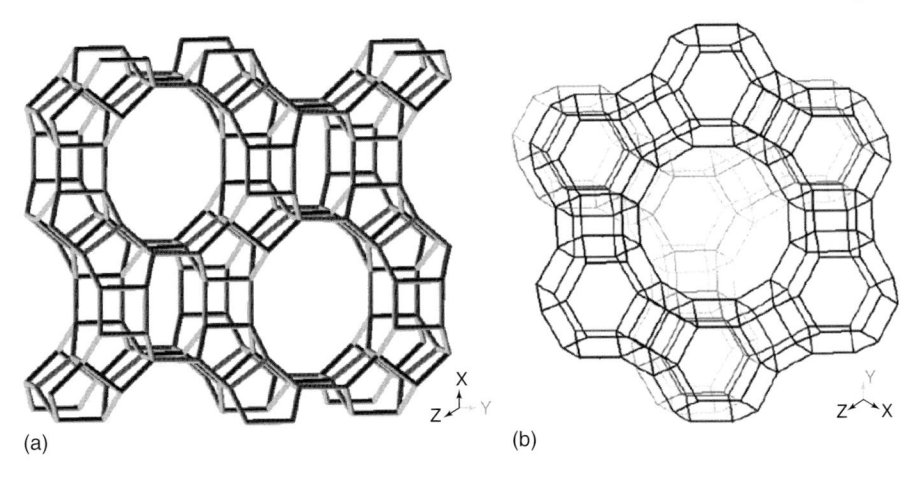

Figure 21.6 Structure of (a) mordenite (MOR) and (b) zeolite Y (faujasite, FAU). In these representations each meeting point of lines represents the position of the metal cation in a TO$_4$ unit. The oxygen atoms are not represented. (Pictures adapted from Ref. [10].)

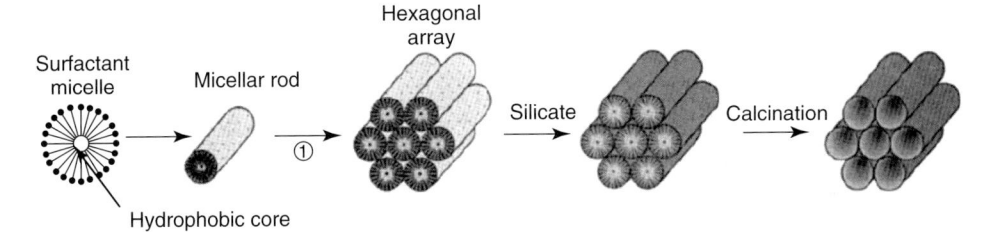

Figure 21.7 The formation of ordered mesoporous silica (MCM-41): at high surfactant concentrations the micelles pack into rods, and the rods pack into regular hexagonal structures. Upon deposition of silica from the aqueous phase (which is excluded from the hydrophobic inner core or the micelles) a nanocomposite is formed. After silica formation the micelles are removed by heat treatment, and a highly porous silica structure remains [13].

electropositive than the Al. Hence, an optimum in Si/Al ratio is typically found. If the lattice does not contain any Al, the zeolite does not contain any hydroxyl groups and can be hydrophobic. However only in a few selected cases, all silica zeolites can be synthesized. The Si/Al ratio in a zeolite can be changed post-synthesis by dealumination, for instance using steaming or acid leaching. Often also additional mesoporosity is introduced, which can alleviate diffusion limitations in the microporous structure [11, 12]. Despite the large number of zeolite structures known, catalysis is dominated by only a few: zeolite Y (FAU), ZSM-5 (MFI), and MOR. Nevertheless new structures and compositions, such as zeolites containing Ti in the framework, have potential for industrial application.

21.6
Ordered Mesoporous Materials

In 1992 a new type of mesoporous materials was reported by Exxon Mobil [13]. Like zeolites they were obtained by condensing silica in aqueous solution in the presence of a template. However, in this case the templates were nanometer-sized surfactant micelles, resulting in amorphous silica with regularly arranged pores of 1.5–10 nm diameter (Figure 21.7).

A prominent example is MCM-41, possessing a one-dimensional, hexagonally arranged pore system. However, due to the thin and amorphous silica walls between the pores, the stability (to steam) of these materials is very limited. In 1998 a similar class of materials was reported, but now templated using block copolymers such as polypropene/polyethene [14]. These materials have larger pores (5–20 nm), but also a higher hydrothermal stability due to the thicker pore walls (typically 1–1.5 nm). The most prominent member of this class of material is SBA-15, with an ordered pore system similar to that of MCM-41, but with microporous connections between the mesopores. The use of both surfactants and polymers allows the design of a range of related materials with different but very well-defined pore sizes and pore geometries.

Ordered mesoporous silica has proven very interesting for fundamental studies of such subjects as catalyst preparation, confinement, and gas adsorption. Using the ordered mesoporous silica itself as a template, other materials such as ordered mesoporous carbon (from sucrose or furfuryl precursors) and metals have been prepared [15]. Functional groups have been incorporated into the silica (by choosing different sol-gel precursors), leading to the so-called Periodic Mesoporous Organosilicas (PMOs). Also applications such as controlled drug release, which involve larger molecules, are widely studied. However, large catalytic industrial applications of ordered mesoporous materials have not yet emerged. Inherent disadvantages are the limited thermal stability and relatively high cost. Hence the challenge is to identify catalyzed reactions in which the order and monodispersity of the pore system are a key advantage.

21.7
Metal-Organic Frameworks

Metal-organic frameworks (MOFs) are microporous crystalline materials that consist of metal ion complexes ('nodes'), linked by organic bridging ligands ('linkers'). Polymeric materials consisting of metal ions bridged by organic groups have been known for some time, but these have always been unstable to thermal treatment or solvent removal. In 1999, for the first time, very rigid and structurally stable MOFs were reported [16]. The most prominent example is MOF-5 (see Figure 21.8) which consists of Zn_4O^{6+} nodes and benzene-1,4-dicarboxylate linkers, forming a cubic structure with 1.2 nm diameter cages that are accessible via a 0.8 nm entrance. MOF pore size and functionality can be tuned by the choice of linker and node,

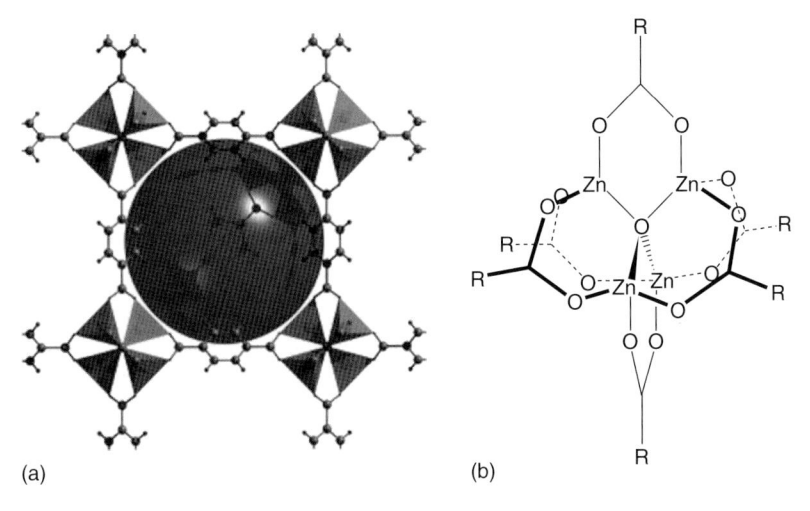

Figure 21.8 (a) Structure of MOF-5, with the central sphere indicating the 1.2 nm cages which are accessible via 0.8 nm windows, and (b) the structure of the Zn_4O connector and carboxylic linkers in more detail.

and the theoretical number of structures that can be formed is huge. MOFs can be surprisingly thermally stable, even to above 300 °C. The porosity can be as high as 2–3 cm³ g^{-1}, and specific surface areas of 3000 m² g^{-1} and even above have been reported. MOFs are currently being proposed for various applications, both in catalysis (with either the metal center, a functional organic group, or a metal cluster included in the micropores as the catalytically active center) and in reversible gas sorption and separation. A practical aspect is the high price, as for most MOFs there are no commercially available linkers.

21.8
Shaping

After preparation, all the materials discussed in this chapter typically consist of fine, micrometer-sized powders. However, handling of these powders on a large scale and packing them into reactors of dimensions measured in meter is not possible. It would lead to unacceptable handling limitations, pressure drops, and difficulties in separating the catalyst from the gas or liquid stream. Hence, in practice, the micro or mesoporous particles are shaped into macroscopic bodies using additional binders, fillers, and other additives. Table 21.2 shows an overview of shaping techniques for catalytic materials [5]. These include the formation of pellets by compression or extrudates by extrusion. The optimum shapes and sizes of the macroscopic support bodies vary, and depend on the type of reactor and reaction. For instance, a fixed-bed reactor requires relatively large bodies to minimize the pressure drop, while a fluidized-bed reactor could use smaller particles to minimize diffusion limitations.

References

1. Iler, R.K. (1979) *The Chemistry of Silica: Solubility, Polymerization, Colloid and Surface Properties and Biochemistry of Silica*, John Wiley & Sons, Ltd.
2. Brinker, C.J. and Scherer, G.W. (1990) *Sol-Gel Science: The Physics and Chemistry of Sol-Gel Processing*, Academic Press.
3. Trimm, D.L. and Stanislaus, A. (1986) The control of pore size in alumina catalyst supports: a review. *Appl. Catal.*, **21** (2), 215–238.
4. Lippens, B.C., Steggerda, J.J., and Linsen, B.G. (1970) *Physical and Chemical Aspects of Adsorbents and Catalysts*, Academic Press, London, New York.
5. Moulijn, J.A., van Leeuwenv, P.W.N.M., and van Santen, R.A. (1993) *Studies in Surface Science and Catalysis*, Elsevier, pp. 309–333.
6. Rodriguez-Reinoso, F. (1998) The role of carbon materials in heterogeneous catalysis. *Carbon*, **36** (3), 159–175.
7. Boehm, H.P. (1994) Some aspects of the surface chemistry of carbon blacks and other carbons. *Carbon*, **32** (5), 759–769.
8. Figueiredo, J.L. et al. (1999) Modification of the surface chemistry of activated carbons. *Carbon*, **37** (9), 1379–1389.
9. Cundy, C.S. and Cox, P.A. (2003) The hydrothermal synthesis of zeolites:history and development from the earliest days to the present time. *Chem. Rev.*, **103** (3), 663–702.
10. International Zeolite Association (IZA) (2009) The Zeolite Structure Database is Kept by the International Zeolite Association (IZA), http://www.iza-structure.org/databases/ (accessed 2012).
11. van Donk, S. et al. (2003) Generation, characterization, and impact of mesopores in zeolite catalysts. *Catal. Rev.*, **45** (2), 297–319.
12. Tao, Y. et al. (2006) Mesopore-modified zeolites: preparation, characterization, and applications. *Chem. Rev.*, **106** (3), 896–910.
13. Beck, J.S. et al. (1992) A new family of mesoporous molecular sieves prepared with liquid crystal templates. *J. Am. Chem. Soc.*, **114** (27), 10834–10843.
14. Zhao, D. et al. (1998) Triblock copolymer syntheses of: mesoporous silica with periodic 50 to 300 angstrom pores. *Science*, **279** (5350), 548–552.
15. Ryoo, R., Joo, S.H., and Jun, S. (1999) Synthesis of highly ordered carbon molecular sieves via template-mediated structural transformation. *J. Phys. Chem. B*, **103** (37), 7743–7746.
16. Li, H. et al. (1999) Design and synthesis of an exceptionally stable and highly porous metal-organic framework. *Nature*, **402** (6759), 276–279.

22
Development of Catalytic Materials
Manfred Baerns

22.1
Introduction

The main challenge in catalyst design is identifying the essential catalytic material. This may consist of a single component or a mixture of components which cause the gas-phase or liquid-phase chemical reaction to proceed to the desired product, either solely or along with some usually undesired products. These materials may be unsupported or supported by an inert material or a material that interacts with the primary catalytic compound affecting the reaction. The support materials may have different functionalities, such as acidity, basicity, or electronegativity. Certain interactions between noble metals and the support material are well known as metal–support interaction (MSI). In general, electronic effects play a paramount role in the elementary steps on the catalytic surface; this is in particular true for the large class of redox reactions occurring on metal oxides with the potential of changing valences. Such redox reactions comprise processes like oxidation, reduction, and hydrogenation.

Against this background, it is not surprising that the discovery and design of new catalysts is usually a rather complex endeavor which is based on both empirical and fundamental knowledge. In the last decade, this knowledge has significantly increased, although empirical experience still remains an essential part of the development work. This work, however, concerns not only the selection of the catalytic compounds, their support materials, and certain promoters which improve their catalytic action, but also the synthesis of the catalytic materials, their conditioning, and their long-term performance, which may be hampered by deactivation. As already mentioned, this chapter deals mainly with identifying the catalytic material that leads with high selectivity (for complex reactions) and activity to the desired product(s). Thus, the development consists of preparing and testing large amounts of materials, which requires a rational screening procedure, the results of which are then implemented in the further development of an optimal catalyst. Long-term performance of new catalytic materials is of high importance, requiring the suppression of deactivation phenomena that lead to losses in activity

---

Catalysis: From Principles to Applications, First Edition. Edited by Matthias Beller, Albert Renken, and Rutger van Santen.
© 2012 Wiley-VCH Verlag GmbH & Co. KGaA. Published 2012 by Wiley-VCH Verlag GmbH & Co. KGaA.

and/or selectivity of new materials. Such phenomena have to be studied with the aim of appropriately modifying the preparation processes.

The present chapter deals with different aspects of catalyst development. The requirement for fundamental and empirical knowledge holds true in conventional as well as in the more recent statistical approaches. Emphasis is put on the latter approaches since they appear to have a high potential for the future. Although, from a scientific point of view, it would be most desirable to base catalyst design solely on fundamental knowledge, this is so far still not possible due to the complexity of most of the catalytic reactions. It should be possible to detect synergistic effects by considering both types of knowledge.

It is important to mention that there are also *a priori* approaches to the design of a catalyst by theoretical modeling (see e.g., Refs. [1–4]). Although these procedures, which have been proven to be successful for some bi- or tri-metallic catalysts, appear to be very promising for the future, they cannot yet be applied for materials or reactions of high complexity. Therefore, these methods are not further considered in the present context.

Catalysts that are more efficient than the present ones are needed for many processes to reduce consumption of feedstocks and energy. This determines the direction of present catalyst development as well as the need for discovering new catalysts for novel reactions. In the development process, emphasis is usually put on three aspects of catalyst performance in the order of selectivity > deactivation > activity.

For process engineering purposes, the type of reactor needed for a specific reaction usually determines the shape and texture of catalytic solid materials, which in turn may influence inter- and intra-particle transport phenomena affecting catalyst performance. Most frequently, packed fixed-bed, fluidized-bed, slurry-phase, and membrane reactors are used, all requiring different particle and pore sizes, shapes, specific surface areas, and crushing and abrasion strengths (e.g., pellets, extrudates, spherical and granular particles, and powders). Although these aspects play a vital role in the final preparation process for use of the catalysts in a pilot plant and later on in a commercial process plant, the reader is referred to the specific literature on these subjects (see, e.g., Ref. [5]). Catalytic reaction engineering principles with respect to mass and heat transfer as well as reactor selection and operation have already been discussed in Chapter 4. The present chapter focuses primarily on the methods of developing a complex catalytic material.

22.2
Fundamental Aspects

The knowledge needed for catalyst development is based on the reactions occurring on the catalytic surface and the type of interactions between reactants and active sites. For illustration, the assumed primary reaction steps in the oxidative dehydrogenation of alkanes on metal oxides determine the selection of catalytic components. Metal oxides can, in principle, catalyze the reaction (see Refs. [6, 7]):

1) by participation of the lattice oxygen of redox metal oxides,
2) by dissociation of oxygen on the surface of metal oxides which then interacts directly with the hydrocarbon reactant, and
3) by abstraction of hydrogen on acidic surfaces, this being subsequently removed from these surfaces by oxidation to water, which is then desorbed.

This knowledge, needed in catalyst development, is based partly on empirical facts but also on theoretical considerations concerning the interactions of reactants and metal oxide surfaces. As catalysis science progresses, the methodology of catalyst development becomes more rational and sophisticated.

In the following, general principles in the design of a catalyst, which comprise its composition (active components, supports), preparation, conditioning, and testing by screening of different materials, are summarized, and the reader is introduced to the challenges associated with catalyst development.

When designing a catalyst, the molecular chemistry [8] which is believed to occur on the catalyst surface should be included in the development process (see Table 22.1). Hypothetical reaction mechanisms have to be set up; moreover, knowledge of the kinetics of the reaction steps can be helpful.

Essentially, all chemical transformations including bond-breaking and bond-forming steps have to be considered, as they require certain surface characteristics like acidity or basicity, also solid-state, electronic, and especially redox properties, particular types of surface sites, and so on. With this in mind and referring to empirical and theoretical pre-knowledge as well as intuition, patents, and literature, the catalyst developer starts carefully to think of possible components and support materials as well as methods of preparation for the catalysts.

On the basis of all the above variables and their interrelationships, appropriate materials compositions have to be defined, prepared, and tested for their functional behavior, that is, their catalytic performance (see Section 22.4). The choice of catalyst compositions (selection of components and their proportions), in particular for complex (multi-step) reactions, which usually involve a larger number of components, becomes a challenging task. This is complicated by the many elements and their compounds that fulfill the fundamental requirements. That is to say, the optimal catalyst composition has to be searched for within a multi-parameter space. This results in a large number of specimens of catalytic materials (up to several thousands), which need to be prepared, tested for their catalytic performance,

Table 22.1 Requirements in the design of heterogeneous catalysts.

Surface chemistry
Assumed reaction mechanism
Kinetics of elementary reaction steps
Surface properties required for bond-breaking and bond-formation steps
Bulk and surface characteristics: elemental composition, acidity, and basicity,
Solid-state properties
Type of active sites.

and finally screened for the best hits as precursors for further optimization. The design of these experiments is based on mathematical optimization methods. Data mining and data analysis form an essential part of the development process, that is, identifying the catalyst compounds as well as physical and physical-chemical properties of the catalytic material, which significantly determine catalytic performance (see Section 22.3). In the author's opinion, these techniques of combinatorial catalyst development can ultimately also contribute to an increased fundamental understanding of the catalysis under invesigation, which, in turn, could deliver new insights for further catalyst improvement.

For illustration, the importance of chemical kinetics and of some relationships between catalytic performance and solid-state properties of the catalytic material as a source of knowledge are summarized below for the oxidative coupling of methane (OCM) to ethane and ethylene (for further details see Ref. [9]).

22.3
Micro-Kinetics and Solid-State Properties as a Knowledge Source in Catalyst Development

Micro-kinetics describes a catalytic reaction scheme in the form of elementary reaction steps which occur on the surface. The determination of micro-kinetics requires that the reactions are not limited by mass- and heat-transport phenomena, by which activity and selectivity might be influenced. It is certainly obvious that reaction-rate schemes including catalytic surface reaction steps are most suited to set up relations between kinetic parameters and the bulk and surface properties of the solid catalytic material. This would also support the theoretical description of its catalytic properties. When deriving micro-kinetics and the respective parameters it is absolutely necessary that the reactions be in no way affected by transport phenomena disguising the intrinsic catalytic reaction (see Section 4.3). This concept of using micro-kinetics is demonstrated in the following case study concerning the OCM reaction.

22.3.1
Reaction Mechanism and Kinetics of the Catalytic OCM Reaction

The OCM reaction occurs via a heterogeneous–homogeneous mechanism, that is, the reaction is initiated on the catalyst surface by generating methyl radicals, which then recombine in the gas phase to yield ethane. In competitive reactions, methyl radicals can react with gas-phase O_2 forming $CH_3O_2^{\bullet}$ or/and $CH_3O^{\bullet}$ radicals, which are precursors of CO_x. Alternatively, these radicals and also the methyl radicals can undergo heterogeneous transformations to CO_x. Consecutive oxidation of ethane and ethylene is the main reaction pathway leading to CO_x at high methane conversions and occurs also via homogeneous and heterogeneous reaction steps, the latter playing the most significant role. These reactions compete with generation of methyl radicals from methane.

$$CH_4 + \text{surface (s)} \longrightarrow \text{s-H} + CH_3(\text{gasphase})$$
$$2CH_3(\text{gasphase}) \longrightarrow C_2H_6 + O_s \longrightarrow C_2H_4$$
$$C_xH_y + \{O_2(\text{gas}) \text{ or } O_x\} \longrightarrow CO_x + H_2O \quad (22.1)$$

As described below, the type of oxygen species formed from gas-phase oxygen on the catalyst surface and the physico-chemical properties of solids are essential selectivity-determining factors. Therefore, establishing the relationships between these properties and individual selective and non-selective reaction pathways of the OCM reaction is highly important for designing novel selective catalytic materials.

22.3.2
Surface Oxygen Species in Methane Conversion

For various heterogeneous oxidation reactions, oxygen activation, that is, the formation of active oxygen species from gas-phase O_2 on the catalyst surface, is an essential reaction step [10–13]. Figure 22.1 shows the diversity of oxygen species which can exist on the catalyst surface and which participate in hydrocarbon oxidation, yielding both selective and non-selective reaction products. However, there is no unanimous agreement on the nature of the reactive oxygen species participating in selective and non-selective reaction pathways. This is due to both the variety of catalysts for the OCM reaction and the experimental limitations for unambiguous identification of active sites at high temperatures (>873 K), which are required for methane activation. Therefore, thorough mechanistic and micro-kinetic analysis is a tool which can be used for deriving insights into the nature of oxygen species and their role in methane conversion.

The importance of O^- oxygen species in the OCM reaction has been repeatedly highlighted by Lunsford and associates [16, 17]. The ability of O_2^{2-} species for methane activation was proven by oxygen-free methane oxidation over Na_2O_2 and BaO_2 [18]. By means of in-situ Raman analysis, O_2^{2-} species were identified over Na/La_2O_3 and Sr/La_2O_3 at 973 K [16] and over BaO/MgO [17] at 973 and 1023 K upon methane oxidation with oxygen. In addition to O_2^{2-} and O_2^- species were also observed over catalytic materials based on rare-earth oxides in a recent in-situ Raman study [19]. It was suggested that O_2^- and O_2^{2-} originate from O_3^{2-}. The latter species is formed via reversible coupling of O_2 with a neighboring O^{2-}.

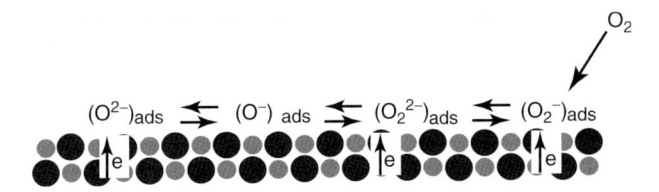

Figure 22.1 Oxygen species stabilized on the oxide catalysts upon activation of gas-phase O_2 [14, 15].

From previous work on the micro-kinetics of O_2 activation over Na/CaO catalysts [20, 21], it has been suggested that mono-atomic oxygen species are responsible for selective oxidation (ethane and ethylene formation), while bi-atomic oxygen species take part in consecutive oxidation of C_2-hydrocarbons to CO_x. By combining the micro-kinetic knowledge with electronic and catalytic properties of Na/CaO, the catalyst's ability for fast transformation of bi-atomic oxygen species into atomic ones was identified as a key catalyst property for achieving high C_2-selectivity. The transformation can be accelerated by creation of anion vacancies in the oxide lattice upon adding low-valence additives into the host oxide matrix. This statement was experimentally proven by doping CaO with Na_2O [22], CeO_2 with CaO [23], and Nd_2O_3 with SrO [24].

22.3.3
Kinetic Analysis

D. Wolf et al. [23] applied a micro-kinetic analysis of the OCM reaction to identify the relationships between the kinetic parameters of individual reaction steps and surface properties of the catalysts. These authors concluded that strong surface OH groups formed upon abstraction of one H atom from CH_4 by active oxygen species block active sites (oxygen species and anion vacancies) and suppress catalytic activity. A decrease in the concentration of anion vacancies results in inhibition of the dissociation of molecular adsorbed oxygen species and therefore in a decrease in C_2-selectivity. The model predicts that the steady-state surface coverage with OH groups, oxygen, and vacant sites of these specific catalysts depends on the CaO content of the mixed metal oxides. However, due to the inferior performance of the earlier computing machines, the authors had to simplify their micro-kinetic models by considering reaction pathways which are far from being elementary in nature. By applying the potential of modern computing machines and software, this approach appears to be attractive for catalyst design.

Based on the above background, the main challenge in the OCM reaction is to suppress the non-selective CO_x production. There are two main reaction pathways leading to a loss of selectivity in the OCM reaction:

1) homogeneous, that is, reactions of $CH_3^{\bullet}$ or $C_2H_5^{\bullet}$ with O_2 yielding $CH_3O_2^{\bullet}/CH_3O^{\bullet}$ or $C_2H_5O_2^{\bullet}$ radicals followed by their further gas-phase transformations to CO_x;
2) heterogeneous oxidation of $CH_3^{\bullet}$ radicals and surface intermediates as well as C_2-hydrocarbons.

Since the non-selective gas-phase reactions are difficult to control, significant improvements in the OCM performance may be only achieved in a first approximation by controlling heterogeneous reaction pathways. For example, CO_x formation can be suppressed if the catalyst's capability for desorption of methyl radicals is decreased in favor of their surface recombination to ethane, which should quickly desorb before its subsequent oxidation to CO_x. Alternatively, the rates of generating methyl radicals, of their desorption, and of the formation of selective oxygen species should be optimized. The idea is to minimize the probability of methyl radicals and gas-phase O_2 to meet each other in the gas. This can be achieved when the rates of chemisorption of gas-phase O_2 yielding surface oxygen species and desorption of methyl radicals are high. Due to the high rate of chemisorption of O_2, its gas-phase concentration should be low, favoring recombination of desorbed methyl radicals in preference to their reaction with gas-phase O_2. Moreover, the catalyst's ability for adsorption of C_2-hydrocarbons has to be reduced.

It should be mentioned that transient kinetic experiments could also contribute to the understanding of the reaction mechanism and to deriving kinetic information (see also Section 4.2). A typical reactor setup (TAP: Temporal Analysis of Products reactor) for such experiments as carried out for the OCM reaction is shown Figure 22.2. More details have been reported in Ref. [25].

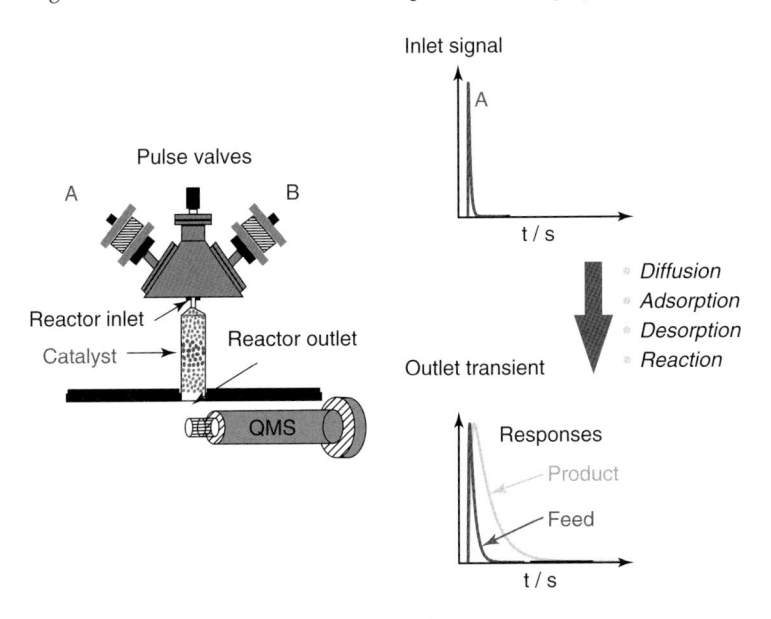

Figure 22.2 TAP reactor setup for transient kinetic experiments.

22.3.4
Physico-Chemical Properties of Catalytic Solid Materials for the OCM Reaction

Physico-chemical properties are of high relevance for achieving high C_2-selectivity; therefore, they have to be taken into account for the rational development of OCM catalyst. These properties are summarized below.

22.3.5
Structural Defects

The importance of the structural defects of catalytic materials for achieving high C_2-selectivities has been repeatedly reported for methane conversion

[11, 23, 26]. The authors in Ref. [27] divided all known OCM catalysts into two groups: (i) multiphase catalysts and (ii) single-phase catalysts. For the former group of catalysts, particular phases appear to be required for good performing catalysts. Oxygen vacancies and ions of impurity transitional metals are the main defects of single-phase catalysts. Oxygen lattice defects promote activation of gas-phase oxygen and the mobility of active oxygen species into the bulk, thereby partly controlling the oxidation state of the catalyst surface. The ability of non-reducible catalytic materials for oxygen activation and stabilization of a certain type of adsorbed oxygen species is also a key factor in the selective transformation of light alkanes [27–29].

However, the defect structure is changed under OCM conditions, as outlined in Refs. [27–29]. These reaction-induced defects influence the catalytic performance. For M/CaO (M = Li or Na < 2 at%), $[MCO_3]^-$, CO_2^{2-}, and O_3^{2-}, defects were identified in the volume of the CaO lattice after the OCM reaction with O_2 took place at 1023 K. An increase in the fraction of the $[M(Li\ or\ Na)CO_3]^-$ centers in doped CaO resulted in a rise in C_2 selectivity using an O_2-containing reaction feed [27]. In the silver-catalyzed OCM reaction, Ag faceting promotes the formation of O_γ (nucleophilic species embedded in the uppermost layer of silver atoms) species as determined by Nagy et al. [29]. These species were suggested to be responsible for improved catalytic performance.

The structural defects influence the type of electrical conductivity of solid materials. Solid materials of n-type conductivity are usually non-selective catalysts, while solids containing both p-type and oxygen-ion conductivity are desirable catalysts [26]. The poor performance of n-type conductors was ascribed to their low ability for generation but high activity for oxidation of methyl radical. Moreover, it also has to be stressed that the band gap should be in the range of 5–6 eV. This is due to the fact that the band gap determines the concentration pf charge carriers. If the band gap is too low, a high concentration of adsorbed oxygen species is expected, which does not favor the formation of selective reaction products.

22.3.6
Surface Acidity and Basicity

Based on past and more recent characterization studies, it has been suggested that catalytic performance of various solid materials for selective oxidative transformation of light hydrocarbons was affected negatively by surface acidity [30, 31] and positively by basicity [32]. These basic properties are suggested to be important for heterolytic breaking of the C–H bond in CH_4.

22.3.7
Redox Properties, Electronic Conductivity, and Ion Conductivity

Furthermore, reducibility, the ability to re-oxidize the active sites in redox-based catalytic materials, and electrical conductivity [26] have an effect on catalytic performance.

22.3.8
Supported Catalysts

The interaction between the oxide support and the deposited active components determines the structure of the resulting surface MeO_x species which influence selective and non-selective reaction pathways. The nature of the reaction intermediates and their adsorption/desorption ability are other important factors influencing product selectivity.

22.3.9
Conclusions

From the above it has become clear that the OCM reaction and its kinetics are influenced by various physico-chemical properties, which, however, have not been previously considered in total for deriving a final conclusion about their relevance for catalyst design. Therefore, a better understanding of the various physico-chemical properties and their effect on reaction mechanism, micro-kinetics and, as a result thereof, on catalytic performance would certainly allow control of the overall process of methane conversion, hence providing useful guidelines for rational catalyst design and development. It remains to be seen whether there are synergistic or possibly even adverse effects of different solid-state properties on the desired catalytic performance.

22.4
Combinatorial Approaches and High-Throughput Technologies in the Development of Solid Catalysts

In combinatorial development of catalytic materials with maximum performance, a large number of catalyst compositions are required (see Section 22.4.1). These compositions have to be prepared and tested by applying high-throughput technologies (see Section 22.4.2). This procedure deviates significantly from conventional approaches, which focus, in an iterative process, on only a few compositional structural variables, which have mostly been identified on a fundamental basis (see e.g., Refs. [33–35]).

The state of the art of combinatorial approaches as a whole but also various specific aspects has been reported in numerous monographs, to which the interested reader is referred [36–38].

22.4.1
Combinatorial Design of Catalytic Materials for Optimal Catalytic Performance

Activity and selectivity of a catalytic material as well as its long-term stability depend on many variables, by which a multi-parameter space is defined. These variables comprise essentially the chemical composition of the material, the mode of its preparation and conditioning (calcination and formation procedure of the final material), and reaction conditions such as temperature, partial pressures, or concentrations of the reactants in the fluid phase (usually gas or occasionally liquid or gas/liquid systems). The large number of variables defines a multi-parameter space in which the best catalytic material, that is, the final catalyst, exists. If one were to try to investigate such a space by conventional means, an extremely high number of experiments would be needed. Taking only the catalyst composition into account, Senkan [39] has calculated the number of possibilities which would exist. Based on 50 chemical elements, which might be considered for designing a catalyst, 1225 binary, 19 600 ternary, and 230 300 quaternary combinations are possible. If different molar fractions of the components, different modes and parameters of preparation, and different reaction conditions are considered, a 'combinatorial explosion' occurs, which means that the number of experiments to be carried out would get out of control.

As has already been pointed out above, a suitable search algorithm has to be applied to find the optimal catalyst. To start any combinatorial development of catalytic materials for a specific reaction, a pool of elements/components has to be established, from which – according to a predefined algorithm – a first generation of materials compositions has to be prepared. If different preparation procedures are applied, these can be also included in such an algorithm. For the sake of simplicity at this point, only compositional changes are considered. If, for example, an alkane is to be oxidatively dehydrogenated (ODH) in the presence of gas-phase oxygen, assumptions have to be made about possible mechanisms of such a reaction. On this basis, a pool of primarily selected components, which are in the present case metal oxides, is set up. In Figure 22.3 this is illustrated for ODH of ethane and propane to their respective olefins.

(a) Redox-mechanism (Mars-van Krevelen)

$$C_nH_{2n+2} \xrightarrow{MeO_x} C_nH_{2n+1} + MeO_xH$$
$$\Downarrow$$
$$C_nH_{2n} + MeO_{x-1} + H_2O$$
$$MeO_{x-1} + 0.5O_2$$

(b) Activation by adsorbed oxygen

$$C_nH_{2n+2} \xrightarrow{MeO_x\text{-}O_{ad.}} C_nH_{2n+1} + MeO_x\text{-}OH$$
$$\Downarrow$$
$$C_nH_{2n} + MeO_x + H_2O$$
$$MeO_x + 0.5O_2$$

(c) Activation by lattice oxygen (no redox-mechanism)

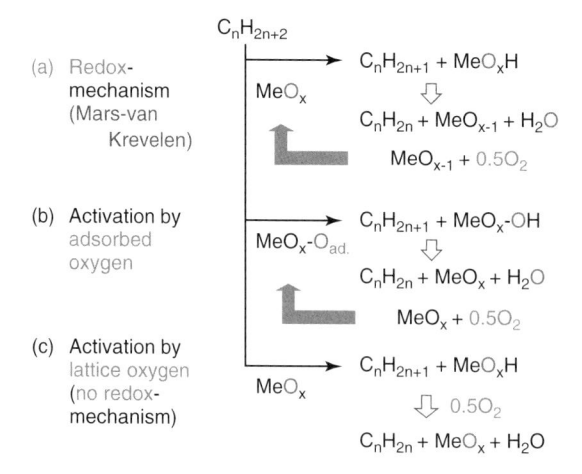

$$C_nH_{2n+2} \xrightarrow{MeO_x} C_nH_{2n+1} + MeO_xH$$
$$\Downarrow 0.5O_2$$
$$C_nH_{2n} + MeO_x + H_2O$$

Figure 22.3 Selection of assumed primary reaction steps of the oxidative dehydrogenation of alkanes on metal oxides as catalyst components.

Table 22.2 Selection of a pool of elements from fundamental and empirical knowledge for preparing catalyst compositions for the oxidative dehydrogenation of ethane and propane according to Refs. [40–42].

Assumed mechanism	Required property	Metal oxides	
		Ethane	Propane
Participation of removable lattice oxygen (Mars–van Krevelen mechanism)	Redox properties (medium M–O binding energy)	Cr_2O_3, CuO, MnO_2, MoO_3, WO_3, Ga_2O_3, CoO, SnO_2	V_2O_5, MoO_3, MnO_2, Fe_2O_3, ZnO, Ga_2O_3, GeO_2, Nb_2O_5, WO_3, Co_3O_4, CdO, In_2O_3, NiO
Activation by adsorbed oxygen	Dissociative adsorption of oxygen	CaO, La_2O_3	La_2O_3
Activation by lattice oxygen	Non-removable lattice oxygen: high Me–O binding energy	ZrO_2	Acidity: B_2O_3 Basicity: MgO

For the three mechanisms, different metal oxides were selected that slightly differed for the two reactions (see Table 22.2). For the first generation, randomized compositions with respect to type of component and its proportion in the material were prepared and tested for their functional properties, that is, catalytic, performance. For further improvements of the performance of the materials tested, the compositions of the materials were changed according to a genetic algorithm; other optimization methods could have been also chosen.

Without going into further detail, the results for catalytic performance, expressed as olefin yield at complete oxygen conversion are shown in Figure 22.4. It is obvious that the yield for the best materials of each generation increases with increasing number of generations until a maximal value is reached (just as the average yield based on all catalytic materials of a generation increases with the number of generations).

Additional improvements would be possible by applying a conventional optimization technique to the catalyst composition in the compositional space of maximal performance. The catalysts could be further improved by taking all experimental performance data into account and fitting them to an artificial neural network to describe the whole multi-dimensional space representing the relationships between catalytic performance and composition (as well as possibly other variables). From such results, even some fundamental insights in catalysis may be eventually derived, as has been shown in Ref. [43]; this, indeed, could be a fruitful application of high-throughput experimentation.

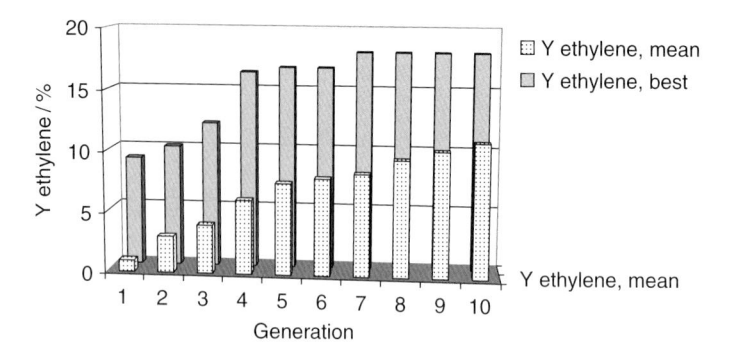

Figure 22.4 Example of catalyst optimization applying a genetic algorithm for the oxidative dehydrogenation of ethane to ethylene: Maximum and average olefin yield as a function of the number of generations in the iterative development process. (After: Ref. [40].)

22.4.2
High-Throughput Technologies for Preparation and Testing of Large Numbers of Catalytic Materials

For covering the whole compositional parameter space, a combinatorial design of catalyst libraries, that is, a full grid search, is required. The combinatorial design takes into account all elements at different randomized levels of concentrations. All catalytic materials from the libraries have to be tested for their catalytic performance, which can be quantified by different objective functions, the main ones being selectivity, activity, and stability.

As already indicated in the preceding section, the required number of materials is usually very high, depending on the number of elements involved and their concentration levels, which have to be defined by the experimenter. To a certain degree, the number of parameters can be restricted by utilizing available knowledge and intuition. The type of material desired determines the method of its preparation and testing. There are basically two forms of materials for catalytic testing, fulfilling two different purposes, namely stage 1 and stage 2 screening (see Ref. [44]). Stage 1 screening features maximum sample throughput but delivers only reduced information, since only target compounds are analytically identified. This method is primarily for discovering new materials, the compositions of which have to be further optimized. This procedure is often based on solid films on supports, which change their composition locally. Stage 2 screening approaches real conditions for particulate materials and reaction conditions; solid particles (grains or micro-pellets) with or without support are frequently used. The mixture of reactants is subjected to a detailed analysis. In this way, continuous improvement is to be expected.

According to the present author's philosophy, the evolutionary strategy incorporates stages 1 and 2 as a whole. However, the only experimental tools described here are for the evolutionary strategy which corresponds to stage 2 screening (for so-called stage 1 sample preparation see, e.g., Ref. [45]).

22.4.2.1 Preparation of Catalytic Materials

For high-throughput preparation of solid catalytic materials, automated robots equipped with synthesis control protocols are used. These robots carry out the various synthesis steps with high speed in parallel or consecutively. The various steps require high accuracy and reproducibility, otherwise no reliable data on the optimal catalyst composition and its best performance are obtained. Moreover, the steps have to be performed in a manner which is transferable to pilot- and technical-scale plant operation. For illustration, an automated synthesis robot is shown in Figure 22.5; this equipment includes such steps as impregnation of support materials, precipitation, filtering, washing, and drying. (Similar devices also exist for conditioning, e.g., calcination.)

22.4.2.2 Testing and Screening of Catalytic Materials

For stage 2 testing of catalytic materials, parallel fixed-bed reactors are used. The requirements for the catalytic materials correspond to those of single reactor operation: internal and external heat- and mass-transfer limitations must be excluded during testing in order to obtain catalytic results which are unbiased by transport phenomena. Furthermore, isothermal operation is needed, which means that radial and axial temperature gradients have to be avoided. Appropriate criteria

Figure 22.5 Equipment for parallel synthesis of catalytic materials including such steps as impregnation of support materials, precipitation, filtering, washing, and drying. (Courtesy of Zinsser Analytic GmbH, Frankfurt am Main, Germany.)

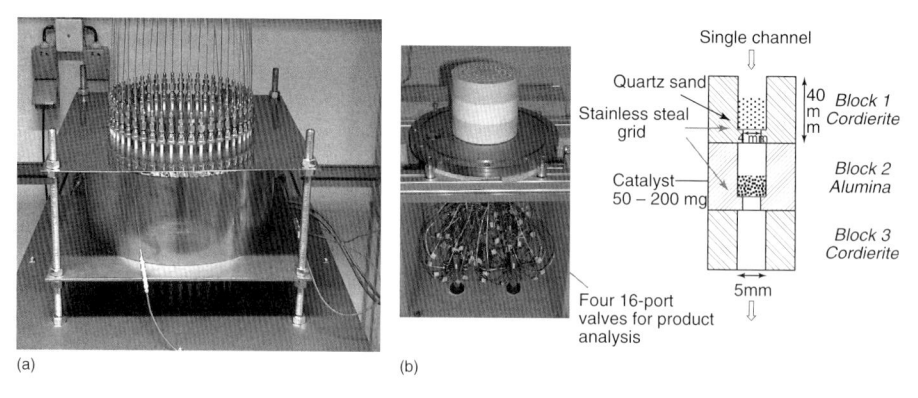

Figure 22.6 (a) Reactor for high-throughput testing of catalytic materials at temperatures up to about 1200 °C (e.g., for HCN formation from methane and ammonia using capillaries to compensate for equal pressure drop in all reactor channels [43]. (Courtesy of Leibniz-Institute for Catalysis at University of Rostock, Germany.) (b) Reactor (insulation not shown) for high-throughput testing of catalytic materials (e.g., catalysts for the oxidative dehydrogenation of alkanes at temperatures up to about 500 °C using quartz sand to compensate for equal pressure drop in all reactor channels. (Courtesy of Leibniz-Institute for Catalysis at University of Rostock, Germany.)

on these phenomena can be found in textbooks dealing with catalytic reaction engineering.

Comparable activity measures for each catalyst in the different parallel reactor channels can be derived from the degree of conversion of a key-feed component after the same contact time (catalyst volume divided by flow rate of feed (volume/time)). Exposure of all the open cross sections of the reactor channels to the total gas volume, will give even distribution to all channels as long as the pressure drops in the individual reactor channels are the same. If this is not the case, steps must be taken to establish the same pressure drop in all channels. Usually an appropriate flow resistance (larger than that caused by the bed of catalyst particles) is placed in front of the reactor channel (Figure 22.6).

Finally, a multi-fold (625 channels) single-bead reactor is shown in Figure 22.7. Only one catalyst bead is hosted in each single channel. The concept of this reactor type is very flexible; it allows the adoption of several analysis techniques for the product gas leaving the channels. The analysis can be either sequential, retrieving gas samples from each individual catalytic-bead channel, typically via a capillary, or via integral techniques over all the samples at a specific time. MS, GC/MS, GC, or dispersive or non-dispersive IR are suitable for fast sequential analysis. Certain integral analysis techniques can also be applied for true parallel analyses, for example, IR-thermo-graphic or photo-acoustic spectroscopy [44].

22.4.3
Data Analysis

In high-throughput preparation and testing of catalysts, large amounts of data have to be dealt with. Assuming that a genetic algorithm is applied for combining

Figure 22.7 Single-bead reactor with 625 parallel channels for testing of solid catalyst beads in gas-phase reactions. (Courtesy of the Aktiengesellschaft, Heidelberg, Germany.)

catalytic components with numerous catalytic materials and for the design of new generations from previous ones with the aim of maximizing their catalytic performance, different aspects of statistics have to be taken into account. The specific algorithms may change from application to application (e.g., type of preparation, type of testing, and closeness to the final optimum composition). For the eventual discovery of optimal compositions, including methods of preparation, conditioning, and so on, and the relationships between the input (catalytic compositions, methods of catalyst preparation, and conditioning) and output variables (selectivity, activity, and its temporal development), specific statistical methods are required. In a recent monograph [6], the present author and M. Holeňa [6] have documented a comprehensive overview of the statistical methods used in combinatorial development of heterogeneous catalysts, supplemented by numerous applications.

References

1. Jacobsen, C.J.H. et al. (2002) Optimal catalyst curves: connecting density functional theory calculations with industrial reactor design and catalyst selection. *J. Catal.*, **205** (2), 382–387.
2. Norskov, J.K. et al. (2009) Towards the computational design of solid catalysts. *Nat. Chem.*, **1** (1), 37–46.
3. Greeley, J. and Norskov, J.K. (2007) Large-scale, density functional theory-based screening of alloys for hydrogen evolution. *Surf. Sci.*, **601** (6), 1590–1598.
4. Andersson, M.P. et al. (2006) Toward computational screening in heterogeneous catalysis: Pareto-optimal methanation catalysts. *J. Catal.*, **239** (2), 501–506.
5. Ertl, G. et al. (2008) *Handbook of Heterogeneous Catalysis*, Wiley-VCH Verlag GmbH & Co. KGA, Weinheim.
6. Baerns, M. and Holeňa, M. (2009) *Combinatorial Development Of Solid Catalytic Materials Design of High-Throughput Experiments, Data Analysis, Data Mining*, Catalytic Science Series, Imperial College Press, London.
7. Rodemerck, U. and Baerns, M. (2004) in *Basic Principles in Applied Catalysis*, Chapter III (ed. M. Baerns), Springer, Berlin, Heidelberg, pp. 259–280.

8. Christensen, C.H. and Nørskov, J.K. (2008) A molecular view of heterogeneous catalysis. *J. Chem. Phys.*, **128**, 182503.
9. Kondratenko, E.V. and Baerns, M. (2008) in *Handbook of Heterogeneous Catalysis* (eds G. Ertl, H. Knözinger, F. Schüth, and J. Weitkamp), Wiley-VCH Verlag GmbH & Co KGA, Weinheim, pp. 3010–3023.
10. Bielanski, A. and Haber, J. (1991) *Oxygen in Catalysis*, Marcel Dekker, New York.
11. Gellings, P.J. and Bouwmeester, H.J.M. (2000) Solid state aspects of oxidation catalysis. *Catal. Today*, **58** (1), 1–53.
12. Moro-oka, Y. (1998) Reactivities of active oxygen species and their roles in the catalytic oxidation of inactive hydrocarbon. *Catal. Today*, **45** (1–4), 3–12.
13. Panov, G.I. (2000) Advances in oxidation catalysis; oxidation of benzene to phenol by nitrous oxide. *CATTECH*, **4** (1), 18–31.
14. Driscoll, D.J. et al. (1985) Formation of gas-phase methyl radicals over magnesium oxide. *J. Am. Chem. Soc.*, **107** (1), 58–63.
15. Tong, Y., Rosynek, M.P., and Lunsford, J.H. (1989) Secondary reactions of methyl radicals with lanthanide oxides: their role in the selective oxidation of methane. *J. Phys. Chem.*, **93** (8), 2896–2898.
16. Mestl, G., Knözingers, H., and Lunsford, J.H. (1993) High temperature in situ Raman spectroscopy of working oxidative coupling catalysts. *Ber. Bunsen-Ges. Phys. Chem.*, **97** (3), 319–321.
17. Lunsford, J.H. et al. (1993) In situ Raman spectroscopy of peroxide ions on barium/magnesium oxide catalysts. *J. Phys. Chem.*, **97** (51), 13810–13813.
18. Sinev, M.Y., Korchak, N.V., and Krylov, O.V. (1986) Highly selective ethane formation by reduction of bao2 with methane. *Kinet. Katal.*, **27** (5), 1110.
19. Zhang, H.B. et al. (2001) Active-oxygen species on non-reducible rare-earth-oxide-based catalysts in oxidative coupling of methane. *Catal. Lett.*, **73** (2), 141–147.
20. Kondratenko, E.V. et al. (1999) Transient kinetics and mechanism of oxygen adsorption over oxide catalysts from the TAP-reactor system. *Catal. Lett.*, **63** (3), 153–159.
21. Kondratenko, E.V., Buyevskaya, O., and Baerns, M. (2000) Mechanistic insights in the activation of oxygen on oxide catalysts for the oxidative dehydrogenation of ethane from pulse experiments and contact potential difference measurements. *J. Mol. Catal. A: Chem.*, **158** (1), 199–208.
22. Kondratenko, E.V., Wolf, D., and Baerns, M. (1999) Influence of electronic properties of Na_2O/CaO catalysts on their catalytic characteristics for the oxidative coupling of methane. *Catal. Lett.*, **58** (4), 217–223.
23. Wolf, D. et al. (2001) Predictions of relationships between catalytic and solid phase properties by kinetic models and their validation. *J. Catal.*, **199** (1), 92–106.
24. Gayko, G. et al. (1998) Interaction of oxygen with pure and SrO-Doped Nd_2O_3 catalysts for the oxidative coupling of methane: study of work function changes. *J. Catal.*, **178** (2), 441–449.
25. Kondratenko, E.V. and Perez-Ramirez, J. (2007) Micro-kinetic analysis of direct N2O decomposition over. *Catal. Today*, **121** (3–4), 197–203(Special Issue).
26. Zhang, Z., Verykios, X.E., and Baerns, M. (1994) Effect of electronic properties of catalysts for the oxidative coupling of methane on their selectivity and activity. *Catal. Rev.*, **36** (3), 507–556.
27. Kondratenko, E.V., Maksimov, N.G., and Anshits, A.G. (1995) Formation and properties of bulk defects in li/cao and na/cao catalysts on the basis of solid-solutions in the oxidative conversion of methane. *Kinet. Katal.*, **36** (5), 658.
28. Anshits, A.G. et al. (1995) The role of the defect structure of oxide catalysts for the oxidative coupling of methane. The activation of the oxidant. *Catal. Today*, **24** (3), 217–223.
29. Nagy, A.J., Mestl, G., and Schlögl, R. (1999) The role of subsurface oxygen in

the silver-catalyzed, oxidative coupling of methane. *J. Catal.*, **188** (1), 58–68.
30. Buevskaya, O.V. *et al.* (1987) Activation of hydrocarbons in the oxidative dimerization of methane over alkaline earth metals. *React. Kinet. Catal. Lett.*, **33** (1), 223–227.
31. Sokolovskii, V.D. *et al.* (1989) Type of hydrocarbon activation and nature of active sites of base catalysts in methane oxidative dehydrodimerization. *Catal. Today*, **4** (3–4), 293–300.
32. Wolf, E.E. (ed.) (1992) *Methane Conversion by Oxidative Processes: Fundamental and Engineering Aspects*, Van No strand Reinhold Catalysis Series, Van Nostrand Reinhold, New York, p. 123.
33. Trimm, D.L. (1980) *Design of Industrial Catalysts*, Chemical Engineering Monographs, Vol. 11, Elsevier Scientific Publishing Co., Amsterdam, Oxford, New York.
34. Richardson, J.T. (1989) *Principles of Catalyst Development*, Plenum Press, New York, London.
35. Hegedus, L.L. *et al.* (1987) *Catalyst Design: Progress and Perspectives*, John Wiley & Sons, Inc., New York.
36. Cawse, J.N. (2003) *Experimental Design for Combinatorial and High Throughput Materials Development*, John Wiley & Sons, Inc., Hoboken, New Jersey.
37. Hagemeyer, A., Strasser, P., and Volpe, A.F. (eds) (2004) *High-Throughput Screening in Heterogeneous Catalysis*, Wiley-VCH Verlag GmbH & Co. KGaA, Weinheim.
38. Baerns, M. (2004) *Basic Principles in Applied Catalysis*, Springer Series in Chemical Physics, Vol. 75, Springer, Berlin, Heidelberg.
39. Senkan, S. (2001) Combinatorial heterogeneous catalysis – a new path in an old field. *Angew. Chem. Int. Ed.*, **40** (2), 312–329.
40. Grubert, G. *et al.* (2003) Fundamental insights into the oxidative dehydrogenation of ethane to ethylene over catalytic materials discovered by an evolutionary approach. *Catal. Today*, **81** (3), 337–345.
41. Buyevskaya, O.V. *et al.* (2001) Fundamental and combinatorial approaches in the search for and optimisation of catalytic materials for the oxidative dehydrogenation of propane to propene. *Catal. Today*, **67** (4), 369–378.
42. Rodemerck, U. and Baerns, M. (2004) in *Basic Principles in Applied Catalysis*, Springer Series in Chemical Physics, Vol. **75** (ed. M. Baerns), Springer, Berlin, Heidelberg. 246–279.
43. Moehmel, S. *et al.* (2008) New catalytic materials for the high-temperature synthesis of hydrocyanic acid from methane and ammonia by high-throughput approach. *Appl. Catal. A: Gen.*, **334** (1–2), 73–83.
44. Schunk, S.A. *et al.* (2004) in *High-Throughput Screening in Heterogeneous Catalysis* (eds A. Hagemeyer, P. Strasser, and A.F. Volpe), Wiley-VCH Verlag GmbH & Co. KGaA, Weinheim, pp. 19–61.
45. Cong, P. *et al.* (1999) High-throughput synthesis and screening of combinatorial heterogeneous catalyst libraries. *Angew. Chem. Int. Ed.*, **38** (4), 483–488.

Part V
Characterization Methods

In the first two chapters in-situ spectroscopic methods for the study of homogeneous and heterogeneous catalytic systems are presented. Several applications are discussed. These two first chapters are followed by a chapter on techniques for porous system characterization. The final chapter contains practical instructions how to properly test a heterogeneous material for catalytic reactivity.

23
In-situ Techniques for Homogeneous Catalysis
Detlef Selent and Detlef Heller

23.1
Introduction

The classical method for following reactions and to measure concentrations as a function of time is to take samples during the course of a reaction and to analyze them by means of GC, HPLC, NMR, or other methods. In doing so, both the sampling in different temperature and pressure regimes and the analysis of the samples can be automated. The advantage of this approach is that intermediates can be individually detected in time, making them accessible for mechanistic and kinetic interpretations. On the other hand, the relatively large effort is a disadvantage. Moreover, this method is less suitable for fast reactions, and an often underestimated problem is that of completely terminating a reaction after sampling and preventing a change in the nature and distribution of catalytically relevant intermediates.

The typically high substrate/catalyst ratio which is characteristic of homogeneous catalytic reactions and the fact that one is working most of the time under inert condition often make the analysis of the catalyst difficult. Moreover, the precatalyst, has to be transformed completely into catalytically active species which are often in equilibrium with other complexes relevant to the catalytic cycle.

Therefore, it is much more elegant to follow a homogeneously catalyzed reaction *in situ*. In this way, stationary states of the catalytic cycle can be characterized and/or variations of substrate and product concentrations in time can be detected directly and kinetically evaluated. Additionally, the large amount of generated data is a clear advantage.

The latin term '*in situ*' essentially means 'at the place of origin.' Conditions are 'quasi stationary' when the temporal variation in concentration is about zero, that is, the concentration is practically constant. In contrast to stoichiometric reactions, quasi stationary conditions are typical for catalytic reactions, especially when very high substrate/catalyst ratios are present. More recently, the term '*operando spectroscopy*' is to be be found in literature. It is used for the commonly used *in-situ* spectroscopic methods with the emphasis on the characterization of the catalyst under real working conditions.

Catalysis: From Principles to Applications, First Edition. Edited by Matthias Beller, Albert Renken, and Rutger van Santen.
© 2012 Wiley-VCH Verlag GmbH & Co. KGaA. Published 2012 by Wiley-VCH Verlag GmbH & Co. KGaA.

This chapter on *in-situ* techniques for homogeneous catalysis focuses on methods for recording gas uptakes and gas formations as well as on selected examples of NMR-, IR-, and UV/Vis spectroscopy. It should, however, be pointed out that a complete overview is not given. For the interested reader we recommend therefore Ref. [1], while Ref. [2] deals especially with *in-situ* NMR spectroscopy.

Other methods, such as, for example, X-ray absorption spectroscopy (XAS) and X-ray fluorescence spectroscopy (XAFS, EXAFS, XANES) have for a long time been standard techniques for heterogeneous catalysis and are more and more being combined with other methods [3]. In homogeneous catalytic reactions, XAS and XAFS can provide access to very fast processes, because the diffraction process is shorter (femtosecond range) than the time necessary for exchanging atomic positions. This method, however, provides only time-averaged information concerning the environment of a coordination center. Applying these methods to the gold-catalyzed oxidative formation of esters from aldehydes and alcohols showed that not only are different gold complexes present but also, for the first time, that gold nanoparticles, which are typically necessary for oxidation reactions, are absent [4]. A more detailed description of X-ray absorption methods, coupled with various other spectroscopic techniques, has been published, together with an investigation of the Michael addition homogeneously catalyzed by iron [5].

Sophisticated measuring techniques which give a deeper insight into catalytic reactions should also be mentioned. One example is PHIP-NMR spectroscopy (PHIP = *para* hydrogen induced polarization) [2]. Via signal amplification, PHIP enables considerably improved detection of intermediates in reactions in which hydrogen is involved. A prerequisite is the pair-wise transfer of the polarized hydrogen molecule as, for example, in the case of the transition-metal mediated hydrogenation reaction [6]. In several cases, this method confirmed the presence of dihydrido rhodium complexes in rhodium-catalyzed hydrogenations [7]. An improved design for the generation of hyperpolarized samples is described in a recent patent, which contains a flow-through cell for the repeated polarization of the sample in a gas mixer [8].

Recent important developments are also described, including principal component analysis and factor analysis. These methods are used especially in the analysis of reaction spectra and are discussed here using the examples of IR- and UV/Vis spectroscopy.

23.2
In-situ Techniques for Homogeneous Catalysis

When choosing a suitable *in-situ* technique it should be kept in mind that with several methods, such as gas-uptake measurements, UV/Vis spectroscopy, and calorimetry, only overall effects are recorded. In contrast, molecule-specific signals are obtained with methods such as NMR and IR spectroscopy. It must also be realized that some methods are more suitable for determining the conversion of the substrate or product, while others are more suitable for characterizing the catalyst. Moreover, signals obtained from *in-situ* techniques can be divided into *integral* measurands (e.g., hydrogen consumption in a hydrogenation reaction) and *differential* measurands (e.g., registration of the rate of hydrogen consumption by means of a flowmeter or via the heat flow obtained by making use of a calorimeter).

23.3
Gas Consumption and Gas Formation

In many homogeneously catalyzed reactions, gaseous reaction partners are involved, either as a starting material (e.g., in hydrogenation, hydroformylation, or carbonylation reactions), or as a product (e.g., the production of hydrogen as an alternative energy carrier).

There are many techniques of following gas uptake or gas formation, and these generally differ according to whether applications are at normal pressure or higher pressure. One can work either under isobaric or isochoric conditions. The isobaric mode has the advantage that the concentration of the gaseous reactant is constant, which generally simplifies the corresponding kinetics.

Following gas uptake under normal pressure is relatively easy, and the general setup of the equipment is fairly standard. Isobaric conditions are guaranteed by adjusting pressure differences resulting from a proceeding reaction, either by decreasing reaction volumes or by supplying the consumed gas from a storage vessel.

If necessary, deviations from the ideal gas behavior have to be taken into account, and pV/p-isotherms give a good overview. The virial development of the thermal equation of state allows us to calculate (according to Eq. (23.1)) precisely the amount of a given gas (n_{gas}) in a storage vessel of known volume and consequently also the gas consumption [9].

$$n_{gas} = \frac{\text{volume of the storage vessel}}{\text{real molar volume}}$$

$$\text{real molar volume} = \frac{R \cdot T}{p} + B + \left(\frac{C - B^2}{R \cdot T}\right) \quad (23.1)$$

R = gas constant, T = temperature, p = pressure
B and C = virial coefficients.

Flowmeters are frequently used for higher operating pressure ranges. One can distinguish between volume and mass flow measurements. A pressure- and temperature-dependent calibration is necessary for volume flow measures in case of non-ideal behavior of the gas. This is also the case for the heat conductivity, which is often used as a measurand in mass flow controllers. The real pressure- and temperature-dependent measurement of a small gas mass flow on the laboratory scale (e.g., by means of the gyrostatic principle [10]) is still currently problematic for gases with a small molecular weight, such as hydrogen.

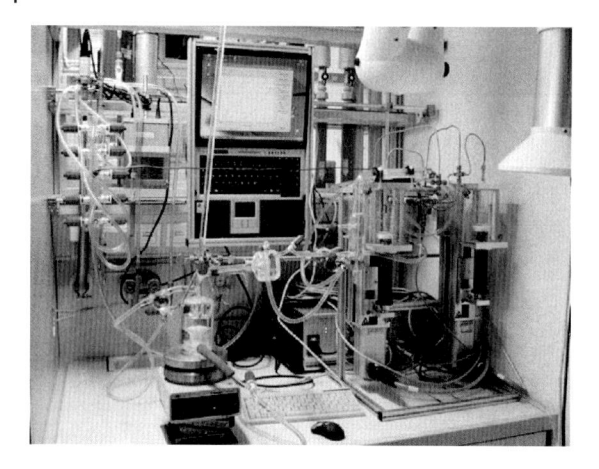

Figure 23.1 Normal pressure equipment for the automated detection of infinite gas consumptions and gas formations under isobaric conditions.

If larger gas consumptions have to be recorded, such as for the polymerization or trimerization of ethylene, the time-dependent loss of weight of a storage vessel due to gas consumption can be determined. This is a simple, yet sufficiently accurate method for following a catalytic reaction in time.

In the following, typical sources of errors and problems are discussed for the equipment we used for the measurement of infinite gas consumptions and gas formations under normal pressure and isobaric conditions (Figure 23.1) The equipment has been developed and constructed in cooperation with the company Messen Nord (www.messen-nord.de) and is commercially available there.

The basic principle for the realization of isobaric conditions is demonstrated for the equipment shown in Figure 23.1. The volume of the gas phase above the reaction solution is altered by means of a gas-tight syringe (not commercially available). In this way, the atmospheric pressure as a reference in the reaction vessel is constantly guaranteed. By using two syringes, which can be switched over by means of magnetic valves infinite gas consumptions and gas formations can be determined.

A general problem is the influence of transport phenomena when following gas-consuming reactions. The gaseous reaction partner has to be transported across the gas–liquid interface toward the catalyst much faster than it is consumed by the chemical reaction. Only then can the measured data be interpreted in terms of kinetics.

There are several possibilities to determine the influence of transport phenomena. Equation (23.2) describes the rate of physical absorption of a gas by a given liquid in the absence of a chemical reaction.

$$\frac{d[C]}{dt} = k_L \cdot a \cdot ([C^*] - [C]) \quad \text{or} \quad [C] = [C*] \cdot (1 - e^{-k_L \cdot a \cdot t}) \tag{23.2}$$

k_L = physical mass-transfer coefficient (liquid side)

a = interfacial area
$[C^*]$ = gas concentration in solution at time $t\infty$ (gas solubility), and
$[C]$ = gas concentration in solution at time t.

As a rule of thumb, the $k_L \cdot a$ value should be approximately 10 times higher than the rate of the gas consuming chemical reaction in order to exclude any effects of diffusion [11].

The frequently used method for the determination of the initial rate of a reaction as a function of the stirring speed should be evaluated with care for two reasons. On the one hand the initial rate of a reaction could be smaller than the rate at higher conversions due to induction periods. On the other hand, an increase in the stirrer speed does not necessarily lead to an increase in the interface.

An easily viable alternative is the following method: a diffusion-determined regime leads to a zeroth order reaction and the rate of the reaction exclusively depends on the constant concentration gradient at the interface. Consequently, reactions following zeroth order in a kinetically controlled regime are especially suitable for the systematic investigation of influences of diffusion phenomena [12]. The pseudo-rate constant is thus a linear function of the applied catalyst concentrations. A continuous increase in catalyst concentration under given reaction conditions (geometry of the reactor, stirrer speed, and size of the stirrer) leads to linear gas uptake curves, independently of the regime (kinetics versus diffusion). Plotting the slopes of the straight lines as a function of the catalyst concentration gives information about the boundaries of the exclusively kinetically controlled regimes.

Another source of error is the vapor pressure of the solvent. While this plays a minor role at high pressures, neglecting the vapor pressure in reactions with gaseous reaction partners at normal pressure is often an underestimated problem. It should be kept in mind, however, that substrates can also contribute significantly to the total pressure of the system even for reactions under pressure. The hydroformylation of 2-pentene is such an example.

The vapor pressure of dichloromethane (CH_2Cl_2) is 580 mbar at $T = 25\,°C$, while that of toluene ($C_6H_5CH_3$) is only 38 mbar. Consequently, more than half of the gas phase consists of solvent vapor in the case of dichloromethane at normal pressure, while it is only approximately 4% in the case of toluene. In other words, the partial pressures of a reactive gas above the solvents toluene and CH_2Cl_2 differ by a factor 2.25 at 1.01 bar/25 °C, and a factor 1.37 at 2.02 bar/25 °C. This has to be taken into account for the serious comparison of catalyst activities in different solvents at low total pressures.

Another problem is the solubility of gases in different common solvents. Table 23.1 illustrates the *effective* solubilities of hydrogen in different solvents at $T = 25\,°C$ and a total pressure of 1.01 bar above the solution [13]. Generally, gas solubilities are given in molar fractions. A small molar fraction basically corresponds to the molar ratio. These values can easily be converted into gas concentrations taking the molar volume and the density of the solvent into account. Consideration of the vapor pressure of the solvent leads to an 'effective' gas solubility, which corresponds to the gas concentration in solution.

Table 23.1 Effective hydrogen solubility (in mmol L^{-1}) for different solvents at 25.0 °C and 1.01 bar total pressure above the reaction medium.

THF	MeOH	i-PrOH	H$_2$O	Toluene
2.62	3.33	3.27	7.56	2.85

These results show that the hydrogen concentrations vary significantly in different solvents at normal pressure, and the effective solubility is affected by both the physical gas solubility and the vapor pressure of the solvent. Although THF and iso-PrOH have under the mentioned conditions an almost identical molar fraction solubility (2.70 × 10^{-3} versus 2.66 × 10^{-3}), the effective hydrogen concentrations differ by 20% due to the different vapor pressure. On the other hand, H$_2$O and toluene have very similar vapor pressures at $T = 25$ °C, but the effective hydrogen concentrations differ by a factor of 2.7 due to the different physical solubilities.

Practical solutions for minimizing the above-mentioned systematic errors, which result from transport phenomena, vapor pressure of the solvent, gas solubilities, and characteristic gas uptake curves and their kinetic interpretation, can be found in Ref. [12].

23.4
NMR Spectroscopy

NMR spectroscopy is an exceptionally important tool for the investigation of homogeneously catalyzed reactions. Next to detailed information concerning substrate/product distributions, this technique provides extensive structural information on catalyst precursors and catalytically active intermediates. NMR measurements can be done time-resolved and at varying temperature. The study of different NMR active nuclei present in the samples under investigation and their interaction is of high diagnostic value and has been termed *Multinuclear NMR Spectroscopy*.

In-situ spectroscopic investigations of homogeneously catalyzed reactions with participation of gases such as CO, H$_2$, CH$_2$=CH$_2$, and O$_2$ are especially valuable. As already discussed under Chapter 23.3 the solubility of a gas in the liquid phase is an important parameter, which effects the interpretation of mechanistic details and kinetic data. Applying gas mixtures, the partial pressure ratio according to a certain stoichiometry in the head space of the reactor is not guaranteed in solution. Thus, a mixture with a composition of approximately CO/H$_2$ = 2 : 1 is present in toluene during the hydroformylation of olefins with CO/H$_2$ = 1 : 1 at $T = 100$ °C, the respective Henry constants being $K_H(H_2) = 24.89 \pm 1.26$ and $K_H(CO) = 11.94 \pm 0.27$ MPa L mol^{-1} [14].

Special requirements apply to the spectroscopic equipment used for the investigation of such reactions. The design should be chosen such that the gas concentration in the sense of a saturated liquid phase is guaranteed constant over a large range of conversion. Any deviation from the saturation concentration can lead to shifts of the equilibria and consequently to changes in the ratio of intermediates. This has to be taken into account when interpreting reaction mechanisms.

The constancy of the hydrogen concentration in an NMR tube can, for example, be checked time-resolved by means of the intensity of the signal at $\delta = 4.44$ ppm (toluene-d8). These investigations can be done under normal pressure and with compressed gases. A normal 10 mm NMR tube can be modified in such a way that a gas can be introduced into the sample without limit. Gas uptakes of up to $0.1\,\text{mL}\,\text{min}^{-1}$ can be tolerated without influencing the rate of a reaction by diffusion [15]. In this way it was proven that the catalyst precursor Rh[(DIPAMP)(COD)]BF_4 (DIPAMP = 1,2- bis[(2-methoxyphenyl)(phenylphosphino)]ethane, COD = 1,5-cyclooctadiene), which was used for the hydrogenation of (Z)-2-(N-acetylamino) cinnamic acid methyl ester (AMe) in methanol, reacts only slowly and is still detectable after full conversion of the substrate next to the catalytically active species Rh[(DIPAMP) (methanol)$_2$]BF_4 (Figure 23.2).

Apparently, the catalytic reaction and the formation of the catalyst are processes which run in parallel for a certain period of time, while only a fraction of the applied rhodium is catalytically active for the asymmetric hydrogenation of the substrate. For a precise kinetic description and the comparison of different catalysts a complete hydrogenation of the catalyst precursor is therefore a prerequisite.

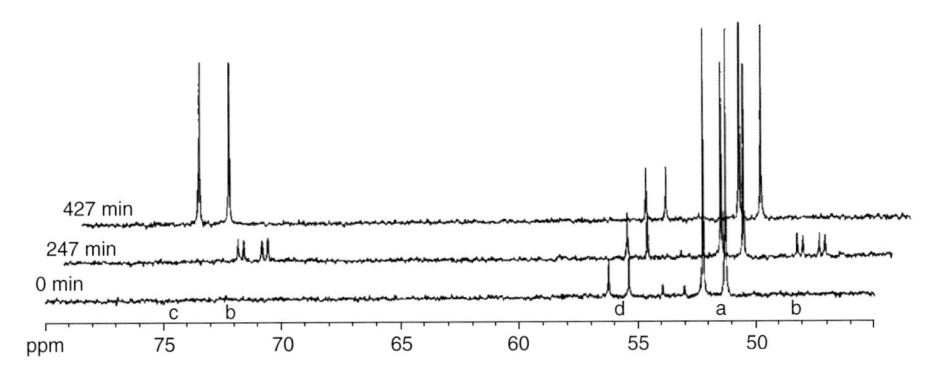

Figure 23.2 ^{31}P NMR spectra recorded during the hydrogenation of AMe: (a) precursor [Rh(DIPAMP)(COD)]BF_4; (b) [Rh(DIPAMP)(AMe)]BF_4; (c) [Rh(DIPAMP)-(methanol)$_2$]BF_4; and (d) [Rh(DIPAMP)$_2$]BF_4 as minor impurities.

Suitable pressure gas flow NMR cells are not commercially available, and consequently in-house developments are common. Until now, their application has been limited to a relatively small number of users, also taking into account that special requirements are necessary. Reviews dealing with the development of the methodology have been published [1, 16]. An elegant and effective solution is the development of an NMR cell which acts as a bubble reactor. In this way, the CO-pressure-dependent dissociation of a carbonyl-bridged hetero bimetallic ruthenium-rhodium complex has been studied in detail (Figure 23.3) [17].

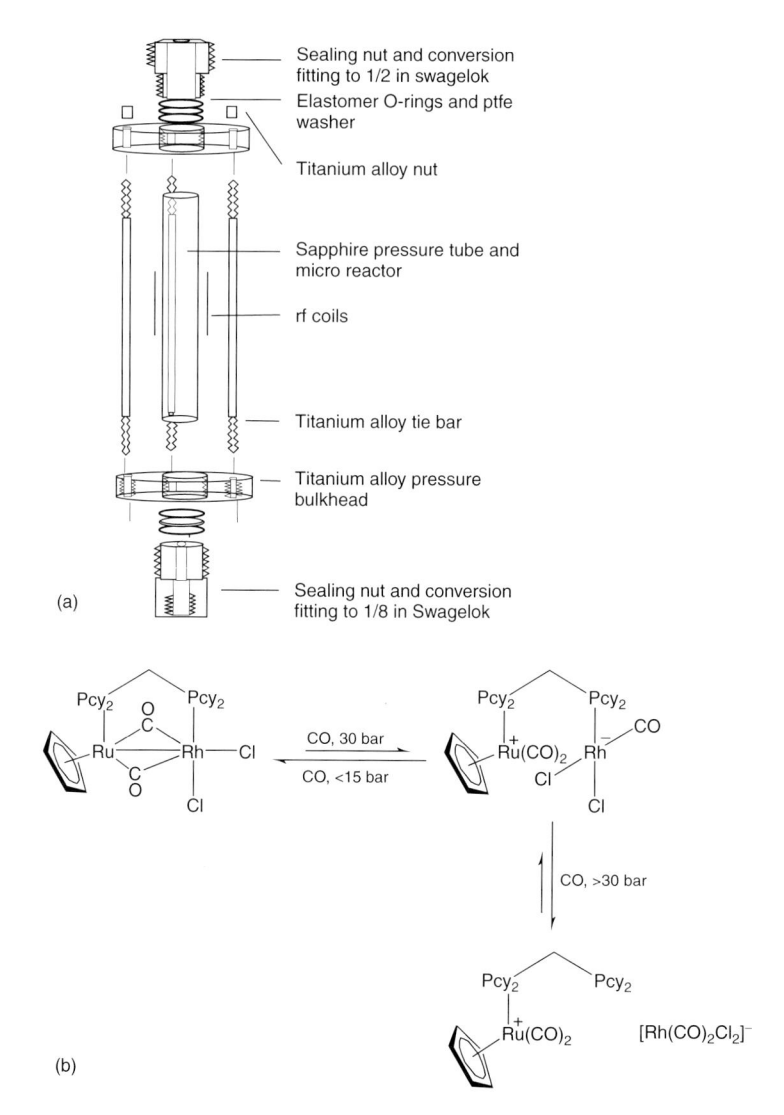

Figure 23.3 (a) Exploded view of an NMR gas flow cell. (b) Pressure induced dissociation of a carbonyl-bridged Ru-Rh heterometallic complex at variable CO pressure >30 bar, as proved by *in-situ* NMR spectroscopy.

The need for a dynamic gas insertion is impressively shown in Figure 23.4. This graph shows the saturation curve of hydrogen in deuterated benzene as a result of diffusion through a static interface between liquid and gas phase at 100 bar within the NMR cell [18].

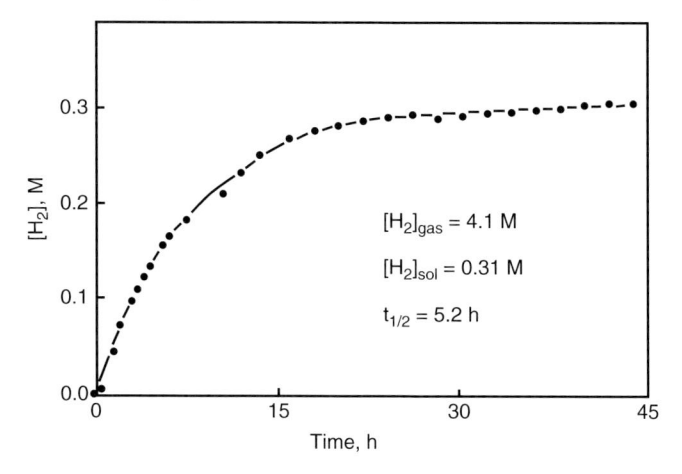

$[H_2]_{gas} = 4.1$ M

$[H_2]_{sol} = 0.31$ M

$t_{1/2} = 5.2$ h

Figure 23.4 Saturation of hydrogen in benzene-d_6 in an NMR cell without stirring; $p = 100$ bar, $T = 22\,°C$.

Even in the commonly used sapphire pressure NMR cells according to the design of Roe, several hours are necessary to generate a saturated solution due to the small specific exchange surface [19]. This, however, is circumvented by pressurizing the NMR cell several hours prior to the actual measurement [20].

More recently, an easy-to-handle gas flow cell with gas insertion via a capillary tube and gas feedback has been designed, this being based on a 10 mm sapphire NMR pressure tube. This cell is especially suitable for investigating reactions under defined pressure, and ensures gas saturation with a small gas uptake per time unit as in the case of subsequent catalyst preformation, or for the characterization of catalysts under controlled conditions. Instead of the commonly used one-way valve with gas connection, this flow cell carries a titanium adapter containing two stainless steel capillaries which fit together coaxially [21, 22]. The inner capillary is made of Teflon and passes through the measuring range of the NMR coil and the solution to the bottom of the NMR tube. The outer capillary is used for the back-flow of the gas. A pumping system, which is positioned next to the NMR magnet, causes gas circulation, while gas flow and pressure can be regulated precisely. This dynamic gas insertion allows the saturation of a 2 mL sample with hydrogen within 2 min under pressures up to 50 bar (Figure 23.5). Moreover, the NMR cell remains in position in case the used gas has to be exchanged.

In contrast to measurements in commercially available 10 mm pressure NMR tubes, this gas flow cell is especially suitable for the determination of realistic preformation conditions for hydroformylation catalysts and consequently also for the comparison of different catalyst precursors. Under static conditions, for

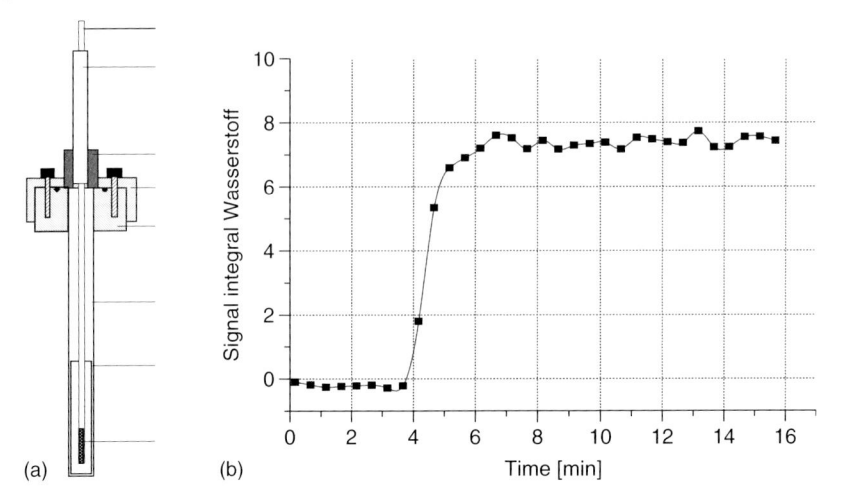

Figure 23.5 (a) Gas flow cell based on a commercially available sapphire NMR tube, Ref. [21]. (b) Saturation kinetics of hydrogen ($\delta = 4.44$ ppm) in toluene-d8 (2 mL) at 25 °C. The hydrogen is added as a gas mixture ($CO/H_2 = 1:1$) at $p_{total} = 19$ bar and a gas flow of 3 mL min^{-1}.

example, the complete hydrogenolysis of the Rh(0) dimer [$Rh_2(\mu\text{-}CO)_2(P\cap P^*)_2$] ($P\cap P^*$ = asymmetrical bidentate diphosphite ligand) at 25 °C and 20 bar syngas required as long as 16 h. By making use of the dynamic gas insertion, the rate of hydrogenolysis is not limited by diffusion, and thus the pure hydrido complex [$HRh(CO)_2(P\cap P^*)$] is present after only 5 min (Figure 23.6). With insertion of gas, the allyl complex [$(\eta^3\text{-}C_3H_5)Rh(P\cap P^*)$] requires 2 h reaction time and a temperature of $T = 60\,°C$ in order to form quantitatively the hydrido complex. In comparison, the Rh(0) dimer is therefore an ideal precursor, which can be rapidly transformed into the catalyst without the formation of by-products [23].

The identification of the structure of hydroformylation catalysts by means of NMR spectroscopy under reaction conditions remains an interesting field of activity. Bidentate ligands based on carbohydrates, calixarenes, and Taddol have been investigated, among others [24].

Two mechanisms – the hydrido and the carboxy cycle – have been proposed for the palladium-catalyzed hydroalkoxycarbonylation of olefins. *In-situ* NMR spectroscopic investigations have shown that the hydrido mechanism dominates for the catalyst system $Pd(II)/1,3\text{-}(^iBu_2P)_2C_3H_6$ (dibpp) (see comparison in Scheme 23.1) [25].

Figure 23.6 (a) Formation of a hydroformylation catalyst from the corresponding Rh(0) dimer containing an unsymmetrically substituted, bidentate diphosphite ligand. Reaction time under static gas feed: 16 h. Time requirement when applying a gas flow cell as shown in Figure 23.11: 5 min. (b) ^{31}P NMR spectrum (proton-coupled) of the Rh(I) hydrido complex (mixture of diastereomers).

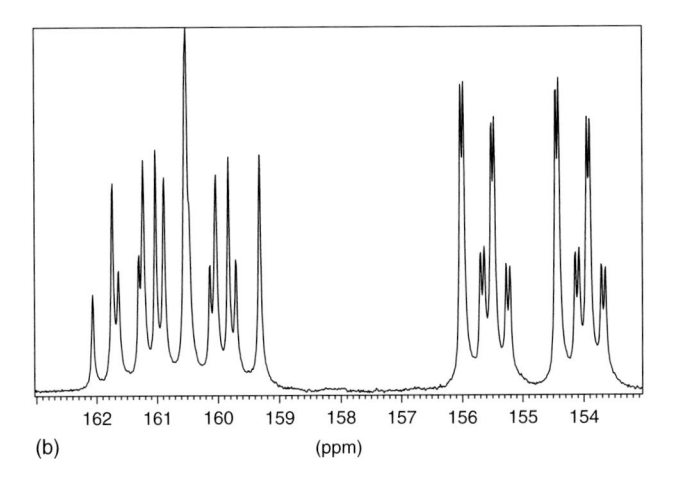

(a)

(b)

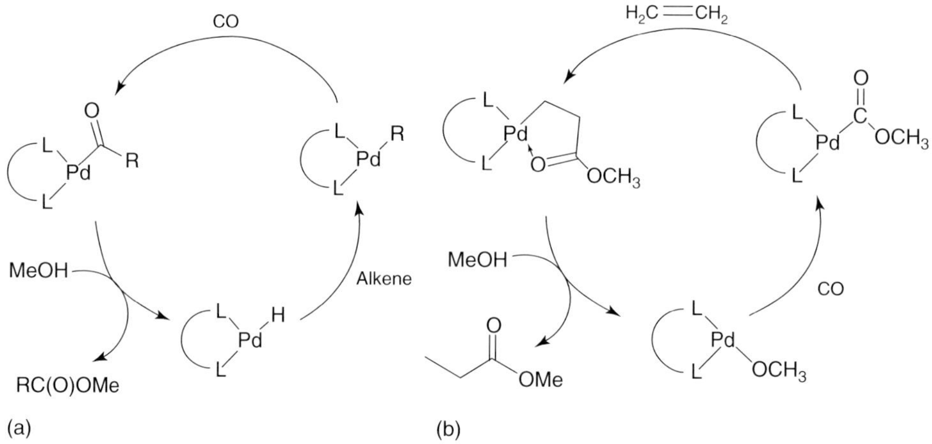

Scheme 23.1 (a) Hydride- and (b) carboxy cycle of palladium-catalyzed hydroalkoxycarbonylation.

In contrast to present interpretations the reason for the minor contribution of the carboxy cycle is not the hindered insertion of the alkene into the Pd-carboalkoxy bond, but the slow methanolysis of the corresponding ß-chelate. Additionally, both cycles involve as a side reaction the alternating insertion of alkene and carbon monoxide, respectively, based on either Pd-acyl or Pd-alkyl intermediates. In this way copolymers are formed as products, which contain either ketone or ester end groups. Figure 23.7 shows the mechanism of the copolymer formation via hydride cycle, and includes several in-situ ^{31}P NMR spectra which result from the reaction of the catalyst [CH$_3$Pd(dibpp)(CH$_3$CN)] with CO and ^{13}CO, respectively, and which document the formation of the intermediates **2** and **3**.

By combining NMR spectroscopy, isotope tracing, kinetic investigations, and isolation of intermediates, plausible mechanisms could also be proposed for several other reactions.

Brookhart and co-workers could, for example, show that the pincer-type Ir(III) complex fragment HIr[C$_6$H$_4$-2,6-(OP^{t}Bu$_2$)$_2$] occurs as an intermediate in several transfer dehydrogenation reactions as well as in the hydrosilylation of carbonyl compounds [26].

Kinetic as well as temperature-dependent in-situ NMR spectroscopic investigations are in line with a non-radical character of the hydrogenation of alkyl halides (Scheme 23.2) [26a,b]. Several substrate complexes of alkyl halides could be characterized spectroscopically.

The asymmetric hydrogenation of prochiral substrates can be catalyzed by cationic rhodium(I) complexes containing chiral ligands. The formation of Rh(III)-dihydrido complexes could be confirmed by means of low-temperature NMR spectroscopy on a complex containing (S,S)-1,2-bis(tert-butylmethylphosphino)ethane (^{t}Bu-Bis-P*) (see Scheme 23.3) [27a].

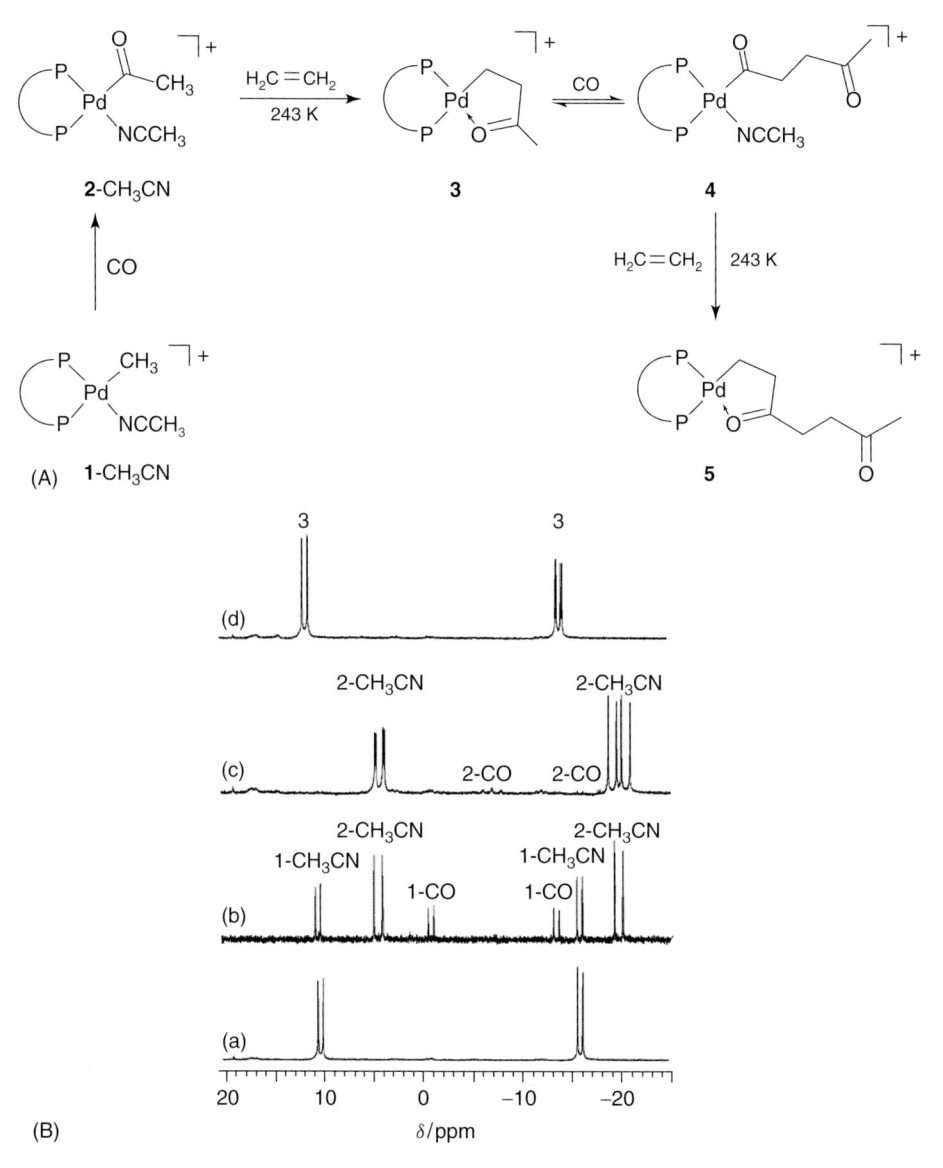

Figure 23.7 (A) Mechanism for the formation of an carbon monoxide ethylene copolymer. The acyl-Pd(II) complex 2-CH$_3$CN is part of the hydride cycle of the hydroalkoxy carbonylation. (B) Detected Pd(II) complexes by means of in situ ^{31}P NMR spectroscopy in CH$_2$Cl$_2$/CH$_3$CN (9 : 1): (a) 1-CH3CN (T = 193 K), (b) after addition of CO at 213 K, (c) after addition of ^{13}CO to a new sample and subsequent purging with N$_2$ at 213 K, and (d) after addition of C$_2$H$_4$ and subsequent purging with N$_2$ at 213 K.

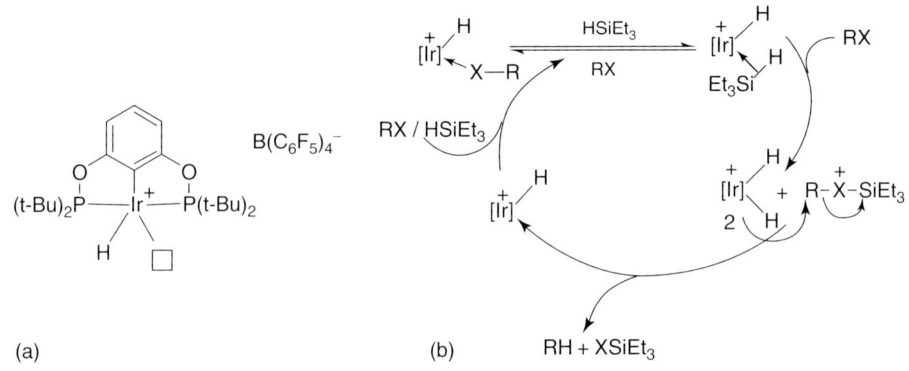

Scheme 23.2 Iridium catalyst (a) for transfer dehydrogenation and hydrosilylation, respectively, and proposed mechanism for the iridium-catalyzed hydrogenation of alkyl halides with triethyl silane (b).

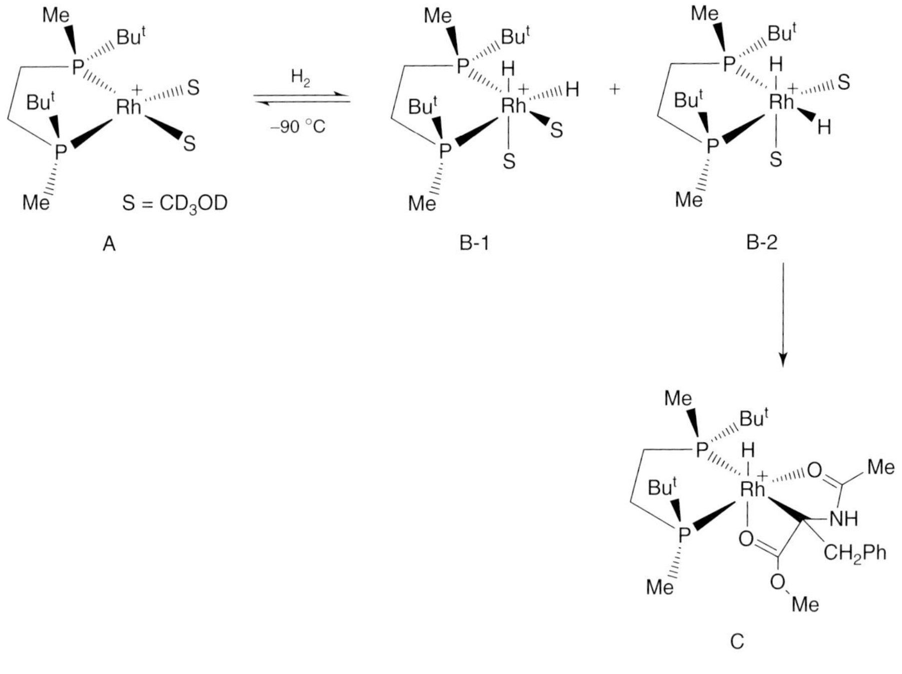

Scheme 23.3 Formation of the hydridoalkyl intermediate from the corresponding dihydrido solvent complex $[H_2Rh(tBu\text{-}BisP^*)(CD_3OD)_2]BF_4$ during the rhodium-catalyzed hydrogenation of AMe.

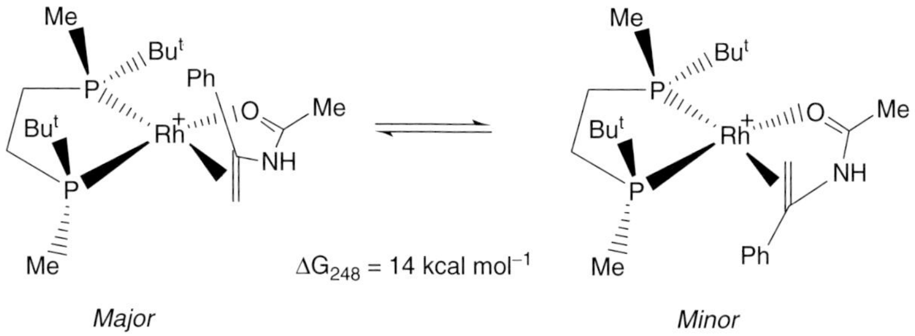

Scheme 23.4 Intramolecular equilibrium between the diasteromeric substrate complexes of [(tBu-BisP*)Rh((N)-(1-phenylvinyl)acetamide]BF$_4$.

The solvent complex A given in Scheme 23.3 is generated initially under a hydrogen atmosphere at $T = -50\,°C$ in deuterated methanol. At $T = -90\,°C$ approximately 20 mol% of the dihydride B containing the major isomer B-1 can be detected. The ^{1}H NMR B-1 shows two signals at $\delta = -7.7$ and -23.0 ppm, respectively. The corresponding protons are exchanging with one another as confirmed by ^{1}H 2D EXSY spectroscopy. In analogy to the literature, the authors discuss a non-detectable Rh(I) complex of molecular dihydrogen, which presumably corresponds to a real minimum on the energy hypersurface and is of similar energy to that of the resulting Rh(III)-dihydride complex [28].

Even though it is not clear whether or not complexes of type B contribute to the hydrogenation reaction at room temperature, it is nevertheless interesting that the reaction of B with AMe as substrate leads to the same alkyl complex as discussed generally for the reaction of the corresponding substrate complex with hydrogen.

The diastereomeric substrate complexes which are formed during the asymmetric hydrogenation of a phenyl enamide undergo both an intra- and intermolecular process of transformation between a *major* and a *minor* diastereomer (Scheme 23.4). According to ^{31}P-^{31}P EXSY-NMR spectroscopic investigations the intramolecular process is only operative at $T = -50\,°C$ (Figure 23.8a), while the transformation by substrate dissociation becomes increasingly important at higher temperatures (Figure 23.8b) [27b].

A deviant behavior from the *major/minor* principle can be observed in the asymmetric hydrogenation of specific β-acyl aminoacrylates. Crucial for this was the assignment of the dominant diastereomer in solution (diasteromeric ratio = 88 : 12) to the crystallographically characterized substrate complex [Rh(S,S-DIPAMP)(Z)-methyl-3-acetamido-3-(4-chlorophenyl)acrylate]BF$_4$ by means of low-temperature NMR spectroscopy [29]. Due to a marginal reactivity difference between the *major* and the *minor* complex ($k_{maj}/k_{min} = 0.3$) the hydrogenation of this β-amino acid precursor via the *major* complex becomes the preferred reaction path, which corresponds to the key-lock principle in enzyme catalysis.

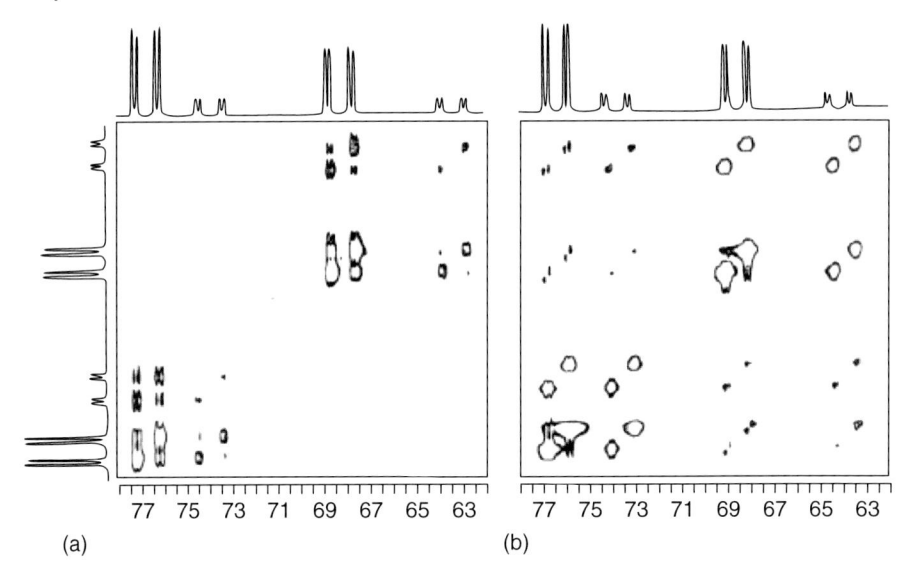

Figure 23.8 ^{31}P-^{31}P EXSY spectra (162 MHz, CD$_3$OD) of the diastereomers shown in Scheme 23.4 at $T = 223$ K (a) and at $T = 248$ K (b). The additionally observed cross peaks at 248 K account also for the presence of an intermolecular way of interconversion of diasteromers under complete substrate dissociation.

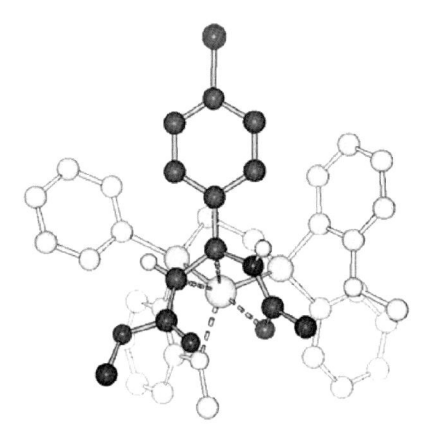

Figure 23.9 Molecular structure (cation) in the crystal of the major-substrate complex of [Rh(S,S-DIPAMP)(Z)-methyl-3-acetamido-3-(4-chlorophenyl)acrylate]BF$_4$, which determines the (S)-configuration of the product.

23.5
IR-Spectroscopy

A deep insight into catalytic reactions can be obtained when concentration profiles of substrates and products can be correlated with the ones of catalysts or intermediates of the corresponding catalytic cycle. The catalyst concentration, however, is often rather low. Moreover, a precise identification of intermediates is often complicated by signal overlap. Next to the sensitive PHIP-NMR method, FT-IR spectroscopy offers the possibility to observe catalysts. It increases the chance to detect short-lived species due to its higher sensitivity in comparison to routine NMR spectroscopy. Similarly to NMR spectroscopy, interesting developments for the investigation of homogeneously catalyzed reactions under pressure and gaseous substrates have occurred. As an example, the high-pressure IR autoclave, developed at the University of Amsterdam, should be mentioned here [1, 30].

Precious tools for processing signals and for the reconstruction of single spectra are principal component analysis and factor analysis, respectively [31]. Factor analysis allows access to single spectra of each component derived from a reaction spectrum measured at different wavelengths and without *a priori* information.

The attenuation of the total light is cumulatively composed of the contribution of each absorbing substance according to Bouguer–Lambert–Beer. In this way, a reaction spectrum can be regarded as a linear superposition of the total absorption of the absorbing species in the form of a matrix. An analysis of the matrix rank then gives the desired information. This analysis is usually already implemented in specific algorithms and programs. Special emphasis should be given to the BTEM-program (*Band-Target Entropy Minimization*), developed by Garland *et al.*, which impressively demonstrated the analysis of mixtures of different rhodium-carbonyl complexes and which allowed for the first time the identification of the hypothetical [HRh(CO)$_4$] species by means of FTIR spectroscopy [32].

Moreover, it was shown that the hydroformylation of 3,3-dimethyl-1-butene by a mixture of Rh$_4$CO$_{12}$ and HMn(CO)$_5$ follows the classical rhodium-catalyzed cycle, but the elimination of the aldehyde is also possible by a bimolecular reaction between the acyl-rhodium intermediate and the hydrido-manganese complex (Scheme 23.5). By detecting 1.5 ppm Rh$_6$CO$_{16}$, the BTEM method further allowed the identification of minor amounts of a carbonyl complex which is not directly involved in the catalytic cycle [33].

A detailed investigation of the hydroformylation of 3,3-dimethyl-1-butene catalyzed by a bulky monophosphite-modified rhodium complex was performed. The time-dependent concentrations of organometallic intermediates and organic components were followed by means of FTIR spectroscopy (Figure 23.10) and an automated sampling unit [34].

It was demonstrated that the hydrido complex (**1**) and the acyl-rhodium complex (**7**) shown in Figure 23.10 exist in parallel over a wide range of conversion. The acyl compound dominates at high olefin concentrations and can thus be regarded as a substrate complex according to Michaelis-Menten kinetics.

Similar behavior was observed by means of FTIR and NMR spectroscopy in the hydroformylation of 1-octene with a rhodium/phenylphosphonic acid diamide

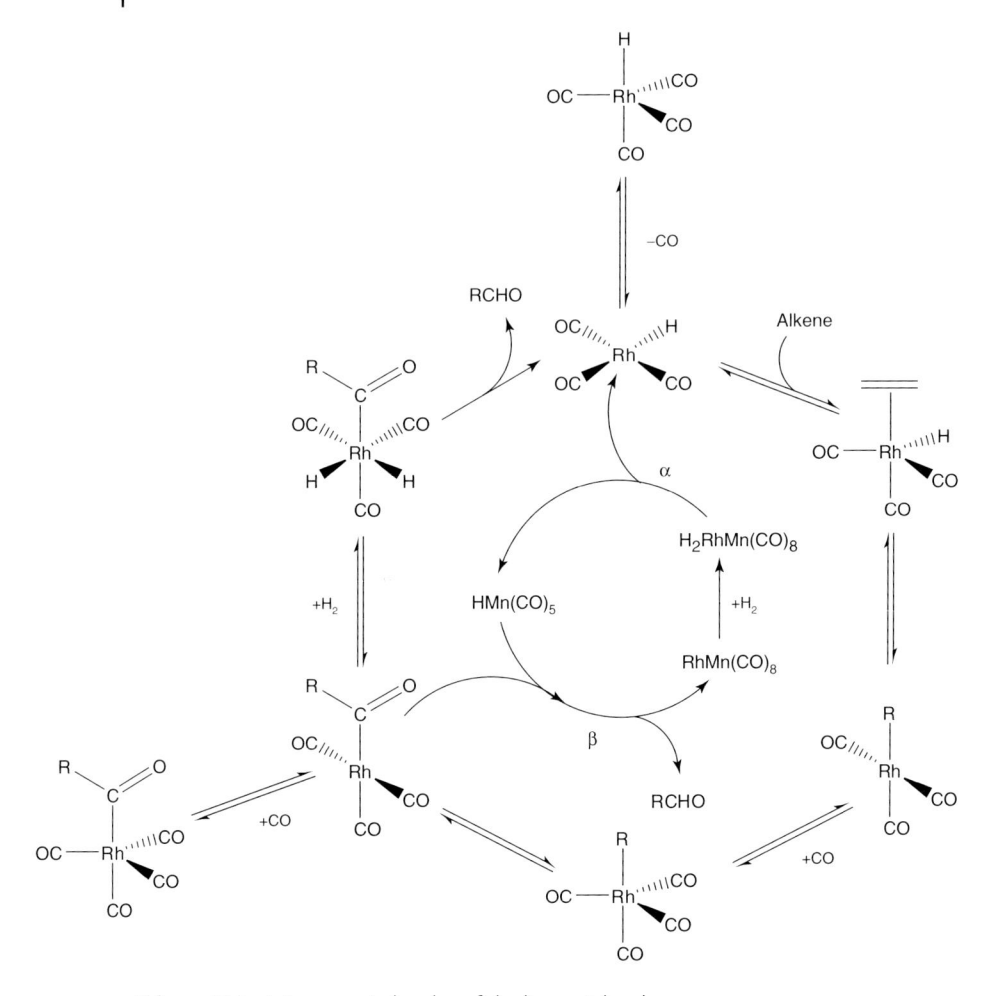

Scheme 23.5 Interconnected cycles of rhodium-catalyzed and Rh/Mn bimetallic hydroformylation of 3,3-dimethyl-1-butene as derived from BTEM analysis of FTIR spectroscopic data.

catalyst [35]. In spite of their bulkiness, two *ortho*-substituted mono-triarylphosphite ligands can coordinate to the hydrido complex (**1**).

In the case of 3,3-dimethyl-1-butene the system undergoes a shift from approximately zeroth order to first order in olefin during the course of the reaction along with an accumulation of the hydrido complex. The observed dynamic behavior of the system at $T = 70\,°C$ is illustrated in Figure 23.11.

The time-dependent profile of the observed organometallic intermediates could be extracted from the superimposed reaction IR spectra. For this purpose the program *Pure Component Decomposition* (PCD) was used as a new algorithm for the extraction of single components from superimposed spectra [36].

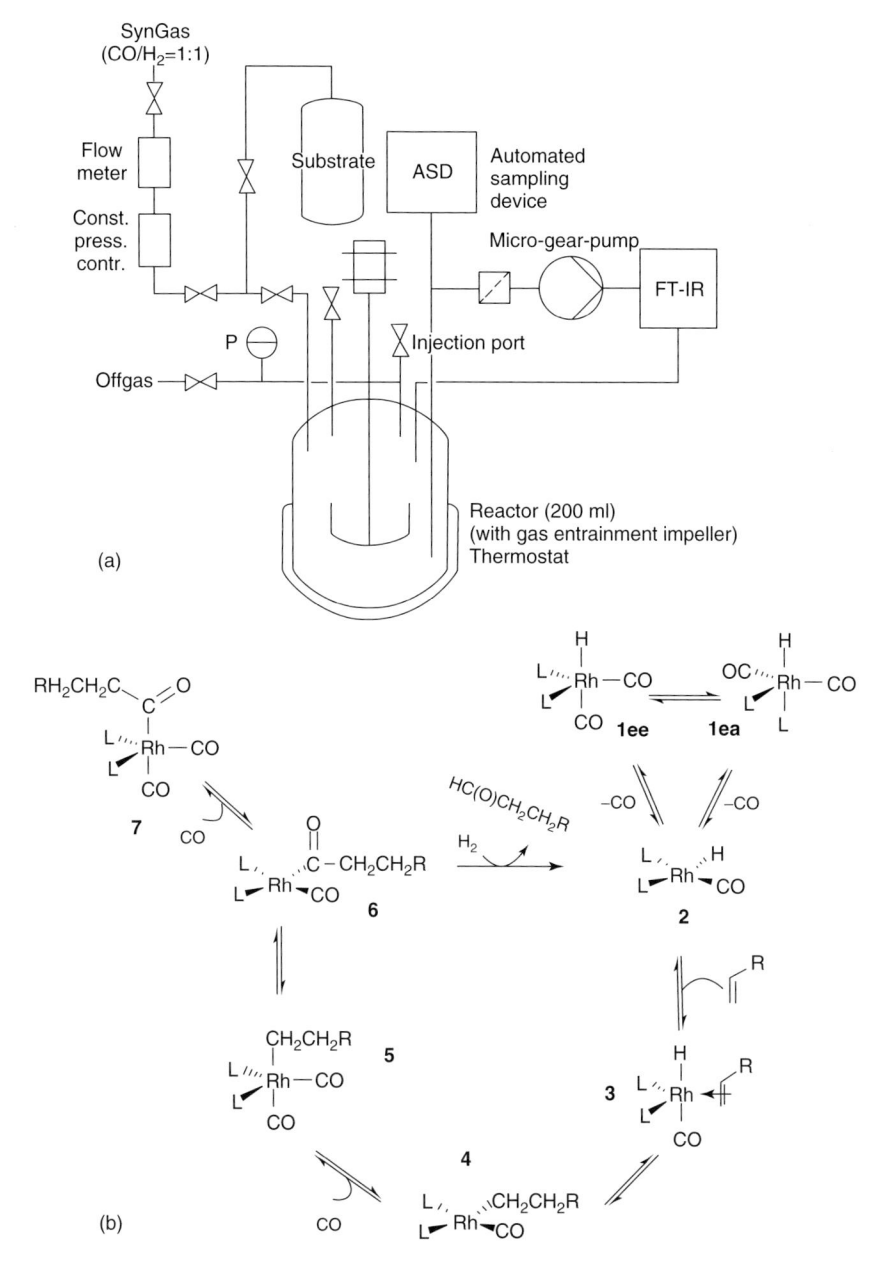

Figure 23.10 (a) Experimental setup used for *in-situ* FTIR spectroscopic study of hydroformylation of 3,3-dimethyl-but-1-ene. (b) Mechanism of the rhodium-catalyzed hydroformylation. Only the path for the formation of the linear aldehyde is shown. The hydrido complex 1 and/or the acyl complex 7 are observed as the resting states of the electronically unsaturated catalyst complex 2.

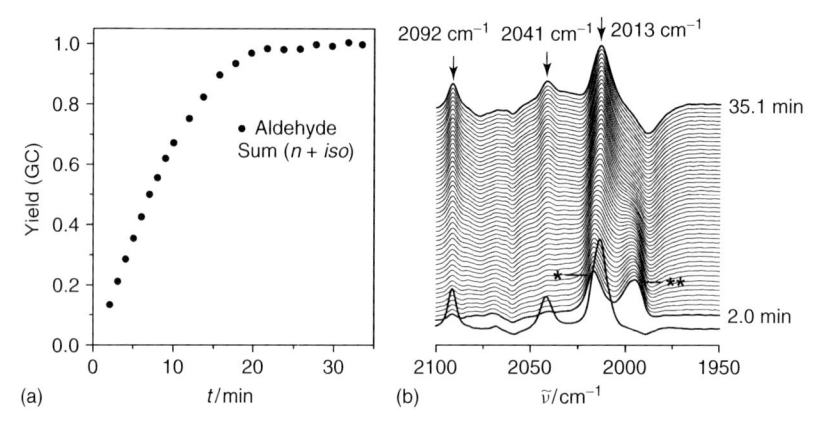

Figure 23.11 Hydroformylation of 3,3-dimethyl-1-butene with a Rh/monophosphite catalyst at 70 °C. (a) Aldehyde yield versus time curve (GC, similar data obtained from FTIR spectroscopy). (b) [HRh(CO)$_3$(phosphite)] after preformation (first spectrum), following series of FTIR spectra taken during the reaction. Signals (*), 2017 cm^{-1}, and (**), 1995 cm^{-1}, were assigned to the acyl complex [(CH$_3$)$_3$C(CH$_2$)$_2$C(O)Rh(CO)$_3$(phosphite)].

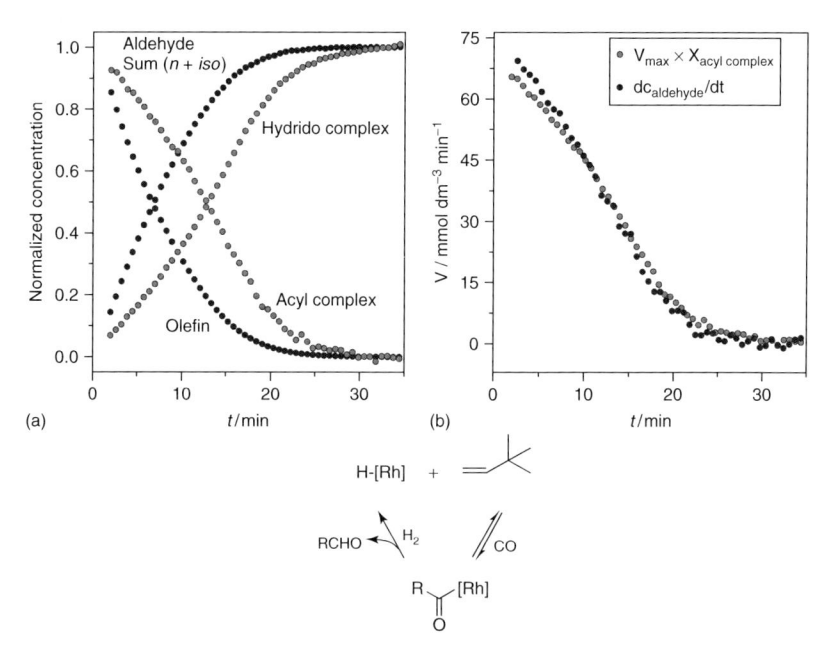

Figure 23.12 Hydroformylation of 3,3-dimethyl-1-butene with a Rh/monophosphite catalyst at 70 °C. (a) Normalized concentration profiles for [HRh(CO)$_3$(phosphite)] and [(CH$_3$)$_3$C(CH$_2$)$_2$C(O)Rh(CO)$_3$(phosphite)] from PCD. (b) Comparison of the rate profile for aldehyde formation, with the product of initial rate and the molar fraction of the acyl rhodium intermediate.

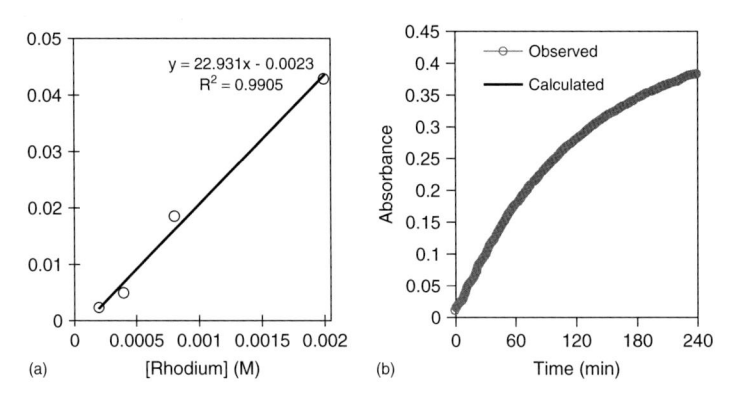

Figure 23.13 (a) Effect of the rhodium concentration on the hydroformylation of styrene with a Rh/(S,S,S)-bisdiazaphos catalyst at 80 °C. (b) The graph shows a comparison of the absorption of formed aldehydes with a calculated curve progression for a first-order dependency with respect to the styrene concentration.

Interestingly, the change in rate for the formation of the aldehyde correlates directly with the concentration profile of the acyl intermediate. This shows that the hydrogenolysis of this complex determines practically the reaction rate during the whole conversion under these reaction conditions (Figure 23.12).

Tracking the aldehyde formation in time is possible by means of both gas chromatography and the change in intensity of the corresponding carbonyl band of the aldehyde. If, as shown above, transmission cells are used for the additional detection of small amounts of rhodium carbonyl complexes, non-linear behavior or total absorption, respectively, in the region of the band maxima around 1740 cm^{-1} can occur during the course of the reaction. Alternatively, the determination of the concentration of the organic component is possible by a less sensitive ATR-IR probe (ATR = attenuated total reflection).

Figure 23.13 shows the hydroformylation of styrene, which follows first-order kinetics with a Rh/(S,S,S)-bis diazaphos catalyst. This reaction was followed with a ReactIR® system for investigating pressure effects on the regio- and enantioselectivity of the catalyst [37].

23.6
UV/Vis Spectroscopy

Nowadays it is possible to stir inside cuvettes to keep them at constant temperature, to use flow-through cuvettes, and to perform spectroscopy on very small samples in microcuvettes. By making use of submersible optrodes in essentially any dimensioned reaction vessel, it is possible to investigate homogeneously catalyzed reactions by means of UV/Vis spectroscopy *in situ* even under inert atmosphere and in different temperature regimes. Additionally it is simultaneously possible to follow other measurands, such as conductivity, pH-value, or gas uptakes under catalytic conditions. Figure 23.14 shows a combination of UV/Vis spectroscopy and hydrogenation.

In this way important information on the catalytic reactions can be obtained from the reaction spectra. The presence of isosbestic points (the wave-length at which the absorption does not change during the reaction) documents a kinetically 'uniform reaction,' and a single reaction variable is necessary for a complete description of the system [38].

In the following, the scope of *in-situ* UV/Vis spectroscopy is briefly discussed using as an example the determination of stability constants of rhodium-arene complexes and the formation of the catalytically active species by hydrogenation of a corresponding cationic rhodium(I) diolefin precursor.

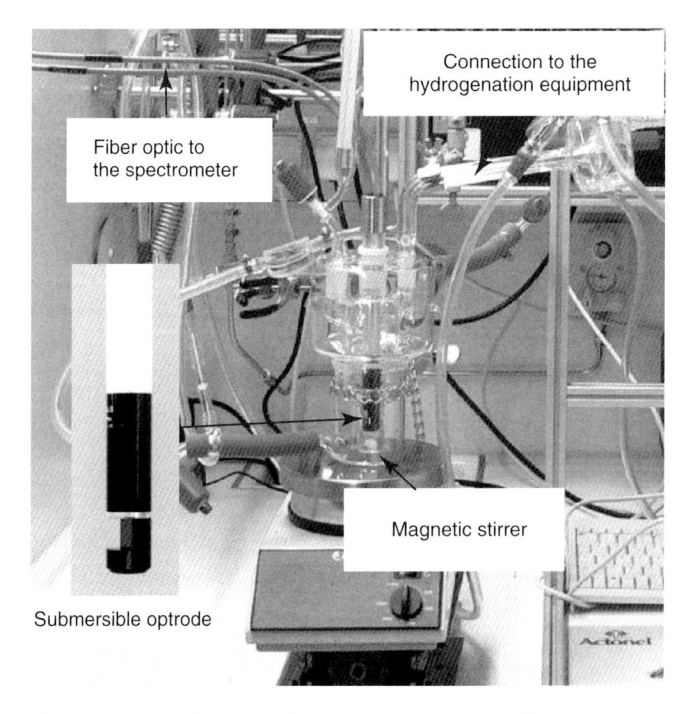

Figure 23.14 Combination of UV/Vis spectroscopy and hydrogenation by making use of a submersible optrode.

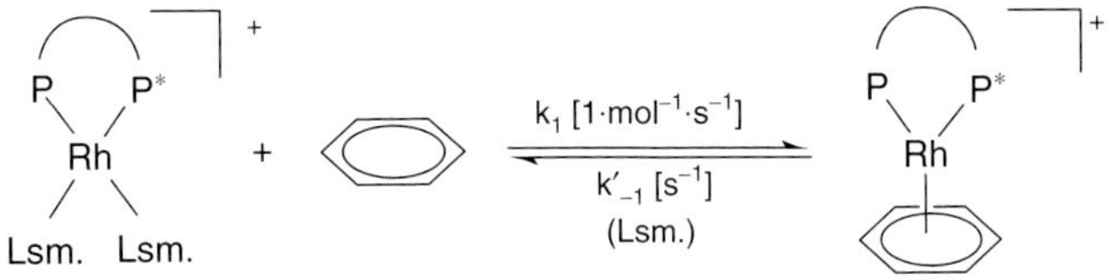

Scheme 23.6 Dynamic equilibrium between the solvent complex, free arene, and arene complex.

Rhodium-arene complexes have been known in the literature for a long time [39]. These complexes are relatively stable and are consequently often responsible for the deactivation of catalytically active cationic rhodium complexes in hydrogenation reactions. Unfortunately, literature data for the corresponding stability constants are rare and are usually based on the classical titration of the arene to the solvent complex. While the established equilibrium is measured directly in this case, the establishment of equilibrium as a function of time can be followed as an alternative. Additionally, this dynamic method allows the determination of the rate constants of the forward and back reactions, the ratio of which corresponds to the stability constant (Scheme 23.6).

Because the equilibria are established rapidly, the combination of a diode array and a stopped-flow technique was chosen. A glove box was used in order to guarantee anaerobic conditions [40].

As an example, the reaction spectrum for the establishment of equilibrium for the system [Rh(Et-DuPhos)(MeOH)$_2$]BF$_4$ and toluene in methanol is depicted in Figure 23.15 for a reaction time of approximately 1 min. The presence of an isosbestic point and the corresponding linear extinction diagrams confirm the existence of a kinetically uniform reaction [38].

The analysis of the spectra results in $k_1 = 1.51$ L mol^{-1} s^{-1} and $k'_{-1} = 1.01 \times 10^{-2}$ s^{-1}, respectively, which gives a stability constant of 149 L mol^{-1}. This value corresponds very well with the stability constant of 132 L mol^{-1} found by classical titration (D. Heller and C. Fischer, unpublished results.).

Reactions with gases as reactants are generally difficult to follow *in situ* in cuvettes under isobaric conditions, and also because argon has to be exchanged at a certain point by the reactive gas, such as hydrogen. This, however, is not a problem by making use of a submersible optrode, which is also available in a pressure-resistant design.

The so-called solvent complexes are the catalytically active species in the asymmetric hydrogenation reaction with rhodium-bisphosphine complexes. The solvent complexes are generated by stoichiometric hydrogenation of the diolefin precursors of the type [Rh(bisphosphine)(diolefine)]anion. Figure 23.16 shows the UV/Vis reaction spectra obtained by following *in-situ* the hydrogenation of COD in the complex [Rh(Me-DuPHOS)(COD)]BF$_4$ in methanol. The spectrum shows four isosbestic points in the depicted regime. They demonstrate the presence of a kinetically uniform reaction, which was confirmed by corresponding extinction diagrams [41].

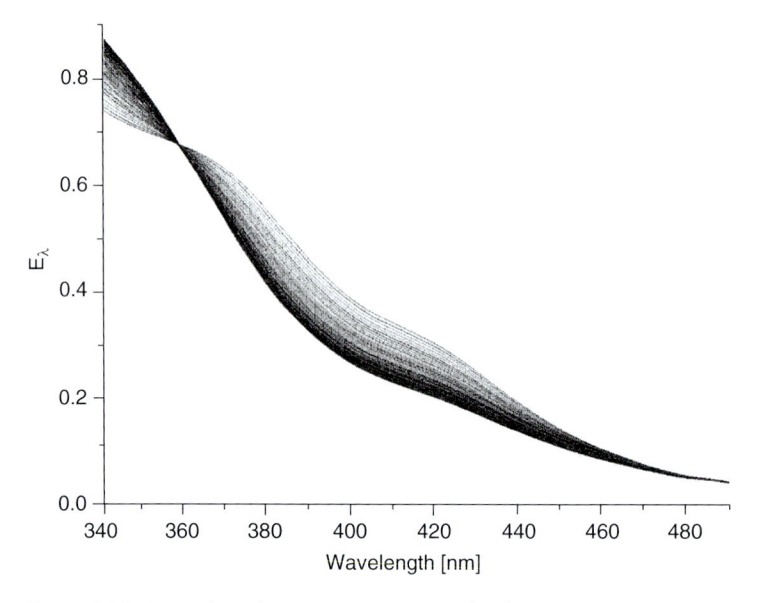

Figure 23.15 Time dependent UV/Vis spectrum for the equilibration of the reaction of [Rh(Et-DuPhos)(MeOH)$_2$]BF$_4$ (5.32×10^{-4} mol L^{-1}) with toluene (4.68×10^{-3} mol L^{-1}) to the corresponding arene complex.

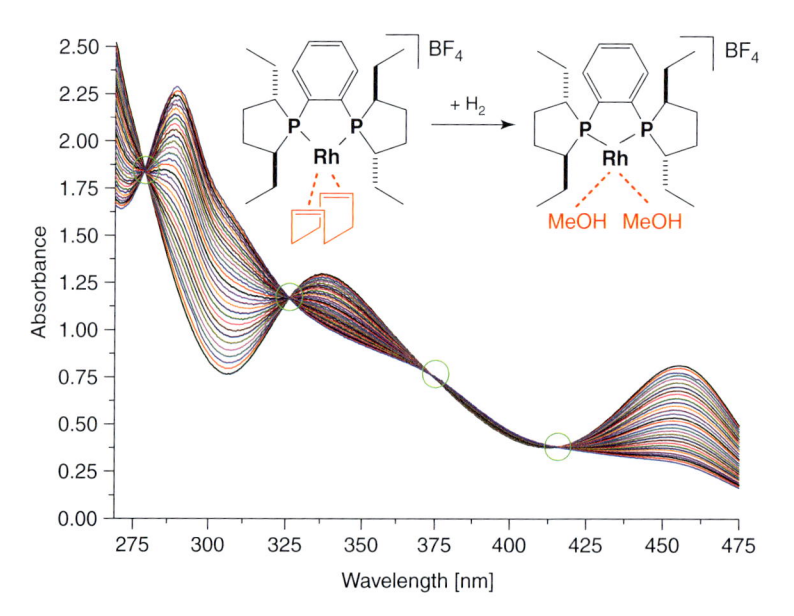

Figure 23.16 Reaction spectrum for the stoichiometric COD-hydrogenation in [Rh(Et-DuPHOS)(COD)]BF$_4$.

23.6 UV/Vis Spectroscopy

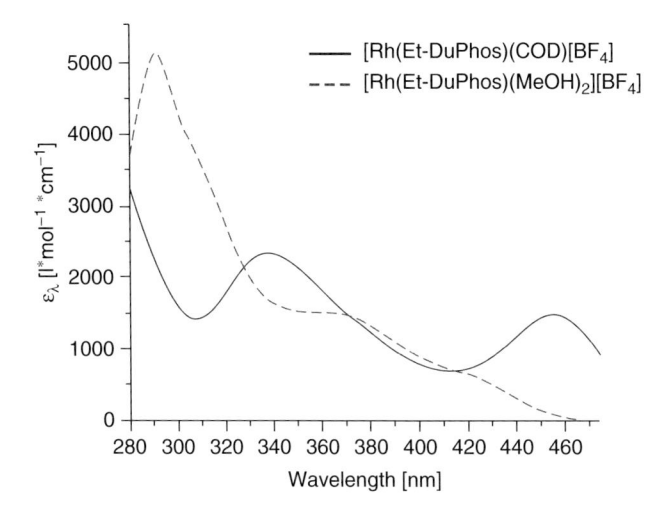

Figure 23.17 Principal component analysis of the reaction spectrum shown in Figure 23.16 calculated with PCD. The calculated molar extinction coefficients are shown as a function of the wavelength.

The analysis of the whole absorption matrix as a pseudo first-order reaction (Specfit/32™ [42],) results in a rate constant of 2.8×10^{-2} min^{-1}. Only this information allows the quantification of the transformation of the diolefin precatalyst into the active species.

Likewise, very fast reactions with gases as reactants can be monitored by means of the diode array/stopped flow technique. The formation of metallacyclopentanes of hafnium and zirconium is described in Ref. [40].

Also in the case of UV/Vis spectroscopy, the factor analysis allows the determination of molar extinction coefficients even of highly unstable intermediates. Once they are known, the corresponding concentration–time profiles can be reconstructed. The power of the factor analysis for UV/Vis reaction spectra is demonstrated for the hydrogenation of the diolefin COD in the complex [Rh(Et-DuPhos)(COD)]BF$_4$ (Figure 23.17). The program of choice was once again PCD [36]. The calculated molar extinction coefficients $\varepsilon_{\lambda,i}$ of the COD complex and the solvent complex are depicted in Figure 23.17.

The isosbestic points of the measured reaction spectrum at 327.4, 374.0 and 415.5 nm (D. Heller and C. Fischer, unpublished results.) match perfectly with the intersections of the calculated single-component spectra. Because the extinction coefficients of both species are thus known, the concentration vectors could be determined from the absorption matrix. The determination of the pseudo reaction rate constants from the concentration–time data, generated with the program PCD without any assumptions, was carried out with a non-linear regression analysis for a first-order reaction. In this way the calculated constant of $k'_i = 2.9 \times 10^{-2}$ min^{-1} is in good agreement with the value of $k'_i = 2.8 \times 10^{-2}$ min^{-1}, which was obtained by full analysis of the whole experimentally obtained absorption matrix with the program Specfit/32™. This promising result documents the power of using the mathematical analysis of principal components without a priori information, also for the interpretation of UV/Vis spectroscopic data.

23.7
Summary

The *in-situ* methods described here not only allow an efficient quantification of homogeneously catalyzed reactions, but also give in-depth insight into mechanisms even in the presence of gaseous reactants.

Gas consumptions and gas formations only characterize the overall product formation or substrate consumption, respectively, and several systematic errors are possible. Nevertheless, the kinetic analysis of such measured curves in combination with characterized intermediates is a particularly important method for the elucidation of reaction mechanisms.

Selected examples of NMR spectroscopic investigations show that much information can be obtained by making use of this technique. If chosen appropriately, even NMR experiments which do not run under real catalytic conditions can provide details on reaction mechanisms. Recording NMR spectra in the presence of gases and at different pressures in gas flow cells is possible, but is not yet as well-established as the technique of performing different catalytic reactions in autoclaves. This is due not only to the necessary investment but also to the fact that such measurements have to be integrated into the routine operation of the NMR facilities and require special skill of the experimentalist.

FTIR spectroscopy and UV/Vis spectroscopy are independent investigative methods capable of providing valuable information. They are based on continuously improving equipment and provide spectra which are subjected to high-quality analysis by suitable available methods. In contrast to NMR spectroscopy, the disadvantage is often a lack of structural information on organometallic intermediates. It should be kept in mind, however, that considerably improved time resolution as well as the possibility of generating spectra of hypothetical species by means of quantum chemical calculations allows comparison with reconstructed experimental single-component spectra.

A future experimental trend in the elucidation of reaction mechanisms is without any doubt the direct combination of different *in-situ* techniques, provided that the distinct reaction conditions applied allow for simultaneous and reliable measurements at one and the same sample. Because of their high potential, chemometric methods will also become more popular.

References

1. Heaton, B. (ed.) (2005) *Mechanism in Homogeneous Catalysis A Spectroscopic Approach*, Wiley-VCH Verlag GmbH.

2. Bargon, J. and Kuhn, L.T. (2007) *In situ NMR Methods in Homogeneous Catalysis*, Springer-Verlag, Berlin, Heidelberg.

3. Tinnemans, S.J., Mesu, J.G., Kervinen, K., Visser, T., Nijhuis, T.A., Beale, A.M., Keller, D.E., van der Eerden, M.J., and Weckhuysen, B.M. (2006) *Catal. Today*, **113**, 3–15.
4. Hashmi, A.S.K., Lothschütz, C., Ackermann, M., Doepp, R., Anantharaman, S., Marchetti, B., Bertagnolli, H., and Rominger, F. (2010) *Chem. – Eur. J.*, **16**, 8012–8019.
5. Bauer, M. and Gastl, C. (2010) *Phys. Chem. Chem. Phys.*, **12**, 5575–5584.
6. Colebrooke, S.A., Duckett, S.B., Lohman, J.A.B., and Eisenberg, R. (2004) *Chem. – Eur. J.*, **10**, 2459–2474.
7. Harthun, A., Kadyrov, R., Selke, R., and Bargon, J. (1997) *Angew. Chem. Int. Ed. Engl.*, **36**, 1103–1105.
8. Hofmann, M., Duckett, S.B., Gianotti, A., Green, G., and George, G.G.R. Nmr measurement apparatus with flow-through probehead DE 10200805931A1 (27.11.2008, to Bruker BioSpin GmbH).
9. Virial coefficients for gases can be found in: Dymond, J.H. and Smith, E.B. (1980) *The Virial Coefficients of Pure Gases and Mixtures*, Clarendon Press, Oxford (Virial coefficients for different temperatures can be determined by interpolation with, for example, cubic spline functions).
10. Profos, P. and Pfeifer, T. (1994) *Handbuch der Industriellen Messtechnik*, 6. Auflage, R. Oldenbourg Verlag, München, Wien.
11. Sun, Y., Wang, J., le Blond, C., Reamer, R.A., Laquidara, J., Sowa, J.R., and Black-Mond, D.G. Jr. (1997) *J. Organomet. Chem.*, **548**, 65–72.
12. Drexler, H.-J., Preetz, A., Schmidt, T., and Heller, D. (2007) in *Handbook of Homogeneous Hydrogenation* (eds H.G. deVries and C. Elsevier), Wiley-VCH Verlag GmbH, pp. 257–293.
13. Fogg, P.G.T. and Gerrard, W. (1991) *Solubility of Gases in Liquids*, John Wiley & Sons, Ltd, Chichester.
14. Jáuregui-Haza, U.J., Pardillo-Fondtevila, E.J., Wilhelm, A.M., and Delmas, H. (2004) *Latin Am. Appl. Res.*, **34**, 71–74.
15. Baumann, W., Mansel, S., Heller, D., and Borns, S. (1997) *Magn. Reson. Chem.*, **35**, 701–706.
16. Frey, U., Helm, L., Merbach, A.E., and Roulet, R. (1996) in *Advanced Applications of NMR to Organometallic Chemistry* (eds M. Gielen, R. Willem, and B. Wrackmeyer), John Wiley & Sons, Ltd, pp. 193–225.
17. Iggo, J.A., Shirley, D., and Tong, N.C. (1998) *New J. Chem.*, **22**, 1043–1045.
18. Rathke, J.W., Klingler, R.J., and Krause, T.R. (1991) *Organometallics*, **10**, 1350–1355.
19. Roe, D.C. (1985) *J. Magn. Res.*, **63**, 388–391.
20. Crause, C., Bennie, L., Damoense, L., Dwyer, C.L., Grove, C., Grimmer, N., Janse van Rensburg, W., Kirk, M.M., Mokheseng, K.M., Otto, S., and Steynberg, P.J. (2003) *Dalton Trans.*, 2036–2042.
21. Selent, D., Baumann, W., and Börner, A. Accessory for the Modification and for Operating a 10 mm Sapphire Pressure NMR Tube is available at Warnow-Hydraulik GmbH, DE 10333143, This accessory is commercially available from Warnow-Hydraulik GmbH, Germany.
22. Horváth, I.T. and Ponce, E.C. (1991) *Rev. Sci. Instrum.*, **62**, 1104–1105.
23. Selent, D., Wiese, K.-D., and Börner, A. (2005) in *Catalysis of Organic Reactions* (ed. J.R. Sowa), Taylor&Francis, Boca Raton, FL, pp. 459–469.
24. (a) Gual, A., Godard, C., Claver, C., and Castillón, S. (2009) *Eur. J. Org. Chem.*, 1191–1201; (b) Sémeril, D., Matt, D., Toupet, L., Oberhauser, W., and Bianchini, C. (2010) *Chem. – Eur. J.*, **16**, 13843–13849; (c) Robert, T., Abiri, Z., Wassenaar, J., Sandee, A.J., Romanski, S., Neudörfl, J.-M., Schmalz, H.-G., and Reek, J.N.H. (2010) *Organometallics*, **29**, 478–483.
25. Liu, J., Heaton, B.T., Iggo, J.A., Whyman, R., Bickley, J.F., and Steiner, A. (2006) *Chem. – Eur. J.*, **12**, 4417–4430.
26. (a) Yang, J. and Brookhart, M. (2009) *Adv. Synth. Catal.*, **351**, 175–187; (b) Yang, J. and Brookhart, M. (2007) *J. Am. Chem. Soc.*, **129**, 12656–12657; (c) Park, S. and Brookhart, M. (2010) *Organometallics*, **29**, 6057–6064.

27. (a) Gridnev, I.D., Higashi, N., Asakura, K., and Imamoto, T. (2000) *J. Am. Chem. Soc.*, **122**, 7183–7194; (b) Gridnev, I.D., Yasutake, M., Higashi, N., and Imamoto, T. (2001) *J. Am. Chem. Soc.*, **122**, 5268–5276.
28. Landis, C.R., Hilfenhaus, P., and Feldgus, S. (1999) *J. Am. Chem. Soc.*, **121**, 8741–8754.
29. Drexler, H.-J., Baumann, W., Schmidt, T., Zhang, S., Sun, A., Spannenberg, A., Fischer, C., Buschmann, H., and Heller, D. (2005) *Angew. Chem. Int. Ed.*, **44**, 1184–1188.
30. Kamer, P.C.J., van Rooy, A., Schoemaker, G.C., and van Leeuwen, P.W.N.M. (2004) *Coord. Chem. Rev.*, **248**, 2409–2424.
31. Malinowski, E.R. (2002) *Factor Analysis in Chemistry*, John Wiley & Sons, Inc., New York.
32. (a) Chew, W., Widjaja, E., and Garland, M. (2002) *Organometallics*, **21**, 1982–1990; (b) Li, C., Widjaja, E., Chew, W., and Garland, M. (2002) *Angew. Chem.*, **114**, 3939–3943.
33. (a) Li, C., Widjaja, E., and Garland, M. (2003) *J. Catal.*, **213**, 126–134; (b) Li, C., Widjaja, E., and Garland, M. (2003) *J. Am. Chem. Soc.*, **125**, 5540–5548.
34. Kubis, C., Ludwig, R., Sawall, M., Neymeyr, K., Börner, A., Wiese, K.-D., Hess, D., Franke, R., and Selent, D. (2010) *ChemCatChem*, **2**, 287–295.
35. van der Slot, S.C., Kamer, P.C.J., van Leeuwen, P.W.N.M., Iggo, J.A., and Heaton, B.T. (2001) *Organometallics*, **20**, 430–441.
36. Neymeyr, K., Sawall, M., and Hess, D. (2010) *J. Chemom.*, **24**, 67–74.
37. Watkins, A.L. and Landis, C.R. (2010) *J. Am. Chem. Soc.*, **132**, 10306–10317.
38. Polster, J. (1995) *Reaktionskinetische Auswertung Spektroskopischer Messdaten*, Friedrich Vieweg & Sohn Verlagsgesellschaft mbH.
39. (a) Halpern, J., Riley, D.P., Chan, A.S.C., and Pluth, J.J. (1977) *J. Am. Chem. Soc.*, **99**, 8055–8057; (b) Heller, D., Drexler, H.-J., Spannenberg, A., Heller, B., You, J., and Baumann, W. (2002) *Angew. Chem. Int. Ed.*, **41**, 777–780.
40. (a) Fischer, C., Beweries, T., Preetz, A., Drexler, H.-J., Baumann, W., Peitz, S., Rosenthal, U., and Heller, D. (2010) *Catal. Today*, **155**, 282–288; (b) Beweries, T., Fischer, C., Peitz, S., Burlakov, V.V., Arndt, P., Baumann, W., Spannenberg, A., Heller, D., and Rosenthal, U. (2009) *J. Am. Chem. Soc.*, **131**, 4463–4469.
41. Preetz, A., Drexler, H.-J., Fischer, C., Dai, Z., Börner, A., Baumann, W., Spannenberg, A., Thede, R., and Heller, D. (2008) *Chem. – Eur. J.*, **14**, 1445–1451.
42. Gampp, H., Maeder, M., Meyer, C.J., and Zuberbuhler, A.D. (1986) *Talanta*, **33**, 943–951, and references therein.

24
In-situ Characterization of Heterogeneous Catalysts
Bert Weckhuysen

24.1
Introduction

In order to design new and better heterogeneous catalysts, it is essential to fundamentally understand their working principles, including phenomena such as molecular diffusion, reactant activation, and catalyst deactivation [1–5]. During the last two decades, many successful efforts were made (i) to unravel in great detail the reaction mechanisms of catalytic solids, (ii) to determine the nature and distribution of their catalytically active sites, and (iii) to reveal the relationships between the structure/composition of catalytic solids and their functions, including activity, selectivity, and stability.

In this scientific endeavor, catalyst characterization methods play a pivotal role, often in combination with modeling and theoretical calculations. It is no wonder that spectroscopy has become the key tool for the characterization of heterogeneous catalysts. Spectroscopy (Latin: spectrum – an appearance and Greek: skopein – to view) can be used to determine the identity, quantity, structure and environment of atoms, molecules, and ions by analyzing the electromagnetic radiation emitted or absorbed by these atoms, molecules, ions and their related ionic, covalent, and hydrogen bonds. There are many ways to obtain such information about catalytic solids. Figure 24.1a presents a general scheme from which almost all characterization techniques can be derived [5]. A characterization method is based on some type of excitation, represented by an ingoing arrow, to which the catalyst material responds, symbolized by the outgoing arrow. For example, the method of X-ray Photoelectron Spectroscopy (XPS) is based on the photoelectric effect and the detection of the energy distribution of electrons (represented by the outgoing arrow) that are emitted from the catalytic solid when irradiated by X-rays (ingoing arrow). In general, characterization techniques applied in the field of heterogeneous catalysis are based on the interaction of solid matter with electromagnetic radiation, ranging from radio- and microwaves through infrared (IR), visible (Vis), and ultraviolet (UV) light to X-rays. In addition, many valuable surface science techniques (e.g., XPS, UV Photoelectron Spectroscopy (UPS) and Low-Energy Electron Diffraction (LEED)), microscopy techniques (e.g., Transmission Electron Microscopy (TEM),

---

Catalysis: From Principles to Applications, First Edition. Edited by Matthias Beller, Albert Renken, and Rutger van Santen.
© 2012 Wiley-VCH Verlag GmbH & Co. KGaA. Published 2012 by Wiley-VCH Verlag GmbH & Co. KGaA.

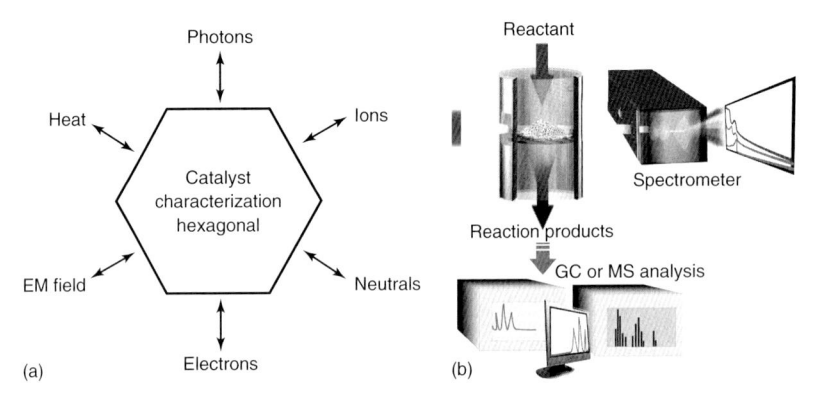

Figure 24.1 (a) Schematic illustrating the different physico-chemical methods to characterize a catalytic solid and (b) Schematic of the operando characterization approach to investigate a catalytic solid by spectroscopic means during reaction.

Scanning Electron Microscopy (SEM), and Atomic Force Microscopy (AFM)) as well as diffraction and scattering techniques (e.g., X-Ray Diffraction (XRD), Small-Angle X-ray Scattering (SAXS), and Wide-Angle X-ray Scattering (WAXS)) can be used to obtain a better understanding of how a heterogeneous catalyst material really operates under reaction conditions.

Characterization studies focus on different stages of the catalyst life cycle, which include the birth of a catalyst (i.e., the synthesis, calcination, and activation), the life of the catalyst (i.e., during reaction), and the death of the catalyst (i.e., during deactivation) [6, 7]. In the conventional approach, a catalyst material is investigated under ambient conditions (atmospheric pressure and room temperature) and, most importantly, outside the reactor tube. Although interesting information can be obtained from examination of spectroscopic data recorded under *static* conditions, no insight into the chemical processes occurring under *dynamic* conditions is revealed. More specifically, information on the heterogeneous catalyst measured before (*pre-natal*) and after reaction (*post-mortem*) does not provide sufficient knowledge on the spatiotemporal behavior of the heterogeneous catalyst in the course of the catalytic action. In order to overcome these limitations, catalytic reactions can be quenched to room temperature before measuring the active catalyst. However, a major drawback of this approach is the possibility that the catalytic solid is restructured during the rapid cooling. Similar arguments have also been made for surface science techniques, which very often require ultra-high vacuum (UHV) conditions.

A solution has been found in the development of *in-situ* characterization techniques, allowing us to continuously monitor physicochemical processes taking place in a catalyst material in real time under working conditions, for example, at elevated pressures and temperatures in the gas and liquid phase. *In situ* is the Latin expression for 'at its place'; it is the converse of *ex-situ*, which describes the measurements mentioned above [6, 7]. Preferably, the *in-situ* characterization approach is accompanied by the simultaneous measurement of catalytic activity and selectivity (e.g., with a mass spectrometer (MS) or by gas chromatography (GC)), enabling the characterization to be linked to the catalytic data, and so leading to useful structure–performance relationships. This way of measuring, known as 'operando' (from Latin: operating or working) characterization or spectroscopy, is schematically illustrated in Figure 24.1b [6–9]. In other words, operando spectroscopic measurements are a subclass of the broader group of *in-situ* characterization studies and allow for proper verification, whether or not the catalyst material placed in the measurement device is functioning properly.

However, it is important to realize that by bringing (often literally) the characterization device (e.g., the spectrometer) to the catalytic reactor (with on-line MS or GC), compromises have to be made with regard to both the characterization (e.g., lower signal-to noise ratio) and catalytic (e.g., less reaction product yield) measurements. In what follows, the phrase '*in-situ* characterization' is used, as on-line activity measurements have not always been used to evaluate the catalytic performances of *in-situ* catalyst 'characterization-testing' cells. In addition, it has to be realized that both the organic (reactant molecules, reaction intermediates, spectator species, product molecules, and deactivation products) and inorganic part (inorganic catalyst material, including information on the molecular structure, cluster size, and oxidation state of, e.g., a supported metal oxide) of the catalytic event have to be probed, and often a particular characterization technique is more sensitive toward either the organic or the inorganic part of an active catalyst system. This implies that in most cases a combination of complementary *in-situ* characterization methods is required to provide sufficient mechanistic insight into catalytic processes.

24.2
Some History, Recent Developments, and Applications

Most probably, the first report on some sort of *in-situ* spectroscopy in the field of heterogeneous catalyst research dates back to 1954 where the group of Eischens used IR spectroscopy to study the interaction of NH_3 with a porous SiO_2-Al_2O_3 solid and the interaction of CO with different supported metals, for example, Pt, Pd, Ni, and Cu [10, 11]. From 1954 to 1980 only a few papers per year appeared on *in-situ* characterization studies in the heterogeneous catalysis literature. However, in the last two to three decades, we have seen a tremendous increase in the number of available physicochemical techniques, potential applications, and publications on *in-situ* catalyst characterization. It is fair to say that nowadays *in-situ* capabilities are becoming more and more natural extensions of research equipment used for catalyst characterization in both academic and industrial laboratories. It even has the promise for controlling the operation of a catalytic solid in an industrial reactor by on-line spectroscopy and experimental protocols are in place for, for example, catalyst regeneration once a critical level of coke deposits is formed on the catalytic solid [12, 13].

Figure 24.2 outlines the three research areas, which have received most attention in recent years by scientists in this field of research [14]; that is, (i) the realization of time-resolved combined spectroscopy under realistic reaction conditions; (ii) the development of *in-situ* characterization tools to characterize catalytic transformations in the liquid-phase, such as water; and (iii) the development of space-resolved and tomographic characterization approaches, ultimately allowing to perform *in-situ* studies at the single molecule, single active site, and/or single catalyst particle level.

In what follows, the potential of *in-situ* characterization methods for investigating catalytic solids during reaction will be illustrated by describing different showcases, which focus on either the characterization of a catalytic solid loaded in a reactor or the characterization of an individual catalyst particle. All examples are from recent work from my own group with the exception of the showcases on single molecule spectroscopy, which originate from two other institutions. For a more exhaustive coverage of the subject the interested reader is referred to a themed issue of *Chemical Society Reviews* on the recent advances in the *in-situ* characterization of heterogeneous catalysts two edited books on *in-situ* spectroscopy [15, 16] as

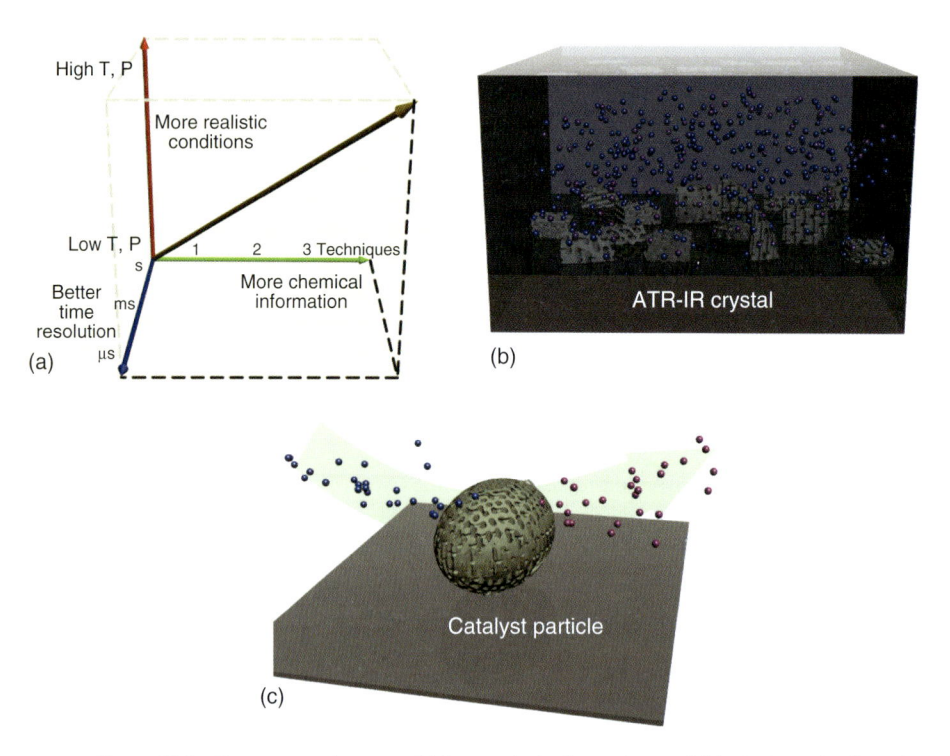

Figure 24.2 Three research areas which have received much attention in recent years in the area of *in-situ* catalyst characterization: (a) combined time-resolved spectroscopy under realistic reaction conditions; (b) spectroscopy in the liquid phase, including water; and (c) single-molecule and single-active-site spectroscopy of catalyst particles. (Reproduced from Ref. [14] with permission from the Royal Society of Chemistry, 2010).

well as the Proceedings of the last three International Congresses on Operando Spectroscopy [17–19].

24.3
In situ Characterization of a Reactor Loaded with a Catalytic Solid

24.3.1
A Reactor Loaded with a Catalytic Solid Probed by One Characterization Method

The first showcase, performed by Iglesias-Juez and co-workers [20], illustrates the use of *in-situ* high-energy resolution fluorescence detected (HERFD) X-ray Absorption Spectroscopy (XAS), currently available at some synchrotron radiation sources (e.g., ESRF, Grenoble, France) to investigate the dynamics of an industrial-like Pt-Sn/Al_2O_3 propane dehydrogenation catalysts during activation (i.e., reduction in H_2) and successive dehydrogenation-regeneration treatment (i.e., a propane treatment at 600 °C, followed by an oxygen treatment and a subsequent reduction treatment in H_2) under realistic reaction conditions. Unique in this *in-situ* investigation is the focus on the catalyst cycling experiments as up to 10 dehydrogenation-regeneration treatments have been characterized.

The advantage of the *in-situ* HERFD-XAS approach is that detailed information can be obtained on the oxidation state of a catalyst material, in this case the state of Pt, while the use of a specially designed *in-situ* cell allows for spectroscopic measurements up to 1000 °C. As an example, Figure 24.3a shows the intensity contour map of the HERFD X-ray Absorption Near Edge Spectroscopy (XANES) data acquired for an industrial-like Pt-Sn/Al_2O_3 catalyst during a heat treatment up to 600 °C in H_2, while Figure 24.3b presents some representative HERFD XANES spectra. From the spectral data it can be concluded that the supported Pt species are initially present as an oxidic Pt phase. During H_2 reduction at elevated temperatures, the Pt L_3 edge shifts to lower energies and the intensity of the rising absorption edge decreases indicating that the reduction of the Pt species takes place at about 140–210 °C leading to the initial formation of metallic Pt. At intermediate temperatures (at about 210–350 °C) a new Pt species is observed. By increasing the temperature, the total reduction of Pt takes place including the insertion of Sn into the structure and the consequent formation of a Pt-Sn alloy. The intensity contour map of the HERFD XANES spectra, acquired during the first two consecutive propane dehydrogenation-regeneration treatments, is given in Figure 24.3c. From the position and shape of the HERFD XANES spectra it is clear that during the process of coke burning, oxygen oxidizes the Sn present in the Pt-Sn alloy, which segregates out of the supported Pt nanoparticles. The subsequent H_2 reduction restores the Pt-Sn alloy.

Interestingly, the Pt nanoparticle size increases after the first regeneration cycles, although this behavior is much less pronounced as for a promoter-free Pt/Al_2O_3 catalyst. Seemingly, the decoration of Pt particles by Sn could prevent the Pt sintering phenomenon. Figure 24.3d illustrates the different stages

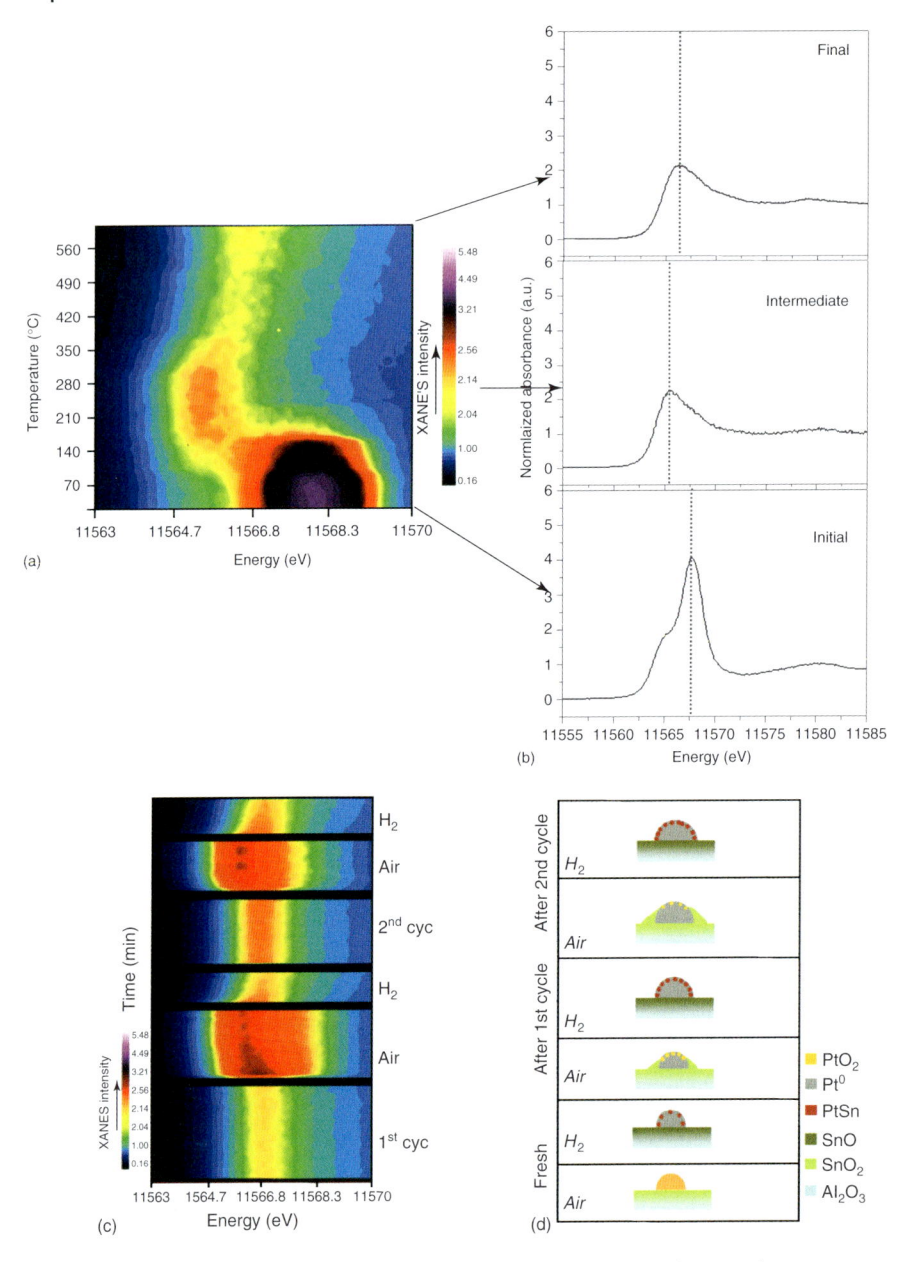

Figure 24.3 (a) Intensity contour map of the HERFD XANES spectra of a Pt-Sn/Al$_2$O$_3$ catalyst acquired during the ramp of temperature up to 600 °C in H$_2$; (b) Selection of HERFD XANES spectra of the initial, intermediate, and final Pt species formed in the catalyst material; (c) Intensity contour map of the HERFD XANES spectra of the Pt-Sn/Al$_2$O$_3$ catalyst acquired at 600 °C during two propane dehydrogenation–regeneration cycles; and (d) Proposed model of the Pt and Pt-Sn species during the different treatment stages for the Pt-Sn/Al$_2$O$_3$ catalyst. (Reproduced from Ref. [20] with permission from Elsevier Science Publishers, 2010).

and forms of the Pt and Pt-Sn entities during the first two successive propane dehydrogenation-regeneration cycles based on the HERFD XANES spectra as well as related XANES and Extended X-ray Absorption Fine Structure (EXAFS) analysis of the Pt L_3 and Sn K edge XAS spectra. The latter studies underline the need of complementary characterization data to arrive at a more detailed picture of the events taking place during catalysis. After 10 propane dehydrogenation-regeneration cycles it was found that a progressive Sn enrichment as well as an increased Sn dilution occurred within the supported Pt-Sn nanoparticles. These effects result in more homogeneous bimetallic entities. Another conclusion was that the presence of Sn keeps the Pt surface sites clean from coke deposits thereby increasing the catalyst stability.

24.3.2
A Reactor Loaded with a Catalytic Solid Probed by Multiple Characterization Methods

In the last decade, many *in-situ* set-ups, in which different spectroscopic techniques are combined in one spectroscopic-reaction cell, have been developed and explored. These advances in *in-situ* instrumentation offer great opportunities to accomplish a more comprehensive understanding of catalytically relevant systems. The key advantage of such multi-technique approaches originates from the acquisition of *in-situ* characterization and catalytic data from the same catalyst system with assured identical reaction conditions. The first techniques combined in one reactor set-uploaded with a catalytic solid are XRD and XAS, as pioneered by the groups of Thomas [21] and Topsøe [22]. In the years thereafter, promising combinations of two or more spectroscopic techniques to study catalytic reactions *in-situ* have been reported in literature. Most combinations involve the use of vibration (IR, Raman) and electronic (UV-VIS-NIR) spectroscopies, but also magnetic resonance (Nuclear Magnetic Resonance (NMR) and Electron Paramagnetic Resonance (EPR)) and synchrotron-based techniques (XAS, SAXS, and WAXS) are employed. Examples include UV-Vis-NIR/EPR [23], Raman/IR [24], Raman/UV-Vis [25], EPR/UV-Vis-NIR/Raman [26], NMR/UV-Vis-NIR [27], Raman/UV-Vis-NIR/XANES [28], and SAXS/WAXS/XANES [29]. Three recent review articles by Tinnemans *et al.* [30], Bentrup [31], and Newton and van Beek [32] summarize the different characterization combinations available for catalyst scientists.

Figure 24.4a illustrates a novel multi-technique set-up, as developed by O'Brien *et al.* [33], which allows obtaining in a combined fashion *in-situ* WAXS, fluorescence, Compton scattering, Raman, and UV-Vis-NIR data of a 15 mm long packed bed of a catalytic solid at reaction temperature as provided by heating guns. On-line activity measurements were made possible by using an on-line coupled MS. As an example, Figure 24.4b shows the spatiotemporal WAXS intensity plots of a MoO_3 catalyst exposed for 240 min to a methanol stream at 345 °C. MoO_3 was initially found to be relatively evenly packed throughout the catalyst bed, but as the experiment progressed the catalytic solid was rapidly consumed as oxygen was stripped from the bulk oxide in a Mars-van Krevelen-type manner under the anaerobic reaction

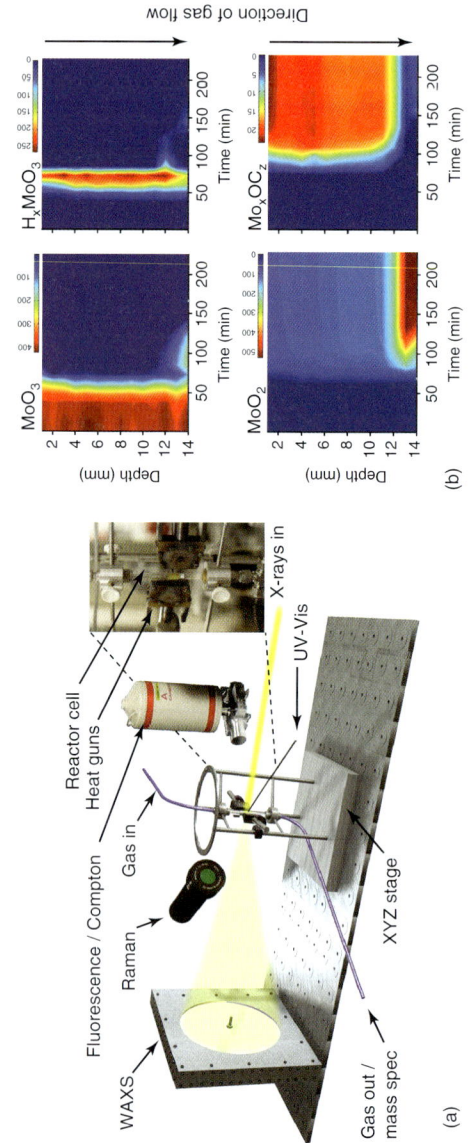

Figure 24.4 (a) Schematic of an *in-situ* multi-technique set-up for performing spatiotemporal characterization studies of catalytic solids at work. Synchrotron radiation-based techniques, that is, WAXS, fluorescence, and Compton scattering, were utilized, and were complemented with Raman, UV-Vis, and on-line mass spectrometry. The reactor consists of a quartz reactor heated by air guns in an 'open' architecture; and (b) Spatiotemporal WAXS intensity plots of phases formed through the depth of the reactor during the reduction of MoO_3 under methanol flow conditions at 345 °C. (Reproduced from Ref. [33] with permission from Wiley-VCH, 2009).

conditions applied. This oxygen depletion process resulted after 50 min of reaction in the formation of new catalyst phases. More specifically, as MoO_3 was consumed, a short lived H_xMoO_3 bronze intermediate was detected by WAXS, which then further converted into more reduced phases. After approximately 100 min MoO_2 was the dominant phase formed, but some additional phases were also detected, including molybdenum oxycarbide (Mo_xOC_z). Unlike MoO_3, all the reduced phases demonstrated some scattering variation through the depth of the catalyst bed. More specifically, MoO_2 had a gradient increase in scattering intensity from top to bottom, while the opposite can be noticed for the Mo_xOC_z phase. These results show that gradients occur along the reactor bed during a reduction process, illustrating that different catalyst phases are exposed to the reactant within a reactor set-up. As a final remark it has to be stressed that complementary *in-situ* Raman and UV-Vis spectroscopy data corroborated these WAXS findings as these methods could reveal the disappearance of the MoO_3 phase as evidenced by the decrease in the typical Raman bands located at 820 and 994 cm^{-1} as well as the formation of Mo^{4+} species, observed in the UV-Vis data by an increasing intensity of the 460 nm absorption band.

Whereas in the first showcase on Pt-Sn/Al_2O_3 catalysts the unique investigation aspect was on the use of high energy resolution fluorescence detection enabling the distinction between different reduction states of a metal (oxide), the second example on MoO_3 catalysts stands out because very high energy X-rays are employed, which allow to investigate large catalyst volumes in much more realistic reactor tubes (i.e., 4 mm inner diameter). Furthermore, the open cell architecture makes it possible to interrogate the catalyst system with a wide variety of complementary characterization methods, underlining the variety of catalytic solids which can be investigated with such *in-situ* characterization set-ups.

24.4
In situ Characterization at a Single Catalyst Particle Level

The second example of MoO_3-based catalyst materials illustrates the presence of spatial heterogeneities along a packed bed of the catalyst material within an operating reactor. Interestingly, such spatiotemporal phenomena also take place at the level of an individual catalyst particle. In order to demonstrate these spatial heterogeneities at the single particle level, *in-situ* micro- and nano-spectroscopy investigations are needed [34].

24.4.1
In situ Micro-Spectroscopy of a Catalytic Solid

In recent years, micro-spectroscopy methods have been fully developed to elucidate the catalytic chemistry and related spatiotemporal heterogeneities within single catalyst particles [34]. The main focus of these studies has been on the investigation of large coffin-shaped H-ZSM-5 zeolite crystals, and an SEM image of such a crystal is shown in Figure 24.5a.

The explored micro-spectroscopic techniques include (confocal) fluorescence microscopy [36, 37], UV-Vis micro-spectroscopy [38], IR micro-spectroscopy [39], interference microscopy [40], hard X-ray micro-spectroscopy [41], Coherent Anti-Stokes Raman Spectroscopy (CARS) [42], and Second Harmonic Generation (SHG) microscopy [43]. The potential of UV-Vis micro-spectroscopy for heterogeneous catalyst research is illustrated in Figure 24.5b,c. The acid-catalyzed oligomerization of styrene derivatives has been used by Kox and co-workers as a probe reaction to visualize the catalytic activity of Brønsted acid sites within large H-ZSM-5 zeolite crystals [35, 44]. Since the carbocation intermediates, formed during the reaction, absorb light in the visible region, they can act as reporter molecules for catalytic activity. In Figure 24.5b, optical microphotographs of the H-ZSM-5 crystals after oligomerization with 15 different styrene derivatives are shown. It is evident that there are distinct differences in coloration between the edges and main body of the crystals as well as between the different styrene derivatives. These spatial differences are due to the intensity variations of different UV-Vis absorption bands (Figure 24.5c), which points to the formation of different (amounts of) carbocation intermediates. In most cases, more carbocation intermediates are formed in the main body of the crystal than at the crystal's edge. The lack of coloration in some crystals indicates diffusion hindrance effects (e.g., for bulky compounds) and provides a direct demonstration of zeolite shape selectivity, that is, reactant selectivity. The UV-Vis band at around 585 nm arises from an allylic dimeric carbocation, while the UV-Vis band at around 635 nm is due to a more extended carbocation species. More specifically, oligomerization starts with the protonation of styrene on the Brønsted acid sites of H-ZSM-5, and this carbocation reacts with a neutral styrene, forming a secondary 1,3-bis(phenyl)-1-butylium cation. The latter is thermally unstable in the acidic environment of the zeolite and reacts further to form the allylic 1,3-bis(phenyl)buten-1-ylium cation, which is readily observable with UV-Vis. Subsequent reactions, for example, further oligomerization, may occur within the acidic environment of the zeolite H-ZSM-5 crystals.

Before the spatial heterogeneities observed can be properly explained, fundamental insight into the crystals' microscopic structure is needed. Based on a detailed investigation of a series of morphologically different H-ZSM-5 crystals, Karwacki and co-workers, by means of confocal fluorescence microscopy (focused ion beam (FIB) milling), electron backscatter diffraction (EBSD) measurements, and TEM labeling, have been able to reveal that the crystals comprise at least six different subunits, which all have pyramidal boundaries with each other, leading to distinct molecular diffusion barriers [45]. For the most complex H-ZSM-5 crystal morphologies, two molecular diffusion barriers can be distinguished, as illustrated in Figure 24.6. These barriers prevent the easy diffusion of the styrene carbocations and lead to spatial heterogeneities between the edge and the main body of the H-ZSM-5 crystals as well as differences between the studied styrene derivatives, as observed in Figure 24.5b,c.

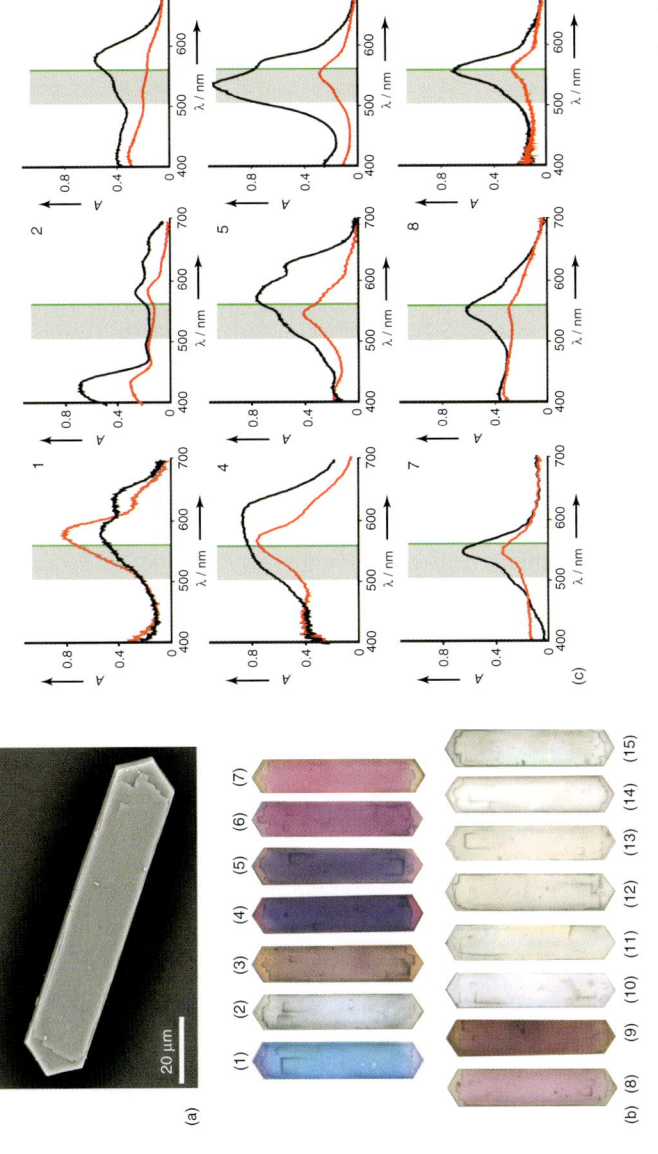

Figure 24.5 (a) SEM image of a H-ZSM-5 crystal; (b) Optical microphotographs of individual zeolite H-ZSM-5 crystals after reaction with 15 different styrene derivatives; and (c) Spatially resolved UV-Vis absorption spectra of individual H-ZSM-5 crystals after reaction with (1–9) recorded at the edges (red) and the center (black). (Reproduced from Ref. [35] with permission from Wiley-VCH, 2007). The Styrene derivatives are: (1) 4-methoxystyrene; (2) 4-ethoxystyrene; (3) 4-methylstyrene; (4) 4-bromostyrene; (5) 4-chlorostyrene; (6) 4-fluorostyrene; (7) Styrene; (8) β-methoxystyrene; (9) trans-β-methylstyrene; (10) 3-chlorostyrene; (11) 2,3,4,5,6-penta-fluoro styrene; (12) α-methylstyrene; (13) 3-trifluoromethylstyrene; (14) 3,4-dichlorostyrene; (15) 2,6-dichlorostyrene.

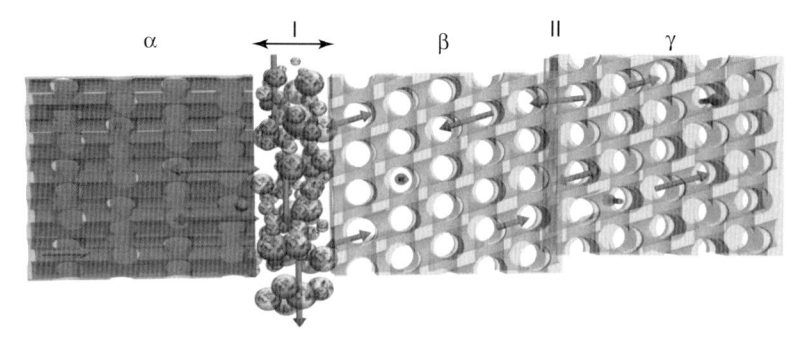

Figure 24.6 Schematic of the internal diffusion barriers separating subunit α, β, and γ within a ZSM-5 zeolite crystal. Barrier I imposes a 90° rotational difference and is formed by the disordered structure along the boundary (indicated by the spheres), whereas barrier II indicates a displacement along the boundary, caused by a mismatch in structure of 0.5–2°. (Reproduced from Ref. [45] with permission from Nature Publishing Group, 2009).

Another powerful method is IR micro-spectroscopy, and some relevant characterization data, as measured by Stavitski *et al.* [46], on the Brønsted acid-catalyzed 4-fluorostyrene oligomerization occurring within a single H-ZSM-5 crystal, are given in Figure 24.7. Figure 24.7a shows the gradual transformation of styrene into the dimeric styrene carbocation intermediate, while Figure 24.7c shows the spatial heterogeneities within an individual zeolite crystal. In line with the findings by UV-Vis micro-spectroscopy a higher amount of carbocation intermediates can be found in the center of the zeolite crystal. Furthermore, thanks to the mediated diffusion of styrene within the large H-ZSM-5 crystal and the minimal external zeolite surface, the carbocation intermediates can be accumulated within micropores in sufficient concentrations to allow their detection by synchrotron-based IR micro-spectroscopy. Strong polarization dependence of the IR spectra due to the molecular alignment of the dimeric styrene carbocations imposed by the zeolite channels, as illustrated in Figure 24.7b, can be rationalized in terms of the vibrational transitions of the intrazeolite carbocations. Based on these findings, a detailed molecular-level picture of the arrangement of the reaction intermediates confined within microporous zeolite matrices has been obtained [39].

24.4.2
Single-Molecule *in-situ* Spectroscopy of a Catalytic Solid

One of the very appealing properties of fluorescent molecules is the possibility to visualize them at the single-molecule level. This interesting property has recently been exploited in the context of *in-situ* heterogeneous catalyst characterization by making use of fluorescence microscopy [34]. In what follows, two

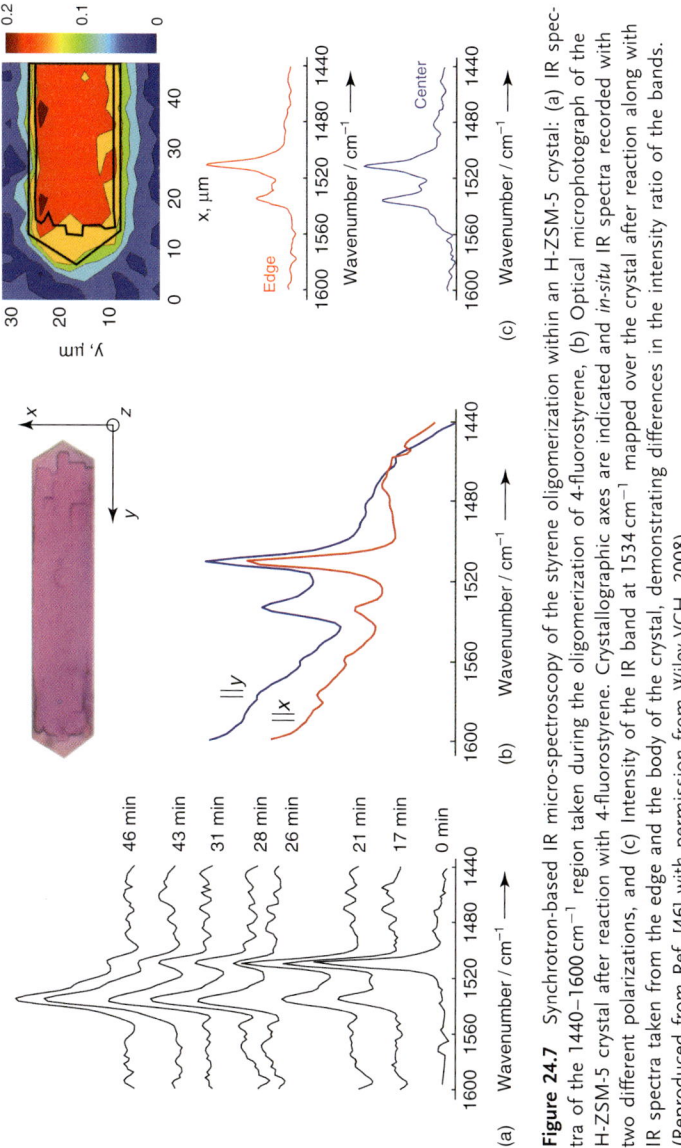

Figure 24.7 Synchrotron-based IR micro-spectroscopy of the styrene oligomerization within an H-ZSM-5 crystal: (a) IR spectra of the 1440–1600 cm^{-1} region taken during the oligomerization of 4-fluorostyrene, (b) Optical microphotograph of the H-ZSM-5 crystal after reaction with 4-fluorostyrene. Crystallographic axes are indicated and in-situ IR spectra recorded with two different polarizations, and (c) Intensity of the IR band at 1534 cm^{-1} mapped over the crystal after reaction along with IR spectra taken from the edge and the body of the crystal, demonstrating differences in the intensity ratio of the bands. (Reproduced from Ref. [46] with permission from Wiley-VCH, 2008).

recent breakthrough studies are briefly discussed. The first single-molecule characterization study involves tracking the diffusion patterns of single molecules in order to assess the pore structure of a catalyst material. Bein, Brauchle, and co-workers have been studying in great detail the diffusion of single fluorescent dye molecules within a thin mesoporous silica film [47–50]. For this purpose, the highly fluorescent terrylene diimide (TDI) dye molecule has been used. By carefully superimposing the fluorescence microscopy and TEM images, the porous structure of the mesoporous silica film can be directly correlated with the diffusion dynamics of the TDI dye molecule. In this way, solid proof was obtained that the molecular diffusion pathway through a porous material correlates with the pore orientation of the two-dimensional hexagonal mesoporous material. Following this approach, the trajectories of individual TDI molecules were measured with high spatial accuracy in various mesoporous environments, which include straight or strongly curved sections as well as regions of less ordered material (dead ends) or a bypass to another main channel. This is shown in Figure 24.8.

The second example of single-molecule fluorescence microscopy involves the monitoring of single-molecule catalytic events. Hofkens, De Vos, Sels, and co-workers have been investigating base-catalyzed reactions over large layered double hydroxide (LDH) crystals [51–53].Two different catalytic reactions, namely [Figure 24.9a] the hydrolysis and the transesterification of 5-carboxyfluorescein diacetate (C-FDA), have been investigated by fluorescence microscopy in order to discriminate between different active-site populations in a single LDH catalyst particle [51]. It is important to note that the C-FDA molecules become emissive only upon hydrolysis or transesterification with, for example, 1-butanol. By exposing the LDH crystals to solutions of C-FDA, bright fluorescent spots are observed which are due to the formation of single molecules of reaction product. The rapid disappearance of the formed fluorescent spots happens by photo-bleaching, which allows the detection of catalytic activity in different regions of the LDH crystals, thereby assessing space-resolved information on their catalytic activity.

When performing the transesterification of C-FDA with 1-butanol it was found that the fluorescent spots appear all over the catalyst surface without any preference for the crystal edges. This is shown in Figure 24.9b,c. Since no fluorescence signal was observed in the solution surrounding the catalyst particle it can be concluded that transesterification is catalyzed by the LDH material only. Instead, when studying the ester hydrolysis by replacing the alcohol solvent by an aqueous solution it was found that the fluorescent spots primarily occurred at the LDH crystal edges and much less at its basal surfaces (Figure 24.9d). Based on a detailed analysis of their data, Roeffaers *et al.* concluded that the catalytic activity of LDH materials is not always associated with the same type of active sites; that is, transesterification primarily occurs at the {0001} plane, whereas hydrolysis requires the {1010} faces of the LDHs, where exchanged OH^- ions at the entrance of the LDH catalyst particle may be the active sites [51]. Furthermore, time-dependent experiments allowed determination of the reaction rate for both catalytic reactions, enabling the heterogeneity in the reaction kinetics at the level of individual crystal faces to be quantified [51–53].

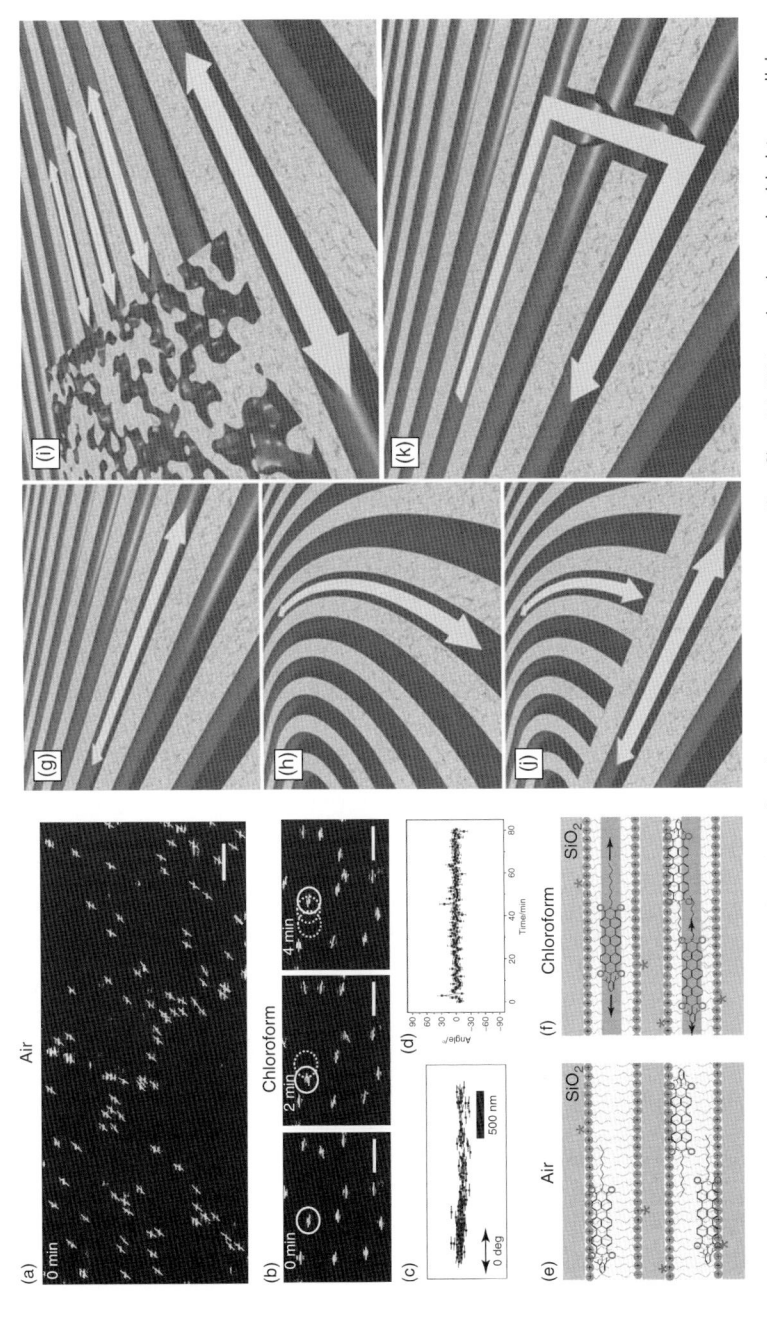

Figure 24.8 Orientation and diffusion of terrylene diimide (TDI) in a mesoporous silica film: (a) TDI molecules embedded in parallel pores in air. Fluorescence image showing oriented TDI molecules (striped patterns); (b) Sequence of fluorescence images showing linear diffusion of TDI molecules in chloroform; (c) The trajectory of a TDI molecule as marked with a white circle in (b); (d) Calculated angular time trajectory of the same TDI molecule; (e) Schematic of TDI molecules immobilized in the mesoporous material in air; and (f) TDI molecules in the mesoporous material in chloroform. The TDI molecules are solvated and diffuse along the channels: their walk is occasionally interrupted by adsorption events; (g) straight segment; (h) curved segment; (i) domain boundaries forcing molecules to turn back; (j) molecular travel stopped at less ordered regions; and (k) lateral motion between neighboring channels. (Reproduced from Refs. [47, 48] with permission from Nature Publishing Group, 2007 and American Chemical Society, 2008).

Figure 24.9 (a) Schematic of the *in-situ* fluorescence microscopy set-up used to measure single catalytic turnovers during the transformation of fluorescein ester over an individual layered double hydroxide (LDH) particle. R = —H for FDA and R = —COOH for C-FDA; (b) and (c) 2D fluorescence images on the transesterification of C-FDA with 1-butanol at 40 nM and 700 nM ester concentration, respectively; (d) 2D fluorescence image of the hydrolysis of 600 nM C-FDA on an LDH crystal, showing the formation of product molecules mainly at the crystal edges and (e) Accumulated spot intensity on the same crystal over 256 consecutive images. (Reproduced from Ref. [51] with permission from Nature Publishing Group, 2006).

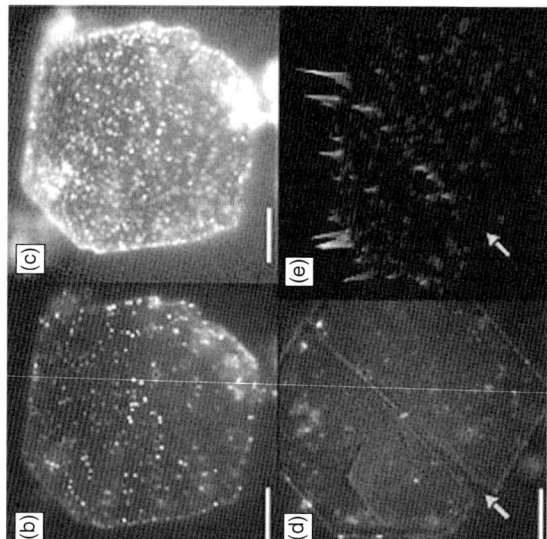

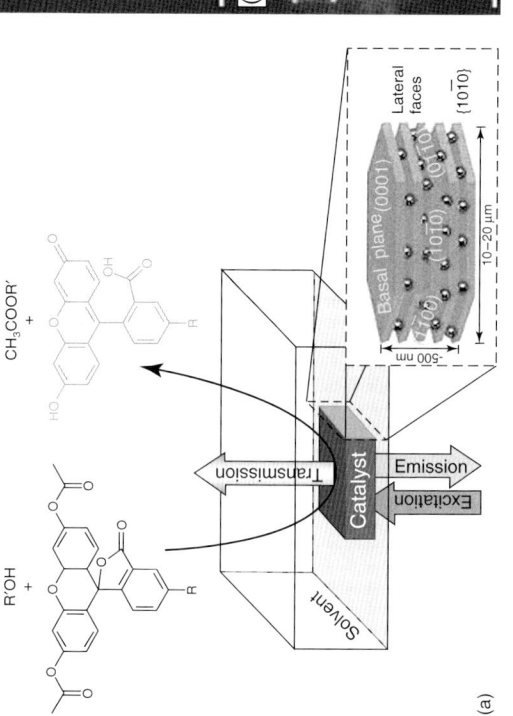

24.4.3
In-situ Nano-Spectroscopy of a Catalytic Solid

As the spatial resolution of the above-mentioned optical microscopy methods is intrinsically determined by the diffraction limits of light, they are – generally speaking – limited to about 0.5 µm (e.g., fluorescence microscopy and CARS), 1–2 µm (e.g., UV-Vis micro-spectroscopy), and 3–10 µm (e.g., IR micro-spectroscopy). X-rays, however, with wavelengths in the nanometer range, should allow in-situ nano-spectroscopic measurements of catalytic solids to be performed. The first example of such a nano-spectroscopy study has been performed by de Smit and co-workers [54]. More specifically, Scanning Transmission X-ray Microscopy (STXM) with wavelengths (X-ray energy) in the range 0.6–6 nm (2000 to 200 eV) has been used in combination with an in-situ nanoreactor to investigate the dynamics taking place during the activation and CO hydrogenation reaction of an Fe-based Fischer-Tropsch Synthesis (FTS) catalyst nanoparticle [54–56]. Figure 24.10a schematically illustrates the nanoreactor developed, which allows heating up to 700 °C via the use of Pt wires and the use of a gas flow [57]. Crucial for its use has been the limited space between the two reactor plates (i.e., ~50 µm), as gas molecules also absorb soft X-rays and could limit the amount of transmitted X-rays.

By using this in-situ cell it has been possible to image an active and fully promoted Fe/Si/K/Cu FTS catalyst in the presence of synthesis gas (CO/H_2) at elevated temperatures. High-quality XAS data concerning the Fe L-, O K-, and C K-edges, with 35 nm spatial resolution, leading to chemical information (i.e., oxidation state and coordination environment) on, for example, Fe during catalytic reaction, have been obtained. Figure 24.10b illustrates this in-situ catalyst characterization approach with a chemical map of the Fe-based FTS catalyst at 250 °C in CO/H_2. Analysis of the C K-edge XAS data revealed that carbon was preferably present in the Fe-rich regions, pointing toward the presence of an active Fe carbide phase (Figure 24.10b, lower region). Furthermore, regions with higher Fe^0 intensity showed a somewhat different C spectrum from that of the regions with lower Fe^0 intensity (Figure 24.10b, upper region). Fe^0-rich regions contain sp^2 hybridized carbon species, whereas regions with lower amounts of Fe^0 (or none) are dominant in sp^3-like carbon species. These observations emphasize the ability of the in-situ STXM method to distinguish between the inorganic phase (i.e., Fe carbide) and the organic phase (e.g., aliphatic and carboxylic carbon deposits) of the catalytic solid during reaction. In a follow-up study [58], de Smit et al. showed that the in-situ STXM method is very powerful in determining the oxidation state of a metal/metal oxide phase in supported or unsupported catalyst formulations. Interestingly, the average oxidation state of a 1 µm-sized Fe-based FTS catalyst particle during H_2 reduction as determined with in-situ STXM correlated very well with bulk reduction behavior of the same catalyst making use of the more conventional Temperature-Programmed Reduction (TPR) method. In other words, the characterization data originating from bulk and nano-spectroscopic methods are in line with each other [58].

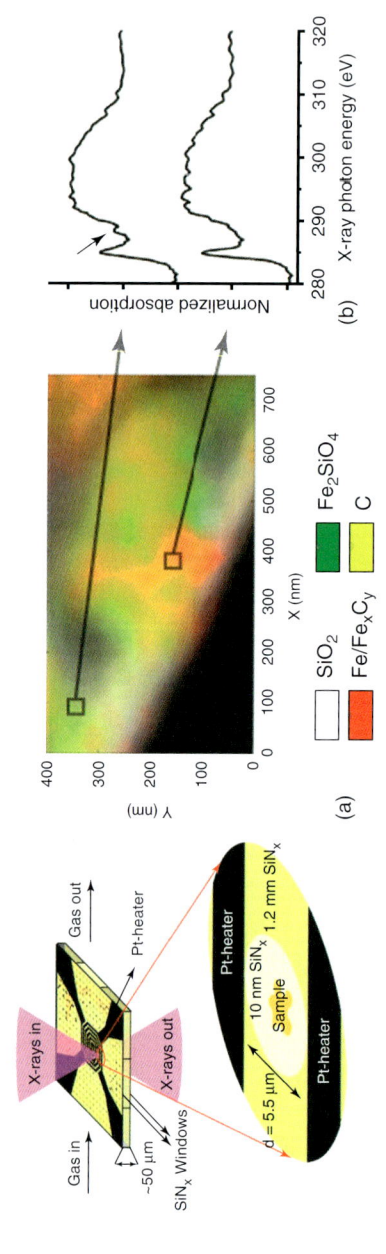

Figure 24.10 (a) Schematic of the *in-situ* STXM nanoreactor including the Pt spiral heater and the $\sim 5 \times 5$ μm^2 windows, and (b) *In-situ* STXM map and related C K-edge XAS spectra of a Fe-based Fischer-Tropsch catalyst at 250 °C in a CO/H$_2$ gas atmosphere. (Reproduced from Ref. [54] with permission from Nature Publishing Group, 2008).

24.5
Concluding Remarks

The catalyst scientist's *in-situ* characterization toolbox for investigating active catalytic solids under realistic reaction conditions has been significantly expanded in the last decade. It is now possible to obtain, in a 'single shot,' (i) multiple sets of *in-situ* characterization data which are highly complementary, (ii) 1D, 2D, and 3D chemical information about a heterogeneous catalyst as a function of reaction time with spatial resolutions measured in micro- or even nanometers, and (iii) single-molecule and single-active-site information, providing insight in the catalyst dynamics. As more laboratories, including beamline stations at synchrotron radiation facilities, are equipped with *in-situ* measurement facilities, it should be clear that the future looks bright for catalyst scientists aiming to unravel mechanistic insights into the working and deactivation principles of catalytic solids. Future developments will most probably include improved time, space, and energy resolutions as well as a further integration of complementary characterization methods, including the development of chemometrics and related software for rapid data analysis and, even more importantly, accurate interpretation of spectral data.

Acknowledgments

B.M.W. acknowledges financial support from NWO-CW (Top research grant) and NRSC-Catalysis.

References

1. Ertl, G., Knözinger, H., Schüth, F., and Weitkamp, J. (eds) (2008) *Handbook of Heterogeneous Catalysis*, Wiley-VCH Verlag GmbH, Weinheim.
2. Somorjai, G.A., and Lin, Y. (2010) *Introduction to Surface Chemistry and Catalysis*, 2nd edn, Wiley-VCH Verlag GmbH, New York.
3. Thomas, J.M. and Thomas, W.J. (1997) *Principles and Practice of Heterogeneous Catalysis*, Wiley-VCH Verlag GmbH, Weinheim.
4. Niemantsverdriet, J.W. (2007) *Spectroscopy in Catalysis: An Introduction*, 3rd edn, Wiley-VCH Verlag GmbH, Weinheim.
5. Niemantsverdriet, J.W. (1998) in *Catalysis: An Integrated Approach*, 2nd Revised and Enlarged edn (eds J.A. Moulijn, P.W.N.M. van Leeuwen, R.A. van Santen, and B.A. Averill), Elsevier, Amsterdam, pp. 489.
6. Weckhuysen, B.M. (2002) *Chem. Commun.*, 97.
7. Weckhuysen, B.M. (2003) *Phys. Chem. Chem. Phys.*, **5**, 4351.
8. Bañares, M.A. (2005) *Catal. Today*, **100**, 71.
9. Guerrero-Perez, M.O. (2002) *Chem. Commun.*, 1292.
10. Mapes, J.E. and Eischens, R.P. (1954) *J. Phys. Chem.*, **58**, 1059.
11. Eischens, R.P., Plisken, W.A., and Francis, S.A. (1954) *J. Chem. Phys.*, **24**, 1786.
12. Bennici, S.M., Vogelaar, B.M., Nijhuis, T.A., and Weckhuysen, B.M. (2007) *Angew. Chem. Int. Ed.*, **46**, 5412.
13. Nijhuis, T.A., Tinnemans, S.J., Visser, T., and Weckhuysen, B.M. (2004) *Chem. Eng. Sci.*, **59**, 5487.

14. Weckhuysen, B.M. (2010) *Chem. Soc. Rev.*, **39**, 4557.
15. Haw, J.F. (ed.) (2002) *In-situ Spectroscopy in Heterogeneous Catalysis*, Wiley-VCH Verlag GmbH, Weinheim.
16. Weckhuysen, B.M. (ed.) (2004) *In-situ Spectroscopy of Catalysts*, American Scientific Publishers, Stevenson Ranch.
17. Weckhuysen, B.M. (2003) *Phys. Chem. Chem. Phys.*, **5**, VI.
18. Bañares, M.A. (2007) *Catal. Today*, **126**, 1.
19. Bruckner, A. (2010) *Catal. Today*, **155**, 155.
20. Iglesias-Juez, A., Beale, A.M., Maaijen, K., Weng, S.C., and Glatzel, P. (2010) *J. Catal.*, **276**, 268.
21. (a) Shannon, I.J., Maschmeyer, T., Sankar, G., Thomas, J.M., Oldroyd, R.D., Sheehy, M., Madill, D., Waller, A.M., and Townsed, R.P. (1997) *Catal. Lett.*, **44**, 23; (b) Couves, J.W., Thomas, J.M., Waller, D., Jones, R.H., Dent, A.J., Derbyshire, G.E., and Greaves, G.N. (1991) *Nature*, **354**, 465.
22. (a) Clausen, B.S., Grabaek, L., Steffensen, G., Hansen, P.L., and Topsøe, H. (1993) *Catal. Lett.*, **20**, 23; (b) Grunwaldt, J.D., Molenbroek, A.M., Topsoe, N.Y., Topsoe, H., and Clausen, B.S. (2000) *J. Catal.*, **194**, 452.
23. Brückner, A. (2001) *Chem. Commun.*, 2122.
24. Le Bourdon, G., Adar, F., Moreau, M., Morel, S., Reffner, J., Mamede, A.S., Dujardin, C., and Payen, E. (2003) *Phys. Chem. Chem. Phys.*, **5**, 4441.
25. Nijhuis, T.A., Tinnemans, S.J., Visser, T., and Weckhuysen, B.M. (2003) *Phys. Chem. Chem. Phys.*, **5**, 4361.
26. Bruckner, A. (2005) *Chem. Commun.*, 1761.
27. Hunger, M. and Wang, W. (2004) *Chem. Commun.*, 584.
28. Beale, A.M., van der Eerden, A.M.J., Kervinen, K., Newton, M.A., and Weckhuysen, B.M. (2005) *Chem. Commun.*, 3015.
29. (a) Beale, A.M., van der Eerden, A.M.J., Jacques, S.D.M., Leynaud, O., O'Brien, M.G., Meneau, F., Nikitenko, S., Bras, W., and Weckhuysen, B.M. (2006) *J. Am. Chem. Soc.*, **128**, 12386; (b) Nikitenko, S., Beale, A.M., van der Eerden, A.M.J., Jacques, S.D.M., Leynaud, O., O'Brien, M.G., Detollenaere, D., Kapteijn, R., Weckhuysen, B.M., and Bras, W. (2008) *J. Synchrotron Radiat.*, **15**, 632.
30. Tinnemans, S.J., Mesu, J.G., Kervinen, K., Visser, T., Nijhuis, T.A., Beale, A.M., Keller, D.E., van der Eerden, A.M.J., and Weckhuysen, B.M. (2006) *Catal. Today*, **113**, 3.
31. Bentrup, U. (2010) *Chem. Soc. Rev.*, **39**, 4718.
32. Newton, M.A. and van Beek, W. (2010) *Chem. Soc. Rev.*, **39**, 4543.
33. O'Brien, M.G., Beale, A.M., Jacques, S.D.M., Di Michiel, M., and Weckhuysen, B.M. (2009) *ChemCatChem*, **1**, 99.
34. Weckhuysen, B.M. (2009) *Angew. Chem. Int. Ed.*, **48**, 4910.
35. Stavitski, E., Kox, M.H.F., and Weckhuysen, B.M. (2007) *Chem. – Eur. J.*, **13**, 7057.
36. (a) Roeffaers, M.B.J., Sels, B.F., Uji-i, H., Blanpain, B., L'hoest, P., Jacobs, P.A., De Schryver, F.C., Hofkens, J., and De Vos, D.E. (2007) *Angew. Chem. Int. Ed.*, **46**, 1706; (b) Roeffaers, M.B.J., Ameloot, R., Baruah, M., Uji-i, H., Bulut, M., De Cremer, G., Muller, U., Jacobs, P.A., Hofkens, J., Sels, B.F., and De Vos, D.E. (2008) *J. Am. Chem. Soc.*, **130**, 5763; (c) De Cremer, G., Sels, B.F., De Vos, D.E., Hofkens, J., and Roeffaers, M.B.J. (2010) *Chem. Soc. Rev.*, **39**, 4543.
37. (a) Karwacki, L., Stavitski, E., Kox, M.H.F., Kornatowski, J., and Weckhuysen, B.M. (2007) *Angew. Chem. Int. Ed.*, **46**, 7228; (b) Kox, M.H.F., Stavitski, E., Groen, J.C., Perez-Ramirez, J., Kapteijn, F., and Weckhuysen, B.M. (2008) *Chem. – Eur. J.*, **14**, 1718; (c) Mores, D., Stavitski, E., Kox, M.H.F., Kornatowski, J., Olsbye, U., and Weckhuysen, B.M. (2008) *Chem. – Eur. J.*, **14**, 11320.
38. Schoonheydt, R.A. (2010) *Chem. Soc. Rev.*, **39**, 5051.
39. (a) Stavitski, E., Pidko, E.A., Kox, M.H.F., Hensen, E.J.M., van Santen, R.A., and Weckhuysen, B.M. (2010) *Chem. – Eur. J.*, **16**, 9340; (b) Stavitski,

E. and Weckhuysen, B.M. (2010) *Chem. Soc. Rev.*, **39**, 4543.

40. (a) Kärger, J., Kortunov, P., Vasenkov, S., Heinke, L., Shah, D.B., Rakoczy, R.A., Traa, Y., and Weitkamp, J. (2006) *Angew. Chem. Int. Ed.*, **45**, 7846; (b) Tzoulaki, D., Heinke, L., Schmidt, W., Wilczok, U., and Kärger, J. (2008) *Angew. Chem. Int. Ed.*, **47**, 3954; (c) Chmelik, C. and Kärger, J. (2010) *Chem. Soc. Rev.*, **39**, 4543.

41. Kox, M.H.F., Mijovilovich, A., Sättler, J.J.H.B., Stavitski, E., and Weckhuysen, B.M. (2010) *ChemCatChem*, **2**, 564.

42. Kox, M.H.F., Domke, K.F., Day, J.P.R., Rago, G., Stavitski, E., Bonn, M., and Weckhuysen, B.M. (2009) *Angew. Chem. Int. Ed.*, **48**, 8990.

43. (a) van der Veen, M.A., van Noyen, J., Sels, B.F., Jacobs, P.A., Verbiest, T., and de Vos, D.E. (2010) *Phys. Chem. Chem. Phys.*, **12**, 10688; (b) van der Veen, M.A., Sels, B.F., de Vos, D.E., and Verbiest, T. (2010) *J. Am. Chem. Soc.*, **132**, 6630.

44. Kox, M.H.F., Stavitski, E., and Weckhuysen, B.M. (2007) *Angew. Chem. Int. Ed.*, **46**, 3652.

45. Karwacki, L., Kox, M.H.F., de Winter, D.A.M., Drury, M.R., Meeldijk, J.D., Stavitski, E., Schmidt, W., Mertens, M., Cubillas, P., John, N., Chan, A., Kahn, N., Bare, S.R., Anderson, M., Kornatowski, J., and Weckhuysen, B.M. (2009) *Nat. Mater.*, **8**, 959.

46. Stavitski, E., Kox, M.H.F., Swart, I., de Groot, F.M.F., and Weckhuysen, B.M. (2008) *Angew. Chem. Int. Ed.*, **47**, 3543.

47. Zurner, A., Kirstein, J., Doblinger, M., Bräuchle, C., and Bein, T. (2007) *Nature*, **450**, 705.

48. Jung, C., Kirstein, J., Platschek, B., Bein, T., Budde, M., Frank, I., Mullen, K., Michaelis, J., and Bräuchle, C. (2008) *J. Am. Chem. Soc.*, **130**, 1638.

49. Kirstein, J., Platschek, B., Jung, C., Brown, R., Bein, T., and Bräuchle, C. (2007) *Nat. Mater.*, **6**, 303.

50. Michaelis, J. and Bräuchle, C. (2010) *Chem. Soc. Rev.*, **39**, 4543.

51. Roeffaers, M.B.J., Sels, B.F., Uji-i, H., De Schryver, F.C., Jacobs, P.A., De Vos, D.E., and Hofkens, J. (2006) *Nature*, **439**, 572.

52. Roeffaers, M.B.J., De Cremer, G., Uji-I, H., Muls, B., Sels, B.F., Jacobs, P.A., De Schryver, F.C., De Vos, D.E., and Hofkens, J. (2007) *Proc. Nat. Acad. Sci. U.S.A.*, **104**, 12603.

53. Roeffaers, M.B.J., Hofkens, J., De Cremer, G., De Schryver, F.C., Jacobs, P.A., De Vos, D.E., and Sels, B.F. (2007) *Catal. Today*, **126**, 44.

54. de Smit, E., Swart, I., Creemer, J.F., Hoveling, G.H., Gilles, M.K., Tylisczak, T., Kooyman, P.J., Zandbergen, H.W., Morin, C., Weckhuysen, B.M., and de Groot, F.M.F. (2008) *Nature*, **456**, 222.

55. de Groot, F.M.F., de Smit, E., van Schooneveld, M.M., Aramburo, L.R., and Weckhuysen, B.M. (2010) *ChemPhysChem*, **11**, 951.

56. Beale, A.M., Jacques, S.D.M., and Weckhuysen, B.M. (2010) *Chem. Soc. Rev.*, **39**, 4656.

57. Creemer, J.F., Helveg, S., Hoveling, G.H., Ullmann, S., Molenbroek, A.M., Sarro, P.M., and Zandbergen, H.W. (2008) *Ultramicroscopy*, **108**, 993.

58. de Smit, E., Swart, I., Creemer, J.F., Karunkaran, C., Bertwistle, D., Zandbergen, H.W., de Groot, F.M.F., and Weckhuysen, B.M. (2009) *Angew. Chem. Int. Ed.*, **48**, 3632.

25
Adsorption Methods for Characterization of Porous Materials

Evgeny Pidko and Emiel Hensen

25.1
Introduction

Porous materials play an important role in many industrial processes. The fields of application include catalysis, chromatography, fluid transport, and gas storage. Porous solids are characterized by a large surface to volume ratio. This means that a large part of their atoms is located close to or at the surface and therefore they become accessible for the surrounding medium. As a result, porous solids exhibit numerous remarkable properties that are useful for catalysis and sorption.

For the development of heterogeneous catalysts based on porous solids, detailed information about the pore structure is important. Most often, catalytic reactions occur on the internal surface area, because it is much larger than the external one. Besides, these solids exert specific types of selectivity on the catalytic reactions occurring in their pores, and this critically depends on the pore size and topology. Adsorption is a key step in heterogeneous catalysis. It is also extensively used to probe the texture of porous solids.

This chapter presents an overview of adsorption-based methods for the characterization of porous materials. The main focus will be on the determination of the surface area and the pore size and volume of heterogeneous catalysts. The information that is available through such experimental techniques ranges from properties such as the total surface area and pore volume to the fine details about the individual pores in the materials. In this chapter we will limit ourselves to the discussion of such methodologies based on the non-specific interactions between adsorbed molecules and solid sorbents.

25.2
Physical Adsorption

Non-specific intermolecular interaction between adsorbed molecules (adsorbate) and a solid material (adsorbent) are at the heart of the phenomenon called *physical adsorption* (physisorption). By definition, *adsorption* is the adhesion of molecules to solid surfaces. Two types of adsorption can be distinguished, namely physical and chemical adsorption. The latter involves the formation of a chemical bond between the adsorbate and the surface atoms of the solid. This process is usually quite exothermic and can be viewed as a true chemical reaction. The heats of chemisorption are rarely less than 80 kJ mol^{-1} and often exceed 400 kJ mol^{-1}. Because this type of adsorption is very specific, it cannot be used as a general method to determine the textural properties of a solid. An example is H_2-chemisorption to determine the metal surface area of supported metal particles: chemical bonds formed between the dissociated H atoms and the metal surface. This method cannot be generally used to determine the surface area of an oxide material. In contrast, physisorption relies on the formation of only weak non-specific bonds between the surface and the adsorbed molecules. It involves rather weak forces of molecular interaction such as permanent dipole, induced dipole, or quadrupole interactions (van der Waals forces, London dispersion forces). As a result, the physisorption energy is usually in the range between 10 and 40 kJ mol^{-1}. By analogy with the attractive forces in real gases, physical adsorption can be viewed as an increase of the concentration at the gas–solid or gas–liquid interface driven by the accumulation of van der Waals forces. At high adsorbate pressures, physisorption becomes almost indistinguishable from the condensation process.

The strength of attractive intermolecular interaction is proportional to the inverse sixth power of the intermolecular distance

$$E_{attr} = -\frac{C}{r^6} \tag{25.1}$$

where C is a coefficient that depends on the nature of the interacting species (e.g., ionization energies, polarizability, and dipole moment) and r is the mean distance between them.

When molecules approach each other closely, the nuclear and electronic repulsions and the rising electronic kinetic energy outweigh the attractive force. Such repulsions increase very rapidly with decreasing intermolecular distance. The repulsive energy between two molecules can be approximated by

$$E_{rep} = \frac{B}{r^{12}} \tag{25.2}$$

where B is an empirical coefficient.

The potential energy between two interacting molecules is then a sum of the attractive and repulsive terms:

$$E(r) = -\frac{C}{r^6} + \frac{B}{r^{12}} \tag{25.3}$$

This equation is often designated as the Lennard–Jones potential (L–J potential). As will be shown later, the interaction potential expressed in such a form usually forms the basis of practical methodologies involving the description of intermolecular interactions. The general forms of the curve $E(r)$ as well as the contributions from the *repulsion* and the *attraction* terms are indicated in Figure 25.1.

It should be noted that the L–J potential is the result of substantial oversimplifications. As a result, the details of the potential energy curve may vary significantly

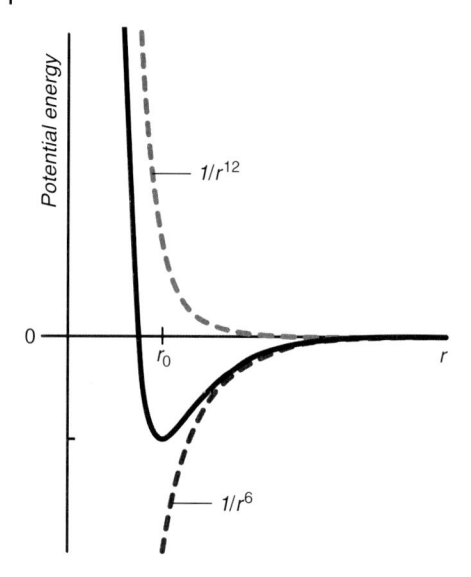

Figure 25.1 The general shape of the Lennard–Jones potential. The dashed lines are the repulsive ($1/r^{12}$) and attractive ($-1/r^6$) contributions.

depending on the nature of the interacting species. Nevertheless, it still provides a reliable description of the adsorption phenomenon if the constants B and C describing the nature of the interacting atoms are properly adjusted for a particular type of interaction.

To describe the adsorption of a gas on a solid surface using this approach, the interactions of the atoms from the surface layers with the atoms of the adsorbed molecule have to be estimated. The nature of the interaction is crucial in this case. As mentioned above, the dispersion interactions dominating physisorption are non-specific and additive. Therefore, the total interaction between the adsorbed molecule and the adsorbent can be calculated by summing the individual pairwise interactions (containing both the attraction and the repulsion terms) of each atom of the gas molecule with each surface atom. In practice, only a limited number of atoms of the solid need to be considered because of the rapid decay of the interatomic potential described by Eq. (25.3) (see Figure 25.1). The use of such an approach for the description of physical adsorption implies that the dominant contribution to the total adsorbate–adsorbent interaction arises from the potential between the neighboring atoms that directly interact with each other, whereas the contribution of the atoms further apart is only minor. Taking into account the additive nature of the interatomic potential, we can conclude that the curvature of the surface will have a crucial influence on the mutual interaction. When the curvature of the surface increases (the pore becomes smaller), a larger number of surface atoms are located in the immediate vicinity of the adsorbed molecule. This results in a larger number of pairwise interatomic interactions and, ultimately, in a higher strength of the total interaction between the adsorbate and the adsorbent.

25.3
Classification of Porous Materials

There are different types of porous systems in solid materials. Very often, porous solids will contain pores within a certain range of sizes. The average width of the pores (the diameter of a cylindrical pore or the distance between the sides of a slit pore) is the most basic and important feature for the adsorption properties of porous materials. A classification of pores according to their average width was originally proposed by Dubinin [1] and has now been officially adopted by the International Union of Pure and Applied Chemistry [2]. The basis of this classification (Table 25.1) is that each pore size range corresponds to characteristic adsorption effects apparent from the isotherms. The interaction potential in micropores is substantially higher than in systems with larger pores, resulting in a larger amount of adsorbed gas at a given pressure. Mesopores are characterized by a specific hysteresis loop in the adsorption isotherm corresponding to capillary condensation, whereas the very large size of macropores make it practically impossible to map out the isotherm in detail. It is, however, important to note that the borders between different pore types are diffuse, because the adsorption behavior depends not only on the pore size, but also on the pore shape and the nature of the adsorbent.

25.4
Adsorption Isotherms

The types of interactions realized between the gas molecules and the solid surface upon adsorption are reflected in the adsorption isotherms. An adsorption isotherm shows the relationship between the amount of molecules taken up by an initially clean solid surface at a constant temperature and the equilibrium pressure of the gas. Normally, the quantity of gas adsorbed is expressed as the mass of adsorbate or the gas volume reduced to STP (standard temperature and pressure), while the pressure is expressed as a dimensionless relative pressure (p/p_0) that is the actual pressure p normalized to the saturation pressure p_0 at the temperature at which the experiment is performed.

For all kinds of solids and adsorbates, adsorption isotherms generally follow one of six forms (Figure 25.2). The classification, including the first five classes (I–V), was originally proposed by Brunauer, Deming, Deming, and Teller (BDDT) [3],

Table 25.1 Classification of pores according to their size.

	Pore size
Micropores	Below ~2 nm
Mesopores	Between ~2 and 50 nm
Macropores	Larger than ~50 nm

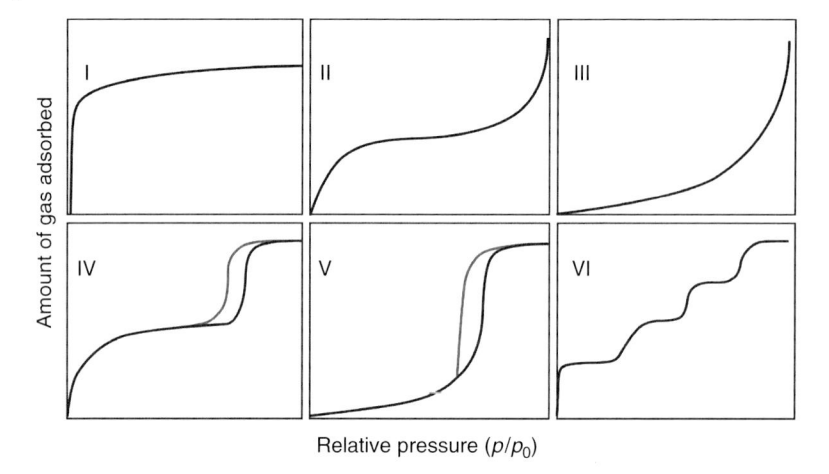

Figure 25.2 The six basic types of adsorption isotherms.

often referred to as *Brunauer classification* [4]. The sixth form is a later addition. Type I is typical for adsorption in micropores and is usually observed in the case of adsorption on molecular sieves or activated carbons. Isotherms II and IV are indicative of multilayer physisorption on either nonporous adsorbents or those having large pores. In addition, the shape of isotherm IV is indicative of capillary condensation in mesopores. This type is often observed in the case of practical heterogeneous catalysts. Isotherms III and V appear when the interaction between the molecules of adsorbent is stronger than that between the adsorbent and the surface of the solid, for example, water adsorption on gold. The type VI isotherm is characteristic of a nonporous solid with an almost perfectly uniform surface. Such isotherms are quite rare and can be observed, for example, in the case of Kr adsorption on graphite at 90 K [5]. In this chapter the application of isotherms of types I, II, and IV for the determination of surface area and pore size distribution (PSD) will be discussed.

25.5
The Application of Adsorption Methods

The adsorption of a gas by a solid can yield valuable information about the specific surface area, pore structure, and size of the solid. In practice, the range of gases suitable for the extraction of such information is quite narrow. The most commonly used adsorbent is nitrogen at its boiling point of 77.4 K and to a lesser extent argon. Nitrogen and argon adsorption and capillary condensation are the best probes for assessing specific surface area and PSDs of mesoporous materials. A type II isotherm is characteristic of a non-porous material, whereas the isotherm IV indicates the presence of mesoporosity in solids. In both cases, specific surface area can be determined from the adsorption experiment. *Specific surface area (A)* is directly related (Eq. (25.4)) to the monolayer capacity of the solid, that is, defined as the amount of adsorbate which can be accommodated in a completely filled dense single molecular layer (a monolayer) on the surface of unit mass of the solid (1 g).

$$A = n_m a_m L \tag{25.4}$$

where n_m is the monolayer capacity expressed in moles of adsorbate per gram of adsorbent; a_m is the average area occupied by a molecule of adsorbate in the completed monolayer, and L is the Avogadro constant. To obtain meaningful information, the dimensions of the adsorbed molecules must be smaller than the diameter of the pores. Usually for determining specific surface area adsorption of N_2 at its boiling point 77.4 K is used. One should note that to obtain the monolayer capacity from the adsorption measurement, the adsorption isotherm must be interpreted in quantitative terms. Selected theories available for these purposes are described later.

In microporous materials, which are characterized by the type I isotherm, the pore size and the size of the adsorbate molecules are similar. As a result, the very concept of specific surface area is not applicable. The values of specific surface area determined for microporous substances have no physical meaning, but only indicate the volume of the adsorbed gas. The average pore size in microporous substances may be readily determined by size exclusion measurements. These involve a series of measurements of the uptake of different sorbates with increasing minimal kinetic diameter. The minimum pore diameter is then defined by the drop in the amount of adsorbed gas with increasing molecular size.

The vast majority of heterogeneous catalysts are solids with high surface area. The catalytic performance of these materials crucially depends on the area accessible by the reactant molecules. This area is related the specific surface of the catalyst and also crucially depends on the PSD, which determines the ease of transport of the molecules from the exterior of the catalytic particles into their internal space where the catalytic sites are accommodated.

25.6
Theoretical Description of Adsorption

There is a dynamic equilibrium between the molecules in the gas phase and those adsorbed on the surface. The variation of surface coverage with pressure at a chosen temperature is reflected in the adsorption isotherm. Theoretical description of such experimental data is required to extract useful information about the surface characteristics of the solid material.

25.6.1
Langmuir Isotherm

The simplest physically sound isotherm was originally put forward by Langmuir almost a century ago [6]. It is based on the following three assumptions:

- Gases form only one monolayer on a solid.
- The surface of the adsorbent is uniform. All adsorptions sites are equivalent.
- Lateral interactions are absent. The ability of molecules to adsorb at a site does not depend on the occupation of neighboring sites. This implies that the adsorbed molecules do not interact with each other.

An adsorption process can be viewed as a chemical reaction

$$A_{gas} + \text{Surface} \rightleftharpoons A - \text{Surface}$$

The rate of the forward reaction – adsorption – (r_{ads}) can then be written as

$$r_{ads} = k_a \left(\frac{p}{p_0}\right) N(1-\theta) \tag{25.5}$$

where p/p_0 is the relative pressure, $N(1-\theta)$ is the number of sites in the uncovered fraction of the surface, and N is the total number of surface sites; k_a is the rate constant for adsorption.

The rate of desorption (r_{des}) is given by

$$r_{des} = k_d N\theta \tag{25.6}$$

where k_d is the rate constant for desorption.

At equilibrium there is no net change of the surface coverage θ, that is, the rates of the forward and backward reactions are equal ($r_{ads} = r_{des}$), and hence

$$k_a \left(\frac{p}{p_0}\right) N(1-\theta) = k_d N\theta \tag{25.7}$$

As the adsorption equilibrium constant K is equal to k_a/k_d, it follows that

$$\theta = \frac{n}{n_m} = \frac{K p/p_0}{1 + K p/p_0} \tag{25.8}$$

where n is the number of moles adsorbed on 1 g of adsorbent and n_m is the monolayer capacity expressed in the same units.

In order to use the Langmuir isotherm for the analysis of the experimental data, Eq. (25.8) may be rewritten in the so-called reciprocal form:

$$\frac{1}{n} = \frac{1}{n_m} + \frac{1}{K n_m} \cdot \frac{1}{p/p_0} \tag{25.9}$$

The plot of $1/n$ against $1/(p/p_0)$ should therefore yield a straight line. The monolayer capacity can be calculated from the intercept $1/n_m$. The equilibrium constant is found from the tangent $\alpha = 1/Kn_m$.

In practice, however, the assumptions underlying the derivation of the Langmuir isotherm are seldom entirely valid. Therefore, adsorption on a practical material can seldom be described solely by a Langmuir equation. In some cases a good straight line is obtained by plotting $1/n$ against $1/(p/p_0)$, whereas in others the line is substantially curved. In most cases, the surface of the adsorbent is not uniform and contains adsorption sites that bind molecules with different strengths. Furthermore, often the heat of adsorption varies with the amount adsorbed, thus implying the presence of mutual interactions between the molecules on the surface. The adsorption then can be treated by a sum of several Langmuir isotherms including corrections for the intermolecular interactions in the adsorbed monolayer.

25.6.2
BET Theory

For small inert molecules the adsorption energy is close to the heat of condensation. This results in adsorption on the uncovered surface creating the first adsorption layer and parallel adsorption on the already adsorbed layers, thus resulting in multilayer adsorption. A major advance in adsorption theory was the development of the BET (from the surnames of its inventors Braunauer, Emmett, and Teller [7]) theory of adsorption, which generalized the Langmuir approach and included the treatment of multilayer adsorption. The fundamental assumption is that the binding in the second and all subsequent adsorption layers is of the same nature as the forces responsible for condensation of gases. Thus, to derive the relationship between the monolayer capacity and the overall amount of gas adsorbed, the following assumptions are made:

- The surface of the adsorbent is uniform. The adsorption energy within the first layer is constant.
- In all layers, except the first, the adsorption energy is equal to the molar heat of condensation, and the condensation–evaporation conditions are identical in these adsorption layers.
- When the pressure p reaches the saturation pressure p_0, the number of layers becomes infinite and the adsorbent condenses to a bulk liquid on the surface of the solid.

Under dynamic equilibrium, the rate of condensation (adsorption) of gas molecules onto an already adsorbed layer is equal to the rate of evaporation (desorption) from that layer. By summing for an infinite number of adsorption layers, one arrives at the BET equation

$$\frac{V}{V_{ml}} = \frac{c\left(\frac{p}{p_0}\right)}{\left(1 - \frac{p}{p_0}\right)\left[1 - (1-c)\frac{p}{p_0}\right]} \qquad (25.10)$$

where V is the total volume of adsorbed gas, V_{ml} is the amount of gas adsorbed in the monolayer, and c is a constant that, in practice, often depends only on the heat of adsorption in the first layer (ΔH_d) and the heat of vaporization (ΔH_{vap}) for a chosen adsorbent:

$$c = e^{(\Delta H_d - \Delta H_{vap})/RT} \qquad (25.11)$$

For convenience of plotting, Equation 25.10 is often rewritten as

$$\frac{p}{V(p_0 - p)} = \frac{1}{cV_{ml}} + \frac{c-1}{cV_{ml}}\frac{p}{p_0} \qquad (25.12)$$

from which a plot of $p/[V(p_0-p)]$ vs p/p_0 Nθ yields a straight line with intercept $1/V_m c$ and slope $(c-1)/V_m c$. From this, the values of V_m and c are readily derived. The BET surface area is then calculated using Eq. (25.4).

BET theory is applicable for the description of type II and type IV isotherms. The BET isotherms fit experimental observation reasonably well only within a limited range of partial pressures (from p/p_0 0.05 to 0.3). At low pressures it tends to underestimate the extent of adsorption, whereas this is overestimated at higher pressures.

It must be emphasized that the BET analysis is not applicable to the description of adsorption in materials containing micropores and small mesopores, because the conditions of multilayer adsorption are not fulfilled here. The values of BET surfaces reported for such systems should rather be interpreted as numbers proportional to the pore volume than specific surface area.

25.6.3
Standard Isotherms and the t-Method

Every adsorbate–adsorbent system is characterized by a unique adsorption isotherm. Nevertheless, the general features and shapes of the isotherm curves are observed for large classes of materials with different surface areas when analyzed by the same adsorbent under the same conditions. Often these isotherms can be superimposed by normalizing the X axes, resulting in the so-called 'standard' isotherms, which are individually applicable to a specific type of materials. The introduction of the concept of standard isotherms allowed the volume of micropores to be determined, the presence of mesopores to be detected, and the thickness of the adsorbed gas on external surfaces and on the walls of meso- and micropores to be quantified.

There are several methods of normalizing the amounts adsorbed of the reference and sample data sets. One is based on normalizing the volume of adsorbed gas V_a by the monolayer capacity V_m, that is, V_a/V_m.

The concept of standard isotherms is used in the so-called t-method of Lippens and de Boer [8] for the determination of specific surface area. This method is based on the t-curve, which uses the statistical thickness of the adsorbed film t as the dependent variable instead of the fraction of monolayer capacity (V_a/V_m). The thickness t is then determined by taking V_a/V_m to be equal to the number of molecular layers in the film and multiplying by the thickness of a single adsorption layer σ. For nitrogen at its boiling point, Lippens et al. [9] assumed $\sigma = 3.54$ Å. The average thickness is then

$$t = \sigma \frac{V_a}{V_m} = 3.54 \left(\frac{V_a}{V_m} \right) \tag{25.13}$$

The amount adsorbed V_a is then plotted against t values instead of p/p_0, resulting in a t-plot. For any value of t there is some V_a value for both the standard and for the test solid. If the shape of the isotherm for the test material is identical to that of the standard, the t-plot will be a straight line passing through the origin. Its slope

b_t is equal to V_m/σ, because the number of adsorbed layers is equal to both t/σ and V_a/V_m.

Therefore,

$$V_a/V_m = t/\sigma \quad \text{or} \quad V_a = V_m \frac{t}{\sigma} = b_t t \tag{25.14}$$

Using Eq. (25.4), the relationship between the slope b_t and the specific surface A can be derived as:

$$A = a_m \sigma L b_t \tag{25.15}$$

Thus, the specific surface is readily available from the t-plot. It should be emphasized that the t-plot does not provide an independent value of the specific surface area, but rather represents a different and somewhat more convenient approach to determining this value than by using the BET theory.

In practice, t-plots often deviate from ideal linear behavior. The deviations in the behavior of t-curves can be used to deduce the nature of the pores in the adsorbent and to determine its micropore volume. The relevant examples are illustrated in Figure 25.3. The linear t-curve 1 can be extrapolated to the origin, which indicates that this curve corresponds to a so-called t-material, whose specific surface area can be readily calculated from the tangent of the curve. In the case of curve 2, the gas uptake rapidly increases at a certain relative pressure p/p_0, resulting in a strong upward deviation of the curve from linearity (curve 2 in Figure 25.3). Such behavior indicates that adsorption is accompanied by capillary condensation starting from t_1. When multilayer adsorption takes place in some narrow pores, this part of the surface becomes unavailable for adsorption. In this case, the t-curve (curve 3 in Figure 25.3) deviates downwards from the standard. Such behavior is observed, for example, in the presence of slit-shaped pores. Curve 4 in Figure 25.3 illustrates adsorption in micropores. At very low partial pressures, large quantities of gas are adsorbed, resulting in a positive intercept of the curve. Such deviation is due to the much stronger adsorption in the micropores. The value of the intercept can be used to estimate the micropore volume (V_{micro}) of the adsorbent. In this case, the surface can still be determined from the tangent of the t-curve.

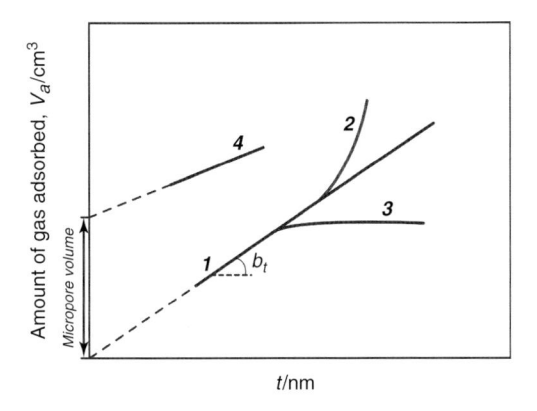

Figure 25.3 Common deviations in t-curves.

25.7
Characterization of Microporous Materials

Pores in real materials often have a non-uniform structure, and are typically represented by various distorted shapes. They twist, turn, or become broader. The analysis of pore systems requires the use of mathematical methods that assume a certain shape of pores in the solid. The pores are often assumed to be cylinders of variable length or slits of infinite size with parallel walls. The information of PSD is important for all types of porous materials, but it is particularly important for the microporous materials in view of their unique role in catalysis and gas separation. Micropores are characterized by pore dimensions below 2 nm. The well-known examples of microporous materials are zeolites and activated carbons. Such materials are characterized by type I isotherms, where large quantities of gas are adsorbed at very low pressures ($p/p_0 \approx 0.15$). In the range of p/p_0 between 0.0 and 0.15 a sharp bend in the isotherm is observed, followed by a horizontal plateau until $p/p_0 = 0.95$. The uptake at the level of the horizontal plateau represents the microporous volume. Extracting the information about the pore distribution requires further analysis of the adsorption data. In this section we will consider a selection of techniques suitable for this analysis.

25.7.1
Dubinin–Radushkevich and Dubinin–Astakhov Methods

Whereas the concept of an adsorbed layer of gas on the surface is valid for large-pore adsorbents, it is not entirely applicable for microporous materials, where the pore dimensions are comparable to the size of the adsorbed molecules. An empirical equation developed by Dubinin and Radushkevich [10] is based on the adsorption potential A, derived by Polanyi [11], which describes the differential molar work of adsorption.

$$A = RT \ln\left(\frac{p_0}{p}\right) = -\Delta G \tag{25.16}$$

where R is the universal gas constant, T is the absolute temperature, p_0/p is the reciprocal relative pressure, and ΔG is the Gibbs free energy of adsorption.

The Dubinin–Radushevich (DR) equation is:

$$W = W_0 e^{[-(A/\beta E_0)^2]} \tag{25.17}$$

where W is the amount of gas adsorbed at a particular p/p_0 and temperature T, W_0 is the limiting micropore volume, E_0 is the characteristic energy of adsorption for benzene vapor as the reference gas, and β is the affinity coefficient that equals 1 when a reference vapor is used as the adsorbent.

Equations (25.16) and (25.17) together give

$$\ln W = \ln W_0 - \left(\frac{RT}{\beta E_0}\right)^2 \left(\ln^2 \frac{p_0}{p}\right) \tag{25.18}$$

The values of ln W are then plotted against $\ln^2(p_0/p)$. In some cases experimental isotherms yield straight lines over a range of pressures, while in other cases substantial deviations are observed. If this procedure yields a straight line, the value of β can be extracted from its slope, while the limiting micropore volume W_0 can be calculated from the y-axis intercept. It should be noted that the W_0 value in this case corresponds to a gas volume, but in order to characterize appropriately the total microporosity, it must be converted into its liquid volume. Usually, straight lines are obtained when benzene and nonane are used as the adsorbates, while sometimes nitrogen is also able to produce straight lines within the DR approach. If the plotted curve deviates from a straight line, the interpretation of this is questionable.

A further development of the DR equation resulted in a more general approach based on the equation derived by Dubinin and Astakhov [12] (DA equation).

$$W = W_0 \, e^{[-(A/\beta E_0)^n]} \qquad (25.19)$$

One can see that the DR equation represents a special case of the DA equation. In this approach the value of n is selected in such a way as to yield the best linear fit. The n values usually range between 1 and 3. In the DA approach the data is analyzed similarly to the manner described above, resulting in a value of the micropore capacity of the adsorbent.

25.7.2
Horvath–Kawazoe (HK) Equation

The Horvath–Kawazoe (HK) method for determining the micropore size distribution [13] is based on the assumption that the relative pressure p/p_0 required to completely fill micropores of a certain size and shape is directly related to the interaction energy between the adsorbate and adsorbent. This implies that with increasing adsorbate pressure the micropores are progressively filled. Therefore, at a given relative pressure, only pores that are smaller than a certain unique size will be filled. Another assumption is that the gas adsorbed in micropores behaves thermodynamically as a two-dimensional gas. Using the HK method, the PSD in the micropore range can be calculated from the data obtained at very low adsorbate pressures. In practice, the HK method entirely relies on high resolution adsorption measurements at very low pressure. This methodology constitutes a very attractive approach because it can be readily combined with the conventional measurements of nitrogen and argon isotherms. It should, however, be noted that in order to obtain reliable results this method requires calibration against a well-characterized microporous material before analyzing microporosity in unknown solids.

The HK model implies that the entropy contribution to the free energy of a system upon adsorption is small compared to the large change in enthalpy. Therefore, the free energy change (ΔG) can be expressed as

$$\Delta G^{\text{ads}} = RT \ln \left(\frac{p}{p_0}\right) = U_0 + P_a \qquad (25.20)$$

where U_0 expresses the interaction between the adsorbate and the adsorbent and P_a is an implicit function of the adsorbate–adsorbate–adsorbent interaction.

The HK method utilizes a model describing the interaction of one adsorbate molecule with two infinite lattice planes of adsorbent separated by a distance, L (Figure 25.4):

$$E(z) = K\varepsilon_0 \left[-\left(\frac{\sigma}{z}\right)^4 + \left(\frac{\sigma}{z}\right)^{10} - \left(\frac{\sigma}{L-z}\right)^4 + \left(\frac{\sigma}{L-z}\right)^{10} \right] \quad (25.21)$$

where E is the potential energy of interaction for the distance z between the adsorbate molecule and surface atom in a slit wall, K is a constant, ε_0 represents the potential energy minimum, and σ is the distance from a surface atom at zero interaction energy.

The volumetric averaging of the potential expressed in Eq. (25.21) results in the average interaction energy, which is further related to the free energy change of adsorption (ΔG^{ads}, Eq. (25.20)). This ultimately yields

$$\Delta G^{\text{ads}} = RT \ln \frac{p}{p_0} = N_a \left(\frac{N_{AS}A_{AS} + N_{AA}A_{AA}}{\sigma^4(L-2d)} \right) \left[\frac{\sigma^4}{3(L-d)^3} - \frac{\sigma^{10}}{9(L-d)^9} - \frac{\sigma^4}{3d^3} + \frac{\sigma^4}{3d^9} \right] \quad (25.22)$$

where N_a is the Avogadro constant, N_{AS} is the number of atoms per unit area of solid surface, N_{AA} is the number of molecules adsorbed per surface area unit, A_{AS} and A_{AA} are the constants in the L–J potential for the adsorbate–adsorbent and adsorbate–adsorbate interactions, and d is the distance between the adsorbed molecule and the solid surface.

In a simplified form, Eq. (25.22) can be rewritten as

$$RT \ln \frac{p}{p_0} = U_0 + P_a = N_a \xi(L) \quad (25.23)$$

Using the HK method, the pore size in microporous materials can be estimated. To calculate the PSD, the relative pressure corresponding with a certain pore width L is calculated first by means of Eq. (25.22). Then, the amount of gas adsorbed at this value of relative pressure is estimated using the experimental adsorption isotherm. Differentiation of the amount adsorbed with respect to the pore width yields the PSD in the microporous range.

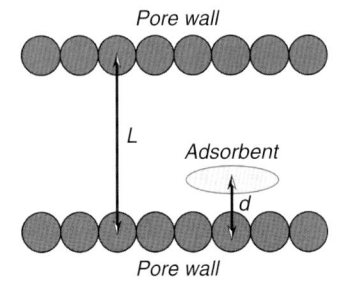

Figure 25.4 A simplified view of adsorption in a slit pore.

25.8
Characterization of Mesoporous Materials

According to the general definition, mesoporous materials are characterized by the presence of pores with diameter between 2 and 50 nm, while larger pores (above 50 nm) are normally classified as macropores. Already below saturated vapor pressures, condensation of the adsorbate takes place in the mesoporous space. Such a capillary condensation in mesopores is associated with a shift in the vapor–liquid equilibrium in the confined space compared to the bulk fluid. The condensation pressure in this case is influenced by such factors as pore size and pore shape as well as by the strength of interaction between the adsorbate and the pore walls.

Capillary condensation is characterized by a step in the adsorption isotherm (type IV). A sharp step associated with this phenomenon is observed for materials with uniform PSDs (Figure 25.5). The adsorption mechanism at low relative pressures resembles that on planar surfaces (stage 1, Figure 25.5). The formation of a dense monolayer is followed by multilayer adsorption (stage 2). After reaching a critical thickness of the adsorbed layer (3), the adsorbate undergoes capillary condensation within the pore (stages 3 and 4). The subsequent plateau section of the isotherm observed at high relative pressures corresponds to the situation when a pore is completely filled with the condensed adsorbate (liquid), which is separated from the gas by a hemispherical meniscus. Upon desorption, the adsorbate is released from the pore by a thinning of the meniscus. At the pressure where the desorption branch coincides with the adsorption branch of the isotherm (end of hysteresis loop, stage 6), there is an equilibrium between a multilayer film on the pore walls and the gaseous adsorbate in the center of the pore [14].

Four types of hysteresis loop are usually distinguished [15], and each of them can be correlated with a particular type of capillary shape of the adsorbent (Figure 25.6). The H1 type is usually associated with porous materials containing well-defined cylindrical pore systems or agglomerates of uniform spheres. The materials that give rise to the H2 type hysteresis are usually disordered and are characterized by a not well-defined PSD. Materials exhibiting the type H3 loop do not show

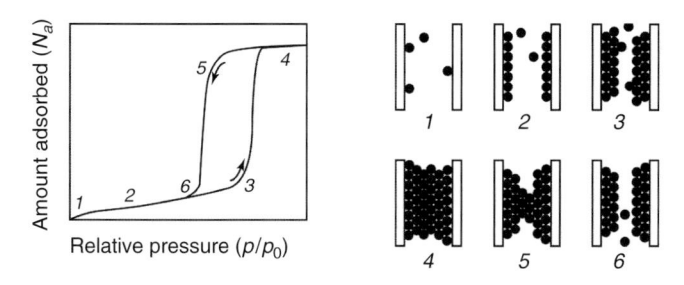

Figure 25.5 An adsorption isotherm in a cylindrical mesopore and a schematic representation of the involved multilayer adsorption and pore condensation processes [14].

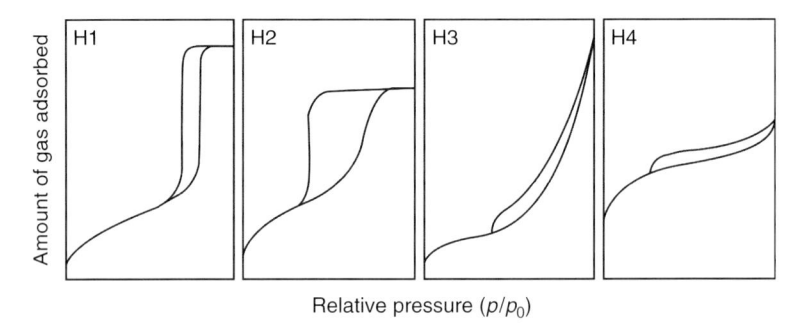

Figure 25.6 Types of hysteresis loops.

any limiting adsorption at high relative pressures. Such behavior is observed for non-rigid aggregates of plate-like particles, resulting in slit-shaped pores. The type H4 is usually associated with the presence of narrow, slit-like pores that may include pores in the micropore region.

One can see that adsorption isotherms can directly provide crucial information about the properties of the mesoporous space. By using the information about the relationship between the critical condensation pressure and the pore size, the mesopore size distribution of the adsorbent can be characterized. In this section we discuss the application of selected classical and more advanced mathematical models that allow a correlation function between these parameters to be established.

25.8.1
The Kelvin equation

The Kelvin equation provides a basis for the description of the condensation process in pores with regular shape, such as ideal slit pores or cylindrical pores. Using this method, the distribution of pore volume as a function of pore size can be calculated from the relative pressures at which the filling and the emptying of pores occurs. This method rests on the macroscopic description of the shift of the gas–liquid phase transition in a confined space from the bulk coexistence into quantities like surface tension and densities of gas and liquid. For the adsorption process in a cylindrical pore, the Kelvin equation for the adsorption branch is usually written as:

$$\ln \frac{p_A}{p_0} = -\left(\frac{\gamma V_L}{RT(r-t)} \right) \qquad (25.24)$$

and for the desorption branch it is given as

$$\ln \frac{p_D}{p_0} = -\left(\frac{2\gamma V_L}{RT(r-t)} \right) \qquad (25.25)$$

where r is the pore radius, p_A/p_0 and p_D/p_0 are the relative pressures at which pore filling and emptying, respectively, start; γ is the liquid surface tension, t is the thickness of an adsorbed multilayer film that is formed prior to the pore condensation.

The Kelvin equation gives a correlation between the pore size and the relative pressure at which pore condensation starts. At each value of p_A/p_0, the value of t can be calculated by using the t-method. The pore radius can then be calculated from the relative pressure at which capillary condensation starts. The Kelvin equation predicts that pore condensation will occur at higher relative pressures for materials with larger pores. This method forms the basis for the traditional methodologies for analysis of mesoporous systems such as the Barret–Joyner–Halada (BJH) method [16].

25.8.2
BJH Method

To calculate PSD using the Kelvin equation, an imaginary process is considered that involves stepwise emptying of the pores occupied with the condensed adsorbate following a decrease of the relative pressure. This approach was originally proposed by Barrett *et al.* [16]. An additional (to those already made for the derivation of the Kelvin equation) assumption in the BJH methodology is that all pores of the adsorbent are completely filled with the condensed adsorbate (fluid) at the initial relative pressure of desorption, close to unity, that in practice is usually close to 99.5% ($p/p_0 = 0.995$). Thus, in the first step of the desorption process, only the capillary condensate is removed from the pores. One should note that already at the next step both the condensate is removed from the filled pores and the depletion of the adsorbed multilayers occurs in the pores that have been partially emptied in the first step.

Let us divide the isotherm into a series of finite equal steps along the decreasing relative pressure: for example, 0.95, 0.90, 0.85, 0.80, and so on. The volume of pores emptied at each of these steps corresponds to the amount of adsorbent (converted to a liquid volume) released at that step. The thickness t of the remaining adsorbed multilayer in the pore is then calculated by using, for example, an equation proposed by de Boer *et al.* [17] for estimating t in the case of N_2 physisorption at 77 K:

$$t = \left(\frac{13.9}{0.034 - \log\left(\frac{p}{p_0}\right)} \right)^{1/2} \tag{25.26}$$

Then the pore radius r is calculated using the Kelvin equation. To determine distributions of radius, area, and volume of the pores, the complete procedure has to be carried out by performing repetitive calculations for the sequence points of steps defined for the isotherm.

Nowadays, the BJH method can be viewed as the standard adsorption-based method for determining the PSD of mesoporous materials. It should be emphasized that its application results in substantial errors. The method is based on a simplified macroscopic description of the capillary condensation process. As a result the microscopic phenomena are completely neglected within this methodology, which makes the method inaccurate for materials with pore sizes in the range of 2–4 nm. For such mesoporous systems, the BJH method may lead to underestimation of pore sizes by up to 1 nm. Recent advances in analyzing adsorption isotherms based on the nonlocal density functional theory (DFT) have dramatically improved the accuracy of calculating PSD in porous materials.

25.8.3
Nonlocal Density Functional Theory (NL–DFT)

The applicability of classical methodologies discussed above for extracting valuable information about porosity from the physical adsorption data is limited to particular types of isotherms and to limited pressure regimes. Whereas the BJH method based on the Kelvin equation is suitable for the analysis of mesopores, it fails to describe adequately the pore-filling mechanism in microporous solids and those containing narrow mesopores. On the other hand, the DR and HK approaches developed for the description of the adsorption in micropores cannot be used to analyze mesoporous adsorbents. Thus, to analyze materials containing both micro- and mesopores, a combination of at least two different methods has to be used. In view of the limited accuracy of such macroscopic approaches, an alternative methodology is required that is capable of accurate analysis of adsorption in materials combining different types of porosity.

Molecular models, as for instance the DFT, that make use of statistical thermodynamics to describe the adsorption isotherms using the microscopic properties of the system under investigation, provide a more comprehensive and realistic microscopic model of the pore-filling process and the phase transition in the pores. These theories are capable of connecting the macroscopic observables, such as the adsorption isotherm, to the microscopic properties of the system, such as the geometry and size of the pores as well as the interaction energy parameters between the different components involved in the adsorption.

The fundamental basis for the DFT methodology originates from the work of Hohenberg and Kohn [18], who formally proved that the ground-state energy for a quantum system is a unique function of its electron density. When the application of the DFT formalism to a classical system involving adsorption in a porous solid is considered, this involves consideration of the equilibrium density profile $\bar{\rho}(r)$ of the inhomogeneous fluid at a surface or within a certain set of pores. The derivation of $\bar{\rho}(r)$ allows a direct calculation of the adsorption isotherm and other thermodynamic properties, for example, the adsorption energy.

Under equilibrium conditions the chemical potential is uniform throughout the system. In the case of a homogeneous fluid, only temperature and pressure determine the chemical potential. This is not the case when the fluid is confined inside the pore. Because of its interaction with the pore walls, the fluid is not homogeneous and its density is not constant. This inhomogeneous fluid is stable only when immersed in the external potential, and this represent a layered adsorbate. The density distribution in this case can be described in terms of density profile, $\rho(r)$, which is a function of the distance from the walls throughout the pore.

The DFT formalism utilizes the statistical mechanical grand canonical ensemble. To establish the density profile $\rho(r)$, two unique free-energy functions of the density

25.8 Characterization of Mesoporous Materials

are introduced: $\Omega[\rho(r)]$ and $F[\rho(r)]$. $\Omega[\rho(r)]$ is the grand thermodynamic potential or grand free energy, and $F[\rho(r)]$ is a function representing the intrinsic Helmholtz free energy. For a multicomponent mixture that is characterized by the grand canonical variables, namely temperature T, volume V, and chemical potentials $\mu_a, \ldots, \mu_i$, we can write:

$$\Omega = F - \Sigma \mu_i N_i \quad (25.27)$$

where F is the Helmholtz free energy and N is the number of species of component i. When an adsorbent gas enters the pore, it is influenced by the presence of the solid surface. The adsorbent is exposed to the attractive potential (due to van der Waals and dispersion interactions with the surface), which causes an increase of the local concentration of the adsorbent near the surface. Thus, for a single-component fluid immersed in the external potential U_{ext} the grand potential functional is given as

$$\Omega[\rho(r)] = F[\sigma(r)] - \int dr\, \rho(r)[\mu - U_{ext}(r)] \quad (25.28)$$

where $\rho(r)$ is the local fluid density at a position r and the integration is performed over the pore volume.

The $F[\rho(r)]$ function can be written as a combination of an ideal gas term (F_{id}), representing a system of noninteracting particles, and the excess intrinsic Helmholtz free energy (F_{ex}) that includes contributions from attractive (F_{att}) and repulsive (F_{rep}) interactions between the adsorbed molecules (fluid–fluid interactions). The interactions between the adsorbent and the pore walls are included in U_{ext} (fluid–solid interactions).

$$F[\rho(r)] = F_{id}[\rho(r)] + F_{ex}[\rho(r)] = F_{id}[\rho(r)] + F_{att}[\rho(r)] + F_{rep}[\rho(r)] \quad (25.29)$$

The ideal gas free energy (F_{id}) can be readily calculated by using an exact expression, whereas the calculation of the excess free energy requires a number of assumptions to be employed. The contribution from the repulsive interactions between the adsorbed molecules (F_{rep}) is commonly represented as the hard-sphere potential (F_{HS}). The free energy due to the attractive interactions (F_{att}) is represented by a mean-field free energy due to the pairwise L–J potentials.

The external potential created by the pore walls depends on the assumed pore geometry and the chemical properties and composition of the adsorbent and the solid. Such a solid–fluid potential U_{ext} is the result of intermolecular interactions, commonly described by the L–J potentials, between the adsorbate and the surface of the solid.

All the interactions that occur within the pore are included in the grand thermodynamic potential Ω. At equilibrium the system is at the minimum of its free energy. The minimization of Ω with respect to $\rho(r)$ yields the equilibrium density profile $\bar{\rho}(r)$ of the system. Because the density profile corresponds to the local density, it must be integrated over the internal volume of the pore to calculate the amount of molecules adsorbed in the pore. The values of chemical potential are directly related to the relative pressure in this case. Therefore, by repeating the integration procedure for different values of μ and, accordingly, p/p_0 the adsorption isotherm can in principle be constructed.

A proper description of the fluid–fluid interactions and accordingly the correct representation of $F[\rho(r)]$ is a challenge. Because the exact form of the hard-sphere free energy functional is not known, additional approximations have to be employed. The free energy of an inhomogeneous fluid depends not only on the density at positions r, but also on the density at other positions near r. To account for such behavior of the confined adsorbate, the so-called nonlocal density functional theory has to be applied (NL-DFT).

Within the NL-DFT approach, the adsorption and desorption isotherms are calculated using the intermolecular potentials for the fluid–fluid and solid–fluid interactions. It is assumed that adsorption occurs within a pore system, where each individual pore has a fixed geometry and is in contact with the bulk adsorbate. The adsorbed fluid in the pores is in equilibrium with the gas phase. The local density of the adsorbate confined within a pore, and therefore exposed to the inhomogeneous external potential U_{ex}, is calculated by the minimization of the grand thermodynamic potential Ω given by Eq. (25.28). The intrinsic Helmholtz free energy is split into the ideal gas, attractive, and repulsive contributions Eq. (25.29)

To describe reliably the repulsions within the confined fluid, several functions for hard-sphere fluids have been developed so far. These include the so-called weighted-density approximation [19], the fundamental-measure theory [20], and the smoothed-density approximation [21]. This last approach is commonly applied in almost all NL–DFT calculations used for pore size characterization. A specific weighted function that takes into account the nonlocal effects is then introduced, resulting in a smoothed density profile. The main concept underlying the smoothed-density or weighted-density approximations is the construction of a smoothed density that represents an average of the true density profile over a local volume.

Thus, the DFT approaches allow a realistic and comprehensive description of physical adsorption in porous solids. By using the DFT methodologies the equilibrium density profile can be calculated, which, after integrating over the pore geometry, forms a theoretical basis for the generation of the realistic adsorption isotherm when the PSD is known or, alternatively, for determining the PSD with high accuracy by using the experimental adsorption isotherm. To compute the PSD, the experimental isotherm is described as a combination of theoretical isotherms in individual pores obtained by means of DFT.

Unlike the classical methods, the application of DFT approaches is not limited to particular ranges of pressures and temperatures. It is important to emphasize, however, that the classical methods for studying adsorption yield analytic functional relationships between the magnitude of adsorption and equilibrium pressures. Since analytic solutions are not possible for DFT approaches, they only produce a numerical result that cannot be generalized for other adsorbate–adsorbent systems aside from the particular one for which the data was calculated.

25.9
Mercury Porosimetry

Another widely-applied method for the determination of the total pore volume and the size distribution of meso- and macropores is mercury porosimetry. This methodology is based on the physical principle that a non-reactive, non-wetting liquid will not enter pores until sufficient external force is applied. Mercury is a non-wetting liquid for almost all substances. The technique involves a progressive increase in the external pressure to force mercury to enter the pores of the solid material. A key assumption of mercury porosimetry is that all pores in the materials have cylindrical geometry. This allows the application of a modified Young–Laplace equation, usually referred to as the *Washburn equation*, which describes the relationship between the applied pressure p and the pore radius r.

$$r = \left(\frac{2\gamma}{p}\right)\cos\theta \tag{25.30}$$

where θ is the contact angle of the mercury on the surface of the solid sample and γ is the surface tension of mercury. In practice, the surface tension of pure mercury and a contact angle of 141° are used for calculations. This allows the determination of pore size from the extent of mercury entering the pores of the solid as a function of the applied pressure. The pressure can usually vary in the range between 0.1 and 2000 bar. Thus, the valid pore range for mercury porosimetry is from 7500 to 3.5 nm. It should be noted that mercury porosimetry does not actually measures the size of the pores, but rather determines the size of the widest connection between the external surface and the pore. Further information about the complex pore structure (e.g., interconnectivity) can be extracted from the hysteresis between the intrusion and extrusion that is almost always observed in mercury porosimetry measurements. This effect is associated with trapping of Hg in the pore system.

25.10
Xenon Porosimetry

Nuclear magnetic resonance (NMR) spectroscopy is a powerful technique to obtain information about the local surrounding of specific nuclei in chemical systems. Obviously, when a molecule is confined within a porous solid, its chemical environment is changed compared to the free state. Even though in the case of physical adsorption the interactions are quite subtle, the nuclei of the atoms in the adsorbed molecules are perturbed by the interaction with the surrounding atoms of the adsorbent. This can result in an observable chemical shift of their NMR, which, in turn, can be correlated with the properties of the porous space.

Ito and Fraissard first introduced the use of NMR spectroscopy to determine the pore size of zeolitic materials from the chemical shift of the adsorbed ^{129}Xe as a probe [22]. The NMR signal of the ^{129}Xe isotope is highly sensitive to its local environment. Xenon is an inert gas with the atomic dimensions of 0.44 nm, and at room temperature it is a supercritical gas. These properties make xenon an excellent probe molecule to characterize microporous materials using the NMR technique. The properties of the porous space (such as the shape and size of micropores as well as their orientation) can in principle be obtained by monitoring the ^{129}Xe chemical shift upon adsorption. However, due to the fast diffusion of the adsorbed gas, the chemical shift observed represents an average of the properties of all pores in the material. To characterize experimentally porous materials at the single-pore level, Telkki, Lounila, and Jokisaari have recently introduced a methodology based on ^{129}Xe NMR spectroscopy of xenon dissolved in a solid or liquid medium confined in a porous material [23]. This approach results in a slowdown of the Xe diffusion and allows the signal of specific Xe atoms located within different parts of the material to be monitored. These signals are characteristics of the local properties of the adsorbent. Using the chemical shifts observed, the distribution of pore size, shape, and orientation in the sample can be deduced.

The sensitivity of conventional NMR techniques based on the use of thermally polarized ^{129}Xe NMR in porous materials is intrinsically limited by the ordinarily low-spin polarization possible even with the strongest NMR magnets. This results in rather low signal-to-noise ratio because of the low concentrations of adsorbed xenon and long relaxation times. A great improvement of the sensitivity of ^{129}Xe NMR can be achieved by transferring angular momentum from a circularly polarized laser to the nuclear spins of xenon. Using these optical pumping methods, hyperpolarized xenon is produced, and the sensitivity of the resulting laser-hyperpolarized (HP) ^{129}Xe NMR method can be increased by several orders of magnitude compared to the conventional technique [24]. The 2D exchange experiments involving HP^{129}Xe NMR have a great potential for detailed characterization of the interconnectivity of micro- and mesoporous systems within solid materials [25].

References

1. (a) Dubinin, M.M. (1960) *Z. Phys. Chem.*, **34**, 959; (b) Dubinin, M.M. (1960) *Chem. Rev.*, **60**, 235.
2. Everett, D.H. *IUPAC Manual of Symbols and Terminology, Appendix 2, Pt. 1, Colloid and Surface Chemistry* (1972) *Pure Appl. Chem.*, **31**, 579.
3. Brunauer, S., Deming, L.S., Deming, W.S., and Teller, E. (1940) *J. Am. Chem. Soc.*, **62**, 1723.
4. Brunauer, S. (1945) *The Adsorption of Gases and Vapours*, Oxford University Press.
5. Amberg, C.A., Spencer, W.B., and Beebe, R.A. (1955) *Can. J. Biochem.*, **33**, 305.
6. Langmuir, I. (1916) *J. Am. Chem. Soc.*, **38**, 2221.
7. Brunauer, S., Emmett, P.H., and Teller, E. (1938) *J. Am. Chem. Soc.*, **60**, 309.
8. Lippens, B.C. and de Boer, J.H. (1965) *J. Catal.*, **4**, 319.
9. Lippens, B.C., Linsen, B.G., and de Boer, J.H. (1964) *J. Catal.*, **3**, 32.
10. (a) Dubinin, M.M. and Radushkevich, L.V. (1947) *Proc. Acad. Sci. USSR*, **55**, 331; (b) Dubinin, M.M. (1965) *Russ. J. Phys. Chem.*, **39**, 697.
11. Polanyi, M. (1914) *Verb. Deutsch. Physik. Ges.*, **16**, 1012.
12. Dubinin, M.M. and Astakhov, V.A. (1971) *Adv. Chem. Ser.*, **102**, 69.
13. Horvath, G. and Kawazoe, K. (1983) *J. Chem. Eng. Jpn.*, **16**, 470.
14. Roque-Malherbe, R.M.A. (2007) *Adsorption and Diffusion in Nanoporous*

Materials, CRC Press, Taylor & Francis Group, New York.
15. Sing, K.S.W., Everett, D.H., Haul, R.A.W., Moscou, L., Pierotti, R.A., Rouquerol, J. and Siemieniewska, T. (1985) IUPAC Recommendations on Reporting Physisorption Data for Gas/Solid Systems. *Pure Appl. Chem.*, **57**, 603.
16. Barret, E.P., Joyner, L.G., and Halenda, P.H. (1951) *J. Am. Chem. Soc.*, **73**, 373.
17. de Boer, J.H., Lippens, B.C., Linsen, B.G., Broekhoff, J.C.P., van den Heuvel, A., and Osinga, Th.J. (1966) *J. Colloid Interface Sci.*, **21**, 405.
18. Hohenberg, P. and Kohn, W. (1964) *Phys. Rev.*, **136**, B864.
19. Curtin, W.A. and Ashcroft, N.W. (1985) *Phys. Rev. A*, **32**, 2909.
20. Kierlik, E. and Rosinberg, M. (1990) *Phys. Rev. A*, **42**, 3382.
21. (a) Taranoza, P. (1985) *Phys. Rev. A*, **31**, 2672; (b) Taranoza, P., Marconi, U.M.B., and Evans, R. (1987) *Mol. Phys.*, **60**, 573.
22. Ito, T. and Fraissard, J. (1987) *J. Chem. Soc., Faraday Trans.*, 1, **83**, 451.
23. (a) Telkki, V.-V., Lounila, J., and Jokisaari, J. (2006) *J. Chem. Phys.*, **124**, 034711; (b) Telkki, V.-V., Lounila, J., and Jokisaari, J. (2005) *J. Phys. Chem. B*, **109**, 24343.
24. Happer, W., Miron, E., Schaefer, S., Schreiber, D., van Wijngaarden, W.A., and Zeng, X. (1984) *Phys. Rev. A*, **28**, 3092.
25. (a) Onfroy, T., Guenneau, F., Springuel-Huet, M.-A., and Gedeon, A. (2009) *Carbon*, **47**, 2352; (b) Springuel-Huet, M.-A., Guenneau, F., Gedeon, A., and Corma, A. (2007) *J. Phys. Chem. C*, **111**, 8268.

26
A Critical Review of Some 'Classical' Guidelines for Catalyst Testing
Frits Dautzenberg

26.1
Introduction

It appears that about every decade the practitioners of catalysis at the bench need to be reminded of the generic values of some of the quality principles involved in catalyst testing. It is not clear exactly why this is necessary, but it has been speculated that the original bench scale scientists may have been promoted to higher responsibilities and have had no time to coach their successors to apply the catalyst testing quality principles of the past. Catalytica Studies Division published the 'A practical guide to catalyst testing' in 1987 [1] with great success, and this was followed by a summary paper in 1989 [2]. An updated version was published as part of the proceedings of a NATO conference in 1999 [3]. Given the current challenging catalyst tasks, this paper provides a new view of these 'classical' guidelines, adjusted in a new arrangement with reformatting and calibration to modern times. The purpose of this paper is to recreate new awareness of the content of the original guidelines, which have never lost any of their intrinsic value.

26.2
Encouraging Effectiveness

The development of new catalysts on a continuous basis is always desirable. Oil refinery operations demand higher product quality and must take into account increasingly severe environmental concerns. For the chemical industry, the quest for improved catalysts will also continue. Biomass-derived feedstocks are entering industry, and many associated R&D projects foresee the use of new catalysts. Progress in catalyst discovery, testing and evaluation, scale-up, and process development, however, is still costly and time consuming.

The funding of catalyst R&D projects comes from the profit of a parent corporation or from government and other outside clients. Whatever the source of funding, the research 'sponsor,' as in any other business, must be kept satisfied, and this is

---

Catalysis: From Principles to Applications, First Edition. Edited by Matthias Beller, Albert Renken, and Rutger van Santen.
© 2012 Wiley-VCH Verlag GmbH & Co. KGaA. Published 2012 by Wiley-VCH Verlag GmbH & Co. KGaA.

Table 26.1 Quality guidelines for catalyst testing.

Encouraging effectiveness	Ensuring efficiency
Specify goals and obtain sponsor support.	Apply effective strategies.
Experimental program (see specifics under Section 26.3)	Collect meaningful data.
Document, evaluate, and report.	Select appropriate reactor.
	Establish ideal flow pattern.
	Ensure isothermal operation.
	Diagnose and minimize transport disguises.
	Asses catalyst stability early.

critical for the survival of any project and sometimes even for the survival of the catalysis R&D team itself!

As in other projects one can distinguish three key process steps: (i) preparation and planning; (ii) execution and data gathering, and (iii) reporting. In order to address sponsor satisfaction needs, the preparation and planning step must receive special attention, bearing in mind the sponsor's special requirements. With these in mind, one can start the development of a specific R&D plan. Extensive industrial experience has shown again and again that one should not start any experimentation before the planning step is in place.

Since sponsor support is usually essential, it follows that the objectives of the envisioned catalyst project and its sub-stages should be defined and agreed upon as clearly as possible during the planning phase of the project. It is extremely important to identify what the sponsor really wants because this in essence becomes the specific goal of the planned catalyst project. Agreeing with the sponsor that sufficient planning aspects are in place should be standard practice and assures an anticipated catalysis project of high quality.

It has also been established that following the executing and data-gathering step it is good practice to include a reporting and documentation phase. During this phase the results are reported in a written document containing, among other information, the conclusions of the R&D work. Without a well-documented report, the value of the R&D may be quickly lost, along with opportunities for future improvements. For these reasons the quality recommendations listed in Table 26.1 under the heading 'Encouraging Effectiveness' have been incorporated.

26.3
Ensuring Efficiency

Once the planning phase has been successfully finalized, one can start the experimental/data-gathering phase, since it has now been established what the 'right' issues are that need to be addressed. This means that the emphasis can be shifted now to 'Ensuring Efficiency' by applying the 'right' experimental protocols, methods, and well-established review procedures. Catalyst testing and evaluation can be rather complex, and it is easy to forget the relatively obvious in favor of the recondite. Accordingly, practicing the following rules may seem obvious, but they are important requirements to ensure reliable experimental results:

- Compare the performance of different catalysts under identical experimental conditions (reactor design; flow rates; catalyst size, shape and quantity; type and degree of agitation; and temperature and pressure regimes).
- Check the reproducibility of results on a regular basis.
- Determine whether the reaction thermodynamics are favorable. In particular, be aware of any limitation on conversion because of the establishment of equilibrium.
- Observe and record, in full, the conditions under which the experiments are carried out and the results obtained.
- Keep possible homogeneous contributions in mind when investigating heterogeneous catalysts and vice versa. Check this mutual interference by changing surface-to-volume ratios in the heterogeneous or homogeneous reactor. Conducting experiments in the absence of the catalyst is not a conclusive proof of the absence of heterogeneously initiated homogeneous reactions. Nevertheless, blank experiments should be used to check the possible reactivity of reactor walls and other internals or inert dilution materials.

Following these test practices is an important step in terms of general quality assurance. In addition, specific attention must be paid to the seven recommendations listed in Table 26.1 under the heading 'Ensuring Efficiency' for the execution of catalyst projects.

26.3.1
Apply Effective Experimental Strategies

During catalyst testing, multiple parameters frequently need to be varied. In such cases, statistically based strategies can offer important advantages, including:

- More information per experiment
- Early isolation of key variables
- Valid conclusion despite experimental uncertainty
- Built-in procedures to check the validity of conclusions
- Detection of interactions among variables
- Significant time savings
- Organized collection and presentation of results
- Up-front estimate of required number of experiments.

User-friendly software is commercially available for factorial designs, regression analysis, and optimization strategies. These are often applied nowadays during high-throughput catalyst experiments.

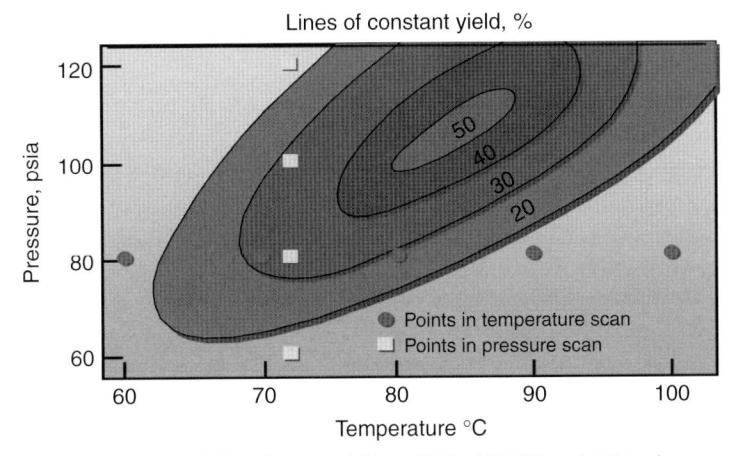

Figure 26.1 Contour map of a two-variables system.

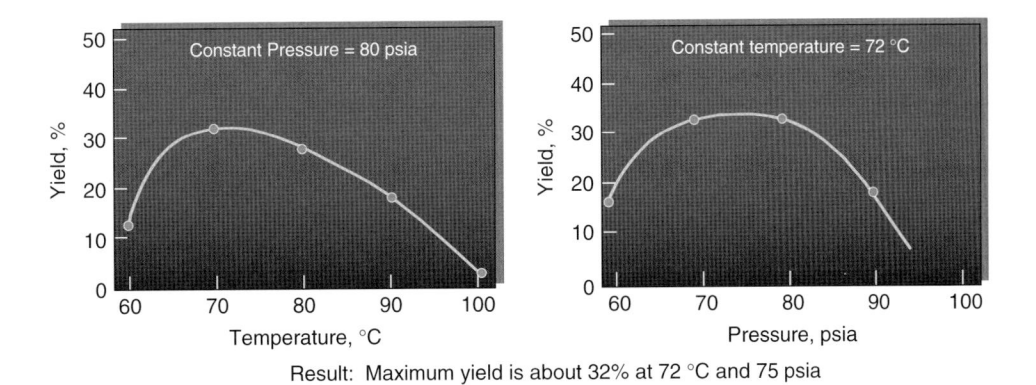

Result: Maximum yield is > 50% at 85 °C and 105 psia

Figure 26.2 Scanning variables one at a time in a two-variables system.

Result: Maximum yield is about 32% at 72 °C and 75 psia

Many catalyst researchers unfortunately still stick to familiar 'one-at-a-time' strategies and by so doing underestimate the influence of experimental uncertainties on the reliability of their conclusions. The following example illustrates what may happen if the performance of a hypothetical catalyst does depend on two variables – pressure and temperature. Figure 26.1 shows the assumed yield contour map of the catalyst in question. Applying a series of 'one-at-a-time' tests, the results are shown in Figure 26.2. This may lead to the wrong conclusion that the catalyst has a maximum yield of 32% at 72 °C and 75 psi. After many more experiments one may have reached the right maximum: a yield of at least 50% at 85 °C and 105 psi. If one had used a statistical factorial design methodology the 'right' maximum would probably have been found with fewer experiments and a higher confidence level. Furthermore, one would have found a possible interaction between the variables, an important piece of information that is usually not found using the classical 'one-at-a-time' approach.

26.3.2
Collect Meaningful Data

(Especially regarding: *activity, selectivity,* and *space–time yield*).

During catalyst testing, one measures **activity** on a regular basis, but unfortunately often without mentioning the definition of activity used, which then creates a lot of unnecessary confusion. Table 26.2 lists five definitions that are used on a regular basis in the open literature. The turnover frequency (or rate) is the preferred measurement of activity, but it requires exact knowledge about the number of active sites. This information is not always available, while the assumptions when estimating the number of active sites are also often not mentioned. The second definition of catalyst activity is more practical, because no assumptions about the number of active sites are made. Since in many cases one needs to rank catalysts, Definition number 3 is readily applicable in many situations. It is strongly recommended not to use definitions 4 and 5, based on the reasoning shown below.

The relative activities of two catalysts can be determined using Eq. (26.1):

$$\frac{a}{\text{WHSV}} = \int_0^{X_B} \frac{dX_B}{R_B M_B / X_{B_0}} = f(X_B, C_{B_0}, T) \qquad (26.1)$$

The right side of Eq. (26.1) is some function f of the achieved conversion of reactant B (X_B), feed concentration of B (C_{B_0}), and reactor Temperature (T). The underlying assumption is that *all catalysts under consideration perform according to the same rate law described by the (often unknown) function f*. If experiments with various catalysts are run at fixed feed composition, at fixed temperature, and up to the same final conversion, function f is fixed and the activity ratio of two catalysts is a function only of the ratio of space velocities (Eq. (26.2)):

$$\frac{a_2}{a_1} = \frac{\text{WHSV}_2}{\text{WHSV}_1} \qquad (26.2)$$

In comparing catalyst activities, the second definition gives a measurable, quantitative result without the need to know the function f. This is not the case for

Table 26.2 Defining catalyst activity.

Turnover frequency (reaction events per site per unit time)
Reaction rate per unit mass or volume of catalyst
Space velocity at which a given conversion is achieved at a specified temperature
Temperature required to achieve a given conversion level
Conversion achieved under specified reaction conditions

---

definitions 4 and 5 (see Table 26.2). If the catalyst comparisons are done at fixed weight hourly space velocity (WHSV), and if the temperature is varied to achieve the desired conversion, applying definition 4 (see Table 26.2) and Eq. (26.1) gives Eq. (26.3):

$$\frac{a_2}{a_1} = \frac{f(X_B, C_{B_0}, T_2)}{f(X_B, C_{B_0}, T_1)} \tag{26.3}$$

A more active catalyst ($a_2 > a_1$) will achieve the desired conversion at lower temperature ($T_2 < T_1$). Equation (26.3), however, cannot provide any quantitative assessment without knowledge of the functional dependence of f on all the variables. The same reasoning can be applied to definition 5 (see Table 26.2), which yields Eq. (26.4):

$$\frac{a_2}{a_1} = \frac{f(X_{B_2}, C_{B_0}, T)}{f(X_{B_1}, C_{B_0}, T)} \tag{26.4}$$

In other words, catalyst activities of similar materials that are likely to have similar kinetic dependences are most readily compared by fixing the temperature and varying the space velocity to obtain a chosen degree of conversion. In modern, high-throughput testing this is very often overlooked, especially since dissimilar materials are also screened.

By definition, **selectivity** is the ratio of the molar amount of a key reactant converted to a desired product relative to the total molar amount of the converted reactant. Selectivity measures how well a catalyst promotes the formation of the desired product as compared to other products. A moderate selectivity improvement of a commercial catalyst can generate large feedstock savings. For this reason the search for more selective catalysts is ongoing because of the potential economic impact. Catalyst selectivity is also important because it can determine the reactor selection. This can be demonstrated by reviewing reactions for exothermic oxidations. When the heat of reaction is low (<25 kcal mol^{-1} of reactant), a simple fixed-bed reactor can be used in a quenched or intercooled configuration. More expensive multi-tubular reactors are used if moderate (50–120 kcal mol^{-1}) amounts of heat need to be removed. For highly exothermic reactions (>125–150 kcal mol^{-1}), fluidized-bed reactors are applied. To obtain sufficient heat transfer capacity, the current maximum heat release is about 300 kcal mol^{-1}.

This became an issue during the development of oxidative coupling of methane [4]. Ideally, the reaction can be represented as follows:

$$2CH_4 + O_2 \rightarrow C_2H_4 + 2H_2O \quad \Delta H_{R,800°C} = -64.4 \text{ kcal mol}^{-1} C_2H_4 \tag{26.5}$$

The reaction, however, is much more complicated, because not only is ethylene formed but so also are other hydrocarbons and carbon dioxide. Taking this into account the overall reaction stoichiometry becomes:

$$400\ CH_4 + 251\ O_2 \rightarrow 90\ C_2H_6 + 70\ C_2H_4 + 64\ CO_2 + 374\ H_2O$$
$$\Delta H_{R,800°C} = -123 \text{ kcal mol}^{-1} C2's \tag{26.6}$$

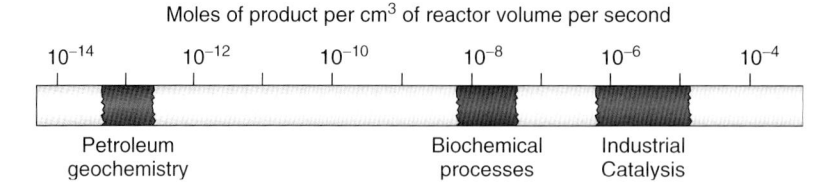

Moles of product per cm³ of reactor volume per second

```
10⁻¹⁴      10⁻¹²      10⁻¹⁰      10⁻⁸       10⁻⁶       10⁻⁴
```

Petroleum geochemistry — Biochemical processes — Industrial Catalysis

Figure 26.3 Is catalyst performance adequate?

This means that one has approached the practical limit for multi-tubular reactors, and one may even consider using a fluidized-bed reaction. To keep the reactor under control, one would need a selectivity exceeding 65%.

As pointed out above, selectivity is important for economical feedstock utilization and reactor selection. One should, however, bear in mind that selectivity will also be a function of the reaction temperature, space velocity, feed composition, reactor geometry, and the degree of conversion and thus heat transfer. Comparing selectivities of different catalysts is therefore only relevant when all these parameters are known as a *function of selectivity*. It goes without saying that this a complex task, especially for small-scale laboratory testing.

Selectivity is sometimes confused with product **yield**, because selectivity is an efficiency measurement. *Yield* is best defined as the ratio of a key reactant converted to the desired product, relative to the total key reactant fed into the reactor system. Therefore:

$$(\text{Selectivity}) \times (\text{Conversion}) = \text{Yield}$$

Based on this definition one can calculate the so-called *'space–time yield'*: the amount of product obtained per unit of time and unit of reactor space, for which one usually takes the reactor volume. A catalytic process becomes viable if the space–time yield is greater than 10^{-7} gmol cm^{-3} s^{-1} and usually lower than 10^{-5} gmol cm^{-3} s^{-1}. These limits are based on the classical papers of Paul Weisz [5–7] and illustrated in Figure 26.3. The so-called Weisz' window has been slightly extended to allow acceptable run lengths based on industrial data (see Table 26.3).

The Weisz' window's upper limit is 'imposed by the achievable rates of heat transfer for the selected reactor, while the lower limit is basically controlled by economics.' If a catalyst is not active enough the reactor residence time becomes too long and the reactor equipment may require too high an investment.

The Weisz' 'window' is an important concept during scale-up and catalyst development. It indicates whether further catalyst improvements are still necessary or whether the catalyst under consideration is already too active. In this case the catalyst needs to be 'tamed,' for instance, by dilution with an inert matrix material. Calculating the space–time yield is a useful exercise in determining the catalyst adequacy.

Table 26.3 Paul Weisz[a] window for some commercial processes.

Technology	Description	Space–time yield (gmol s^{-1} cm^{-3})
Ethylbenzene [8, 9]	Aromatic alkylation	6.8×10^{-7}
Styrene [10–12]	Dehydrogenation	6.9×10^{-7}
Maleic anhydride [13]	Oxidation	2.1×10^{-7}
Isomerization [14]	Paraffin/isomerization	7.0×10^{-6}
Catofin [15]	Paraffin dehydrogenation	9.7×10^{-7}
lummus-crest (LC) fining [16]	Residue hydrocracking	1.2×10^{-7}

[a] References [5–7].

26.3.3
Select the Most Appropriate Laboratory Reactor

Industrial reactors are most of the time adiabatic and designed for effective heat transfer and operating conditions leading to viable space–time yields. In the laboratory one should resist the temptation to use a reduced replica of the industrial reactor that the process designers have chosen. The laboratory reactor must, especially for catalyst screening and early stage process development, ensure operating conditions that lead to (i) isobaricity, (ii) isothermality, (iii) an 'ideal' flow profile, and (iv) no significant concentration profiles. This can best be done in either continuously stirred tank reactors (CSTRs) or in plug flow reactors (PFRs), applying so-called differential conditions which can be achieved at low conversion, isobaricity, and isothermality. The two preferred laboratory reactors are shown in Figure 26.4.

For laboratory catalyst evaluation one should try to stay away from batch, fluidized-bed, bubble column, and trickle-bed reactors. These reactors have been studied extensively in the past 20 years and can now be used if certain precautions are met. This will be further discussed in Appendix A.

Batch reactors, although still popular in many laboratories, are not well suited for kinetic investigations. It is impossible to uncouple the main reaction kinetics from deactivation, and it is difficult to determine the actual reaction time. The complex hydrodynamics in fluidized-bed and bubble-column reactors do not permit accurate assessment of intrinsic catalyst behavior. Expert assistance is required to interpret process conditions.

It can be stated that as a laboratory catalyst test instrument, the laboratory CSTR is by far the preferred reactor type for good reasons, including:

- Gradientless operation
- Simple mathematical treatment of data
- Separation of reaction kinetics and deactivation parameters
- Uniform catalyst deactivation.

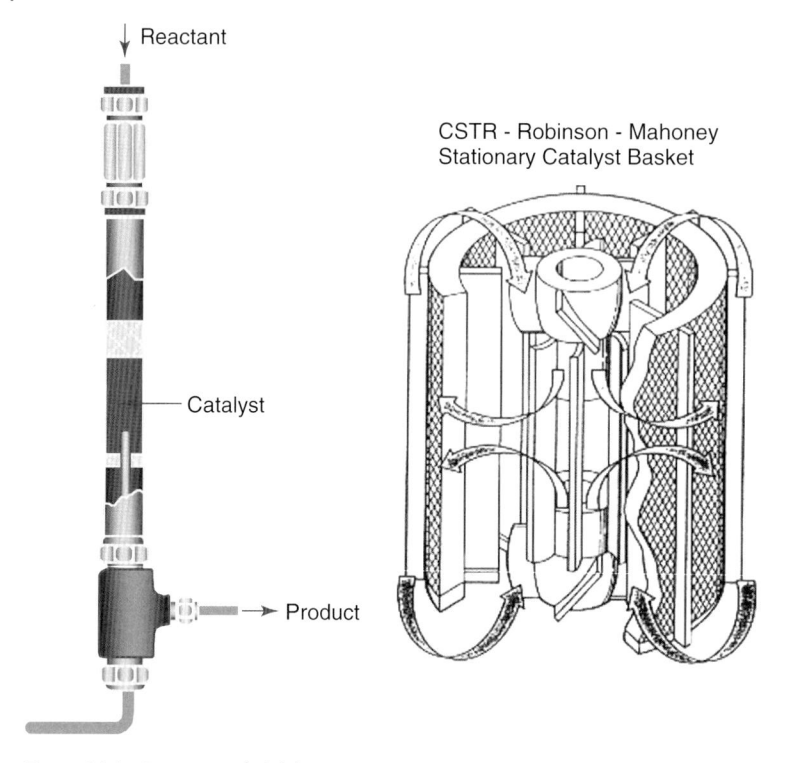

Figure 26.4 Recommended laboratory reactors.

From an ease of operation point of view the laboratory CSTR is sometimes considered to be too difficult in the laboratory, but there are several companies that presently offer laboratory reactors of this type.

Without doubt, the PFR laboratory reactor is the most popular. The PFR can be used efficiently for solid catalyst screening in a single fluid phase. It can also be used during later development activities, especially if one tries to develop a new process with a fixed-bed reactor. Many times this is the first choice for the commercial reactor if at all possible.

The theory of similarity teaches that, for the laboratory and the industrial fixed-bed reactors, the Damkohler number (N_{Da}), the Sherwood number (N_{Sh}), and the Thiele modulus (ϕ) should be kept constant during scale-up (Figure 26.5). This ensures that heat transfer phenomena, fluid-dynamic regime, and catalyst utilization are basically kept constant. The experimental reactor must therefore have the *same length* as the envisioned commercial reactor. Scale-up is done by increasing the diameter of the reactor for fixed-bed applications or increasing the number of tubes in multi-tubular reactors. This example further illustrates that experimental development reactors are not necessarily small in size.

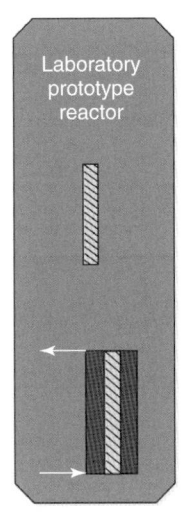

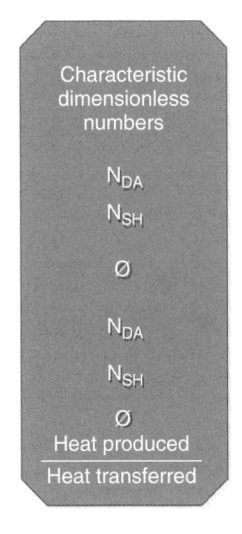

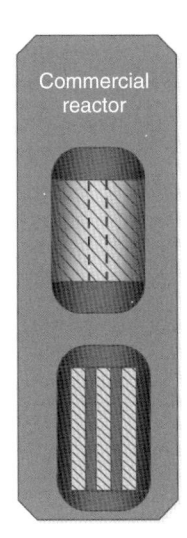

Figure 26.5 Schematic of prototype tube.

26.3.4
Establish Ideal Flow Pattern

In order to use laboratory reactor rate data straightforwardly, two ideal reactor flow patterns are possible: (i) PFR or (ii) completely mixed flow for CSTR. In these cases, the radial and axial dispersion terms in the continuity equation become negligible, and, assuming continuous reactor operation ($\delta p/\delta t = 0$), the rate analysis of a CSTR and a PFR become, respectively:

$$R = \frac{Q}{V}(C_1 - C_0) \quad (26.7)$$

$$R = Q \times \frac{dc}{dV} \quad (26.8)$$

Any other flow pattern requires much more complicated mathematical procedures. Laboratory reactors with non-ideal flow patterns should therefore be avoided for rate measurements.

It is well known that the diameter of a laboratory PFR should be at least 10 times the catalyst particle diameter to ensure the required ideal flow pattern. This is presumed to eliminate the influence of the reactor wall on the flow pattern. However, actual gradients may still exist, depending on the degree of conversion. For fixed-bed reactors, it is often assumed that if the ratio of bed length (L) to particle diameter (d_p) is greater than 50 ($L/d_p > 50$), this is a sufficient requirement to avoid axial gradients, which is usually indeed the case if the particle Reynolds number (Re_p) is 10 or above. In laboratory experiments, the particle Reynolds numbers, however, are mostly much smaller, and the usual situation is $Re < 0.1$. To take this into account it is recommended to operate at low conversion and selectivity and correct the ratio of the reactor bed length (L) to the catalyst particle diameter (d_p), as

Table 26.4 Calculation of the required minimum L/d_p.

Acceptable deviation from plug flow can be ensured if:

$$N_{Pe} > N_{Pe_{min}} \quad (26.9)$$

The Peclet number can be calculated using:

$$N_{Pe} = (0.034) N_{Rep}^{0.53} \left(\frac{L}{d_p}\right) \text{ for liquid - phase operation} \quad (26.10)$$

$$N_{Pe} = (0.087) N_{Rep}^{0.23} \left(\frac{L}{d_p}\right) \text{ for gas - phase operation} \quad (26.11)$$

The minimum Peclet number follows from:

$$N_{Pe_{min}} = 8n \ln \frac{1}{1-x} \quad (26.12)$$

The required minimum L/d_p follows then from:

$$\frac{L}{d_p} > (235.3) N_{Rep}^{-0.53} n \ln \frac{1}{1-x} \text{ for liquid - phase operation} \quad (26.13)$$

$$\frac{L}{d_p} > (92.0) N_{Rep}^{-0.23} n \ln \frac{1}{1-x} \text{ for gas - phase operation} \quad (26.14)$$

In Eqs. (26.14) and (26.15), the particle Reynolds number is applied:

$$N_{Rep} = \frac{u d_p \rho}{\mu} \quad (26.15)$$

shown by the leading work of Gierman [17] in 1988. There is a quick procedure that has been shown to work effectively for calculating the required L/d_p for laboratory PFRs (see Table 26.4).

Typical results using this quick calculation are shown in Table 26.5. Note that the Re_p is very low and that L/d_p is much greater than 50!

26.3.5
Ensure Isothermal Conditions

Since most reactions are either endothermic nor exothermic, achieving 'true' isothermal conditions in reactors is basically impossible without taking special precautions. In laboratory situations, one often wishes to measure intrinsic kinetics, which requires (quasi-) isothermal conditions. Relatively small differences in temperatures and temperature gradients affect reaction rates, selectivities, and

Table 26.5 Effect of reactor scale on nitrogen removal experiments (gas hourly space velocity (GHSV) = 0.67 N in feed − 700 ppm) [17].

Reactor type	Microreactor	Bench-scale reactor
Catalyst content (cm^3)	16	200
Catalyst bed length (cm)	27.4	67.8
Reactor diameter (cm)	0.864	1.939
Catalyst particle diameter (cm)	0.036	0.036
Conversion	0.65	0.71
N in product (ppm)	245	200
Linear velocity (cm s^{-1})	0.51×10^{-2}	1.26×10^{-2}
N_{Rep}	0.061	0.151
Required L/d_p	1090	790
Actual L/d_p	760	1880

Table 26.6 Activation energy and temperature dependence (temperature rise needed to increase the rate of reaction by about 30%).

°C	ΔE (cal mol^{-1})		
	10 000	30 000	50 000
0	4	1	<1
100	7	2	1
200	12	4	2
400	24	8	5
500	32	11	6

also deactivation rather significantly. Temperature sensitivity is greater at lower temperatures (see Table 26.6). Consequently the requirement to control the temperature more precisely becomes more important in determining intrinsic kinetics of high-activity catalysts.

As indicated in Figure 26.6, the axial and radial intra-reactor temperature gradients are nearly always more severe than the interface temperature gradients, which in turn are generally more severe than intra-particle gradients. Moreover, the intra-particle temperature gradients are less significant because the effective thermal conductivity of the catalyst material is much greater than that of the surrounding fluid.

Within the catalyst particle, conductivity through the solid phase determines the heat transfer. The extent to which catalyst activity measurements are disturbed by the various heat transfer processes can be assessed by evaluating

- For small experimental reactors:
  (Low flow rates and conversion to avoid axial and radial gradients)

  | Interphase temperature effects | > | Intraparticle concentration effects | > | Interphase concentration effects | > | Intraparticle temperature effects |

- For commercial reactors:
  (high flow rates)

  | Intraparticle concentration effects | > | Intraparticle temperature effects | > | Interphase temperature effects | > | Interphase concentration effects |

Figure 26.6 Relative importance of various gradients.

Table 26.7 Criteria for isothermal operation [18].

Intrareactor $\dfrac{|\Delta H| R_v r_t^2}{k_b T_w} < 0.2 \dfrac{R T_w}{E}$

Interphase $\dfrac{|\Delta H| R_m \rho_p r_p}{h T_b} < 0.15 \dfrac{R T_b}{E}$

Intraparticle $\dfrac{|\Delta H| R' \rho_p r_p^2}{k_p T_s} < 0.75 \dfrac{R T_s}{E}$

experimental catalyst performance data using the various mathematical criteria given in Table 26.7.

The mathematical criteria compare functions A and B containing observable or measurable parameters. If A<B it can be assumed that the relevant temperature gradient may cause a <5% deviation of the reaction rate. Consideration of the above mathematical criteria together with practical laboratory experience indicate what laboratory experience reactor system features can be manipulated to achieve quasi isothermal control. Based on these, the following recommendations can be applied effectively in a laboratory setting:

- Use CSTR if possible
- Work at low conversion levels
- Use small catalyst particles
- Decrease bed voidage by using small inert diluent particles
- Select catalyst supports with high thermal conductivity, if possible
- Add feed diluents with high thermal conductivity (H_2, steam, and He above 500 K)
- Apply high flow rates.

26.3.6
Diagnose and Minimize Effects of Transport

Heat and mass transfer effects can cause intra-reactor, inter-phase, or intra-particle gradients as discussed above, and these can distort ('disguise') the true results. These effects must be eliminated, which can be done by adjusting the experimental conditions. In a well-stirred CSTR, intra-reactor gradients will be absent, but inter-phase and intra-particle gradients may still be present. Conversely, in a fixed-bed PFR with small catalyst particles, intra-particle gradients may be eliminated, although intra-reactor gradients still occur.

For any catalyst project, the following tests are highly recommended to ensure that indeed intrinsic catalyst properties are measured, free from spurious effects.

- In a PFR the linear flow rate can be varied while the space velocity is kept constant (see Figure 26.7). The influence of the inter-phase and inter-reactor effects maybe assumed to be insignificant if the conversion remains constant.
- A similar test can be done with a laboratory CSTR. In this case the absence of inter-phase and inter-reactor effects can be assumed if the reaction rate is independent of the stirring speed. One basically increases the velocity of the fluid through the catalyst bed, which leads to the elimination of the inter-phase and inter-reactor gradients.
- In a CSTR with a heterogeneous catalyst as a slurry, increasing the stirrer speed maybe insufficient to eliminate inter-phase gradients, as the catalyst particles are moving around at the same speed as the reactor fluid. It is for this reason that a CSTR as shown in Figure 26.4 is recommended for laboratory testing.
- Changing catalyst particle sizes is an absolute must and can be used to test inter-particle effects (see Figure 26.8). If there is no change of catalyst activity with change in particle size (assuming that the exposed surface area of active catalyst is constant), the catalyst is considered to be free of inter-particle gradients.

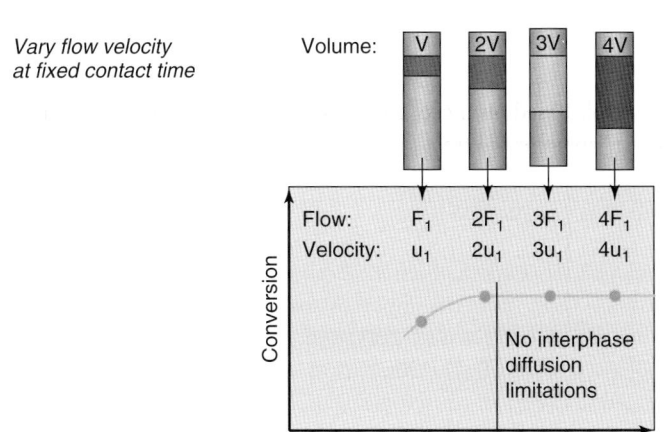

Figure 26.7 Diagnostic tests for interphase (external) transport disguises.

Vary catalyst particle size

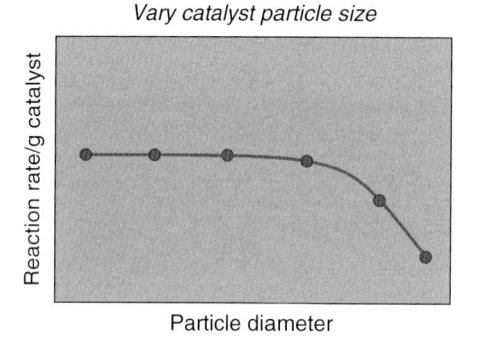

[Graph: Reaction rate/g catalyst vs Particle diameter, showing flat response at small diameters decreasing at larger diameters]

Figure 26.8 Diagnostic tests for intraparticle (internal) transport disguises.

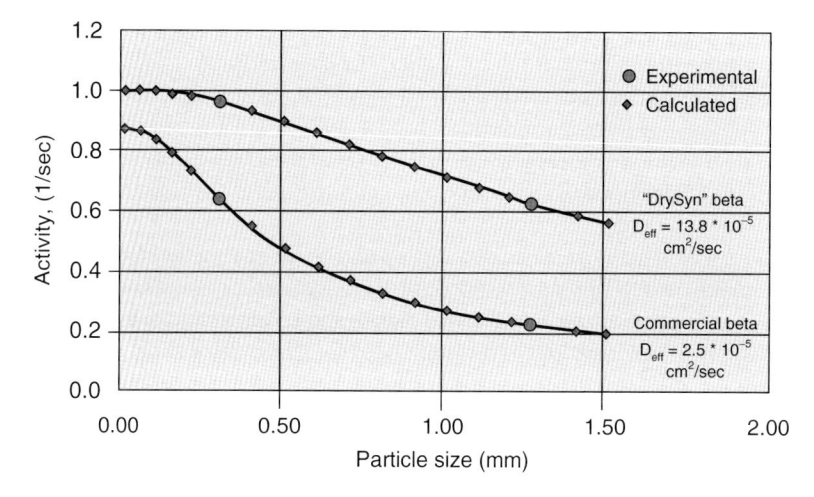

[Graph: Activity (1/sec) vs Particle size (mm), comparing "DrySyn" beta ($D_{eff} = 13.8 * 10^{-5}$ cm²/sec) and Commercial beta ($D_{eff} = 2.5 * 10^{-5}$ cm²/sec), with Experimental and Calculated data points]

Figure 26.9 Performance of 'DrySyn' Beta vs commercial Beta.

Inter-particle diffusion limitation may be unavoidable, especially if one tests fast reactions or uses highly active catalysts. In this case and in cases where commercially sized catalysts must be tested, the quick procedure shown in Table 26.7 can be used to estimate the degree of catalyst utilization. Figure 26.9 shows an application of this method [19].

As discussed previously there are mathematical criteria to assess the influence of the various gradients (see Table 26.8). It is important to realize that some parameters used in these mathematical criteria (i.e., activation energy, reaction order, and reaction rate) are not available during the initial experiments. Therefore it is always strongly recommended to use small catalyst particles, moderate temperatures, and low conversions in the preliminary stages of catalyst testing projects. Using the iterative approach described in the above set of experimental tests establishes a so-called 'disguise-free' regime.

Table 26.8 How to determine $k(o)$ = catalyst k at a particle size = 0 and D = molecular diffusivity.

Determine k (app) at two catalyst particle sizes
It follows that:

$$\frac{k(\text{app})(1)}{k(\text{app})(2)} = \frac{\eta(1)}{\eta(2)} = \frac{\frac{3}{\Phi_1}\left[\frac{1}{\tanh \Phi_1} - \frac{1}{\Phi_1}\right]}{\frac{3}{\Phi_2}\left[\frac{1}{\tanh \Phi_2} - \frac{1}{\Phi_2}\right]} \qquad (26.16)$$

In which $\Phi_i = r_p(i) \times \sqrt{\frac{k(o)}{D}}$ \qquad (26.17)

and $\eta(i) = \frac{k(\text{app})(i)}{k(o)}$ \qquad (26.18)

Apply iterative approximation to establish $k(o)/D$ to satisfy Eq. (26.16)
For each value of r_p one now can calculate ϕ_i and thus $k(o)$ and D

26.3.7
Assess Catalyst Stability Early

Although treated in this paper as the last quality guideline, characterization of catalyst stability is a very important activity and should preferably be addressed during the early phase of a catalyst development project. Stability is a key catalyst attribute that deserves careful attention. It does more than determine how much catalyst must be installed to obtain a desired run length. As indicated in Table 26.9, catalyst stability may actually determine the type of reactor technology that must be applied to obtain a viable commercial process.

The decrease in catalyst activity with time may be extremely fast (as in fluidized-bed catalytic cracking), or it can occur over a much longer time scale as in residue desulfurization [20]. In this case, deactivation appears to occur by a process of metal deposition in the outer core of the catalyst particles, while

Table 26.9 Design of reactors with deactivating catalysts.

Catalyst deactivation	Cycle length	Reactor system	Examples
Minor	Six months or more	Fixed bed and tubular	Naphtha HT Ethylene oxide
Moderate	Days/weeks	Swing	Fully Regenerative Reactors Reforming
Fast	Hours up to one day	Continuous addition and withdrawal of catalyst	Continuous Catalytic Reforming (CCR) LC-fining Hycon
Very fast	Seconds/minutes	Fluid bed	FCC

desulfurization and hydro cracking occur mainly in the inner core of the catalyst particles. This understanding has opened the path to various industrial solutions for the treatment of high-metal feedstocks [21].

The above is a well-known illustration of why one needs to address catalyst stability during the early phases of a catalyst development program. Not only can one avoid pursuing a wrong direction but one may also make new discoveries and recognize new opportunities. This important guidelines for catalyst testing are demonstrated by many industrial patents and examples in the open literature. Many catalyst publications, however, do not even mention catalyst stability at all.

26.4
Concluding Remarks

Reviewing these principal guidelines on a regular basis is a highly recommended activity, the following guidelines being of particular importance:

- Prior to starting the experimental testing it is vital to agree with the sponsors what are the goals of the project. This is probably the most important guideline to ensure that the 'right' issues will be addressed.
- Select a laboratory PFR or CSTR for the catalyst testing and data gathering program phase, especially during the early testing and screening of potential catalyst candidates. Remember that laboratory reactors are not usually small replicas of commercial reactors.
- The two most important experimental conditions are an ideal flow pattern and isothermal or quasi-isothermal operation. An ideal flow pattern, either plug flow or perfectly mixed, is essential for a straightforward interpretation of kinetic data. Quasi-isothermal operation is critical to ensure that reliable qualitative kinetic data are obtained.
- Do not forget to test the influence of catalyst size and check catalyst stability.

Appendix A: Three-Phase Trickle-Bed Reactors

Trickle-bed reactors are three-phase reactors used in many important industrial units for applications such as: hydro-desulfurization, hydro-denitrogenation of gas oils, fluid catalytic cracking (FCC) hydro-treating, hydro-cracking of oils, and so on. Trickle-bed reactors have some important advantages, but also some disadvantages (see Tables 26.10 and 26.11).

As a consequence of these, many studies have been carried out in the last decades [24]. In general, attempts were made to obtain authentic data from small-scale experiments that could be used to predict performance on an industrial scale. For laboratory trickle-bed reactors, the following six aspects need to be closely monitored and controlled:

1) The homogeneity of the catalyst bed
2) The wetting of the catalyst particle
3) Liquid and gas axial dispersion

Table 26.10 Advantages of trickle-bed operation (see Ref. [22]).

Close to plug flow behavior for liquid and gas phases
Changes in gas and liquid velocities
Small liquid phase hold-up
Permits operation at high pressure and temperature
No problem of flooding in co-current flow of gas and liquid
No mobile phase
No losses of catalyst
Large reactors possible
Simple construction and easy operation
Limited costs, etc.

Table 26.11 Disadvantages of trickle-bed operation (see Ref. [22]).

Mass and heat radical dispersion
Poor efficiency of the catalyst
Large influence of the hydrodynamics on the efficiency of the reactor
Large pressure drop
Low liquid-to-solid mass transfer rate
Counter-current operation provides flooding (but also a higher level of conversion)
Impossible to use with a liquid of high viscosity
Particle size should usually be not less than 1 mm
Maldistribution, channeling, and incomplete catalyst wetting at low velocities
Limited capacity in the case of fast deactivation of the catalyst
May act as a filter, complicating use in the case of a feedstock containing suspended solids
Difficulty of construction for multi-tubular reactors, etc.

4) Non-preferential flow, especially wall effects and channeling
5) Isothermal behavior
6) Mass transfer limitations.

Based on a very recent, important review paper [22], criteria have now been developed for each of these factors with the exception of the homogeneity of the catalyst bed. However, with perfect packing, no unforeseen phenomena occur due to catalyst maldistribution caused by channeling, hot spots, and so on.

The mathematical criteria are shown in Table 26.12. Eq. (26.19) is applied to assess catalyst wetting. If the criteria are fulfilled, then good catalyst wetting and contact with the reactants occur. To review the importance of actual dispersion, Eqs. (20–23) can be used. If no actual dispersion is observed, the reactor behavior is close to plug flow. To avoid channeling, especially close to the wall of the reactor, Eqs. (26.24) and (26.25) are applied. For isothermal behavior Eqs. (26–29) are recommended. In this case, one assesses whether important temperature gradients inside the reactor are significant. Whether mass transfer is limiting the kinetics can be determined from Eqs. (26.30) and (26.31).

Table 26.12 Criteria for trickle-bed reactors [22].

Axial mixing
Catalyst wetting

$$\eta_L U_L / \rho_L d_p^2 g > 5 \times 10^{-6} \quad (26.19)$$

Axial mixing

$$L/d_p > 100 \quad (26.20)$$

$$D/d_p > 10 \quad (26.21)$$

$$L/d_p > [20n/Bo_L] \times \ln(C_o/C_f) \quad (26.22)$$

$$L/d_p > [20n/Bo_g] \times \ln(C_o/C_f) \quad (26.23)$$

Channeling

$$D/d_p > 25 \quad (26.24)$$

$$d_{pi} > d_p/10 \quad (26.25)$$

Isothermality

$$D/d_p > 100 \quad (26.26)$$

$$1.6 R T_w / E > \Delta_r H.r.D^2 / k_c T_w \quad (26.27)$$

$$0.30 R T_o / E > \Delta_r H.r.d_p / h_p T_o \quad (26.28)$$

$$0.6 R T_o / E > \Delta_r H.r.d_p^2 / \lambda T_o \quad (26.29)$$

Mass transfer

$$r(1-\varepsilon) \times (d_p/2)^2 / D_e C < 1 \quad (26.30)$$

$$k_{est} > (10 d_p/C).r.(1-\varepsilon) \quad (26.31)$$

In the review paper [22], the criteria of Table 26.12 are applied in a study of a hypothetical, but very representative hydrodesulfurization example. The parameters of this example case are listed in Tables 26.13 and 26.14.

Based on these values, the numerical values according to the various criteria are listed in Table 26.15. These indicated where the criteria were *not* satisfied. Since the original paper [22] did contain a few minor errors, the general conclusions of this comparison are slightly different (see Table 26.16) compared to the review paper [22]. The example indicates that a longer laboratory reactor might have given no axial liquid dispersion issue, which also could have enabled the external mass transfer limitation to be avoided. It also demonstrates that using inert fines in laboratory experiments for three-phase reactors is good practice, as originally recommended by van Klinken and van Dongen in 1980 [23].

Table 26.13 Trickle-bed input data (see Ref. [22]).

Main parameters	Reactor	
	Laboratory	Industrial
L, length of the bed (m)	0.4	5
D, diameter of the reactor (m)	5×10^{-2}	1.5
d_p, diameter of the catalyst particle (mm)	1.6	3.0
d_{pi}, diameter of the inert fines (mm)	0.2	
R, catalyst to inert ratio (mL/mL)	1	
U_L, velocity of the liquid (m s^{-1})	5×10^{-5}	1×10^{-3}
U_G, velocity of the gas (m s^{-1})	1×10^{-3}	5×10^{-2}
Other parameters		
Type of reaction		HDS
Regime		Downflow
Type of catalyst		NiMo/$\gamma - Al_2O_3$
Type of inert	Silicon	
ε, bed voidage		0.3
T, temperature (K)		900
r, rate of the reaction (mol s^{-1}m^{-3})		1
n, order of the reaction considered		1
X, conversion level	0.8	0.95
C, concentration of the limiting component (mol m^{-3})		500
ρ^L, density of the liquid (kg m^{-3})		750
η_L, dynamic viscosity of the liquid (Pa s^{-1})		5×10^{-4}
ρ_G, density of the gas (kg m^{-3})		5
η_G, dynamic viscosity of the gas (Pa s^{-1})		1×10^{-4}
E, energy of activation (kJ mol^{-1})		100
ΔH_r, heat of reaction (kJ mol^{-1})		250
h_p, particle to fluid heat transfer coefficient (W m^{-2} K^{-1})	1000	10 000
k_g, bed effective radial conductivity (W m^{-1} K^{-1})	10	100
λ, effective thermal conductivity of the particle (W m^{-2} K^{-1})	5	5
D_e represents the molar diffusivity (m^2 s^{-1})		1×10^{-8}
k_{exp} coefficient of mass transfer (m s^{-1})	1×10^{-5}	1×10^{-3}

Table 26.14 Trickle-bed data (see Ref. [22]).

Correlated results	Lab reactor	Industrial reactor
v_L, kinematic viscosity of the liquid (m^2 s^{-1})		6.67×10^{-7}
v_G, kinematic viscosity of the gas (m^2 s^{-1})		2×10^{-5}
u_L, superficial liquid velocity (kg/m^2 s^{-1})	3.75×10^{-2}	0.75
U^G, Superficial gas velocity (kg/m^2 s^{-1})	5×10^{-3}	0.25
Reynolds Number Re_L for the liquid phase	0.12	4.5
Reynolds Number Re_G for the gas phase	8×10^{-2}	7.5
Bodenstein Number Bo_L for the liquid phase	1.18×10^{-2}	7.60×10^{-2}
Bodenstein Number Bo_G for the gas phase	10.5	0.417

Table 26.15 Application of criteria for three-phase trickle-bed reactors.

(a)

Data		Laboratory	Industrial	
ηl	Dynamic viscosity of the liquid	5.00E−04	5.00E−04	Pa s^{-1}
Ul	Liquid velocity	5.00E−05	1.00E−03	m s^{-1}
ρl	Density of the liquid	7.50E+02	7.50E+02	kg/m^3
d_p	Diameter of catalyst particle	1.60E−03	3.00E−03	m
g	Gravity force	9.80E+00	9.80E+00	m s^2
L	Length of bed	4.00E−01	5.00E+00	m
d_p	Diameter of catalyst particle	1.60E−03	3.00E−03	m
D	Diameter of reactor	5.00E−02	1.50E+00	m
n	Order of the reaction	1.00E+00	1.00E+00	–
Bo_L	Bodenstein number of liquid	1.18E−02	7.60E−02	–
C_o	Concentration of feed	5.00E+02	5.00E+00	mol/m^3
C_f	Concentration of effluent	1.00E+02	2.50E−01	mol/m^3
Bo_g	Bodenstein number of gas	1.05E+01	4.17E−01	–
D	Diameter of reactor	5.00E−02	1.50E+00	m
d_p	Diameter of catalyst particle	1.60E−03	3.00E−03	m
d_{pi}	Diameter of inert particle	2.00E−04	0.00E+00	m
D	Diameter of reactor	5.00E−02	1.50E+00	m
d_p	Diameter of catalyst particle	1.60E−03	3.00E−03	m
R	Gas constant	8.31E+00	8.31E+00	J K^{-1} mol^{-1}
T	Temperature	9.00E+02	9.00E+02	K
E	Activation energy	1.00E+02	1.00E+02	kJ mol^{-1}
ΔH_r	Heat of reaction	2.50E+02	2.50E+02	kJ mol^{-1}
r	Rate of reaction	1.00E+00	1.00E+00	mol s^{-1} m^{-3}
k_e	Effective radial bed conductivity	1.00E+01	1.00E+02	W m^{-1} K^{-1}
h_p	Particle to fluid heat transfer coefficient	1.00E+03	1.00E+04	W m^{-2} K^{-1}
gl	Effective thermal conductivity of the particle	5.00E+00	5.00E+00	W m^{-2} K^{-1}
r	Rate of reaction	1.00E+00	1.00E+00	mol s^{-1} m^{-3}
ε	Bed voidage	3.00E−01	3.00E−01	–
d_p	Diameter of catalyst particle	1.60E−03	3.00E−03	M
De	Molar diffusivity	1.00E−08	1.00E−08	m^2 s^{-1}
C_o	Concentration of feed	5.00E+02	5.00E+02	mol m^{-3}
k_{ext}	Coefficient of mass transfer	1.00E−05	1.00E−03	m s^{-1}

Table 26.15 (continued)

(a)

Data		Laboratory	Industrial	
(b)	Laboratory Left side	Laboratory Right side	Industrial Left side	Industrial Right side
Catalyst wetting				
Equation (26.19)	1.33E−06	5.00E−06	7.56E−06	5.00E−06
Axial dispersion				
Equation (26.20)	2.50E+02	1.00E+02	1.67E+03	1.00E+02
Equation (26.21)	3.13E+01	1.00E+01	5.00E+02	1.00E+01
Equation (26.22)	2.50E+02	2.73E+03	1.67E+03	7.88E+02
Equation (26.23)	2.50E+02	3.07E+00	1.67E+03	1.44E+02
Non-preferential flow				
Equation (26.24)	3.13E+01	2.50E+01	n.a	n.a
Equation (26.25)	2.00E−04	1.60E−04	n.a	n.a
Isothermicity				
Equation (26.26)	3.13E+01	1.00E+02	5.00E+02	1.00E+02
Equation (26.27)	1.20E−01	6.94E−02	1.20E−01	6.25E+00
Equation (26.28)	2.24E−02	4.44E−04	2.24E−02	8.33E−05
Equation (26.29)	4.49E−02	1.42E−04	4.49E−02	5.00E−04
Mass transfer				
Equation (26.30)	8.96E−02	1.00E+00	3.15E−01	1.00E+00
Equation (26.31)	1.00E−05	2.24E−05	1.00E−03	4.20E−05

Table 26.16 Comparing laboratory scale and industrial trickle-bed reactors.

Scale of operation	Laboratory reactor	Industrial trickle-bed
Catalyst wetting	Poor, but probably better with inert fines	Good wetting
Axial dispersion	May occur in liquid phase; inert fines may help; gas phase is OK	No axial dispersion effects
Preferential flow	Wall effect may occur; inert fines can help	No preferential flow
Isothermicity	OK, but wall effect may occur; inert fines can help	Not isothermal, but no wall effect
Mass transfer	No internal limitations but external limitations are possible	No mass transfer limitations

List of Symbols and Abbreviations

A	Catalyst activity multiplier (dimensionless)
C	Concentration (mol volume^{-1})
C_o	Inlet concentration (mol length^{-3})
C_i	Concentration of species (mol length^{-3})
D_r	Radial diffusion coefficient (length^2 time^{-1})
D_x	Axial diffusion coefficient (length^2 time^{-1})
d_p	Solid particle diameter (length)
E	Activation energy (energy mol^{-1})
F	Molar flow rates (mol time^{-1})
ΔH	Heat of reaction (energy mol^{-1})
H	Heat transfer coefficient (energy length^{-1} time^{-1} temperature^{-1})
k_b	Effective thermal conductivity of catalyst bed (energy length^{-1} time^{-1} temperature^{-1})
k_p	Thermal conductivity of catalyst particle (energy length^{-1} time^{-1} temperature^{-1})
L	Catalyst bed length (length)
M_B	Molecular weight of species B (mass mol^{-1})
N_{Pe}	Peclet number (dimensionless)
N_{Rep}	Reynolds number based on particle diameter (dimensionless)
N_{Sh}	Sherwood number (dimensionless)
N_{Da}	Damkohler number (dimensionless)
N	Number of moles (mol)
Q	Volumetric flow rate (volume time^{-1})
R	Gas constant (energy mol^{-1} temperature^{-1})
R_a	Kinetic expression for dependence of reaction rate on temperature and concentration (determined by applicable rate units)
R'	Rate of reaction per unit mass of catalyst (mol time^{-1} mass^{-1})
R_v	Rate of reaction per unit volume of catalyst (mol time^{-1} length^{-3})
R	Radial distance (length)
r_p	Radius of catalyst particle (length)
r_t	Radius of reactor (length)
T	Absolute temperature (temperature)
T_b	Temperature of bulk fluid (temperature)
T_s	Temperature of catalyst surface (temperature)
T_w	Temperature of reactor wall (temperature)
T	Time (time)
U	Superficial fluid velocity
V	Reactor volume (length^3)
WHSV	Weight hourly space velocity (h^{-1})
X	Fraction of reactant consumed via chemical reaction (dimensionless)
X_B	Fractional conversion of B (dimensionless)
Z	Distance along reactor axis (length)
μ	Fluid viscosity (mass length^{-1} time^{-1})
ρ	Fluid density (mass length^{-3})
ρ_p	Catalyst particle density (mass length^{-3})
ϕ	Thiele modulus (dimensionless)

References

1. Boudart, M., Dautzenberg, F.M., De deken, J.C., and Schlatter, J.C. (1987) Catalytica: A Practical Guide to Catalyst Testing, Catalytica Studies Division, Mountain View, California.
2. Dautzenberg, F.M. (1989) in *Ten Guidelines for Catalyst Testing*, ACS Symposium Series, Vol. 411 (eds S.A. Bradley, M.J. Galhuso, and R.J. Bertolacini), Characterization and Catalyst Development, Los Angeles.
3. Dautzenberg, F.M. (1999) Quality principles for catalyst testing during process development. Combinatorial Catalysis and High Throughput Catalyst Design and Testing Conference, Vilamoura, Quarteira (Algarve), NATO Advances Study Institute, Portugal, July 19–24, 1999.
4. Dautzenberg, F.M., Schlatter, J.C., Fox, J.J., Rostrup-Nielsen, J.R., and Christiansen, L.J. (1992) Catalyst and reactor requirements for the oxidative coupling of methane. *Catal. Today*, **13**, 503–509.
5. Weisz, P.B.Z. (1957) Diffusivity of Porous Particles. *Phys. Chem.*, **11**, 1.
6. Weisz, P.B. (1982) The science of the possible. *Chem. Technol.*, **12** (7), 424–425.
7. Weisz, P.B. and Miale, J.N. (1965) Super active crystalline alumina silicate hydrocarbon catalysts. *J. Catal.*, **4**, 527.
8. Gandhi, D., Mortimer, T., and Wilcox, R. (1997) Superior ethylbenzene presented technologies: lummus/UOP liquid phase EB process and CDTECH EB ® process. Paper presented at ABB Lummus Global Conference, Mumbai, India, December 3–5, 1997.
9. Narsolis, F., Woodle, G., Gajda, G., and Gandhi, D. (1997 3rd Q), High Performance Catalyst for Liquid Phase EB Technology. *Petroleum Technology Quarterly*, 77–81.
10. Gandi, D., Romatier, J., and Parker, R. (1991) Revamp/expansion of existing styrene plant using integrated SM process. Paper presented at Achema 1991, International Meeting on Chemical Engineering and Biotechnology, Frankfurt, June 12, 1991.
11. Mukherjee, U., Gandi, D., and Sardina, H. (1993) Proven EB/SM technology of ABB lummus crest. *J. Jpn. Aromat. Ind. Assoc.*, **45**, 35–42.
12. Dautzenberg, F.M. and Mukerjee, M. (1993) Current challenges in styrene technology. Paper presented at Large Chemical Plants-10-Conference, Antwerp, Belgium.
13. Arnold, S.C., Suciu, G.D., Verde, V., and Neri, L. (1985) Use fluid bed reactor for maleic anhydride from butane. *Hydrocarbon Process*, **9**, 123.
14. de Boer, M., Nat, P., and van Broekhoven, E.H. (1997) New, high performance catalysts for the isomerization of nC4 and nC5/C6. Paper presented at the National Petroleum Refiners Association Annual Meeting, San Antonio, Texas, March 16–18, 1997.
15. Feldman, R.J., Dufallo, L., Tucci, E.L., and Balogh, P. (1992) Commercial performance of the Houdry Catofin ® process. Paper presented at the National Petroleum Refiners Association Annual Meeting, New Orleans, March 22–24, 1992.
16. Trambrouze, P., van Landeghem, H., and Wanquer, J.P. (1984) *Les Reacteurs Chimiques*, Editions Technip, Paris, pp. 360.
17. Gierman, H. (1988) Design of laboratory hydrotreating reactors: Scaling down of trickle-flow reactors. *Appl. Catal.*, **43** (2), 277–286.
18. Mears, D.E. (1971) The role of axial dispersion in trickle-flow laboratory reactors. *Chem.Eng.Sci.*, **26**, 1361.
19. Dautzenberg, F.M. (2004) New synthesis and multifunctional reactor concepts for emerging technologies in the process industry. *Catal. Rev.*, **46** (3–4), 1–30.
20. Dautzenberg, F.M., van Klinken, J., Pronk, K.M.A., Sie, S.T., and Wijffels, J.B. (1978) Catalyst deactivation through pore mouth plugging during residue desulfurization. Paper presented at the 5th International Symposium on Chemical Reaction

Engineering, Houston, Texas, March 13–15, 1978.
21. Dautzenberg, F.M., George, S.E., Ouwerkerk, C., and Sie, S.T. (1982) Advances in the catalytic upgrading of heavy oils and residues. Paper presented at the Advances in Catalytic Chemistry Symposium, Salt Lake City, Utah, May, 1982.
22. Mary, G., Chaouki, J., and Luck, F. (2009) Trickle-bed laboratory reactors for kinetic studies. *Int. J. Chem. Reactor Eng.*, **7**, R2.
23. van Klinken, J. and van Dongen, R.H. (1980) Catalyst dilution for improved performance of laboratory trickle-flow reactors. *Chem. Eng. Sci.*, **35** (1–2), 59–66.
24. Sie, S.T. and Krishna, R. (1998) Process development and scale-up: III Scale-up and scale-down of trickle bed processes. *Rev. Chem. Eng.*, **14** (3), 203–252.

Part VI
Catalytic Reactor Engineering

Catalysis: From Principles to Applications, First Edition. Edited by Matthias Beller, Albert Renken, and Rutger van Santen.
© 2012 Wiley-VCH Verlag GmbH & Co. KGaA. Published 2012 by Wiley-VCH Verlag GmbH & Co. KGaA.

27
Catalytic Reactor Engineering
Albert Renken and Madhvanand N. Kashid

27.1
Introduction

The chemical reactor is the most important part of every chemical production process. The choice of a suitable reactor for catalytic reaction systems depends strongly on the type of catalysis involved. We distinguish between homogeneous (catalyst is in the same as reactant phase) and heterogeneous (catalyst is in another phase) systems. The heterogeneous systems are further subdivided into fluid-solid (generally one mobile phase), fluid-fluid (generally two mobile phases), and three phase reactions (generally two mobile and one fixed phase). Chemical reactor influences the quality of the product and the economy of the entire process. Because of the large variety and complex relationships of chemical reactions, the design and mode of operation of chemical reactors are also highly diversified.

For single-phase reactions, mixing of the reaction partners and an efficient heat transfer are the main parameters for the reactor design. In heterogeneous systems, mass transfer between the different phases is a further important issue. Mass transfer may strongly influence or even limit the effective transformation rate, thus reducing the reactor performance and the product yield and selectivity as described in detail in Chapter 4.

Besides the distinction between single- and multi-phase systems, reactors can be operated in batch, semi-batch, or continuous modes.

The batch or discontinuous operation is characterized by the fact that the reaction partners together with the catalysts and solvents are charged into the reactor, where they remain for a defined reaction time under preselected reaction conditions. The composition of the reaction mixture changes with operation time, while efficient stirring prevents the occurrence of local concentration and temperature differences. The reactor operation is non-stationary. Batch operations are still widely used in fine chemicals and pharmaceutical production due to their high flexibility, that is, use of the same reactor for different products or product specifications. The disadvantages of batch operation are the occurrence of down times during filling,

Catalysis: From Principles to Applications, First Edition. Edited by Matthias Beller, Albert Renken, and Rutger van Santen.
© 2012 Wiley-VCH Verlag GmbH & Co. KGaA. Published 2012 by Wiley-VCH Verlag GmbH & Co. KGaA.

emptying, heating, and cooling of the reactor, and the sometimes difficult control of the non-stationary process.

In continuously operated reactors, the reaction partners are fed into the reactor at constant flow rates. The reaction mixture is similarly removed from the reactor. All reaction parameters: pressure, temperature, throughput, and inlet concentrations of the reactants are held constant with time. The reactor operation is thus stationary. The advantages of continuous operation are constant and reproducible product quality as a result of constant operating conditions, the higher performance compared to batch operation, and the automation of the process.

Conventionally, continuous reaction processes are used in mass production. But, with the development of micro-structured reactors (MSRs) continuous operation for small chemical productions becomes attractive. Micro-structured catalytic reactors provide high mass and heat transfer performances, thus enabling chemical syntheses under unconventional conditions like high temperature and pressure. This in turn allows considerable process intensification. In addition, high process safety is attained. MSRs are discussed in more detail in Section 27.5.

27.2
Types of Catalytic Reactors

The reactors used for catalytic reactions can be broadly divided into four types, depending on the number of phases involved: single-phase, fluid–solid, fluid–fluid (gas–liquid, liquid–liquid), and three-phase reactors (gas–liquid–solid, gas–liquid–liquid).

In this section, the characteristic criteria and the peculiarities of different types of reactors used for catalytic reactions will be mentioned, but a complete description is not intended. Rather, an attempt is made to introduce the basic concepts and tools of catalytic reactor engineering.

Besides the number of phases present, the choice and the design of catalytic reactors depend on:

- chemical kinetics (stoichiometry, rate of main and side reactions)
- enthalpy of the reactions
- chemical equilibrium and its dependence on pressure and temperature.

27.2.1
Single-Phase Reactors

27.2.1.1 Stirred-Tank Reactor
The stirred tank is the classical reactor for homogenous liquid-phase reactions. It can be operated either discontinuously (batchwise) or continuously. The advantage of a batch reactor is the well-defined and sometimes very long reaction time. In addition, reaction conditions such as temperature, pH value, or catalyst concentrations can be varied and optimized during the reaction period. However, the product

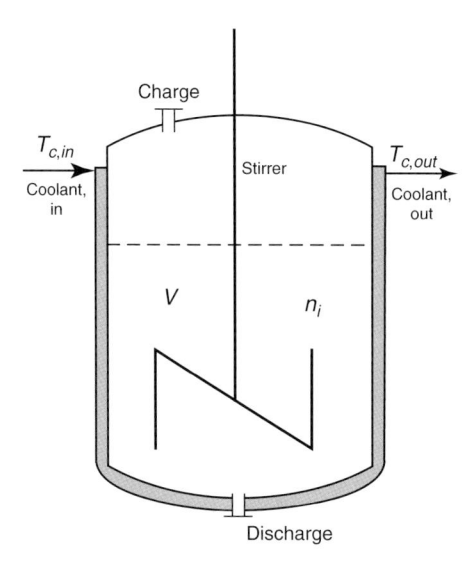

Figure 27.1 Stirred tank reactor.

quality depends essentially on the precise maintenance of the operation procedure, and therefore some efforts are required to control such a non-stationary process.

A typical stirred tank is illustrated in Figure 27.1. Stirred tanks are available in a large number of standard sizes and dimensions [1, 2]. Whenever possible, standard dimensions and materials are chosen, and expensive special constructions are avoided for economic reasons.

Mixing One of the main tasks of a stirred tank reactor is the homogenization of two, mutually soluble solutions down to the molecular level. In addition, heat exchange through the tank walls and sometimes integrated cooling coils play a major role in the reactor design.

A whole series of stirrer types have been developed to fulfill the mixing and/or heat transfer requirements in different media; these are illustrated schematically in Figure 27.2. They are classified principally according to the flow patterns they cause and the viscosity ranges in which they can be employed. In order to choose the stirrer type which guarantees the desired degree of homogeneity in the required time t_m with the minimum power dissipation, its mixing and power characteristics must be known [3, 4].

The mixing is especially important when the reaction participants are fed into the vessel separately. If the mixing is insufficient and/or too slow, concentration gradients can arise which, in the case of complex reactions, can lead to a change in the product distribution, resulting in decreasing product yield. In the case of very rapid reactions or reactions in highly viscous media, for example, polymerizations, concentration gradients can also occur due to the chemical transformation which must be compensated by intense mixing.

The so-called mixing time t_m can be taken as a criterion for evaluating the mixing efficiency. This is the time required to mix a soluble tracer completely with the

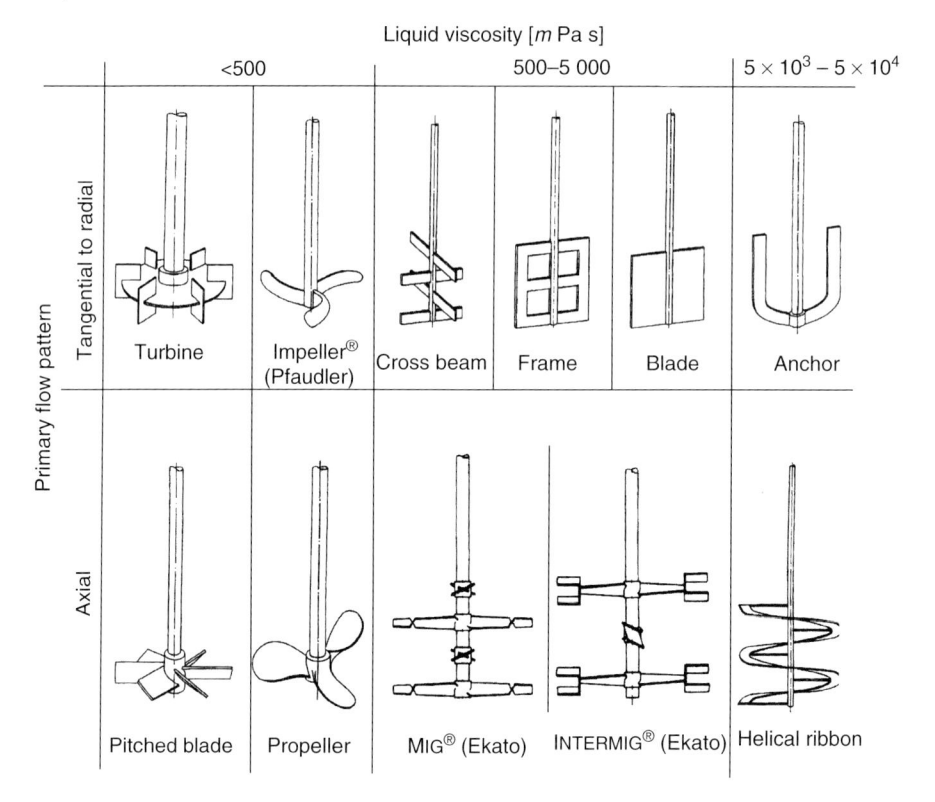

Figure 27.2 Classification of stirrers according to the flow pattern and the viscosity range for their use [5].

reaction mass. A series of physical and chemical methods have been developed for this purpose [5].

In order to ensure a homogeneous reaction mixture in the tank, the mixing time t_m should be much shorter than the characteristic reaction time t_r. Under practical conditions t_m should be at least 10 times smaller than t_r. The characteristic reaction time is referred to the reaction rate (r_0) under the inlet or initial conditions and is defined as follows:

$$t_r = \frac{c_{1,0}}{r_0} \tag{27.1}$$

In continuously operated reactors, the incoming flow of reactants must be homogeneously mixed with the reaction mixture present in the reactor as quickly as possible. It is thus required that the mixing time t_m is small compared to the space time, τ:

$$\tau = \frac{V}{\dot{V}_0} \tag{27.2}$$

where V represents the volume occupied by the reaction mass and $\dot{V}_0$ is the volumetric inlet flow rate. For design purposes, the relationship $t_m \leq 0.1\tau$ is used.

The mixing time depends on the viscosity of the liquid and the power dissipated in the reactor. In general the following relationship holds [6]:

$$t_m \sim \left(\frac{v}{\varepsilon_m}\right)^{1/2} \tag{27.3}$$

with v the kinematic viscosity and ε_m, the specific power dissipation (W kg^{-1}).

Thus, to reduce the mixing time by a factor of 2, the power dissipation must be increased by a factor of 4. Therefore, for the economic operation of the reactor, it is essential that the technically needed degree of homogeneity is achieved without unnecessary stirring power being dissipated and hence wasted.

Heat Transfer The convective heat transfer in a stirred tank can be described with sufficient accuracy by the following expression relating the Nusselt number (Nu) with the Prandtl (Pr) and Reynolds (Re) number [4]

$$Nu = C \cdot Re^a \cdot Pr^b \left(\frac{\mu_{fl}}{\mu_w}\right)^c$$

with $\quad Nu = \dfrac{h \cdot d_{tank}}{\lambda}$ or $Nu = \dfrac{h \cdot d_{coil}}{\lambda} \quad Pr = \dfrac{v}{\alpha} = \dfrac{\mu \cdot c_p}{\lambda} \quad Re = \dfrac{n \cdot d_{stirrer}^2}{v}$
$$\tag{27.4}$$

where d_{tank} is the tank diameter (heat exchange by double jacket), d_{coil} the tube diameter (integrated cooling coil), $d_{stirrer}$ the stirrer diameter, n the stirrer revolutions per second, and μ_{fl} and μ_w are the dynamic viscosities of the fluid in the bulk and near the wall of the heat exchanger, respectively. Lambda is the heat conductivity and cp the specific heat.

In the flow range $10^2 \leq Re \leq 10^6$, the heat transfer characteristics can be represented by: $a \simeq 2/3, b \simeq 1/3, c \simeq 0.14$.

Different types of stirrers and tank geometries can be taken into account by changing the constant C. It should be noted that the heat transfer coefficient h increases with increasing stirrer speed according to

$$h = \text{constant} \cdot n^{2/3} \tag{27.5}$$

while the power requirements increase much more strongly with n, as indicated in Eq. (27.6)

$$P = \text{constant} \cdot n^2, \quad \text{laminar flow}$$
$$P = \text{constant} \cdot n^3, \quad \text{turbulent flow} \tag{27.6}$$

This may cause problems as the increasing heat transfer performance with n is accompanied by an important increase in the power dissipated by the stirring, which will heat up the reaction mixture.

27.2.1.2 Tubular Reactors

In a discontinuously operated, ideally mixed stirred tank reactor, conversion proceeds as a function of time at the same reactor position whereas, in a plug flow reactor (PFR), it changes with the length of the reactor. Under plug flow conditions and at a constant density of the reaction mixture, the reaction time, t_R, necessary for getting a required conversion in a batchwise operated tank corresponds to the space time $\tau = V/\dot{V}_0$ in the 'ideal' tubular reactor.

Because of the high ratio of surface area to reactor volume, the tubular reactor is especially suitable for strongly endothermic or exothermic reactions.

In general, tubular reactors are operated in the turbulent flow regime ($Re > 10^4$). The ratio of tube length to tube diameter should be $L/d > 50$ in order to minimize the influence of axial dispersion in the reactor.

For slow reactions, the use of tubular reactors is limited due to the long residence times required resulting in low flow rates. Controlled reaction conditions are difficult to attain for $Re < 1000$. In low-viscosity media, secondary flows may arise due to concentration and temperature gradients leading to unpredictable behavior. In highly viscous media, for example, in polymerizations, a laminar velocity profile is developed which gives rise to a very wide residence time distribution and to poor product quality. In the region of low Re numbers, the tube must be equipped with internal devices such as, for example, static mixers or packed beds [7, 8]. In addition to the suppression of undesired axial dispersion, improved radial concentration and temperature homogenization are achieved by these measures. Furthermore, a two- to fivefold increase in heat transfer is reported.

27.2.2
Fluid–Solid Reactors

In multi-phase reaction systems, the transfer of one or more reactants from one phase into anther phase or to its surface precedes the chemical reaction. The physical transfer phenomena can considerably influence the conversion and the selectivity for the desired product (see Chapter 4). The rate of mass transfer and the size of the interfacial area are thus decisive parameters which influence the choice of the reactor type.

Fixed-bed and fluidized-bed reactors are the most commonly used types of reactors for heterogeneous catalyzed reactions.

27.2.2.1 Fixed-bed Reactors

The simplest type of catalytic fixed bed consists of a random packing of catalyst particles in a tube. Different shapes of particles may be used like spheres, cylinders, rings, or even crushed materials. The randomly packed bed reactor is suitable for reactions requiring relatively high amounts of catalyst as the catalyst volume can be up to 60–65% of the reactor volume. The technical design is mainly determined by the desired temperature profile in the reactor. A rough distinction can be made between adiabatic and polytropic operation of fixed-bed reactors (Figure 27.3). If adiabatic reactor operation is possible it is preferred because of its simplicity. Since no heat has to be removed through the outside wall, no radial temperature profiles can arise. However, the axial temperature increases with increasing conversion. If the reaction temperature needs to be limited for reasons of kinetics or thermodynamics, the reactor is divided into different stages. Reaction heat is evacuated by interstitial cooling or by injection of cold gas between the

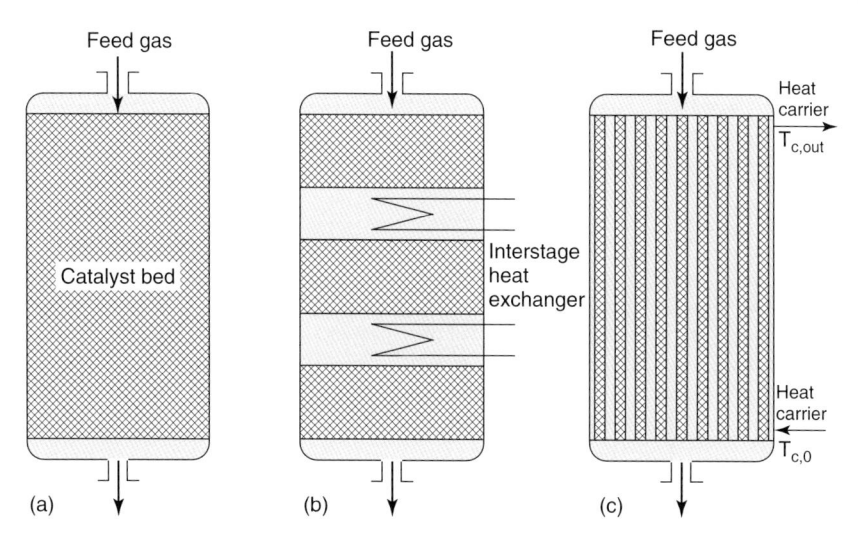

Figure 27.3 Basic types of fixed-bed reactors. (a) Adiabatic reactor, (b) staged fixed-bed, and (c) multitubular reactor.

stages. Examples of the use of adiabatic fixed-bed reactors are oxidation of sulfur dioxide and syntheses of methanol and ammonia.

If an intensive heat exchange is necessary, the reaction is performed in a multi-tubular fixed-bed reactor. Reactors of this type can contain up to 10 000 parallel tubes and are employed for strongly exothermic or endothermic reactions. Examples of the industrial use of multitubular fixed-bed reactors are partial oxidation of hydrocarbons, and Fischer–Tropsch synthesis.

With regard to the flow characteristics, multitubular reactors correspond to PFRs with narrow residence time distribution (see Section 27.3).

The pressure drop in fixed beds can be estimated by the Ergun equation [9]:

$$\frac{\Delta p}{L} = 150 \frac{(1-\varepsilon)^2}{\varepsilon^3 d_p^2} \cdot \mu_f \cdot u_0 + 1.75 \frac{1-\varepsilon}{\varepsilon d_p} \cdot \rho_f \cdot u_0^2, \qquad (27.7)$$

ε : porosity, u_0 : superficial velocity

Since the pressure drop increases strongly with decreasing particle diameter, particles with mean diameters in the range of $2\,\text{mm} \leq d_p \leq 10\,\text{mm}$ are used industrially.

Pressure drop and fluid/solid mass transfer considerations led to the introduction of structured and arranged catalysts like monoliths, membranes, fibers, or foams [10]. In particular, monolith catalysts are widely used for catalytic gas purification (selective catalytic reduction of NO_x, automotive exhaust purification) (Figure 27.4).

27.2.2.2 Fluidized-bed Reactors

Fluidized beds are widely used for gas–solid reactions [11]. Examples of their use in heterogeneous catalytic gas–solid reactions include catalytic cracking, production of acrylonitrile, ethylene dichloride, and o-phthalic anhydride. Liquid–solid fluidized

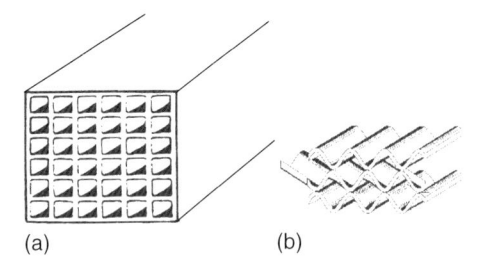

Figure 27.4 Shapes of monolith catalysts. (a) Square-channel monolith, and (b) corrugated plate packing.

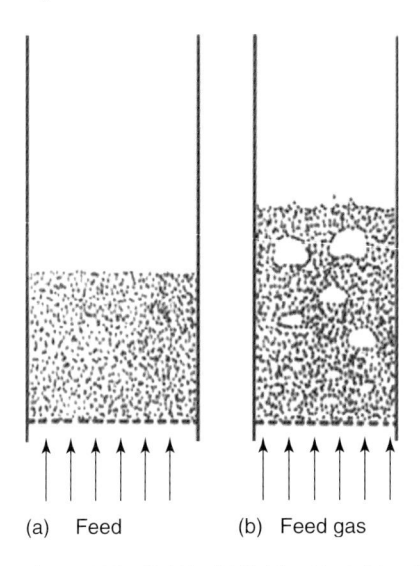

(a) Feed   (b) Feed gas

Figure 27.5 Fluid/solid fluidized-bed (a) uniform and (b) gas bubbles formation.

beds are characterized by homogeneous fluidization (Figure 27.5). They are employed in electrochemical processes and to an increasing extent in biotechnological processes involving immobilized enzymes and cells on inert supports.

The main advantages of fluidized beds are the mobility of the solid, allowing easy handling and transportation, use of small solid particles ($d_P \approx 10$–$800 \ \mu m$), consequently large external fluid–solid surface area, no limitation of internal mass transfer, intense radial and axial solid mixing, thus uniform temperatures, and large heat transfer coefficients, thus suitable for reactions with high reaction enthalpies. The disadvantages of gas–solid fluidized beds are broad residence time distribution of the gas due to axial dispersion and bypassing bubbles, erosion of the vessel, and abrasion of the solid particles. In addition, scaling-up of fluidized beds is quite difficult.

Minimum Fluidization Velocity and Pressure Drop   When a layer of solid particles is subjected to a flow of fluid from the bottom upwards and the linear flow velocity of the fluid is increased above the so-called minimum fluidizing velocity (u_{mf}), the individual particles float, that is, they are suspended homogenously in the fluid. In gas–solid fluidized-beds, increase of the throughput up to about two or three times the minimum fluidizing velocity gives rise to inhomogeneity. In fact, particle-free gas bubbles are formed which rise up through the fluidized bed and increase in size by coalescence. In the case of tall and thin fluidized beds the bubbles can fill the entire cross-section of the reactor. This behavior can often be counteracted by the incorporation of internals (e.g., guiding plates). On further increase of the throughput the fluid flow rate will finally reach the free-fall velocity of the particles and the solid will be transported out of the reactor.

When the pressure drop is calculated by Ergun's empirical relationship (Eq. (27.7)) the minimum fluidization velocity, u_{mf}, is given by Eq. (27.8) [11].

$$u_{mf} = 7.14 v\,(1 - \varepsilon_{mf})\, S_v \left[\sqrt{1 + 0.067 \frac{\varepsilon_{mf}^3 \cdot g \cdot (\rho_P - \rho_f)}{(1 - \varepsilon_{mf})^2\, v^2 \rho_f} \cdot \frac{1}{S_v^3}} - 1 \right] \quad (27.8)$$

The minimum fluidization velocity depends on the kinematic viscosity, v, the density, ρ_f, of the fluid, the density of the particles, and the volume-specific surface area S_v.

$$S_v = \frac{\text{surface area of all particles in the bed}}{\text{volume of all particles in the bed}}$$

27.2.3
Fluid–Fluid Reactors

Gas–liquid reactions such as partial oxidation, hydrogenation, chlorination, and aerobic fermentation belong to this large group of chemical processes. Examples of liquid–liquid reactions in immiscible liquids are nitrations and sulfonations of organic products and extractive reactions in which, for example, the intermediate product is extracted into a phase which is not miscible with the reaction phase in order to avoid subsequent reactions. In general, it can be assumed that the reaction proceeds in only one of the phases, hence the actual reaction is preceded by the transport of one of the reactants into this reaction phase.

Examples of catalytic fluid–fluid reactors are sparged stirred tanks, bubble columns, Buss loop reactors, and static mixer reactors. The contacting principles are bubbling, filming, or spraying of one fluid into the other. For gas–liquid–liquid and gas–liquid–solid reactions, the equipment used is similar (see Figure 27.6: suspension reactors: (a) sparged stirred tank, (b) Buss loop reactor, (c) slurry bubble column).The operating ranges for typical fluid–fluid reactors are summarized in Table 27.1.

Special attention has to be paid to the mass transfer efficiency in the reactor. Because the reaction takes place in the catalyst-containing liquid phase, the reactants must first be transferred from the gas and/or the second liquid phase into the reaction phase. The mass transfer rate between the different phases depends on the area of the interface and the mass transfer coefficient. Whether the reaction

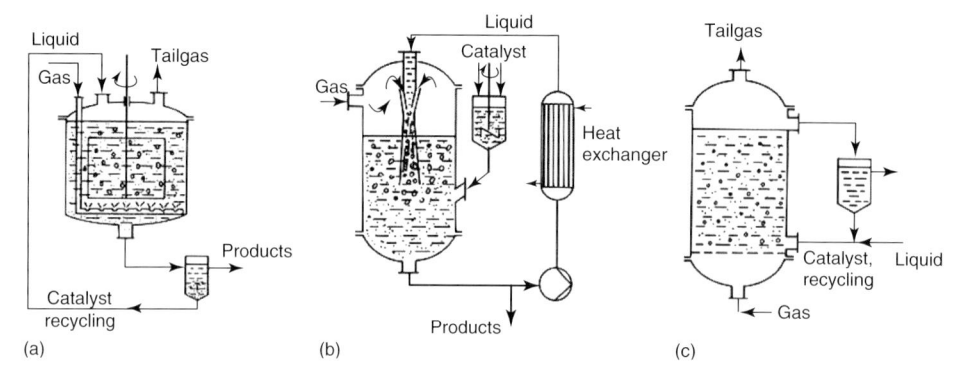

Figure 27.6 Suspension reactors; (a) sparged stirred tank, (b) Buss loop reactor, and (c) slurry bubble column.

Table 27.1 Operating ranges for various gas–liquid reactors [12].

Reactor type	ε_g^* (%)	$10^{-1} k_g$ (mol cm^{-2} s^{-1} Pa^{-1})	10^2 k_l (cm s^{-1})	a (cm^2 cm^{-3} reactor)
Packed column				
Countercurrent	75–98	0.03–2	0.4–2	0.1–3.5
Co-current	5–98	0.10–3	0.4–6	0.1–17
Plate column				
Bell plates	5–90	0.5–2	1–5	1–4
Sieve plates	5–90	0.5–6	1–20	1–2
Bubble column	2–40	0.5–2	1–4	0.5–6
Stirred tank	5–30	–	0.3–4	1–20
Jet nozzle reactor	1–6	–	0.15–0.5	0.2–1.2
Jet scrubber	–	–	–	1–20
Venturi scrubber	70–95	2–10	5–10	1.6–25

ε_g^* = gas fraction in reactor.

will take place in the bulk of the reaction phase or near the interface depends on the ratio between the characteristic reaction time (t_r) to the characteristic time for mass transfer (t_{tr}). This ratio is known as the *Hatta number* (Ha) (see Chapter 4).

$$\mathrm{Ha} = \sqrt{t_r/t_{tr}} \tag{27.9}$$

The discussion can be facilitated by basing it on the film model and by supposing a first-order irreversible reaction in the reaction phase [2, 13, 14]. Under these conditions the following relationships result:

$$\mathrm{Ha} = \sqrt{k_r' D_{i,\mathrm{II}}} \big/ k_{L,\mathrm{II}} \tag{27.10}$$

k_r, the reaction rate constant, is a function of the catalyst concentration ($k'_r = k_r \cdot c_{cat}$), the diffusion coefficient for the compound 1 in the second liquid phase ($D_{i,II}$), and the mass transfer coefficient in this liquid phase ($k_{L,II}$). Increasing values of Ha lead to a decreasing reactant concentration in the bulk of the reacting phase and, as a consequence, the available volume of the reacting phase is decreasingly used. As a logical consequence, fast reactions have to be carried out in reactors ensuring a high interfacial area and high mass transfer coefficient.

27.2.3.1 Liquid–Liquid–Gas System

An important example of a liquid–liquid–gas triphasic system is the hydroformylation of olefins (oxo synthesis) developed by Roelen in 1938 at Ruhrchemie [15]. Two immiscible liquids are used, one containing the homogeneous catalytic metal complex and the other containing the unconverted reactant and the products. Both liquids can be easily separated after the reaction and the catalyst is recycled in the reactor. This can be done without any thermal or chemical treatment. As the reaction is carried out in the presence of the homogeneously dissolved catalyst, the advantages of homogeneous catalysis are fully preserved. In addition, possible consecutive reactions of the product are avoided or diminished, since the reaction products are preferentially present in the second inert liquid phase. The basic flow scheme of the hydroformylation process is shown in Figure 27.7 [16].

27.2.4
Three-Phase Gas–Liquid–Solid Systems

Reactions in which the three phases gas, liquid, and solid participate are encountered frequently in the chemical industry. In general, the catalyst is in solid form while the reactants are distributed in the gas and liquid phases.

Industrially used three-phase reactors differ mainly in the way in which the phases are moved. The most important types are fixed-bed and suspension reactors.

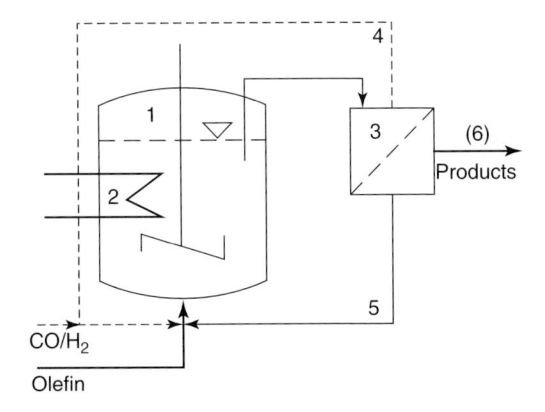

Figure 27.7 Basic flow sheet of the oxo-process. (1) reactor, (2) heat exchanger, (3) separator, (4) gas recycle, (5) catalyst recycle, and (6) product stream.

27.2.4.1 Fixed-Bed Reactors

In fixed-bed reactors the catalyst is arranged in the form of random packing; the particle size ranges from 3 to 7 mm. With regard to the mode of operation, one can distinguish between a packed bubble column (Figure 27.8a) (three-phase fixed-bed) and a trickle bed reactor (Figure 27.8b).

In the first case, the liquid is normally fed in at the bottom and flows upwards to form the continuous phase in which the gas is dispersed. Practical examples are the amination of alcohols and the selective hydrogenation of acetylene or olefins. In general, a packed bubble column is preferred when long residence times are required and/or the liquid is to be reacted with a relatively small amount of gas.

Trickle bed reactors operating in co-current have found wide practical application. Examples are hydro-desulfurization and hydrocracking. The full-scale reactors have a total volume of up to 200 m^3. The height amounts to 10–30 m and the diameter 1–4 m. In general, the reactors are operated adiabatically, that is, without heat removal. Attention must be paid to uniform distribution of the liquid over the cross-section of the reactor and complete wetting of the catalyst. Non-uniform liquid distribution leads to broadening of the residence time distribution and consequently diminishes the reactor performance and product yield. Incomplete wetting of the solid catalyst can result in local overheating associated with a loss of catalytic activity and selectivity.

27.2.4.2 Slurry–Suspension Reactors

Stirred tanks and bubble columns are employed as suspension reactors. Typical examples inlude selective hydrogenation of organic intermediates, fat hydrogenation, and Fischer–Tropsch synthesis. The liquid always constitutes the continuous phase in which the gas and solid are dispersed. Particle size of the solid is usually less than 100 μm, and the solid concentration is typically 1–5%. Suspension reactors behave in a very similar manner to gas–liquid systems.

Hydrogenation reactions are highly exothermic; therefore heat removal is an important issue and may even limit the reactor performance. To increase the heat exchange performance, loop reactors with external heat exchanger are commonly used to control the reactor temperature and to avoid thermal runaway. A further advantage of this reactor type is the high gas/liquid interfacial area generated in the venturi injector.

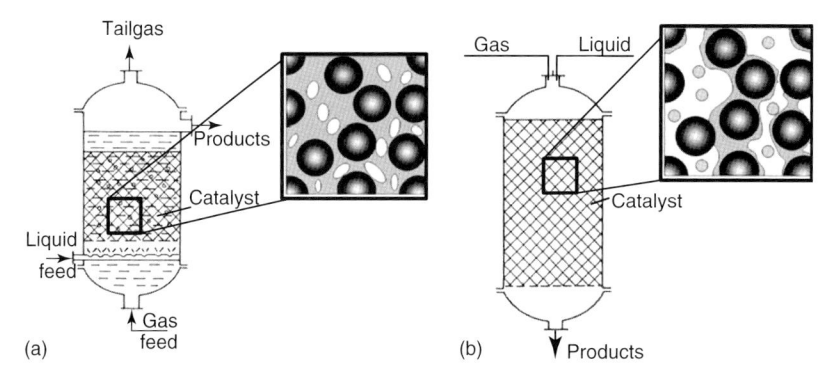

Figure 27.8 Randomly packed bed reactors; (a) packed bubble column and (b) trickle bed reactor.

For small particles, only a low relative velocity (u_{ls}) is to be expected between solid and liquid. In the extreme case, the particles move with the same speed as the liquid. Therefore, the liquid–solid mass transfer can be estimated supposing the particles to be surrounded by a stagnant liquid. The estimated value based on the Frössling equation [6] gives the minimum external mass transfer coefficients.

$$Sh_{L-S} \simeq 2, \text{ and } k_m = 2 \cdot \frac{D_l}{d_p} \quad \text{Dl: diffusion coefficient; dp: particle diameter} \quad (27.11)$$

In contrast, due to the small particle size, internal mass transfer resistances are in most cases negligible (see Chapter 4).

27.2.4.3 Structured Catalysts for Multiphase Reactions

The performance of catalytic multiphase processes is strongly dependent on the catalyst activity/selectivity and the interaction of chemical kinetics with mass transfer. To avoid internal and external mass transfer limitations, catalyst particles in the micrometer range are required, leading to the use of suspension reactors. The size of the catalysts is, however, limited due to catalyst handling, such as solids charging, filtration, and discharging, which often pose safety and environmental problems.

Therefore, considerable effort has been dedicated to the application of structured catalytic materials or catalysts arranged in multiphase processes. Apart from the fact that problems due to catalyst handling are effectively avoided, the main advantage of these reactors is that the shape of the catalysts can be designed in full detail according to the system requirements.

Structured catalysts are characterized by a non-randomly ordered fixed bed of catalysts. Important examples are monoliths, foams, and fibers. Monolith catalysts consist of small parallel channels in the millimeter range. For gas/liquid systems the reactor is operated in the slug flow regime. This regime is characterized by high gas–liquid–solid mass transfer. The temperature of the adiabatic catalytic bed has to be limited, the liquid is recycled, and the reaction heat must be evacuated in an external heat exchanger as shown in Figure 27.9 [17].

While the external recycle allows sufficient temperature control, it comes at the expense of total backmixing of reaction products, counteracting the plug flow characteristics in the catalytic bed. In complex reaction systems like the selective hydrogenation of alkynes, complete backmixing can lead to considerable losses in product yield at high conversions.

Bubble column reactors consisting of structured catalyst layers in combination with integrated heat transfer exchangers can overcome the mentioned disadvantage [18].

27.3
Ideal Reactor Modeling/Heat Management

Most chemical reactors used in practice can be classified according to some common criteria and assigned to so-called basic or ideal reactor types. Based on

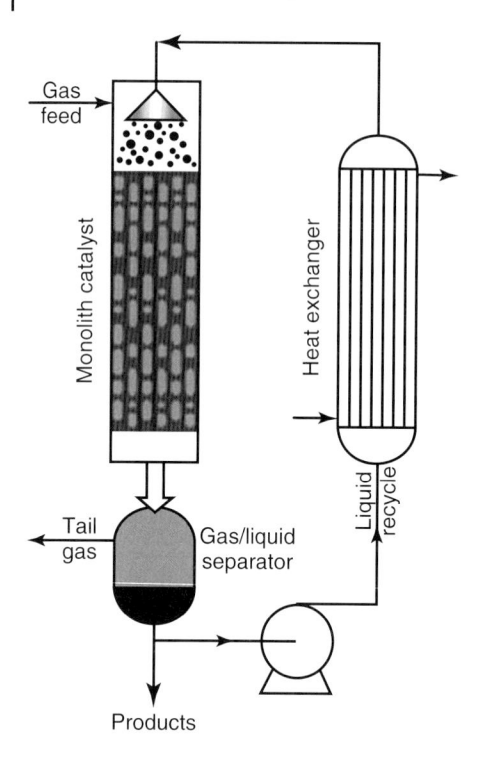

Figure 27.9 Monolith reactor with external liquid recycle.

the characteristics of ideal reactors, the complex interactions of chemical reaction, mass, heat, and impulse transport can be discussed in a general way. The behavior of many reactors actually used approaches that of the ideal types, so that their fundamental relationships can be applied, at least for a first reactor design. In other cases, the reactor behavior of real systems has to be described with the help of models often containing the ideal reactors as individual elements.

For ideal reactors, highly simplifying assumptions are used as a starting point, for example, ideal mixing down to the molecular level or plug flow. We distinguish between:

- the ideally mixed batch-operated stirred tank reactor (BSTR),
- the ideally mixed continuously operated stirred tank reactor (CSTR), and
- the ideal PFR.

27.3.1
Mass and Energy Balances

The interactions between chemical reaction and the simultaneously occurring transport processes for mass, energy, and impulse can be treated mathematically by the fundamental conservation laws. For this, a so-called system volume is defined in which the changes with time of specific conditions of state are described. The changes arise from the difference between the flows entering and leaving this system volume as well as the generation or consumption inside this volume.

For an unambiguous description, it is necessary to select the control volume in such a way that the conditions of state in it can be considered as constant.

The mass balance for a component i can then be formulated as follows:

$$\left\{\begin{array}{c}\text{accumulation}\\\text{of }A_i\text{ in}\\\text{the system}\end{array}\right\} = \left\{\begin{array}{c}\text{rate of flow}\\\text{of }A_i\text{ into}\\\text{the system}\end{array}\right\} - \left\{\begin{array}{c}\text{rate of flow}\\\text{of }A_i\text{ out of}\\\text{the system}\end{array}\right\} + \left\{\begin{array}{c}\text{transformation}\\\text{rate of }A_i\text{ within}\\\text{the system}\end{array}\right\}$$

$$\frac{dn_i}{dt} = \dot{n}_{i,0} - \dot{n}_i + L_{p,i} \qquad (27.12)$$

where n_i represents the number of moles of species A_i at time t, $\dot{n}_i$ is the molar flow rate, and $L_{p,i}$ is the rate of A_i production. The rate of A_i production corresponds to the product of the reaction volume, V, and the rate of A_i transformation:

$$L_{p,i} = R_i \cdot V \qquad (27.13)$$

The rate R_i is given by the sum of the rates of reactions (r_j) in which A_i participates.

$$R_i = \sum v_{i,j} r_j \qquad (27.14)$$

For the following two reactions occurring in the reactor volume, the transformation rates are given in Eq. (27.16).

$$j = 1 : v_{1,1}A_1 + v_{2,1}A_2 \rightarrow v_{3,1}A_3 + v_{4,1}A_4$$
$$j = 2 : v_{3,2}A_3 + v_{4,2}A_4 \rightarrow v_{1,2}A_1 + v_{2,2}A_2 \qquad (27.15)$$
$$R_1 = v_{1,1}r_1 + v_{1,2}r_2; \; R_3 = v_{3,1}r_1 + v_{3,2}r_2 \qquad (27.16)$$

With the exception of special cases, it is generally sufficient for chemical reaction engineering to limit the energy balance to a heat balance. Other forms of energy, such as, for example, the kinetic energy of the reaction mass, can generally be considered separately since they have no influence on the chemical kinetics.

$$\left\{\begin{array}{c}\text{accumulation}\\\text{of energy within}\\\text{the system}\end{array}\right\} = \left\{\begin{array}{c}\text{rate of energy}\\\text{added by}\\\text{convection}\\\text{to the system}\end{array}\right\} - \left\{\begin{array}{c}\text{rate of energy}\\\text{leaving the}\\\text{system by}\\\text{convection}\end{array}\right\}$$

$$+ \left\{\begin{array}{c}\text{rate of energy}\\\text{exchanged with}\\\text{the surroundings}\end{array}\right\} + \left\{\begin{array}{c}\text{rate of energy}\\\text{production}\end{array}\right\}$$

$$\frac{dQ}{dt} = (\dot{m}c_p T)_0 - \dot{m}c_p T + UA(T_c - T) + V \sum_j r_j (-\Delta H_{r,j}) \qquad (27.17)$$

27.3.2
Batchwise-Operated Stirred-Tank Reactors

In the case of a discontinuously (batchwise) operated, ideal stirred tank reactor (BSTR), complete mixing down to the molecular level is assumed (Figure 27.1). The reaction mixture is homogeneous, that is, neither temperature nor concentration gradients exist. Since reactants are neither added nor removed during the reaction time, the mass balance reduces to two terms:

$$\left\{\begin{array}{c}\text{rate of reactant}\\\text{accumulation}\end{array}\right\} = \left\{\begin{array}{c}\text{rate of reactant}\\\text{transformation}\end{array}\right\}$$

$$\frac{dn_i}{dt} = VR_i = V\sum_j v_{ij} r_j \qquad (27.18)$$

The volume V occupied by the reaction mass may change slightly, if the density of the reaction mixture varies during the reaction time as a result of changing product composition and physical processes like heating and cooling.

The BSTR can exchange heat (in contrast to mass) through the reactor wall with the surroundings.

It follows for the general heat balance:

$$(\overline{C}_w + m\overline{c}_p)\frac{dT}{dt} = U \cdot A \cdot (T_c - T) + V\sum_j r_j(-\Delta H_{r,j}). \qquad (27.19)$$

with U, the global heat transfer coefficient, A, the heat exchange surface area in the reactor, T_c, the mean temperature of the cooling/heating medium, and T, the temperature of the reaction mixture.

The total heat capacity of the reactor is designated as $\overline{C}_w$ and is assumed to be independent of temperature. The same holds for the average specific heat $\overline{c}_p$ of the reaction mixture, for which it is additionally assumed that it does not change with the product composition. Equations (27.18) and (27.19) are used to describe the behavior of the reactor during the reaction period and to determine the reactor performance.

The *reactor performance* L_p is defined as the amount of product A_i produced per unit time. In batch-operated reactors, L_p depends on the entire reaction cycle t_c. The cycle consists of the reaction time t_R required to reach a desired degree of conversion and the shut-down time t_a needed for charging, emptying, cleaning, heating, and cooling of the reactor.

$$L_p = \frac{n_i - n_{i0}}{t_a + t_R} = \frac{n_i - n_{i0}}{t_c} \qquad (27.20)$$

The term n_{i0} corresponds to the product present at the start of the cycle.

The reaction time t_R, which is needed to achieve the desired degree of conversion, is obtained by integrating Eq. (27.18). For a single reaction ($R_i = v_i \cdot r$), X, the degree of conversion of the key compound A_1, is given by

$$X = \frac{n_{1,0} - n_1}{n_{1,0}} \tag{27.21}$$

$$t_R = n_{1,0} \int_{X_0}^{X} \frac{dX'}{R_1 \cdot V}. \tag{27.22}$$

For the example of a simple first-order irreversible reaction:

$$R_1 = -kc_1 = -kc_{10}(1-X); \quad \frac{dX}{dt} = k(1-X) \tag{27.23}$$

In the case of strongly exothermic or endothermic reactions a strict isothermal reactor operation is difficult to realize. Thus, for the reactor design, mass and heat balances must be solved simultaneously.

With the Arrhenius law it follows for an nth order reaction:

$$\frac{dX}{dt} = k_0 \exp\left(-\frac{E}{RT}\right) c_{10}^{(n-1)}(1-X) \tag{27.24}$$

$$\frac{dT}{dt} = \frac{U \cdot A}{\bar{C}_w + m \cdot \bar{c}_p}(T_c - T) + \Delta T_{ad}\frac{dX}{dt}. \tag{27.25}$$

with $\Delta T_{ad} = \frac{V(-\Delta H_R) \cdot c_{10}}{(\bar{C}_w + m \cdot \bar{c}_p)}$ the adiabatic temperature rise.

The temperature profile in the reactor is governed by the following parameters:

- the adiabatic temperature rise ΔT_{ad}
- the rate of heat removal.
- which is determined by the overall heat exchange coefficients (U), the heat exchanger surface (A), and the average temperature difference between the heating or cooling medium and the reaction mixture ($T_c - T$),
- the rate of heat generation by the reaction and its temperature dependence.

In the case of constant coolant temperature, the removal of heat increases linearly with the temperature in the reactor, whereas the heat generation increases exponentially according to the Arrhenius equation. This can lead to extremely high temperature peaks (hot spots).

Particular attention has to be paid to the maximum temperature T_{max} which must be limited for safety reasons or for reasons concerning the product yield and selectivity. Side reactions often have higher activation energies than the desired catalyzed reaction. As a consequence, the formation of by-products is favored by increasing temperature, leading to losses in product yield. In addition, the exponential increase of heat production with temperature can lead to thermal runaway of the reactor (see also Section 27.3.4).

27.3.3
Continuously Operated Ideal Stirred Tank Reactors

In an ideal CSTR, no concentration or temperature gradients occur. The reactants fed to the reactor are instantaneously mixed up to the molecular level at the reactor entrance (Figure 27.10).

Since the reaction mixture is homogeneous, the balance equation can be extended over the entire volume, and the mass balance for the component i can be formulated as follows:

$$\frac{dn_i}{dt} = \dot{n}_{i0} - \dot{n}_i + V \sum_j v_{ij} r_j. \tag{27.26}$$

When the molar flow $\dot{n}_i$ is replaced by the volume flow $\dot{V}$ and the concentration c_i, we obtain:

$$V\frac{dc_i}{dt} = \dot{V}_0 c_{i0} - \dot{V}_{out} c_i + V \sum_j v_{ij} r_j. \tag{27.27}$$

The volume occupied by the reaction mixture is designated as V. In general, V corresponds to about 3/4 of the nominal reactor volume.

The ratio of the reaction volume to the volumetric inlet flow $\dot{V}_0$ is known as the *space time* (Eq. (27.2)).

The reciprocal value of τ is often designated as the *space velocity* or, in biotechnology, the *dilution rate*.

After a transient period, which corresponds to about five times the space time, the reactor operates at steady state, that is, the composition of the reaction mixture

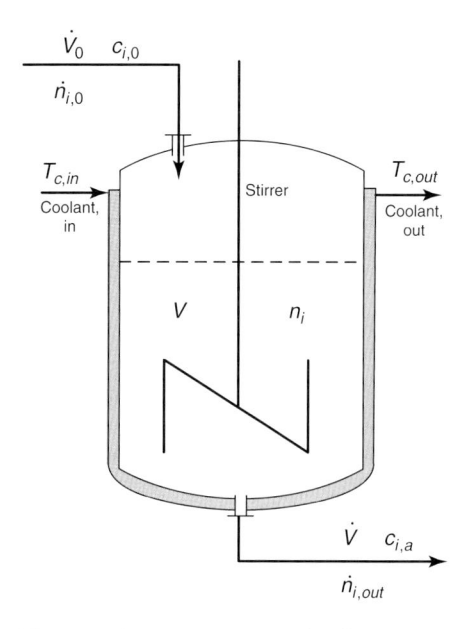

Figure 27.10 Continuous stirred tank reactor.

is time invariant. The mass balance is reduced to a simple algebraic expression.

$$\dot{V}_0 c_{i0} - \dot{V}_{out} c_{i,out} + V \sum_j v_{ij} r_j = 0 \tag{27.28}$$

Introduction of the degree of conversion for the key reactant A_1, leads to

$$X = \frac{\dot{n}_{1,0} - \dot{n}_{1,out}}{\dot{n}_{1,0}} = \frac{-V \sum_j v_{ij} r_j}{\dot{V}_0 c_{1,0}} = \frac{-V R_1}{\dot{V}_0 c_{1,0}}. \tag{27.29}$$

The space time necessary to achieve a required conversion is:

$$\tau = \frac{c_{1,0} X}{-R_1} \tag{27.30}$$

For the heat balance at steady state:

$$\dot{V}_0 \rho_0 c_{p0} T_0 - \dot{V} \rho c_p T + U \cdot A(T_c - T) + V \sum_j r_j (-\Delta H_{r_j}) = 0. \tag{27.31}$$

For the simple case of a single, stoichiometrically independent reaction, we obtain with the mass balance for the key component A_1

$$-R_1 V = r \cdot V = X c_{1,0} \dot{V}_0.$$

$$\dot{V}_0 \rho_0 c_{p0} T_0 - \dot{V} \rho c_p T + U \cdot A(T_c - T) + \dot{V}_0 c_{1,0} X(-\Delta H_R) = 0 \tag{27.32}$$

or, at constant specific heat capacity ($c_p = c_{p0}$)

$$\dot{V}_0 \rho_0 c_{p0} (T - T_0) + U \cdot A(T - T_c) = \dot{V}_0 c_{1,0} X(-\Delta H_R). \tag{27.33}$$

Together with the mass balance, Eq. (27.33) serves for the design of the reactor, that is, to determine the operating parameters for a required reactor performance. In addition, the temperature T_c and the heat exchange area A can be determined for known inlet temperatures.

27.3.4
Ideal Plug Flow Reactor

Under stationary conditions, the mass flow in a continuously operated tubular reactor is independent of time and identical at every point. In contrast, the composition of the reaction mixture changes with increasing distance from the reactor inlet. For an ideal PFR we assume constant concentration and temperature over the entire cross-section. In addition, any axial mass and heat dispersion is excluded.

The axial concentration profile in an ideal PFR is based on the mass balance for a differential volume element as shown in Figure 27.11. For tubes with constant cross section A_{cs}:

$$-\frac{d(c_i \dot{V})}{dV} + \sum_j v_{ij} r_j = -\frac{d(c_i u)}{dz} + R_i = 0 \tag{27.34}$$

Any change in the inlet flow travels with linear velocity u through the reactor. Therefore, a novel stationary axial concentration profile is reached after the space time $\tau = L/u$ (constant density).

27 Catalytic Reactor Engineering

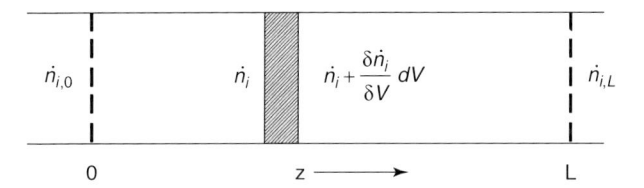

Figure 27.11 Mass balance in plug-flow reactor.

For a single, stoichiometrically independent reaction we obtain

$$\frac{d\dot{n}_i}{dV} = \frac{d(c_i u)}{dz} = v_i r = R_i. \tag{27.35}$$

After introduction of the conversion for the key component A_i, it follows that:

$$\frac{dX}{dV} = \frac{-R_1}{\dot{n}_{1,0}} = \frac{-R_1}{\dot{V}_0 \cdot c_{1,0}}. \tag{27.36}$$

The reactor volume required to achieve a certain conversion can be calculated by integration. In general, the conversion is a function of the ratio between space time and characteristic reaction time:

$$X = f(DaI), \quad DaI = \tau/t_r, \text{ 1st Damköhler number} \tag{27.37}$$

The space time in the reactor for getting a final conversion X_L is given by

$$\tau = \frac{V}{\dot{V}_0} = c_{1,0} \int_{X_0}^{X_L} \frac{dX}{-R_1}. \tag{27.38}$$

The space time necessary for a required conversion corresponds to the hatched gray surface under the curve $c_{1,0}/(-R_1) = f(X)$ shown in Figure 27.12. The space time necessary for getting the same conversion in a CSTR is represented by the whole gray surface. Due to the low reactant concentration the space time in the latter is considerably higher, leading to poor reactor performance.

If the density of the reaction mixture does not change throughout the reactor, the linear velocity of the reaction mixture remains constant and Eq. (27.35) can then be transformed with $d\tau = dz/u$ to

$$u\frac{dc_i}{dz} = \frac{dc_i}{d\tau} = \sum_j v_{ij} r_j = R_i.; \quad \text{with } u = \frac{\dot{V}_0}{A_{cs}} = \text{constant.} \tag{27.39}$$

The derived expressions for the ideal tubular reactor are the same as those for an ideal BSTR. The reaction time t_R corresponds to the space time τ, that is, the conversion achieved in a batch reactor is identical to that in an ideal PFR when the reaction time t_R and the space time τ are equal. However, this comparison is no longer valid when the reaction is accompanied by a density change, which leads to variations in the linear velocity.

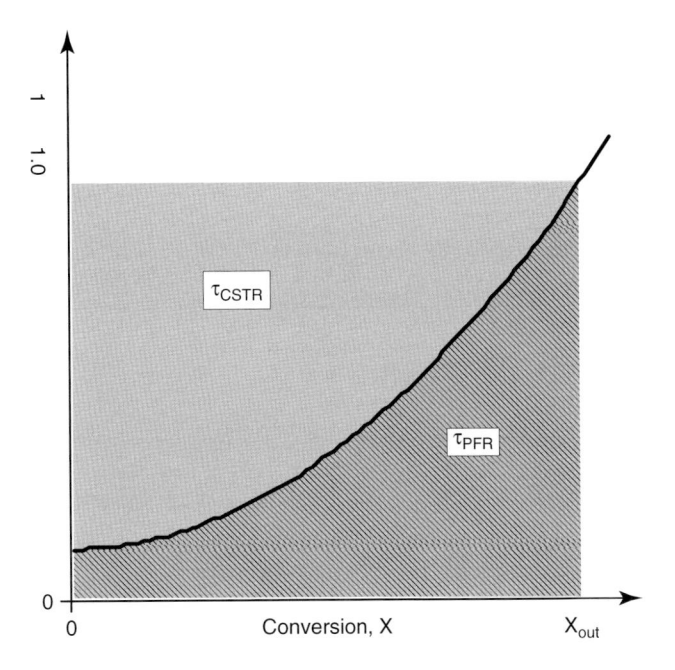

Figure 27.12 Space time in CSTR and PF reactor.

The **heat balance** for a volume element in an ideal PFR at steady state can be formulated as follows:

$$\bar{m}\bar{c}_p \frac{dT}{dV} = \sum_j r_j(-\Delta H_{R_j}) + U(T_c - T)\frac{dA}{dV}. \tag{27.40}$$

In accord with the previously made assumptions, the temperature over the cross-section is constant and only a function of the axial position. The term (dA/dV) corresponds to the reactor surface per volume element (specific surface area, a). For circular tubes with a constant diameter d_t it follows that:

$$a = \frac{dA}{dV} = \frac{4}{d_t} \tag{27.41}$$

In general, the heat exchanged through the tube wall will be different from the heat generated or consumed by the reaction at the same axial position. As a consequence, an axial temperature profile develops.

To determine the axial temperature and concentration profiles, heat and mass balances must be solved simultaneously. For a single, stoichiometrically independent reaction and under stationary conditions, we obtain for the key component A_1

$$\frac{dX}{dV} = \frac{-R_1}{\dot{V}_0 \cdot c_{1,0}} = \frac{r}{\dot{V}_0 \cdot c_{1,0}}$$

$$\frac{dX}{dZ} = \frac{-R_1 \cdot \tau}{c_{1,0}} \tag{27.42}$$

$$Ua(T_c - T) + r(-\Delta H_r) - \bar{m}\bar{c}_p \frac{dT}{dV} = 0$$

$$\frac{dT}{dZ} = \frac{U \cdot \tau}{\rho_0 \cdot \bar{c}_p} \left(\frac{dA}{dV}\right)(T_c - T) + \Delta T_{ad} \frac{r \cdot \tau}{c_{1,0}} \tag{27.43}$$

$Z = z/L$ (relative length of the reactor)

For a tube with constant diameter d_t, we obtain from Eq. (27.43)

$$\frac{dT}{dZ} = \underbrace{\frac{U \cdot \tau}{\rho_0 \cdot \bar{c}_p} \cdot \frac{4}{d_t}(T_c - T)}_{A} + \underbrace{\Delta T_{ad} \frac{r \cdot \tau}{c_{1,0}}}_{B}$$

with: $\Delta T_{ad} = \dfrac{c_{1,0}(-\Delta H_r)}{\rho_0 \bar{c}_p}$, the adiabatic temperature rise (27.44)

The behavior of an ideal tubular reactor is determined by the rate of heat exchange and the heat capacity of the mixture (term A, Eq. (27.44)) and by the rate of heat generation and its dependence on the temperature (term B, Eq. (27.44)). Due to the exponential increase of the reaction rate, and thus the rate of heat generation with temperature, high local temperature peaks may occur in the case of exothermic reactions. Under certain operating conditions, small variations of the reaction parameters such as cooling temperature or inlet concentration can lead to extremely large changes in the temperature and concentration profiles and thermal runaway of the reactor. This behavior is known as *parametric sensitivity*. To predict the thermal behavior of PFRs and to avoid the dangerous domain of high parametric sensitivity, we follow the general considerations of Semenov and Barkelew [19].

For an irreversible reaction of nth order, Eq. (27.44) can be rewritten in a dimensionless form:

$$\frac{d(\Delta T')}{dZ} = \left[S' \exp\left(\frac{\Delta T'}{1 + \Delta T'/\gamma}\right)(1 - X)^n - N\Delta T'\right] \text{DaI}(T_c)$$

with: $\Delta T' = \dfrac{T - T_c}{T_c} \cdot \dfrac{E_a}{RT_c}$, $\text{DaI}_{T_c} = \tau \cdot k(T_c)c_{10}^{n-1}$ (27.45)

The first Damköhler number $\text{DaI}(T_c) = \dfrac{\tau}{t_r(T_c)} = \tau \cdot k(T_c)c_{10}^{(n-1)}$ in Eq. (27.45) corresponds to the ratio of space time τ to the characteristic reaction time t_r determined at the temperature of the cooling medium.

The temperature profile in the reactor is determined by the following parameters:

- $N = \dfrac{t_r(T_c)}{t_c} = \dfrac{U \cdot a}{\rho_0 \bar{c}_p} \cdot \dfrac{1}{k(T_c)c_{10}^{(n-1)}}$, the ratio between the characteristic reaction time $t_r(T_c)$ at cooling temperature and the characteristic cooling time, t_c.
- $S' = \Delta T_{ad} \dfrac{E_a}{RT_c^2}$, the heat production potential
- $\gamma = \dfrac{E_a}{RT_c}$, the Arrhenius number.

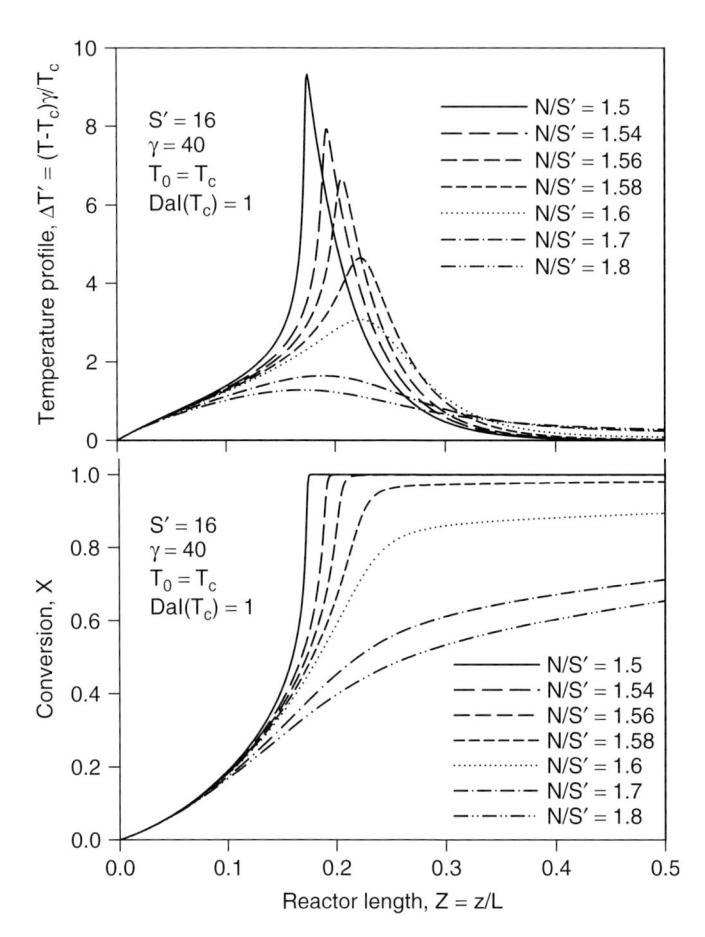

Figure 27.13 Axial temperature profiles (top) and conversion (bottom) in a cooled plug flow reactor with first-order reaction and different values (N/S′).

As the Arrhenius number is generally high; the term $\Delta T'/\gamma$ in Eq. (27.45) can often be neglected, and we obtain:

$$\frac{d(\Delta T')}{dZ} \simeq \left[S' \exp(\Delta T')(1-X)^n - N\Delta T'\right] \mathrm{DaI}(T_c) \tag{27.46}$$

Of primary importance for the temperature and conversion profiles is the value of N/S', which corresponds to the inverse Semenov number $\frac{N}{S'} = \frac{1}{S_e}$.

Typical temperature profiles are shown in Figure 27.13 for various values of N/S' with S' as parameter. The initial temperature is assumed to be identical to the temperature of the cooling medium ($T_0 = T_c$).

For a given heat production potential, S', the axial temperature profile is strongly influenced by the ratio between characteristic reaction and cooling time, N. The

27 Catalytic Reactor Engineering

minimum ratio, N_{min}, for stable reactor operation as a function of S' can be estimated from the following empirical relationship [19]:

$$\left(\frac{N}{S'}\right)_{min} = 2.72 - \frac{B}{\sqrt{S'}} \qquad (27.47)$$

The parameter B is dependent on the reaction order n and is found to be: $B = 0$ ($n = 0$), $B = 2.6$ ($n = 0.5$), $B = 3.37$ ($n = 1$), $B = 4.57$ ($n = 2$).

Highly exothermic reactions are characterized by large adiabatic temperature rises resulting in high values of S'. Safe reactor operation can then only be realized when the characteristic cooling time is reduced. This is in general only possible by increasing the specific heat exchange surface, for example, by reducing the diameter of the tubular reactor.

A number of papers have been published on the problem of parametric sensitivity of chemical reactors. The object was to provide simple criteria to estimate whether the reactor is operating in a safe range or whether the occurrence of dangerous situations leading to a 'runaway' need to be considered. The results from the literature are summarized in a stability diagram where the minimal N/S' to insure safe operation versus the heat generation potential S' is given (Figure 27.14).

In Figure 27.14, regions of high parametric sensitivity are separated by stable regions. As the criteria used by several authors are different, it can only be presented as a gray band and not as a sharp line. The proposed simple relationship for estimating the regions of high parametric sensitivity (equation (27.47)) are added to the diagram. The reactor sensitivity decreases with increasing reaction order.

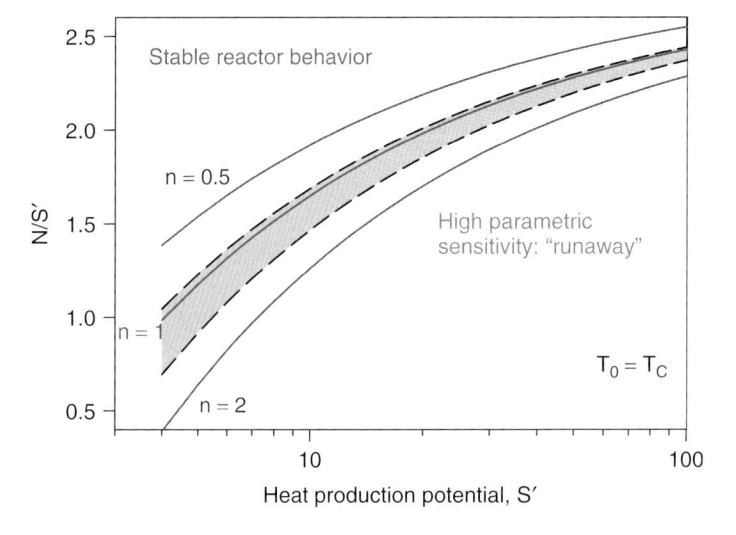

Figure 27.14 Stability diagram.

27.4
Residence Time Distribution

For the ideal reactors discussed in Section 27.3, their hydrodynamic behavior is well defined: plug flow should prevail in ideal tubular reactors, where axial dispersion is excluded.

The behavior of real reactors often deviates from that of the ideal type. In tubular reactors, radial velocity and concentration profiles may develop at laminar flow. In turbulent flow, velocity fluctuations can lead to an axial dispersion. In packed beds, irregular flow with the formation of channels may occur while stagnant fluid zones (dead zones) may develop in other parts of the reactor. Incompletely mixed zones and thus inhomogeneities can also be observed in CSTR, especially in the cases of viscous media.

The above-mentioned phenomena lead to a non-uniform residence time of the fluid elements in the reactor, which cannot be described by the model equations presented for the ideal reactors. The influence of the non-ideality on the reactor performance and on the product yield and selectivity will be discussed in the following sections.

The residence time distribution of a fluid can be characterized by the age distribution of volume elements which have left the reactor. A typical residence time distribution $(E(t))$ is shown in Figure 27.15. The function $E(t)$ indicates the probability that a fraction of the amount (n_0) entering the reactor at $t = 0$ has left the reactor after the time t. The dimension of $E(t)$ is s^{-1}.

$$E(t) = \frac{\dot{n}(t)}{n_0} = \frac{\dot{V}c(t)}{\int_0^\infty \dot{V}c(t)dt} \tag{27.48}$$

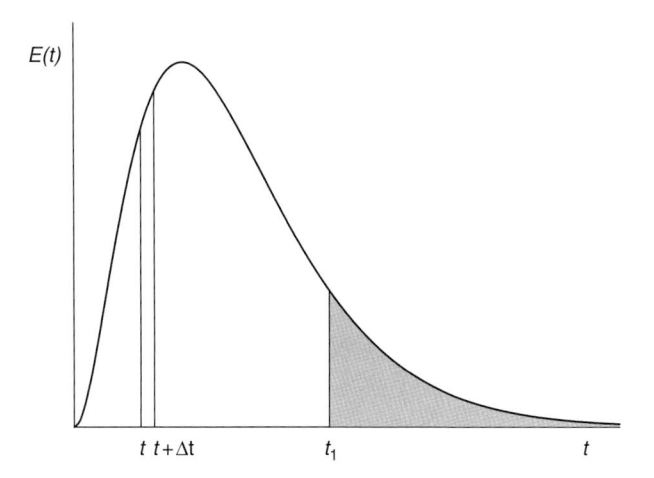

Figure 27.15 Residence time distribution or exit age distribution curve, $E(t)$.

After an infinitely long observation time, the probability that all volume elements fed into the reactor at the time $t = 0$ have left it is equal to one.

$$\int_0^\infty E(t)dt = 1 \tag{27.49}$$

The fraction of the fluid in the exit stream which is younger than t_1 is given by

$$\int_0^{t_1} E(t)dt, \tag{27.50}$$

and that which is older is

$$\int_{t_1}^\infty E(t)dt = 1 - \int_0^{t_1} E(t)dt. \tag{27.51}$$

It is often advantageous to relate the distribution functions to the mean residence time in order to be able to compare reactors of different sizes and with different throughputs. Thus, a dimensionless time referred to the mean residence time is introduced.

$$\theta = t/\bar{t} \tag{27.52}$$

We then obtain the following relationships for the distribution functions

$$E(\theta) \equiv E = \bar{t} \cdot E(t) \tag{27.53}$$

For practical purposes, it is beneficial to characterize distribution functions by a few characteristic terms such as the mean residence time and the variance around the mean. The mean residence time ($\bar{t}$) corresponds to the first moment of the distribution density function.

$$\mu_1 = \bar{t} = \int_0^\infty t \cdot E(t)dt \tag{27.54}$$

If the density of the fluid does not change in the reactor ($\dot{V}_0 = \dot{V}_{\text{out}}$), the mean residence time corresponds to the space time.

$$\bar{t} = \tau = V/\dot{V}_0; \rho = const. \tag{27.55}$$

The variance of the distribution is obtained from the second moment of the distribution function.

$$\sigma^2 = \mu_2 - \mu_1^2 = \int_0^\infty t^2 \cdot E(t)dt - \bar{t}^2 = \int_0^\infty (t - \bar{t})^2 E(t)dt \tag{27.56}$$

The higher moments of the residence time distribution are of little practical interest since they are difficult to obtain experimentally with the required accuracy.

27.4.1
Experimental Determination of the Residence Time Distribution

To determine the residence time distribution experimentally a tracer is introduced at the inlet of the reactor. The response of the system to the imposed inlet function is obtained by measuring the tracer concentration at the outlet as a function of time. It is important that the tracer does not change the physical properties of the reactor contents, that is, the viscosity and density of the fluid must remain constant. In addition, the tracer should not take part in the reaction process or be adsorbed at parts of the reactor; furthermore, it should be easy to measure even in low concentrations.

Usually the tracer is injected in the form of a well-defined function: as a step or impulse function, in the form of a sine function, or as a random signal with known properties. The first two mentioned functions are the most used and will be discussed in detail.

27.4.1.1 Step Function
At the inlet of the reactor, the concentration of a tracer is abruptly changed at time $t = 0$. The response of the system at the reactor outlet is measured. The momentary tracer concentration $c(t)$ is referred to the constant inlet concentration c_0. The response curve is thus dimensionless and is designated as an F curve after Danckwerts [13]; it thus has values between 0 and 1 (see Figure 27.16). It corresponds to the cumulative curve of the residence time distribution.

$$F(t) = \frac{c(t)}{c_0} \tag{27.57}$$

The following relationship between the external residence time distribution and the F curve holds:

$$F(t) = \int_0^t E(t')dt' = \int_0^\theta E d\theta' = F \tag{27.58}$$

$$E(t) = \frac{dF(t)}{dt} = \frac{dF}{dt} \tag{27.59}$$

The mean residence time in the reactor is obtained from the F curve according to

$$\bar{t} = \int_0^\infty t \cdot E(t)dt = \int_0^1 t \cdot dF \tag{27.60}$$

Or with discrete measurement points

$$\bar{t} \simeq \sum_i t_i \cdot \Delta F_i. \tag{27.61}$$

In analogy, the variance is obtained with.

$$\sigma^2 = \int_0^\infty (t - \bar{t})^2 E(t)dt = \int_0^1 (t - \bar{t})^2 dF \tag{27.62}$$

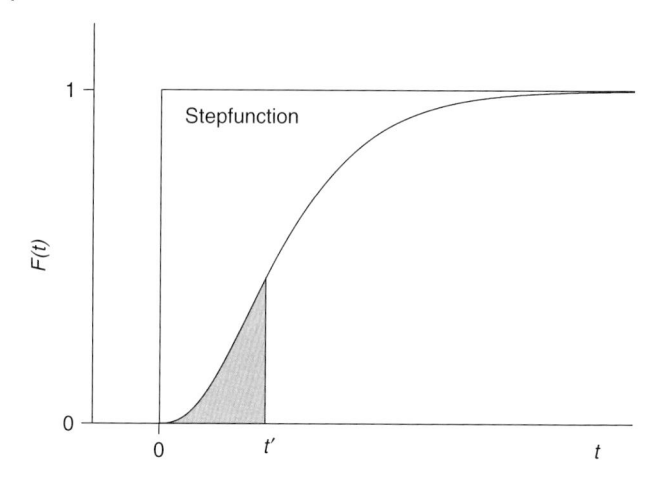

Figure 27.16 Response to a stepfunction at the reactor inlet.

Or with discrete measured points

$$\sigma^2 \simeq \sum_i (t_i - \bar{t})^2 \Delta F_i \qquad (27.63)$$

27.4.1.2 The Pulse Function

The entire amount of the tracer is fed to the reactor inlet within a very short time to approach the Dirac delta function as closely as possible. The Dirac function has the following properties:

$$\begin{array}{ll} t = 0 & \delta(t) = \infty \\ t \neq 0 & \delta(t) = 0 \end{array} \qquad (27.64)$$

$$\int_{-\infty}^{+\infty} \delta(t) dt = 1$$

In practice, the input time Δt should be small compared to the space time ($\Delta t \leq 0.01\tau$). The response of the system at the outlet to the pulse-like inlet function is called the *C curve*. The values of this curve are dimensionless since the measured outlet concentration of the tracer is referred to the formal mean concentration c_0 in the reactor. If n_0 moles of the tracer are introduced over the time period Δt, the mean concentration is defined as:

$$c_0 = n_0/V \qquad (27.65)$$

Experimentally, the amount of tracer injected can also be determined by measuring its concentration at the reactor outlet.

$$n_0 = \int_0^\infty \dot{V} c(t) dt. \qquad (27.66)$$

The relationship between the measured $C(t)$ curve and the residence time distribution $E(t)$ is given by the following:

$$E(t) = \frac{\dot{V} \cdot c(t)}{n_0} = \frac{1}{\bar{t}} \frac{c(t)}{c_0} = \frac{1}{\bar{t}} C(t) . \quad C(t) = c(t)/c_0 \qquad (27.67)$$

27.4.2 RTD for Ideal Reactors

Before we consider the residence time distribution in real systems, we will first examine the behavior of the ideal reactors presented in Section 27.3.

27.4.2.1 Ideal Plug Flow Reactor

The ideal PFR acts solely as a delaying element without changing the form of the input signal. In the case of an impulse function at the inlet, the same pulse function is obtained at the outlet after a time delay corresponding to the average residence time $\bar{t}$.

$$E(t) = \delta(t - \bar{t}) \qquad (27.68)$$

The same is true for the step function and its response $F(t)$.

27.4.2.2 Ideal Continuously-Operated Stirred Tank Reactor

If the amount n_0 of a tracer is added in form of a pulse to a completely mixed stirred tank, the maximum average concentration is established instantaneously

$$c_0 = \frac{n_0}{V}, \quad t = 0. \qquad (27.69)$$

The concentration–time course can then be predicted from the mass balance by integration

$$V \frac{dc(t)}{dt} = -\dot{V} c(t)$$

and

$$C(t) = \frac{c(t)}{c_0} = \exp\left(-\frac{t}{\bar{t}}\right) = \exp(-\theta) \qquad (27.70)$$

With Eq. (27.67) we obtain for the residence time distribution (RTD):

$$E(t) = \frac{1}{\bar{t}} C(t) = \frac{1}{\bar{t}} \exp\left(-\frac{t}{\bar{t}}\right)$$

and

$$E = \exp\left(-\frac{t}{\bar{t}}\right) = \exp(-\theta). \qquad (27.71)$$

From this it follows by integration for the cumulative residence time distribution (curve $F(t)$):

$$F(t) = \int_0^t E(t') dt' = 1 - \exp\left(-\frac{t}{\bar{t}}\right). \qquad (27.72)$$

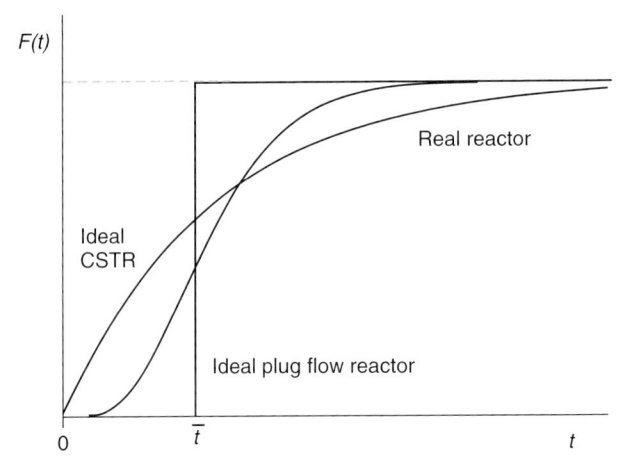

Figure 27.17 Cumulative RTD of ideal and real reactors.

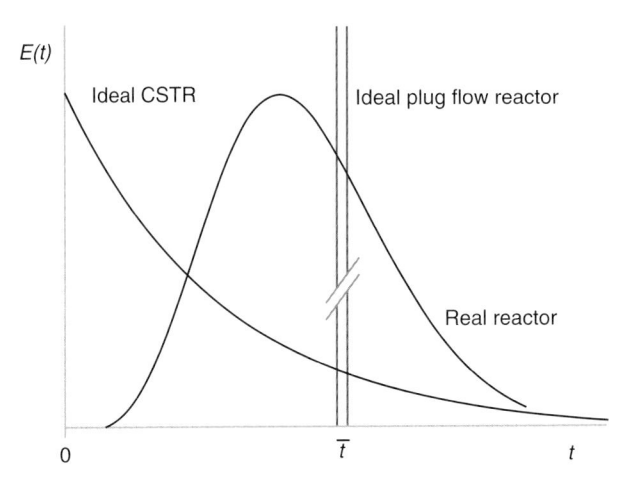

Figure 27.18 RTD curves of ideal and real reactors.

In Figures 27.17 and 27.18 the residence time distributions of ideal reactors are presented together with the RTD of a real reactor. The ideal, continuously-operated stirred tank reactor has the broadest residence time distribution of all reactor types. The most probable residence time for an entering volume element is $t = 0$. After a mean residence time $\bar{t}$, 37% of the tracer injected at time $t = 0$ is still present in the reactor. After 5 mean residence times, a residue of about 1% still remains in the reactor. This means that at least 5 mean residence times must pass after a change in the inlet conditions before the reactor effectively reaches its new stationary state.

27.4.2.3 Cascade of Ideally Stirred Tanks

The cascade consists of a series of ideal stirred tank reactors connected one after the other (Figure 27.19). The outlet function of one CSTR is thus simultaneously the

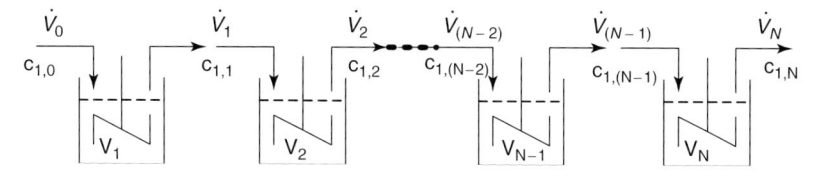

Figure 27.19 Cascade of continuous stirred tank reactors.

input function of the next. Since the transfer functions of each reactor are identical and known, the distribution function of a cascade of N tanks can be determined by successive convolution.

For a cascade of N tanks of equal space time $\bar{\tau}_i$ we obtain for the residence time distribution:

$$E(t) = \frac{1}{\bar{t}_i} \left(\frac{t}{\bar{t}_i}\right)^{N-1} \cdot \frac{1}{(N-1)!} \exp\left(-\frac{t}{\bar{t}_i}\right). \qquad (27.73)$$

With $\bar{t} = N \cdot \bar{t}_i$ as mean residence time of the cascade:

$$\bar{t} E(t) = E = \frac{N(N \cdot \theta)^{N-1}}{(N-1)!} \exp(-N\theta) \qquad (27.74)$$

The cumulative residence time distribution curve can be calculated from this by integration.

$$F(t) = \int_0^t E(t')dt' = \frac{c(t)_N}{c_0}$$

$$= 1 - \exp(-N\theta) \left[1 + N \cdot \theta + \frac{(N \cdot \theta)^2}{2!} + \cdots \frac{(N \cdot \theta)^{N-1}}{(N-1)!}\right] \qquad (27.75)$$

The residence time distributions of ideal cascades with different numbers of tanks are given in Figures 27.20 and 27.21. With increasing subdivision of the entire reactor volume into ideally mixed individual elements, the residence time becomes more and more uniform and the residence time distribution curves more symmetrical. The residence time distribution of the cascade approaches that of an ideal PFR and becomes identical with this when N approaches infinity.

27.4.2.4 Laminar Flow Reactor

Laminar flow is characterized by a parabolic velocity profile according to the Hagen–Poiseuille law:

$$u(r) = u_{max}\left(1 - \frac{r^2}{R^2}\right) = 2\bar{u}(1 - y^2) \cdot y = \frac{r}{R} \quad \begin{array}{l}\text{radial distance}\\ \text{tube radius}\end{array} \qquad (27.76)$$

with u_{max} the velocity in the center of the tube and the average value over the cross-section. If diffusion processes are neglected, the residence time of a volume element depends on the radial position in the tube.

$$t = \frac{L}{u} = L[u_{max}(1-y^2)]^{-1} = \frac{t_{min}}{(1-y^2)} \qquad (27.77)$$

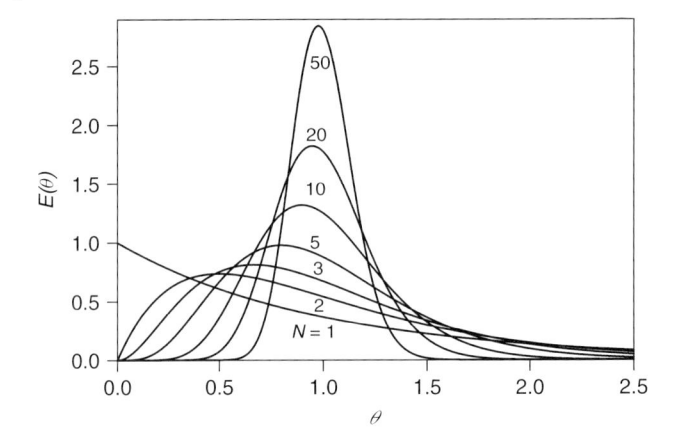

Figure 27.20 Residence time distribution in a cascade of stirred tanks, parameter: number of tanks.

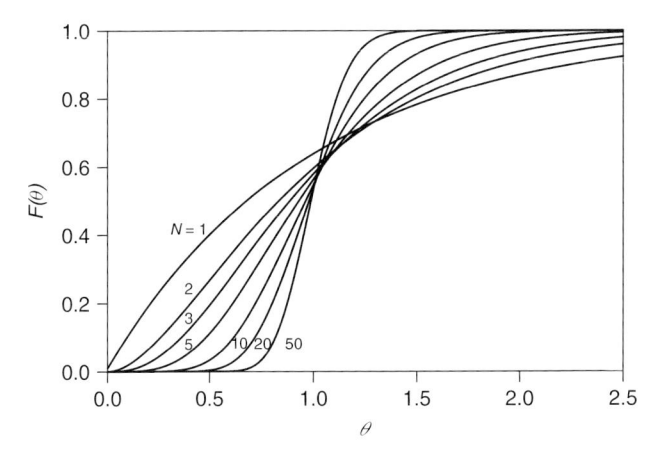

Figure 27.21 Cumulative residence time distribution curves of a cascade of stirred tanks, parameter: number of tanks.

With $\bar{t} = 2 \cdot t_{min}$.

The fraction of the total liquid which is in the position y and thus has a residence time t can be deduced from the area of a circle with the radius R.

$$\frac{d\dot{V}}{\dot{V}} = \frac{u \cdot 2\pi \cdot R \cdot dR}{\pi R_0^2 \bar{u}} = \frac{2 \cdot u \cdot R \cdot dR}{\bar{u} \cdot R_0^2} = E(t)dt \tag{27.78}$$

With $t_{min}/t = 1 - y^2$, it follows that

$$E(t)dt = \frac{2t_{min}^2}{t^3}dt = \frac{\bar{t}^2}{2t^3}dt \tag{27.79}$$

and in dimensionless form

$$E = \tau \cdot E(t) = 0.5\theta^{-3} \tag{27.80}$$

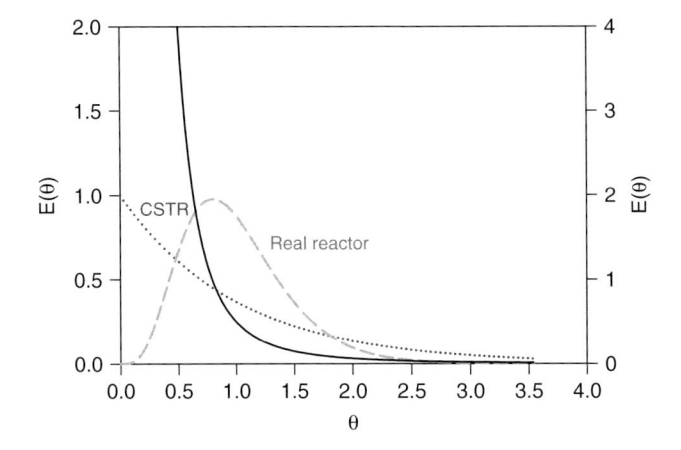

Figure 27.22 Residence time distribution in a laminar flow reactor (without radial diffusion).

The F curve is obtained by integration

$$F = \int_{t_{min}}^{t'} E(t) dt = 1 - \left(\frac{\tau}{2t}\right)^2 = 1 - \frac{1}{4\theta^2} \qquad (27.81)$$

The residence time distribution in a laminar flow reactor without radial diffusion is shown in Figure 27.22. The first volume elements reach the reactor outlet after $\bar{t}/2$ ($\theta = 0.5$) and the tracer concentration slowly approaches 0.

27.4.3
RTD Models for Real Reactors

The experimental determination of the residence time distribution in real reactors has two purposes: to characterize the reactor and to compare the behavior with that of an ideal system. Unwanted short-circuit flows or dead zones are recognized and can sometimes be eliminated by constructional modifications. First of all, a real reactor is classified according to the degree of back-mixing. The degree of back-mixing lies between that of the ideal PFR and the ideal stirred tank. The proposed RTD model serves, in combination with the kinetic model, for the prediction of the reactor performance and the achievable product selectivity and yield.

27.4.3.1 Dispersion Model

Starting from an ideal PFR, a term taking into account the effective axial dispersion is added to the model. The axial back-mixing does not take place solely through molecular diffusion, which is usually negligibly small, but mainly through deviations from ideal plug flow caused by turbulent velocity variations and eddies. Since all of these processes are linearly dependent on concentration gradients, they can be lumped together and treated in analogy to Fick's law. The dispersion processes are described by the equation

$$J = -D_{ax}\frac{dc}{dz} \tag{27.82}$$

with D_{ax} the axial dispersion coefficient.

The response of a real tubular reactor to a pulse injection of a tracer is given by the general mass balance:

$$\frac{\partial c}{\partial t} = -u\frac{\partial c}{\partial z} + D_{ax}\frac{\partial^2 c}{\partial z^2} \tag{27.83}$$

Regrouping of the variables leads to:

$$\frac{\partial c}{\partial \theta} = \frac{D_{ax}}{u \cdot L}\frac{\partial^2 c}{\partial Z^2} - \frac{\partial c}{\partial Z} = \frac{1}{Bo}\frac{\partial^2 c}{\partial Z^2} - \frac{\partial c}{\partial Z} \tag{27.84}$$

with $\theta = t/\tau = t \cdot u/L$, and $Z = z/L$.

The model parameter in the above expressions is the axial dispersion coefficient D_{ax} and the dimensionless group $Bo = t_D/\tau = u \cdot L/D_{ax}$, known as the *Bodenstein number*. The Bo number represents the ratio of the characteristic axial dispersion time $(t_{dis,ax} = L^2/D_{ax})$ to the space time $(\tau = L/u)$ in the reactor. In the case of negligible dispersion $(t_{dis,ax} \gg \tau)$ the Bo number tends to infinity $(Bo \to \infty)$. The reactor behavior approaches that of an ideal PFR. In practice, dispersion in tubular reactors can be neglected for $Bo \geq 100$.

If the axial mixing is very large $(Bo \to 0)$ the axial concentration profile disappears and the reactor behaves as an ideally mixed CSTR.

27.4.3.2 Cell Model

The discussion of residence time distribution in a cascade of identical ideal stirred tank reactors resulted in a narrower residence time distribution with increasing

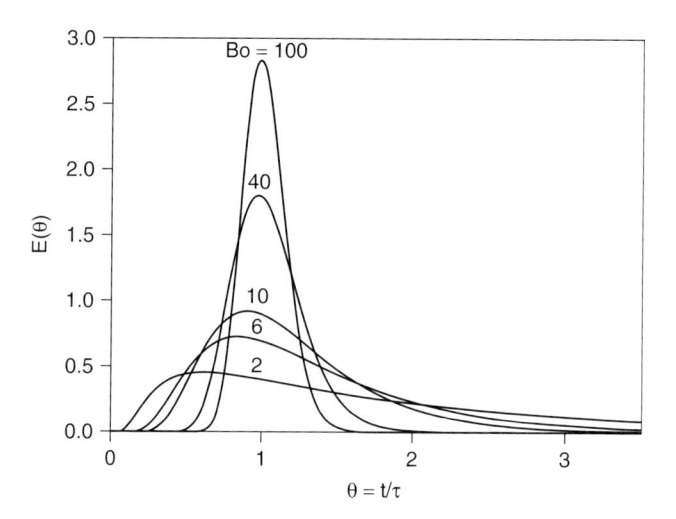

Figure 27.23 Residence time distribution according to the dispersion model.

number of tanks in series, and we showed that for N approaching infinity the RTD corresponds to that of an ideal PFR. It is, therefore, possible to describe the RTD of a real system by the imaginary subdivision of the total volume into N identical, completely mixed cells.

The residence time distribution according to the cell model is described by the equation

$$E = \frac{N(N \cdot \theta)^{N-1}}{(N-1)!} \exp(-N\theta) \qquad (27.85)$$

The only model parameter is the number N of cells in series, which can be determined by a direct fitting of the measured and calculated distribution curves or from the moments of the distribution. The relationship between N and the variance according to the cell model is given by

$$\sigma_\theta^2 = \frac{1}{N} \qquad (27.86)$$

$$\bar{t} = \tau(\rho = \text{constant}) \qquad (27.87)$$

When, however, the dispersion is low (Bo > 50), the distribution curves calculated by the dispersion and the cell models coincide. The following equivalence between the model parameters results:

$$N \approx \frac{\text{Bo}}{2} (\text{Bo} > 50) \qquad (27.88)$$

27.4.4
Estimation of the Residence Time Distribution in Tubular Reactors

For the design of tubular reactors, *a priori* estimation of the axial dispersion is indispensable. The dispersion in tubular reactors depends on the flow regime, characterized by the Reynolds number, Re, and the physical properties of the fluid, characterized by the Schmidt number, Sc. In addition, the presence of internal packing influences the flow behavior and, in consequence, the axial dispersion of the fluid.

In the literature, a great number of experimental data are available correlating the axial Péclet number (Pe) with the Reynolds and Schmidt numbers [20]. The axial Péclet number has as characteristic length the tube diameter, or the particle diameter in packed bed reactors.

$$\text{Pe}_{\text{ax}} = \frac{u \cdot d_t}{D_{\text{ax}}} \text{ (tube) } \text{Pe}_{\text{ax}} = \frac{u \cdot d_p}{D_{\text{ax}}} \text{(packed bed); } u: \text{ superficial velocity} \qquad (27.89)$$

The relationship between the Bodenstein number characterizing the dispersion in the chemical reactor and the Péclet number becomes:

$$\text{Bo} = \text{Pe}_{\text{ax}} \frac{L}{d_t} \qquad (27.90)$$

Correlations between Pe_{ax} and Re or Re · Sc are summarized in Table 27.2 together with the definitions of the model parameters. The presented correlations are

27 Catalytic Reactor Engineering

Table 27.2 Estimation of axial dispersion [19].

Definitions: $Pe_{ax} = \dfrac{u \cdot d_p}{\varepsilon \cdot D_{ax}}$; $Re_p = \dfrac{u \cdot d_p}{v}$; $d_p = 6\dfrac{V_p}{A_p}$; $Re = \dfrac{u \cdot d_t}{v}$; $Sc = \dfrac{v}{D_m}$;

Empty tube, laminar flow:

$$D_{ax} = D_m + \chi \dfrac{u^2 d_t^2}{D_m}$$

$$\dfrac{1}{Pe_{ax}} = \dfrac{1}{Re \cdot Sc} + \dfrac{Re \cdot Sc}{\chi}; \quad \dfrac{L}{d_t} > 0.04 \dfrac{u \cdot d_t}{D_m}; \quad \chi = \dfrac{1}{192} \text{ for circular tubes} \quad (27.91)$$

Empty tube, turbulent flow:

$$\dfrac{1}{Pe_{ax}} = \dfrac{3 \cdot 10^7}{Re^{2.1}} + \dfrac{1.35}{Re^{1/8}}; \quad Pe_{ax} = \dfrac{u \cdot d_t}{D_{ax}} \quad (27.92)$$

Packed bed, gas flow:

$$\dfrac{1}{Pe_{ax}} = \dfrac{0.3}{Re_p \cdot Sc} + \dfrac{0.5}{1 + \dfrac{3.8}{Re_p \cdot Sc}}; \quad \dfrac{d_p}{d_t} > 15; \quad \begin{array}{c} 0.008 < Re_p < 400 \\ 0.28 < Sc < 2.2 \end{array} \quad (27.93)$$

Packed bed, liquid flow:

$$\varepsilon \cdot Pe_{ax} = 0.2 + 0.011 \cdot Re_p^{0.48}; \quad \dfrac{d_p}{d_t} > 15 \quad 10^{-3} < Re_p < 10^3 \quad (27.94)$$

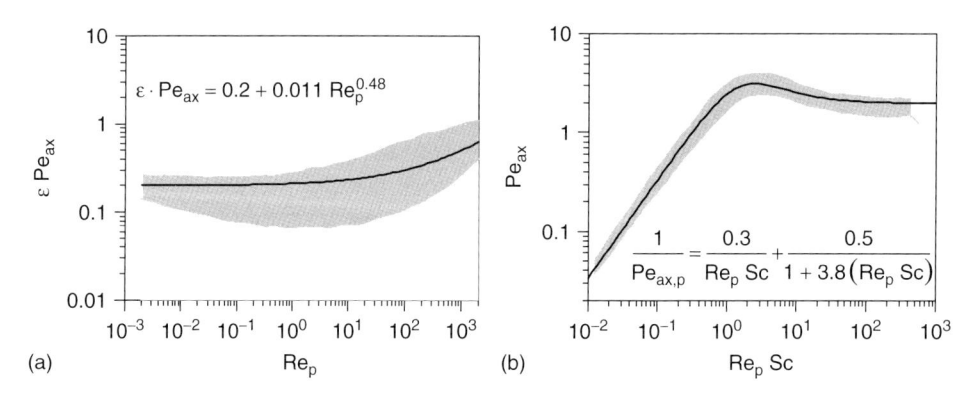

Figure 27.24 Axial dispersion in fixed-bed reactors: (a) liquid flow and (b) gas flow [19].

compared with experimental results, indicated as gray area, in Figures 27.24 and 27.25.

In general, axial dispersion decreases with increasing values of Re and Re · Sc. An exception is the behavior of empty tubes under laminar flow conditions. For laminar flow a parabolic velocity profile develops. Under these conditions, molecular diffusion in axial and radial directions plays an important role in RTD. The diffusion in the radial direction tends to diminish the spreading

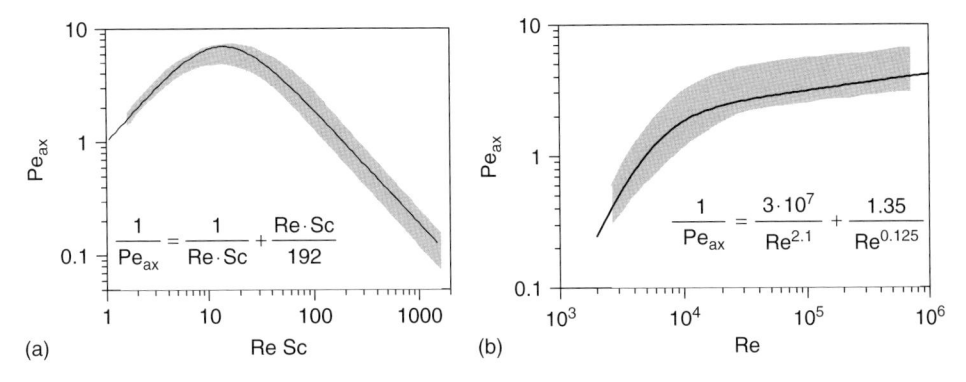

Figure 27.25 Axial dispersion in tubular reactors: (a) laminar flow and (b) turbulent flow [19].

effect of the parabolic velocity profile, while in the axial direction the molecular diffusion increases the dispersion. As a result the axial dispersion passes through a minimum (Pe_{ax} passes through a maximum) as a function of $Re \cdot Sc = u \cdot d_t/D_m$ at $Re \cdot Sc = \sqrt{\chi}$.

27.4.5
Influence of RTD on Performance of Real Reactors

In the case of identical mean residence times, the conversion and selectivity of a reaction will depend on the residence time distribution in the reactor. With increasing backmixing intensity, the reactor will approach the behavior of a continuously-operated, ideally mixed reactor. Accordingly, for example, the reactor performance will decrease with increasing residence time distribution at a constant mean residence time for reactions with formally positive reaction orders.

Backmixing in a tubular reactor has a direct influence on the axial concentration profile. With decreasing axial dispersion time compared to the space time (decreasing Bo) the concentration profile flattens, and finally a uniform concentration results (Bo→0). This is demonstrated in Figure 27.26 for an irreversible first-order reaction at $DaI = k \cdot \tau = 3$.

For reactions with positive reaction order, the flattening concentration profiles diminish the mean reaction rate in the tubular reactor and the conversion will decrease at constant space time and constant DaI.

Based on the dispersion model, the following mass balance for a small volume element results:

$$u\frac{dc_1}{dz} - D_{ax}\frac{d^2c_1}{dz^2} - R_1 = 0 \tag{27.95}$$

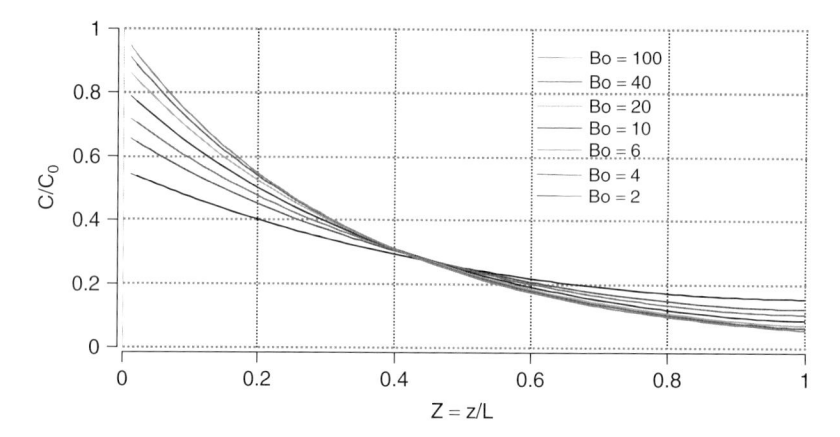

Figure 27.26 Influence of Bo on the axial concentration profile (first order reaction, DaI = 3).

or, in dimensionless form:

$$-\frac{dX}{dZ} - \frac{\tau(R_1)}{c_{1,0}} + \frac{D_{ax}}{u \cdot L} \frac{d^2 X}{dZ^2} = 0 \quad (27.96)$$

Applying Danckwerts' boundary conditions [19], Eq. (27.96) can be solved for an irreversible first-order reaction (Eq. (27.97)).

$$1 - X = \frac{4a \exp(Bo/2)}{(1+a)^2 \exp(aBo/2) - (1-a)^2 \exp(-aBo/2)}$$
with $a = \sqrt{1 + 4\mathrm{DaI}/\mathrm{Bo}}$ \quad (27.97)

The conversion is a function of DaI and the axial dispersion characterized by Bo, as shown in Figure 27.27. With decreasing Bo the conversion diminishes at constant DaI (constant space time). At DaI = 5 a conversion of $X = 0.99$ is attained in a PFR (Bo = ∞), whereas the conversion drops to $X = 0.83$ for Bo = 0 (continuous stirred tank reactor).

In complex reaction systems axial dispersion will also influence the product yield and selectivity attainable in real tubular reactors. This will be demonstrated for first order consecutive reactions.

$$A_1 \xrightarrow{k_1} A_2 \xrightarrow{k_2} A_3; \quad \kappa = \frac{k_2}{k_1}$$

The yield of the intermediate product $Y_2 = \frac{c_2}{c_{1,0}}$ depends on the space time and the ratio of the rate constants, κ. It follows that with DaI_1, the Damköhler number referred to the first reaction:

$$Y_2 = \frac{1}{\kappa - 1}[\exp(-\mathrm{DaI}_1) - \exp(-\kappa \mathrm{DaI}_1)] \quad (27.98)$$

The product yield first increases up to a maximum value ($Y_{2,\max}$) and decreases to 0 for $\mathrm{DaI}_1 \to \infty$ (Figure 27.28). The maximum is attained at the optimal time,

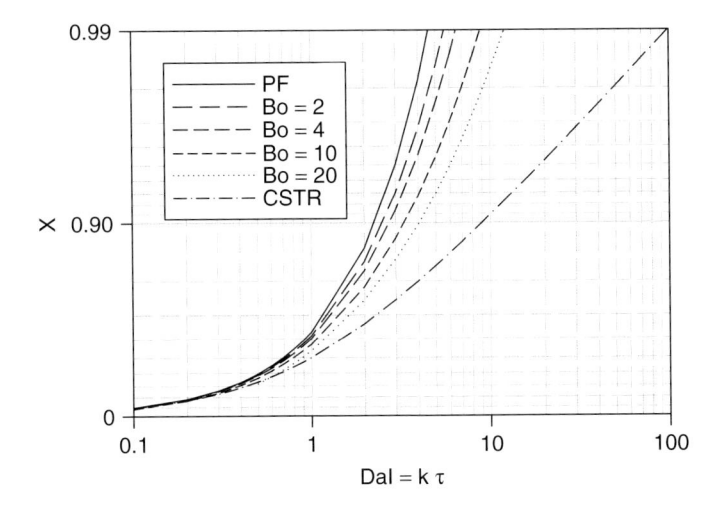

Figure 27.27 Conversion as function of Dal and Bo ($n = 1$).

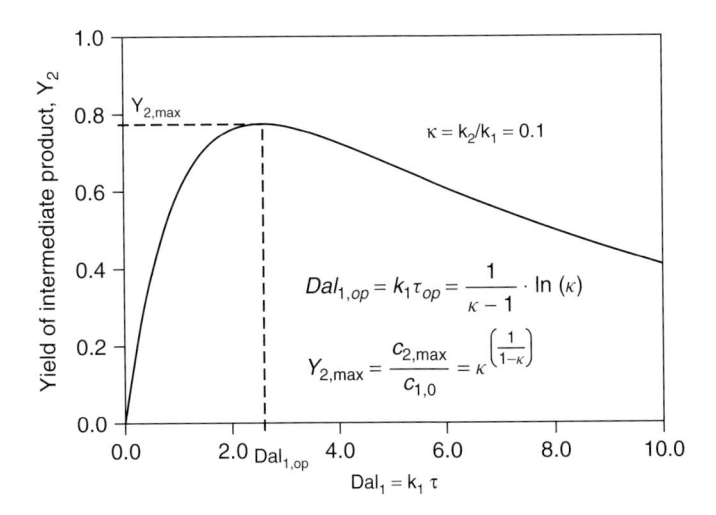

Figure 27.28 First-order consecutive reactions. Yield as function of space time.

$$Dal_{1,op} = k_1 \tau_{op} = \frac{1}{\kappa - 1} \cdot \ln(\kappa)$$

$$Y_{2,max} = \frac{c_{2,max}}{c_{1,0}} = \kappa^{\left(\frac{1}{1-\kappa}\right)}$$

and the Da-number is $Dal_{1,op}$. At higher or lower values the yield diminishes. Therefore, it is evident that the highest yield can only be reached in an ideal PFR with a space time corresponding to $Dal_{1,op}$. Any residence time distribution in real tubular reactors will never reach the maximum yield of the intermediate. This is demonstrated for three different values of $\kappa = k_2/k_1$ in Figure 27.29. The real tubular reactor is modeled with the cell model.

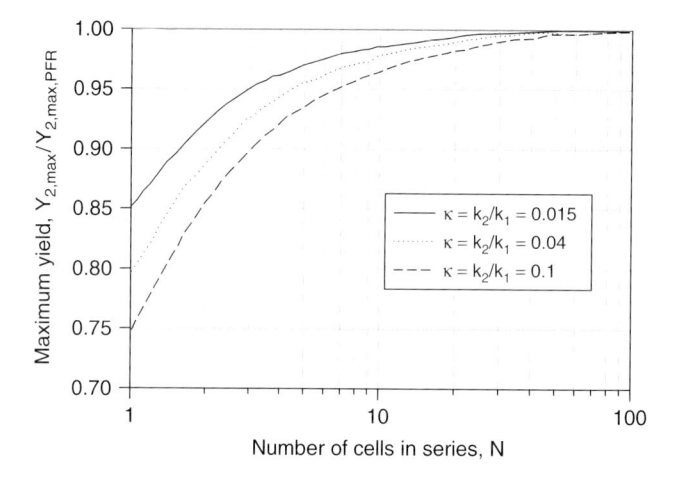

Figure 27.29 Maximum yield of intermediate product referred to the max yield in PFR as function of axial dispersion (cell model).

27.5
Microreaction Engineering

MSRs, or micro-structured reactors, are innovative reactors with dimensions in the submillimeter range along one or two axes, where chemical transformation takes place [21]. The main characteristic features of these reactors are their high surface to volume ratio, in the range $10\,000–50\,000$ $m^2 \cdot m^{-3}$. For comparison, the specific surface of conventional laboratory and production vessels is usually ~ 100 $m^2 \cdot m^{-3}$ and seldom exceeds 1000 $m^2 \cdot m^{-3}$. Therefore, MSR presents mass and heat transfer performances one order of magnitude higher than that of traditional reactors and heat exchangers, leading to process intensification [22].

High rates of heat and mass transfer reduce or eliminate transport resistances, thus improving effective reaction rates and product selectivities and enabling the reaction to achieve its 'chemical' potential in the optimal temperature and concentration window. A typical MSR is shown in Figure 27.30.

27.5.1
General Criteria for Reactor Selection

The characteristic time of chemical reactions, t_r, which is defined by intrinsic reaction kinetics, can vary from hours for slow organic or biological reactions to milliseconds for high-temperature partial oxidation reactions.

The characteristic times of physical transport processes in conventional reactors range from a few seconds to milliseconds, depending on reactor configuration and physical properties of the materials involved (Figure 27.31). This means that relatively slow reactions ($t_r \gg 10$ s) are carried out in the kinetic regime, and the global performance of the reactor is controlled by the intrinsic reaction kinetics. The

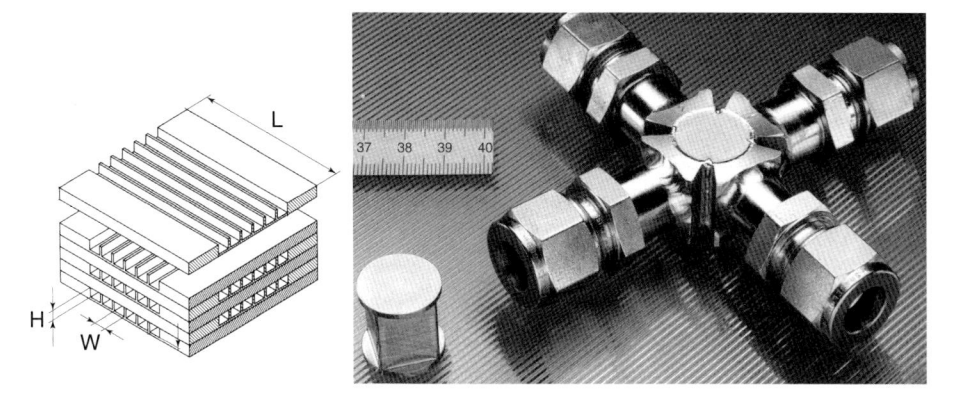

Figure 27.30 Schematic representation (left) and photograph (right) of a micro structured reactor (channel length L = 14 mm, W = 100 μm, and H = 78 μm [23].

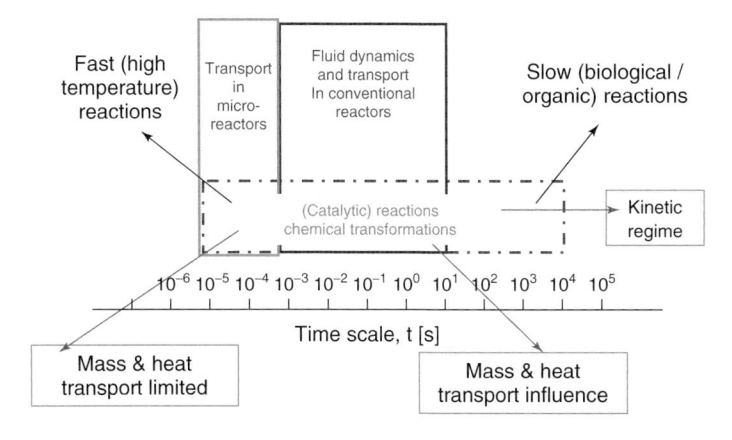

Figure 27.31 Time scale of chemical and physical processes [37].

chemical reactor is designed and dimensioned to get the required product yield and conversion of the raw material. The attainable reactant conversion in the kinetic regime depends on the ratio of the space time in the reactor, τ, to the characteristic reaction time, that is, the first Damköhler number, DaI. Depending on the kinetics and the type of the reactor, τ should be several times higher than t_r (DaI $\gg$ 1) to get conversions >90% [13, 24]. For fast chemical reactions, the characteristic reaction time is of the same order of magnitude as the characteristic time of the physical processes. The performance of a conventional reactor is influenced in this case by mass and/or heat transfer. For very fast reactions, the global transformation rate may be completely limited by transfer phenomena. As a result, the reactor performance is diminished as compared to the maximal performance attainable in the kinetic regime, and the product yield is very often reduced.

To eliminate mass transfer resistances in practice, the characteristic transfer time should be roughly one order of magnitude smaller than the characteristic reaction time. As the mass and heat transfer performance in an MSR is up to two orders of magnitude higher than that in conventional tubular reactors, the reactor performance can be considerably increased, leading to the desired intensification of the process. In addition, consecutive reactions can be efficiently suppressed by strict control of residence time and narrow RTD. Therefore, fast reactions carried out in MSRs show higher product selectivity and yield.

27.5.2
Types of Microstructured Reactors

MSRs are characterized based on the number of phases involved, that is single-phase, fluid–solid, fluid–fluid, and three-phase reactors. The detailed descriptions of reactors and examples of reactions presented in the following sections illustrate the choice of reactors.

27.5.2.1 Single-Phase MSR

Mixing Mixing of the reaction partners is an important task in homogeneous single-phase reactors and has a decisive impact on the performance of MSR and the product yield attainable. Mixing in micro-channels takes place by diffusion and convection, depending on the flow geometry and the operating conditions used [25, 26]. The short diffusion radial lengths decrease the mixing time in micro-channels even at low flow rates in the laminar flow regime. The mixing time can be further reduced in convective mixing by creating eddies. These convective effects increase the concentration gradients and material surfaces, which increases mixing rate in both turbulent and laminar flows. A commonly used method to enhance the mixing in micro-channels is to distribute the flow into compartments and reduce diffusion paths beyond the geometric dimensions of the channel. For instance, splitting and recombining the feed streams or injecting substreams via a special microstructure can break the laminar profile and enable better mixing to be achieved [27]. In both cases, pressure drop is increased, resulting in higher power consumption. Thus, the choice of mixing device should always be governed by the tradeoff between characteristic mixing time (defined by the chemical system) on the one hand and energy consumption on the other. Different mixing principles used in the MSR are depicted in Figure 27.32.

Assuming laminar mixing in microchannels, the theoretical mixing time (t_m) is given as a function of Péclet number (Pe: ratio of mixing by convection to molecular diffusion) by the following equation [25]:

$$t_m = \frac{(d_t^2/D_m)}{8\text{Pe}} \ln(1.52\text{Pe})$$

$$\text{Pe} = \text{Re} \cdot \text{Sc} = \frac{\rho u d_t}{\mu} \frac{\mu}{\rho D_m} = \frac{u d_t}{D_m} \quad (27.99)$$

In laminar liquid mixing applications, the range of variation of the Pe is between 10^3 and 10^6.

27.5 Microreaction Engineering

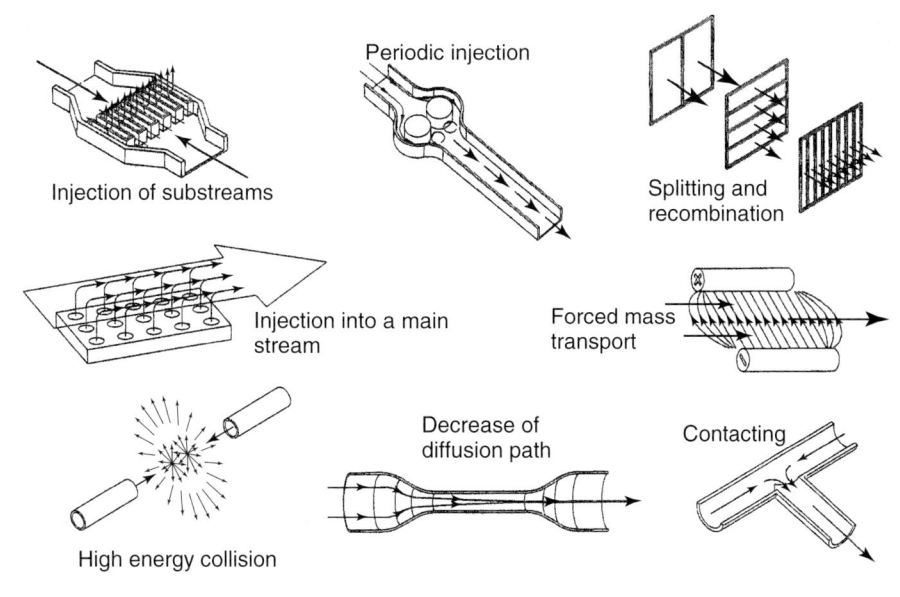

Figure 27.32 Different mixing principles used in MSR [28].

The pressure drop through open channels with laminar flow is given by the Hagen–Poiseuille equation [29].

$$\Delta p = 32\zeta \frac{\mu u}{d_t^2} L_t \qquad (27.100)$$

where ζ is a geometric factor, which is 1 for a circular tube and depends on the height (H) to width (W) ratio for a rectangular channel (Figure 27.30). The correction factor becomes 0.89 for quadratic channels and assumes the asymptotic value 1.5 when the ratio goes to zero. An empirical correlation is given by the following expression [30]:

$$\zeta = 0.8735 + 0.6265 \exp\left(-3.636 \frac{H}{W}\right) \qquad (27.101)$$

A micromixer with structured internal surfaces shows a different trend compared to the other micro-channels, namely quadratic dependence of the pressure loss on the flow rate [26].

Specific power dissipation ε (W · kg^{-1}) is an important parameter for benchmarking technical reactors. It is calculated from the pressure loss and the flow rate [25]:

$$\varepsilon = \frac{\dot{V} \Delta P}{\rho V} \qquad (27.102)$$

where $\dot{V}$ is the volumetric flow rate and V is the reactor volume.

If the above mixing time in different MSRs is compared with the power dissipation, the following theoretical relationship is obtained for all sizes of

27 Catalytic Reactor Engineering

Figure 27.33 Evolution of the mixing time in different micromixers as a function of the specific power dissipation (adapted from ...)

microchannels.

$$t_m = 0.0075\varepsilon^{-0.5} \tag{27.103}$$

However, when the mixing time was investigated experimentally in all micro mixers, related to specific power dissipation, the following relationship was obtained as shown in Figure 27.33:

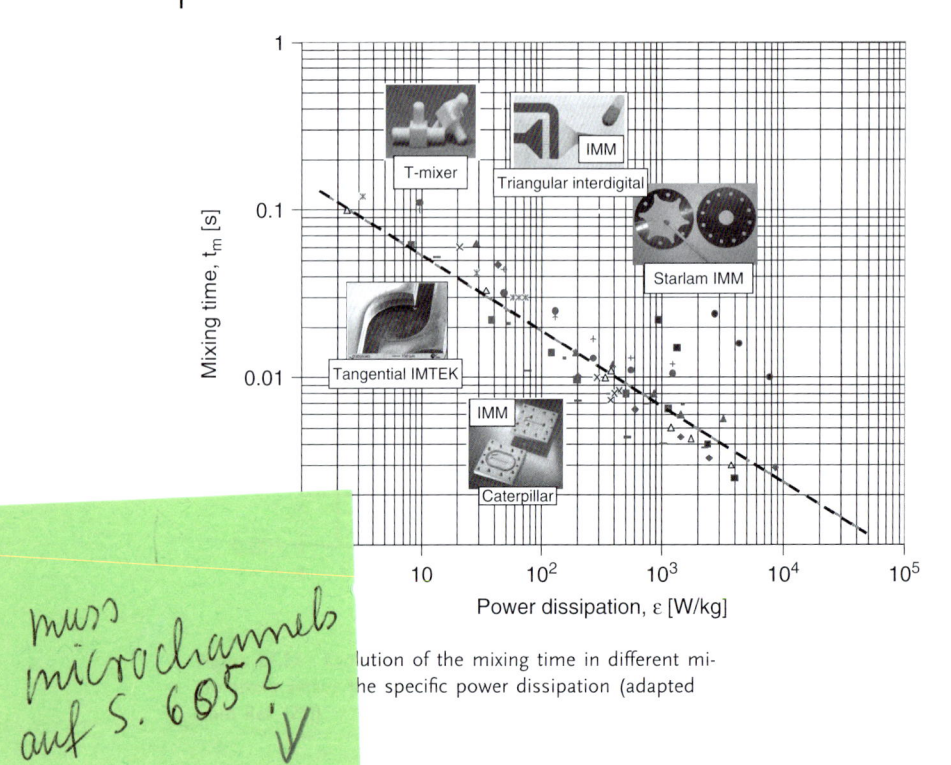

$$t_m = 0.15\varepsilon^{-0.45} \tag{27.104}$$

By comparing Eq. (27.103) with Eq. (27.104) it can be seen that only about 5% of the power dissipated is effectively used for mixing. Mixing efficiency in conventional reactors is even lower.

Residence Time Distribution Flow in micro-channels with diameters between 10 and 500 μm is mostly laminar and has a parabolic velocity profile. Therefore, the molecular diffusion in axial and radial directions plays an important role in RTD. As pointed out in Section 27.4.4, diffusion in the radial direction tends to diminish the spreading effect of the parabolic profile, while in the axial direction the molecular diffusion increases the dispersion. As a result, the axial Péclet number passes through a pronounced maximum corresponding to a minimal axial dispersion.

The Bodenstein number in microchannels can be determined from Eqs. (27.90) and (27.91):

$$\frac{1}{Bo} = \frac{D_m}{L^2} \frac{L}{u} + \frac{1}{192} \frac{d_t^2}{D_m} \frac{u}{L} = \frac{D_m L}{L^2 u} + \frac{1}{192} \frac{4 \cdot u}{D_m L}$$

$$\frac{1}{Bo} = \frac{\tau}{t_{D,ax}} + \frac{1}{48} \frac{t_{D,rad}}{\tau}; \text{ with } t_{D,ax} = \frac{L^2}{D_m}; t_{D,rad} = \frac{R_t^2}{D_m} \quad (27.105)$$

The first term in Eq. (27.105) corresponds to the ratio between space time and the characteristic axial diffusion time. The diffusion coefficient lies within the range 10^{-5}–10^{-9} m^2 s^{-1} for gases and liquids, respectively. Typical lengths of MSR are several centimeters, and the space time is of the order of seconds. Therefore, the axial dispersion in microchannels is mainly determined by the second term in Eq. (27.105), and the Bodenstein number can be estimated from Eq. (27.106).

$$Bo \cong 48 \cdot \frac{\tau}{t_{D,rad}} \cong 50 \cdot \tau \cdot \frac{D_m}{R_t^2} \quad (27.106)$$

It follows that axial dispersion can be neglected (Bo $\geq$ 100) if the space time is at least twice the radial diffusion time. The characteristic radial diffusion times for different channel diameters are summarized in Table 27.3. Accordingly, axial dispersion of gases in microchannels can be neglected if their diameters are less than 1000 μm and the space time is longer than 0.1 s. This could also be proved experimentally [31, 32].

27.5.2.2 Fluid–Solid MSR

Two types of heterogeneous catalytic MSR have been described in the literature for different applications: the micro fixed-bed reactor and the catalytic wall reactor.

Fixed-bed MSRs accommodate a large amount of catalyst particles of size ~10–250 μm, depending on the reactor diameter, offering a large specific surface area for chemical reaction. The micro packed bed is mainly used for catalyst screening. The advantage of packed-bed MSR stems from the fact that the developed catalyst used in traditional reactors can be applied. To avoid flow maldistribution in

Table 27.3 Characteristic radial diffusion time as a function of channel diameter.

Channel diameter d_t (μm)	Characteristic radial diffusion time $t_{D,rad}$ (s) Gas: $D_m = 10^{-5}$ m^2 s^{-1}	Characteristic radial diffusion time $t_{D,rad}$ (s) Liquid: $D_m = 10^{-9}$ m^2 s^{-1}
2000	100×10^{-3}	1000
1000	25×10^{-3}	250
500	6.3×10^{-3}	63
200	1×10^{-3}	10
100	0.25×10^{-3}	2.5
50	0.06×10^{-3}	0.6
20	0.01×10^{-3}	0.1

the bed, the particle diameter should not be greater than a 0.1 of the tube diameter ($d_p \leq d_t/10$) and the channel length should be greater than 50 particle diameters ($L_{bed} > 50 \cdot d_p$). This can lead to a relatively high pressure drop in the MSR, which can be estimated with the Ergun equation ((Eq. (27.7))).

The microstructured multichannel reactors with catalytically active walls are by far the most used devices for heterogeneous catalytic reactions (Figure 27.34). Advantages are low pressure drop, high external and internal mass transfer performance, and quasi-isothermal operation.

In most cases the reactors are based on micro heat exchangers with typical channel diameters in the range of 50–500 µm and a length between 20 and 100 mm. About 200–2000 channels are assembled in 1 unit.

In fluid–solid systems, the reaction takes place on the catalyst surface. Prior to this, the reactant molecules have to reach the catalyst surface, and therefore the rate of mass transfer is an important operational parameter (see Chapter 4). To avoid internal mass transfer influences, the following criterion must be fulfilled:

$$\delta_{cat,max} \leq b \sqrt{\frac{D_{eff} c_s}{r_{eff}}} \tag{27.107}$$

where $\delta_{cat,max}$ is the maximal thickness of the porous catalytic wall layer, and D_{eff}, and r_{eff} are the effective diffusion coefficient and the observed reaction rate, respectively. The coefficient b depends on the formal reaction order and has a value of 0.8, 0.3, and 0.18 for zero-, first-, and second-order reactions, respectively.

In the case of strongly exothermic or endothermic reactions, the reactions may give rise to a temperature profile within the catalytic layer which is dependent on reaction enthalpy (ΔH_R), activation energy (E), and thermal conductivity of the porous catalytic material (λ_{eff}). For quasi-isothermal behavior, the observed rate, r_{eff}, should not differ from the rate that would be observed at constant temperature by more than 5%, and thus the resulting criterion for effectively isothermal catalytic wall behavior is given by:

$$\delta_{cat,max} \leq 0.3 \sqrt{\frac{R}{E} \frac{\lambda_{eff} T_s^2}{|\Delta H_R| r_{eff}}} \tag{27.108}$$

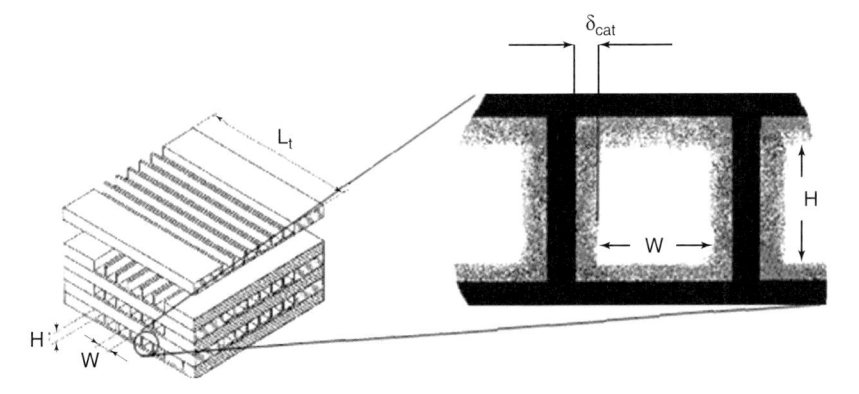

Figure 27.34 Micro-structured catalytic wall reactor.

where T_s corresponds to the temperature of the catalyst surface and R is the gas constant.

In general, the thickness of the catalytic layer is kept sufficiently small to avoid the influence of internal heat and mass transfer on the kinetics. Therefore, only the transfer of the reactants from the bulk to the catalytic wall must be considered. The effective reaction rate is determined by the ratio of the characteristic mass transfer time, t_{tr} and the characteristic reaction time t_r, which gives the second Damköhler number (Chapter 4):

$$\text{DaII} = \frac{t_D}{t_r} = \frac{k_r c_{1,b}^{n-1}}{k_m a_p} \tag{27.109}$$

Due to the small channel diameters in MSR, laminar flow can be considered. The radial velocity profile in a single channel develops from the entrance to the position where a parabolic velocity profile is established. The length of the entrance zone depends on the Re number and can be estimated from the following empirical relation [33, 34]:

$$L_e \leq 0.06 \cdot \text{Re} \cdot d_t \tag{27.110}$$

Within the entrance zone the mass transfer coefficient diminishes, reaching a constant value. The dependence can be described in terms of Sherwood numbers, [35, 36]: $\text{Sh} = k_m d_t / D_m$

$$\text{Sh} = B \left(1 + 0.095 \frac{d_t}{L} \text{Re} \cdot \text{Sc} \right)^{0.45} \tag{27.111}$$

The constant B in Eq. (27.111) corresponds to the asymptotic Sh number for constant concentration at the wall, which is identical to the asymptotic Nusselt number, Nu, characterizing the heat transfer in laminar flow at constant wall temperature. The constant B depends on the geometry of the channel as summarized in Table 27.4.

If the entrance zone in the tube can be neglected, the mass transfer is constant and given by B. It follows that for a circular shaped reactor:

$$\text{Sh}_\infty = 3.66; \text{ for } L \geq 0.05 \text{Re} \cdot \text{Sc} \cdot d_t \quad \text{(constant wall concentration)} \tag{27.112}$$

Table 27.4 Values of B for different channel geometries [35].

Geometry	B
Circular	3.66
Ellipse (width/height = 2)	3.74
Parallel plates	7.54
Rectangle (width/height = 4)	4.44
Rectangle (width/height = 2)	3.39
Square	2.98
Equilateral triangle	2.47
Sinusoidal	2.47
Hexagonal	3.66

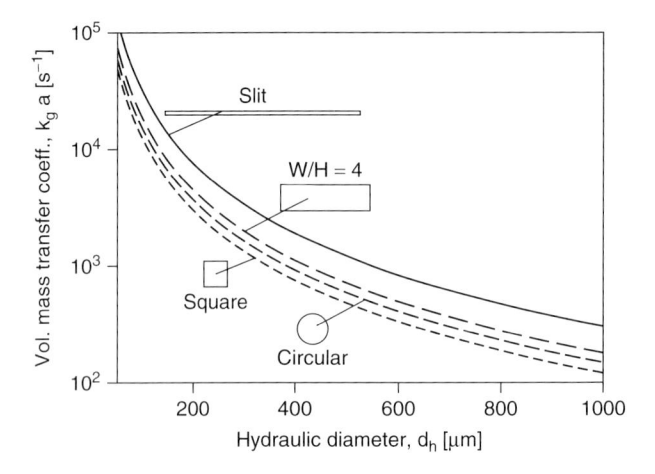

Figure 27.35 Volumetric mass transfer coefficient as function of the hydraulic diameter in micro-structured channels ($D_m = 10^{-5}$ m$^2 \cdot$s^{-1})

The specific performance of the MSR under mass transfer limitations depends on the volumetric mass transfer coefficient ($k_m \cdot a$), which determines the maximal reactor performance for very fast catalytic reactions. Its value increases with $1/d_h^2$ with $d_h = \frac{4 \cdot A_{cs}}{l_c}$, the hydraulic channel diameter, A_{cs}, the cross section area, and l_c the circumference. The volumetric mass transfer coefficient as a function of the channel diameter is shown in Figure 27.35.

27.5.3
Fluid–Fluid MSR

Fluid–fluid multiphase MSRs exist for gas–liquid and liquid–liquid systems. They generally take advantage of the large interfacial area, fast mixing, and reduced transfer limitations, providing an enhanced performance relative to conventional bench-scale systems due to reduced diffusion times and increased mass transfer across phase boundaries.

Fluid–fluid MSRs can be roughly classified into three types based on the contacting principles [37]:

- Micromixer: a device in which chaotic mixing occurs due to high flow rates or static internals, for example, micromixer settler, cyclone mixer, interdigital mixer
- Microchannel: a channel in which mixing occurs by both convection and diffusion, for example, microchannel with partial overlap, microchannel with membrane or metal contactor, microchannels with inlet Y-shaped or T-shaped contactor, microchannel with static internals bed, microstructured packed-bed reactor, parallel microchannels with internal redispersion units
- Falling film microchannel.

The first two types of MSR are used for all fluid–fluid applications, while the micro-structured falling film reactor is used only for gas–liquid systems. The micromixers are used for applications requiring intense mixing, as a very high specific interfacial area can be reached within less than 1 s, being up to fivefold greater than microchannels [21].

27.5.3.1 Gas–Liquid Systems

In gas–liquid systems, mass transfer takes place from the gas phase to the liquid phase as well as in the reverse direction, and chemical reactions can occur in the gas and/or the liquid phase. Examples of catalytic gas–liquid reaction include nitration of naphthalene using N_2O_5 [38] and asymmetric hydrogenation of Z-methylacetamidocinnamate (mac) with rhodium chiral diphosphine complexes [39].

The continuous feeding of gas and liquid into a small channel in the millimeter or sub-millimeter range leads to various flow regimes: dispersed bubble flow, slug flow (Taylor flow), churn flow, and annular flow [40, 41]. The various flow regimes develop as a result of the gas and liquid properties, the superficial velocities, and the channel diameter. The overall gas-liquid holdup, the pressure drop, and the mass transfer performance depend on the flow regime that is established. Figure 27.36 shows the various flow regimes in a single capillary.

Laminar flow is observed for low to moderate superficial velocities, and the predominant pattern is that of segmented flow, also known as *slug* or *Taylor flow* (see Figure 27.37 a and b). Taylor flow is characterized by gas bubbles that are too large to retain a spherical shape and are deformed to fit inside the channel. Depending on the gas and liquid flow rates and fluid properties, the bubbles often have hemispherically shaped tops and flattened tails [42]. This flow regime appears at low capillary numbers, $Ca < 0.01$, where interfacial forces dominate shear stresses, and the dynamics of break-up are dominated by pressure drop across the emerging droplet [3, 43].

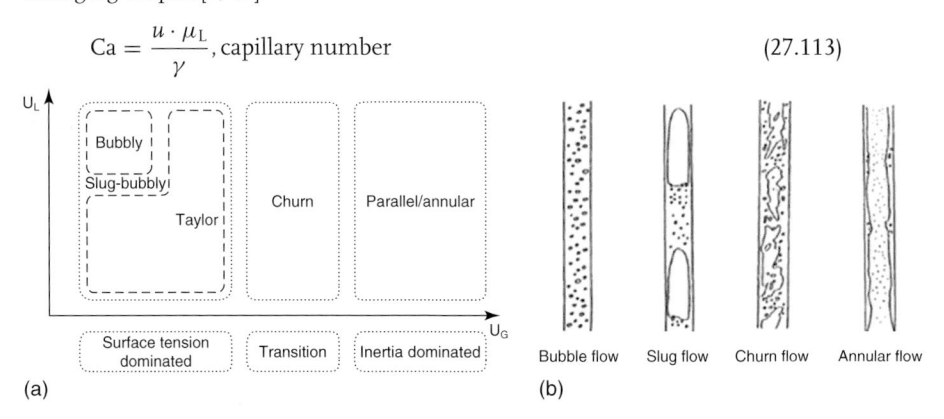

$$Ca = \frac{u \cdot \mu_L}{\gamma}, \text{capillary number} \tag{27.113}$$

Figure 27.36 Observed flow regimes in capillary tubes. ((a) Adapted from Shao et al. [41]. (b) Adapted from Ref. [40].) (Adapted with permission from Elsevier).

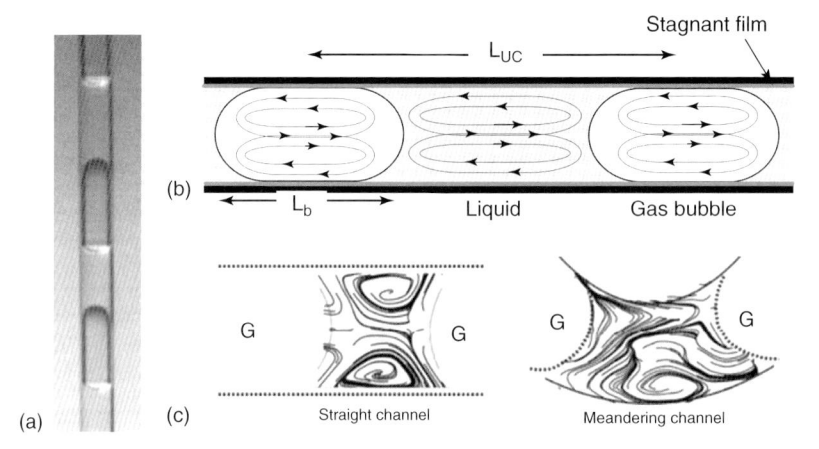

Figure 27.37 Experimental snapshots and schematic presentation of Taylor flow in different configurations. (a) Taylor flow in vertical capillary. (Adapted from Liu et al. [42].). (b) Schematic presentation of Taylor flow in horizontal capillary (L_b – length of bubble and L_{UC} – unit slug length). (c) Comparison between flow behavior in the liquids slugs of Taylor flow in straight and meandering channel. (Adapted from Günther et al. [45]).

The wide application of this regime for mass transfer comes from its stability and ability to provide well-defined high specific interfacial area. The recirculation within the liquid slugs improves heat and mass transfer from liquid to wall and interfacial mass transfer from gas to liquid [44]. Thus, due to the combined advantage of reduced axial dispersion and enhanced radial mixing, the slug flow regime attracts a variety of industrial applications. In addition to this, the radial mixing in Taylor flow can be enhanced using a meandering channel, as shown in Figure 27.37c.

Gas–liquid MSRs are particularly suitable for fast chemical reactions, where the reaction takes place mainly in the liquid film near the gas–liquid interphase. This situation is characterized by Hatta numbers (Ha) greater than 3 (Chapter 4).

In Taylor flow regimes, the liquid usually exhibits good wettability of the channel wall, forming a thin wall film over which the gas bubbles glide. The wall film provides a lubricating action to the enclosed bubbles, and, as a consequence, bubbles flow at higher velocity than the liquid. The interfacial gas–liquid area thus comprises two parts: the lateral part (that of the wall film) and the perpendicular part (between the bubble and the adjacent liquid plug). Often the length of the bubble is several times greater than the channel diameter; so that the lateral part of the interfacial area is greater than the perpendicular part. The thin liquid wall film may become saturated with the absorbed component or may be exhausted of the liquid-phase reactant. In such cases, the lateral part of the interfacial area will become inactive, and therefore it is important in the design of an MSR to determine conditions under which this part of interface remains active.

Pohorecki [46] developed criteria for keeping the lateral wall film active. The derived relationship corresponds to a modified Fourier number as shown in

Eq. (27.114):

$$\text{Fo}' = \frac{L_b D_m}{u_b \delta_{film}^2} \ll 1 \qquad (27.114)$$

where D_m corresponds to the molecular diffusion coefficient of the reactant in the liquid phase.

The above equation shows that reagent exhaustion in the liquid film can be avoided by decreasing the length of the bubble, L_B, or by increasing the flow velocity, u_B.

A simplified mode of mass transfer in Taylor flow was proposed by Vandu et al. [42] taking film contribution into account:

$$k_L a = 4.5 \sqrt{\frac{D_m u_G}{L_{UC}}} \frac{1}{d_t} \quad \text{for} \quad \sqrt{((u_G + u_L)/L_S)} > 3 \qquad (27.115)$$

The agreement between the model and the experiment is reasonably good for both circular and square capillaries, showing dependence of $k_L a$ on capillary diameter (d_t). Use of the above correlation requires the unit cell lengths (L_{UC}) and also the length of the liquid slug (L_s) to check the validity of the correlation. They can be related by assuming the negligible volume of liquid in the film between the gas bubble and capillary wall to be

$$L_s = L_{UC}(1 - \varepsilon_G) \qquad (27.116)$$

where, ε_G is the gas hold-up, which is estimated from the following equation [42]:

$$\varepsilon_G = \frac{u_G}{u}\left[1 - 0.61 \text{Ca}_L^{0.33}\right] \qquad (27.117)$$

The same authors proposed the following relationship between two-phase flow velocity and slug length:

$$\frac{u}{\sqrt{L_s}} = 0.088 \text{Re}_G^{0.72} \text{Re}_L^{0.19} \qquad (27.118)$$

The mass transfer performances obtained in different types of MSR as well as in conventional gas–liquid contactors are listed in Table 27.5. When the mass transfer performance of an MSR was compared with conventional contactors, it showed that liquid side $k_L a$ and interfacial area in MSR are at least one or two orders of magnitude higher than those in conventional contactors such as bubble columns or packed columns.

27.5.3.2 Liquid–Liquid Systems

In the chemical industry, a large number of reactions which are strongly exothermic and involve two-phase liquid–liquid systems are carried out on a large scale,. Thus, the intensification of heat and mass transfer is an important issue for reactor design. The highly exothermic and fast nitration reactions constitute an important field of application for microstructured liquid–liquid reactors. Examples are given in Table 27.6.

Similarly to gas–liquid flow, different flow regimes based on various operating conditions and flow properties are observed such as drop, slug, slug-drop, deformed

Table 27.5 Comparison of MSR performance with conventional contactors for gas–liquid mass transfer.

Type of contactor	$k_L \times 10^5$ (m s^{-1})	a (m$^2 \cdot$m^{-3})	$k_La \times 10^2$ (s^{-1})
Bubble columns	10–40	50–600	0.5–24
Couette-Taylor flow reactor	9–20	200–1200	3–21
Impinging jet absorbers	29–66	90–2050	2.5–122
Packed columns, concurrent	4–60	10–1700	0.04–102
Packed columns, countercurrent	4–20	10–350	0.04–7
Spray column	12–19	75–170	1.5–2.2
Static mixers	100–450	100–1000	10–250
Stirred tank	0.3–80	100–2000	3–40
Tube reactors, horizontal and coiled	10–100	50–700	0.5–70
Tube reactors, vertical	20–50	100–2000	2–100
Gas–liquid microchannel	40–160	3400–9000	30–2100

Adapted from Yue et al. [47].

Table 27.6 Examples of liquid–liquid reactions studied in the MSR [37].

Reaction	Reactor
Nitration of benzene	Stainless steel and PTFE capillary MSR Borosilicate glass microreactor
Nitration of toluene	PTFE capillary MSR
Trans-esterification – production of alkyl esters	PTFE capillary MSR
Production of β-ionone (vitamin precursor)	Multichannel MSR
Hydrolysis of p-nitrophenol acetate	PMMA microreactor
Tandem diazotation/Heck reaction sequences	PMMA microreactor
Nitration of dialkyl-substituted thioureas	Silicon multitubular MSR with
Villermaux–Dushman – instantaneous neutralization and rapid redox reactions	PDMS MSR
Bromination of styrene	Thiolene-based resin MSR
Extraction and detection of carbaryl derivative	Two Pyrex glass MSRs
Degradation of p-chlorophenol	Pyrex glass MSR
Kinetics of ribonuclease A (RNase A)	PDMS MSR
Hydrolysis of p-nitrophenyl acetate 1	PTFE MSR
Phase-transfer alkylation of β-keto esters	Microchip connected to a Teflon tube

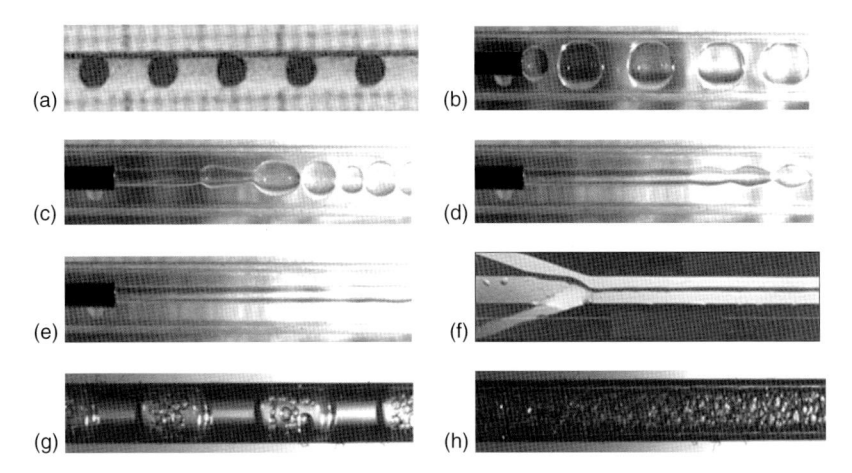

Figure 27.38 Flow regimes observed for liquid–liquid systems in MSR. (a) Drop flow, (b) slug flow, (c) slug-drop flow, (d) deformed interface flow, (e) annular flow, (f) parallel flow in a wedge-shaped microchannel (adapted from Ref. [48]), (g) slug-dispersed flow, and (h) dispersed flow.

interface, annular, parallel, and dispersed flow regimes (Figure 27.38). When the stability of liquid–liquid flow regimes is compared with that of gas–liquid flow, the former are seen to be relatively stable, as the interface in the latter undergoes severe deformations and break-up due to large viscosity and density differences between the two fluids.

The most commonly used liquid–liquid two-phase flows are 'slug flow' and 'parallel flow.' In the case of slug flow, two mechanisms are known to be responsible for the mass transfer between the two liquids: (i) internal circulation within each slug and (ii) concentration gradients between adjacent slugs. In the case of the parallel pattern, the flow is laminar, and the mass transfer between the two phases is thought to occur only by diffusion.

The mass transfer in liquid–liquid MSR can be estimated with acceptable accuracy using following empirical correlation:

$$Da' = k_L a \cdot \tau = \left(\frac{k_L a \cdot L}{u}\right) = a \cdot (Ca)^b \cdot (Re)^c \cdot \left(\frac{d_t}{L}\right)^d \qquad (27.119)$$

Ca and Re are calculated with the mean properties of the mixture using volumetric flow fraction (ε) as follows:

$$\rho_M = \varepsilon_1 \rho_1 + (1 - \varepsilon_1) \rho_2; \quad \mu_M = \varepsilon_1 \mu_1 + (1 - \varepsilon_1) \mu_2; \quad \varepsilon_1 = \left(\frac{\dot{V}_1}{\dot{V}_1 + \dot{V}_2}\right) \qquad (27.120)$$

The constant a and exponents b–d in Eq. (27.119) are adjustable parameters that are determined by fitting experimental data to the correlation. The above correlation can be considered as flow regime independent and the values of adjustable parameters for different microchannels obtained are listed in Table 27.7.

Table 27.7 Fitting parameter for correlation on mass transfer in liquid–liquid systems.

MSR	a	b	c	d
T-square [49]	0.47	5.50	−5.65	−8.34
T-trapezoidal [49]	0.65	4.81	−5.10	−7.35
Y-square [49]	3.76	−4.81	4.70	6.07
Concentric [49]	4.51	5.11	−5.14	−11.45
Caterpillar [49]	3.86	−3.97	3.99	3.36

Table 27.8 Mass transfer coefficients and effective interfacial area obtained in different liquid–liquid contactors.

Contactor	a (m^2 m^{-3})	$k_L a$ (s^{-1})
Agitated contactor	32–311	$(48-83) \times 10^3$
Packed bed column (Pall/Raschig ting, Intalox saddles)	80–450	$(3.4-5) \times 10^3$
RTL extractor (Graesser raining bucket)	90–140	$(0.6-1.3) \times 10^3$
Air operated two impinging jet reactors	350–900	0.075
Two impinging jets reactor	1000–3400	0.28
Capillary microchannel (ID = 0.5–1 mm)	830–3200	0.88–1.67

Adapted from Ref. [50].

The volumetric mass transfer coefficients found in the liquid–liquid MSR at various flow rates were compared with those for conventional equipment in Table 27.8. It can be seen that the mass transfer coefficients found in MSRs are way above those of conventional contactors.

Energy input is an important parameter for benchmarking technical reactors. For a continuous process, it is calculated as amount of energy required per unit volume of liquid fed to the reactor. The energy input in MSRs is compared with various liquid–liquid contactors in Table 27.9. The comparison shows that the capillary MSR requires much less power than the alternatives to provide such a large interfacial area.

27.5.3.3 Three-Phase Reactors

Gas–Liquid–Solid Different types of fluid–solid MSR have been developed, using various contacting principles [51]. These principles can be classified as follows:

- Continuous-phase contacting, where the fluid phases are separated. Examples are microstructured falling-film and mesh reactors.

Table 27.9 Energy input requirement for different liquid/liquid contactors.

Contactor	Energy input (kJ·m^{-3} of liquid)
Agitated extraction column	0.5–190
Mixer-settler	150–250
Rotating disk impinging streams contactors	175–250
Impinging streams	280
Impinging streams extractor	35–1500
Centrifugal extractor	850–2600
Capillary microchannel (ID = 0.5–1 mm)	0.2–20

Table 27.10 Examples of gas–liquid–solid reactions studied in MSR.

Reaction	Reactor	Reference
Hydrogenation of cyclohexene	Micro packed-bed reactor and microchannel with staggered arrays columns	Losey et al. [63, 64]
Hydrogenation of nitrobenzene	Falling film MSR	Yeong et al. [53]
Hydrogenation of p-Nitrotoluene	MSR made of stack of aluminum wafers	Födisch et al. [65]
Hydrogenation of α-methylstyrene	Micro packed bed reactor	Losey et al. [66]
Hydrogenation of α-methylstyrene	Micro-structured mesh contactor	Abdallah et al. [39]

- Dispersed-phase contacting, obtained when one of the fluid phases is dispersed in the other. Examples are micro-bubble columns and micro-packed-bed reactors. Examples are listed in Table 27.10.

Continuous-Phase Contacting In *falling-film contactors* a thin film is created by a liquid falling under the influence of gravity. The liquid flows over a solid support, which is normally a thin wall or a stack of pipes. In conventional falling-film devices, a film with a thickness of 0.5–3 mm is generated [51]. This rather thick liquid film results in a significant mass transfer resistance for gaseous reactants or products that need to diffuse to and away from the solid catalyst deposited on the reactor wall. Furthermore, the film flow becomes unstable at high throughputs, and it may break up into rivulets, fingers, or droplets.

These problems can be overcome by microstructuring the solid wall. The microstructured falling-film reactor consists of open microchannels, which are typically 300 µm wide, 100 µm deep, and about 80 mm long. The channels are separated by 100-µm wide walls (Figure 27.39). Inflow and outflow of the liquid occur through boreholes which are connected via one large slit to numerous small orifices at the top of the channels. A structured heat-exchanger plate is inserted

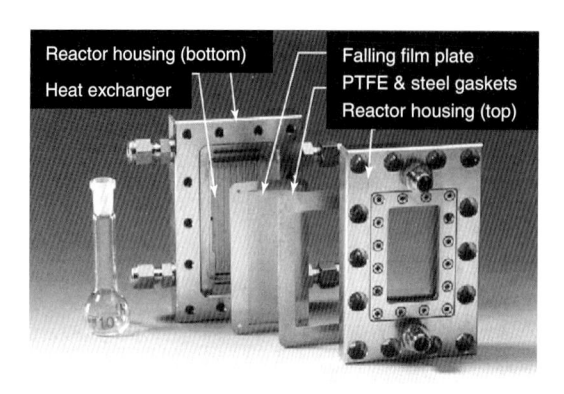

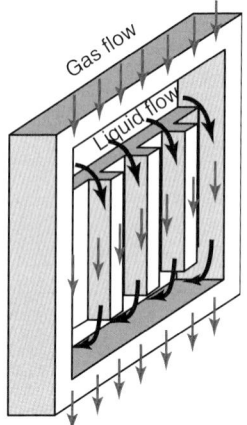

Figure 27.39 Components (left) and schematic drawing (right) of the microstructured falling film reactor (64 parallel channels, 300 µm wide, 100 µm deep, and 78 µm long) [53]. (Adapted with permission from Elsevier).

beneath the falling-film plate for heat removal, and nearly isothermal operation can be achieved even for highly exothermic reactions. The top part of the housing has a transparent window that allows use of the reactor for photochemical reactions [52].

The temperature control in such reactors is excellent, and any local or global overheating can be avoided, which is essential if thermally unstable products need to be volatilized.

A further advantage is the small holdup and short contact time. Klemm et al. [70] applied microstructured film contactors for the evaporation of hydrogen peroxide that was then used for the partial oxidation of propene to methyloxirane.

Yeong et al. [53, 54] used a microstructured film reactor for the hydrogenation of nitrobenzene to give aniline in ethanol at a temperature of 60 °C, an H_2 partial pressure of 0.1–0.4 MPa, and residence times of 9–17 s. Palladium catalysts were deposited as films or particles on a microstructured plate. Confocal microscopy was used to measure the liquid film thickness, which increased from 67 to 92 µm as flow rates were increased from 0.5 to 1.0 cm^3 min^{-1}. The value of $k_L a$ characteristic of this system was estimated to be 3–8 s^{-1} at an interfacial surface area (per reactor volume) of 9000–15000 m$^2 \cdot$ m^{-3}. Conversion was found to be affected by both liquid flow rate and H_2 partial pressure, and the reactor operated between the kinetic and mass transfer controlled regimes.

The main drawback of the microstructured falling-film reactor is the short residence time of the liquid in the channels, which typically varies from 5 to 20 s, depending on the physical properties of the liquid and the operating conditions. The residence time can be increased by lengthening the channels or by decreasing the angle of descent, which can be achieved with a helicoidal microchannel falling-film reactor. The residence time was found to be increased by a factor of about 50 in a microchannel when the angle of descent was decreased from 90 to 7.5° [51].

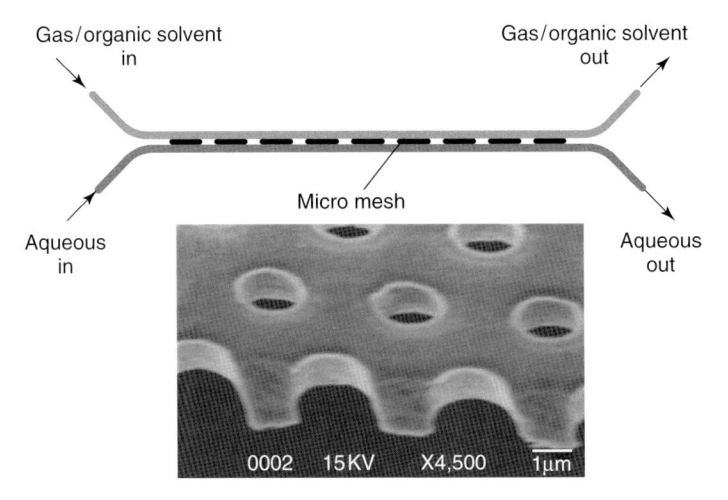

Figure 27.40 Principle of two-phase microcontactor and SEM image of the micromesh [55]. (Adapted with permission from the Royal Society of Chemistry).

In a *mesh microcontactor*, gas and liquid flow through separate compartments and are brought into contact by a partially porous wall between the compartments. Typically, the wall consists of a 5 μm thin plate with well-defined openings, into which the fluid penetrates [55] (Figure 27.40). The mesh-to-wall distances were typically about 100 μm. Pore widths of about 5 μm provided for adequate stability of the interface. The pore length was approximately equal to the pore width and ensured a low resistance to diffusion through the stagnant liquid in the mesh. The mesh was fabricated from nickel by use of photolithography and a two-stage electroplating method.

The shape of the meniscus at the interface between the two phases defines the available area for mass transfer and is a function of contact angle, pore geometry, and the pressure difference between the phases. The open area of the micromesh contactor is about 20–25%, which leads to a gas–liquid interfacial area of 2000 $m^2 \cdot m^{-3}$. This value is significantly higher than those typical of stirred tank reactors. The combination of a high gas–liquid interfacial area with a thin fluid layer results in high volumetric mass transfer coefficients. Abdallah *et al.* [39, 56] estimated the volumetric mass transfer coefficients $k_l a$ during the very fast hydrogenation of α-methylstyrene on a Pd/γ-alumina catalyst. The measured global gas–liquid–; solid volumetric mass transfer coefficient $k_L a$ was in the range of 0.8–1.6 s^{-1}. These values are well above those predicted by the film model and those obtained from a CFD (3D model) simulation. In addition, mesh contactors are characterized by a low pressure drop, and there is no danger of flooding as it exists in bubble or packed columns.

Dispersed Phase Contacting Micro bubble reactors (microchannels with slug or bubble/drop flow) and micro packed-bed reactors come under the class of dispersed phase contacting. Regular flow patterns are provided by the segmented flow in a single capillary or in multi-channel microreactors. In segmented-flow gas–liquid–solid reactors, the liquid usually flows over the solid surface while the gas flows through the liquid in the central core.

Miniaturized packed-bed microreactors follow the paths of classical reactor engineering by enabling trickle-bed or packed bubble column operation. The hydrodynamic characteristics of three-phase reactors, such as pressure drop and residence time distribution, can be determined similarly to those of fluid–solid and fluid–fluid reactors.

In *segmented flow gas–liquid–solid reactors*, the liquid usually flows over the solid surface while the gas flows through the liquid in the form of bubbles or as an annular flow, depending on the MSR geometry and the catalyst arrangement.

The difference between the gas–liquid and gas–liquid–solid systems is that due to the reaction at the surface of the catalyst, there is always a concentration gradient in the liquid phase in the latter case. In contrast to gas–liquid reactions, it is always important to saturate the liquid film with the gaseous component. To assure constant saturation, the following criterion has to be satisfied:

$$\frac{L_b D_m}{u_b \delta^2} \gg 1 \qquad (27.122)$$

The global transformation rate of a gas–liquid reaction catalyst by a solid catalyst is influenced by the mass transfer between the gas and the liquid and between the liquid and the solid. Mass transfer and surface reaction are in series, and, for fast chemical reactions, mass transfer will influence the reactant concentration on the catalytic surface and, as a consequence, influence the reactor performance and the product selectivity. For the gaseous reactant, three mass transfer steps can be identified [57]: (i) the transfer from the bubble through the liquid film to the catalyst ($k_{GS}a_{GS}$), (ii) the transfer from the caps of the gas bubbles to the liquid slug ($k_{GL}a_{GL}$), and (iii) the transfer of dissolved gas to the catalytic surface ($k_{LS}a_{LS}$). Steps (ii) and (iii) are in series, and in parallel with respect to step (i). The following expression describes the overall mass transfer:

$$k_{ov}a = k_{GS}a_{GS} + \left(\frac{1}{k_{GL}a_{GL}} + \frac{1}{k_{LS}a_{LS}}\right)^{-1} \qquad (27.122)$$

Various attempts were made to determine the mass transfer coefficients separately in non-reactive systems. However, the concentration profiles in the liquid surface film and in the slugs are strongly affected by fast chemical reactions, and the results must be interpreted with caution.

As hydrogenations are fast and highly exothermic reactions, the heat of reaction must be effectively evacuated to ensure isothermal operation and to avoid reaction runaway. An efficient device for this purpose was proposed by Hessel *et al.* [59]. The authors arranged the microchannels in parallel in between cooling channels as shown in Figure 27.41. Each channel works under segmented flow conditions. The main problem to overcome is to achieve uniform distribution of both gas and liquid flow over the microchannels to ensure identical flow behavior and residence time.

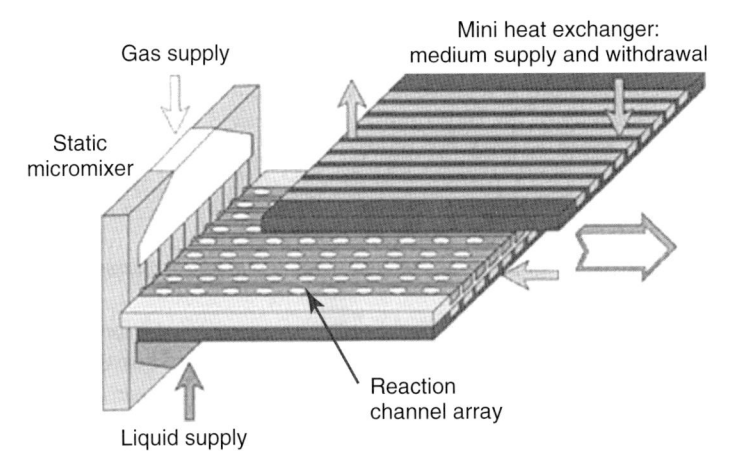

Figure 27.41 Micro bubble column with integrated cooling channels [59].

Many reported microreactors used for gas–liquid–solid reactions are *micro packed beds*. An advantage of microstructured packed beds for hydrogenation processes stems from the fact that active and selective catalysts are commercially available. In addition, the particle size of these catalysts, used in suspension reactors, is in the micrometer range and is well suited for their use in microchannels. However, proper design of the reactor is required to maintain an acceptable pressure drop. To avoid an excessive pressure drop, Losey *et al.* [66] constructed a microchemical system consisting of a microfluidic distribution manifold and a microchannel array. Multiple reagent streams (specifically, gas and liquid streams) were mixed on-chip, and the fluid streams were brought into contact by a series of interleaved, high-aspect-ratio inlet channels. These inlet channels deliver the reactants continuously and cocurrently to 10 reactor chambers containing standard catalytic particles in a diameter range of 50–75 µm. Flow regimes in the microreactor were characterized visually for different flow rates and gas-to-liquid flow ratios. For low liquid and gas velocities, bubbles were formed at the entrance and were carried by the liquid through the packed bed. Under these conditions the hydrogenation of cyclohexene was studied and used as a model reaction to measure the mass transfer resistances. Overall mass transfer coefficients ($k_{ov}a$) were measured and found to range from 5 to 15 s^{-1}, which is nearly two orders of magnitude larger than values reported in the literature for standard laboratory-scale reactors.

Increasing flow velocities in the microreactor led to pulsations and the formation of segmented flow. The different flow pattern observed in microstructured packed beds were studied in detail by van Herk *et al.* [60]. They confirm the segregated flow pattern at high gas fractions.

More recently, the catalytic hydrogenation of *o*-nitroanisole in a microstructured packed bed reactor was studied by Tadepalli *et al.* [61]. The reactor had an inner diameter of 0.775 mm and was filled with Pd/zeolite catalyst with particle diameter in the ranges of 45–75 and 75–150 µm. The length of the catalytic bed could be varied between 60 and 80 mm. It is stated that segmented gas–liquid flow was observed, but further hydrodynamic studies are lacking.

To increase mass transfer in solid-catalyzed gas–liquid hydrogenations, the reactions are often operated at high pressures. Silicon/glass microreactors present a possibility of safely handling high pressure and provide optical access into the reaction channel for flow investigations. Trachsel *et al.* [62] have described a silicon/glass microreactor with soldered microfluidic connections for high-pressure and high-temperature applications. Mechanical testing of the device using tensile and pressure tests showed no failure in continuous operation at 14 MPa and 80 °C. The microreactor design was applied for the well-described solid-catalyzed exothermic hydrogenation of cyclohexene at operating conditions up to 5.1 MPa and 71 °C.

Gas–Liquid–Liquid Gas–liquid–liquid reactions have several applications such as hydroformylation, carbonylation, hydrogenation, oligomerization, polymerization, hydrometallurgical applications, biochemical processes, and fine chemicals manufacturing. Developments in homogenous catalysis have made these reaction systems increasingly attractive in recent years. Gas–liquid–liquid systems are encountered in reaction systems that comprise three phases of two (or more) immiscible reactants, reaction products, or catalyst [67]. In some cases, the three reactants are supplied from three different phases (e.g., Koch reaction). It is also possible to intensify the mass transfer in a liquid–liquid system by means of an additional gas phase to enhance mixing and augment the interfacial area.

The flow patterns of gas–liquid–liquid flow in the MSR depend on the volume fraction of each phase within the reactor. At low gas volume fractions, the gas remains in one of the liquids while both liquids flow in the form of slugs. If the reaction is mass transfer limited, the overall reaction rate is strongly dependent on the interfacial liquid–liquid mass transfer. By reducing the capillary diameter, the specific interfacial area increases and leads to an intensified process.

Önal *et al.* [71] carried out selective hydrogenation of α, β-unsaturated aldehydes in aqueous solution using PTFE capillary of diameter from 500 to 1000 µm. The gas–liquid–liquid flow observed in the reactor is depicted in Figure 27.42. It shows the alternate flow of two liquid phases with organic forming a wall film due to its affinity toward the capillary material and hydrogen in the form of small bubbles in the organic phase. There was a significant effect on global reaction rates, with a threefold increase on reducing the channel diameter from 1000 to 500 µm.

27.5.4
Heat Management in Microstructured Reactors

MSRs are characterized by high surface-to-volume ratio, offering unique transport capabilities for enhanced heat and mass transfer. The use of MSRs allows new chemical routes with reaction conditions that are not feasible in batch vessels or in conventional continuous tubular reactors, thus opening new process windows for

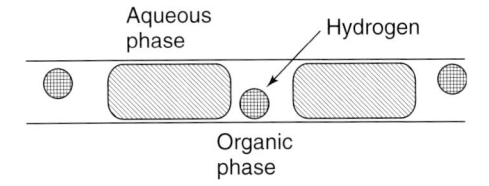

Figure 27.42 Schematics of gas–liquid–liquid flow patterns observed in the selective hydrogenation of α, β-unsaturated aldehydes.

chemical production. Mainly high temperatures and high pressures are applied to speed up the chemical reaction and, in consequence, the reactor performance. However, the advantages can only be realized if the reactor can be properly controlled to ensure safe and stable operation even under harsh non-usual conditions.

Temperature control of intensified chemical reactions is an important issue even in MSRs [22, 23, 66–68]. Fast exothermic reactions lead to the development of important axial temperature profiles, as discussed in Section 27.3.4. This in turn has an often detrimental impact on the product quality or a dramatic loss in product selectivity due to undesired side reactions. In addition, flow reactors can become very sensitive to small variations in the operation parameters such as cooling temperature, inlet temperature and concentration, or pressure. This so called high parametric sensitivity depends on the heat production potential and the ratio of the characteristic reaction time to the characteristic cooling time. In this situation the specific reactor performance is limited by the characteristic cooling time:

$$t_c = \frac{\rho \cdot c_p}{h \cdot a} = \frac{\rho \cdot c_p \cdot d_t}{h \cdot 4} \tag{27.123}$$

For laminar flow with developed velocity and temperature profiles the Nusselt number reaches a final value (Nu_∞). As a consequence, the characteristic cooling time depends on the thermal conductivity of the fluid, λ, and the square of the channel diameter:

$$t_c = \frac{\rho \cdot c_p \cdot d_t^2}{Nu_\infty \lambda \cdot 4} \tag{27.124}$$

As pointed out in Section 27.3.4, a maximal characteristic cooling time is required for safety reasons, avoiding reactor runaway ($t_c \leq t_r/N_{min}$). This maximum value depends on the characteristic reaction time at cooling temperature, the adiabatic temperature increase, and the sensitivity of the reaction rate to temperature variations. The estimated characteristic cooling times as a function of the channel diameter is shown in Figure 27.43 for water, gasoline, and air. Physical properties were assumed for $T = 293$ K and $p = 0.1$ MPa.

As the reaction rate increases exponentially with temperature, high wall temperatures are preferred to obtain high reactor performances. For safety reasons the characteristic cooling time has to be adapted. This is only possible by reducing the diameter of the micro channels. In Figure 27.44 the relative performance of

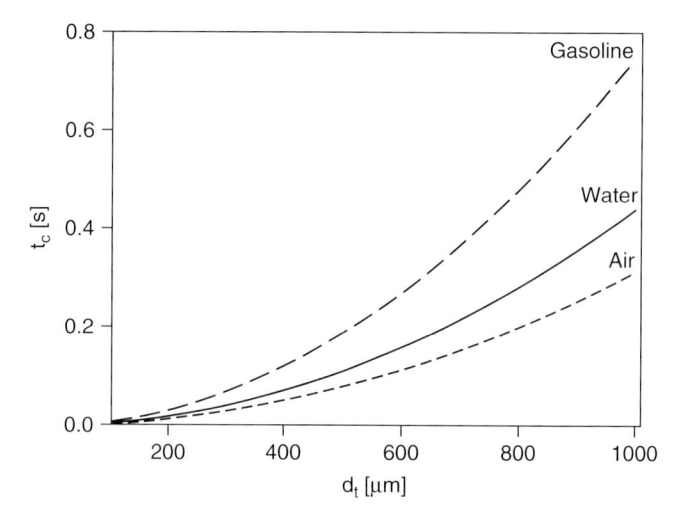

Figure 27.43 Characteristic cooling time in a micro-structured reactor for different fluids.

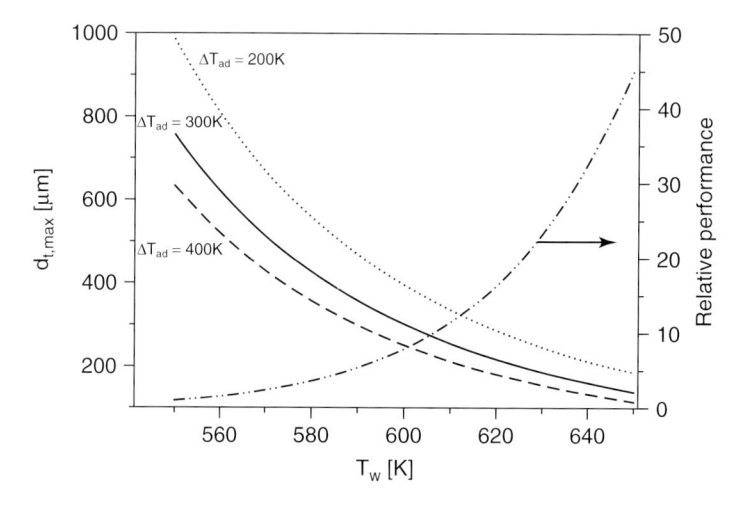

Figure 27.44 Relative reactor performance and maximum channel diameter for safe operation.

the reactor $\left(L_{P,T_w}/L_{P,550K}\right)$ and the maximum channel diameter for safe operation are plotted as a function of the wall temperature ($T_w = T_c$) for three different adiabatic temperatures. One can conclude from the shown data that for effective and safe reactor operation a channel diameter of about 200 μm is required when carrying out the reaction, with a ΔT_{ad} up to 400 K and wall temperatures up to 620 K [69].

References

1. Baerns, M., Hofmann, H., and Renken, A. (1999) in *Lehrbuch der Technischen Chemie*, vol. Bd. 1, 3rd edn (eds J. Falbe, M. Baerns, F. Fetting, H. Hofmann, W. Keim, and U. Onken), Georg Thieme Verlag, Stuttgart, New York, 444 p., 215 figures; 41 tables.
2. Trambouze, P. and Euzen, J.-P. (2002) *Les Réacteurs Chimiques*, Editions Technip, Publications de l'Institut Français du Pétrole.
3. Falk, L. and Commenge, J.-M. (2009) Characterization of mixing in homogeneous flow systems, in *Micro Process Engineering* (eds V. Hessel, A. Renken, J.C. Schouten, and J. Yoshida), Wiley VCH Verlag GmbH, Weinheim, Vol. 1, pp. 147–173.
4. Zlokarnik, M. (2001) *Stirring*, Wiley VCH Verlag GmbH, Weinheim, online ISBN 9783527306732.
5. Zlokarnik, M. (2005) Stirring, *Ullmann's Encyclopedia of Industrial Chemistry*, John Wiley & Sons, Inc.
6. Villermaux, J. (1993) *Génie de la Réaction Chimique*, TEC & DOC Lavoisier, Paris.
7. Flaschel, E., Nguyen, K.T., and Renken, A. (1985) *Proceedings of the 5th European Conference on Mixing*, Würzburg, BHRA, Bedford, pp. 549–554.
8. Nguyen, K.T., Streiff, F., Flaschel, E., and Renken, A. (1990) Motionless mixers for the design of multitubular polymerization reactors. *Chem. Eng. Technol.*, **13**, 214–220.
9. Eigenberger, G. (2008) Catalytic fixed-bed reactors, in *Handbook of Heterogeneous Catalysis* (eds G. Ertl, H. Knözinger, F. Schüth, and J. Weitkamp), Wiley VCH Verlag GmbH, Weinheim, Vol. 4, pp. 2075–2105.
10. Cybulski, A. and Moulijn, J.A. (2006) *Structured Catalysts and Reactors*, 2nd edn, Taylor & Francis, London.
11. Werther, J. (2008) Fluidized-bed reactors, in *Handbook of Heterogeneous Catalysis* (eds G. Ertl, H. Knözinger, F. Schüth, and J. Weitkamp), Wiley VCH Verlag GmbH, Weinheim, Vol. 4, pp. 2106–2131.
12. Laurent, A. and Charpentier, J.C. (1981) Role and use of experimental laboratory-scale models for predicting the performance of an industrial gas-liquid reactor (Anwendung Experimenteller Labormodelle Bei Der Voraussage Der Leistung Von Gas/Fluessigkeits-Reaktoren). *Chem. Ing. Tech.*, **53** (4), 244–251.
13. Baerns, M., Hofmann, H., and Renken, A. (2002) *Chemische Reaktionstechnik*, 3rd edn, Wiley VCH Verlag GmbH, Weinheim.
14. Levenspiel, O. (1999) *Chemical Reaction Engineering*, 3rd edn, John Wiley & Sons, Inc., New York.
15. Bahrmann, H. and Bach, H. (2010) Oxo Synthesis, *Ullmann's Encyclopedia of Industrial Chemistry*, Wiley VCH Verlag GmbH, Weinheim.
16. Renken, A. (2010) Chemical reaction engineering aspects for heterogenized molecular catalysis, in *Heterogenized Homogeneous Catalysts for Fine Chemical Production. Materials and Processes* (eds P. Barbaro and F. Liguori), Springer, pp. 247–282.
17. Guettel, R., Kunz, U., and Turek, T. (2008) Reactors for Fischer-Tropsch synthesis. *Chem. Eng. Technol.*, **31** (5), 746–754.
18. Grasemann, M., Renken, A., Kashid, M., and Kiwi-Minsker, L. (2010) A novel compact reactor for three-phase hydrogenations. *Chem. Eng. Sci.*, **65**, 364–371.
19. Baerns, M., Behr, A., Brehm, A., Gmehling, J., Hofmann, H., Onken, U., and Renken, A. (2006) *Technische Chemie*, 4th edn, Wiley VCH Verlag GmbH, Weinheim, 733 p.
20. Wen, C.Y. and Fan, L.T. (1975) *Models for Flow Systems and Chemical Reactors*, Marcel Dekker, New York, Basel.
21. Ehrfeld, W., Hessel, V., and Löwe, H. (2000) *Microreactors*, Wiley-VCH Verlag GmbH, Weinheim.
22. Renken, A., Hessel, V., Lob, P., Miszczuk, R., Uerdingen, M., and Kiwi-Minsker, L. (2007) Ionic liquid synthesis in a microstructured reactor for process intensification. *Chem. Eng. Process.*, **46** (9), 840–845.

23. Bier, W., Keller, W., Linder, G., Seidel, D., Schubert, K., and Martin, H. (1993) Gas to gas heat transfer in micro heat exchangers. *Chem. Eng. Process.*, **32** (1), 33–43.
24. Levenspiel, O. (1999) *Chemical Reaction Engineering*, John Wiley & Sons, Inc.
25. Falk, L. and Commenge, J.M. (2010) Performance comparison of micromixers. *Chem. Eng. Sci.*, **65** (1), 405–411.
26. Kockmann, N., Kiefer, T., Engler, M., and Woias, P. (2006) Convective mixing and chemical reactions in microchannels with high flow rates. *Sens. Actuator. B: Chem.*, **117** (2), 495–508.
27. Hessel, V., Hofmann, C., Löwe, H., Meudt, A., Scherer, S., Schönfeld, F., and Werner, B. (2004) Selectivity gains and energy savings for the industrial phenyl boronic acid process using micromixer/tubular reactors. *Org. Process Res. Dev.*, **8** (3), 511–523.
28. Löwe, H., Ehrfeld, W., Hessel, V., Richter, T., and Schiewe, J. (2000) Micromixing technology. 4th International Conference on Microreaction Technology (IMRET4), Atlanta.
29. Verein Deustscher Ingenieure (2002) *VDI-Wärmeatlas*, 9th edn, Springer, Berlin, Heidelberg, New York.
30. Kiwi-Minsker, L. and Renken, A. (2008) in *Handbook of Heterogeneous Catalysis* (eds G. Ertl, H. Knözinger, F. Schüth, and J. Weitkamp), Wiley VCH Verlag GmbH, Weinheim, pp. 2248–2264.
31. Rouge, A. and Renken, A. (2001) in *Reaction Kinetics and the Development and Operation of Catalytic Processes* (eds G.F. Froment and K.C. Waugh), Elsevier Science B.V., Amsterdam, pp. 239–246.
32. Rouge, A., Spoetzl, B., Gebauer, K., Schenk, R., and Renken, A. (2001) Microchannel reactors for fast periodic operation: the catalytic dehydrogenation of isopropanol. *Chem. Eng. Sci.*, **56**, 1419–1427.
33. Hoebink, J.H.B.J. and Marin, G.B. (1998) Modeling of monolithic reactors for automotive exhaust gas treatment, in *Structured Catalysts and Reactors* (eds A. Cybulski and J.A. Moulijn), Marcel Dekker, Inc., New York.
34. Sherony, D.F. and Solbrig, C.W. (1970) Analytical investigation of heat or mass transfer and friction factors in a corrugated duct heat or mass exchanger. *Int. J. Heat Mass Transfer*, **13** (1), 145–146.
35. Cybulski, A. and Moulijn, J.A. (1994) Monoliths in heterogeneous catalysis. *Catal. Rev. Sci. Eng.*, **36** (2), 179–270.
36. Hayes, R.E. and Kolaczkowski, S.T. (1994) Mass and heat transfer effects in catalytic monolith reactors. *Chem. Eng. Sci.*, **49** (21), 3587–3599.
37. Kashid, M.N., Agar, D.W., Renken, A., and Kiwi-Minsker, L. (2009) in *Micro Process Engineering* (eds V. Hessel, A. Renken, J.C. Schouten, and J. Yoshida), Wiley VCH Verlag GmbH, Weinheim, pp. 395–440.
38. Antes, J., Tuercke, T., Kerth, J., Marioth, E., Schnuerer, F., Krause, H.H., and Loebbecke, S. (2001) Use of microreactors for nitration processes. 4th International Conference on Microreaction Technology (IMRET 4), Atlanta, GA.
39. Abdallah, R., Meille, V., Shaw, J., Wenn, D., and De Bellefon, C. (2004) Gas-liquid and gas-liquid-solid catalysis in a mesh microreactor. *Chem. Commun.*, **10** (4), 372–373.
40. Mishima, K. and Hibiki, T. (1996) Some characteristics of air-water two-phase flow in small diameter vertical tubes. *Int. J. Multiphase Flow*, **22** (4), 703–712.
41. Shao, N., Gavriilidis, A., and Angeli, P. (2009) Flow regimes for adiabatic gas-liquid flow in microchannels. *Chem. Eng. Sci.*, **64** (11), 2749–2761.
42. Liu, H., Vandu, C.O., and Krishna, R. (2005) Hydrodynamics of Taylor flow in vertical capillaries: flow regimes, bubble rise velocity, liquid slug length, and pressure drop. *Ind. Eng. Chem. Res.*, **44** (14), 4884–4897.
43. Garstecki, P., Fuerstman, M.J., Stone, H.A., and Whitesides, G.M. (2006) Erratum: formation of droplets and bubbles in a microfluidic T-junction – scaling and mechanism of break-up. *Lab Chip – Miniaturisation for Chem. Biol.*, **6** (5), 693–694. (*Lab Chip – Miniaturisation for Chem. Biol.* (2006) **6** (437) doi: 10.1039/b510841a).

44. Berčič, G. and Pintar, A. (1997) The role of gas bubbles and liquid slug lengths on mass transport in the Taylor flow through capillaries. *Chem. Eng. Sci.*, **52** (21–22), 3709–3719.
45. Günther, A. and Kreutzer, M.T. (2009) in *Micro Process Engineering* (eds V. Hessel, A. Renken, J.C. Schouten, and J. Yoshida), Wiley VCH Verlag GmbH, Weinheim, pp. 3–40.
46. Pohorecki, R. (2007) Effectiveness of interfacial area for mass transfer in two-phase flow in microreactors. *Chem. Eng. Sci.*, **62** (22), 6495–6498.
47. Yue, J., Chen, G., Yuan, Q., Luo, L., and Gonthier, Y. (2007) Hydrodynamics and mass transfer characteristics in gas-liquid flow through a rectangular microchannel. *Chem. Eng. Sci.*, **62** (7), 2096–2108.
48. Burns, J.R. and Ramshaw, C. (1999) Development of a microreactor for chemical production. *Chem. Eng. Res. Des.*, **77** (A3), 206–211.
49. Ghaini, A., Kashid, M.N. and Agar, D.W. Effective interfacial area for mass transfer in the liquid-liquid slug flow capillary microreactors. *Chem. Eng. Process.: Process Intensificat.*, **49**, 358–366.
50. Kashid, M.N. (2007) *Experimental and Modelling Studies on Liquid-Liquid Slug Flow Capillary Microreactors*, University of Dortmund, Dortmund.
51. De Bellefon, C., Lamouille, T., Pestre, N., Bornette, F., Pennemann, H., Neumann, F., and Hessel, V. (2005) Asymmetric catalytic hydrogenations at micro-litre scale in a helicoidal single channel falling film micro-reactor. *Catal. Today*, **110** (1–2), 179–187.
52. Jähnisch, K., Hessel, V., Löwe, H., and Baerns, M. (2004) Chemistry in microstructured reactors. *Angew. Chem. Int. Ed.*, **43** (4), 406–446.
53. Yeong, K.K., Gavriilidis, A., Zapf, R., and Hessel, V. (2003) Catalyst preparation and deactivation issues for nitrobenzene hydrogenation in a microstructured falling film reactor. *Catal. Today*, **81** (4), 641–651.
54. Yeong, K.K., Gavriilidis, A., Zapf, R., and Hessel, V. (2004) Experimental studies of nitrobenzene hydrogenation in a microstructured falling film reactor. *Chem. Eng. Sci.*, **59** (16), 3491–3494.
55. Wenn, D.A., Shaw, J.E.A., and Mackenzie, B. (2003) A mesh microcontactor for 2-phase reactions. *Lab Chip - Miniaturisation Chem. Biol.*, **3** (3), 180–186.
56. Abdallah, R., Magnico, P., Fumey, B., and De Bellefon, C. (2006) CFD and kinetic methods for mass transfer determination in a mesh microreactor. *AIChE J.*, **52** (6), 2230–2237.
57. Heiszwolf, J.J., Kreutzer, M.T., Van Den Eijnden, M.G., Kapteijn, F., and Moulijn, J.A. (2001) Gas-liquid mass transfer of aqueous Taylor flow in monoliths. *Catal. Today*, **69** (1–4), 51–55.
58. Aartun, I., Silberova, B., Venvik, H., Pfeifer, P., Görke, O., Schubert, K., and Holmen, A. (2005) Hydrogen production from propane in Rh-impregnated metallic microchannel reactors and alumina foams. *Catal. Today*, **105** (3–4), 469–478.
59. Hessel, V., Ehrfeld, W., Golbig, K., Haverkamp, V., Löwe, H., Storz, M., and Wille, C. (1999) in *Proceedings of the 3rd International Conference on Microreaction Technology (IMRET3)* (ed. W. Ehrfeld), Springer, Berlin, pp. 526–540.
60. van Herk, D., Kreutzer, M.T., Makkee, M., and Moulijn, J.A. (2005) Scaling down trickle bed reactors. *Catal. Today*, **106** (1–4), 227–232.
61. Tadepalli, S., Halder, R., and Lawal, A. (2007) Catalytic hydrogenation of o-nitroanisole in a microreactor: reactor performance and kinetic studies. *Chem. Eng. Sci.*, **62** (10), 2663–2678.
62. Trachsel, F., Hutter, C., and von Rohr, P.R. (2008) Transparent silicon/glass microreactor for high-pressure and high-temperature reactions. *Chem. Eng. J.*, **135** (Suppl. 1), S309–S316.
63. Losey, M.W., Jackman, R.J., Firebaugh, S.L., Schmidt, M.A., and Jensen, K.F. (2002) Design and fabrication of microfluidic devices for multiphase mixing and reaction. *J. Microelectromech. Syst.*, **11** (6), 709–717.
64. Losey, M.W., Schmidt, M.A., and Jensen, K.F. (2001) Microfabricated multiphase packed-bed reactors:

characterization of mass transfer and reactions. *Ind. Eng. Chem. Res.*, **40** (12), 2555–2562.

65. Födisch, R., Hönicke, D., Xu, Y., and Platzer, B. (2001) in *International Conference on Microreaction Technology (IMRET 5)*, France Springer-Verlag, Berlin, Strasbourge, pp. 470–478.

66. Losey, M.W., Schmnidt, M.A., and Jensen, K.F. (2000) *International Conference on Microreaction Technology (IMRET 3)*, Springer-Verlag, Frankfurt, Germany, pp. 277–286.

67. Kaur, R., Machiraju, R., and Nigam, K.D.P. (2007) Agitation effects in a gas-liquid-liquid reactor system: methyl ethyl ketazine production. *Int. J. Chem. React. Eng.*, **5**, 1–19.

68. Renken, A. (2009) Micro-structured reactors and catalysts for the intensification of chemical processes. Proceeding of the 7th International ASME Conference on Nanochannels, Microchannels and Minichannels, 2009, CNMM-2009(Vol. B), pp. 1361–1369.

69. Renken, A. and Kiwi-Minsker, L. (2006) in *Micro Process Engineering* (ed. N. Kockmann), Wiley VCH Verlag GmbH, Weinheim, pp. 173–201.

70. Klemm, E., Mathivanan, G., Schwarz, T., and Schirrmeister, S. (2011) Evaporation of hydrogen peroxide with a microstructured falling film. *Chem. Eng. Proc. / Process Intensification*, **50**, 1010–1016.

71. Önal, Y., Lucas, M., and Claus, P. (2005) Insatz eines Kapillar-Mikroreaktors für die selektive Hydrierung von alpha, beta ungesättigten Aldehyden in der wässrigen Mehrphasenkatalyse. **77** (1–2), 101–106.

Index

a

acid/base catalysis 154–157
– bimolecular, acid catalyzed reaction 155
– bimolecular, base catalyzed reaction 156
– Brønsted–Lowry theory 155
– RNA hydrolysis catalyzed by enzyme RNase-A 157
acids 132–143, *See also* solid acids and bases
Acinetobacter calcoaceticus 181
activated carbon supports 439
activation, catalyst 58–64
– induction periods as catalyst activation 59–60
active sites 269–273
activity of catalyst, in homogeneous catalysis 54–58
acylases 185–186, 256
adaptive chemistry methods 370
adiabatic reactor 569
adsorption entropy 27–30
adsorption methods for porous materials characterization 514–534
– adsorption isotherms 517–518
– application of adsorption methods 518–519
– Barret–Joyner–Halada (BJH) method 529–530
– classification of porous materials 517
– mercury porosimetry 533
– mesoporous materials characterization 527–532
– – capillary condensation 527
– – Kelvin equation 528–529
– microporous materials characterization 524–526
– – Dubinin–Astakhov methods 524–525
– – Dubinin–Radushkevich (DR) methods 524–525
– – Horvath–Kawazoe (HK) method 525–526
– – Nonlocal Density Functional Theory (NL–DFT) 530–532
– physical adsorption 514–516
– theoretical description of adsorption 519–523
– – BET theory of adsorption 521
– – Langmuir isotherm 519–521
– – standard isotherms 522–523
– – t-method 522–523
– xenon porosimetry 533–534
Aerosil® process 435
alcohols
– carbonylations of 245–247
– synthesis 176–177
aldolases 190–193
alkene metathesis 402–406
altered surface reactivity 38–39
alumina (Al_2O_3) and other oxides 436–438
amidases 185–186
amidases catalyzed reactions 256
amino acid dehydrogenases (AADHs) 177, 253
amino acids synthesis 177–178
6-aminopenicillanic acid (6-APA) 256
7-aminocephalosporanic acid (7-ACA) 256
ammonia synthesis 114–119, 289–299
– activation energies 116
– based on natural gas 290
– catalysis 292–295
– composition dependence 114–119
– mechanism of reaction 114
– process optimization 295
– structure sensitivity 114–119
– technology development 291–292 emkin kinetic expression 118

Catalysis: From Principles to Applications, First Edition. Edited by Matthias Beller, Albert Renken, and Rutger van Santen.
© 2012 Wiley-VCH Verlag GmbH & Co. KGaA. Published 2012 by Wiley-VCH Verlag GmbH & Co. KGaA.

ammonia synthesis (*contd.*)
– Temkin Pyzhev kinetics   292
– ultra-high-vacuum techniques   292
Anderson–Schulz–Flory curve   313
ansa-zirconocene-based olefin polymerization catalysis   271–272, 275
anti-lock and key model   28, 167
apparent activation energy   69
aqueous ammonium heptamolybdate (AHM)   425
Arrhenius law   49, 69
Arrhenius number   85
Arthrobacter globiformis   255
aryl halides, carbonylations of   245–247
aryl–X derivatives, carbonylation of   246
asymmetric synthesis   174–175
atom utilization   17
atomic force microscopy (AFM)   494

b

Baeyer–Villiger oxidation   179
– Baeyer-Villiger Monooxygenases (BVMOs)   180–181
– enzyme-catalyzed, mechanism   182
'band gap'   218
Barret–Joyner–Halada (BJH) method   529–530
bases   132–143, *See also* solid acids and bases
basicity and nucleophilicity, relation between   154
batchwise-operated stirred-tank reactors (BSTR)   578–579
Bayer process   436
bimolecular catalytic reactions   76–77
biocatalysis   171–194, 250–259, *See also* hydrolases; lyases; oxidoreductases
– applications   253–257
– best route, choosing   250–253
– – enzymatic routes   251–252
– biocatalysis cycle   173
– case studies   257–259
– – lipitor building blocks synthesis   257–259
– development   172
– reaction strategy choice   174–175
– – kinetic resolution or asymmetric synthesis   174–175
– reaction systems, choice   175–176, 250–259
biotransformations   171
biphasic fluid/fluid systems, homogenous catalysis in   103–108
Bodenstein number   596–597, 607

Boltzmann constant   34
Braunauer, Emmett, and Teller (BET) theory   6, 521
Brønsted acid/base catalysis   132–143, 154, *See also* acid/base catalysis
Brønsted–Evans–Polanyi (BEP) relationship   14, 22
Brønsted–Lowry theory   155
Brunauer classification   518
Brunauer, Deming, Deming, and Teller (BDDT)   517
Burkholderia plantarii   254
Butler–Volmer setting   382
butylbranching mechanism   329

c

Candida cylindracea   254
Carberry number   83, 85
carbon materials   438–440
carbonylation reactions   234–247
– of alcohols and aryl halides   245–247
– – aryl–X derivatives   246
– – Ibuprofen   246
– – lazabemide   247
– – Rhodium-catalyzed carbonylation of methanol   245
– hydroformylation   235–239
– of olefins and alkynes   239–244
– – 1,3-butadiene   243
– – palladium-catalyzed carbonylation   242
5-carboxyfluorescein diacetate (C-FDA)   506–507
catalytic partial oxidation (CPOX)   357
catalytic reactor engineering   563–624, *See also* fluid–solid reactors; ideal reactor modeling/heat management; microreaction engineering; residence time distribution (RTD); single-phase reactors
– fixed-bed reactors   574
– fluid–fluid reactors   571–573
– liquid–liquid–gas system   573
– principles   67–108
– – formal kinetics of catalytic reactions   68–77, *See also* individual entry
– – homogenous catalysis in biphasic fluid/fluid systems   103–108
– – mass and heat transfer effects   77–103, *See also* individual entry
– slurry–suspension reactors   574–575
– structured catalysts for multiphase reactions   575
– three-phase gas–liquid–solid systems   573–575
– types   564–575

catalytic selective oxidation 341–363
– complexity/issues of 354, 356–358
– consolidated technologies 341–363
– in continuous development of more-sustainable industrial technologies 355–356
– development of industrial process, challenges 353–355
– directions for innovation 361–363
– dream oxidations 359–361
– fundamentals 341–363
– main features 341–353
– for organic compounds synthesis in petrochemical industry 342–352
– – oxidation process PHASE reactor 342–353
C–C bond cleavage 126–132
chain growth 314–315
chain walking 279
chemocatalysis, equivalence of 30–32
Chilton–Colburn analogy 84, 100
clariant process 247
classical chlorohydrin route 18
'classical' guidelines for catalyst testing 536–557
– appropriate laboratory reactor, selecting 546–548
– catalyst stability, assessing 554–555
– data collection 543–546
– effective experimental strategies 541–543
– – two-variables system 542
– encouraging effectiveness 536–540
– ensuring efficiency 540–555
– ideal flow pattern, establishing 548–549
– isothermal conditions, ensuring 549–551
– quality guidelines 540
– selectivity 544
– space–time yield 545
– transport, diagnosing and minimizing the effects of 552–554
– – 'DrySyn' Beta vs commercial Beta performance 553
– – diagnostic tests for interphase (external) transport 552
– – diagnostic tests for intraparticle (internal) transport 553
– yield 545
cobalt 312
– cobalt catalyst formation 321–322
– cobalt promoter 394–395
co-catalysts 59
co-catalyzed carbonylation 241
cofactor regeneration 178

Coherent Anti-Stokes Raman Spectroscopy (CARS) 502
combinatorial approaches in solid catalysts development 453
– for optimal catalytic performance 453–456
compensation effect 44–46
– exocyclic methylation reaction 45
– pairing reaction 45
complex reaction 73
consecutive first-order reactions 82
consecutive reactions 92
constant selectivity relationship 159
constrained geometry catalysts 279
Continuously Operated Ideal Stirred Tank Reactors (CSTR) 580–581
– space time in 583
continuously stirred tank reactors (CSTRs) 546–547
continuous-phase contacting 616–618
– falling-film contactors 617
Corynebacterium glutamicum 253
coupling of catalytic reaction 42
crystalline catalysts design 222–223
cumylhydroperoxide (CHP) 362
cycle, catalytic 20–27

d
Damköhler number 81–82, 547
de Donder concept 32
deactivation, catalyst 58–64
– due to irreversible reactions 65–64
– due to multinuclear complexes formation 61–62
– due to non-reactive complexes formation 61
dehydrogenases 176–178
– alcohols synthesis 176–177
– amino acids synthesis 177–178
– formate dehydrogenase (FDH) 178
– polyethylene glycol (PEG) 178
– Prelog-rule 177
dehydrogenation 126–132
– cyclohexane dehydrogenation 131
– mechanism of 126
– olefin hydrogenation kinetics 126–127
density functional theory (DFT) 209, 367, 395–397
deposition precipitation 427–428
desulfurization 291
development of catalytic materials 445–459
– catalytic OCM reaction, reaction mechanism and kinetics of 448–449
– combinatorial approaches in 453–459
– fundamental aspects 446–448

development of catalytic materials (contd.)
- high-throughput technologies in 456–459
- kinetic analysis 450–451
- micro-kinetics and solid-state properties as knowledge source 448–453
- – electronic conductivity 452
- – ion conductivity 452
- – redox properties 452
- – structural defects 451–452
- – supported catalysts 453
- – surface acidity and basicity 452
- surface oxygen species in methane conversion 449–450

development of catalytic processes 11–13
- history 11–13
- future 11–13

dibenzothiophene 392
dihydroxyacetone phosphate (DHAP) 191–192
- DHAP-dependent aldolases 192
dilution rate 580
direct alkane activation 139–141
direct desulfurization (DDS) 392
dispersed phase contacting 619–622
- gas–liquid–liquid 622
- micro packed beds 621
- segmented flow gas–liquid–solid reactors 620
dispersion model 595–596
'dormant' sites 269–273
dry impregnation 425
Dubinin–Astakhov methods 524–525
Dubinin–Radushkevich (DR) methods 524–525
Dusty Gas Model 382
dynamic kinetic resolution (DKR) 175

e

efficiency factor 105
electric double layer 204–207
- diffuse double-layer 206
- inner layer 206
electrocatalysis 201–213
- electric double layer 204–207
- – diffuse double-layer 206
- – inner layer 206
- electrochemical potentials 203–204
- equivalence of 30–32
- theory 203–207
electrochemical potentials 203–204
electrolyte-reservoir 205
electron backscatter diffraction (EBSD) measurements 502

electrophilic catalysis 154–157
enantioselective ring-closing metathesis 405
enzyme catalysis 8–9, 155
ethane hydrogenolysis mechanism 127–132
ethylene/propylene rubber (EPR) 264
Evonik-Uhde process 361–362
exocyclic methylation reaction 45
extended X-ray absorption fine spectroscopy (EXAFS) 400, 499
external and internal transfer resistances, combination 96–101
- external and internal temperature gradient 100–101
- in isothermal pellets 96–98
- mass transfer implication on temperature dependence 98–99
external mass and heat transfer 78–85

f

falling-film contactors 617
'fingerprinting' 267
Fischer–Tropsch (FT) synthesis 40, 301–323, 569
- catalysts, general 311–313
- – cobalt 312
- – iron 312–313
- – nickel catalysts 312
- – Ruthenium 312
- gas chromatogram of 303
- operation ranges 306–308
- processes and product composition 308–311
- – commercial FT-synthesis 308–311
- – high-temperature Fischer–Tropsch synthesis 310–311
- – hydrocracking 309
- – hydroisomerization 309
- – isomerization 309
- – low-temperature synthesis 309–310
- – multi-tubular fixed-bed reactors 309
- – slurry-phase bubble column reactors 309
- – slurry reactors 310
- – synthesis gas 311
- rate equations 306
- reaction fundamentals 313–322
- – ideal polymerization model 313–322
- stoichiometry 304–305
- thermodynamic aspects 305–306
Fischer-Tropsch Synthesis (FTS) catalyst 509
fixed-bed reactors 568–569, 574
- adiabatic reactor 569
- multitubular reactor 569
- staged fixed-bed 569
fluid catalytic cracking (FCC) process 391

Index

fluid–fluid MSR 610–612
- falling film microchannel 610
- microchannel 610
- micromixer 610

fluid–fluid reactors 571–573
fluidized-bed reactors 569–571
fluid–solid MSR 607–610
fluid–solid reactors 568–571
- fixed-bed reactors 568–569
- fluidized-bed reactors 569–571
- minimum fluidization velocity and pressure drop 570–571

fluorescence microscopy (focused ion beam (FIB) 502
formal kinetics of catalytic reactions 68–77
- general definitions 69–70
- heterogeneous catalytic reactions 70–71
- Langmuir adsorption isotherms 72–73
- reaction mechanisms 73–77

formate dehydrogenase (FDH) 178, 254
fractional surface coverage 71
framework-substituted redox ions 335–339
- Ti-catalyzed epoxidation 335–338
fumed silica 433–436
fundamental catalysis in practice 13

g

gas–liquid–liquid reactors 622
gas–liquid–solid reactors 616
gas–liquid systems 611–613
glycerolphosphate oxidase (GPO) 192
Gouy–Chapman diffuse double layer 206
Gouy–Chapman model 205
Gouy–Chapman–Stern model 205
'growing chain orientation' mechanism 268

h

Haber–Bosch process 3, 291
Hagen–Poiseuille law 593, 605
half-titanocene precatalyst structures 281
Hammett acidity function 154
Hatta number (Ha) 103, 572
heat balance 583
heat management in microstructured reactors 622–624
heat transfer 85–96, See also mass and heat transfer effects
Helmholtz model 205
heterogeneous catalysis 4–7, 113–150, 273–278, See also kinetics of heterogeneous catalytic reactions; reducible oxides; solid acids and bases; transition metal catalysis
- catalyst preparation methods 7
- catalyst's performance and its composition and structure 4–7
- – Brunauer-Emmett-Teller (BET) technique 6
- – characterization tools 6
- – dynamic Monte Carlo methods 7
- – Langmuir–Hinshelwood–Watson–Hougson (LHWH) expressions 6
- – Michaelis–Menten expression 6
- *in situ* spectroscopic measurements 5
- – T-plot techniques 6
- – X-ray scattering techniques 6
- formal kinetics 70–71
- molecularly defined systems in 399–413, See also individual entry
- monomer insertion 274
- 'replication' phenomenon 274, 277
- steps involved in 68
- Ziegler–Natta catalysts 274

heterogeneous catalysts 493–510, See also *in-situ* characterization of heterogeneous catalysts
- design of, requirements 447

heterogeneous chemistry, of high-temperature catalysis 365–385
- heterogeneous reaction mechanisms 367–368

heterogeneous photocatalysis 216–228
- case studies 220–228
- – crystalline catalysts design 222–223
- – energy conversion 222–225
- – microreactors 227–228
- – supported chromophores 223–225
- – visible light-sensitive systems 223
- – water purification 220–222
- – Z-scheme process 224
- photocatalysis 216–217
- from surface chemistry to reactor design 216–228

Hevea brasiliensis (HbHNL) 189, 257
High Density PolyEthylene (HDPE) 262
high-energy resolution fluorescence detected (HERFD) X-ray Absorption Spectroscopy (XAS) 497
High Throughput Experimentation (HTE) 280
Highest Occupied Molecular Orbital (HOMO) 217
high-temperature catalysis 365–385
- applications 372–378
- – formulation of an optimal control problem 375–378

high-temperature catalysis (*contd.*)
– – olefin production by high-temperature oxidative dehydrogenation of alkanes   373–378
– – synthesis gas from natural gas by high-temperature catalysis   373
– – turbulent flow through channels with radical interactions   372–373
– fundamentals   366–372
– – coupling of chemistry with mass and heat transport   369–370
– heterogeneous chemistry role   365–385
– homogeneous chemistry role   365–385
– hydrogen production from logistic fuels by   378–380
– mathematical optimization of reactor conditions and catalyst loading   372
– radical chemistry role   365–385
– in solid oxide fuel cells   380–385
high-temperature Fischer–Tropsch synthesis   310–311
high-temperature fuel cells   381
high-throughput technologies in solid catalysts development   456–459
– catalytic materials preparation   457
– data analysis   458–459
– screening of catalytic materials   457
– testing of catalytic materials   457
historic review   3–19
– development of catalytic processes   11–13
– – history and future   11–13
– fundamental catalysis in practice   13
– history of catalysis science   3–12
– – chemical engineering   3
– – organic chemistry   3
– – nineteenth century   3
– – synthetic organic chemistry   4
– – hysical chemistry   4
– – inorganic chemistry   3–4
– important scientific discoveries   9–11
– – new industries/new catalytic processes   10
– process choice   17–19
– reactor choice   16–17
Hofmann-type β-hydrogen elimination   391
Hombikat UV   100, 220
homogeneous catalysis   8–9, 152–169, 278–280, 465–490, *See also* acid/base catalysis; *in-situ* techniques for homogeneous catalysis; kinetics in homogeneous catalysis
– in biphasic fluid/fluid systems   103–108
– characteristics   152
– – isolation of products and recovery and recycle of catalyst   153
– – ligands for   153–154
– – metal-catalyzed reactions   153
– – nature of compounds   152
– – reaction conditions   152
– – reactions mediated by transition metal complexes   153
– – reactions phase   153
– enzyme catalysis   155
– nucleophilic and electrophilic catalysis   154, 157–159
– – acid/base catalyzed RNA hydrolysis   155
– organometallic complex catalysis   157–160
– transition metal-centered homogeneous catalysis   155, 159–169, *See also individual entry*
homogeneous chemistry, of high-temperature catalysis   369
Horvath–Kawazoe (HK) method   525–526
hydantoinases   187–188
hydride transfer   141
hydroaminomethylation reaction   238
hydrocarbon/oxygen (C/O) ratio   377
hydrocarboxylation   240
hydrocracking reaction, acid catalysis   309, 325–332
– cracking selectivity dependence   326–328
– hydrocarbon chain length, activity on   326–328
– symmetric versus asymmetric cracking patterns   328–332
– – butylbranching mechanism   329
– – pore size   328–332
– – side-chain elongation mechanism   329
– – stereo-selective behavior   330
– – stereoselectivity   328–332
– – topology dependence   328–332
hydrodenitrogenation   397
hydrodesulfurization   390–398
– C-X bond-breaking mechanism   393
– sulfidic catalyst, structure   393–397
– – DFT calculations   395–397
– – Mo structure   393–394
– – promoter structure   394–395
– surface sites determination   398
hydroesterification   240
hydroformylation   235–239
– co-catalyzed carbonylation   241
– rhodium-catalyzed hydroformylation   236
hydrogen cyanide (HCN)   171
hydrogen evolution reaction (HER)   202

hydrogenation 126–132
– hydrogenolysis of isopentane, single-center route for 130
– mechanism of 126
– route 392
hydrogenolysis of grafted hydrocarbyl-containing systems 409
hydroisomerization 309, 327–328
hydrolases 182–188
– acylases 185–186
– amidases 185–186
– hydantoinases 187–188
– lipases 182
– nitrilases 186–187
– nitrile hydratases 186–187
– peptidases 185–186
hydroxynitrile lyases (HNLs) 171, 188–190, 257

i

ibuprofen 246
ideal continuously-operated stirred tank reactor 591–592
ideal plug flow reactor 591
ideal polymerization model of FT synthesis 313–322
– alcohols in 316
– alternative reactions on growth site 315
– Anderson–Schulz–Flory curve 313
– branching 315–316
– chain growth 314–315
– desorption (olefins/paraffins) 316–319
– – primary reactions 317
– – secondary reactions 317
– *in situ* catalyst formation 319–322
– – cobalt catalyst formation 321–322
– – iron catalyst formation 319–321
– – self-organization of FT regime 319
ideal reactor modeling/heat management 575–586
– batchwise-operated stirred-tank reactors 578–579
– continuously operated ideal stirred tank reactors (CSTR) 580–581
– ideal plug flow reactor 581–586
– – heat balance 583
– – highly exothermic reactions 586
– – mass balance in 582
– – parametric sensitivity 584
– – stability diagram 586
– mass and energy balances 576–577
in situ catalyst formation in FT synthesis 319–322

in situ characterization at a single catalyst particle level 501–510
– single-molecule *in-situ* spectroscopy of a catalytic solid 504–508
in situ generation of organo-catalyst 42–44
in situ hydroformylation 238
in situ micro-spectroscopy of catalytic solid 501–504
in situ techniques 59, See also individual entries
induction periods as catalyst activation 59–60
inhomogeneous site distribution 40–42
inorganic chemistry 4
– inorganic solid chemistry 42
in-situ characterization of heterogeneous catalysts 493–510
– applications 495–497
– dynamic conditions 494
– gas chromatography (GC) 495
– history 495–497
– mass spectrometer (MS) 495
– reactor loaded with catalytic solid 497–501
– – Extended X-ray Absorption Fine Structure (EXAFS) 499
– – *in-situ* HERFD-XAS approach 497
– – probed by multiple characterization methods 499–501
– – probed by one characterization method 497–499
– – XANES 497
– recent developments 495–497
– static conditions 494
– ultra-high vacuum (UHV) conditions 494
in-situ high-resolution transmission electron microscopy (HRTEM) 296
in-situ nano-spectroscopy of a catalytic solid 509–510
in-situ techniques for homogeneous catalysis 465–490, See also IR-spectroscopy; NMR spectroscopy; UV/Vis spectroscopy
– gas consumption and gas formation 467–470
instantaneous or point selectivity 92
interfacial activation 183
internal mass and heat transfer 85–96
– isothermal pellet 87–94
– non-isothermal pellet 94–96
Internally Illuminated Monolith Reactor (IIMR) 227
ion adsorption 423–424
iron 312–313
iron middle pressure synthesis 309

irreversible reactions, catalyst deactivation due to 65–64
IR-spectroscopy 481–485
– hydroformylation 484–485
– *in-situ* FTIR spectroscopic study 483
isomerization catalysis 141–143
isothermal pellet 78–84, 87–94
– consecutive reactions 92
– external concentration profile 80
– instantaneous or point selectivity 92
– isothermal yield and selectivity 82–84, 91
– parallel reactions 91
– porous catalysts, concentration profiles in 79
isothermal yield and selectivity 82–84, 91
– consecutive first-order reactions 82
– parallel reactions 84
isotherms 71

k

Kelvin equation 528–529
kinetic resolution 174–175
– dynamic kinetic resolution (DKR) 175
kinetics 4–7, 68
– microkinetics 7
– rate-limiting step 6
kinetics in homogeneous catalysis 48–64
– activation and deactivation, catalyst 58–64
– – co-catalysts 59
– – *in situ* techniques 59
– – induction periods as catalyst activation 59–60
– – spectator ligands 59
– catalyst activity 54–58
– kinetic description 48–54
– principles of catalyst 48–54
kinetics of heterogeneous catalytic reactions 20–46
– altered surface reactivity 38–39
– equivalence of electrocatalysis and chemocatalysis 30–32
– materials gap 39–42
– microkinetics, rate-determining step 32–34
– physical chemical principles 20–27
– – activation energy 26
– – catalytic cycle 20–27
– – rate-limiting step 21
– pressure gap 36–39
– surface reconstruction 37–38

l

lactate dehydrogenase (LDH) 194
laminar flow reactor 593–595
Langmuir adsorption isotherms 72–73
Langmuir-Hinshelwood kinetics 74–75, 327
Langmuir–Hinshelwood–Watson–Hougen (LHWH) equation 6, 21, 25
Langmuir isotherm 519–521
Laser Doppler Anemometry/Laser Doppler velocimetry (LDA/LDV) 371
laser-induced fluorescence (LIF) 371
lazabemide 247
Lennard–Jones potential (L–J potential) 515–516
leucine dehydrogenase (Leu-DH) 178, 253
Lewis acid–lewis base catalysis 132–143, 332–333
– hydrocarbon activation 332–333
ligand-accelerated catalysis (LAC) 167
Linum usitatissimum 189
lipases 182, 253–255
lipitor building blocks synthesis 257–259
liquid–liquid systems 613–616
liquid–liquid–gas system 573
liquid loading (α) 107
lock and key model 27–30, 168
– anti-lock and key model 28
Low Density PolyEthylene (LDPE) 262–263
Low-Energy Electron Diffraction (LEED) 493
Lowest Unoccupied Molecular Orbital (LUMO) 217
low-pressure oxo processes (LPO) 236
lyases 188–193
– aldolases 190–193
– hydroxynitrile lyases 188–190
– pyruvate/phosphoenolpyruvate-dependent aldolases 192–193

m

major/minor concept 168, 479
Manihot esculenta 190
mass and heat transfer effects 77–103
– external and internal transfer resistances, combination 96–101
– external mass and heat transfer 78–85, *See also* internal mass and heat transfer
– isothermal pellet 78–84, *See also individual entry*
– non-isothermal pellet 84–85
– transport effects, estimation criteria for 101–103
mass Biot number 97
mass transfer implication on temperature dependence 98–99

materials gap 39–42
– catalyst activation or deactivation 40
– inhomogeneous site distribution 40–42
– structure sensitivity 39
maximum rate (r_{max}) 53
mercury porosimetry 533
mesh microcontactor 619
mesoporous materials characterization 527–532
metal-organic frameworks (MOFs) 442–443
methane reforming 120–125
– activation energies and reaction energies 123
– composition dependence 120–125
– mechanism of reaction 120
– structure sensitivity 120–125
methyl methacrylate (MMA) 240
Michaelis–Menten equation 6, 27, 50–55, 166
micro packed beds 621
microkinetics 7
– in catalysts development 448–453
– rate-determining step 32–34
microporous materials characterization 524–526
microreaction engineering 602–624, See also three-phase reactors
– criteria for reactor selection 602–604
– fluid–fluid MSR 610–612
– fluid–solid MSR 607–610
– gas–liquid systems 611–613
– heat management in microstructured reactors 622–624
– liquid–liquid systems 613–616
– micro-structured catalytic wall reactor 608
– residence time distribution 606–607
– single-phase MSR 604–607
– types of 604–610
microreactors 227–228
micro-spectroscopic techniques 502
micro-structured reactors (MSRs) 564
minimum fluidizing velocity (u_{mf}) 571
modern pertochemical route 18
molecular basis of catalysis 5
molecular vs surface chemistry 401
– characterization tools 401
molecularly defined systems in heterogeneous catalysis 399–413
– bridging the gap with classical heterogeneous systems 406–408
– molecular vs surface chemistry, characterization tools 401
– – chemical characterization (reactivity) 401
– – elemental analysis 401
– – ESR spectroscopy 401
– – IR (Raman) spectroscopy 401
– – mass spectrometry 401
– – NMR spectroscopy 401
– – UV spectroscopy 401
– – X-ray crystallography 401
– single sites
– – on border between homogeneous and heterogeneous catalysis 400–412
– taking homogeneous catalysis to heterogeneous phase 402–406, See also single-site alkene metathesis catalysts
– toward new reactivity 408–411
– – alkane metathesis 409
– – cross-metathesis of methane and higher alkanes 409–410
– – H/D reaction of D2/H2 or D2/RH mixtures 409
– – hydrogenolysis of alkanes 409
– – methanol-to-olefin (MTO) process 413
– – non-oxidative coupling of methane 410
– – titanium-substituted silicate-1 (TS-1) 413
monolithic catalysts 370–371
– DETCHEM^{MONOLITH} 370–371
– shapes
– – corrugated plate packing 570
– – square-channel monolith 570
monomer insertion 274
Monte Carlo methods 7
Montsanto process 245
mordenite (MOR) 441
most abundant surface intermediate (masi) approximation 75–76
multinuclear complexes formation, catalyst deactivation due to 61–62
multinuclear NMR spectroscopy 470
multitubular reactors 309, 569

n

N-acetylneuraminate (NeuAc) 192
natural gas-based ammonia plant 290–291
– desulfurization 291
– secondary reforming 291
– steam reforming 291, 295–299
nicotinamide adenine dinucleotide (phosphate)(NAD(P)H) 251
nitrilases 186–187
nitrile hydratases 186–187
NMR spectroscopy 470–480
– carbon monoxide ethylene copolymer formation 477
– hydridoalkyl intermediate formation 478
– hydroformylation catalyst formation 474

NMR spectroscopy (*contd.*)
– Iridium catalyst 478
– major/minor principle 479
– multinuclear NMR spectroscopy 470
– palladium-catalyzed hydroalkoxycarbonylation 476
non-isothermal pellet 84–85, 94–96
– internal and external mass transport in 96–98
nonlocal Density Functional Theory (NL–DFT) 530–532
non-oxidative coupling of methane 410
non-reactive complexes formation, catalyst deactivation due to 61
non-specific adsorption 204
nucleophilic/electrophilic catalysis 154, 157–159
– bimolecular reaction 158

o

olefin hydrogenation kinetics 126–127
olefin polymerization process technology 264–280, *See also* heterogeneous catalysis; homogeneous catalysis
– 'growing chain orientation' mechanism 268
– propene polymerization catalysts 266
– reactivity 264–269
– structure 264–269
Operando spectroscopy 465
ordered mesoporous materials 442
organo-catalyst, *in situ* generation of 42–44
organometallic complex catalysis 152, 157–160
– major/minor concept 159
– 'Reppe chemistry' 157
Ortho-F effect 282
overall effectiveness factor 97
oxidation reactions 148–150
– propane ammoxidation mechanism 149
– propane oxidation 150
– selective oxidation of propylene 148–150
oxidative addition 162–163
– in transition metal-centered homogeneous catalysis 159–169
– – of non-polar addenda in apolar solvents 171
– – via radical chain mechanism 163–164
oxidative coupling of methane (OCM) 448
– catalytic solid materials for OCM reaction
– – physico-chemical properties of 451
– kinetics of 448–449
– reaction mechanism 448–449

oxidoreductases 176–182, *See also* oxygenases
– dehydrogenases 176–178
– – alcohols synthesis 176–177
oxo synthesis (hydroformylation) 103
oxygen reduction reaction (ORR) on Pt(111), application 202, 207–212
– electrochemical ORR mechanisms 210
– HOOH pathway 209
– O_2 pathway 208
– OOH pathway 209
oxygenases 178–182
– Baeyer-VilligerMonooxygenases (BVMOs) 180–181
– P450 mono-oxygenases 179–180
– P450 superfamily 180

p

P450 mono-oxygenases 179–180
pairing reaction 45
palladium-catalyzed carbonylation 242
parallel reactions 84, 91
parametric sensitivity 584
Pareto plot 15–16
Pauling valency concept 133
Pausen-Khand reaction 244
Péclet number 597
Penicillin G amidase (PGA) 256
peptidases 185–186
phenyl alanine dehydrogenase (Phe-DH) 178, 254
photocatalysis 216–217, *See also* heterogeneous photocatalysis
– applications of 219
– in practice, reactor considerations 225–228
– principle of 217–219
physical adsorption 514–516
physical chemistry 4
pig liver esterase (PLE) 171
planar laser-induced fluorescence (PLIF) 371
plug flow reactors (PFRs) 546
Point of Zero Charge (PZC) 422–423, 436
polyethylene glycol (PEG) 178
polymerase chain reaction (PCR) 253
polymer-electrolyte (or proton-exchange) membrane fuel cell (PEMFC) 207
polymerization 261–285
– kinetics
– – active sites 269–273
– – 'dormant' sites 269–273
– – 'triggered' sites 269–273
– latest breakthroughs 280–285
– – half-titanocene precatalyst structures 281

- olefin polymerization process technology 264–273, See also individual entry
- polyolefins 262–264
- Ziegler–Natta-type olefin polymerizations 261
polyolefins 262–264
porcine pancreatic lipase (PPL) 184
pore size distribution (PSD) 518
pore volume impregnation (PVI) 425
porous catalysts, concentration profiles in 79
porous materials as catalysts and catalyst supports 431–444, See also adsorption methods for porous materials characterization
- activated carbon supports 439
- alumina (Al_2O_3) and other oxides 436–438
- carbon materials 438–440
- fumed silica 433–436
- general characteristics 431–433
- ordered mesoporous materials 442
- shaping 443–444
- sol-gel method of preparation of 433–436
- - Aerosil® process 435
- zeolites 440–441
Prelog-rule 177
pressure gap 36–39
primary proton attachment 137
Proactinomyces erythropolis 181
process choice 17–19
- classical chlorhydrin route 18
- modern pertochemical route 18
propane oxidation 150
propene polymerization catalysts 266
proton activation by zeolites 135–139
Pseudomonas chlororaphis B23 256
Pseudomonas putida 181
pyruvate decarboxylase (PDC) 194
pyruvate/phosphoenolpyruvate-dependent aldolases 192–193

q

α-quartz 433
quasi-surface equilibrium approximation 75

r

radical chemistry, of high-temperature catalysis 365–385
- experimental evaluation of models 371
rate-controlling step or slow step 33
rate-determining step 34
rate-limiting step 6, 21
reaction engineering principles, See catalytic reactor engineering
reaction mechanisms 70, 73–77

- bimolecular catalytic reactions 76–77
- complex reaction 73
- Langmuir–Hinshelwood model 74–75
- 'masi' approximation 75–76
- quasi-surface equilibrium approximation 75
reaction order 69
reaction rate 69
reactor choice 16–17
reactor engineering, See catalytic reactor engineering
reactor performance (L_p) 578
redox catalysis 333–335
reducible oxides 143–150
- heats of formation 145
- oxidation reactions, mechanism 148–150
- - propane ammoxidation mechanism 149
- - propane oxidation 150
- - selective oxidation of propylene 148–150
- relative stabilities, comparison 143–145
- structure sensitivity 145–148
residence time distribution (RTD) 587–602, 606–607
- experimental determination of 589–591
- - pulse function 590–591
- - step function 589–590
- residence time distribution in tubular reactors, estimation 597–599
- RTD for ideal reactors 591–595
- - cascade of ideally stirred tanks 592–593
- - ideal continuously-operated stirred tank reactor 591–592
- - ideal plug flow reactor 591
- - laminar flow reactor 593–595
- RTD influence on performance of real reactors 599–602
- RTD models for real reactors 595–597
- - cell model 596–597
- - dispersion model 595–596
reversible reaction 69
rhodium-catalyzed carbonylation of methanol 245
rhodium-catalyzed hydroformylation 236
Rhodococcus rhodochrous 256
Ruhrchemie/Rhône-Poulenc process 237
ruthenium 312

s

Sabatier Principle 21–23
Sabatier's catalytic reactivity principle 14
Scanning Electron Microscopy (SEM) 494
Scanning Transmission X-ray Microscopy (STXM) 509
Schmidt number 85

second Damköhler number (DaII) 81–82, 98
Second Harmonic Generation (SHG) 502
secondary proton attachment 137
segmented flow gas–liquid–solid reactors 620
selection, catalyst 13–16
– Brønsted–Evans–Polanyi (BEP) relationship 14
– computational approach 13
– Sabatier's catalytic reactivity principle 14
selective oxidation 333–335
selectivity 544
shaping, porous material catalysts 443–444
Shell Higher Olefin Process (SHOP) 103
Sherwood number (NSh) 547
side-chain elongation mechanism 329
silica (SiO_2) 433
single-molecule in-situ spectroscopy of catalytic solid 504–508
single-phase MSR 604–607
– mixing 604
single-phase reactors 564–568, See also stirred-tank reactor
– tubular reactors 567–568
single-site alkene metathesis catalysts 402–406
– enantioselective ring-closing metathesis 405
slurry-phase bubble column reactors 309
slurry reactors 310
slurry–suspension reactors 574–575
sol-gel method 433–436
solid acids and bases 132–143, 440
– Brønsted acid or base 132
– Lewis acid or base 132
– mechanistic considerations 139–143
– – direct alkane activation 139–141
– – hydride transfer 141
– – isomerization catalysis 141–143
– primary proton attachment 137
– proton activation by zeolites 135–139
– secondary proton attachment 137
– stereochemical effects 139
– van der Waals interactions 139
solid oxide fuel cells (SOFC)
– high-temperature catalysis in 380–385
solid-state properties, in catalysts development 448–453
solvent complexes 487
Sorghum bicolor 189–190
space time 580, 583
space–time yield 545
space velocity 580

specific adsorption 204
specific surface area (A) 518
spectator ligands 59
staged fixed-bed 569
standard isotherms 522–523
steam reforming, in ammonia sysnthesis 291, 295–299
– catalysis 296–297
– secondary phenomena 297–299
– technology 295–296
– tubular reformer 296
stirred-tank reactor 564–567
– heat transfer 567
– mixing 565
structural defects 451–452
structured catalysts for multiphase reactions 575
sulfidic catalyst 393–397, See also under hydrodesulfurization
supported catalysts 453
– preparation 420–429
– – deposition precipitation 427–428
– – drying 425–427
– – impregnation 425–427
– – ion adsorption 423–424
– – outer-sphere complex formation 423
– – selected catalysts, applications 421
– – selective removal 420
– – support surface chemistry 422–423
– – thermal treatment 428–429
Supported Liquid-Phase Catalyst (SLPC) 106
supported transition-metal hydrides 408–411
surface oxygen species in methane conversion 449–450
surface reactions 34–36
– elementary rate constant expressions for 34–36
surface reconstruction 37–38
suspension reactors 572
symmetric versus asymmetric cracking patterns 328–332
synthesis gas 311
synthetic organic chemistry 4

t
Temkin kinetic expression 118
Temkin Pyzhev kinetics 291–292
temperature-programmed reduction (TPR) method 509
Temporal Analysis of Products reactor (TAP) 451
Terrylene diimide (TDI) dye molecule 506
Tetrahydrofuran (THF) 57, 176

thermal reaction   216
thermal treatment   428–429
Thiele modulus   88, 95, 98, 547
Thomas chemistry   339
three-center (M-C-H) transition states   128
three-phase gas–liquid–solid systems   573–575
three-phase reactors   616–622
– continuous-phase contacting   616–618
– dispersed phase contacting   619–622
– gas–liquid–solid   616
– mesh microcontactor   619
– trickle-bed reactors   555, See also individual entry
Ti-catalyzed epoxidation   335–338
Time-Resolved Microwave Conductivity (TRMC)   220
t-method   522–523
Topsøe radial flow converter   292
T-plot techniques   6
TPPTS (tris sodium salt of meta-trisulfonated triphenylphosphine)   237
transaminases (TAs)   193–194
transition metal catalysis   114–132
– ammonia synthesis   114–119, See also individual entry
– C–C bond cleavage   126–132
– dehydrogenation   126–132
– ethane hydrogenolysis mechanism   127–132
– hydrogenation   126–132
– methane reforming   120–125, See also individual entry
transition metal-centered homogeneous catalysis   155, 159–169
– hydrometalation of alkenes   165
– kinetic activity   167
– ligand substitution
– – associative pathway   160
– – dissociative pathway   160
– – limiting mechanisms   160
– migratory insertion   164–165
– oxidative addition   162–163
– reductive elimination   164
– substrates and reagents, activation of   159
– Ziegler–Natta polymerization   164
Transmission Electron Microscopy (TEM)   494
transport effects, estimation criteria for   101–103
trickle-bed reactors   552
– advantages   553

– application of   556
– criteria for   554
– – axial mixing   554
– – channeling   554
– – isothermality   554
– – mass transfer   554
– disadvantages   553
– input data   555
– laboratory scale versus industrial trickle-bed reactors   557
tubular reactors   567–565
tubular reformer   296
turnover frequency (TOF)   55
turnover number (TON)   54–55

u

UV Photoelectron Spectroscopy (UPS)   493
UV/Vis spectroscopy   486–490
– principal component analysis   489

v

van der Waals interactions   139
visible light-sensitive systems, quest for   223

w

Washburn equation   533
weight hourly space velocity (WHSV)   544
Weisz module   90, 102
Weisz' window   545
wet impregnation   425
white biotechnology   171
Wilkinson's catalyst   166

x

xenon porosimetry   533–534
X-ray Absorption Near Edge Spectroscopy (XANES)   497
x-ray absorption spectroscopy (XAS)   466
x-ray fluorescence spectroscopy (XAFS)   466
X-ray Photoelectron Spectroscopy (XPS)   493
x-ray scattering techniques   6

y

yield   545

z

zeolite catalysis   325–339
– framework-substituted redox ions   335–339
– hydrocracking reaction, acid catalysis   325–332, See also individual entry
– Lewis acid–lewis base catalysis   332–333

zeolite catalysis (*contd.*)
– redox catalysis   333–335
– selective oxidation   333–335
– single-site versus two-center Fe oxycations reactivity   334–335
– Thomas chemistry   339

zeolites   7, 28–30, 440–441
– proton activation by   135–139
Ziegler–Natta-type olefin polymerizations   164, 261–262, 266–267, 274–283
Z-scheme process   224